51st Annual Loss Prevention Symposium 2017 (LPS)

Topical Conference at the 2017 AIChE Spring Meeting and 13th Global Congress on Process Safety

San Antonio, Texas, USA
26 - 30 March 2017

ISBN: 978-1-5108-4364-6

Printed from e-media with permission by:

Curran Associates, Inc.
57 Morehouse Lane
Red Hook, NY 12571

Some format issues inherent in the e-media version may also appear in this print version.

Printed by Curran Associates, Inc. (2017)

For permission requests, please contact AIChE
at the address below.

AIChE
120 Wall Street, FL 23
New York, NY 10005-4020

Phone: (800) 242-4363
Fax: (203) 775-5177

www.aiche.org

Additional copies of this publication are available from:

Curran Associates, Inc.
57 Morehouse Lane
Red Hook, NY 12571 USA
Phone: 845-758-0400
Fax: 845-758-2633
Email: curran@proceedings.com
Web: www.proceedings.com

TABLE OF CONTENTS

482030 AGILE Award Keynote Address

Abstract:

Factors in Selecting AGILE Honoree

1. Overall sustained excellent industry leadership in businesses that employ chemical engineers

2. Recognized for efforts in innovation and creating strategies that ensure success

3. Value of the keynote to the meeting attendees and the members of the Institute Schedule

Event	Time	Location
Continental Breakfast	7:00AM - 8:00AM	Henry Gonzalez Convention Center, Lila Cockrell Theatre
AGILE Keynote Address	8:00AM - 9:15AM	Henry Gonzalez Convention Center, Lila Cockrell Theatre

The 2017 AIChE Spring Meeting and 13th GCPS AGILE Keynote Address will be presented by AGILE award recipient John Y. Televantos, Partner, Arsenal Capital Partners.

John Televantos is a Partner at Arsenal Capital Partners, a middle market private equity firm focused on the specialty industrial, healthcare and specialty financial service sectors. Before joining Arsenal in 2006, Dr. Televantos was President of Aqualon Company, owned by Hercules, Inc.; President and Founder of Helios Chemical Company; and Chief Executive Officer for Foamex International.

Dr. Televantos began his career with Union Carbide in 1977 in silicones research. There he developed two key processes for the manufacture of Methyl Chloride and Siloxane polymers that were commercialized. These processes not only improved efficiency and reduced cost but also dramatically reduced waste. He moved through a variety of management positions there and later at ARCO Chemical, where he became vice president of strategy and development.

At ARCO John co-invented and led a team that developed and commercialized a breakthrough technology in polyol manufacture (a key raw material for polyurethanes) eliminating a whole process step along with waste and emissions, while at the same time opening the door to the creation of new higher-performance polyurethanes. The technology has since been licensed by most of the major global polyol manufacturers with significant waste reduction worldwide.

At Arsenal Capital, Dr. Televantos serves as co-head of the specialty industrial practice. During his tenure, he has been part of several acquisitions and successful exits in the chemicals and materials sectors. In addition to serving on the boards of several Arsenal-owned companies, he is lead director on the board of Balchem Corp.

481480 Data visualization Connecting Big Data to Integrate Every Level of the Organization- Messaging, Action and Opportunities!

Abstract:

In an effort to improve data transparency, identify true costs, and most importantly, capitalize on opportunities to improve the safety culture throughout the enterprise, United developed a data visualization program, to communicate information clearly through graphical means including Geo Mapping. By centralizing big data from various departments and organizations, we can mine big data to a level that provide universal messaging through visualization. The ultimate goal was to reduce the frequency and severity of employee injuries and implement the daily use of data to address trends and decrease the number events impacting the health and safety of our employees as well as protecting our aircraft from damage.

United's Corporate Safety team in a collaborative agreement is working with the University of New Haven in changing the way data is processed, analyzed and communicated. Our approach is allowing 24/7 universal access to data and streamlining analyst processing time to a fraction of the time while being more comprehensive in what the data means!

United was able to apply the same logic to employee injuries and aircraft events globally across the network, integrating safety and operational data to allow end users to analyze events down to a specific location (such as a specific gate at an airport); the type of equipment involved; the type of injury; frequency; and impact of such events. We are now applying the same logic to how facilities and equipment are maintained and operated in way to be more predicative.

United Airlines has seen success as frontline leaders continue to rely on the data visualization dashboards to actively manage the health and safety of our employees. These techniques are being utilized by operational division to eliminate waste and improve efficiency throughout the airline.

Leadership in Catastrophic Incident Prevention: Using Organizational Data to Assess and Respond to High Reliability Loss Prevention Precursors

Michael Snyder
DEKRA Insight
113 Campus Drive
Princeton, NJ 08540
Mike.snyder@dekra.com

Prepared for Presentation at
American Institute of Chemical Engineers
2017 Spring Meeting and 13th Global Congress on Process Safety
San Antonio, TX
March 26 – 29, 2017

Leadership in Catastrophic Incident Prevention: Using Organizational Data to Assess and Respond to High Reliability Loss Prevention Precursors

Michael Snyder
DEKRA Insight
113 Campus Drive
Princeton, NJ 08540
Mike.snyder@dekra.com

Keywords: leadership, catastrophic incident prevention, highly reliable organization, early warning signals

Abstract

Catastrophic Occupational & Process Safety Incidents usually occur after a series of early warning signals, often subtle, are overlooked. Some organizations have been very successful in avoiding catastrophic events, even in environments with high risk factors and complex operations. Known as Highly Reliable Organizations (HROs), these operations have developed cultures that embrace the concepts of Anticipation, Inquiry, Execution, and Resilience – leading to significant improvement in the detection of and response to early warning signals.

Contemporary research has identified methods, which allow for identification of incident precursors to Serious & Catastrophic Potential Events, along with diagnostic tools to assess the maturity of an organization's "High Reliability". These tools provide powerful insights, and allow leaders to develop a comprehensive approach to catastrophic incident prevention through the effective use of Organizational Data.

1 Introduction

Four years before BP's Deepwater Horizon became global news, ExxonMobil walked away from a $187 million investment in a similar deep sea well. The well had yet to produce a single drop of oil when engineers detected geologic instability in the seabed. After consulting with geologists and drillers on the project, leaders decided that the best course of action was to cap the well and move on. Industry analysts at the time accused the organization of being too risk averse; after all, no incident had actually occurred. [1, 2]

That narrative changed in 2010, when operators of the Deepwater Horizon lost control of a wellhead that had been experiencing similar disturbances. This time there was no speculation of whether the risk was "real" or not. The resulting explosion and fire killed 11 workers and created an environmental and public relations nightmare. For leaders across industry, the tragedy of the Deepwater Horizon was compounded by the fact that many of the lessons it had to offer had already been "learned" before.

After any serious process safety incident has occurred, a diligent investigator will typically find a long history of less consequential incidents that preceded it. The goal of High Reliability Safety is to understand why these "early warning signals" are overlooked and what can be done in today's manufacturing environment to improve the identification of and response to these signals.

2 Emergence of the High Reliability Organization (HRO)

Throughout the industrial age, human-made complex systems have advanced to provide the world with materials, clothing, fuel, food, and resources. However, these systems are susceptible to catastrophic events that result in multiple fatalities and major property damage. Since the 1980s, academics have worked to define what they call the High-Reliability Organization (HRO). The high-reliability organization typically defined as an enterprise that executes consistent and sustainable operations by creating a culture that anticipates and minimizes variation and implements high-quality decision making and controls.

The HRO paradigm seeks to categorize the characteristics of organizations that stay safe despite operating in high- risk conditions (Figure 1). Much of the HRO organizational research has studied U.S. military organizations, wildfire incident management teams, and nuclear plant operations. In theory, a high-reliability organization would routinely make the same kinds of safety-based decisions that ExxonMobil did back in 2006.

3 Four Disciplines of High Reliability Organizations

Through study of both theoretical models of high performing safety organizations and incident investigations of actual process safety events we have recognized that there are **four aspects of organizational performance** that are key to safe operations (Figure 2). Our studies have identified that successful operations have developed cultures that embrace the concepts of Anticipation, Inquiry, Execution, and Resilience – leading to significant improvement in the detection of and response to early warning signals, resulting in fewer process safety incidents.

3.1 Anticipation

Fostering systems and behaviors that are sensitive to "weak signals" that may be indicative of increased risk of catastrophic events.

Anticipation is about recognizing the early warning signs that help you head off incidents. An organization strong in Anticipation will have mechanisms to capture information from a variety of sources that may be meaningful early indicators of change to exposure. Examples could include

Figure 1: Framework for a High-Reliability Organization (HRO) adapted from Sutcliffe & Weick [3].

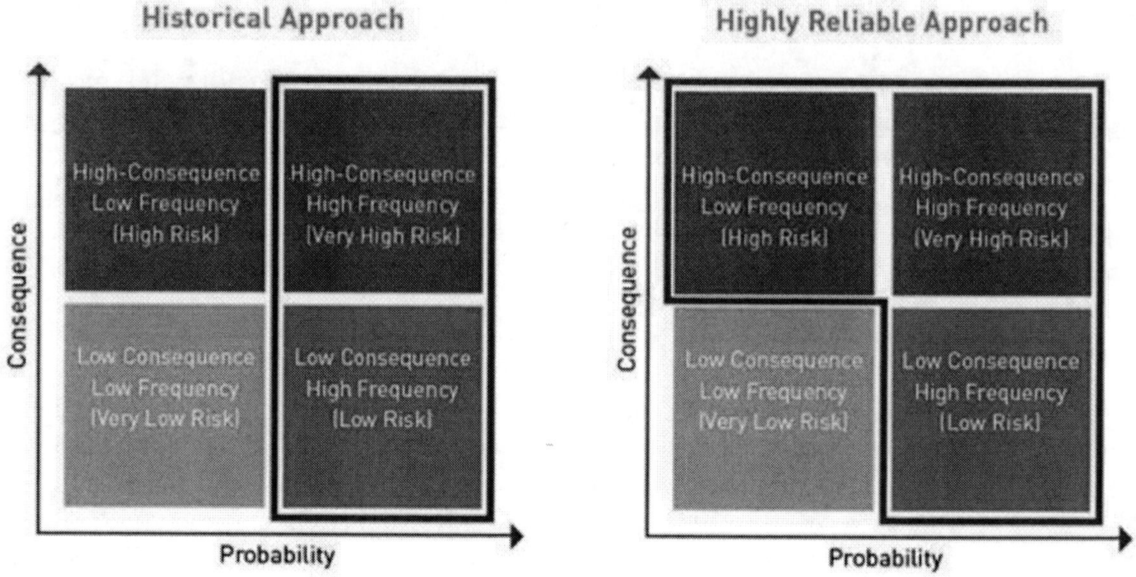

HISTORICAL FOCUS	FUTURE FOCUS
Only high frequency events	**Also high-consequence, low-frequency events**
Lagging metrics	**Leading and "In Process" metrics**
Report only incidents	**Report near misses & other significant events**
Investigate only Recordable events	**Investigate all events with high learning value**
Causal factors: Technical & operational	**Causal factors: Organizational & management system**
Assume past performance predicts future	**Assume the worst case is indeed possible**
Learning environment: Primarily internal	**Learning environment: Includes external view**

process deviations, unusual maintenance requests, and even frontline workers detecting differences in sounds. Strong Anticipation is created by a culture that is always seeking indicators of problems. This means:

- People are encouraged to report and the value of reporting is reinforced
- Data is acted upon and people know that. They won't bother if they perceive that information disappears into a black hole and nothing ever happens
- Leaders understand and accept that there will be lots of information that is not important but it's worth sorting through to be sure we don't miss the information that is critical
- We can identify key leadership behaviors that help create this culture. When those behaviors are practiced we build an organization with strong Anticipation

Organizations with strong cultures have leaders who visibly value the search for early warnings and reinforce the analysis of these indicators, even when this does not result in identification of

Figure 2: The AIER Model

serious risk after it is fully explored. These leaders understand that supporting the detection and investigation of many false positives is worth it, if it results in just one catastrophic event avoided.

3.2 Inquiry

Making effective use of information to analyze, understand, and plan mitigation of risks, while making a conscious effort to overcome bias.

The Inquiry discipline is about preserving the integrity of decision-making and action, specifically by protecting teams from the natural biases innate in all humans and groups. Traditional Process Safety Management (PSM) includes a number of elements (such as process hazard analysis, pre-startup safety review, management of change) designed to evaluate and plan for control of hazards and risks. However, there are common (but often undetected) cultural characteristics that can undermine the effectiveness of these efforts and leave the organization vulnerable.

Cognitive bias refers to the tendency we all have to rely on intuitive, rather than analytical, thinking in order to process information efficiently. Our knowledge and experience allow us to reach conclusions and make decisions quickly and efficiently in many circumstances. However, this can also trap us in poor decisions.

Cognitive bias can also get us into trouble in formal hazard analysis – we can too readily accept what we think we know, e.g., that a particular condition or situation or deviation is not problematic. The tendency to go along with a group's predominant opinion can make this worse.

Biases can include:

- **Confirmation bias.** The tendency to search for or interpret information in a way that confirms one's preconceptions.
- **Normalcy bias.** The refusal to plan for, or react to, a disaster that has never happened before.
- **Availability bias.** The tendency to predict based on how easily an example can be brought to mind.
- **Status quo bias.** The tendency for people to like things to stay relatively the same.
- **Groupthink.** The tendency to do (or believe) things because many other people do (or believe) the same.
- **Risk seeking/risk aversion.** The tendency to make risk-averse choices if the expected outcome is positive, but make risk-seeking choices to avoid negative outcome.

The only way to guard against the insidious effects of cognitive bias is through effective culture and leadership interventions. There are specific leadership behaviors (for example, encouraging the voicing of dissenting opinions) that promote a culture that produces more reliable decision-making. There are also specific skills involved in asking the right question in the right way to get the right data. Leaders must promote and measure the use of these leadership behaviors and skills.

3.3 Execution

Monitoring, reinforcing, and verifying program execution, while staying true to the process.

The Execution discipline is about assuring consistent and reliable use of safety programs and processes. Even the best-designed hazard identification and control efforts are subject to failure or underperformance if the programs and practices are not followed, as intended.

While many organizations use periodic audits to provide a check on implementation, the key to assuring consistent and ongoing activity is developing leaders who frequently monitor, reinforce, and verify effective program execution. Key aspects of these leadership activities are summarized below:

- **Monitoring** involves regularly acquiring information on what subordinates are doing, how they are progressing toward achievement of goals, and what issues or problems they may be encountering. This is not micromanaging; rather it is assuring that the leader has sufficient information to meaningfully recognize good performance, provide support when subordinates need it, and then providing corrective feedback on those (hopefully rare) occasions when subordinates fail to fulfill their responsibilities. A leader's monitoring behavior may take many forms. Depending on the situation and the leader, these may include walking around and observing, informal conversations at the front line, periodic meetings with subordinates to review progress, use of written progress reports, review of appropriate metrics, etc.

- **Reinforcement** involves providing feedback that recognizes good performance. This communicates the importance and priority of the catastrophic event prevention activities and maintains focus on consistent execution. Effective reinforcement is based on effective monitoring, which provides the leader with specific data on which to base reinforcement, avoiding the vague and ineffective "good job" type of feedback to subordinates.
- **Verification** is similar to monitoring, but where monitoring focuses on the performance of subordinates, verification focuses on activities and programs. Assuring that audit findings are resolved in a timely manner is an example of verification, and leaders who do this effectively are more likely to have organizations in which consistent execution is valued as part of the culture.

3.4 Resilience

Developing and exercising the ability to react in ways that prevent upset conditions from becoming catastrophic events — and then learning from the experience.

The Resilience discipline is about developing the agility to recognize and quickly respond to exposures in real time. Upset conditions occur from time to time in any system. A resilient organization is able to react quickly in ways that prevent upset conditions from becoming catastrophic events, and then learning from the experience. This has a major influence on ultimate results. Even where automated control systems are designed to handle upset conditions, it is important that workers understand when and how to intervene, and are not only able but also willing to make appropriate interventions early. An organization that is strong in Resilience is more likely to prevent a small process disruption from becoming a major incident. There are three main questions to consider when looking for aspects for effective Resilience:

- **Knowledge.** ***Do people at various levels have a broad enough understanding of the operation so that they can make good judgments in case of emergency?*** Some organizations approach this through use of extensive sets of rules and procedures. That approach intends to assure consistency and avoid having to rely on technical knowledge at the operating level. However, two problems arise with that approach. First, the range of possibilities that plans for results in a proliferation of procedures and rules that become impractical for anyone to know. Second, this approach assumes that all possibilities (with all variations in every scenario) are identified in advance — something that is unlikely to be true. The alternative is to develop an organization in which people are knowledgeable and taught to make good judgments based on their knowledge and the information at hand.
- **Willingness.** ***Are people more or less likely to take action on their own initiative?*** The answer directly relates to culture. People are less likely to act if they are not confident that the organization will support them. Perceptions of the culture's support for resilience form over a long period and are based on many small actions taken and not taken by

leaders. An organization desiring strong catastrophic event prevention will be sensitive to this and intentionally create the culture that supports resilience.

- **Continuous Learning.** ***Do leaders establish both the expectation that learning is an ongoing process that never stops — as well as promoting the systems and processes to support such continuous learning?*** Resilient organizations make learning routine rather than waiting for an incident to trigger learning mode. They create formal mechanisms (e.g., cross-functional teams, advanced training in incident investigation) to continually capture and respond to risk data. They put the focus on generating new knowledge, not finding fault. Great learning organizations are also characteristically self-aware. They make sure that even the newest employee knows the story of where the organization came from and where it is going. Continuously learning organizations are vigilant about preserving organizational memory and assuring that hard lessons are never forgotten — or repeated.

4 Building a High Reliability Organization

So we come back to our initial question: How can leaders develop a high-reliability organization?

Fundamentally, high reliability is a blended cultural and technical process -- driven by leaders.

Leaders must establish the expectation that attention to exposure is an ongoing learning process that never stops. At the same time, they must also assure that organizational and technical systems, as well as enabling systems, support such continuous learning. Several practices can help an organization move toward high-reliability:

- **Cultural Assessment.** The first step is identifying how the organization is performing now. Rather than look at incident rates (which are poor indicators of actual exposure). The AIER scales identified above are readily measured (Table 1), and can be compared against industry norms. They may be used alongside validated culture indicators shown to predict general safety outcomes.
- **Technical Assessment.** Cultural assessments can be immensely powerful and have been demonstrated to predict safety outcomes. To provide a more holistic picture of organizational reliability, it is advisable to conduct a technical assessment that evaluates the effectiveness of process safety systems, including the identification of risk, robustness of protective barriers, and competency of personnel responsible for system operations.
- **Develop a process safety improvement roadmap**. Based on the findings from the Cultural and Technical assessments, an organization's capability can be assessed against one of several internationally recognized Process Safety Management models. Improvement plans can be developed and paced to meet organizational risk resolution timelines and resource capacity.

- **Adopt a learning orientation.** High reliability organizations are characteristically self-aware. They make sure that even the newest employee knows the story of where the organization came from and where it is going. Continuously learning organizations are vigilant about preserving organizational memory and assuring that hard lessons are never forgotten — or repeated.

Table 1: Sample criteria to evaluate AIER scales

ANTICIPATION	INQUIRY	EXECUTION	RESILIENCE
Always seeking	**Always questioning: Avoiding bias**	**Maintaining control while staying true to process**	**Preparation, response, and learning**
Priority is given to leading measures and weak signals Systems and culture support problem-seeking Input from the front line is given high value High sensitivity to change Reinforcement of reporting	Rigorous analysis systems are in place and constantly used The organization actively works to overcome bias The risk acceptance process is clearly defined and followed Recommendations are applied widely	Leaders set clear expectations Roles and responsibilities are well defined and understood Routine monitoring and checking occurs Follow-up assures findings are addressed in a timely manner	Teams have rehearsed responding to a variety of upset conditions Leadership supports and reinforces employee intervention to minimize impact Post upset activity focuses on learning and putting lessons into action widely

5 Identifying Data to Serve as Early Warning Indicators

While it is intuitive that organizations demonstrating higher performance on the four characteristics (Anticipation, Inquiry, Resilience and Execution) related to excellent catastrophic incident prevention have fewer process safety incidents, research efforts have begun to validate this hypothesis [4]. While this analysis is straightforward for a company with multiple sites using common operational definitions for Process Safety incidents, challenges remain.

In general, comparison among different companies is constrained by the lack of common reporting definitions for Process Safety Incidents, particularly those with much larger potential than the outcome actually experienced. In addition, the use of Tier 1 and Tier 2 metrics from API 754, Process Safety Performance Indicators for the Refining and Petrochemical Industries, generally provides a very limited count of events. While intended for internal company analysis, increased standardization of Tier 3 and Tier 4 metrics may help advance the comparative evaluation of the AIER scales for larger datasets.

Specific performance weakness on an individual AIER scale component have been used to improve focus on the types of leading and in process metrics used to aid in identification of catastrophic incident precursors. Several examples are listed in Table 2 below.

Table 2: Example incident precursor metric focus provided by AIER scales

Incident Attribute	Definition	Example Metrics
ANTICIPATION	**Always seeking**	• **Percentage Process Safety Events (including near misses) that were identified by PHA or Hazard Evaluation.** • **Number of Hazard Reports Filed (Focus on Process Systems – for example, "Valve Inspection is Out of Date")**
INQUIRY	**Always questioning: Avoiding bias**	• **Near Miss / Close Call Reporting (Ratio in respect to actual incidents)** • **Number of Stop Work or Job Pauses Executed** • **Number of Safety Critical Objects that Fail to Pass Preventative Inspection/Test in "As Found" Condition.**
EXECUTION	**Maintaining control while staying true to process**	• **Results of Procedure & Work Process Audits** • **Percentage Process Safety Events (including near misses) that were predicted by PHA or Hazard Evaluation, but had one or more barrier failures (categorize barrier failures)** • **Corrective & Preventive Actions Completed on Site (including audit that actions are effectively implemented)**
RESILIENCE	**Preparation, response, and learning**	• **Local incident learnings applied throughout site or corporation.** • **Emergency Drills on Facility Major Accident Hazards (including follow-up corrective actions)** • **Near Misses with Significant Potential Mitigated by Proactive Employee Action.**

6 Conclusion: Moving Towards HRO Performance

In the high-hazard manufacturing environment, safety expectations have grown well beyond the early charter of reducing costs and injuries. The promise of high-reliability organizations is that they define a path to consistent safe performance, even at the frontiers of industry and technology. As with all innovation, the key to adoption is being able to replicate the conditions and factors that create the phenomenon reliably and predictably. The practical and measureable nature of the AIER scales opens a new way to leaders wishing to create high-reliability organizations of their own.

7 References

[1] S. LeVine. "Exxon: Juggernaut or Dinosaur?, *BusinessWeek*, February 5, 2009, https://www.bloomberg.com/news/articles/2009-02-05/exxon-juggernaut-or-dinosaur

[2] J. Mouawad, "New Culture of Caution at Exxon After Valdez", *New York Times*, July 12, 2010

[3] K. M. Sutcliffe and K. E. Weick "Managing the Unexpected: Assuring High Performance in an age of Complexity", San Francisco: Jossey-Bass, 2001

[4] *The AIER Assessment: Technical Report.* DEKRA Insight/BST, Oxnard, CA, 2015.

13TH GLOBAL CONGRESS ON PROCESS SAFETY

Addressing the Dissonance between Corporate and Individual Process Safety Drivers

Angus Keddie
Process Safety Matters Ltd
23 Somerville Rise
Bracknell
RG12 7LW
UK
angus.keddie@processsafetymatters.com

ID# 476110

Prepared for Presentation at
American Institute of Chemical Engineers
2017 Spring Meeting and 13th Global Congress on Process Safety
San Antonio, TX
March 26 – 29, 2017

Addressing the Dissonance between Corporate and Individual Process Safety Drivers

Angus Keddie
Process Safety Matters Ltd
23 Somerville Rise
Bracknell
RG12 7LW
UK
angus.keddie@processsafetymatters.com

ID# 476110

Keywords: Psychology, Human Factors, Learning, Behaviour

Abstract

Conventional wisdom holds that the Process Safety Goals of the Organization and the Individual are roughly in tune. This paper will argue, however, that the drivers of the individual's Brain and the Organization are sufficiently incoherent so as to merit attention. It will explain why this is and suggest some innovative ways to drive alignment.

1 Corporate Process Safety Drivers

What drives Process Safety in Corporations? My 30 years as a Chemical Engineer working in diverse roles in the international Oil and Gas sector leads me to believe that corporate drivers for Process Safety are two-fold: Money and Reputation. Corporations are and should be driven to maximise a return to their shareholders. If corporations didn't do this, they may become uncompetitive and possibly perish. All the benefits that are attributed to them: useful products, employment, shareholder profit would be lost. Additionally, in some cultures they also have an obligation (financial and/or psychological) to their employees and other stakeholders.

Thus I believe, their approach to Process Safety (PS) should be underpinned by these tenets:

- to minimise PS incidents, which cost, on average, 10000+ times the cost of measures which would have prevented them
- to maximise their PS Return on Investment
- to maximise their uptime, which is negatively correlated with Lagging Indicators such as breaching the PS envelope (e.g. popping PSVs)

Consequently, they strive to align the PS related behaviour of their employees with these goals. The following table gives examples of Current/Recent Methods Employed by Corporations to promote PS within workforce:

Company	Method	Workforce Impact
A	Chronic Unease – workforce encouraged to be vigilant at all times	This is unsustainable and, on its own, can lead to increased levels of stress and higher risk of violation
B	Operator Care – workforce is empowered to tour the facility and be alert to anything out of the ordinary	The worker is acknowledged thus making it more likely that he/she will persevere with any behaviour change such that habits form
C	Safety Day - A day (or at least a significant part thereof) is devoted to considering and discussing an key element of Process Safety	If it is led sincerely and competently from the top, individuals will be impressed by the 'talk walking'.
BP (Texas City 2005)	Focus on Lagging Indicators (days without lost time incident) exacerbated by celebrating milestones	Workers able to maintain unconscious competence mode (avoiding trips, slips and falls) while generating psychological sunken costs ahead of a milestone (not reporting lost time incidents)

2 Individual Process Safety Drivers

What drives Process Safety in People? On operating sites, the tenets are different from those of corporations. I would argue that they are complex, but would certainly include protecting self and colleagues.

Why is it that Individuals and a group of Individuals (a corporation) have different drivers? I believe that individuals generally respond to the drivers of their own brains. Human brains are among the most complex organisms in existence. They are very energy draining (25% of body's energy usage vs 2% weight). Consequently, they have evolved to be very energy efficient and will drive to minimise activities which are energy sapping:

- Conflict
- Uncertainty
- Conscious thought

They have also developed senses to protect us against atavistic threats such as attack by wild animals:

I think it is useful here to look at the Learning Paradigm: how we master a new skill. Firstly, we are not aware that we don't have (or need) the skill - for example a child doesn't know (or need to know) how to drive a car. This first phase is known as Unconscious Incompetence. In the Process Industry, a school leaver becoming a plant operator would be in this mode. The Risk for the corporation is Ignorance leading to Mistakes. The remedy is classroom and 'on the job' training.

Secondly, when we attempt to engage with a new skill, we are aware of our incompetence. We are Consciously Incompetence. Here the Risk for corporation is Nervousness leading to Lapses. The remedy is Buddying or Mentoring.

Thirdly, we start to become competent at the new skill. However, we are still aware of our execution actions. Here we are in the realm of Conscious Competence, which is ideal for corporations. Unfortunately, the brain will drive relentlessly to the final mode - Unconscious Competence, as this is the low energy consumption (ideal) state for the brain. Here again there are risks for the corporation Complacency leading to Violations. How are we to counter these?

Let's examine the key individual Process Safety related behavioural drivers. These are:

- Minimising conflict – risk of groupthink
- Minimising uncertainty – fear of change
- Minimising conscious thought – risk of human factor errors: lapses, mistakes, violations (normalisation of deviation)
- Use of senses (sight, hearing, smell, touch, taste) has evolved to protect us from atavistic threats. Less useful when having to manage PS incidents (when nothing happens for a long time)

3 Alignment Tools

How can we address these issues to drive alignment and improve process safety? The following table illustrates potential alignment tools:

Individual Driver	Impact	Alignment Tool
Minimising Conflict	Groupthink – unwillingness to report unsafe situations	Empower employees and lead by example. Encourage external perspective.
Minimising Uncertainty	Resistance to necessary change (especially in culture which rewards success)	Incentivise the hassle to develop new habit. Reward endeavour over success.
Minimising conscious thought	Increased risk of human factor errors: lapses, mistakes, violations (normalisation of deviation)	Consider use of 'safe surprises' e.g. carry out 'on the spot' site audits.
Use of senses (sight, hearing, smell, touch, taste) has evolved to protect us from atavistic threats	Less useful when having to manage PS incidents (when nothing happens for a long time)	Empower employees to tour the facility and be alert to anything out of the ordinary (applying their 6th sense).

Addressing Conflict Minimisation

Groupthink is alive and well in our industry. Look no further than the BP Deepwater Horizon Disaster. The uncontrolled subsea blowout in the Gulf of Mexico in 2011 led to fire and explosion resulting in 11 deaths. It was the biggest US environmental disaster for 20 years. The cost to BP - $50bn. It was partly caused by collective psychological issues including groupthink.

One BP operator reported that when one of the men carried a handful of rubber material to a superior concerned that the rubber seal down in the well had been damaged, he was told, "No, that can't be. We always get that kind of material coming up."

BP had already sunk millions of dollars into the oil rig's production. Changing direction at that point – in other words stopping production because the rubber seal might be broken - was out of the question. BP had become wilfully blind: the evidence of barrier transgression didn't fit with the collective view that the rig was safe and was therefore ignored.

So how can we best counter Groupthink:

1. Anticipate Groupthink in your Risk Plan. While it might sound like planning to fail, ignoring the potential for Groupthink is a failure to plan for a very real risk. And like any risk plan, there must be processes for monitoring and mitigating emerging Groupthink.
2. Size counts. Limit the typical team size to less than 10 and ensure that there are well-defined boundaries for inclusion. Porous team boundaries and widespread casual involvement on teams breeds dysfunction, including pressure towards consensus for the wrong reasons.
3. Invite external perspectives at various stages of the process. Of course, you've got to have the procedures in place to both protect external viewpoints and to find ways to incorporate them into the group's thinking and plans.
4. Lengthen the discussion phase. Use structured discussion to focus on vetting the issues.
5. Develop a second solution. Challenge your team to assume that management will reject their first solution. Develop an alternative and very different second solution and be prepared to defend it.
6. Invite the Devil's Advocate to the party. While a designated Devil's Advocate is a contrived role and everyone knows it, at least someone will be throwing rocks at the groups beautiful picture. Rules on respecting and vetting the DA's perspective are critical to benefitting from this approach.

Addressing Uncertainty Minimisation

How could we start addressing Resistance to Change? Change is uncomfortable and therefore instinctively avoided. We do start off well however. I remember encouraging the effort both my sons made when they started to walk as babies.

Unfortunately, at some point, we seem to segue to only praising skill. One of the consequences of this change is that, as we grow, we are more likely to develop a 'fixed mindset' (one that believes that ability is inate) rather than a 'growth mindset' (one that believes ability can be improved by practice). As Carol Dwerck argues in her book 'Mindset', people with a fixed mindset are more likely to lie and normalise deviation to counter the discomfort they feel at the prospect of making mistakes.

Over half a century ago. Mick Jagger uttered the famous mantra 'I can't get no satisfaction' in the eponymous song. It seems to me the lyric encapsulates a key drive of human behaviour, where ever the end goal is on the Maslow Hierarchy of Needs. If the purpose of life is simply more life, then perhaps the goal of life should be individual satisfaction. For most of the key tenets of our lives, we know and target the scenarios which lead to consumption gratification.

And mostly that works fine. We buy and consume food and drink, we engage with other people for social and sexual benefit, we sleep when we are tired. All of which is likely to make us feel good if they are achieved. It seems our evolution has created rewards, often in the form of the hormones dopamine and oxytocin, which are released to encourage us to abandon the status quo and engage with the barriers between us and these tenets. Often the more significant the tenet and the higher the barrier, the greater the associated reward: for example, you may risk competition and rejection in pursuit of a mate.

However, in the arena of industrial hazards, this relationship becomes very tenuous. In terms of risk reduction, it benefits the group if near misses are reported and acted upon. Statistically, there are many times more near misses than the accidents they may have become. And as the antecedents of both are often the same - for example, the cause of a dropped load leading to an injury or a 'lucky escape' - the learning will reduce the likelihood of the same or similar cause reoccurring. So, it would seem to make sense to encourage the reporting of near misses.

However, this is not always the case. Currently, many operating companies set milestones for lost time incidents. As the milestone is approached, who wants to be the one to report a minor injury which would scupper the celebrations? Also, many companies have incentive schemes which are aligned to productivity, which is superficially negatively correlated with downtime. Furthermore, in companies with a strong safety culture, where near miss reporting is encouraged and the corresponding reduction in incidents means that nothing untoward happens for increasing lengths of time - which is a good thing, but not a satisfying thing.

So what to do? How can we make the promulgation of less incidents be satisfying to the front line worker? One idea I have is take an idea that Hollywood has hijacked and re-imagine it. I don't think I'm alone is watching action movies where heroes prevent bad things happening. Tom, Arnie et al battling to prevent the explosion, toxic cloud, wild beast from doing their worst. We are experiencing the vicarious echo of the actual experience. Risk and Reward. Why not apply the same principle to rewarding front line workers when they report a near miss which could have led to an injury or fatality.

An example I heard recently from an operating company was a propane leak in the vicinity of the air intake for a compressor. 'There was no explosion but for luck and the quick reaction and reporting of the shift supervisor'. In this case the supervisor was thanked. How about enhancing this thanks with the reward of a simulation of the incident his (or her) quick and robust intervention prevented. I could be shown at an end of year event and a copy presented to the worker. You have just created a lasting psychological reward which will reap benefits long after any monetary offering has been forgotten.

In an interview with Entrepreneur.com in 2012, James Dyson, the billionaire engineer and inventor, said 'we have to embrace failure and almost get a kick out of it.' Before successfully developing his eponymous bag-less vacuum cleaner, he had a large number of failures: 5216 to be precise. If he had become discouraged at, say prototype 3682, for example, he wouldn't have ultimately succeeded. How can we harness this spirit in our industry? By incentivising the hassle to develop new habit and rewarding endeavour over success.

Addressing Conscious Thought Minimisation

What about the Normalisation of Deviation? In his 1997 paper, published in the Journal Safety Science, Jens Rasmussen argues that in the presence of strong cultural and psychological forces (Corporate, Safety, Individual) human behaviour are likely to migrate towards the boundary of acceptable performance.

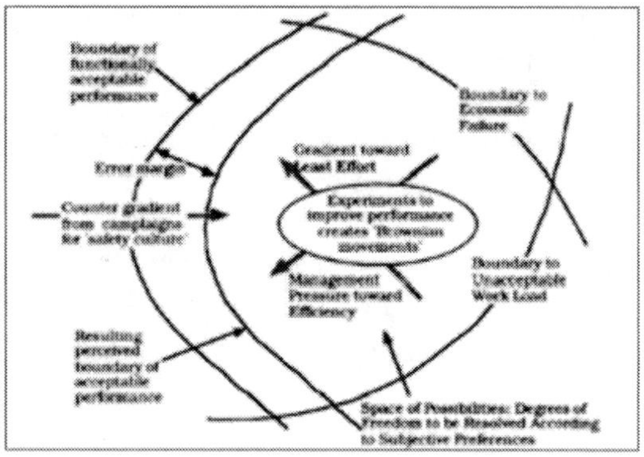

He argued that Experiments to improve performance creates 'Brownian movements' and that individual and corporate drivers will push the operations envelope towards the perceived boundaries of acceptable performance. If we take individual employees, they want to make their working lives as comfortable as possible, generating a gradient of least effort.

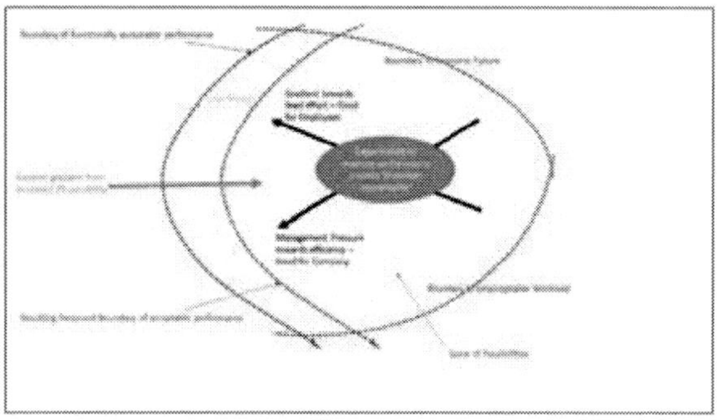

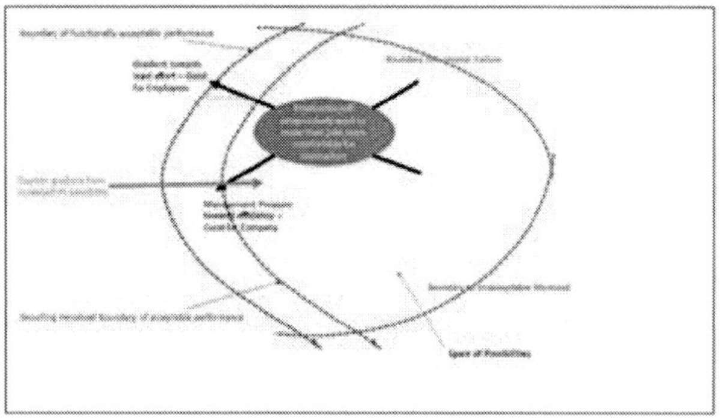

In doing so they may transgress the perceived boundary for acceptable performance.

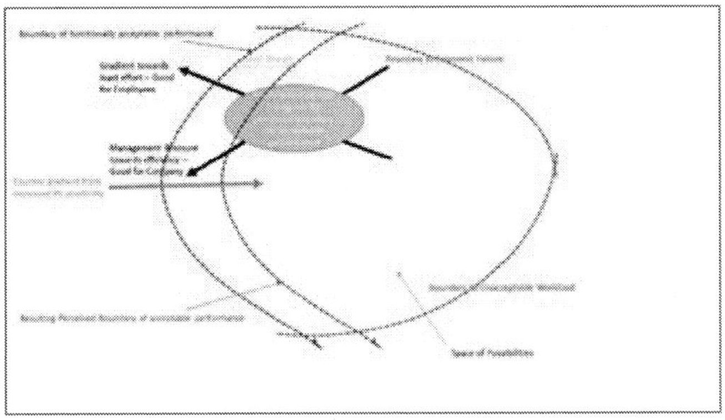

If PS sensibility isn't strong, this may lead to transgressions being normalised.

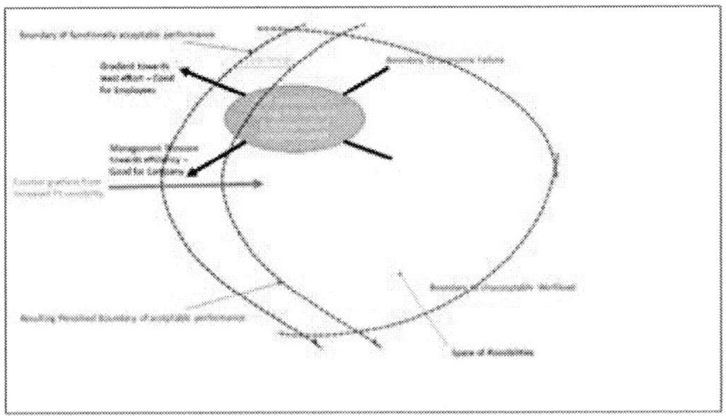

The Perceived Boundary of acceptable performance may recede towards and even through the actual boundary

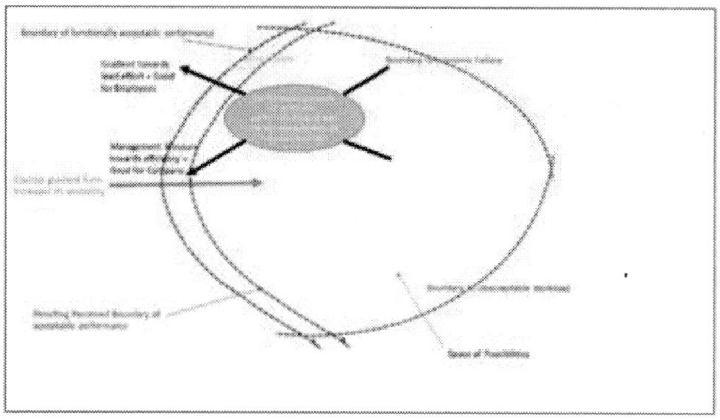

Reducing and eliminating the error margin and leading inexorably to breeching the actual boundary:

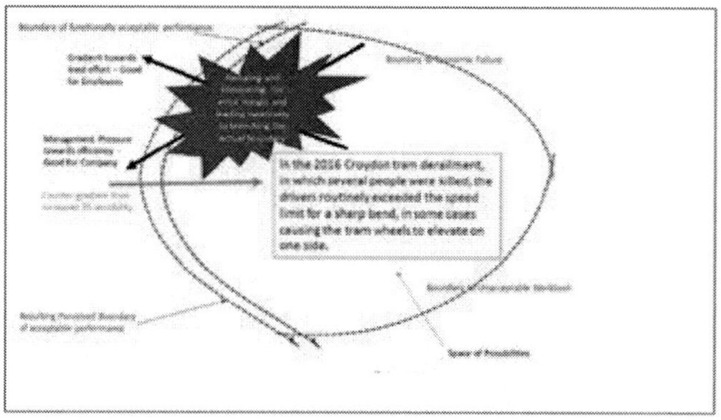

In the 2016 Croydon tram derailment, in which several people were killed, the drivers routinely exceeded the speed limit for a sharp bend, in some cases causing the tram wheels to elevate on one side. The accident took place in the dark and during heavy rain, on a sharp left curve. The curve comes almost immediately after the line emerges from a tunnel. It has been described as a "sharp bend" and has a 20 km/h (12 mph) speed restriction. The tram entered the curve at a speed of 70 kilometres per hour (43 mph) and overturned onto its right side, falling outside the curve on which it was travelling and trapping several people inside. There were seven fatalities with 58 other people injured.

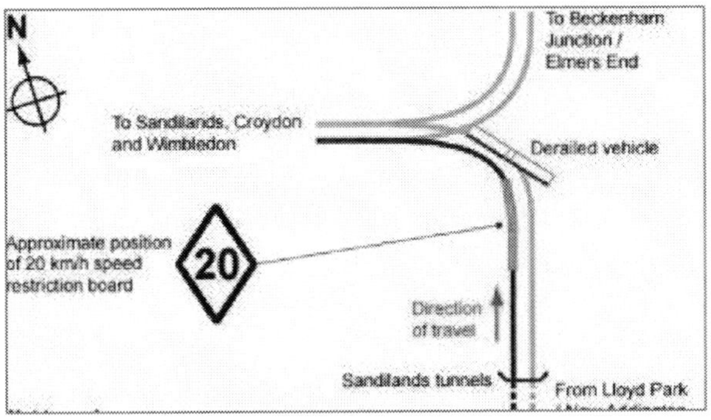

Images from the Aftermath:

However, if we make use of Safe Surprises and Unplanned Alerts, we can help to reset the Perceived Boundary. The Error Margin between Perceived and Actual Boundary of Acceptable Performance is thus re-established.

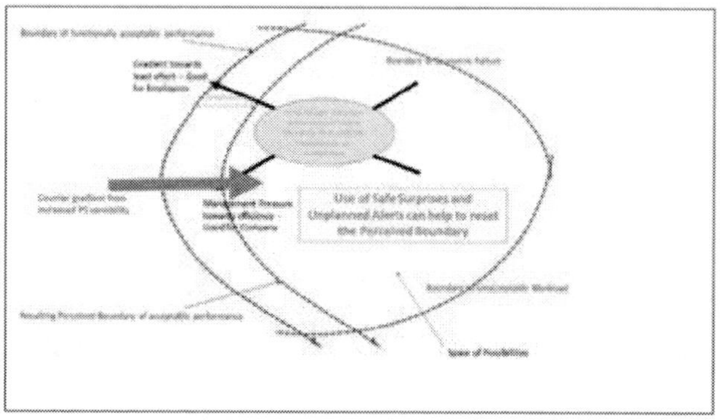

Addressing 'Senseless' Danger

Finally, how can we augment the front line operator's senses to improve PS outcomes? By Providing 6th Sense Tools such as 'Google Glass'. We could upload Key Process Parameter and Condition Based Monitoring Data. The glasses would be able to sense dynamic plant conditions, such as temperature or pressure spikes. The software could then trigger triaged warnings:

- Red – immediate plant shutdown
- Orange – investigate and risk assess
- Yellow – Periodic monitoring

4 Conclusions

So to summarise, this paper has demonstrated that the drivers for Process Safety of individuals and organisations are troublingly distinct, with worrying implications for both employee safety and company profits. The paper has shown that it is possible to better align the drivers and provided a number of tools to effect this, including:

- Groupthink - Develop a second solution (and compare it to the rock damaged initial one)
- Fear of Change - Find ways of rewarding endeavour over success e.g. generate simulations of prevented incidents
- Drive towards unconscious competence - Apply a mix of planned and surprise alerts
- 'Senselessness' of PS Hazards - Purchase technology to enhance front line workers 6th sense
- Use CBT to generate customised message for your workforce

Hopefully, this will lead to a Successful Outcome

5 References

Carol Dwerck. Mindset: How You Can Fulfill Your Potential. ISBN1780332009. 2012.

Nadia Goodman. James Dyson on Using Failure to Drive Success. Entrpreneur.com. November the 5th 2012.

Jens Rasmussen. Risk Modelling in a Society: A Modelling Problem. Safety Science Vol 27No 2/3. Pp 183-213. 1997.

481540 Dan Crowl on Fire - Basics of Flammability *Dan Crowl*

Abstract:

Flammable materials are common in industrial facilities - either as a feedstock, solvent, product or fuel. It is important to understand the flammability characteristics of these materials in order to handle them safely. This talk will provide an introduction to flammability. This will include the nature of fires and explosions, the properties of flammable materials, and how to use these properties to handle flammable materials safely

481541 How Do Flame Arresters Work?

Abstract:

Processes involving the storage, transport, separation, & destruction of flammable vapors are ubiquitous in the chemical & petrochemical industries. These operations present inherent explosion risks that are often addressed through the application & use of Flame Arresters. However, many practicing engineers & recent graduate engineers are unfamiliar with Flame Arresters, which can lead to dangerous misapplications. This introductory presentation details the basic fundamental principles of flame arresters, the industry standards governing these devices, & common flame arrester applications.
Key technical concepts related to flame arresters, such as the MESG, the dynamics of propagating explosions, & arrester element design will be described. The critical regulatory & industry requirements of NFPA 67, NFPA 69, & 33 CFR Part 154 will be cited. A generous cross section of typical flame arrester applications will be provided.

481542 BLEVE Basics, A Primer on Boiling Liquid Expanding Vapor Explosions

Abstract:

Boiling Liquid Expanding Vapor Explosions (BLEVEs) are a special case of bursting pressure vessels in which the pressure vessel is filled with a liquid that is above its normal atmospheric boiling point. When the vessel fails, the liquid boils and adds significant energy to the bursting vessel event. The thermodynamics of these events can be very complicated and interesting and the consequences of these events can be catastrophic. This session will provide new practitioners with a working knowledge of BLEVE phenomena, potential consequences, prediction methodologies, and some sample event sequences that can lead to a BLEVE. The focus will be on developing fundamentals, discussing common misconceptions, and presenting how to deal with the most common scenarios and the session will include a worked example of a BLEVE calculation to illustrate the techniques.

Large Scale Flammability and Explosivity Testing of Low Burning Velocity Gases: Validity of Prediction Tools and Impact On Siting Studies and Risk Assessments

Scott G. Davis
Gexcon US
4833 Rugby Avenue Suite 100
Bethesda, Maryland 20814
Scott.G.Davis@gexcon.com

John L. Pagliaro, Tom F. Debold
Gexcon US
Bethesda, MD USA

Matthijs van Wingerden, Kees van Wingerden
Gexcon AS
Bergen, Norway

Prepared for Presentation at
American Institute of Chemical Engineers
2017 Spring Meeting and 13th Global Congress on Process Safety
San Antonio, TX
March 26 – 29, 2017

Large Scale Flammability and Explosivity Testing of Low Burning Velocity Gases: Validity of Prediction Tools and Impact On Siting Studies and Risk Assessments

Scott G. Davis
Gexcon US
4833 Rugby Avenue Suite 100
Bethesda, Maryland 20814
Scott.G.Davis@gexcon.com

John L. Pagliaro, Tom F. Debold
Gexcon US
Bethesda, MD USA

Matthijs van Wingerden, Kees van Wingerden
Gexcon AS
Bergen, Norway

Keywords: flammability properties, explosivity properties, low burning velocity gases, ammonia, mildly flammable gases, refrigerant flammability

Abstract

Understanding the explosion risk of flammable and combustible fluids is a key factor when performing risk assessments and siting studies for facilities. Past efforts to understand explosion risk have primarily focused on traditional flammable fluids used in industry (e.g. propane and methane), having laminar burning velocities (LBV) typically around 40 cm/s.

Recently enacted policies to reduce human-induced climate change have led to an increase in the use of mildly flammable compounds such as ammonia and other low global warming potential (GWP) working fluids that have LBVs less than 10 cm/s. The lack of available data necessary to assess risk and evaluate potential large-scale explosion consequences associated with these compounds has resulted in two contrasting approaches. The conservative approach is to assume the explosion risks are close to those of methane. The other approach is to assume there is no risk at all from such compounds. In order to properly evaluate risk, the flammability characteristics of mildly flammable compounds must be thoroughly understood and their explosion consequences must be examined at full scales through experimentation and accurate numerical modeling.

To quantify the explosion consequences of mildly flammable compounds, premixed ammonia/air clouds were ignited in a large-scale 50 m^3 rig equipped with pressure transducers and high speed video monitoring. Methane/air mixtures were also considered to provide a direct comparison of deflagration consequences. Tests were performed with the rig partially confined with two degrees of obstacle density (congestion) along the flame path. Supplementing the large-scale results, fundamental flammability properties of ammonia were measured, including: upper and lower flammability limits; auto-ignition temperature; minimum ignition energy; explosion severity index; peak pressure rise; and burning velocity as a function of equivalence ratio. The present study compares the results against the limited data available in literature and to those of methane. To test numerical modeling capabilities, the 50 m^3 rig experiments with ammonia/air were modeled with the CFD package FLACS and predicted deflagration overpressures agreed very well with the measurements. The combined results from the large-scale consequence testing, fundamental flammability and ignitibility experiments, and CFD modeling enable a more accurate assessment of risks, which can be compared directly to the risks for extremely flammable compounds, and which can serve as the basis for proper mitigation measures.

1 Introduction

Understanding the risk of explosions involving flammable and combustible fluids is a key factor when performing risk assessments and siting studies for facilities. However, there are a number of fluids such as ammonia and other new climate-preserving working fluids that have been characterized as mildly flammable, whereby the laminar burning velocity (LBV) is less than 10 cm/s. In comparison, more traditional, extremely flammable gases (e.g., propane and methane) have LBVs of approximately 40 cm/s. A very conservative approach to assess risk and evaluate potential explosion consequences of mildly flammable compounds is to assume the explosion properties are close to those of methane. This will, however, grossly over-predict severity as post-ignition flame speeds and overpressures are directly correlated to the LBV of the fuel. At the opposite end of the spectrum, another approach is to essentially assume that there is no risk from such compounds, which may artificially under-predict the associated consequences of such gases. The actual explosion consequence of ammonia, and other mildly flammable substances, lies somewhere between these two bounds.

To understand the actual large-scale risks associated with mildly flammable compounds, and to design effective mitigation measures, fundamental flammability characteristics must be well defined and consequences must be explored at representative scales. In addition, risk assessments and mitigation design can be improved if consequence tools such as computational fluid dynamics (CFD) models can accurately simulate events involving low LBV compounds ($\leq$ 10 cm/s). Once a CFD model is validated against existing test data, it can then be used to explore the sensitivity of event severity to variables such as geometry, confinement, ventilation conditions, fuel reactivity and heat release, fuel quantity and leak rate, and safety mitigation measures.

To explore the deflagration consequences of mildly flammable compounds, premixed ammonia/air clouds were ignited in a large-scale 50 m^3 rig. Tests were performed with the rig partially confined with two degrees of obstacle density (e.g. empty and congested) along the flame path. Tests were also performed with methane/air to provide a direct comparison of the resulting deflagration consequences. Supplementing the large-scale results, the present study also measures fundamental flammability properties of ammonia, including: upper and lower flammability limits; auto-ignition temperature; minimum ignition energy; explosion severity index; peak pressure rise; and burning velocity as a function of equivalence ratio. The results are compared against the limited data available in literature and to those of methane. The experiments in the 50 m^3 rig are then modeled with the CFD code FLACS to provide preliminary validation of overpressure prediction accuracy when simulating deflagrations involving low burning velocity compounds.

2 Experimental

2.1 Lab-scale

Combustion properties of ammonia and methane were measured using four experiments. Experiments following EN-1839 [1] provided the lower and upper explosion limits (LEL and UEL). Mixtures were prepared in a 2 L cylindrical vessel equipped with a pressure transducer and small polycarbonate windows for visual monitoring of ignition and flame propagation. Test mixtures were subjected to a 15 kV oscillating spark across centrally located electrodes. Once a flammable mixture was identified, tests were repeated at higher and lower fuel concentrations until the pressure rise for a given mixture was negligible (below the 50 mbar threshold).

Experiments following EN-14522 [2] provided the auto-ignition temperature (AIT). Fuel/air mixtures were injected into a 200 mL narrow-necked Erlenmeyer flask placed in a temperature-controlled oven. The test operator monitored the flask for 5 minutes to see if a flame established, and the lowest oven temperature for which a flame did established was recorded as AIT.

Experiments following ASTM E582 [3] provided the minimum ignition energy. Tests were performed in a 2 L cylindrical vessel equipped with a small polycarbonate viewing window. For each test, ignition sparks were generated using one of two circuits, depending on the expected MIE. When the expected MIE was low, a standard capacitive discharge circuit was used. The spark energy was calculated from the capacitance of the circuit and the measured voltage, and for this circuit, the energy could be set as low as 0.1 mJ, which is on the order of the MIE for common hydrocarbons such as methane and propane [4]. When the expected MIE was higher (which is the case for mildly flammable compounds), ignition was attempted with an alternative ignitor capable of generating energies greater than 1.6 mJ. The ignition system used a moderately high voltage power supply (30 kV) and a transformer which created a trigger spark that broke down the spark gap. Mixtures were subjected to sparks with increasing energy until ignition was observed through the viewing window.

Experiments in a 20 L semi-spherical vessel provided the explosion pressure (constant-volume) as a function of time. Further processing of the data yielded the maximum pressure rise p_{max}, the deflagration index K_g, and the laminar burning velocity S_u. The test apparatus is a replica of the Bureau of Mines apparatus described in ref. [5], and used in numerous explosibility studies [6, 7].

Mixtures were prepared in the chamber using the partial pressure method. Once prepared, time was given for the mixture to settle and become quiescent before ignition was attempted (typically around 5 minutes). A 15 kV oscillating spark ignitor (similar to the ignitor used for the explosion limit tests) created a 0.2 s duration spark across a gap between two electrodes at the center of the chamber. The explosion pressure was recorded with two piezo-electric pressure transducers (Kistler 701A) located on opposite sides of the chamber. The signals were averaged, the maximum pressure rise Δp_{max} was recorded, and the maximum rate of pressure rise dp/dt_{max} was determined by numerically differentiating the pressure trace. Once dp/dt_{max} was known, the deflagration index K_g was determined using Eq. 1, where V is the vessel volume (20 L):

$$K_g = dp/dt_{max} * V^{1/3} \tag{1}$$

Burning velocities S_u were determined from the pressure trace by applying the expression derived by Lewis & von Elbe [8],

$$S_u = (R/3)\left[1-(1-y_b)(p_0/p)^{1/\gamma_u}\right]^{-2/3}(p_o/p)^{1/\gamma_u}(dy_b/dt) \qquad (2)$$

where R is the chamber radius, p_0 is the initial pressure, p is the instantaneous chamber pressure, γ_u is the specific heat ratio, and y_b is the instantaneous mass fraction of burned gas in the chamber. Instead of assuming a linear relationship between y_b and p (proposed in ref. [8]), the present work applies the thermodynamic model described in refs. [9, 10], which is similar to the models applied in refs. [11-16].

2.2 50 m³ rig

Large-scale experiments are performed in a 50 m^3 rectangular rig. For each test, the rig is filled with premixed reactants at concentrations resulting in the maximum burning velocity (i.e., slightly rich conditions). The gases are ignited and the severity of the ensuing deflagration is monitored via pressure measurements and high speed video. As illustrated in Figure 1, the rig is 8 m long, 2.5 m tall, and 2.5 m wide with a clear polycarbonate panel along the front so that the deflagration can be filmed. The frame, floor, ceiling, back wall, and one end wall (west end) are made of steel, and the east end is left open for venting. Tests are performed with the partially confined rig with two levels of object density: (1) empty where the only obstructions inside the rig were four horizontal beams located every 2 m at mid-height to provide structural support; and (2) congested where additional objects were added to module as shown in Figure 2. An AC transformer generates a spark to initiate reaction. As seen in Figure 1 and Figure 2, the ignition transformer is located 1 m from the closed end, 1.25 m from the front wall (centered along the width), and 1.35 m above the floor.

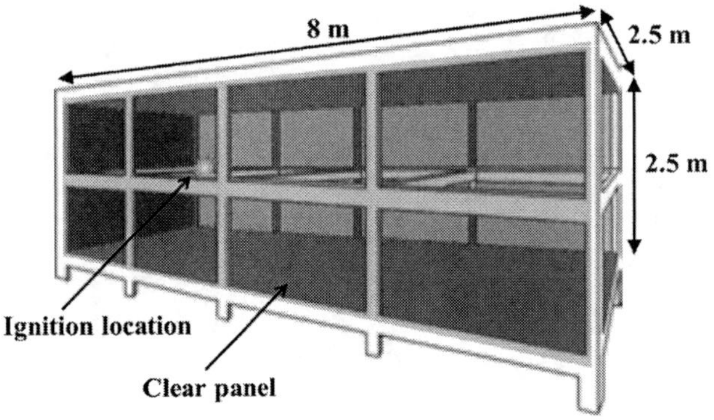

Figure 1: Empty 50 m³ test rig schematic.

Figure 2: Congested 50 m³ test rig.

2.2.1 Instrumentation

Internal pressures are monitored with eight piezo-electric pressure transducers (Kistler 7261) that have a 13 kHz response and are connected to a charge amplifier (Kistler 5011A). Four sensors form a row along the floor near the back wall, and four sensors form a row along the ceiling down the center of the rig (mid-depth). The external pressure along the vent path (i.e. the blast wave pressure) is monitored with three skimmer plate transducers that are connected to a separate charge amplifier (Kistler 4601). The sensors are spaced every 4 m forming a row outside the rig at mid-depth and mid-height. The sensor locations are illustrated in Figure 3, with the x, y, and z directions defined. The signals from each sensor are filtered with a low pass filter at 1000 Hz to remove extremely short pressure peaks that are likely caused by reflections inside the sensors [17].

Figure 3: Pressure sensor locations shown by the green circles.

Experiments are recorded with two high-speed digital cameras (Edgetronic), each having a maximum frame rate of 1800 fps. The cameras are triggered just prior to ignition, along with a light mounted on the outside of the rig to provide ignition timing in the videos.

2.2.2 Procedure

For each test, the rig is filled with a homogeneous mixture of fuel and air. In order to prepare the mixture in the rig, the open end is first sealed with a thin plastic sheet. Fuel is added at five different locations to help ensure uniform distribution, and a re-circulation system provides mixing with a centrifugal fan. During mixture preparation, a multi-channel suction system removes small samples that are analyzed using an infrared gas analyzer (Servomex 4200) and an oxygen analyzer (Servomex Xendos 225) to check the gas concentrations. Once the desired concentrations are reached and mixing is complete, two 8" butterfly valves are closed to isolate the mixing system from the rig, the thin plastic sheet covering the vent is released, and the data acquisition and ignition are triggered.

3 Numerical Methods

The experiments in the 50 m^3 rig were simulated with FLACS, which is a computational fluid dynamics (CFD) tool predominantly used in the Oil & Gas and chemical industries to predict the consequences of hydrocarbon deflagrations. FLACS can predict several transient features of a deflagration including overpressure, pressure impulse, flame dynamics, and flame temperature. These physical phenomena are simulated in a three-dimensional domain while taking into account the interaction between the fluid flow and obstacles, obstructions, and topography.

FLACS solves the compressible Reynolds-Averaged Navier–Stokes (RANS) equations on a 3D Cartesian grid using a finite volume method. Incorporated in the model are the conservation equations for mass, impulse, enthalpy, turbulence and species, with the equations closed by invoking the ideal gas law and the standard k-ε turbulence model [18, 19]. FLACS uses a flamelet-based combustion sub-model with one-step reaction kinetics characterized by the laminar burning velocity of the fuel/air mixture, which is specified in FLACS as a function of equivalence ratio, temperature, and pressure. To model combustion and explosion dynamics, a flame growth sub-model describes how the local flame speed changes with the above-mentioned parameters as well as the turbulence characteristics of the flow field. Then, using sub-grid models, the real flame area is properly described and corrected for curvature at smaller scales than the grid. Furthermore, FLACS accounts for flame acceleration due to flame instabilities, flame-folding by obstacles, and turbulent mixing [20, 21].

FLACS captures the effect of small obstacles while remaining computationally tractable by implementing a "distributed porosity concept" [22]. Large geometries are modeled on a Cartesian grid with larger objects and walls represented on-grid, and smaller objects represented sub-grid. Each grid volume has an assigned volume porosity, area porosity, and drag coefficient based on the congestion contained within the grid cell. The drag from partially blocked cells is then accounted for in the RANS equations through a modified turbulence generation rate term in the k-ε turbulence model. With this approach, small objects contribute to flow resistance, turbulence generation, and flame folding without needing to be fully resolved by the grid.

Figure 4 provides the FLACS geometries for the empty and congested rig simulations. Pressure monitoring points were specified where the pressure transducers were located in the experiment. A uniform 10 cm grid maps the computational domain within the rig and out toward the last far field sensor (12 m from the open end). A coarser grid maps the remainder of the computational domain, which extends 10 m in each direction to minimize the influence of non-physical boundary reflections. PLANE_WAVE boundary conditions are specified to further reduce boundary interactions as recommended by Hansen et al. [17].

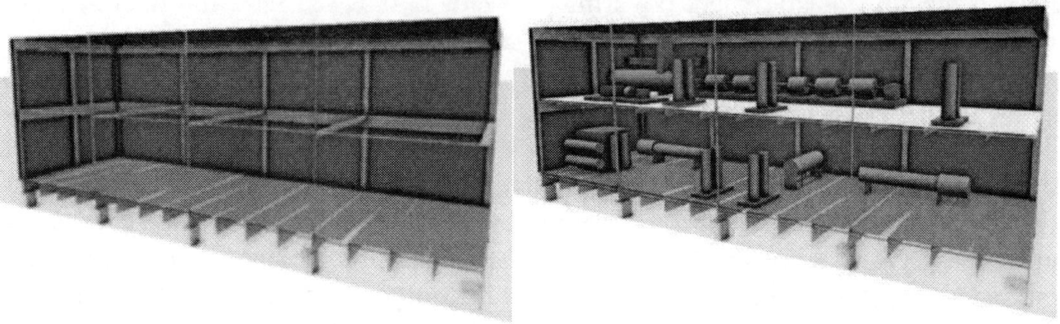

Figure 4: FLACS model of the empty (left) and congested (right) 50 m³ rig test configuration.

4 Results and Discussion

4.1 Flammability and ignitability properties of ammonia

Table 1 provides the flammability properties for ammonia/air determined in the present study and presented previously in literature [23-27]. The present LEL, UEL, AIT, and P_{max} values are close to previously reported results. Larger scatter exists in the MIE data, with the present value close to the lower end of the experimentally determined range. The present K_g value is somewhat lower than reported in NFPA 68. Overall, the flammability properties for ammonia are fairly well established. Additional work is required to reduce the uncertainty in the MIE, which is known to typically have the largest measurement uncertainty considering the sensitivity to test setup [27] and the difficulty in quantifying the actual energy transferred to the flammable mixture as opposed to discharged from a capacitor or even measured across electrodes [28].

Table 1: Flammability and ignitability properties of ammonia.

Source	LEL %v/v	UEL %v/v	AIT (°C)	MIE (mJ)	ΔP_{max} (bar)	K_g (m*bar/s)
Present Study	15.8	32.4	637 ± 5	15-20	4.4	6
Airgas SDS	15	28	651			
Minor et al. 2008	15	28		100		
Kondo et al. 2006	15.2	30				
Ronney et al. 1985	14	32.2				
Verkamp et al. 1967				14		
Takizawa 2015 (calculated)				45		
NFPA 68 1998 Edition					4.4	10

4.1.1 General LBV observations

Figure 5 shows the measured burning velocities for ammonia/air as a function of equivalence ratio Φ, together with previously published data [29-32]. The results are presented as-determined and shifted to account for experimental observations reported in ref. [29]. A shift in the present data is required to compensate for the adsorbing nature of ammonia on the inner surfaces of the test vessel. Takizawa et al. [29] reported that the partial pressure (i.e., volume fraction) decreases to roughly 95% the initial filling pressure 1 hour after filling. Therefore, the "shifted" results in Figure 5 are estimated only and based on equivalence ratios corresponding to 95% of the recorded ammonia volume fraction for each test. The measured maximum burning velocity $S_{u,max}$ for ammonia/air was 7.5 cm/s, which was very close to $S_{u,max}$ reported in previous studies [29, 30]. For ammonia/air, the burning velocity peaks at rich conditions ($\Phi \approx 1.1$) and drops to about 80% of the peak value at $\Phi = 1.0$ and 1.25 (for the shifted results). In general, the present burning velocities are within the scatter of previously published results and the data comparison shows sufficient agreement for $S_{u,max}$.

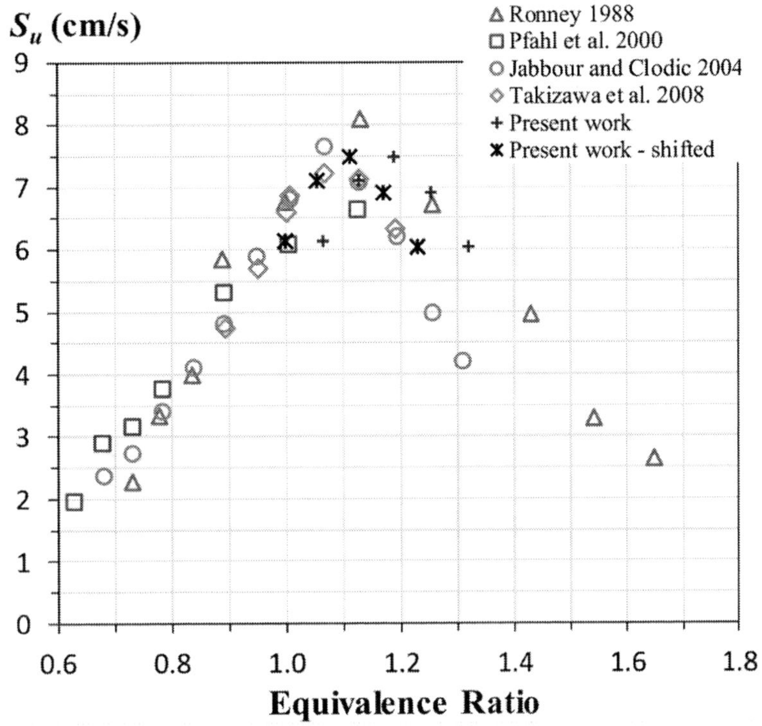

Figure 5: Comparison of present and previously reported burning velocities of ammonia/air as a function of equivalence ratio.

4.1.2 Risk probability based on flammability properties

The flammability properties LEL, AIT, and MIE are used to assess the probability of an ignition event. The LEL establishes the likelihood of forming a flammable mixture (i.e., the higher the LEL the less likely a flammable cloud will form) and AIT and MIE establish the likelihood of ignition in addition to which types of ignition sources are sufficient. The properties P_{max}, K_g, and S_u prefigure the consequence of an event. S_u and K_g are direct and indirect measures of reactivity respectively, and P_{max} is a measure of the energy release and maximum pressure realizable during a constant volume event. By comparing these properties for ammonia and methane, we can begin to differentiate between the risks associated with each.

Table 2 compares the flammability properties of methane and ammonia measured in the present study. The higher LEL for ammonia gives it a lower probability of forming a flammable mixture as compared to methane. The higher MIE and AIT also give it a lower likelihood of ignition. The lower P_{max}, K_g, and $S_{u,max}$ for ammonia suggest less severe consequences from a resulting deflagration event. Presented next are large-scale tests performed to evaluate the actual consequences of methane and ammonia. These results will be correlated back to the lab scale tests.

Table 2: Measured lower and upper flammability limit (*LEL* and *UEL*), minimum ignition energy (*MIE*), auto-ignition temperature (*AIT*), peak explosion pressure (P_{max}), deflagration index (K_g), and maximum burning velocity ($S_{u,max}$) for the ammonia/air and methane/air.

Reactant	LEL (% vol)	UEL (% vol)	MIE (mJ)	AIT (°C)	$\Delta P_{max,Quies}$ (bar)	$K_{g,Quies}$ (m*bar/s)	$S_{u,max}$ (cm/s)
Methane	4.6	15.4	0.24-0.27	537	7.6	65	37.0
Ammonia	15.8	32.4	15-20	637	4.4	6	7.5

4.2 50 m³ rig tests

Experiments in a partially confined 50 m³ test rig provided consequence data for methane/air and ammonia/air post-ignition events. A high speed camera filmed the events and pressure transducers recorded overpressures generated during the events. The following sections discuss the primary differences observed during these experiments.

4.2.1 Flame dynamics

Figure 6 shows the initial flame development of stoichiometric ammonia/air (top) and methane/air (bottom) for tests without congestion. Note that the background is darker in the methane test because the recording frame rate was higher and the test was done later in the day. In Figure 6, the left-most column of images shows the initial flame after ignition, and the time from this point on is provided in the bottom corner of all subsequent images. While both flames start off nearly spherical, the ammonia flame propagates much slower than the methane flame. For methane/air, the diameter of the flame sphere grows much faster than for ammonia/air and the center remains roughly stationary in the z-direction. The high reactivity of methane/air results in flame front velocities that are much faster than the buoyancy driven flow, hence the flame dynamics are dominated by the convective "push" of the expanding burned gases behind the flame. In contrast, the ammonia/air flame begins to rise early on and a dimple forms at the base of the sphere as a result of buoyancy. When the ammonia flame reaches the ceiling, the dimple has extended nearly half-way through the flame and the flame shape is closer to a mushroom cap instead of a sphere.

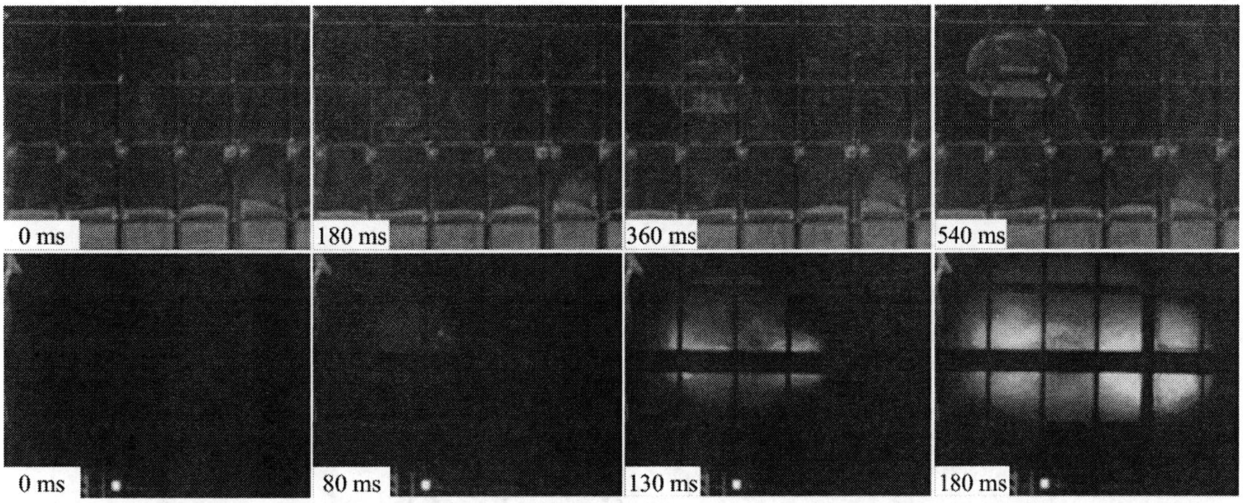

Figure 6: Flame dynamics of ammonia/air (top) and methane/air (bottom) during the initial flame development in the tests without congestion. The time for the flame to progress to the current state, in relation to the initial state seen in the first row of images is provided in the bottom corner.

The flame dynamics of high burning velocity fuels such as methane ($S_u \approx 40$ cm/s) and low burning velocity fuels such as ammonia ($S_u < 10$ cm/s) were not only different during the initial flame development, but also during the latter stages of propagation. Figure 7 shows methane/air (left frames) and ammonia/air (right frames) flames as they propagate throughout the empty rig. The methane/air flame starts off spherical and smooth, and because of the higher burning velocity, spherically expands in all directions until the flame front reaches the back, top and bottom walls. The expanding burned gases continue to convectively "push" the flame front horizontally towards the open end as it consumes the entire reactant mixture, further contributing to the heat release and horizontal expansion.

The flame dynamics are much different for a stoichiometric mixture of ammonia/air, which has a burning velocity ≈ 7 cm/s. The flame sphere is driven upward and contacts the ceiling when the flame diameter is about 2 m. The flame then spreads along the ceiling and its movement is controlled by the buoyancy driven flow of the burned gas. Once the leading edge of the flame reaches the vent opening, the flame begins to slowly propagate downward as a very smooth and well-defined flame surface.

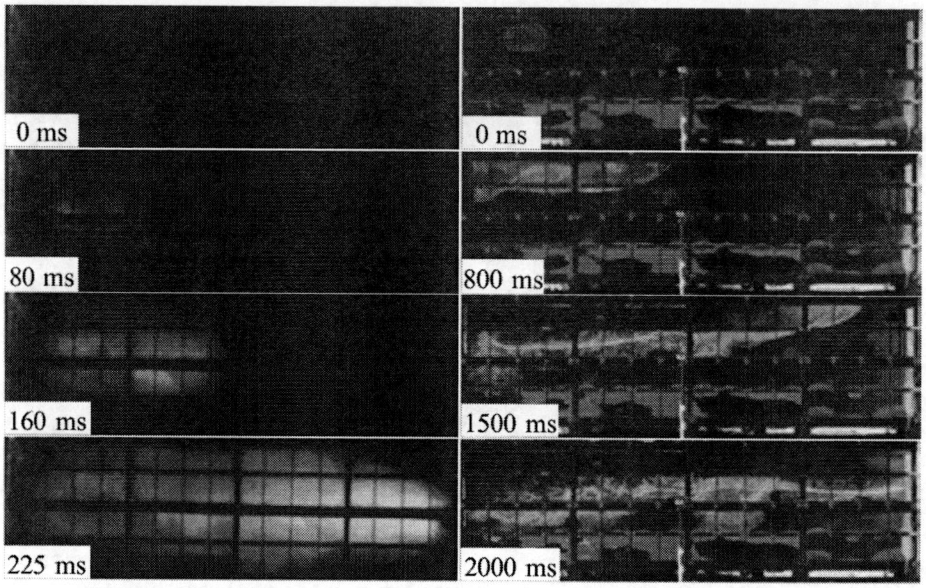

Figure 7: Flame dynamics of methane/air (left) and ammonia/air (right) in the 50 m^3 rig <u>without</u> added congestion.

Images of the methane/air (left frames) and ammonia/air (right frames) tests in the congested rig are shown in Figure 8. Even with added congestion in the rig, the ammonia/air flame starts off buoyancy driven. Notice that the ammonia flames in the first two images are similar to the flames observed during the test with no congestion. The flame rises and propagates along the ceiling until enough congestion induced turbulence is generated ahead of the flame front to sufficiently increase the flame speed. While the methane/air flames in the congested and empty tests are both convectively driven flows, the additional congestion creates turbulence and the flame front is much more wrinkled in the congested case as it propagates through the rig (compare images in Figure 7 and Figure 8).

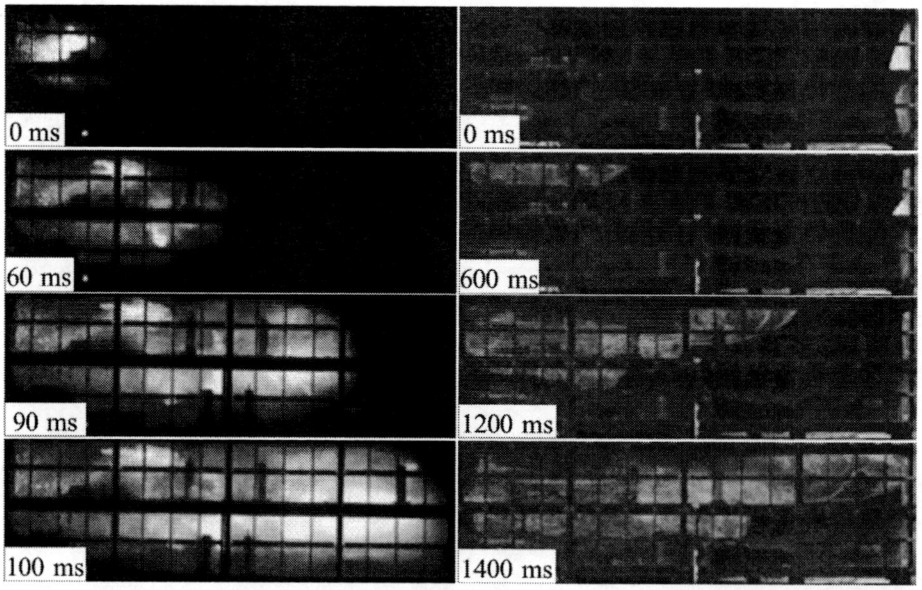

Figure 8: Flame dynamics of methane/air (left) and ammonia/air (right) in the 50 m^3 rig <u>with</u> added congestion.

When low burning velocity fuels are premixed in air, flame propagation is initially dominated by buoyancy and if no confinement is present, may result in negligible overpressures. However, if confinement and congestion are present the consequences from low burning velocity fuels can be orders of magnitude higher. Therefore, methods of accurately predicting the consequences of events involving these fuels must therefore not only take into account buoyancy effects during early flame propagation but also the flame acceleration effects in the event confinement and congestion are present. As shown in an upcoming section, preliminary comparisons of FLACS predictions with experimental results show that FLACS accurately predicts the flame dynamics of slow burning flames, and the explosion consequences can differ by orders of magnitude depending on the actual plant specifics.

4.2.2 Flame front propagation rates

To further illustrate the differences in flame propagation for the different fuels and for the different congestion configurations, the high speed videos were post-processed using in-house software to yield the flame propagation velocity for the leading edge of the flame fronts. The red lines in Figure 9 shows the height at which the flame front was tracked for the methane and ammonia tests with and without congestion (only empty configurations shown since heights do not vary). For methane the flame was tracked near mid-height and for ammonia the flame was tracked along the ceiling.

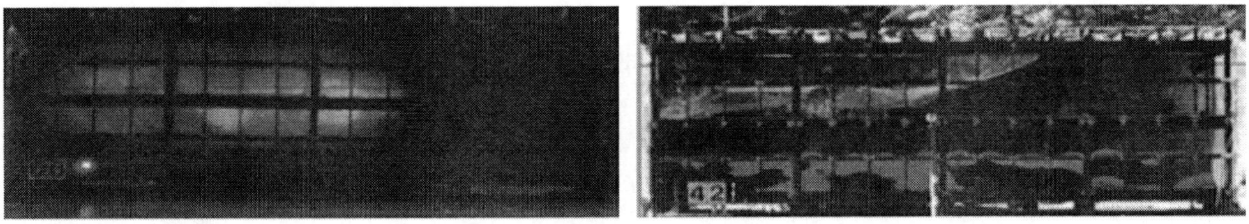

Figure 9: Flame front tracking location in the empty methane (left) and empty ammonia (right) tests shown by the red line.

Flame front tracking is accomplished by cropping a line of pixels in each frame at the height specified by the red line in Figure 9. The individual horizontal lines of pixels are then rotated 90° counterclockwise and stacked in sequential order. By specifying the distance of the red line, and the frame rate of the high-speed video, the combined lines form a position versus time plot which is then digitized and used to determine the flame front position as a function of time (x-t plot). To illustrate, Figure 10 shows the x-t plots for methane and ammonia in the empty test configuration. Notice the different x-axis scales in the plots, and how the ammonia figure shows that the initially rising flame reaches the tracking height around 500 ms before propagating both toward the closed and open ends of the rig.

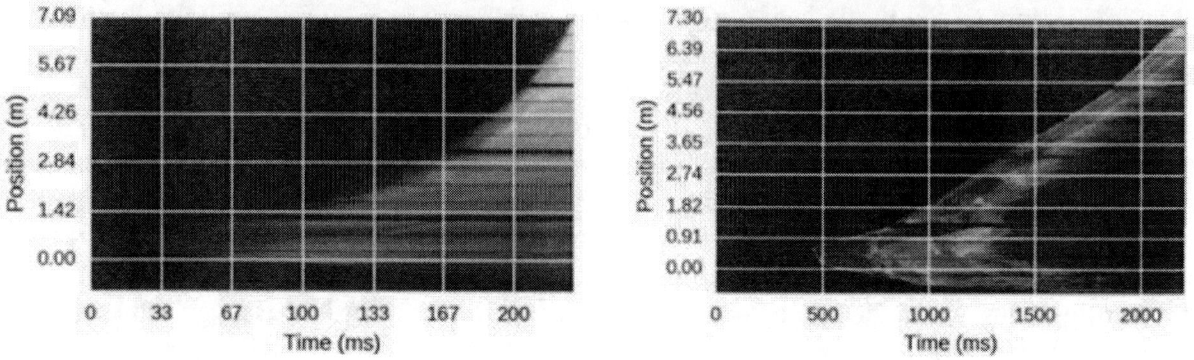

Figure 10: Flame position versus time for methane (left) and ammonia (right) in the empty test configuration.

The flame front position versus time data extracted from x-t plots (illustrated in Figure 10) is differentiated to yield the flame front velocity as a function of either position or time. Figure 11 shows the data versus position, with 0 m and 8 m on the x-axis corresponding to the closed and open ends of the rig. The methane flame speeds are on the order of 100 m/s when approaching the end of the rig whereas ammonia flame speeds are closer to 10 m/s. Congestion clearly causes faster flame acceleration, with velocities for the methane and ammonia tests about a factor of two higher in the congested configuration compared to the empty configuration.

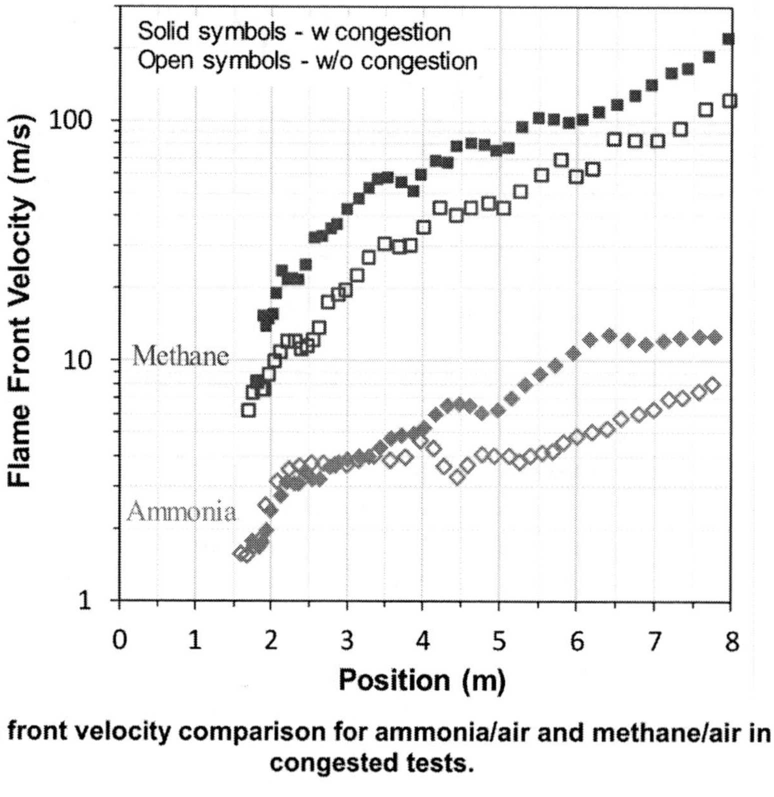

Figure 11: Flame front velocity comparison for ammonia/air and methane/air in the empty and congested tests.

4.2.3 Overpressures

A primary hazard of deflagrations is the overpressure they create. Figure 12 compares overpressure generated from methane and ammonia deflagrations in the empty and congested rig. In general, overpressures generated in the ammonia tests are 1-2 orders of magnitude less than overpressures generated in the methane tests. Peak overpressures within the rig were around 0.1 kPa and 1 kPa for the empty and congested test with ammonia, respectively (see Figure 13). Methane tests in the empty rig yielded peak internal pressures near 10 kPa whereas methane tests in the congested rig yielded peak pressures around 25 kPa. Decaying external overpressures were registered during the ammonia test in the congested rig which were again orders of magnitude lower than registered during the methane tests. No external overpressures were registered during the ammonia test in the empty rig, and were due to pressures waves either dissipating before reaching the first external sensor or they reached the sensors after the data acquisition system stopped recording (set to record for 3 s after ignition).

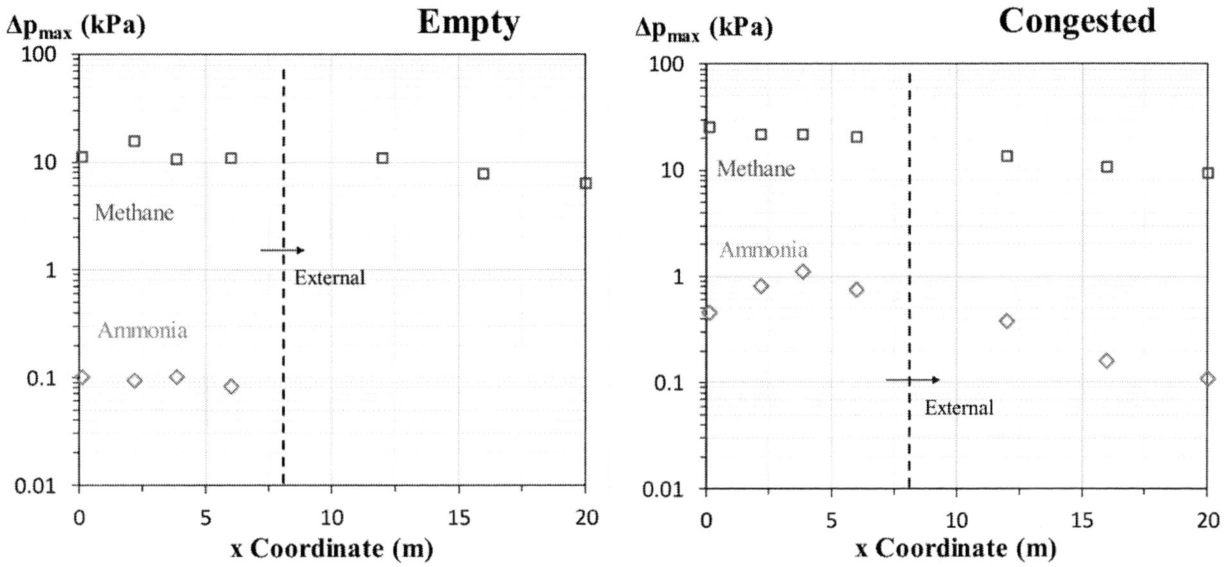

Figure 12: Deflagration overpressures measured within and outside the test rig during tests with methane and ammonia without (left image) and with (right image).

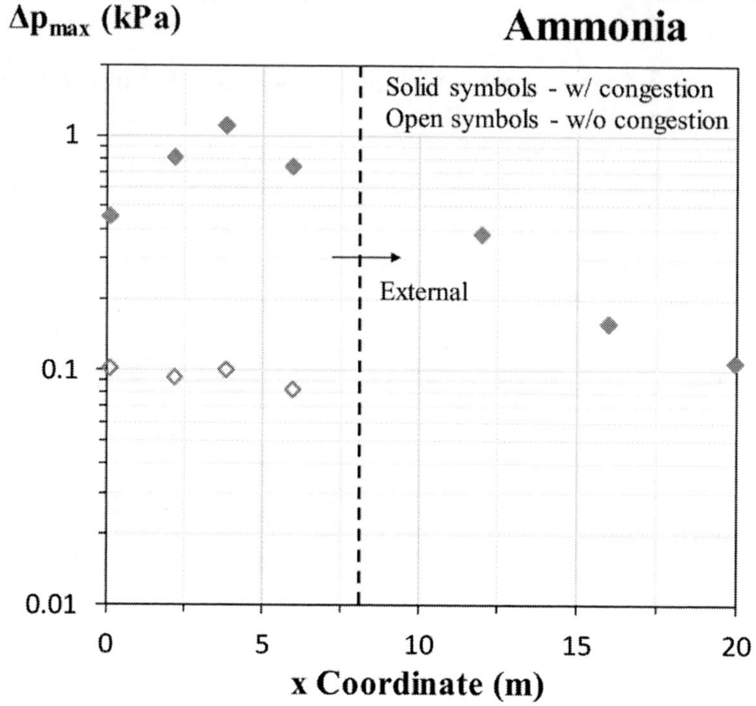

Figure 13: Deflagration overpressures within and outside the test rig during tests with ammonia with and without congestion

4.3 FLACS modeling

Ammonia tests in the 50 m^3 rig were modeled in FLACS and the predicted flame dynamics and overpressures were compared to experimental observations. It should be noted that methane has been extensively validated in FLACS and will not be discussed further [17]. Figure 14 compares the flame dynamics for the ammonia test in the empty rig. The FLACS images show gases at or above 2000 K (i.e. the burned gases), therefore the outer edges of the colored cloud are the flame front. In each row of images is the elapsed time from the first row of images. The evolution of the predicted ammonia flame is similar to what was observed during the experiment. Buoyancy drives the initial flame upward and then the flame expands across the ceiling of the rig. Once the flame front exits the rig, it begins to burn slowly downward throughout the rig. While the flame propagation is slightly faster in the FLACS simulation (through comparison of image time stamps), FLACS captures buoyancy aspects that become increasingly important for low burning velocity fuels.

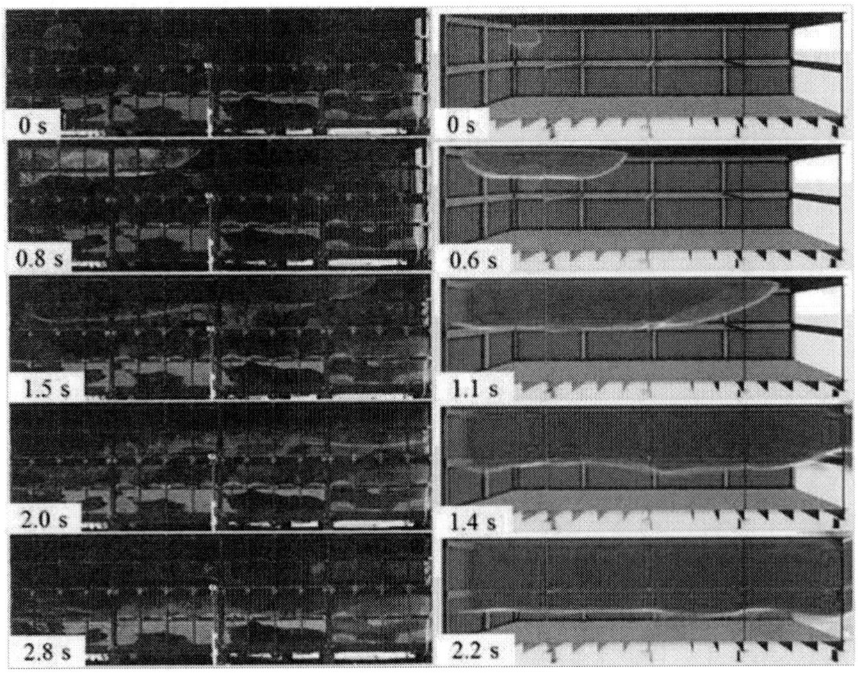

Figure 14: Comparison of flame dynamics observed in experiments and predicted in FLACS for ammonia tests in the <u>empty</u> rig.

Figure 15 compares the FLACS predictions to experimentally observed flame dynamics during the ammonia test in the congested rig. Agreement between the two is similar to what was observed for tests in the empty rig. Images of the experiment show that the flame is more wrinkled as a results of the interaction with the various obstacles in the rig. The flame is also more wrinkled in the FLACS simulation compared to the simulations results for the empty rig test. The time stamps on each image show that the flame in the FLACS simulation propagates slightly faster than the flame in the experiment.

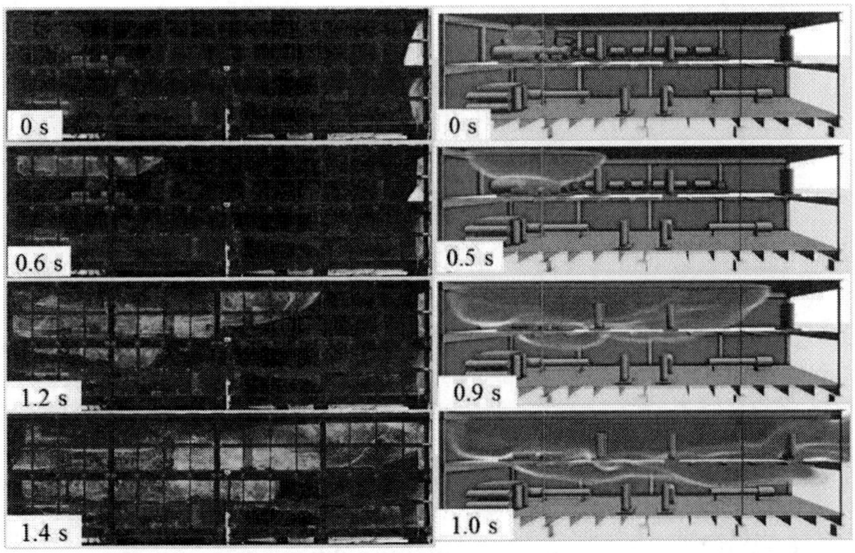

Figure 15: Comparison of flame dynamics observed in experiments and predicted in FLACS for ammonia tests in the <u>congested</u> rig.

Figure 16 presents the predicted overpressures in both the uncongested and congested rig configuration as compared to the experimentally determined values. FLACS predictions were generally good compared to the experimental values, with overpressures being slightly over-predicted in the empty configuration and under-predicted in the congested configuration. As FLACS accurately captures both the flame dynamics (e.g., initial buoyancy dominated flame propagation followed by flame accelerations due to the confinement/congestion) as well as the observed overpressures, the consequences associated with an unintended release of ammonia were evaluated in an actual facility setting.

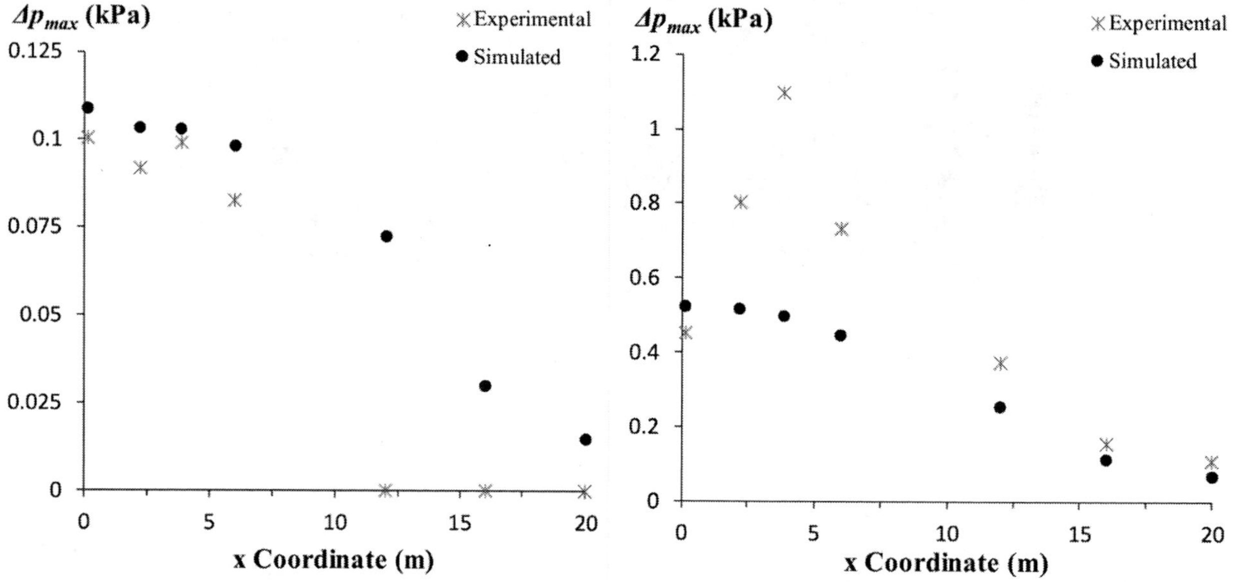

Figure 16: FLACS overpressure predictions compared to experimental results in the empty (left image) and congested (right image) test configuration.

4.4 Practical considerations

Practically speaking, the severity of an ammonia deflagration resulting from unintentional release in a typical process facility could range from negligible overpressures of 0.1 kPa to those orders of magnitude higher. The consequences will be directly related to any confinement present in the area of the vapor cloud and the detailed layout of the facility. Therefore, when evaluating the explosion consequences of an unintentional release of a low burning velocity fuel such as ammonia, it would not be prudent to assume that there is no risk from such compounds. Conversely, it may be overly conservative to assume the explosion risks of such compounds are close to those of methane. As such, it is necessary to have a tool that can accurately predict both the slow, buoyancy dominated flame propagation as well as the faster flame acceleration caused by confined and congested facilities.

To further illustrate the differences in explosion consequences of ammonia in both a relatively open, yet congested environment as compared to a partially confined and congested layout, we ran FLACS explosion simulations with large ammonia vapor clouds in a generic process facility. Frames (a), (b), and (c) of Figure 17 show the unconfined scenario, the confined scenario, and the ammonia/air vapor cloud size and location, respectively. Simulations with ammonia were performed with and without an upper deck which provided some level of confinement compared to the case without the upper deck to demonstrate the vastly different explosion consequences that can be expected for the different configurations. In addition, methane vapor clouds were also ignited to demonstrate that assuming methane explosion consequences for ammonia may be overly conservative.

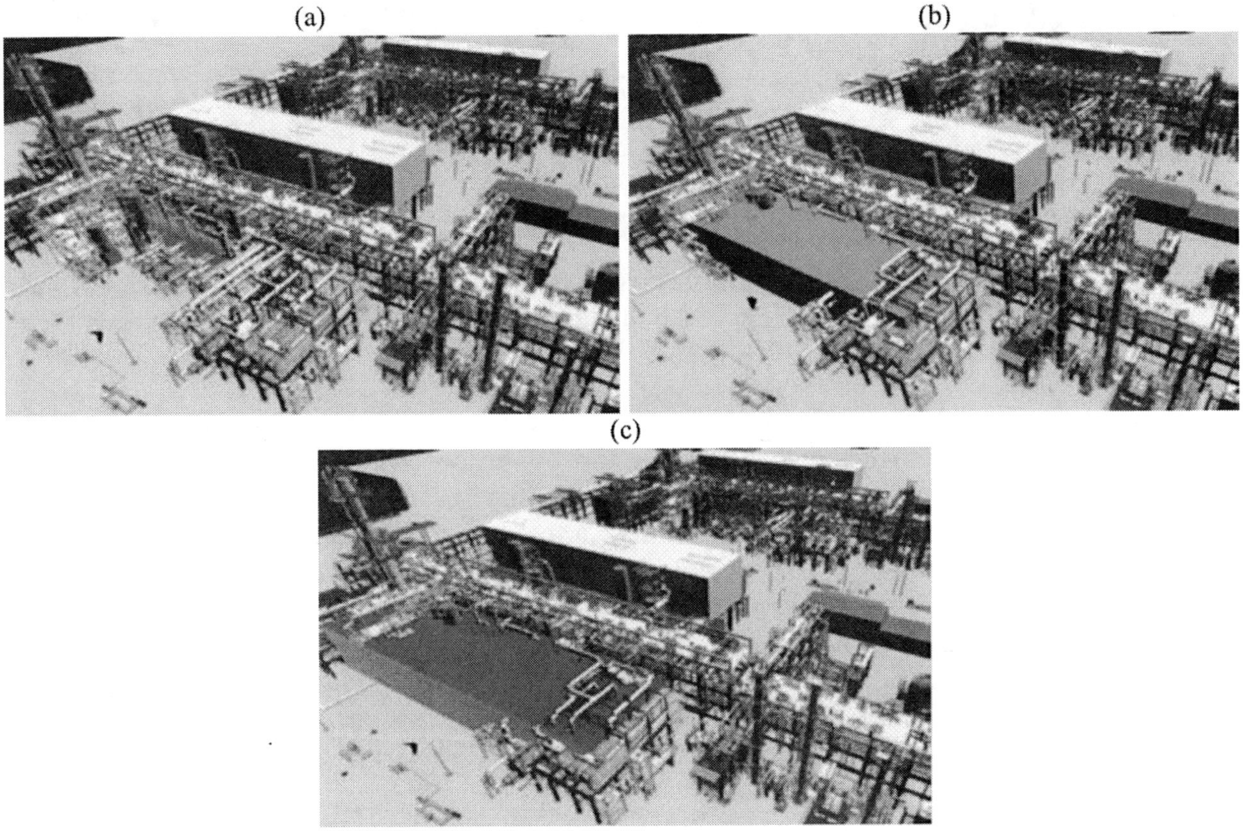

Figure 17: FLACS simulations of ammonia/air and methane air in a generic process facility.

Frames (a), (b), and (c) of Figure 17 show the unconfined scenario, the confined scenario, and the vapor cloud size and location, respectively. Ignition of the ammonia/air cloud in the unconfined scenario resulted in overpressures on the order of 0.01 barg. For the scenario with added confinement, peak overpressures were two orders of magnitude higher and close to 0.3 barg. Figure 18 compares the resulting peak overpressures in FLACS for ammonia/air (left frame) compared to methane/air (right frame). The results show that methane/air vapor cloud explosions have a much higher severity as compared to ammonia/air vapor cloud explosions.

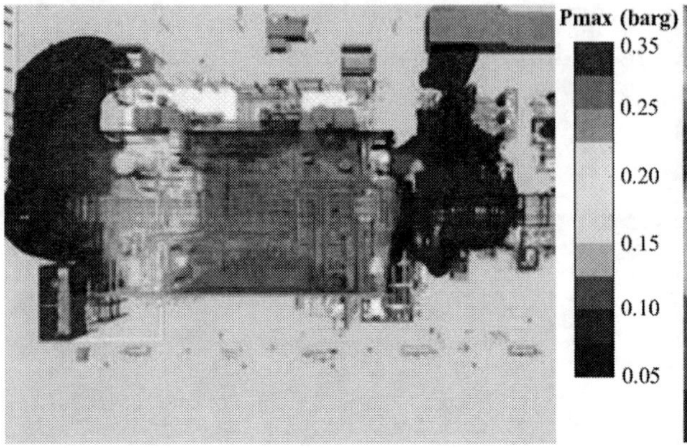

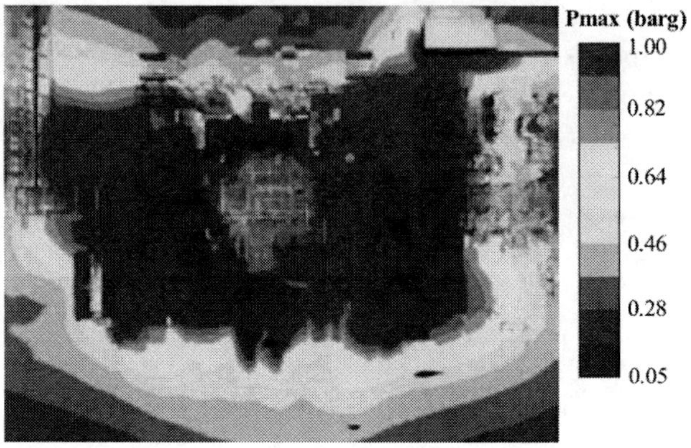

Figure 18: Resulting peak overpressures in FLACS for ammonia/air (left) and methane/air (right) vapor cloud explosions.

5 Conclusions

The present study examined the flammability and explosion characteristics of ammonia as a representative low burning velocity compound and compared the results to those of methane. Small-scale tests provided flammability limits, auto-ignition temperature, minimum ignition energy, peak constant-volume pressure rise, deflagration index, and laminar burning velocity as a function of equivalence ratio. These properties provided the basis for qualitatively evaluating risk for high- and low burning velocity compounds by comparing the likelihood of ignition and the relative differences in associated consequences. Large-scale tests in a 50 m^3 rig quantified the difference in severity associated with ignition of premixed ammonia/air and methane/air. The ammonia tests in the 50 m^3 rig were modeled with FLACS and the results were compared to experimental observations and measurements.

Fundamental combustion properties determined through lab-scale testing were generally in agreement with previously published data, with the exception of MIE, for which large scatter in the data already existed. The LEL, AIT, and MIE of ammonia show that it has a lower likelihood of forming a flammable cloud and of igniting if a flammable cloud forms.

In the 50 m^3 tests, flame front propagation speeds were much lower for ammonia compared to methane. The flame dynamics were also much different, with ammonia flame propagation being buoyancy driven due to the lower laminar burning velocity. In the empty and congested tests, deflagration overpressures were orders of magnitude lower for ammonia/air compared to methane/air. This was consistent with the lower measured values for S_u, K_g, and P_{max}.

FLACS simulations captured the buoyancy dominated flame dynamics for ammonia/air and also predicted overpressures close to what was observed in the experiments. Additional simulations of ammonia clouds in a generic process facility showed that, under certain confined conditions, ammonia/air vapor cloud explosions can create appreciable overpressures. The results also demonstrated that assuming methane explosion consequences for ammonia will be overly conservative.

The results of the present study show that consequences of low burning velocity fuels, such as ammonia, are much lower than methane, but not negligible. Risk assessments must be based on this concept and should incorporate advanced tools to ensure risks are fully apparent and properly mitigated when designing facilities.

6 Acknowledgements

The authors gratefully acknowledge the financial support from Daikin Industries, LTD.

7 References

[1] EN 1839 (2003). *Determination of Explosion Limits of Gases and Vapours.* European Committee for Standardisation, Brussels, Belgium.
[2] EN 14522 (2005. *Determination of the Auto Ignition Temperature of Gases and Vapours.* European Committee for Standardisation, Brussels, Belgium.
[3] ASTM E582 (2007). *Standard Test Method for Minimum Ignition Energy and Quenching Distance in Gaseous Mixtures.*
[4] I. Glassman, Combustion, Academic Press, 1997.
[5] K.L. Cashdollar and M. Hertzberg, 20-l explosibility test chamber for dusts and gases, Review of Scientific Instruments 56 (1985) 596-602.
[6] K.L. Cashdollar, Coal dust explosibility, Journal of Loss Prevention in the Process Industries 9 (1996) 65-76.
[7] K.L. Cashdollar, I.A. Zlochower, G.M. Green, R.A. Thomas and M. Hertzberg, Flammability of methane, propane, and hydrogen gases, Journal of Loss Prevention in the Process Industries 13 (2000) 327-340.
[8] B. Lewis and G. von Elbe, Combustion, Flames, and Explosions of Gases, 1961.
[9] J.L. Pagliaro, G.T. Linteris, P.B. Sunderland and P.T. Baker, Combustion inhibition and enhancement of premixed methane–air flames by halon replacements, Combust. Flame 162 (2015) 41-49.
[10] S.G. Davis, J.L. Pagliaro, T.F. Debold, M. van Wingerden and K. van Wingerden, Flammability and explosion characteristics of marginally flammable refrigerants, in: 11th International Symposium on Hazards, Prevention, and Mitigation of Industrial Explosions, Dalian, China, 2016,
[11] M. Metghalchi and J.C. Keck, Laminar Burning Velocity of Propane-Air Mixtures at High Temperature and Pressure, Combust. Flame 38 (1980) 143-154.
[12] T. Iijima and T. Takeno, Effects of Temperature and Pressure on Burning Velocity, Combust. Flame 65 (1986) 35-43.
[13] K. Saeed and C.R. Stone, Measurements of the laminar burning velocity for mixtures of methanol and air from a constant-volume vessel using a multizone model, Combust. Flame 139 (2004) 152-166.
[14] K. Takizawa, A. Takahashi, K. Tokuhashi, S. Kondo and A. Sekiya, Burning velocity measurement of fluorinated compounds by the spherical-vessel method, Combust. Flame 141 (2005) 298-307.
[15] P.G. Hill and J. Hung, Laminar Burning Velocities of Stoichiometric Mixtures of Methane with Propane and Ethane Additives, Combust. Sci. Technol. 60 (1988) 7-30.

[16] K. Eisazadeh-Far, A. Moghaddas, F. Rahim and H. Metghalchi, Burning Speed and Entropy Production Calculation of a Transient Expanding Spherical Laminar Flame Using a Thermodynamic Model, Entropy 12 (2010) 2485-2496.
[17] O.R. Hansen, P. Hinze, D. Engel and S. Davis, Using computational fluid dynamics (CFD) for blast wave predictions, Journal of Loss Prevention in the Process Industries 23 (2010) 885-906.
[18] B.H. Hjertager, Computer simulation of turbulent reactive gas dynamics, Modeling, Identification and Control 5 (1984) 211.
[19] F.H. Harlow and P.I. Nakayama, Turbulence transport equations, Physics of Fluids (1958-1988) 10 (1967) 2323-2332.
[20] R. Abdel-Gayed, D. Bradley and M. Lawes, Turbulent burning velocities: a general correlation in terms of straining rates, Proceedings of the Royal Society of London A: Mathematical, Physical and Engineering Sciences 414 (1987) 389-413.
[21] K.N.C. Bray, Studies of the turbulent burning velocity, Proceedings of the Royal Society of London A: Mathematical, Physical and Engineering Sciences 431 (1990) 315-335.
[22] D. Bjerketvedt, J.R. Bakke and K. Van Wingerden, Gas explosion handbook, Journal of hazardous materials 52 (1997) 1-150.
[23] B. Minor and M. Spatz, HFO-1234yf low GWP refrigerant update. *International Refrigeration and Air Conditioning Conference*, 2008; Prudue University, Paper 937.
[24] S. Kondo, Y. Urano, K. Takizawa, A. Takahashi, K. Tokuhashi and A. Sekiya, Flammability limits of multi-fluorinated compounds, Fire Safety Journal 41 (2006) 46-56.
[25] P.D. Ronney and H.Y. Wachman, Effect of Gravity on Laminar Premixed Gas Combustion I - Flammability Limits and Burning Velocities, Combust. Flame 62 (1985).
[26] F. Verkamp, M. Hardin and J Williams, Ammonia combustion properties and performance in gas-turbine burners. *Symposium (International) on Combustion*, 1967; Elsevier, 985-992.
[27] K. Takizawa, N. Igarashi, S. Takagi, K. Tokuhashi and S. Kondo, Quenching distance measurement of highly to mildly flammable compounds, Fire Safety Journal 71 (2015) 58-68.
[28] K. Eisazadeh-Far, F. Parsinejad, H. Metghalchi and J.C. Keck, On flame kernel formation and propagation in premixed gases, Combust. Flame 157 (2010) 2211-2221.
[29] K. Takizawa, A. Takahashi, K. Tokuhashi, S. Kondo and A. Sekiya, Burning velocity measurements of nitrogen-containing compounds, Journal of hazardous materials 155 (2008) 144-52.
[30] T. Jabbour, Flammable refrigerant classification based on the burning velocity. Ecoles de Mines de Paris, 2004.
[31] P.D. Ronney, Effect of chemistry and transport properties on near-limit flames at microgravity, Combust. Sci. and Technol. 59 (1988) 123-141.
[32] U.J. Pfahl, M.C. Ross and J.E. Shephard, Flammability Limits, Ignition Energy, and Flame Speeds in H2-CH4-NH3-N2O-O2-N2 Mixtures, Combust. Flame 123 (2000) 140-158.

Hydrogen Jet Vapor Cloud Explosion: A Model for Predicting Blast Size and Application to Risk Assessment

Simon Jallais
Air Liquide R&D
1 Chemin de la Porte des Loges - Les Loges-en-Josas 78354 JOUY-EN JOSAS
Simon.Jallais@AirLiquide.com

Elena Vyazmina
Air Liquide R&D

Derek Miller
Air Products

J. Kelly Thomas
Baker Engineering and Risk Consultants (BakerRisk)

Prepared for Presentation at
American Institute of Chemical Engineers
2017 Spring Meeting and 13th Global Congress on Process Safety
San Antonio, TX
March 26 – 29, 2017

Hydrogen Jet Vapor Cloud Explosion: A Model for Predicting Blast Size and Application to Risk Assessment

Simon Jallais
Air Liquide R&D
1 Chemin de la Porte des Loges - Les Loges-en-Josas 78354 JOUY-EN JOSAS
Simon.Jallais@AirLiquide.com

Elena Vyazmina
Air Liquide R&D

Derek Miller
Air Products

J. Kelly Thomas
Baker Engineering and Risk Consultants (BakerRisk)

Keywords: jet explosions, TNO ME index, hydrogen, CFD modeling, blast size prediction, risk assessment

Abstract

Releases of hydrogen at elevated pressures form turbulent jets which may pose vapor cloud explosion (VCE) as well as jet fire hazards. The turbulence induced by the jet release can lead to flame speeds sufficient to produce damaging blast loads if the release is not immediately ignited, even in the absence of confinement or congestion. The VCE hazard posed by such high pressure hydrogen releases is not well-recognized.
The authors have previously presented test data and computational fluid dynamic (CFD) modeling analyses which characterize high pressure hydrogen releases and the associated VCE hazard. The current paper provides an overview of this VCE hazard, presents a simplified method for predicting the potential resultant blast loads, and discusses the implications of such events for building siting analyses and risk assessments.

1 Introduction

Releases of hydrogen at elevated pressures form turbulent jets which may pose vapor cloud explosion (VCE) as well as jet fire hazards. The turbulence induced by the jet release can lead to flame speeds sufficient to produce damaging VCE blast loads if the release is not immediately ignited, even in the absence of confinement or congestion. The ignition of hydrogen released as a jet has resulted in accidental VCEs [1]. Hydrogen jet release VCE scenarios may therefore be relevant to consequence assessment studies, particularly in the absence of alternative release scenarios involving large congested volumes. However, the VCE hazard posed by high pressure hydrogen releases is not well-recognized, and there is no well-established methodology to predict the associated VCE blast loads.

A number of moderate to large hydrogen jet release VCE tests have been performed, including those reported by Miller et al. [2], Daubech et al. [3], Grune et al. [4], Willoughby and Royle [5], Takeno et al. [6] and Chaineaux [7]. The blast load data from these tests have been compared to the results obtained by computational fluid dynamics (CFD) modeling using the FLACS code [8], and reasonable representations for the peak pressure have been obtained using tailored approaches, although these models have not been able to reproduce the blast wave shape or duration [2, 9].

This paper describes an alternative approach for predicting the blast loads resulting from ignition of hydrogen jet releases. The mass of hydrogen in the concentration range which would contribute to the VCE is calculated using a simplified engineering analysis model. The VCE blast load strength is correlated to the jet release rate using simplified blast curve methods, such as the TNO Multi-Energy or Baker-Strehlow-Tang (BST) VCE blast load prediction methods [10]. This approach is validated by comparison against available hydrogen jet release VCE tests.

2 Description of experiments

2.1 INERIS experiment

INERIS together with Air Liquide, and AREVA Energy Storage in the frame of a Joint Industrial Program EXJET performed medium scale experiments on delayed explosions of high pressure hydrogen jets in a free field environment in 2013. Tests were performed at the INERIS Montlaville experimental site (abandoned rock quarry). For a detailed description of the test setup and results see Daubech et al [3].

2.1.1 Dispersion Tests

The hydrogen was released through an orifice of 12 mm diameter in horizontal direction at an elevation of 1.48 meters above the ground into a flat open area. Temperature and pressure were also measured at the release point.

The H_2 concentrations in the cloud, radial and stream wise velocities were measured at different axial positions (1.25, 2, 3, 4.5, 7.5 and 10 m from the release point). Concentrations in the cloud were measured using paramagnetic O_2 analyzers. Velocities were measured using Pitot probes (also called McCaffrey probe). These probes are equipped with rapid pressure sensors allowing the measurement of turbulence intensity and also the integral lengths.

The mass flow considered for simulations was 250g/s in Daubech et al [3]. Experiments were performed in an abandoned rock quarry, where there is no perceptible wind.

2.1.2 Explosion Tests

For explosion experiments, aerial overpressures were measured using 3 high speed piezo-resistive sensors placed in a lens allowing measurements of the incident overpressure. The sensor L_1 was placed at 20 cm from the release point. The sensors L_2 located on the axis of the jet at 2 m downstream from the igniter and L_3 was perpendicularly to the jet axis at 2.5m from the igniter.

A high speed camera was also used to monitor the flame propagation. It was placed at 10 m from the setup. The image was centered on the igniter allowing visualization of 3 m around.

For all the experiments, the ignition occurs 5 seconds after the start of the release. All the experiments in the same conditions were duplicated.

Daubech et al [3] used a powerful ignition (8500 J) produced by the explosion of a stoichiometric H_2/O_2 mixture in a tube of 5.5 cm diameter and of 50 cm length. The extremity of the tube was placed 20 cm below the release axis to obtain the ignition strictly on the release centerline. The ignition position was chosen at the location corresponding to a 30% concentration of H_2 on the centerline (1.8m downstream of the release point).

2.2 DNV GL experiment

Air Products commissioned DNV GL (formerly Advantica Ltd.) to perform a series hydrogen jet release tests in 2008, targeted primarily at measuring the thermal flux from ignited open air hydrogen jets. The tests were performed at the DNVGL test facility in Spadeadam, UK. For a detailed description of the test setup and results see Miller et al [2].

2.2.1 Test Description

The hydrogen jets were created by depressurizing a high-pressure hydrogen reservoir via a control manifold that recorded supply flows, temperatures, and pressures. The hydrogen was released through a horizontal vent pipe at an elevation of 3.25 meters into a flat open area. The hydrogen employed was greater than 99.9% pure.

Pipe diameters of ¾-inch and 2-inches were tested at a release pressure of approximately 60 barg, giving initial flow rates of roughly 1 and 8 kg/s, respectively. For both release sizes, a single test was performed where the ignition system was not activated until approximately 2 seconds after the release had been initiated (±0.5 seconds), such that the resulting flame propagated into a turbulent flammable gas cloud of limited volume. The ignition system was an incendiary device positioned close to the hydrogen release point. The release conditions for these two tests are summarized in Table 1. The wind was oriented in approximately the same direction as the release for both tests.

Table1. Summary of Test Conditions

Parameter	¾-inch Test	2-inch Test
Pipe Inner Diameter (mm)	20.9	52.5
Air Temperature (°C)	6.3	6.8
Wind speed (m/s)	3.4	2.6
Initial Release Rate (kg/s)	1	8

Blast loads were measured by pressure transducers located at 10 meters and 20 meters off the jet centerline, at axial distances along the jet centerline of 11 meters (¾-inch test) and 23 meters (2-inch test).

2.2.2 Test Results

The measured blast overpressure histories are shown in Figure 1 (¾-inch) and Figure 2 (2-inch); the pressure histories shown in these plots have been smoothed by time averaging. The time axes in these figures are not synchronized with ignition of the flammable gas cloud. The estimated cloud length prior to ignition, measured peak blast overpressures, and estimated blast wave velocity (Mach number) are summarized in Table 1. The 2-inch test yielded significant overpressures (e.g., approaching ½ bar at 10 meters from the centerline), while the ¾-inch test gave much lower overpressures. The blast wave from the 2-inch test has very little rise time and appears to be a shock wave, whereas that from the ¾-inch test has a pronounced rise time and would be classified as a pressure wave.

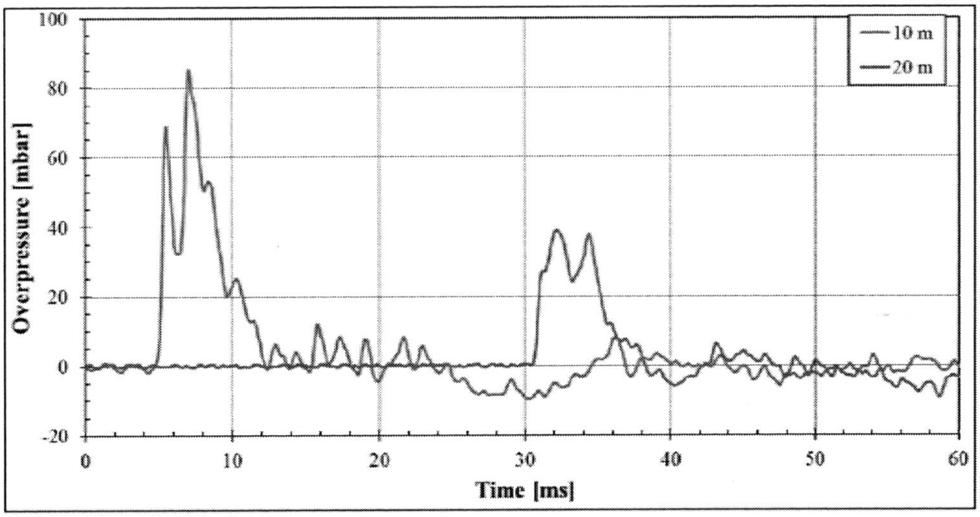

Figure 1. Blast Pressure History for ¾-Inch Release Test

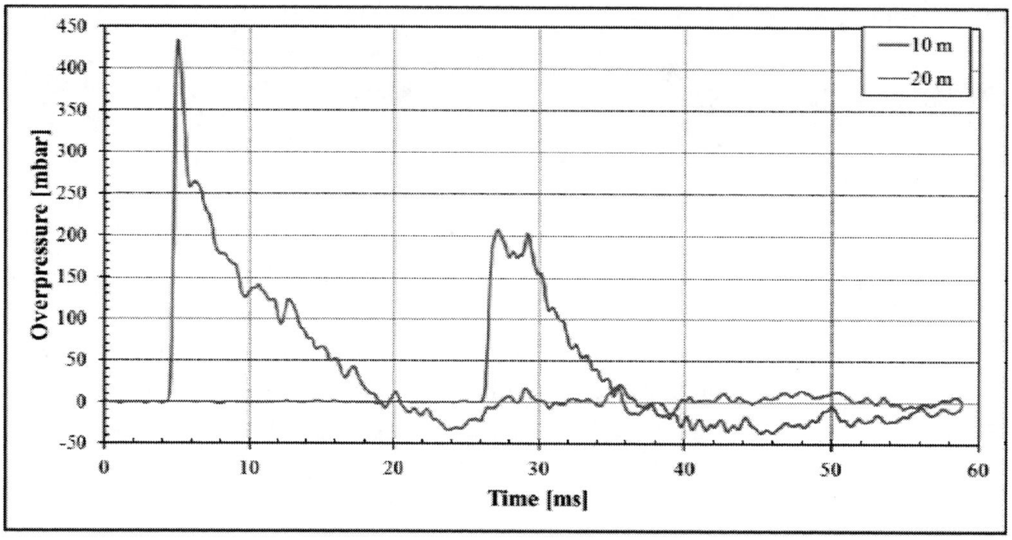

Figure 2. Blast Pressure History for 2-Inch Release Test

Table 1. Summary of Test Results

Parameter	¾-inch Test	2-inch Test
Unignited cloud length (m)	20	35
Peak overpressure @ 10 m (bar)	0.085	0.43
Peak overpressure @ 20 m (bar)	0.039	0.21
Blast wave velocity (Mach number)	1.0	1.4

3 Modelling

3.1 CFD model description

For CFD modeling a commercial code FLACS v10.4 is used. FLACS is developed by GexCon to simulate gas explosions in offshore oil and gas platforms. It solves the compressible Navier-Stokes equations on a 3-D Cartesian grid using a finite volume method.

FLACS uses RANS (Reynolds-Averaged Navier-Stokes) approach for fluid mechanics and therefore a turbulence model is required. For turbulence FLACS uses the k-eps model Harlow & Nakayama [11]. The SIMPLE pressure correction algorithm is applied [12]. There are additional source terms for the compression work in the enthalpy equation for correct modeling of compressible flows. Basic equations of conservation used in the FLACS model can be found in Hjertager [13, 14] and Hjertager et al [15].

A second-order scheme is used to solve for diffusive fluxes and a second-order ''kappa'' scheme (hybrid scheme with weighting between 2nd order upwind and 2nd order central differences, with limiters) is applied for the convective fluxes [16].

The flame in FLACS combustion model is considered as a collection of flamelets with one-step kinetic reaction. The laminar burning velocity is pre-defined in tables from experimental measurements. For turbulent burning velocity Bray's expression [17] is used. The reaction zone in a premixed flame is thin compared to grid resolutions in FLACS. Hence numerically the flame zone is thickened by increasing the diffusion by a factor and simultaneously reducing the reaction rate by a factor $1/\beta$. β is chosen such that the flame thickness becomes 3-5 grid cells.

For modeling presented in the paper the following strategy is used:

- at the first step, dispersion unsteady simulations are performed to obtain flammable steady cloud
- at the second step, the cloud is used to initialize explosion simulations.

3.1.1 Equivalent source

Since the target is to model the jet dispersion and explosion in the region located more than 10 diameters downstream of the orifice, for the modelling of releases from the high pressure reservoirs an equivalent source for jets can be used. An equivalent (in terms of mass flow) source has sub-sonic velocities and density at ambient pressure. For this purpose a jet model program integrated into FLACS (starting from v9.1 and later) is used, this model is based on the isentropic expansion from the reservoir conditions to the jet exit; see Birch et al. [18] and FLACS user's guide [16].

Equivalent sources conditions for experiments are given in the table below.

Table 3. Description of the equivalent sources.

	Variable	INERIS	DNV GL, ¾-inch	DNV GL, 2-inch
	Mass flow (kg/s)	0.25	1.04	8.18
Real source conditions	D (mm)	12	20.9	52.5
	P (barg)	36	60	70
Equivalent source	Eq D (mm)	70.55	147.6	413.84
	Eq T (°C)	6.5	4.7	6.6

For turbulence length scale 10% of the original source diameter is used. The release velocity is approximately of 700 m/s in all cases.

3.1.2 Boundary Conditions

The equivalent jet source is located at the same height as the original orifice: 1.5m and 3.25m from the ground for Daubech experiment [3] and Miller experiment [2] correspondingly. The release is done in the streamwise direction. The wind profile or other weather conditions are not taken into account in simulations. The output parameters from the Jet equivalent source are used as the conditions for jet dispersion. The open boundary conditions ("pressure wave", see [16] are imposed on lateral, upstream, downstream and upward boundaries, no-slip (ground effect) are imposed at the ground boundary. The boundary conditions for the explosion simulations remain the same as in the dispersion simulations.

Solution independence on the size of simulation domain is verified for each experiment. The typical sizes of simulation domain are presented in the table below:

Table 4. Size of computational domain.

Size of the domain	INERIS	DNV GL, ¾-inch	DNV GL, 2-inch
stream-wise direction X	[-10; 70]	[-10; 90]	[-10; 90]
cross-stream direction Y	[-15; 15]	[-40; 40]	[-40; 40]
vertical direction Z	[0; 16.5]	[0; 40]	[0; 40]
Position of the release	X=0, Y=0, Z=1.5	X=0, Y=0, Z=3.25	X=0, Y=0, Z=3.25

3.1.3 Grid Choice for dispersion simulations

There are special recommendations to set up dispersion simulations in FLACS [16] including a grid refinement close to the release point in both directions perpendicular to the release directions (cross-stream and vertical). The area of the expanded jet (at ambient pressure) must be resolved using one grid cell $A_{leak} < A_{cv} < 1.25\ A_{leak}$ for hydrogen. Here A_{cv} is the area of the control volume (the grid-cell) and A_{leak} is the area of the leak. These guidelines are followed for all simulation presented in the paper

The solution independence of spatial resolution is verified for each case using at least two meshes, see figure 4, where 3 different spatial resolutions are compared.

The comparison of simulation results obtained with four different grids showed a good match. The grids used for modeling of experiments of Miller are described in the Table 5. The detailed description of mesh for INERIS experiment can be found in Daubech et al [3]. For other cases under consideration, the same guidelines are followed and solution independence on spatial resolution was also verified, see [3] and [9].

Table 5. Description of the meshes used for dispersion simulations.

Experiment		DNV GL, ¾-inch		DNV GL, 2-inch	
X direction, downstream direction	dx_{min}(m)	0.13	0.15	0.37	0.41
	dx_{max}(m)	0.8	0.6	0.37	0.42
	size (m)	[-4.7; 44.2]	[-4.7; 44.7]	[-10; 90]	[-10; 90]
	number of cells	294	294	273	244
Y direction, cross stream direction	dy_{min}(m)	0.13	0.15	0.37	0.41
	dy_{max}(m)	0.2	0.23	0.37	0.41
	size (m)	[-22; 22]	[-22; 22]	[-40; 40]	[-40; 40]
	number of cells	331	297	217	195
Z direction, vertical direction	dz_{min}(m)	0.13	0.15	0.37	0.41
	dz_{max}(m)	0.69	0.63	0.38	0.44
	size (m)	[0; 22.8]	[0; 22.4]	[0; 40]	[0; 40]
	number of cells	119	113	108	97
Total number of cells		11 580 366	9 866 934	6 398 028	4 615 260

3.1.4 Grid Choice for explosion simulations

The region corresponding to the position of the pressure detectors in the experimental tests is of special interest. To correctly compute pressure wave propagation a homogeneous cubic grid must be used in these regions. Since the leakage is not stopped before the ignition, the control volume should be a cell with $A_{leak} < A_{cv} < 1.25\ A_{leak}$. Table 6 gives a brief description of meshes used for modeling of the experimental setup, the cross-comparison of simulation results showed that the minimum possible spatial resolution must be used. However even the coarse resolution gives acceptable results.

For other cases under consideration, the same guidelines are followed and solution independence of spatial resolution was also verified, results, showed close agreement with an error of up to 5% between two grids.

Table 6: Description of the meshes used for explosion simulations.

Experiment	INERIS	DNV GL, ¾-inch		DNV GL, 2-inch	
Grid size, cm	8.5	13	15	37	41
Total number of cells	5 240 862	11 580 366	9 866 934	6 211 350	4 615 260

3.2 *Simple Engineering model*

Houf et al. [19] based on the previous work of Birch et al. [18] propose a jet dispersion engineering model.

This simple model allows user to calculate:

- gas velocity and mass release at the outlet nozzle,
- distances to flammability limits: LFL and UFL (Lower and Upper Flammability Limit),
- maximum flammable volume and flammable mass.

The mass flow rate through the exit orifice can be calculated assuming isentropic flow and an infinitely-large reservoir where stagnation conditions do not vary with time:

$$m = C_d . \pi d_0^2 \sqrt{P_0 \rho_0 \gamma \left(\frac{2}{\gamma+1}\right)^{\frac{\gamma+1}{\gamma-1}}}$$

Where d_0, P_0 and ρ_0 correspond respectively to the diameter, pressure and density upstream of the leak, C_d is the discharge coefficient, γ is the specific heat ratio (γ=1.4).

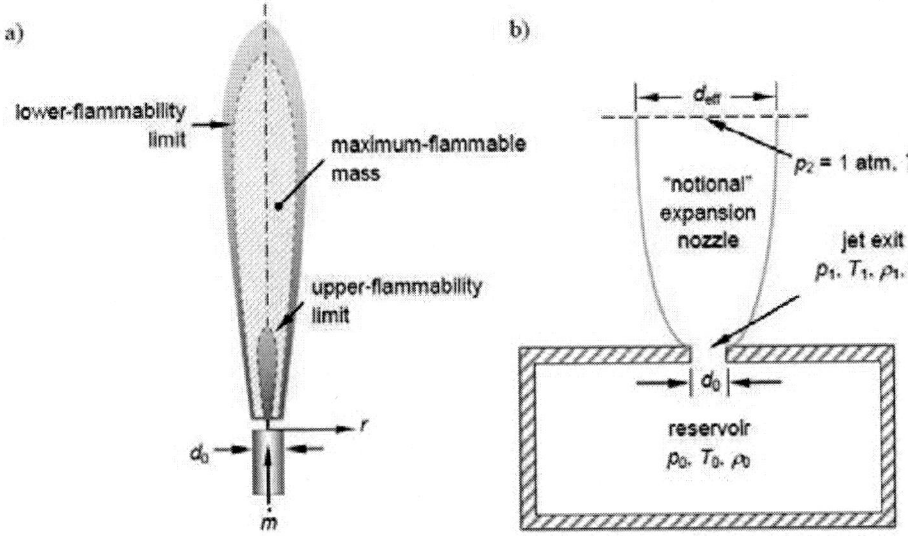

Figure 3. Characteristics of a flammable gas cloud: a) lower and upper flammability limits in a jet from a choked orifice; b) expansion of a supercritical release

Downstream of the choked orifice, an under-expanded jet is formed as hydrogen rapidly expands to ambient pressure via a series of oblique-shock waves. Concentration gradients in both axial and radial directions are generated within the jet as the entrainment and turbulent mixing occurs with the ambient air. In order to bypass the complex region of wave interactions, an equivalent isentropic expansion through a "notional" nozzle with an "effective" diameter d_{eff} can be defined follow Birch [18] (see Fig. 3). Classical relations for concentration decay in a turbulent jet can be used for a supercritical-under-expanded jet and the flammable mass of hydrogen can be calculated once the concentration distribution is known (in both the axial and radial directions), see [19] for calculation details.

4 Results

4.1 Validation of CFD and engineering model for dispersion

Experimental results of INERIS experiment (Daubech et al.[3]) for dispersion are compared with FLACS simulation. As shown in Fig. 4 experiments and FLACS v10.4 simulations are in quite good agreement for the hydrogen concentration decay vs. the distance:

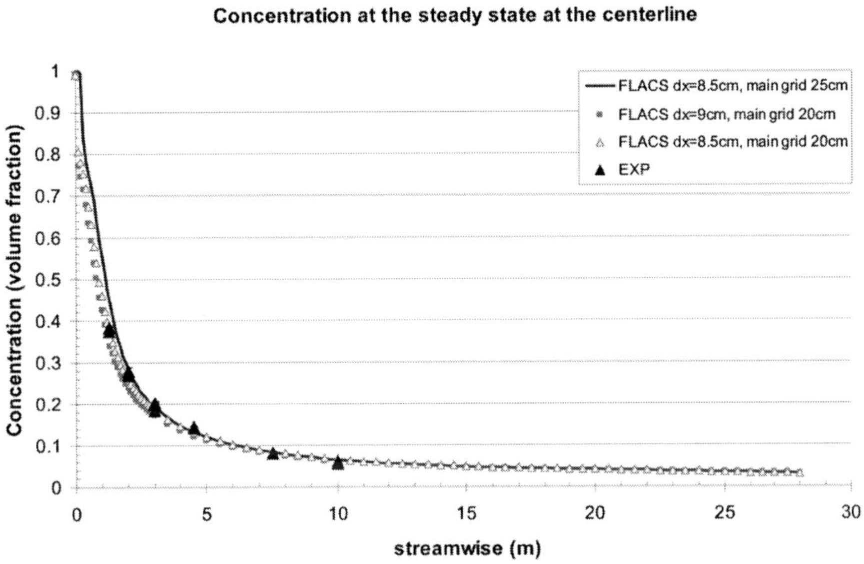

Figure 4: Concentration decay on the centerline vs distance: INERIS experiment

Figures 5 and 6 show the centerline velocity decay with the distance from the source and in the radial direction: FLACS simulations vs. experimental measurements. The simulations/measurements comparison shows that FLACS results are in reasonable agreement with experimental data. Due to quite high error bar for experimental investigations, one can conclude that simulations correctly reproduce the physical phenomenon of jet dilution in the streamwise and in the radial directions.

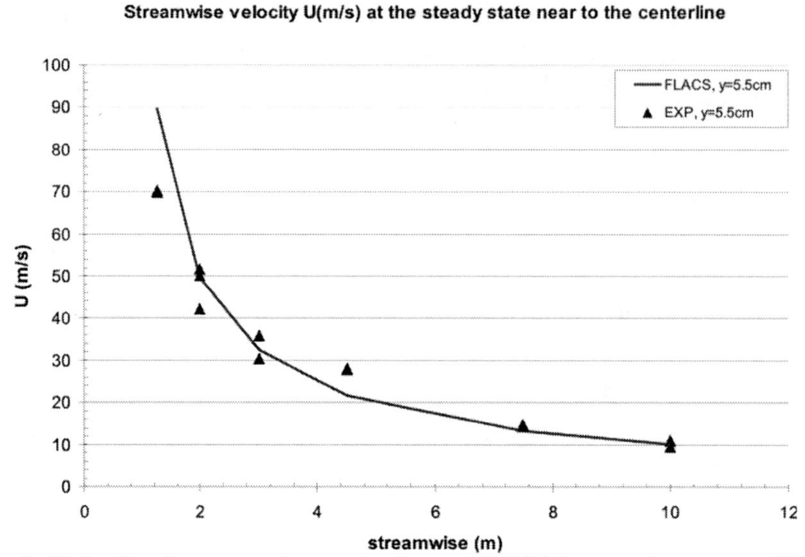

Figure 5: Velocity decay on the centerline – INERIS experiments vs. FLACS.

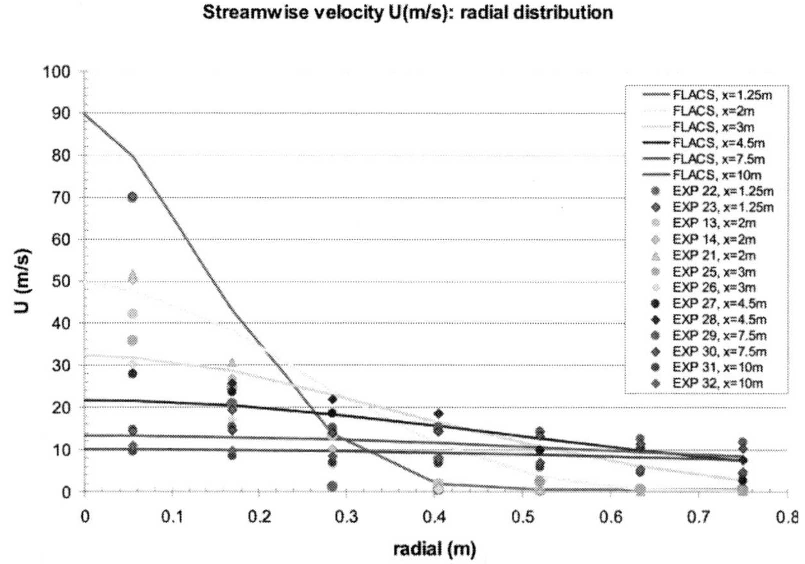

Figure 6: Velocity decay in radial direction – INERIS experiments vs. FLACS.

Hence it can be concluded that CFD simulations correctly reproduce the physical phenomenon of jet dilution.

Figure 7 demonstrates the comparison of the engineering model results with INERIS experiments and FLACS simulations of DNV GL, 2-inch experiment.

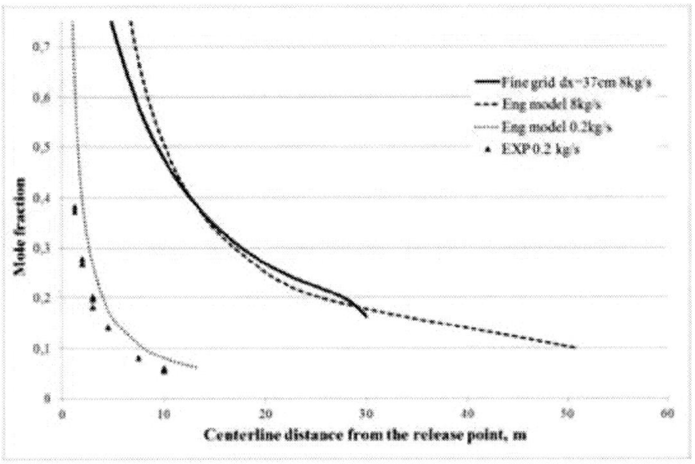

Figure 7: Comparison of the engineering model and FLACS modeling for the 8 kg/s release (DNV GL, 2-inch) and INERIS experiment for the H_2 concentration decay on the centerline.

Fig.7 shows good agreement for the comparison of the engineering model and FLACS modeling for the 8 kg/s release [2] and Daubech et al. [3] for the hydrogen concentration decay on the release centerline.

4.2 Validation of CFD model for explosion

In the Daubech and al. experiments [3], ignition of the dispersed flammable cloud was performed by a pilot hydrogen - oxygen flame. FLACS simulations are performed with a volumetric ignition; see Daubech and al. [3] for more details.

The comparison of simulation results at 3 overpressure monitoring points is shown in Fig.8: L1 sensor is located near the release point (20 cm); the L2 sensor is located on the axis of the jet 2 m downstream from the igniter; the L3 sensor is located at 2.5m from the ignition perpendicularly to the jet axis.

Fig.8 demonstrates a close agreement between simulations and experimental data for this case.

In the case of Miller et al. [2] experiments, ignition of the dispersed cloud was performed by an incendiary device positioned close to the hydrogen release point. In simulations a volume ignition is used (0.1m by 0.1m by 0.1m).

Simulation comparison with experimental data of Miller et al. [2] at two monitor points is presented in Fig.9 for 1 kg/s and for Fig.10 8 kg/s releases. Overpressure monitoring sensors were located at 10 meters and 20 meters off the jet centerline and at 11 m (for 1kg/s flow rate) and at 23 m (for 8 kg/s flow rate) downstream of the release point.

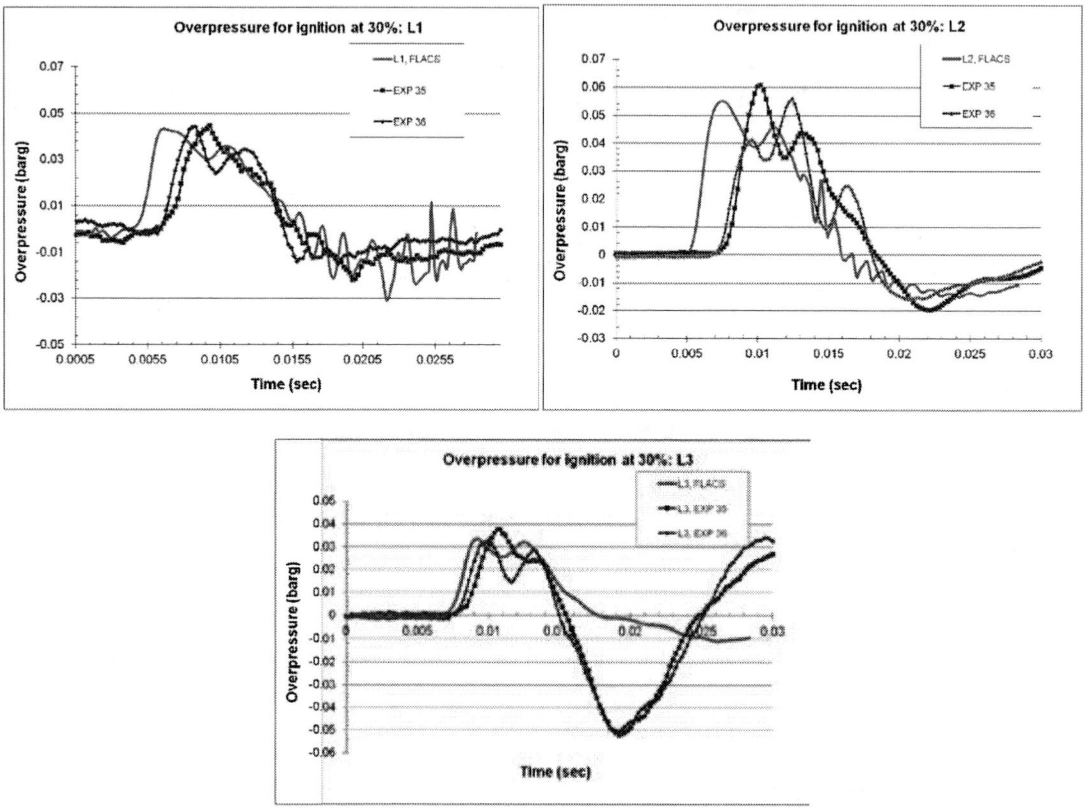

Figure 8: Comparison of simulation results with INERIS experiment (Daubech and al. [3]) at 3 monitor positions (mass flow is 0.245 kg/s).

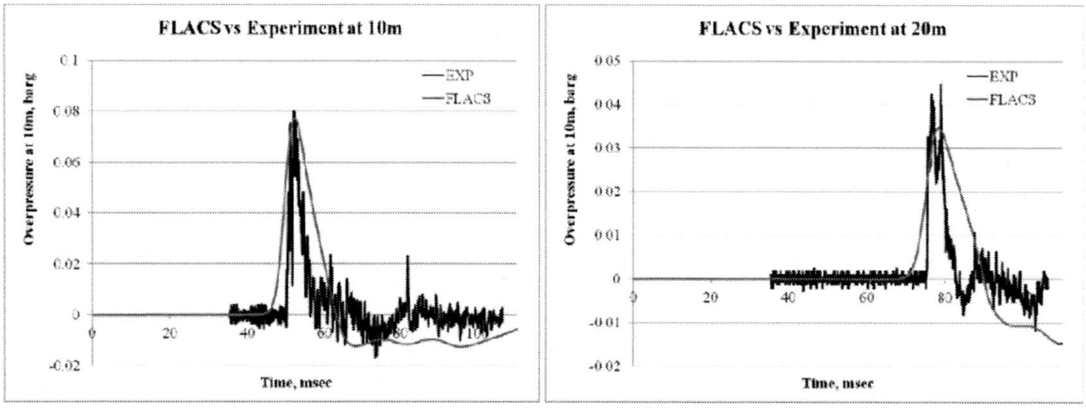

Figure 9: Comparison of FLACS simulation results with experimental data GNV GL, ¾-inch inch (1 kg/s) at 2 monitor positions.

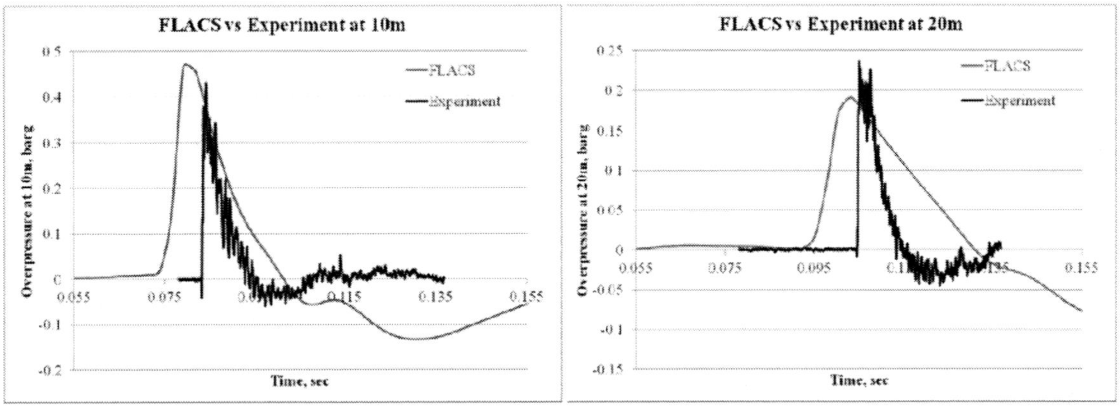

Figure 10: Comparison of FLACS simulation results with experimental data GNV GL, 2-inch (8 kg/s) at 2 monitor positions.

Simulations are in quite good agreement with experimental results for the overpressure magnitude. However, due to the numerical diffusion the calculated pressure signal is much wider than the measured one. This also leads to a significant overestimation of the corresponding positive impulses by CFD. To obtain the correct shape of the signal a much finer spatial resolution must be used, however this would require much more CPU time.

4.3 Parametric study

A parametric study has been performed to find the ignition position corresponding to the maximum overpressure, i.e. to define the worst case ignition scenario.

A parametric study is performed with FLACS, where the ignition position is varied along the centerline. Ignition points correspond to the local concentration from 10% up to 75%. As shown on Fig.11, the overpressure is monitored along the centerline and at locations L2 (2m downstream of the ignition position).

As expected, the ignition in the lean region leads to very low overpressures. Surprisingly, the worst ignition is not in the most reactive region of the cloud (30-40% H_2). The maximum overpressure is obtained for ignition at 65% H_2. This can be explained by a high turbulence of the jet which increases the flame velocity before reaching the most reactive region.

Parametric study to define the worst case scenario is performed for releases with mass flow from 0.1 kg/s up to 8 kg, see figure 12.

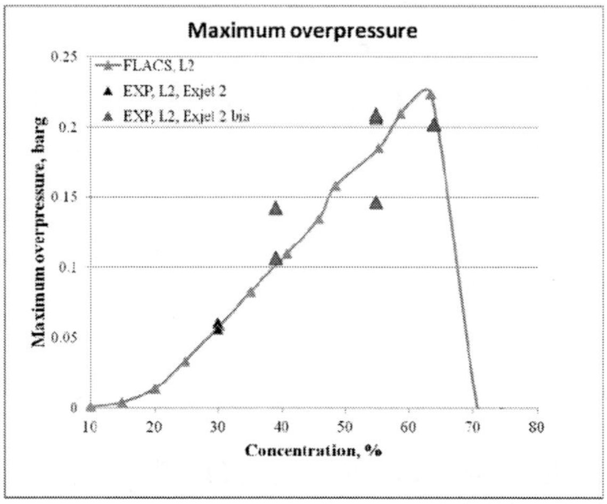

Figure 11: Parametric study: dependence of the overpressure on ignition positions (at different concentrations) and comparison with experiments of Daubech and al. [3]. Black triangles old experiments, red triangles recent experiments.

According to simulations, in all cases it is found that the maximum overpressure corresponds to the ignition at concentration between 55-65% H_2.

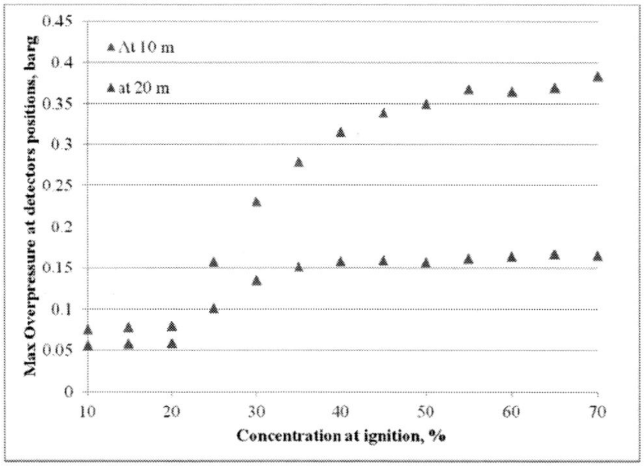

Figure 12: Parametric study: dependence of the overpressure on ignition positions (at different concentrations) for 8kg/s. Overpressure is measured at 10m and 20m perpendicular to the jet axis, 23m downstream of the release point, 2m and 3m from the ground respectively.

5 Discussion

5.1 Worst case ignition point

Based on experiments and modeling, this paper aims to suggest an engineering approach to assess the blast loads of delayed ignition of high pressure hydrogen releases.

It is found that FLACS is able to correctly describe high pressure releases of H_2 from small releases of 250 g/s INERIS) up to high flow rates of 1 and 8kg/s (Miller et al. [2]).

Explosion simulations preformed with FLACS also gives a reasonable agreement for both experiments (INERIS and DNV GL). For Daubech et al. [3] measurements FLACS gives the correct shape of the pressure signal and the positive impulse is correctly estimated. The pressure signals obtained for DNV GL experiments are much wider, leading to a significant overestimation of the positive impulse.

A parametric study for various releases (from 100g/s up to 8kg/s) by means of FLACS is performed for various ignition positions. It is shown that the maximum overpressure is obtained for ignition at the location corresponding to 55-65% concentration. This conclusion is partly validated by experiments of INERIS and data from DNV GL experiment (experimental the ignition is located at 65% H_2).

5.2 A new adapted correlation

Dorofeev et al. [20] revealed that the expansion ratio (σ: ratio of density of reactants to the density of reaction products) could be used to predict the fast flame occurrence by flame acceleration. They demonstrated that there is a critical expansion ratio σ* separating the slow flame mixtures from the fast flames mixtures. For hydrogen air mixtures at atmospheric conditions, this criterion is σ*=3.67 corresponding to a mixture with 11%vol of hydrogen. Daubech et al [3] also observed experimentally that the flames do not propagate in the regions of the flammable cloud with concentrations smaller than 10% of hydrogen.

Hence, it is recommended to consider the hydrogen mass contributing to fast deflagration between 10 and 75% vol of hydrogen.

Using this fast burning mass for different experiments and CFD results, the multi-energy strength index can be defined. Figure 13 represents the TNO multi energy (ME) strength index vs. mass flow rates of hydrogen release for ignition at 30%.

However, it is important to emphasize that this approach is valid only for high pressure releases of hydrogen from 10 bars up to 400 bars.

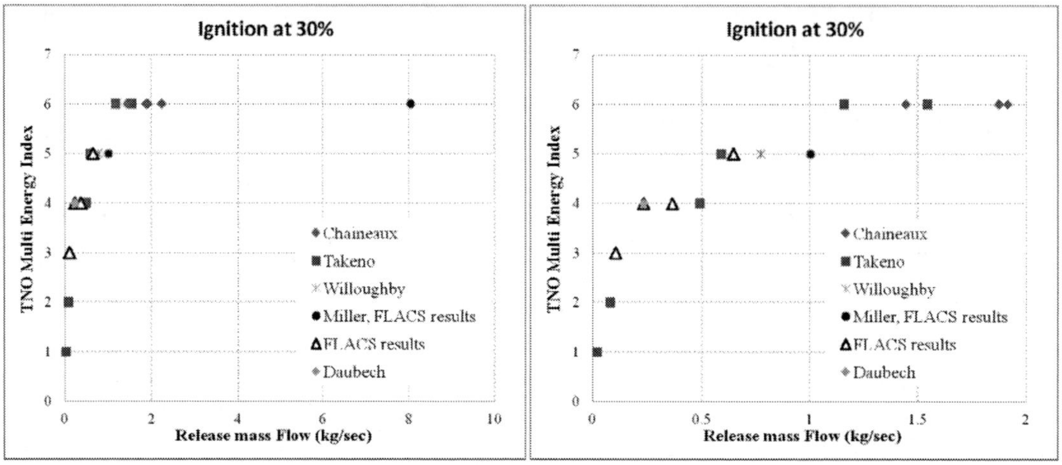

Figure 13: TNO multi energy strength index vs mass flow rate for ignition at 30%: on the left - mass flow rates up to 8kg/s, on the right – close-up on mass flow rate up to 2kg/s.

According to Figure 13, for ignition at 30% it is recommended to use:

- A TNO ME strength index of 4 (corresponding to a maximum overpressure of 100 mbar) for mass flow below 500 g/s
- A TNO ME strength index of 5 (Pmax = 200 mbar) for mass flow between 500 and 1000 g/s
- A TNO ME strength index of 6 (Pmax = 500 mbar) for mass flow between 1 and 10 kg/s.

It should be noted that the ME strength index values are related to the flame speed in the flammable cloud. The ME strength index 4, 5 and 6 correspond respectively to a flame speed of 100, 140 and 240 m/s.

Therefore, this method is easy adaptable for the Baker Strehlow Tang (BST) approach.

Similar investigation is performed for ignition at 55%-65% of hydrogen. Figure 14 represents the TNO multi energy strength index vs mass flow rates at ignition at 55% -65% of hydrogen. Pressure releases are between 5bars-70bars.

According to Figure 14, for ignition at 55-65% it is recommended to use:

- A TNO ME strength index of 4 (corresponding to a maximum overpressure of 100 mbar) for mass flow below 100 g/s
- A TNO ME strength index of 5 (Pmax = 200 mbar) for mass flow between 100 and 1000 g/s
- A TNO ME strength index of 6 (Pmax = 500 mbar) for mass flow between 1 and 10 kg/s.

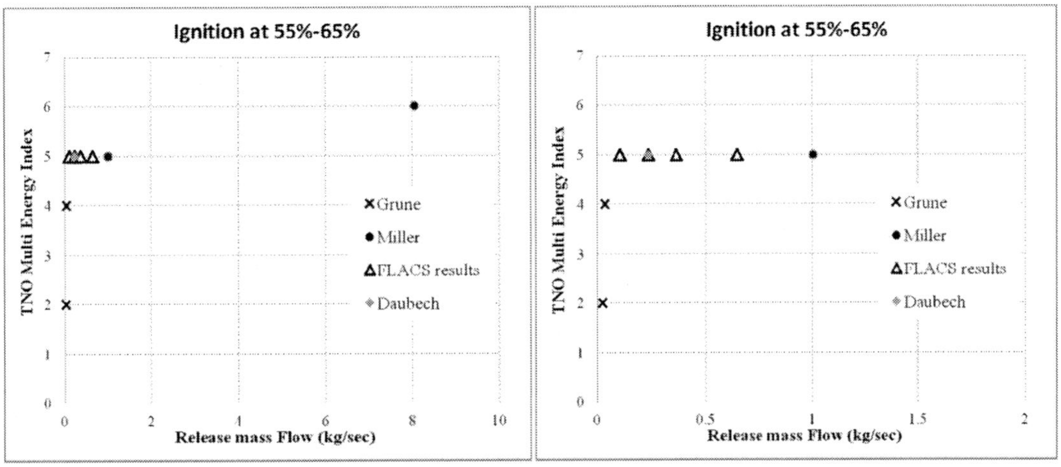

Figure 14: TNO multi energy strength index vs mass flow rate for ignition at 55% - 65%: on the left - mass flow rates up to 8kg/s, on the right – close-up on mass flow rate up to 2kg/s.

A difference in the TNO ME strength index is important only for low mass flow (lower than 500g/s). Thus for such flow rates the TNO ME strength index must be adapted according to the ignition position.

According to simulations results included in graphs in Figure 13 and14 for larger releases (more than 500g/s) the ignition position does not impact the choice of TNO ME strength index. However, this conclusion must be checked against future experimental tests.

Since in a real release it is not known where an ignition source is, it is recommended to use the TNO ME strength index corresponding to the more severe case of ignition at ~ 60%. However for the center of blast, it is suitable to assume it is at H_2=30% which will be further out than H_2=60% but this position is a little more conservative.

5.3 Examples of engineering model application

Risk analysis associated with accidental pressurized hydrogen releases normally involves addressing potential consequences such as jet fires and delayed ignition vapor cloud explosions in confined and/or congested areas. The phenomenon of delayed ignition hydrogen jet explosions in unconfined / uncongested areas has traditionally not been considered but may introduce additional risks. In order to better understand the relative impact of this newly characterized risk, it is useful to compare the hazard with existing risks. For this purpose, two simple examples will be described using engineering models presented in the paper and from [21].

5.3.1 <u>High pressure hydrogen tube trailer hose leak</u>

Tube trailer maximum operating pressure is 165 barg. Hose is 0.5" internal diameter. Typical leak size is assumed to be 3% of hose cross sectional area (by NFPA 2, [22]). Calculated release rate is 0.04 kg/s. Compare consequences thermal radiation from jet fire with delayed ignition jet explosion. Assume a horizontal release at 2m elevation and its impact at head height also at 2m elevation.

Hydrogen jet fire.

The thermal radiation hazard is analyzed using the recently published Air Products Flame model, [21], which is a new simple engineering jet fire model, developed specifically for hydrogen and syngas mixtures, based on extensive data collection and analysis of jet fire tests of these gases. For details of all required equations, see reference [21]. Thermal radiation hazard levels are taken from API 521 [23]: exposure of a few seconds, emergency escape, 9.46 kW/m^2; exposure of 30 seconds, short term emergency action, 6.31 kW/m^2; longer term emergency action, 3 minutes, 4.73 kW/m^2. These levels include allowance for solar background radiation and a standard level of 0.95 kW/m^2 is assumed in the results presented in Figure 15, which is output from the Air Products Flame model for the hose failure release. The graph axes are: *x* (horizontal distance from leak source) and *y* (vertical distance from grade).

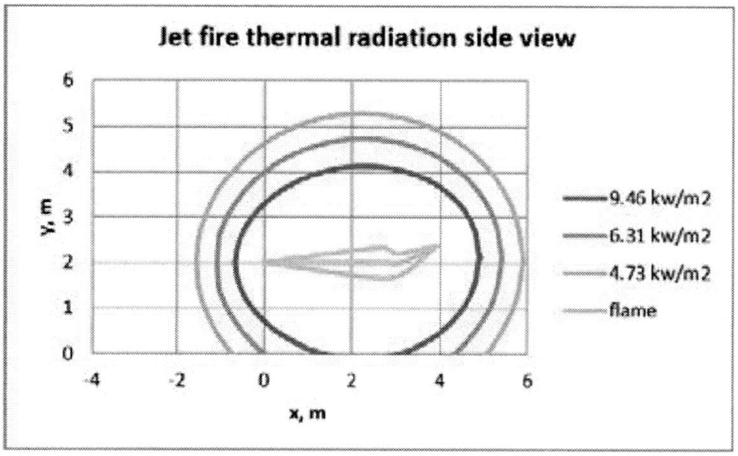

Figure 15: Thermal radiation side view for hydrogen jet fire: example tube trailer hose leak.

Delayed hydrogen jet explosion.

The same hose leak release conditions can also be assessed for blast over-pressure, using the new simple engineering hydrogen jet explosion model presented in this paper, along with standard blast over-pressure hazard levels taken from NASA Glenn Safety Manual [24]: knock a person over, 70 mbar; shatter glass 35 mbar. Figure 16 shows the calculated distances to these two limits. The graph axes are: x (horizontal distance from leak source) and y (vertical distance from grade).

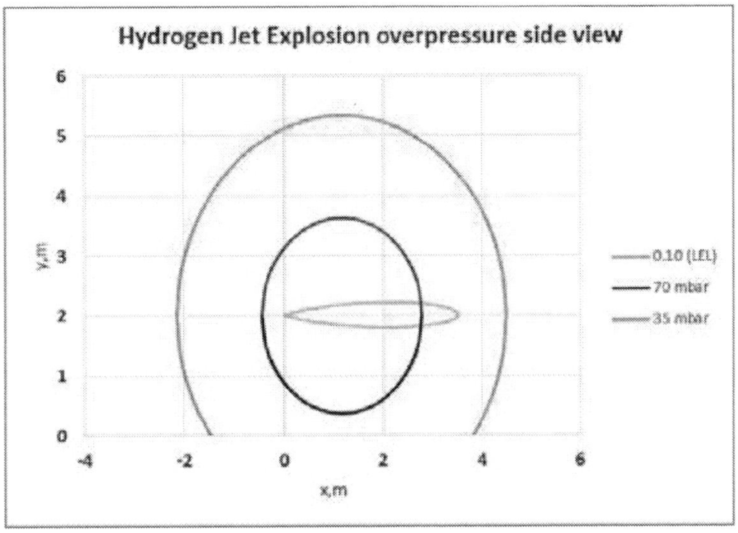

Figure 16: Overpressure side view for hydrogen jet explosion: example tube trailer hose leak.

For this example, the risk from the jet explosion (Figure 16) is comparable to the risk from the jet fire (Figure 15).

5.3.2 2" large leak of hydrogen in process plant at a pressure 30 barg

Compare consequences of explosion resulting from congested volume VCE (vapor cloud explosion) with a jet explosion in an uncongested area of the site. Calculated release rate is 4 kg/s.

Congested VCE.

Assume a stoichiometric mixture in a range of congested volumes typically found in process plants: 50m^3, 100 m^3 and 200 m^3. Use TNO ME method, strength level 7, for typical process plant congestion. Release rate is not considered in this analysis - it is assumed that the release is sufficiently large to fill the confined volume.

Jet explosion.

The explosion is modeled for the stated hole size and calculated release rate. Release is assumed to be horizontal and at an elevation of 2m.

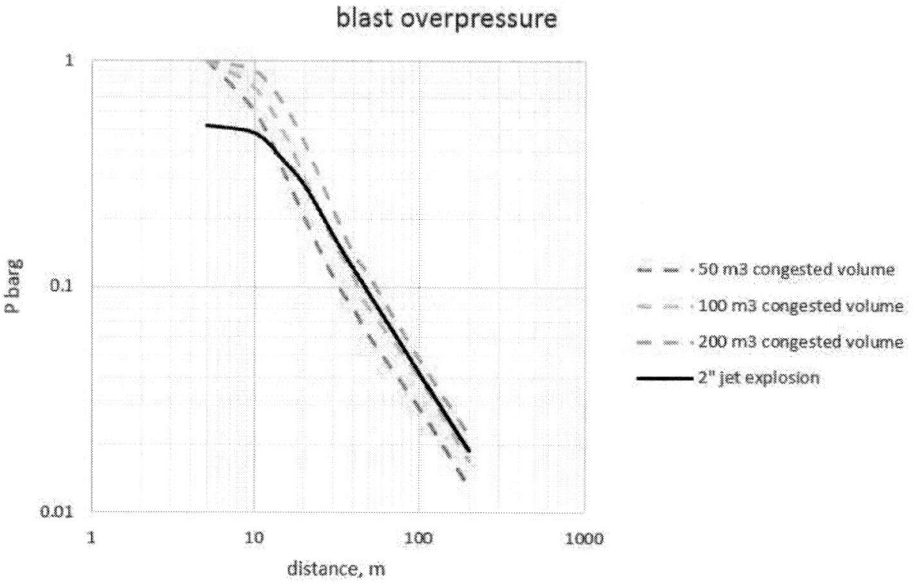

Figure 17: Overpressure vs distance for VCE and jet explosion: example 2" large leak in process plant at a pressure 30 barg.

For this example, the risk from the jet explosion without any confinement or congestion is comparable to the risk from vapor cloud explosions in typical congested volumes found in process units, see Figure 17, at least in the far field. This analysis excludes consideration of any potential domino effects in the near field.

6 Conclusion

Hazards associated with hydrogen jet explosions are comparable with other scenarios typically considered today in risk studies, such as jet fires and vapor cloud explosions. It is, therefore, important that delayed ignition jet explosion is included as one of the scenarios to be considered when assessing the potential consequences of accidental hydrogen releases. For this purpose two different approaches can be used CFD simulations or a new methodology for determining blast strength based on the TNO multi-energy model with strength index correlated to the mass flow rate of the accidental release. For low mass flow (lower than 500g/s) the TNO ME strength index must be adapted according to the ignition position: 30% or 55%-65%. For larger releases (more than 500g/s) the ignition position does not impact the choice of TNO ME strength index. However, this conclusion must be checked against future experimental tests. Since in a real release it is not known where an ignition source is, it is recommended to use of TNO ME strength index corresponding to the more severe case of ignition at ~ 60%, however for the center of blast for both positions (30% and 55%-65%), it is suitable to assume it is at H_2=30% which will be further out than H_2=60% but this position is more conservative.

7 References

[1] J.K. Thomas, C.D. Eastwood and M.L. Goodrich, Are Unconfined Hydrogen Vapor Cloud Explosions Credible?, Process Safety Progress, 34(1) (2015), 36-43 (10th Global Congress on Process Safety, AIChE Annual Meeting, New Orleans, LA, March 30 - April 2, 2014).

[2] D.Miller , C.D. Eastwood and J.K. Thomas, Hydrogen Jet Vapor Cloud Explosion: Test Data and Comparison with Predictions, 11th Global Congress on Process Safety, AIChE Annual Meeting, Austin, TX, 26-30 April 2015.

[3] J. Daubech, J. Hebrard, S. Jallais E. Vyazmina, D. Jamois and F. Verbecke, Un Ignited and Ignited High Pressure Hydrogen Releases: Concentration – Turbulence Mapping and Overpressure Effects, Journal of Loss Prevention in the Process Industries, 36 (2015), 439-446.

[4] J.Grune, , K. Sempert, M. Kuznetsov and W. Breitung, Experimental Study of Ignited Unsteady Hydrogen Releases Into Air, Intl. Journal of Hydrogen Energy, 36(3) (2011), 2497-2504.

[5] D.B. Willoughby and M. Royle, The Interactions of Hydrogen Jet Releases with Walls and Barriers, Intl. Journal of Hydrogen Energy, 36(3) (2011), 2455-2461.

[6] K.Takeno , K. Okabayashi, A. Kouchi and T. Nonaka, K. Hashiguchi and K Chitose, Dispersion and Explosion Field Tests for 40 MPa Pressurized Hydrogen, Intl. Journal of Hydrogen Energy, 32(13) (2007), 2144-2153.

[7] J. Chaineaux, Merge Project : INERIS Report, (1995).

[8] J.R. Bakke, D. Bjerketvedt and M. Bjørkhaug, FLACS as a Tool for Safe Design Against Accidental Gas Explosions, IChemE Conference 'Piper Alpha - Lessons for Life Cycle Safety Management'. 26-27 September 1990, London.

[9] E. Vyazmina, S. Jallais and D. Miller, Effects of the Ignition Position on the Overpressure Originated from a Delayed Ignition of High Pressure Releases of Hydrogen, 11th Intl. Symp. on Hazards, Prevention, and Mitigation of Industrial Explosions (ISHPMIE), 24-29 July 2016, Dalian, China.

[10] Q.A. Baker, et al., Guidelines for Vapor Cloud Explosion, Pressure Vessel Burst, BLEVE and Flash Fire Hazards, Second Edition, Center for Chemical Process Safety (CCPS) & John Wiley and Sons, New York, NY, 2010.

[11] F. H. Harlow and P. I. Nakayama, Turbulence transport equations. Physics of Fluids, 10 (1967), 2323–2332.

[12] S. V. Patankar, Numerical Heat Transfer and Fluid Flow. Taylor & Francis, 1980.

[13] B. H. Hjertager, Computer simulation of turbulent reactive gas dynamics. Modeling Identification and Control, 5(4) (1985), 211–236.

[14] B. H. Hjertager, Three-dimensional modeling of flow, heat transfer, and combustion. Chapter 41 in Handbook of Heat and Mass Transfer, Gulf Publishing Company, Houston, 1986.

[15] B. H. Hjertager, K. Fuhre, & M. Bjørkhaug, Gas explosion experiments in 1:33 and 1:5 scale offshore separator and compressor modules using stoichiometric homogeneous fuel/air clouds. Journal of Loss Prevention in the Process Industries, 1(1988), 197-205.

[16] FLACS User's Manual, GexCon, Bergen, 2015.

[17] K. N. C. Bray, Studies of the turbulent burning velocity. *Proceedings of the Royal Society of London, Series A*, 431(1990), 315-335.

[18] A. D. Birch, D. J. Hughes and F. Swaffield, Velocity Decay of High Pressure Jets, *Combust. Sci. and Tech.*, 52 (1987), 161-171.

[19] W Houf and R Schefer, Predicting radiative heat fluxes and flammability envelopes from unintended releases of hydrogen, *Int. J. Hydrogen Energy*, 32 (2007), 136-151.

[20] S. B. Dorofeev, M. S. Kuznetsov, V. I. Alekseev, A. A. Efimenko and W. Breitung, Evaluation of limits for effective flame acceleration in hydrogen mixtures, *Journal of Loss Prevention in the Process Industries*, V 14, Issue 6 (November 2001), 583-589.

[21] D. Miller, New Model for Predicting Thermal Radiation from Flares and High Pressure Jet Fires for Hydrogen and Syngas, Global Conference on Process Safety, Spring meeting 2016.

[22] "Pressure – relieving and Depressuring Systems", API Standard 521, 6th Edition (2014).

[23] API Standard 521 Pressure-relieving and Depressuring Systems, SIXTH EDITION, JANUARY 2014, 248 PAGES, PRODUCT NO. C52106

[24] "Glenn Safety Manual", NASA GLM – QS- 1700.1.7 Rev. B (2012).

13TH GLOBAL CONGRESS ON PROCESS SAFETY

Aerosol generation approach and aerosol combustion behavior study

Shuai Yuan
Mary Kay O'Connor Process Safety Center
Artie McFerrin Department of Chemical Engineering, Texas A&M University
College Station, Texas 77843-3122, USA
sy2516@tamu.edu

Yogesh Koirala[a,b], Chad V. Mashuga[a,b], Zhengdong Cheng[b] and M. Sam Mannan[a,b]
[a]Mary Kay O'Connor Process Safety Center
[b]Artie McFerrin Department of Chemical Engineering, Texas A&M University
College Station, Texas 77843-3122, USA

Prepared for Presentation at
American Institute of Chemical Engineers
2017 Spring Meeting and 13th Global Congress on Process Safety
San Antonio, TX
March 26 – 29, 2017

Aerosol generation approach and aerosol combustion behavior study

Shuai Yuan
Mary Kay O'Connor Process Safety Center
Artie McFerrin Department of Chemical Engineering, Texas A&M University
College Station, Texas 77843-3122, USA
sy2516@tamu.edu

Yogesh Koirala[a,b], Chad V. Mashuga[a,b], Zhengdong Cheng[b] and M. Sam Mannan[a,b]
[a]Mary Kay O'Connor Process Safety Center
[b]Artie McFerrin Department of Chemical Engineering, Texas A&M University
College Station, Texas 77843-3122, USA

Keywords: Aerosol; Flame speed; Combustion; OpenFOAM

Abstract

The hazard of high flash point liquid is often overlooked by people because the flammable hazard of liquid is conventionally classified by their flash point. However, numerous incidents revealed that high flash point liquid can be ignited and can explode when it is in the form of aerosol. The aerosol usually forms in the situations where the liquid is released under elevated pressure and temperature, or rapid condensation of vapor. Although a flammable aerosol is composed of liquid droplets, the aerosol behaves very differently from the bulk liquid. The small droplets diameter promotes mass and heat transfer because of larger surface area compared with bulk liquid. Moreover, the diffusion of droplets makes the aerosol easily ignited by ignition source. In this study, the combustion behavior of the aerosol with different droplet size generated by electrospray was investigated. Furthermore, the aerosol generation process was simulated by OpenFOAM. The Sauter Mean Diameter (SMD) of droplets obtained from simulation is compared with experimental results.

1. Introduction

The hazards of hydrocarbons are often classified by their flash points in standards and regulations (*e.g.*, NFPA and GHS) [1, 2]. Usually, the liquid with lower flash point (less than 93 °C) is defined as flammable liquid. However, there are many instances of fires and explosions involving high flash point liquid aerosol throughout the process industry [3, 4]. The reason behind those incidents may be due to the misconception that it is safe when liquids are below their flash points. The property of aerosol is significantly different from the bulk liquid. One potential explanation for this phenomenon is a larger surface area of aerosol droplets promotes the evaporation of liquid which would provide flammable gas within the flammable limits. Therefore, the conventional category would make people ignore the high flash point liquid hazards and there is lack of appropriate regulations and measures to handle the hazards related to aerosol.

The aerosol could form via mechanical breakup under high pressure and high temperature conditions or rapid condensation of vapor. There are many severe incidents (*e.g.*, Milliken Carpet factory fire, Buncefield explosion and Flixborough disaster) involving the combustion and explosion of aerosol [3, 5]. Eichhorn has also pointed out that mist (aerosol) can explode in 1955 [6]. As shown in Fig. 1, Eichhorn used the concept of flammability limits in flammable gases as reference and introduced aerosol flammability limits. Researchers have widely studied the upper and lower flammable limits of various gases. But the aerosol flammability regions for different liquids are still unclear.

Although many researchers have dedicated considerable effort to the aerosol flammability limits and aerosol flame speed studies in the past decades, there is still no systematic approach to investigate the aerosol flammability limits and consequence of aerosol flame. In the laboratory, researchers mainly studied the influence of existing droplets on laminar flame. However, the aerosol flames observed in process industries are normally turbulent flame. As a result, there are contradictory conclusions on the trend of aerosol flame propagation speed with respect to droplet diameter. Ballal *et al.* studied isooctane, diesel oil and heavy fuel oil aerosols and reported that the flame speed decreases as droplet diameter increases [7]. However, Burgoyne *et al.* pointed out that the flame propagation speed of tetralin aerosol first increases with respect to droplet size and then decreases [8]. Polymeropoulos used a model to predict a transition range of flame propagation speed, usually from 10 μm to 30 μm, in which the flame speed of aerosol is faster than that of vapor [9].

In this study, different droplet size aerosols generated by electrospray were ignited to obtain the flame speed in order to validate the trend of aerosol flame speed with respect to droplet diameter. Besides, an open source CFD software, OpenFOAM, was applied to simulate the aerosol generation process. The SMD of droplets in the simulation was compared with the experimental results to validate the feasibility of the models used in the simulation. The aerosol combustion and flame propagation also was investigated by the simulation.

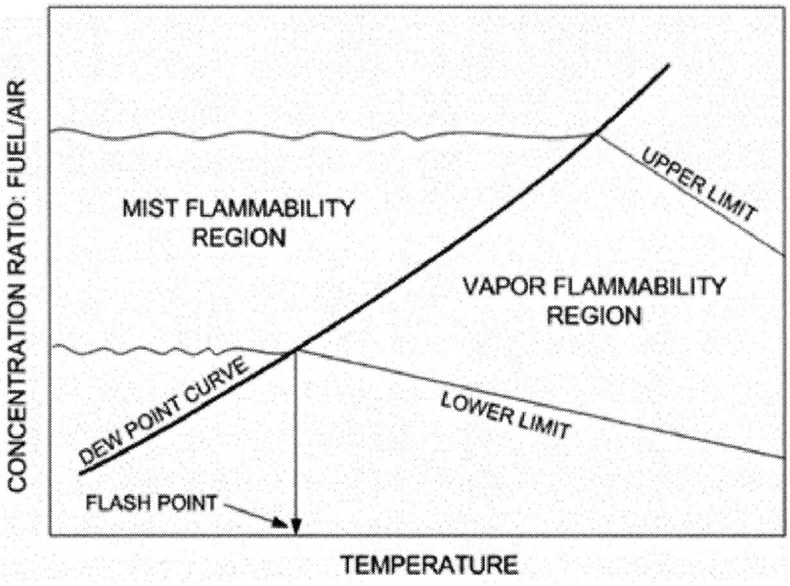

Fig.1 Flammability diagram with respect to concentration and temperature [6]

2. Methodology

2.1 Experiment

The experimental equipment is similar to that used in previous studies done by researchers from the Mary Kay O'Connor Process Safety Center [3-5, 10]. The basic layout of the electrospray aerosol generation device is displayed in Fig. 2. The candidate fluid (D-600) is stored in ten syringes which are dispensed by a syringe pump (KDS 220). D-600 is a commercial heat transfer fluid and the properties are listed in Table 1. The ten syringes are connected to ten stainless steel nozzles (2.54×10^{-4}m i.d. and 5.08×10^{-4}m o.d.) by rubber tubes. The nozzles are placed between a metal plate and a metal mesh. The metal plate is connected to a high voltage source (HV1) which is manipulated between 5-8 kV and the metal mesh is connected to another high voltage source (HV2) which is manipulated between 3-5 kV. The voltage difference between the two is set up around 2-3 kV depending on the desired aerosol droplet size. Both the high voltage sources are generated through function generator (Stanford Research System, DS-345) and amplifier (Trek Inc. 610E). The liquid stream will break up into small droplets under the effect of intensive electric field.

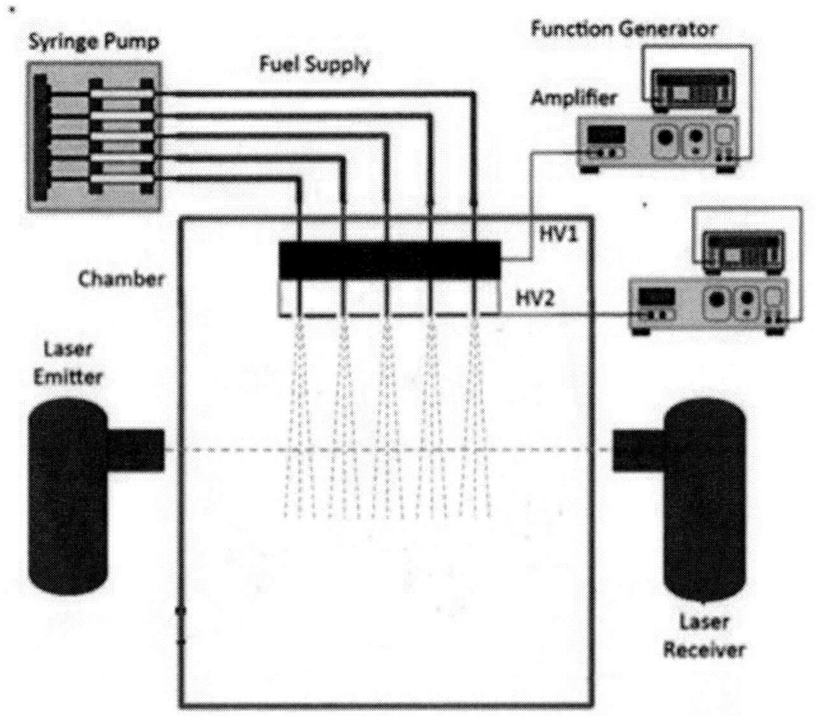

Fig.2 Schematic of electrospray aerosol generator adapted from Lian *et al.* [5]

Table 1. Properties of D-600

Properties	
Appearance	Slight yellow tint liquid, very slight hydrocarbon odor
Composition	Paraffinic hydrocarbon
Average molecular weight	372
Vapor pressure	<0.003 kPa@25°C
Flash point	224°C (435°F)
Auto-ignition temperature	360°C
Specific gravity	0.85 @24°C
Viscosity	70.08 cp@24°C

The diameter and concentration of aerosol droplets are characterized by Laser Diffraction Particle Analyzer (Spraytec, Malvern Instruments Ltd). The Laser Diffraction Particle Analyzer (LDPA) contains a 2 mW He-Ne laser source and a diode detector, which will emit a collimated monochromatic laser beam with a 632.8 nm wavelength and 10 mm diameter. The advantage of LDPA is that the information about the droplet diameter and concentration will be obtained instantaneously without the sampling process. The Sauter Mean Diameter (SMD) is used to denote the diameter of the polydisperse droplets and the definition of SMD is

$$\text{SMD}\ \left(D_{[3][2]}\right) = \frac{\sum n_i d_i^3}{\sum n_i d_i^2} \tag{1}$$

Where
$D_{[3][2]}$ is the Sauter Mean Diameter (SMD)
d_i is the mean diameter of droplets in size-class i
n_i is the number of droplets in size-class i

The concentration is the volumetric concentration of droplets, which is calculated based on Beer-Lambert Law. The intensity of the laser beam obeys Beer-Lambert Law expressed below

$$T = \frac{I}{I_0} = e^{-\alpha b} \tag{2}$$

Where
T is the relative transmission of the laser beam
I_0 is the intensity of laser beam as it enters the aerosol field
I is the intensity of laser beam at a distance b in the aerosol field
α is the absorbance

The absorbance is related to the liquid properties, droplet diameter and droplet concentration. Substituting the expression of absorbance into Beer-Lambert Law then the volumetric concentration of aerosol droplets is

$$C_v(\text{ppm}) = \frac{-2000\ln(T)}{3b\sum\frac{Q_i v_i}{d_i}} \tag{3}$$

Where
b is the distance in the aerosol field in mm
d_i is the mean diameter of droplets in size-class i in μm
Q_i is the mean of efficiency of light extinction for size-class i obtained from Mie theory
v_i is the relative volume in size-class i with mean diameter d_i

After the diameter and concentration of droplets generated by the electrospray becomes stable, the aerosol cloud is ignited by a propane flame with a length of 1.5 cm. The flame expands and moves upward and this process is recorded by a high speed camera (Phantom 4.2, Vision Research Inc.). The flame speed is obtained by analyzing the flame propagation video.

2.2 Simulation

Typically, researchers separate the aerosol generation process into three stages, atomization, primary breakup and secondary breakup. After the liquid stream breaks up into small droplets, the droplets will fall and interact with the surrounding gas vapor. The process is simulated through the

Eulerian-Lagrangian method [11-14]. The gas phase behavior is described by Eulerian gas phase and the liquid droplets are tracked by Lagrangian manner. The continuity equation, momentum conservation equation, energy conservation equation and species equation are shown below

$$\frac{\partial \rho}{\partial t} + \frac{\partial \rho u_j}{\partial x_j} = S_{ev} \tag{4}$$

$$\frac{\partial \rho u_i}{\partial t} + \frac{\partial}{\partial x_j}\left(\rho u_i u_j - \tau_{ij}\right) = -\frac{\partial p}{\partial x_i} + S_{i,mo} \tag{5}$$

$$\frac{\partial \rho h}{\partial t} + \frac{\partial}{\partial x_j}\left(\rho u_j h - \frac{\rho v}{Pr}\frac{\partial h}{\partial x_j}\right) = \frac{\partial p}{\partial t} + \frac{\partial p u_j}{\partial x_j} + S_{he} \tag{6}$$

$$\frac{\partial \rho X_f}{\partial t} + \frac{\partial}{\partial x_j}\left(\rho u_j X_f - \rho D\frac{\partial X_f}{\partial x_j}\right) = S_{ev} \tag{7}$$

Where

ρ is the density of gas phase mixture

u is the velocity of gas phase mixture

p is the gas phase pressure

τ is the stress tensor

v is the kinematic viscosity

h is the specific enthalpy

X_f is the mass fraction of the hydrocarbon in gas phase

D is the diffusion coefficient of the hydrocarbon vapor

$Pr = \frac{\mu C_P}{k}$ is the Prandtl number

μ is dynamic viscosity of gas phase

C_P is the specific heat capacity of gas phase

k is the thermal conductivity of gas phase

S_{ev}, $S_{i,mo}$ and S_{he} are source terms contributed from evaporation in mass conservation, momentum conservation and energy conservation respectively.

The motion of each droplet is calculated based on Newton's second law [15]. In this study, the electric field force and the interaction between droplets are neglected. Therefore, the mathematical equation is

$$\sum \vec{F} = \vec{F}_D + \vec{F}_g = m_d \frac{du_d}{dt} \tag{8}$$

Where

$\vec{F}_D$ is the drag force acted on droplets by the gas

$$\vec{F}_D = m_d \frac{18\mu}{\rho_d d_d^2}\frac{C_D Re}{24}\left(\vec{u} - \vec{u}_d\right) \tag{9}$$

m_d is the mass of single droplet
ρ_d is the droplet density
d_d is the droplet diameter
u_d is the velocity of single droplet
C_D is the drag force
Re is the Reynolds number

$$Re = \frac{\rho d_d |\vec{u}_d - \vec{u}|}{\mu} \tag{10}$$

$\vec{F}_g$ is the gravitational force of droplets which also takes buoyancy into consideration.

$$\vec{F}_g = m_d \vec{g} \left(\frac{\rho_d - \rho}{\rho_d} \right) \tag{11}$$

$\vec{g}$ is the gravitational constant.

Evaporation is viewed as the contribution to the difference between aerosol and bulk liquid. The evaporation of D-600 is not significant because the boiling point of D-600 is greater than 315 °C. In this study, the evaporation of D-600 droplets is negligible in generation process, but the evaporation is considered in the combustion process. In the injection model, the droplet size distribution is represented by Rosin-Rammler size distribution. The reason for using the Rosin-Rammler size distribution instead of normal distribution is because Rosin-Rammler size distribution has a narrower span, which is closer to the data obtained from experiments. The expression of the Rosin-Rammler size distribution is shown below

$$Y_d = 1 - \exp\left(-d/\bar{d}\right)^n \tag{12}$$

Where
Y_d is the cumulative fraction of droplets less than diameter d
n is the spread parameter of droplets
$\bar{d}$ is the characteristic droplet size which means 63.2% droplets are smaller than this value

Besides the models described above, the simulation of the aerosol droplets generation process also involves dispersion model, breakup model and turbulence model. The droplets dispersion is simulated via Discrete Random walk method. The breakup model used in this study is the Reitz and Diwakar model and the turbulence model used is standard $k - \varepsilon$ model. For the combustion process, it also includes evaporation model and combustion model. The evaporation model applied is liquid evaporation boiling model and the combustion model is Partially stirred reactor (PaSR) model.

3. Results

3.1 Droplets size measurement

The droplets size generated by electrospray is mainly controlled by the liquid flowrate through the nozzle and the electric field intensity. From previous studies, Lian *et al.* found that the droplet SMD will decrease with the increase of the voltage difference or with the decrease of the liquid flowrate through the nozzle. In order to determine the relationship between flammability and droplet size, monodisperse droplets are desired. Although the droplet size generated by electrospray is polydisperse, the droplet size of aerosol generated from the Mary Kay O'Connor Process Safety Center devices has a narrower distribution. Table 2. displays the experimental setup (liquid flowrate in syringe and applied voltage on HV1 and HV2). Fig. 3 shows the droplet size distributions of the D-600 aerosol cloud with 75 μm SMD and 116 μm SMD. The quality of the dispersity of droplets is quantified by span ratio.

$$\text{span ratio} = \frac{D_v(90) - D_v(10)}{D_v(50)} \tag{13}$$

Where

$D_v(x)$ is the diameter that $x\%$ droplets are smaller than this diameter

Table 2. Experimental setup used to generate droplets with various diameters

Number	Flow rate (mL/h)	HV1 (kV)	HV2 (kV)	SMD (μm)
1	1	6	3	75
2	1.2	7	5	116

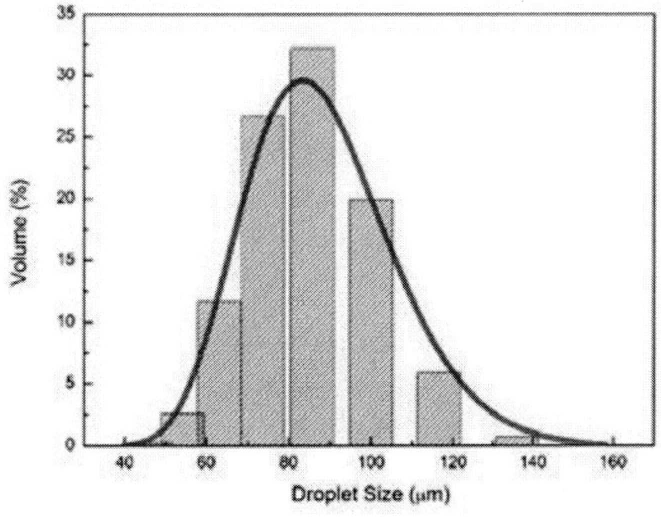

(a)

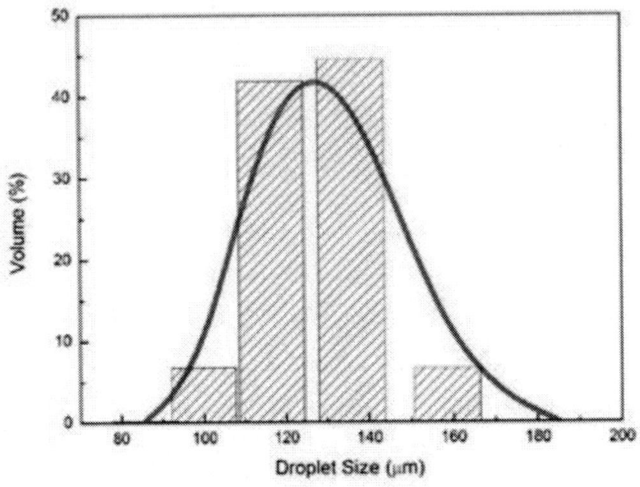

(b)

Fig.3 D-600 aerosol droplet size distribution for a SMD of (a) 75 μm and (b) 116 μm

The span ratios obtained from LPDA for D-600 droplets with a SMD of 75 μm and with a SMD of 116 μm are 0.463 and 0.206 respectively.

3.2 Flame speed measurement

In a previous study, Lin *et al.* measured the flame speed for D-600 aerosol droplets with a SMD of 60 μm, 69 μm, 90 μm, 126 μm and 151 μm. In this study, the flame speed for D-600 aerosol droplets with a SMD of 75 μm and 116 μm are measured in order to validate the trend of flame speed with respect to droplet size. The volumetric concentration in this study is maintained at 0.6 ppm. The high speed camera record 200 frames per second. The center position of flame in each frame was obtained and then was plotted with respect to time. Fig. 4 shows an example method to obtain the flame speed for D-600 aerosol with a SMD of 116 μm. The high linear correlation coefficient indicates that the flame will develop to a stable speed.

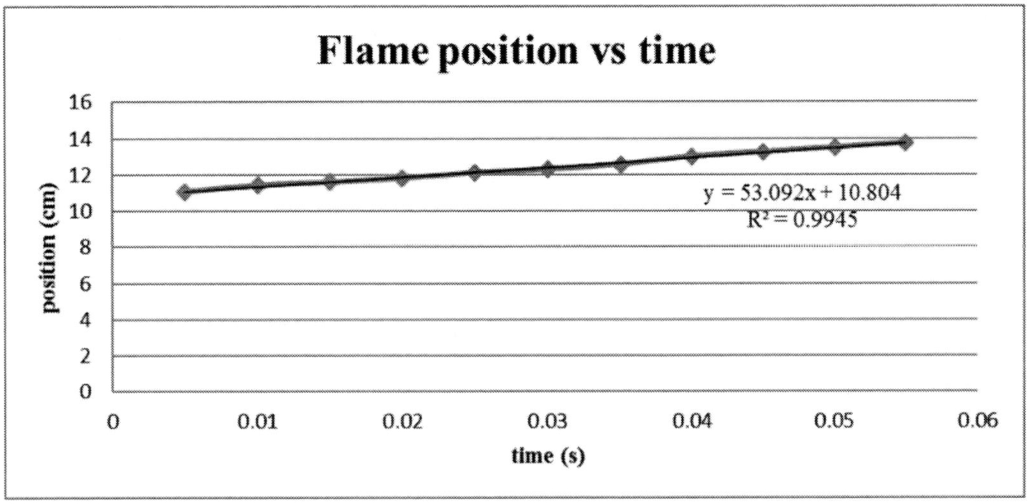

Fig.4 Flame position vs. time for D-600 aerosol with a SMD of 116 μm

The flame speed results for D-600 aerosol with a SMD of 75 μm and 116 μm along with Lin's work are shown in Fig.5. The flame speed for D-600 aerosol with a SMD of 75 μm is higher than the interpolation value of Lin's work. While the flame speed for a SMD of 116 μm is lower than that predicted by Lin's work. Both studies indicated that the flame speed measured for each droplet size has a large deviation. This means the flame speed is effected by several factors, such as droplet concentration and turbulence level, besides droplet diameter. The deviation of current study has overlap with that of previous study and both studies found the same trend that the flame speed will decrease with respect to the increase of droplet size when the droplet size is larger than 60 μm. It is difficult to generate droplets smaller than 60 μm with current experimental setup because when the voltage difference between HV1 and HV2 increases to a limiting value, a spark will be observed. Meanwhile, the decrease of liquid flowrate will lower the aerosol droplets concentration, which will result in the ignition failure of aerosol cloud. Future studies will focus on the improvement of experimental devices to generate droplets with diameters in the predicted transition range. Moreover, the redesign of ignition source will quantify the ignition energy and minimize the influence of turbulence during the ignition process.

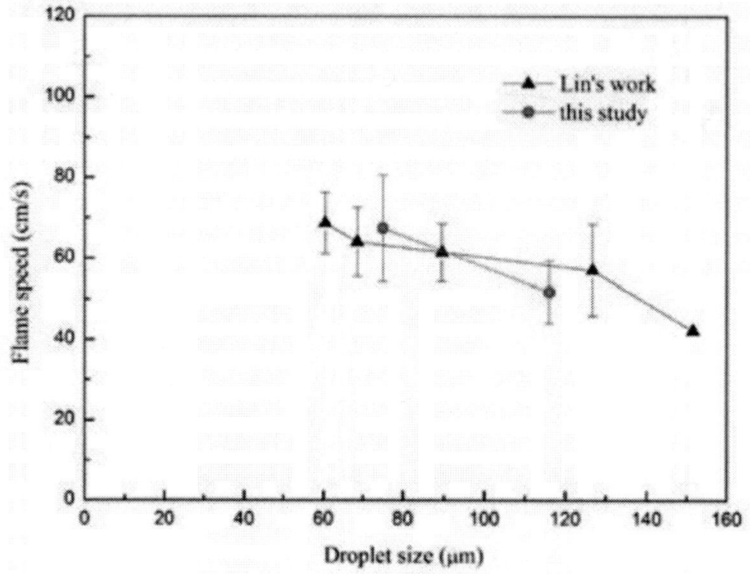

Fig.5 Flame speed vs. droplet size for D-600 aerosol

3.3 Simulation results

3.3.1 Aerosol generation process

Based on the experimental results for aerosol droplets with a SMD of 75 μm, the spread parameter and characteristic droplet size of Rosin-Rammler distribution are calculated. Fig.6 shows the diagram of the Rosin-Rammler distribution fitting plot for 75 μm SMD experiment result. The characteristic droplet size $\bar{d}$ calculated is 82.153 μm and the spread parameter n is 6.599.

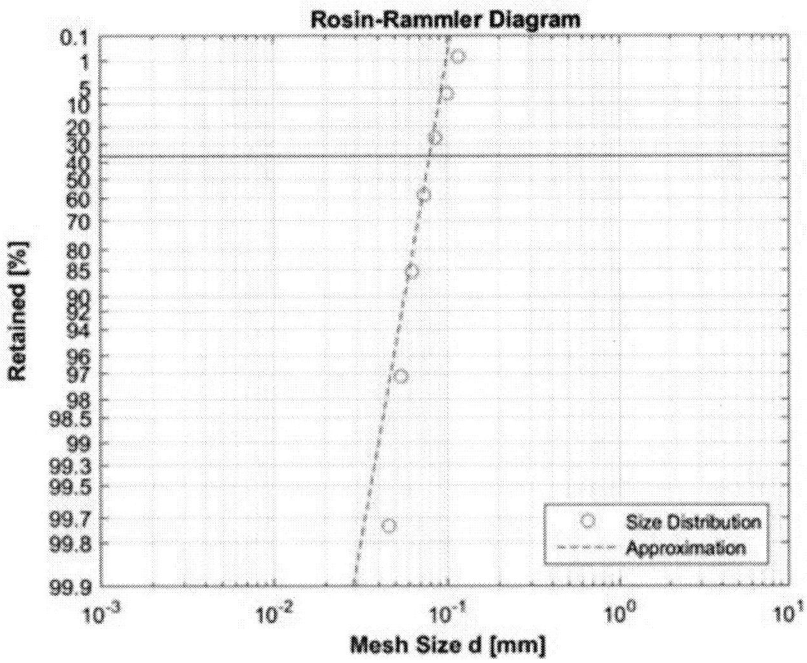

Fig.6 Rosin-Rammler distribution fitting plot for droplets with a SMD of 75 μm

In this study, the mass flow rate of the liquid is 2.17×10^{-7}kg/s which is calculated based on 1mL/h flow rate. The droplet size distribution is based on the Rosin-Rammler fitting. The injection model is pressure driven type injector and the injection pressure is adjusted to make the droplet initial velocity close to 3m/s. The Weber number is 6 in the breakup model. Fig.7(a)-(c) show the position and the number of droplets after 0.05s, 0.2s and 1s respectively, after the start of the injection respectively. This result indicates that the heavier the droplets, the more likely the droplets will fall. The lighter droplets will suspend in the air.

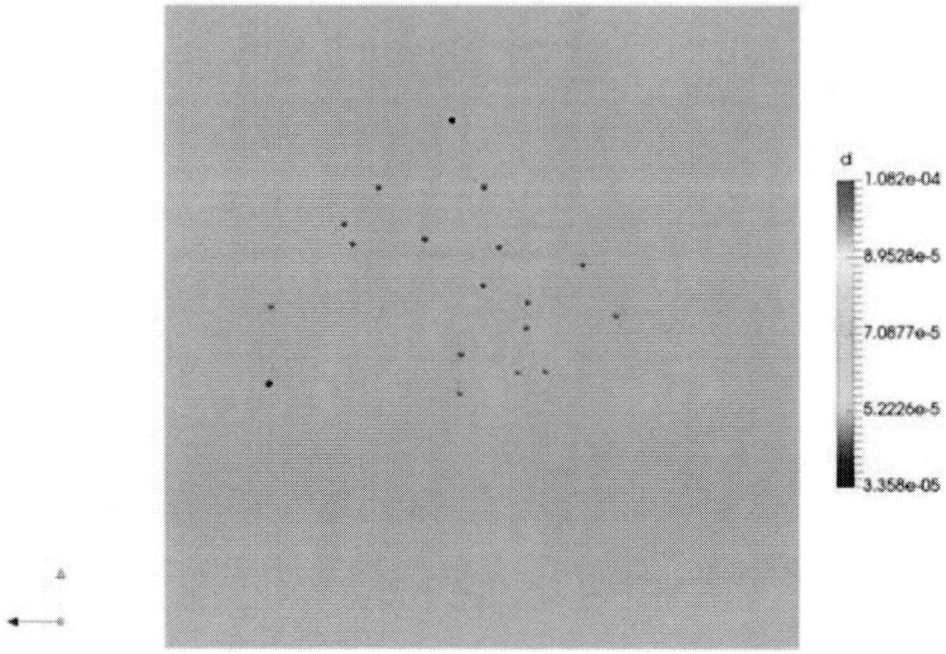
d
1.082e-04
8.9528e-5
7.0877e-5
5.2226e-5
3.358e-05

(a) 0.05s

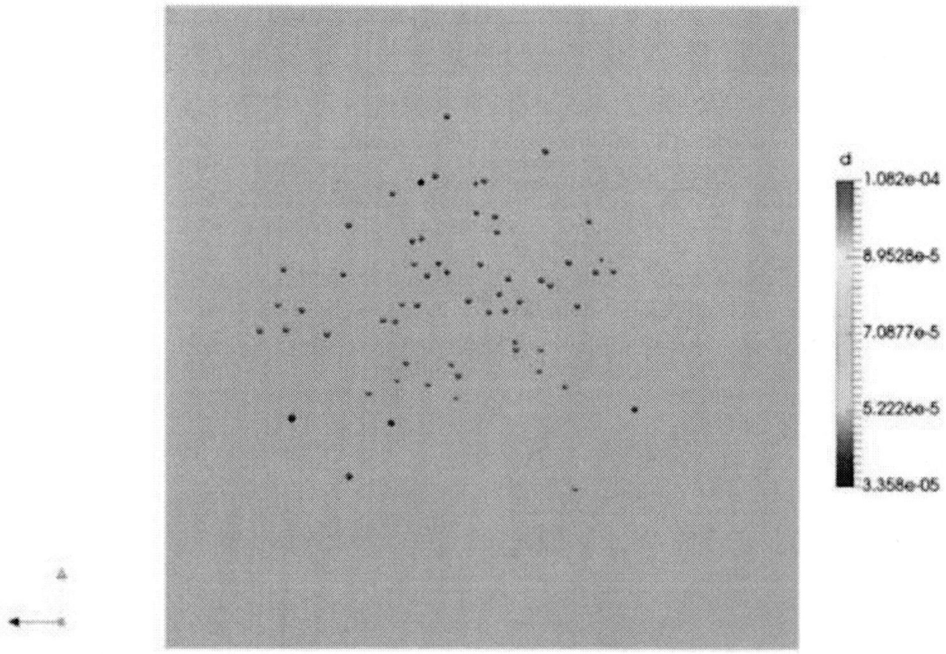
d
1.082e-04
8.9528e-5
7.0877e-5
5.2226e-5
3.358e-05

(b) 0.2s

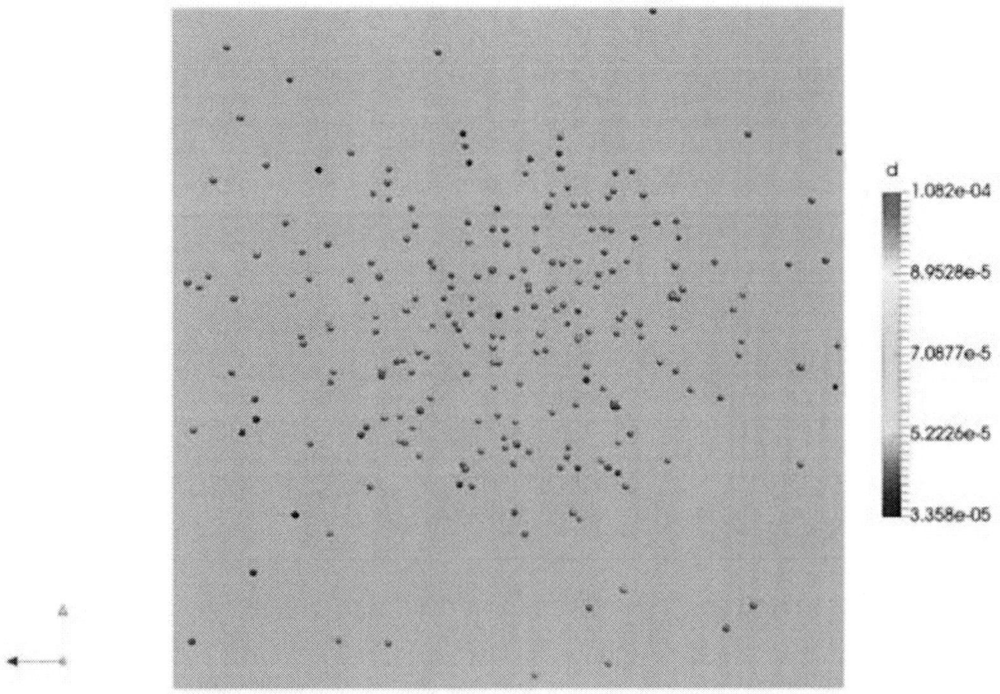

(c) 1s

Fig.7 OpenFOAM simulation results of droplets spray at time (a) 0.05s (b) 0.2s (c) 1s

A Matlab code is applied to extract the SMD value from OpenFOAM log file. Fig. 8 compares the SMD of droplets after the start of injection obtained from simulation and experiment. From the experimental data, the droplets size will continuously increase and becomes stable at around 8s. Fig. 9 shows the trend of droplets size before it reaches the stable size in the simulation. Initially, the SMD of droplets from simulation is larger than the stable droplets size. However, the droplets size will increase and then decrease rapidly to achieve the stable size within 0.05s. The stable droplets size predicted by the simulation has a good fit with the experimental results. This means the breakup model and injection model used can predict the droplets behavior before the ignition.

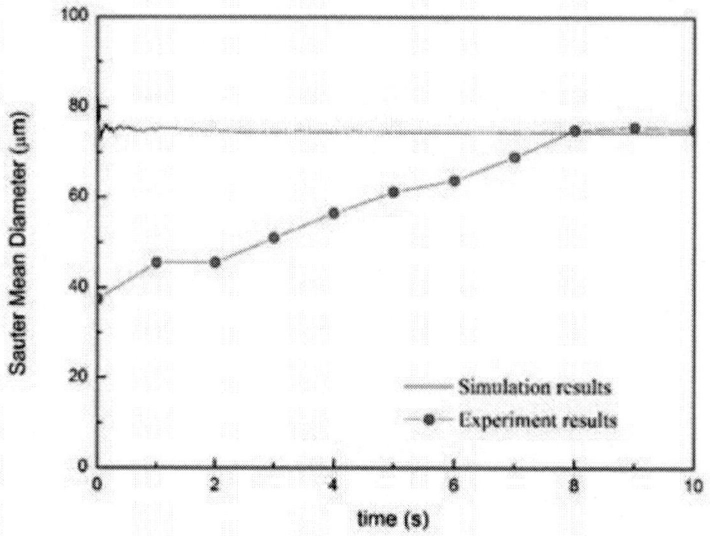

Fig.8 Comparison of SMD vs time between experiment and simulation

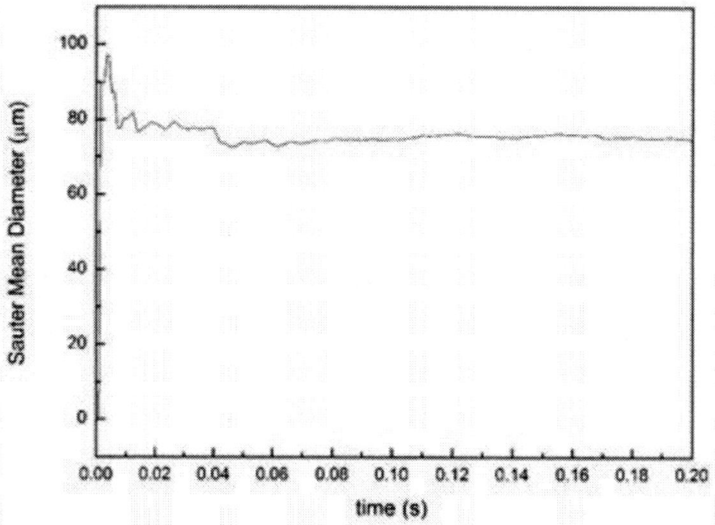

Fig.9 Simulation results of SMD vs time (0-0.2s)

3.3.2 Aerosol combustion process

The ignition of the aerosol is performed by set up the ignition source at the position that is 15 cm below the nozzle. The ignition temperature is set to 2265 K (red color) that is the propane flame temperature. The ignition flame size is 1.5 cm radius and the during is 0.5s. Fig.10 shows the temperature profile of gas phase after the ignition at 0.01s, 0.1s, 0.2s, 0.3s, 0.4s and 0.5s respectively. The green color represents the flame position, which shows the flame moves upward.

Fig.10 (e) indicates that the flame reaches the top of the chamber at 0.4s after ignition. That means the flame travels 15 cm within 0.4s. Therefore, the flame propagation speed is estimated to 37.5 cm/s. Compare to the experimental result, the models used in the simulation underestimate the flame speed.

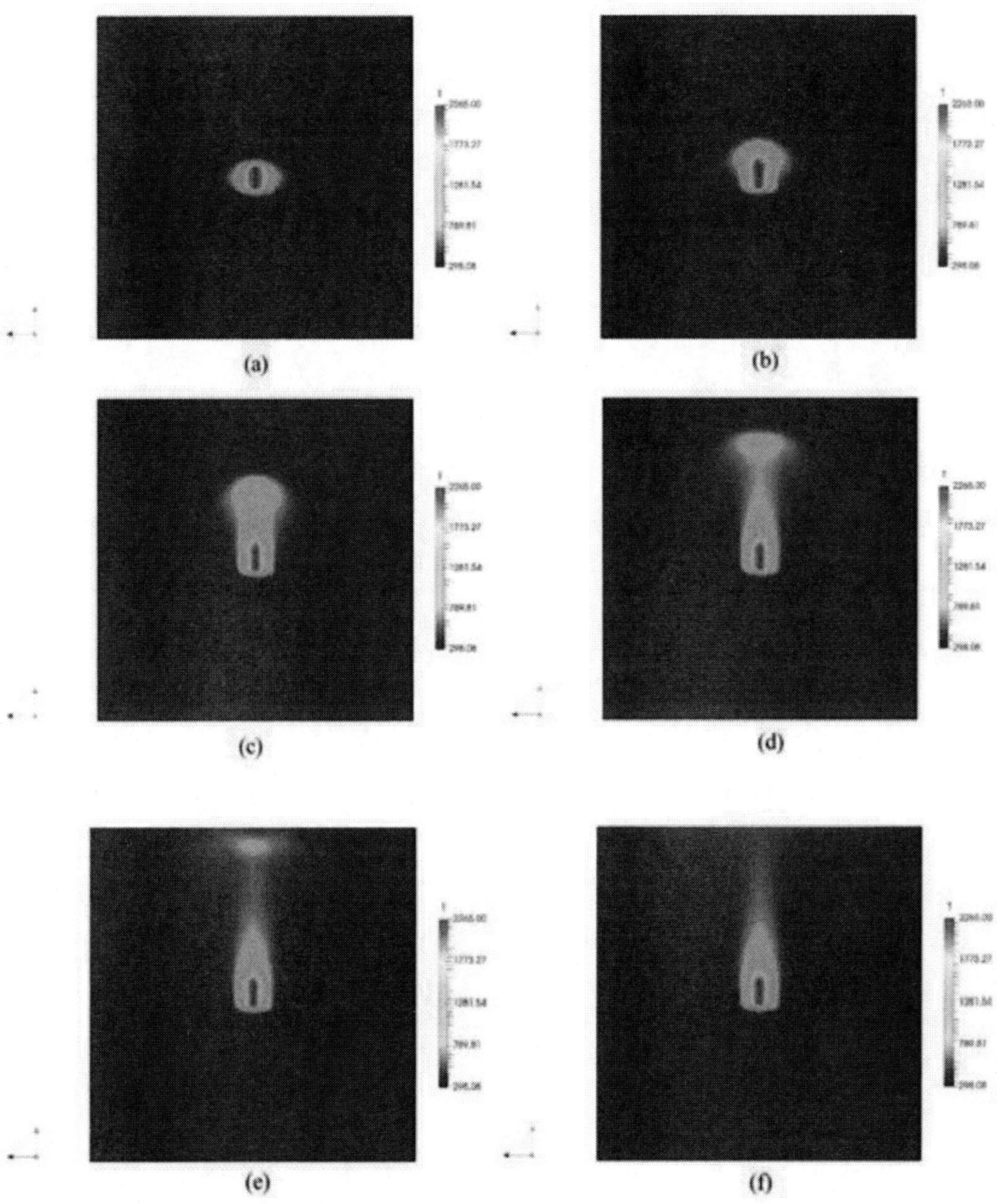

Fig.10 OpenFOAM simulation results of aerosol combustion after the ignition at time (a) 0.01s (b) 0.1s (c) 0.2s (d) 0.3s (e) 0.4s (f) 0.5s

4. Conclusions

In the study, the flame speed of D-600 obtained via experiment can validate the trend of flame speed with respect to droplet size when the droplet size is large. For the simulation of aerosol droplet generation process, the breakup model and injection model can achieve the task because the droplet size obtained from the simulation and the distribution of droplet position fit the experimental results. For the simulation of aerosol combustion process, the flame propagation speed is underestimated when compared with the experimental results. The potential explanation is the ventilation and the moving of ignition source will create more turbulence, which increases the flame propagation during experiment. This study opens up the possibility to use simulation to investigate the ignition and combustion of aerosol cloud.

5. Reference

[1] NFPA 30, *Flammable and Combustible Liquids Code, 2015 edition*, National Fire Protection Association: Quincy, MA.

[2] *Globally Harmonized System of Classification and Labelling of Chemicals (GHS)*, 2013, New York and Geneva: United Nations

[3] Huang, Szu-ying, Xinrui Li, and M. Sam Mannan. *The characterization of heat transfer fluid P-NF aerosol combustion: Ignitable region and flame development.* Journal of Loss Prevention in the Process Industries, 2013. **26**(6): p. 1134-1144.

[4] Lin, Yan-Ru, Hao Chen, Chad Mashuga, and M. Sam Mannan. *Improved electrospray design for aerosol generation and flame propagation analysis.* Journal of Loss Prevention in the Process Industries, 2015. **38**: p. 148-155.

[5] Lian, Peng, Andres F. Mejia, Zhengdong Cheng, and M. Sam Mannan. *Flammability of heat transfer fluid aerosols produced by electrospray measured by laser diffraction analysis.* Journal of Loss Prevention in the Process Industries, 2010. **23**(2): p. 337-345.

[6] Eichhorn, J., *Careful! Mist can explode.* Petroleum Refiner, 1955. **34**(11): p. 194-196.

[7] Ballal, D. R., and Lefebvre, A. H.. *Flame propagation in heterogeneous mixtures of fuel droplets, fuel vapor and air.* Symposium (International) on combustion. Vol. 18. No. 1. Elsevier, 1981.

[8] Burgoyne, J. and L. Cohen. *The effect of drop size on flame propagation in liquid aerosols.* Proceedings of the Royal Society of London A: Mathematical, Physical and Engineering Sciences, 1954. **225**(1162): p. 375-392.

[9] Polymeropoulos, Constantine E. *Flame propagation in aerosols of fuel droplets, fuel vapor and air.* Combustion Science and Technology, 1984. **40**(5-6): p. 217-232.

[10] Lian, Peng. *Flammability and combustion behaviors in aerosols formed by industrial heat transfer fluids produced by the electrospray method.* PhD dissertation., Texas A&M University, 2011.

[11] Costa, Anthony B., and R. Graham Cooks. *Simulation of atmospheric transport and droplet–thin film collisions in desorption electrospray ionization.* Chemical Communications, 2007 **38**: p. 3915-3917.

[12] Gjesing, Rasmus, Jesper Hattel, and Udo Fritsching. *Coupled atomization and spray modelling in the spray forming process using open foam.* Engineering Applications of Computational Fluid Mechanics, 2009. **3**(4): p. 471-486.

[13] Huang, Chen, and Andrei Lipatnikov. *Modelling of gasoline and ethanol hollow-cone sprays using OpenFOAM.* No. 2011-01-1896. SAE Technical Paper, 2011.

[14] Frackowiak, Bruno, Gérard Lavergne, Cameron Tropea, and Alain Strzelecki. *Numerical analysis of the interactions between evaporating droplets in a monodisperse stream.* International journal of heat and mass transfer, 2010. **53**(7): p. 1392-1401.

[15] Jamali, Seyed Hossein. *Computational Modeling of Turbulent Ethanol Spray Flames in a Hot Diluted Coflow.* Master of Science in Mechanical Engineering Thesis, Delft University of Technology, 2014.

13TH GLOBAL CONGRESS ON PROCESS SAFETY

Elongated VCE Blast Waves and Structural Damage

Jihui Geng, Ph.D.
Thomas Mander, P.E.
J. Kelly Thomas, Ph.D.
Quentin Baker, P.E.
Baker Engineering and Risk Consultants, Inc. (BakerRisk)
3330 Oakwell Court, Suite 100
San Antonio, TX 78218
JGeng@BakerRisk.com

Prepared for Presentation at
American Institute of Chemical Engineers
2017 Spring Meeting and 13th Global Congress on Process Safety
San Antonio, TX
March 26 – 29, 2017

Elongated VCE Blast Waves and Structural Response

Jihui Geng, Thomas Mander, J. Kelly Thomas & Quentin Baker
Baker Engineering and Risk Consultants, Inc. (BakerRisk)
3330 Oakwell Court, Suite 100
San Antonio, TX 78218
JGeng@BakerRisk.com

Keywords: Vapor cloud explosion, VCE, elongated congestion, blast wave shape, structural response

Abstract

Elongated congested volumes are common at chemical processing and petroleum refining facilities due to the arrangement of processing units, but there have been relatively few evaluations reported for the blast loads produced by elongated vapor cloud explosions (VCEs). The accidental VCE that occurred at the Buncefield, UK facility in 2005 involved an elongated congested volume formed by the trees and undergrowth along a portion of the site boundary. Some of near-field damage indicators present at the Buncefield site could not be reasonably explained using existing standard VCE blast load prediction techniques that are based on an assumption that the congested volume filled with flammable gas cloud is hemispherical and located at grade level.

This paper summarizes recent work to define the characteristics associated with elongated congested volume VCEs and identify differences relative to standard VCEs involving compact congested volume geometries. The main conclusions from this work with regard to the blast wave shape for an elongated congested volume deflagration are: 1) the blast wave behaves as an acoustic wave along the long axis, 2) the blast wave has a very quick transition from the positive phase peak pressure to the negative phase peak pressure (i.e., quick relative to the positive phase duration), and 3) the magnitude of the pressure drop between the peak positive and negative pressures diminishes quickly with distance outside the congested volume. These observations are not consistent with the behavior of a compact congested volume geometry VCE blast wave.

Deflagration and deflagration-to-detonation transition (DDT) regimes were also identified for unconfined elongated congested volume VCEs as a function of the normalized flame travel distance and flame speed. These regimes were verified with existing test data, including data from the on-going RPSEA test program. These observations and the DDT regime identification provide a frame of reference to develop a better understanding of elongated congested volume VCEs.

Generic structures, with properties typical of conventional construction, were analyzed to illustrate the effect of the elongated congested volume VCE blast loads on structural response. The results show that, for an elongated congested volume VCE, a high flame speed deflagration may result in more severe structural response in the near-field than for a detonation (i.e., due to a DDT). These results provide an alternative near-field damage indicator analysis approach for the investigation of elongated congested volume VCE incidents.

1 Introduction

The accidental VCE which occurred at the Buncefield, UK facility in 2005 involved an elongated congested volume. It was concluded[1] that the trees and undergrowth along a plant boundary at the Buncefield site, which formed an elongated congested volume, caused flame acceleration up to a velocity of several hundred m/s. This elongated congestion played a key role in the flame acceleration, resulting blast loads and subsequent observed structural damage. In particular, the blast loads from the Buncefield incident were reported to show overpressures up to about 2 barg at the cloud/congestion edge, which then diminished rapidly with distance.

Blast waves from explosion sources like a VCE, pressure vessel burst or high explosive exhibit both positive and negative phases, and the relative magnitude of the positive and negative phases varies among these types of explosion sources and the specific source characteristics[2]. VCEs can be categorized into two modes, deflagrations and detonations, according to propagation mechanisms. In a vapor cloud deflagration, the flame propagates through the unburned fuel-air mixture at a flame velocity less than the speed of sound in the reactant gas mixture, while a detonation propagates at supersonic velocity. The blast curves for simplified VCE blast load prediction methodologies are normally based on an assumption that the flammable gas cloud is hemispherical and located at grade level[3]. Numerous studies[19], [6] have shown that as the flame speed increases, the rate of pressure rise and the peak blast overpressure both increase. VCE blast waves can be characterized in three regimes[7]: acoustic wave, pressure wave and shock wave. These regimes can be overlaid on Baker-Strehlow-Tang (BST) scaled overpressure curves[8].

Although elongated congested volumes are common, there have been few evaluations reported for the blast loads produced by elongated VCEs[9] and the blast loads from the elongated VCEs have not been fully addressed in VCE prediction methods. The authors recently performed an evaluation[10], [11] to characterize elongated VCEs and to identify the resultant difference in blast wave shape compared to those predicted by well-known VCE blast prediction methodologies. BakerRisk's Blast Wave Target Interaction (BWTI™) computational fluid dynamics (CFD) code[11] was used to investigate the VCE blast loads resulting from elongated flammable gas clouds. The BWTI™ predictions[10] were validated against test data from the recent Buncefield JIP[1]. A blast wave diagram was developed[10] to illustrate typical waves generated during the flame acceleration process and their subsequent interaction. The evaluation showed that the blast wave shape for a deflagration in an elongated congested volume is similar to that for an acoustic wave in the near-field along the long-axis direction. An elongated VCE blast wave has a very quick transition from the positive phase peak pressure to the negative phase peak pressure, relative to the positive phase duration. The magnitude of the applied negative pressure on a building face depends strongly on the transition time between the positive and negative phase peak pressures, and this applied negative phase can be important to structural response under certain conditions[8].

The influence of the normalized flame travel distance and the flame speed on the resultant blast wave shape was further investigated [11], and it was determined that the blast wave shape in the near-field around the congested cloud boundary can be significantly impacted by the cloud aspect ratio and the flame speed attained.

This evaluation extends the previous work[11], examining how the attained flame speed impacts the resultant blast wave shapes and corresponding blast loads immediately downstream of the cloud end along the long-axis direction. These blast loads were further used to analyze the structural response of conventional structures.

2 Terminology

2.1 Elongated VCE Parameters

Figure 1 depicts an elongated flammable cloud with a length L, a width W and a height H. It is assumed that the cloud covers a congested volume (i.e., that none of the cloud occupies an empty uncongested space), so that the cloud and the congested volume have the same dimensions. The parameters shown in Table 1 are utilized to characterize the cloud geometry and the resultant blast field.

The "*Free Vent Distance*" (L_{FV}) is defined as the minimum of the cloud height and half the cloud width. The "*Cylindrical Distance*" (L_{Cyl}) is introduced to measure the maximum distance a cylindrical flame front can travel before reaching the edge of the cloud. The "*Flame Travel Distance*" (L_f) is the distance from ignition to the flame front, with a maximum value equal to the distance to the edge of the cloud; the ignition source could be located at the cloud center, cloud end or anywhere in the cloud. The "*Target Standoff Distance*" (L_{SD}) is the shortest distance measured from the congested cloud boundary to the target of interest. The circle with an "X" marked through it in Figure 1 denotes the location of a target of interest. The last two parameters shown in Table 1 are normalized parameters used to characterize flame travel and standoff distance.

For the purposes of this paper, a flammable cloud is considered to be "elongated" if the ratio of the cloud length to the free vent distance (L/L_{FV}) is larger than 10 and the ratio of the cylindrical distance to the free vent distance (L_{Cyl}/L_{FV}) does not exceed 2. These conditions are summarized in Table 2, and ensure that the maximum normalized flame travel distance is at least $Ln_f = 5$ if a cloud is ignited at its center.

The elongated cloud is characterized by two parameters: (1) the aspect ratio *(L/W/H)* and (2) a characteristic distance represented by either the free vent distance (L_{FV}) or the equivalent radius (R_{Eq}). The four cloud parameter sets specifically considered in this work are given in Table 3. The "Example Case" cloud is a parametric example case, whereas the "BakerRisk", "Buncefield" and "SRI RPSEA (GexCon)" clouds correspond to existing experimental test rigs. The "Example Case" cloud listed in Table 3 has dimensions of L=37 m, W=7.32 m and H=1.83 m and has (1) an aspect ratio of 20/4/1 and (2) a free vent distance (L_{FV}) of 1.83 m and an equivalent radius (R_{Eq}) of 2.92 m (along with a cylindrical distance of 3.66 m). This cloud is referred to as "20/4/1 – 1.83 m" or "20/4/1 – R 2.92 m" within the context of this paper. The BakerRisk test rig[13], [15] has dimensions of L=14.6 m (48 ft), W=3.66 m (12 ft) and H=1.83 m (6 ft) and is referred to as "8/2/1 – 1.83 m" or "8/2/1 – R 2.06 m". The Buncefield JIP Test 3.2 #4 configuration[1] has dimensions of L=30 m, W=4.5 m and H=3 m is referred to as "10/1.5/1 – 2.25 m" or "15/1.5/1 – R 2.93 m". The SRI RPSEA (GexCon) test #5 configuration[16] has dimensions of L=48 m, W=8 m and H=4 m is referred to as "12/2/1 – 4 m" or "15/1.5/1 – R 4.5 m".

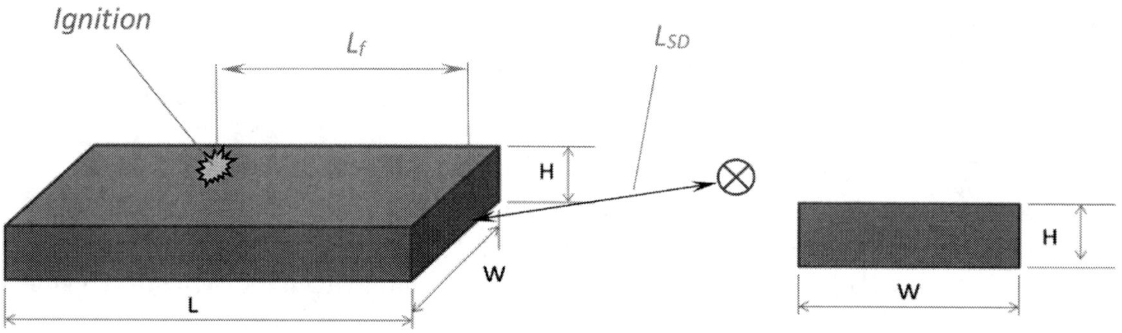

Figure 1. Dimensions of Elongated Vapour Cloud (L/L_{FV}>10)

Table 1. Cloud Geometry and Blast Field Parameter Definitions

Parameter	Symbol	Definition
Free Vent Distance	L_{FV}	$\mathrm{MIN}\left(H, \frac{W}{2}\right)$
Cylindrical Distance	L_{Cyl}	$\mathrm{MAX}\left(H, \frac{W}{2}\right)$
Equivalent Radius of Cross Section	R_{Eq}	$\sqrt{\frac{2 \cdot H \cdot W}{\pi}}$
Flame Travel Distance	L_f	*n/a*
Target Standoff Distance	L_{SD}	*n/a*
Normalized Flame Travel Distance	Ln_f	L_f / L_{FV}
Normalized Target Standoff Distance	Ln_{SD}	L_{SD} / L_{FV}

Table 2. Conditions for Elongated Cloud

Condition	Expanded Condition	If H > W/2	If H < W/2
$L/L_{FV} > 10$	$L > 10 \cdot \mathrm{MIN}\left(H, \frac{W}{2}\right)$	$L > 5W$	$L > 10H$
$L_{Cyl}/L_{FV} < 2$	$\mathrm{MAX}\left(H, \frac{W}{2}\right) \leq 2 \cdot \mathrm{MIN}\left(H, \frac{W}{2}\right)$	$H < W$	$W < 4H$

Table 3. Elongated Vapor Cloud Characteristic Parameters

Test Rig	Dimensions (m)						Aspect Ratio (L/W/H - L_FV)	Volume (m^3)
	Length	Width	Height	L_FV	L_Cyl	R_Equiv		
Example Case	36.6	7.32	1.83	1.83	3.66	2.92	20/4/1 - 1.83 m	490
BakerRisk	14.6	3.66	1.83	1.83	1.83	2.06	8/2/1 - 1.83 m	98
Buncefield JIP	30	4.5	3	2.25	3	2.93	10/1.5/1 - 2.25 m	405
SRI RPSEA (GexCon)	48	8	4	4	4	4.51	12/2/1 - 4 m	1,536

2.2 Elongated VCE Characteristic Parameters

The blast field generated by an elongated cloud depends on several characteristic parameters. The three characteristic parameters examined in this work are as follows:

1) Characteristic time (t_n), which is defined as the time for a rarefaction wave to travel from the cloud edge to the cloud center or ignition location.
2) Radial expansion factor (F_{Exp}), which is defined in terms of the expansion of unburned gas ahead of the flame front.
3) Energy release rate (dV/dr), which is expressed in terms of the volume of flammable gas cloud consumed per unit flame travel distance (V is volume and r is the flame front position).

The characteristic time (t_n) is defined by:

$$t_n = \frac{R_{Eq}}{C} \quad (1)$$

where C is the ambient sound speed (taken to be that for air, 334 m/s = 1.1 ft/ms).

The constant-volume adiabatic combustion (i.e., combustion in a fully enclosed chamber without heat loss) of a stoichiometric hydrocarbon fuel-air mixture will yield a final to initial pressure ratio of approximately 8. Similarly, constant-pressure combustion (i.e., flame moves at a very slow speed in an unconfined space) of a stoichiometric hydrocarbon fuel-air mixture will yield a final/initial volume ratio (i.e., the volume expansion factor) of approximately 8 for a stoichiometric hydrocarbon fuel-air mixture. Hence, for an unconfined hemispherical fuel-air cloud on the ground which is ignited at its center and has a low flame speed, the radial expansion factor (F_{Exp}) will be approximately 2 (i.e., cube root of the final/initial volume ratio).

In reality, if obstacles are present, turbulence will be generated in the combustion-induced flow field ahead of the flame. This turbulence will increase the local burning rate by increasing both the flame surface area and enhancing mass and energy transport. The increased burning rate will produce higher unburned gas flow velocities. This feedback loop results in an acceleration of the propagating flame. During this transient process, the radial expansion factor (F_{Exp}) is a function of flame speed, and will decrease with increased flames speed. Since the volume expansion factor for the constant-pressure combustion of a fuel-air mixture is around 8, the radial expansion factor for an accelerated flame will be between 1 and 2, depending on flame speed.

The burning of a fuel-air mixture within a moving flame front releases combustion energy. Characterizing the energy release rate as *dV/dr* assumes a constant energy density in the cloud, which matches the conditions of the test cases evaluated. Of course, *dV/dr* would actually be the energy release rate only when multiplied by the flame speed and combustion energy released per unit volume. The expression for this parameter depends on the location of the flame front within the cloud. Consider, for example, the case of the "Example Cloud" cloud in Table 3 ("20/4/1 – 1.83m") with ignition at the cloud center. Figure 2 shows four different flame front positions and corresponding times; expansion and distortion of the cloud due to the passage of the flame front is not reflected in this figure for the sake of simplicity.

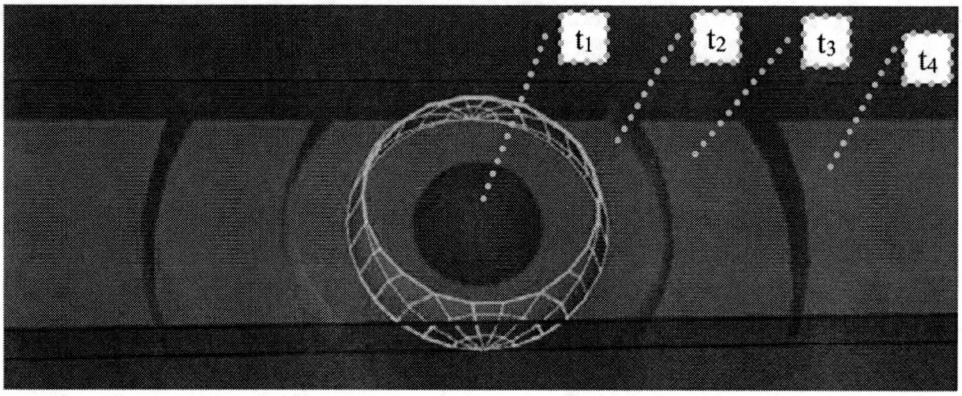

Figure 2. Schematic of Flame Shape in Elongated Cloud (20/4/1 -1.83 m Cloud)

At time t_1 the flame reaches the top of cloud (i.e., $r = L_{FV}$), the flame is spherical, and the energy release rate up to this time is:

$$dV/dr_{r<L\_FV} = 2\pi r^2 \tag{2}$$

Between t_1 and t_2 the flame travels to the side of the cloud/congestion (i.e., at t_2, $r = L_{Cyl}$), at which point the flame front shape is neither spherical nor cylindrical. The energy release rate during this time period (t_1 to t_2) can be approximated as:

$$dV/dr_{L_{FV}<r<L_{Cyl}} = 2\pi L_{FV} r \tag{3}$$

Time t_4 is defined as that time at which the flame front reaches $2L_{Cyl}$, and there after the flame front is almost planar and the energy release rate can be expressed as:

$$dV/dr_{r>2L_{Cyl}} = WH \tag{4}$$

Between t_2 and t_4, the energy release rate can be linearly extrapolated from a rate of $2\pi L_{FV} r$ to *WH*.

Figure 3 provides an example of a normalized energy release rate distribution along the long-axis of the cloud, both without and with cloud expansion taken into account, as shown in the top and bottom charts, respectively. The energy release rates were normalized by LW, the cloud cross sectional area, so that the energy release rate is unity at long flame travel distances ($>2L_{Cyl}$). The flame travel distance was normalized by the free vent distance (L_{FV}), so that the total cloud length would be approximately 20 (i.e., 37/1.83 = 20.2) and the half-length of the cloud would be approximately 10. A radial expansion factor (F_{Exp}) of 1.5 was used for illustrative purposes. The normalized energy release rate distribution without cloud expansion (top chart) matches the schematic description shown in Figure 2. However, the normalized energy release rate distribution with cloud expansion (bottom chart) better explains the blast field resulting from an elongated VCE, as discussed in the following section.

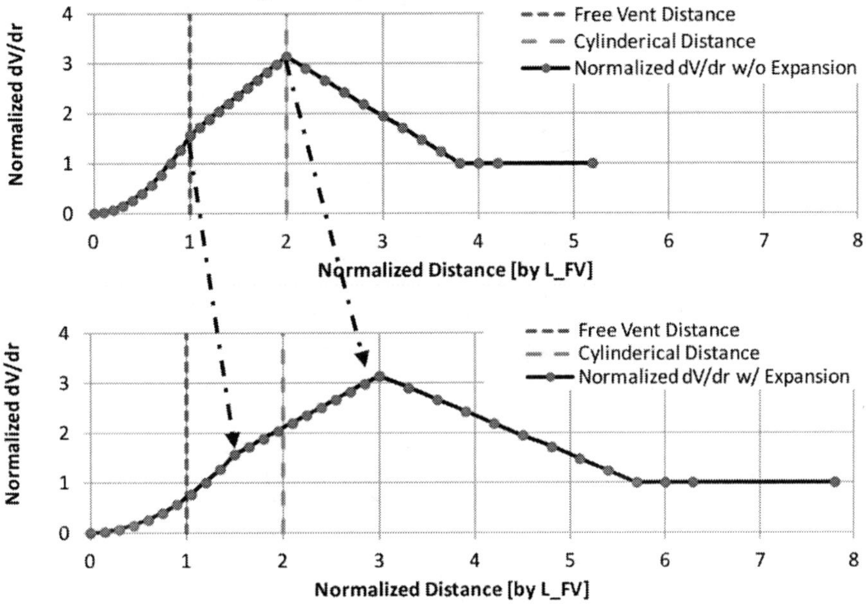

Figure 3. Energy Release Rate Distribution without (top) and with (bottom) Cloud Expansion (20/4/1 -1.83 m Cloud)

This "Example Case" cloud of 20/4/1 – 1.83 m was also modeled in Reference [10] with the FLACS CFD tool[17]-[18]. Three phases of the flame acceleration were observed: (1) one during the time when the flame front is spherical, (2) a second when the flame front is cylindrical, and (3) the latter portion of the flame propagation, where the flame speed is approximately steady (constant) when the flame front is planar. The first two phases of the flame acceleration are consistent with the energy release rate distribution versus the normalized distance (with cloud expansion) shown in Figure 3.

2.3 Impact of Ignition Location

As given in Table 3, the "Example Case" cloud of 20/4/1 – 1.83 m has a cylindrical distance of L_{Cyl} =3.66 m, 2 times the free vent distance (i.e., $L_{Cyl} / L_{FV} = 2$), while BakerRisk test rig and SRI RPSEA (GexCon) test rig have a $L_{Cyl} / L_{FV} = 1$. Thus, the flame acceleration for these two test rigs will be primarily produced by a hemi-spherical flame front. This primary flame acceleration also depends on ignition location as shown in Figure 4 and summarized in Table 4 for a "20/2/1 – 4 m" cloud.

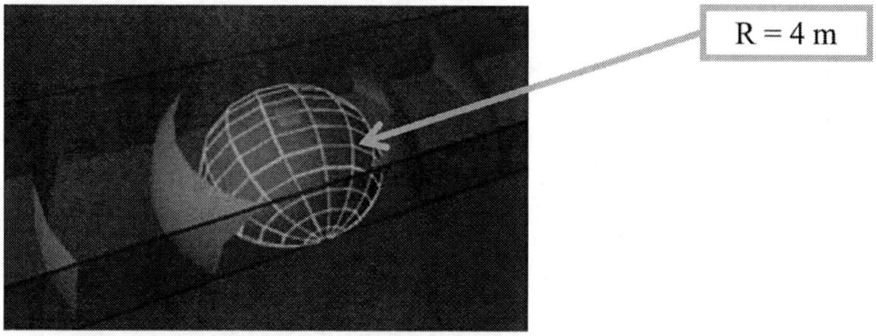

Central Ignition at Grade Level

Central Ignition at Mid-level

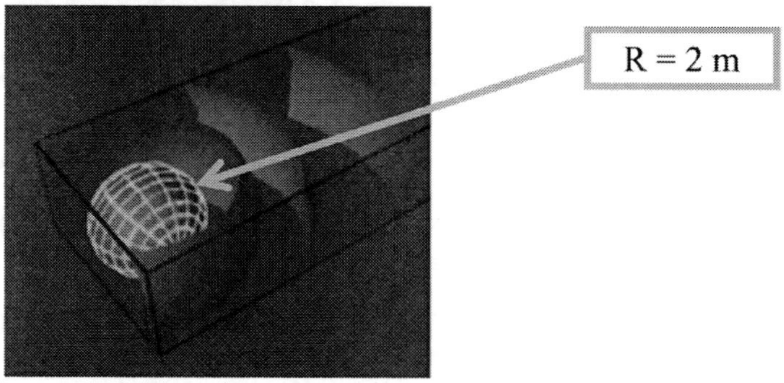

End Ignition at Mid-level

Figure 4. Impact of Ignition Location on Initial Energy Release (12/2/1- 4 m Cloud)

Table 4. Impact of Ignition Location on the Initial Energy Release

Case No	Aspect Ratio	Ignition Location		L_{FV} at Beginning	Overall Energy Release Rate	Distance for DDT
	L/W/H - $L_{\_FV}$	Footprint	Elevation			
1	20/2/1 - 4 m	Central	Ground	4 m	Largest	Shortest
2		Central	Mid-level	2 m	Moderate	Middle
3		End	Mid-level	2 m	Smallest	Longest

For Case 1 in Figure 4, the cloud of "20/2/1 – 4 m" is ignited at the center at grade level. The free vent distance is L_{FV} = 4 m for this case, and the hemi-spherical flame front for this case can travel 4 m. For Case 2 in Figure 4, the same cloud is ignited at the center at mid-level (i.e., at an elevation of 2 m), and the spherical flame front can only travel about 2 m before the flame reaches the top of cloud. The free vent distance at the beginning for this case is therefore 2 m, which is half of that for Case 1. For Case 3 in Figure 4, the same cloud is ignited at end location (2 m inside cloud) at mid-level, and the spherical flame front can only travel 2 m before the flame reaches both the top of cloud and one end of cloud. The free vent distance at the beginning for this case is therefore 2 m, which is half of that for Case 1. Moreover, there is no more energy release on the ignition side afterwards for this case once the flame reaches the end of the cloud and, as a result, the overall energy release rate for Case 3 is the lowest among these three cases. The highest energy release is for Case 1, which will result in the shortest distance for DDT occurrence for Case 1 (Central Ignition, on ground), provided that a DDT would occur. The longest distance for DDT would be for Case 3 (End Ignition, at mid-level). These observations are summarized in Table 4.

3 Characteristics of Elongated VCEs[10]

3.1 BWTI's Simulation and Validation[10] against Buncefield JIP Test Data

The elongated cloud configuration (15/1.5/1 – 2.25 m, 15/1.5/1 – R 2.93 m) for Test 3.2 #6 from the Buncefield JIP test program was modeled in Reference [10]. The cloud, which covered a congested volume made up of vegetation, was 45 m long, 4.5 m wide and 3 m high. Ignition was at one end of the cloud, 1 m inside the cloud edge. The cloud length ratio (L/L_{FV}) is greater than 15. BakerRisk's BWTI™ CFD code[12] was selected to model the blast loads due to the large length ratio, using a prescribed flame speed that matched the measured flame speed, which ensured that the constant flame speed after the initial acceleration phase in the cloud was modeled reasonably. BWTI™ does not directly simulate turbulent combustion; rather, an energy wave approach[4] with a prescribed flame speed is used in BWTI™. The blast wave shape predicted by BWTI is in reasonable agreement with the test data, including the double peaks. The agreement at the end of the cloud is particularly good for both the positive and negative overpressure.

3.2 Blast Wave Diagram for Elongated VCEs

A diagram was also developed to illustrate typical waves generated during the flame acceleration process and their subsequent interaction in Reference [10]. It includes a) a flame front, b) a leading acoustic wave due to any disturbance induced by flame movement, c) primary peak pressure due to the initial flame acceleration, d) secondary peak pressure due to the isolated high pressure core generated by the flame front after initial flame acceleration, and e) the start of the negative phase. The same cloud configuration as the Buncefield Test 3.2 #6 (15/1.5/1 – 3 m or 15/1.5/1 – R 2.93 m) was employed to generate distance time (x-t) wave diagrams and associated pressure histories at selected locations in order to examine the impact of the flame speed on the resultant blast wave shape. A constant flame acceleration distance of 6 m (i.e., L_{Acc}/L_{FV} =2.7) was assumed, with a steady flame speed for the remaining cloud region. Three flame Mach numbers, 0.3, 0.6 and 0.9, were investigated to determine the impact of the flame speed attained at the end of the initial flame acceleration phase on the pressure history developed inside the congested volume.

Figure 5 shows the resultant pressure histories[10] at selected normalized distances for peak flame speeds of (a) $M_f = 0.3$, (b) $M_f = 0.6$ and (c) $M_f = 0.9$. It should be noted that normalized time (ratio of actual time to characteristic time, $t_n = 2.93$ m / 334 m/s = 8.8 ms) is used for the horizontal axis. Relevant observations based on Figure 5 are as follows:

- For $M_f = 0.3$ in Figure 5 (a), the low flame speed attained at the end of the initial flame acceleration phase allows the primary and second pressure waves to decouple beyond a normalized distance (L_f/L_{FV}) of approximately 6, such that the first pressure peak rapidly decays while the second peak remains at about 1.0 psig
- For $M_f = 0.6$, the pressure histories at normalized distances (L_f/L_{FV}) between 4.0 and 12.9 are given in Figure 5 (b). As can be seen by comparing Figure 5 (a) and (b), increasing the flame speed slows the decoupling between the primary and second pressure peaks, with the second peak remaining constant beyond an L_f/L_{FV} of approximately 10.
- For $M_f = 0.9$ (Figure 5 (c)), the energy release rate is sufficiently high that it provides positive feedback to the primary blast wave, which therefore experiences a slower decay (i.e., as compared to the decay rate at lower flame speeds). The peak pressure remains constant at about 10 psig beyond an L_f/L_{FV} of approximately 10.

The following conclusions were drawn based on the observations discussed above. For an unconfined elongated VCE with a uniform fuel concentration and congestion pattern, an initial rapid phase of flame acceleration occurs within the normalized distance (L_f/L_{FV}) region between 2 and 4. If a deflagration-to-detonation transition (DDT) does not occur within this range, an approximately constant subsonic flame speed will be reached within the normalized distance (L_f/L_{FV}) region between 4 and 6. During the constant flame speed phase, an isolated high pressure core travels with the flame front, followed by a negative phase. The negative phase is generated by the primary blast wave developed during the initial acceleration phase.

The blast wave shape within the elongated cloud and in the near-field exhibits a behavior belonging to the acoustic wave regime no matter how high the flame speed as long as the flame Mach number is less than unity (i.e., subsonic flame), while the magnitudes of the peak positive and negative phases are larger than with a traditional acoustic wave.

At a sufficiently high flame speed ($M_f > 0.6$), the blast wave outside the cloud/congestion (along the long-axis direction) experiences a significant pressure drop from the peak positive pressure to peak negative pressure within one characteristic time (t_n). The magnitude of the pressure drop between the peak positive and negative pressures diminishes quickly with distance. For the case where the flame speed reaches Mach 0.9, the pressure at the end of the cloud drops from the peak positive pressure (9.0 psig) to the peak negative pressure (-10 psig) within the characteristic time ($t_n = 8.8$ ms). The difference between the peak positive and negative pressure is 19 psig (1.3 barg). The drop between the peak positive and negative pressures diminishes quickly with distance. This localized pressure drop is due to the high pressure core traveling with the flame front (i.e., in the region beyond an Ln_f of 6) in an elongated VCE (see Figure 5).

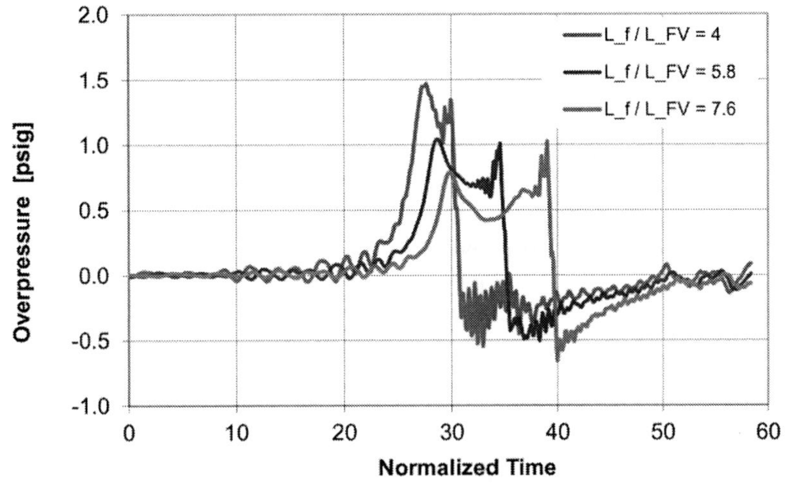

(a) $M_f = 0.3$

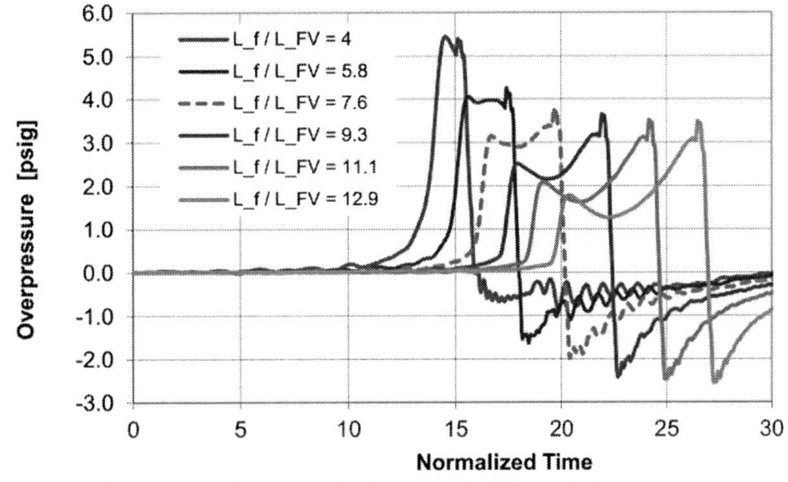

(b) $M_f = 0.6$

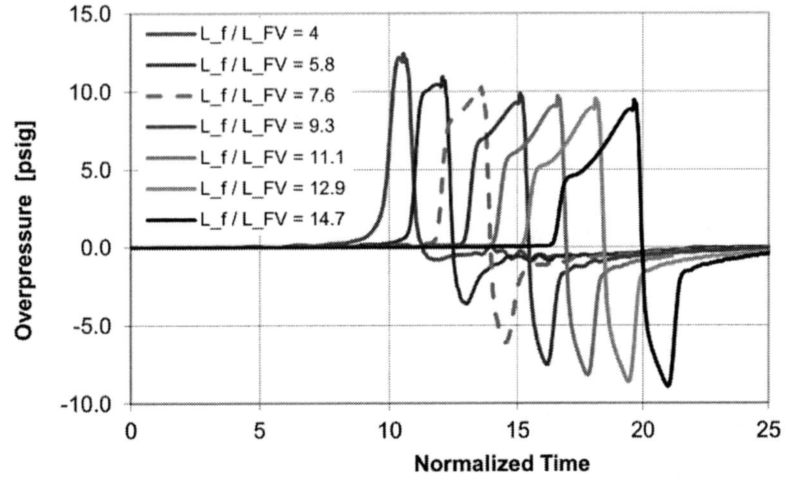

(c) $M_f = 0.9$

Figure 5. Impact of Flame Speed on Blast Wave Shape at Selected Locations

4 Deflagration versus DDT Regime

The previous work[10] concluded that if the flame speed does not exceed Mach 0.6 (i.e., M_f<0.6) within a normalized distance (L_f/L_{FV}) of 6, the slow speed of the steady flame allows the primary and second pressure waves to decouple after this distance. The first peak then rapidly decays, while the second peak remains at a constant value (depending on the steady flame speed achieved).

If a flame speed of Mach 0.9 is attained within a normalized distance (L_f/L_{FV}) of 2 and 4, the decoupling between the primary peak and the secondary peak beyond an L/L_{FV} of 4 is diminished. A DDT may occur if a positive-feedback disturbance (e.g., congestion pattern change) within the normalized distance range of 4 to 6 is present.

At a flame speed of between Mach 0.6 and 0.9, the decoupling between the primary and secondary peaks is not significant in the normalized distance (L_f/L_{FV}) region between 4 and 6, and hence flames traveling at this speed would be most susceptible to a DDT in this range. Of course, the disturbance required for a Mach 0.6 flame to DDT would be much greater than with a Mach 0.9 flame.

Figure 6 illustrates deflagration versus DDT regimes as a function of normalized distance and flame speed based on these observations. DDTs in BakerRisk unconfined test rig were reported for both ethylene[13], [14] and hydrogen[15] mixtures, as shown in Figure 7 (a). A DDT resulted for ethylene within a normalized distance (L_f/L_{FV}) of 4 for an approximately stoichiometric mixture. With a hydrogen mole fraction of less than 18%, no DDT was observed and the deflagration flame speed was less than Mach 0.6. With a hydrogen mole fraction of about 22%, a DDT occurred within a normalized distance (L_f/L_{FV}) of around 3.

Figure 7 (b) was generated based on the Buncefield JIP test data[1]. Propane-air mixtures with an approximately stoichiometric concentration were used for all eight tests, with changing congestion patterns in terms of congestion levels and free vent distances (L_{FV}). Two DDT data sets and one deflagration data set are shown in Figure 7 (b). With the worst congestion pattern (3 trees/m^2, L_{FV}=2.25 m, Test3.2 #2), a DDT occurred within a normalized distance (L_f/L_{FV}) of 4. With a moderate congestion pattern (2 trees/m^2, L_{FV}=2.25 m, Test3.2 #4) a DDT occurred within a normalized distance (L_f/L_{FV}) of around 6. With a less severe congestion pattern (1.5 trees/m^2, L_{FV}=2.25 m, Test3.2 #3), no DDT was observed, resulting in a deflagration with a flame speed less than Mach 0.6.

Figure 7(c) was generated based on the SRI RPSEA (GexCon) test data[16]. Propane-air mixtures with an approximately stoichiometric concentration were used for two tests with a cloud of "12/2/1 – 4 m". Both tests evidenced DDT occurrence. The flame velocity data shown was digitized from the x-t diagram given in [16]. As discussed in Section 2.3, an end-ignition at mid-level (as used in RPSEA test program) resulted in DDT occurrence at a relatively long distance around L_f/L_{FV} = 5 and 6 for Test # 5 and #6, respectively. It is expected that a DDT will occur in a shorter distance if the cloud is ignited at a central location at grade level. The deflagration and DDT regimes in the SRI RPSEA tests are in general agreement with the observed DDTs and deflagrations in unconfined vapor clouds discussed above.

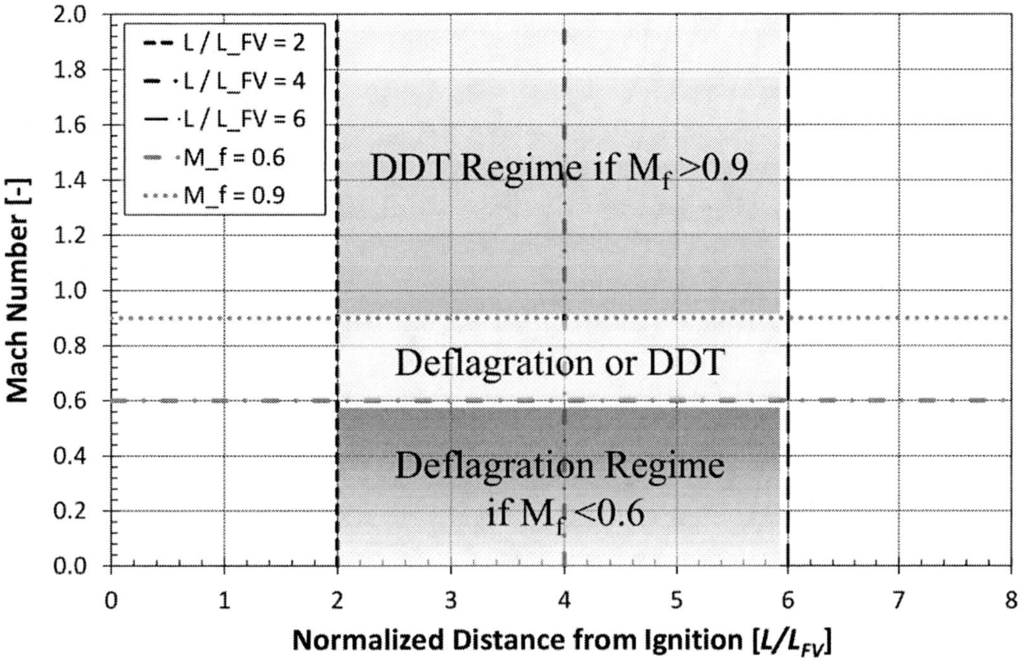

Figure 6. Deflagration vs. DDT Regimes

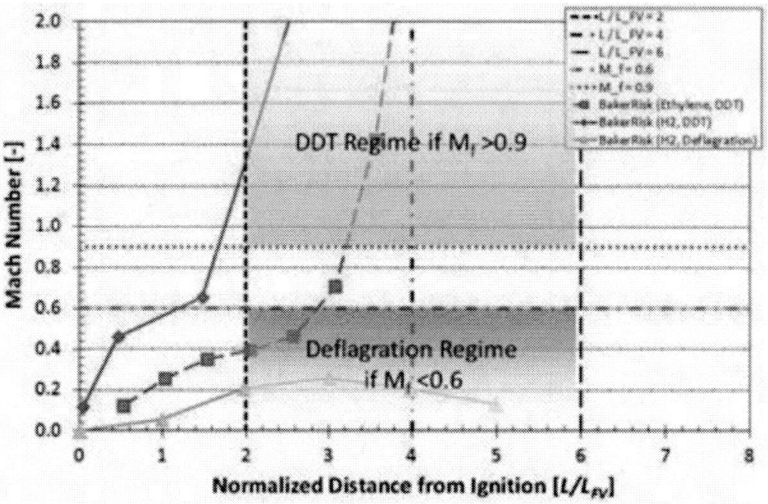

(a) BakerRisk Test Rig (8/2/1 – 1.83 m)

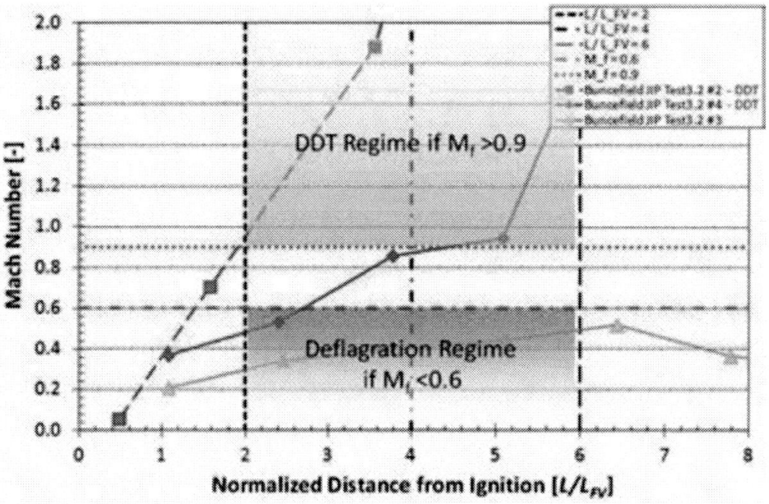

(b) Buncefield JIP Test Rig (10/1.5/1 – 2.25 m)

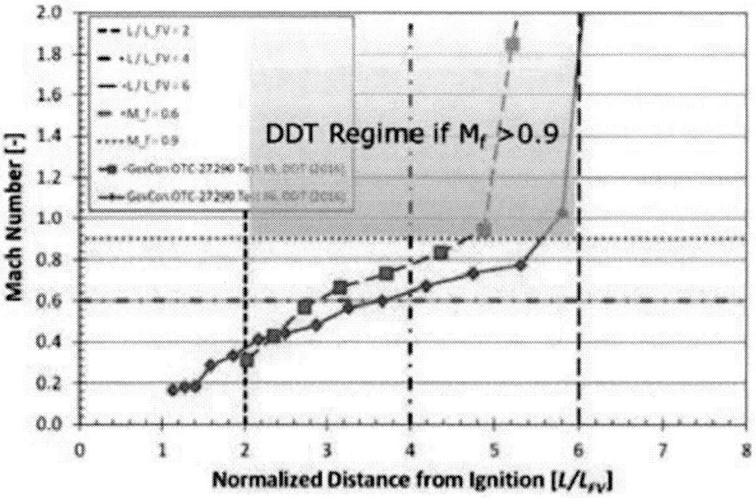

(c) SRI RPSEA (GexCon) Test Rig (12/2/1 – 4 m)

Figure 7. Observed Test Results

5 Blast Loads from Elongated VCEs with Higher Flame Speeds

The previous discussion is based on the assumption that the VCE occurs in a fully unconfined elongated congestion with a uniform fuel concentration and congestion pattern. It was found that for an initial rapid phase of flame acceleration occurs within the normalized flame travel distance ($Ln_f = L_f\, L_{FV}$) region between 2 and 4. An approximately constant subsonic flame speed will be reached within the normalized distance (Ln_f) region between 4 and 6 if a DDT does not occur. During the constant flame speed phase, an isolated high pressure core travels with the flame front, followed by a negative phase. The negative phase is generated by the primary blast wave developed during the initial acceleration phase.

As discussed in Section 2.2, the initial flame acceleration phase in a fully unconfined congested cloud mainly depends on the free vent distance (L_{FV}) and the cylindrical distance (L_{Cyl}) provided that a uniform cloud with a fixed fuel concentration and a constant congestion pattern are considered. On the other hand, flame speeds attained during the initial acceleration phase and the following constant flame speed phase are strongly influenced by a change a fuel concentration gradient (due to dispersion), confinement, congestion level/severity and/or free vent distance, often are present in a real world facility. For example, in another test (Test #8) of Buncefield JIP Test Program[1], increasing congestion level and decreasing free vent distance produced a high flame velocity (>200 m/s, $M_f > 0.6$) at a relatively large flame travel distance ($Ln_f > 20$), but no DDT. When a degree of confinement (due to thick plastic cover on top) is added to a methane-air cloud of 15/2/1 – 4 m[20], an extremely high and constant flame speed ($M_f \approx 2.5$) is achieved, but without a DDT.

BakerRisk's BWTI code was used to perform a parametric study to evaluate blast wave shape outside the cloud along the long-axis as a function of the flame speed attained. Flame speeds of $M_f = 0.7$, 0.9, 1, 1.5, 2 and 5.2 were run for a "12/2/1 - 4 m" cloud. $M_f = 5.2$ represents a DDT, where a detonation wave propagates through the cloud to the end of the cloud. These blast loads were employed to analyze the structural response of conventional structures, as discussed in Section 6.

Figure 8 shows the BWTI model employed. Targets to record the blast pressure history are distributed along the long-axis. Targets 1 through 14 are inside cloud, Target 15 at the cloud end, and Targets 16 through 20 are outside the cloud. The cloud was ignited near the left-hand end of the cloud, as denoted in the figure. The flame speed was increased linearly from the ignition location to the specified constant flame speed within a normalized flame travel distance (Ln_f) of 2.

The calculated blast pressure histories at the last to target locations inside the cloud and for targets outside of the cloud at normalized standoff distances (Ln_{SD}) of up to 3.8 are given in Figure 9 ($M_f = 0.7$), Figure 10 ($M_f = 2.0$) and Figure 11 ($M_f = 5.2$); the normalized standoff distance is defined in Section 2.1 (Table 1). Note that the scales used for the axes in these three figures differ. Direct comparison of pressure histories generated by different flame speeds ($M_f = 0.7$ up to 5.2) at selected normalized standoff distances of $Ln_{SD} = 0.8$, 1.5 and 2.3 are given in Figure 12, Figure 13 and Figure 14, respectively; again, the scales used for the axes in these three figures differ.

As can be seen, the detonation case ($M_f = 5.2$) blast wave decays faster than at lower flame speeds. There is a significant difference in blast wave shape between the supersonic ($M_f > 1.0$) and subsonic ($M_f < 1.0$) cases. The supersonic cases exhibit a normal shock front (i.e., zero rise time), while the subsonic cases exhibit an acoustic wave shape. Within and just outside the cloud, the subsonic case acoustic wave shape has a rapid decline from the positive peak pressure to the negative phase compared to its positive duration (i.e., this transition is rapid).

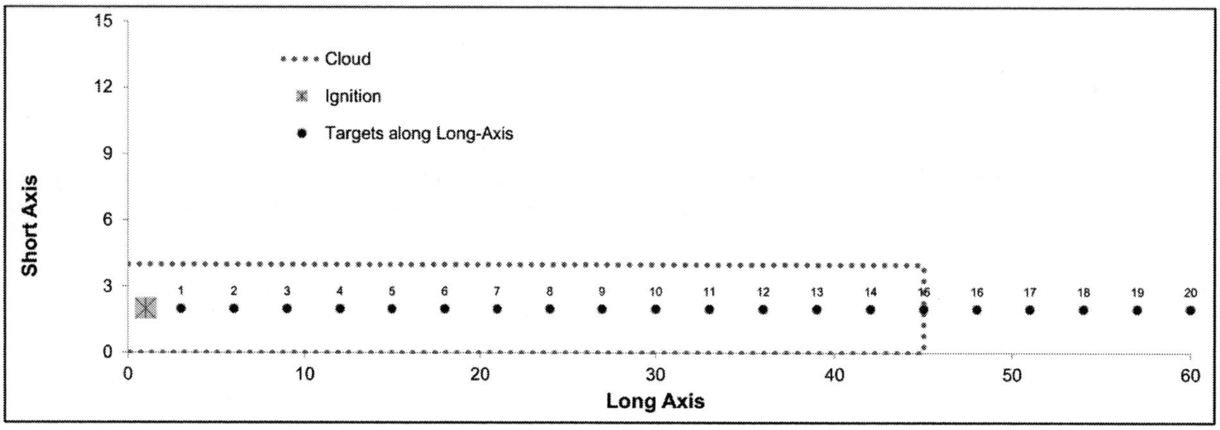

Figure 8. Target Distribution for a Cloud of 12/2/1 – 4 m

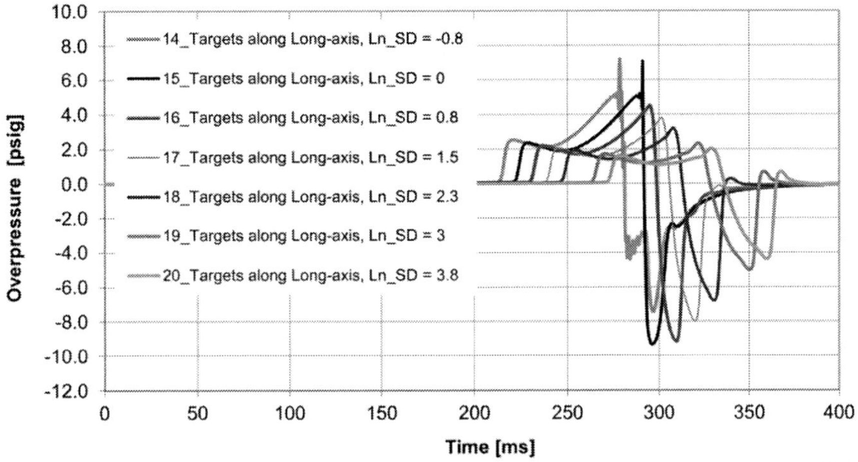

Figure 9. Near-field Blast Waves at Selected Locations (M_f = 0.7)

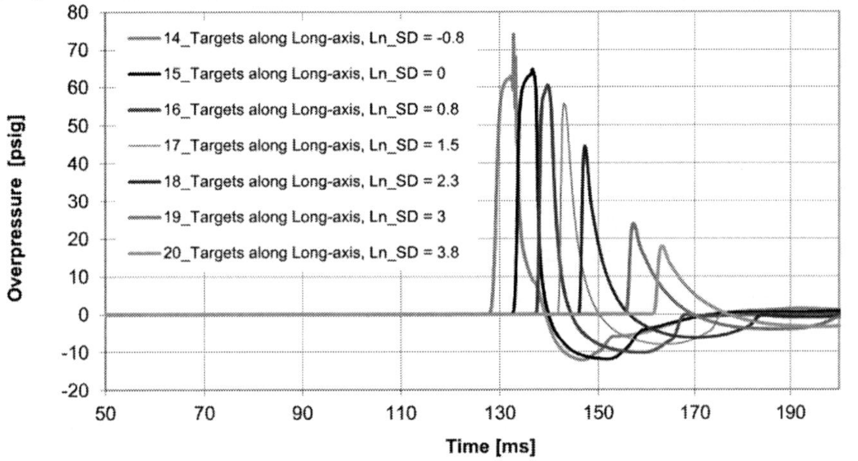

Figure 10. Near-field Blast Waves at Selected Locations (M_f = 2.0)

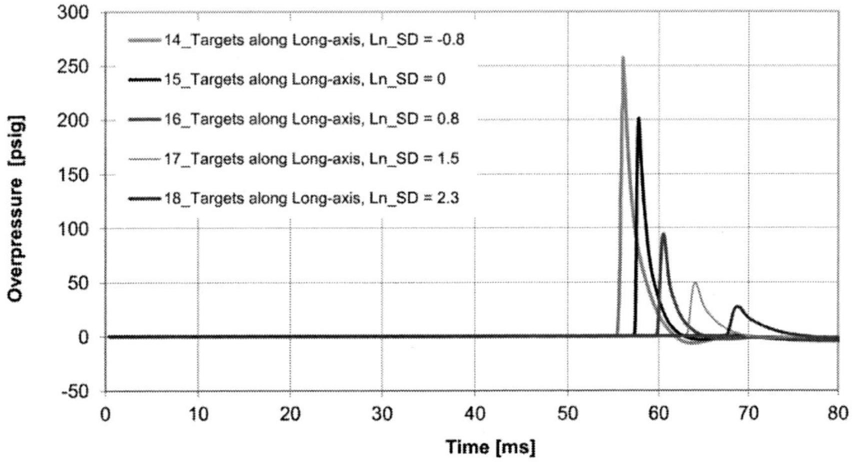

Figure 11. Near-field Blast Waves at Selected Locations (M_f = 5.2)

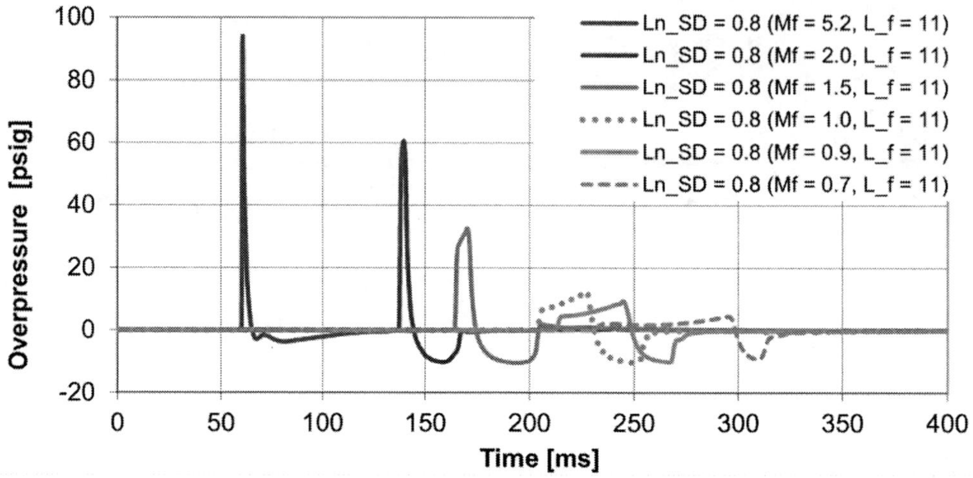

Figure 12. Comparison for Selected Flame Speeds (Ln_{SD}= 0.8)

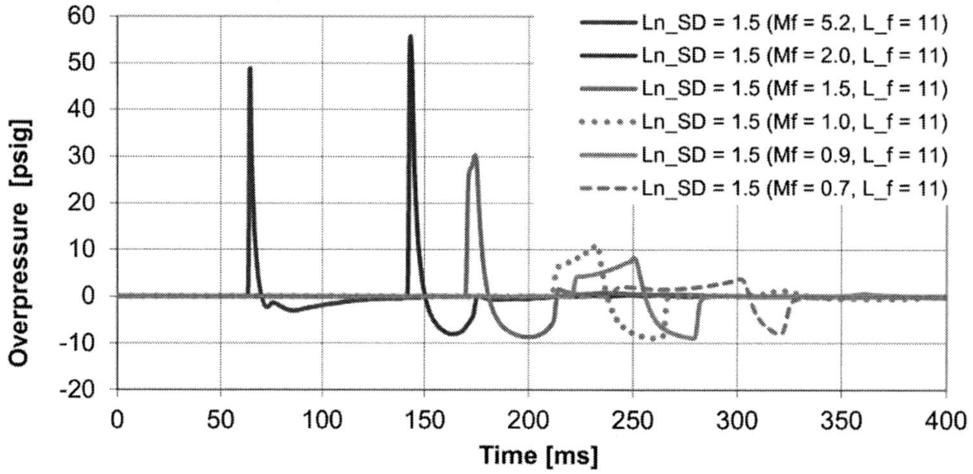

Figure 13. Comparison for Selected Flame Speeds (Ln_{SD}= 1.5)

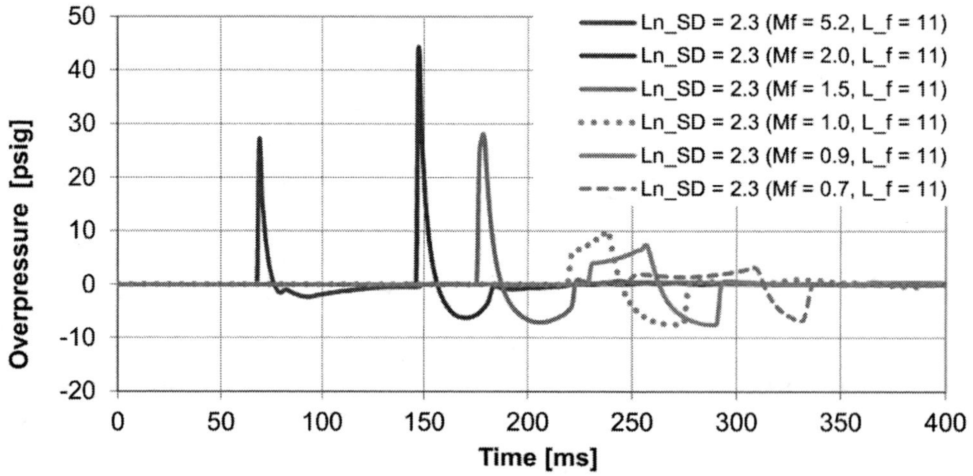

Figure 14. Comparison for Selected Flame Speeds (Ln_{SD}= 2.3)

6 Structural Analysis

Buildings and other structures are designed to withstand established conventional loads for strength and serviceability limit states. Strength limit states require structural members, and components to safely carry a factored static load or quasi-static overload without failure, such as snow or wind loads. Serviceability requirements restrict the deflection of components from everyday use. The former defines the strength, or resistance, of a structural component, while the stiffness is the primary concern for serviceability limits. These two quantities, resistance and stiffness, together with the structural mass, dictate the dynamic properties of a structural member, and how it will respond to blast loading. The structural response of a blast loaded component is also dependent on the applied blast overpressure magnitude, duration and waveform. This section investigates the dynamic response for a range of general structural components using the different blast waveforms presented in this paper.

The complexity of modeling blast loaded structures can be simplified by transforming individual components to equivalent single-degree-of-freedom (SDOF) oscillators. For example, the two-way plate shown in Figure 15(a) deforms as shown from an applied uniform pressure, with the maximum deflection occurring at the plate center. The maximum stresses within the plate can be used to determine the resistance, while the point of maximum deflection is used to compute the stiffness. The deformed shape is used to compute the effective mass that contributes to the inertial resistance under dynamic loading. As illustrated in Figure 15(b), these properties are used to covert the problem into an equivalent SDOF mass-spring-damper system.

When ductile components, such as a steel plate, are loaded beyond their elastic deflection limit, yielding occurs, and the load-carrying capacity is constant (i.e., as shown in Figure 15(c)). Consequently, the component experiences plastic deformation, which is analogous to damage. The plastic deformation is usually expressed in terms of displacement ductility, defined as the ratio of the peak deflection to the yield deflection. As illustrated in Figure 15(d), a component with a yield displacement of 0.5 inches and a peak dynamic displacement of 2.25 inches would have a peak ductility of 4.5.

The blast damage, measured by ductility, was used to compare the response of various SDOF components when loaded with the blast loads shown in Figure 12 to Figure 14. Low resistance values of 0.5 psi and 1.0 psi, typical of conventional structural components not designed specifically for blast loads, were analyzed over a range of different natural periods. For each normalized target standoff distance and flame speed, the peak SDOF ductility was calculated. Results from these analyses are shown in Figure 16. The damage decreases with increased normalized standoff for all cases analyzed.

The graphs of Figure 16 demonstrate some important findings for the particular blast loads developed in this paper. First, for a fixed natural period, a higher flame speed associated with a detonation ($M_f = 5.2$) generates less damage than a deflagration ($M_f = 0.7$), which is counter-intuitive. For example, the chart in Figure 16(a) shows that for a SDOF with a natural period of 200 ms and a resistance of 0.5 psi, the detonation ($M_f = 5.2$) causes a ductility of 6.2, while a deflagration ($M_f = 0.7$) results in a ductility of about 12. This trend holds true for all of the natural periods and resistances studied.

While it is shown that the deflagration loads in this paper result in greater structural damage than the detonations, a variation in this trend is observed for flame speeds around the deflagration or DDT transition ($M_f = 0.7$, 0.9 and 1.0). For these loads, lower flame speeds result in less structural damage for the same SDOF model. This damage trend is primarily due to the impulse (integral of pressure over time) applied to the SDOF oscillator for the natural periods considered, rather than the peak overpressure.

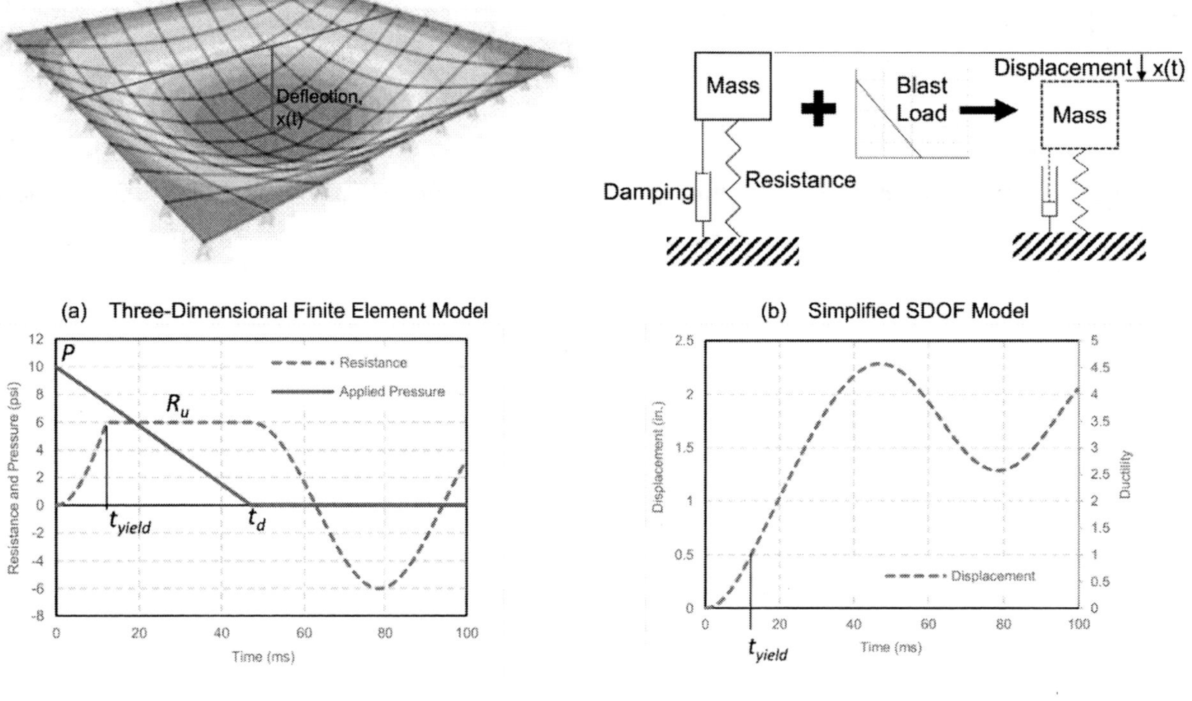

(c) Example SDOF Resistance vs Time

(d) Example SDOF Displacement

Figure 15. Non-Linear Structural Single-Degree-of-Freedom Modeling

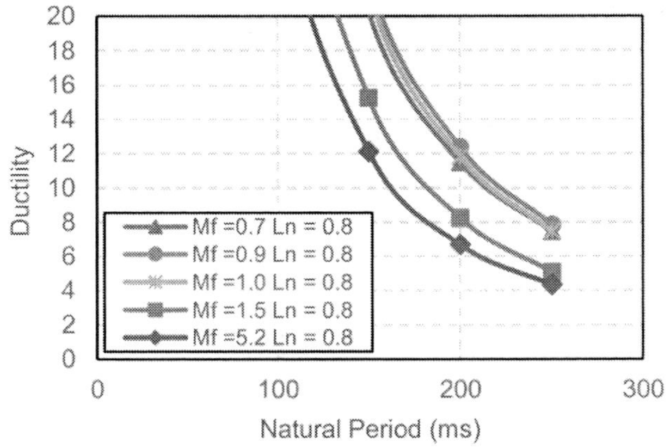

(a) R = 0.5 psi, Ln = 0.8

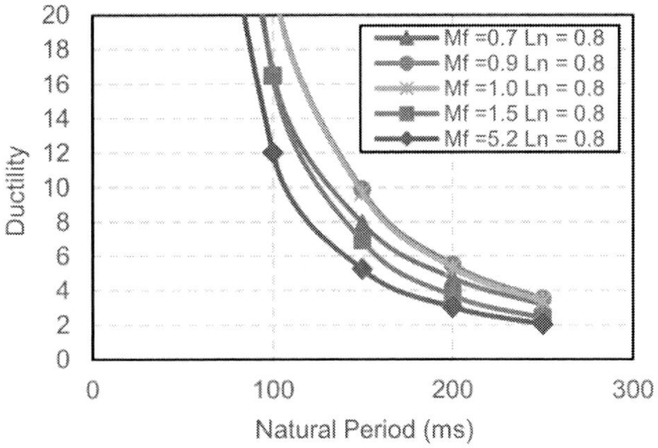

(b) R = 1.0 psi, Ln = 0.8

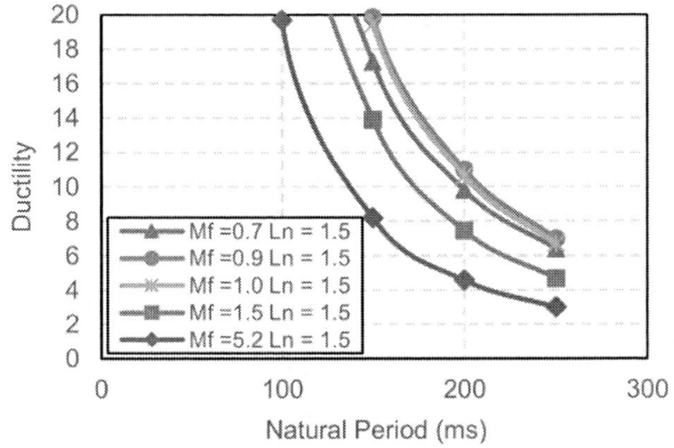

(c) R = 0.5 psi, Ln = 1.5

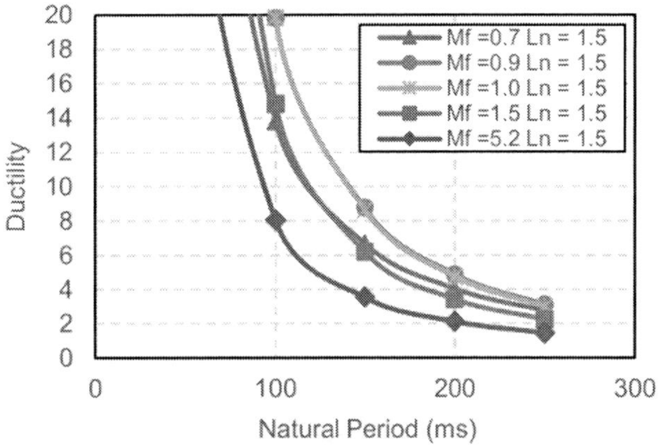

(d) R = 1.0 psi, Ln = 1.5

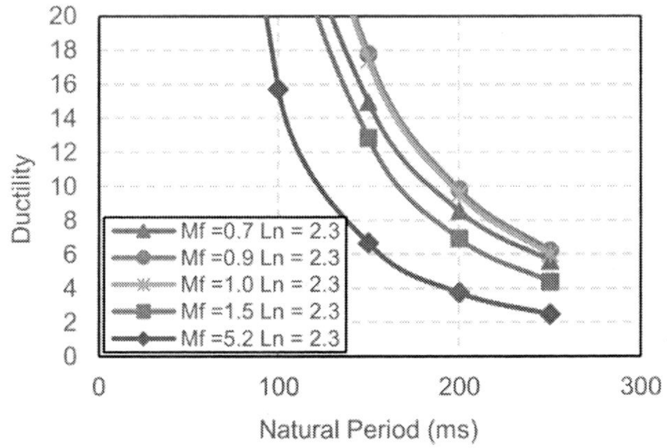

(e) R = 0.5 psi, Ln = 2.3

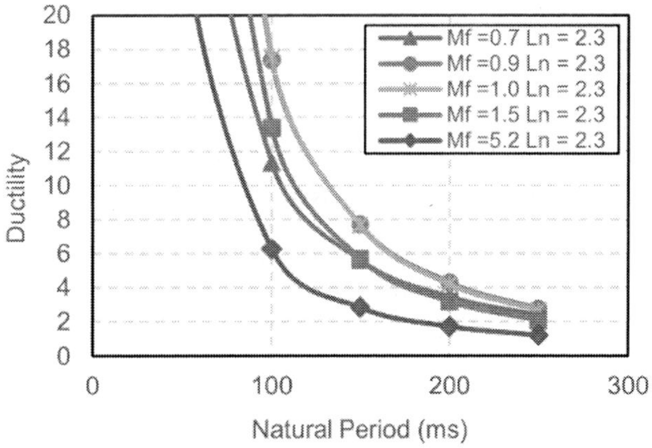

(f) R = 1.0 psi, Ln = 2.3

Figure 16. Structural Response under Different Blast Load Conditions

7 Conclusions

The following conclusions were reached as a result of this evaluation:

- The blast wave shape within the elongated cloud and in the near-field along the long-axis exhibits a behavior belonging to the acoustic wave regime (defined for a standard hemispherical VCE as a flame speed less than Mach 0.35) as long as the flame Mach number in the elongated cloud is less than unity (i.e., subsonic flame), while the magnitudes of the peak positive and negative phases are larger than with a traditional acoustic wave.
- The identified deflagration and DDT regimes for unconfined elongated VCEs as a function of normalized flame travel distance (Ln_f) and flame speed attained at a specified Ln_f were found to be in reasonable agreement with published data.
- There is a significant difference in blast wave shape between the supersonic flame speed case (M_f >1.0) and the subsonic flame speed cases (M_f <1.0), The former exhibits normal shock front (with zero rise time) and the latter the acoustic wave shape which has a rapid decline time from the positive peak pressure to the negative phase, compared to its positive duration.
- Downstream of the cloud end along the long-axis direction, the blast wave for the detonation case ($M_f = 5.2$) decays faster than another other cases with lower flame speeds.
- For representative structural components, it was shown that lower pressure deflagration blast waves result in greater damage than higher pressure detonation waves. This demonstrates the importance of considering blast wave shape, rather than simply peak overpressure, in the prediction of blast loaded structures. This also indicates that high damage levels to conventional structures due to an accidental VCE are not necessarily indicative of a detonation.

8 References

[1] B. A. Burgan, Dispersion and Explosion Characteristics of Large Vapour Clouds (Volume 2 – Appendices), SCI DOCUMENT ED024, 2014.

[2] Baker, W.E., Explosion in Air, University of Texas Press, Austin, TX, 1973

[3] Center for Chemical Process Safety, Guidelines for Vapor Cloud Explosion, Pressure Vessel Burst, BLEVE and Flash Fire Hazards, American Institute of Chemical Engineers, John Wiley & Sons, Hoboken, NJ, 2010

[4] Baker, W.E., Cox, P.A., Westine, P.S., Kulesz, J.J. and Strehlow, R.A., Explosion Hazards and Evaluation, Elsevier, 1983

[5] van den Berg, A.C., "The Multi-Energy Method - A Framework For Vapor Cloud Explosion Blast Prediction", *J. of Hazard Materials*, **12**, 1-10, 1985

[6] Lenoir, E.M., and Davenport, J.A., "A Survey of Vapor Cloud Explosions-Second Update", 26th Loss Prevention Symposium, AIChE, 1992

[7] Tang, M.J. and Baker, Q.A., "A New Set of Blast Curves for Vapor Cloud Explosions," Center for Chemical Process Safety (CCPS)/AIChE, 33rd Loss Prevention Symposium, 1999

[8] Geng, J., Mander, T.J. and Baker, Q.A., "Blast Wave Clearing Behavior for Positive and Negative Phases" *J. of Loss Prevention in the Process Industries* 37: 143-151. 2015

[9] Pickles, J.H. and Bittleston, S,H., "Unconfined Valor Cloud Explosions – The Asymmetrical Blast from an Elongated Cloud", *Combustion and Flame*, 51: 45-53, 1983

[10] Geng, J., Thomas, J.K., T. and Baker, Q.A., "Characteristics of Elongated Vapour Cloud Explosions" *Hazards 25-126, Edinburgh, May 13-15,* 2015

[11] Geng, J., Thomas, J.K. and Baker, Q.A., "A Study of the Blast Wave Shape from Elongated VCEs" *J. of Loss Prevention in the Process Industries* 44: 614-625. 2016

[12] Geng, J. and Thomas, J.K., "Simulation and Application of Blast Wave-Target Interaction," Proceedings of the 41st Annual Loss Prevention Symposium, Houston, Texas, 2007

[13] Thomas, J.K., Pierorazio, A.J., M. Goodrich, Kolbe, M., Baker, Q.A. and Ketchum, D.E. "Deflagration to Detonation Transition in Unconfined Vapor Cloud Explosions," Center for Chemical Process Safety (CCPS) 18th Annual International Conference & Workshop, Scottsdale, AZ, 23-25 September 2003

[14] Thomas, J.K., Goodrich, M.L. and Duran, R.J. "Propagation of a Vapor Cloud Detonation from a Congested Area into an Uncongested Area: Demonstration Test and Impact on Blast Load Prediction," Process Safety Progress, 32(2): 199-206, 2013

[15] Thomas, J.K., Duran, R.J. and Goodrich, M.L. "Deflagration to Detonation Transition in a Lean Hydrogen-Air Unconfined Vapor Cloud Explosion," Mary Kay O'Connor Process Safety International Symposium," College Station, TX, October 27, 2010

[16] Davis, S.G., Engel, D and van Wingerden, K., "OTC27290-MS Large Scale Testing – Development of Advanced CFD Tools for the Enhanced Prediction of Explosion Pressure Development and Deflagration Risk on Drilling and Production Facilities". Offshore Technology Conference, Houston, TX, May 2-5, 2016

[17] Bakke, J.R., Bjerketvedt, D. & Bjørkhaug, M. "FLACS as a Tool for Safe Design against Accidental Gas Explosions". *IChemE Conference 'Piper Alpha - Lossons for Life Cycle Safety Management'.* 26-27 September 1990, London: 12 pp, 1990

[18] van Wingerden, K., Arntzen, B., Storvik, I., Teigland, R., Bakke, J.R., Sand, I.Ø. & Sørheim, H.R. "FLACS-93: A New Explosion Simulator". *ERA-Conference: Offshore Structural Design against Extreme Loads,* 3-4 November, London, UK: 5.2.1-5.2.14, 1993

[19] Strehlow, R.A., Luckritz, R.T., Adamczyk, A.A. and Shimp, S.A.,"The Blast Wave Generated By Spherical Flames", *Combustion and Flame*, 35: 297-310, 1979

[20] Davis, S.G., Merilo, E., Engel, D and van Wingerden, K., "Large Scale Detonation Testing: New Findings in the Prediction of DDTs at Large Scales". 19th Annual Intern. Symposium of MKOPSC, College Station, TX, October 25-27, 2016

The Nature of Ammonium Nitrate Decomposition and Explosions

Ronald J. Willey, P.E., Ph.D.
Northeastern University
Dept. of Chemical Engineering 313 SN
Northeastern University
Boston, MA 02115
r.willey@northeastern.edu

Prepared for Presentation at
American Institute of Chemical Engineers
2017 Spring Meeting and 13th Global Congress on Process Safety
San Antonio, TX
March 26 – 29, 2017

Paper 43b

The Nature of Ammonium Nitrate Decomposition and Explosions

Ronald J. Willey, P.E., Ph.D.
Northeastern University
Dept. of Chemical Engineering 313 SN
Northeastern University
Boston, MA 02115
r.willey@northeastern.edu

Keywords: Case Histories, Runaway Reactions, Modeling of Runaways

Abstract

Ammonium nitrate explosions can cause significant destruction as shown with examples in West, Texas (2013) and Tianjin, China (2015). Yet, ammonium nitrate is available in nearly every garden center, and serves as one of the major fertilizers used throughout the world. Why is it safe to store in a garden center, and yet not safe when stored in wooden framed buildings? This presentation explores the nature of ammonium nitrate decomposition steps using a process simulator where adiabatic and constant pressure scenarios are evaluated. Answers are simple, yet complex. The first decomposition step is endothermic. In other words, there is a natural barrier that needs to be overcome before further decomposition. Further explanations are offered within the manuscript.

1 Introduction

This manuscript is an extension of work presented in a CSB report on the 2013 West Texas Incident (15 people killed) [1], several articles in Process Safety Progress about the incidents in West Texas [2], and the 2015 incident in Tianjin, China (173 killed) [3,4,5,6,7], and a presentation made by the author in Kunming China on October 18, 2016 [8]. References to previous ammonium nitrate based industrial accidents are nicely summarized on the web in Wikipedia [9]. Notable events include the 1921 explosion at BASF plant in Oppau, Germany (561 people killed) [10], the 1947 Texas City, TX disaster (581 people killed) [11], the 2001 Azote de France fertilizer factory explosion, Toulouse, France (31 people killed) [12,13].

This manuscript focuses on the engineering understanding of why ammonium nitrate is considered nonhazardous by some groups, and extremely hazardous by other groups. For example there is a sentence in the WikiPedia article that says "Pure, compact AN is stable and very difficult to initiate. However, there are numerous cases when even impure AN did not explode in a fire." .. "Ammonium nitrate decomposes in temperatures above 210°C (410°F). Pure AN is stable and will stop decomposing once the heat source is removed..." [9]. This observation is based on reports such as the 1947 report by Davis [13] and the 1966 Bureau of Mines report by Van Dolah et al. [14]. A paper published by Meyer et al. [15] demonstrates that pure ammonium nitrate "does not initiate" when packed into a "witness tube" even under the influence of a blasting cap (Figure 1a); however, add 2.5% dodecane and very different results occurs with the destruction of the witness tube (Figure 1b). Ammonium nitrate is extremely moisture sensitive, it attracts moisture, and has an equilibrium moisture content such that granules become hardened solid chunks if exposed to moisture for any length of time. This further mitigates the explosive potential.

2 What is ammonium nitrate?

The Chemical Reactivity Worksheet 4 (CRW4), presently available from the Center for Chemical Process Safety for free [16], provides significant background on ammonium nitrate. Summarized in Table 1 (directly below) and shown in a screen capture (Figure 2) is the information on the individual chemical: Ammonium Nitrate as provided by CRW4.

Table 1. Chemical Information from the CRW4 for Ammonium Nitrate

Chemical info

Ammonium Nitrate: CAS 6484-52-2, UN Number 1942, Chemical Formula $H_4N_2O_3$

DOT Hazard Label

Oxidizer

Reactive group

Nitrate and Nitrite Compounds, Inorganic; Salts, Acidic

Reactivity alert

Explosive; Strong Oxidizing Agent

General Description

Ammonium nitrate is commercially available both as a colorless crystalline solid and processed into prills for specific applications. **Soluble in water**. Does not readily burn but will do so if contaminated with combustible material. Accelerates the burning of combustible material. Produces toxic oxides of nitrogen during combustion. Used to make fertilizers and explosives, and as a nutrient in producing antibiotics and yeast.

Reactivity profile

The hazards of AMMONIUM NITRATE have been well studied because of several extremely severe explosions [Chem. Eng., 1962, 70, 91; Bretherick, 5th Ed., 1995]. Mixtures with alkyl esters may explode, owing to the formation of alkyl nitrates. Mixtures with phosphorus, tin(II) chloride or other reducing agents may react explosively [Bretherick 1979 p. 108-109]. A mixture with aluminum powder (also zinc, cadmium, copper, magnesium, lead, cobalt, nickel, bismuth, chromium, and antimony) can be used as an explosive. A number of explosions in which ammonium nitrate and aluminum were mixed with carbon or hydrocarbons, with or without oxidizing agents have occurred [Mellor 5:219 1946-47]. A mixture with acetic acid ignites when warmed, especially if concentrated [Von Schwartz p. 322 1918]. Causes the decomposition of sodium hypochlorite within a few seconds [Mellor 2 Supp. 1:550 1956].

Air and water reactions

Water soluble. Hot aqueous solutions of the nitrate above 50% conc., under confinement may decompose explosively. This process is aided catalytically with such substances as nitric acid and chloride ion, [Chem. Abs., 1982, 97, 78074].

Health hazards

Water soluble. Hot aqueous solutions of the nitrate above 50% conc., under confinement may decompose explosively. This process is aided catalytically with such substances as nitric acid and chloride ion, [Chem. Abs., 1982, 97, 78074].

Potentially incompatible absorbents

Cellulose-Based Absorbents
Expanded Polymeric Absorbents
Missing Carbon!!!

Observation and comments: It is true that ammonium nitrate by itself will not burn. It is one of the most important chemicals to our current life-style in that it is used as a fertilizer in approximately 12% of the nitrogen based fertilizers is ammonium nitrate (calculation done by author – not confirmed). Eliminating this chemical, eliminates or degrades the food chain to the detriment of the welfare of the people who depend upon these fertilized crops for good health.

2.1 Compatibility matrix check using CRW 4

CRW4 is proving to be invaluable to chemical engineers. The example discussed below is not to discourage engineers from turning to it for ease of compatibility checks. I was curious to see what happens if the combination of carbon and ammonium nitrate are analyzed. Unfortunately, the output is wrong for ammonium nitrate and carbon. Figure 3 below shows that the combination of carbon (activated) and ammonium nitrate are compatible. It is actually just the opposite, the addition of carbon, a source of fuel, to the ammonium nitrate, an oxidizer, creates a mixture that is extremely explosive. The stoichiometric amount of carbon needed to ensure full combustion to nitrogen gas, water, and carbon dioxide, is about 7 % by weight.

3 The chemistry of the decomposition process

An extraordinary amount of research has gone into the decomposition of ammonium nitrate. Only a few key references are reviewed below. Most notable is the work of Marcellin Berthelot in the 1860's to 1880's on explosives [17]. His work is summarized from a Davis'

1944 publication below [13]. It is admirable work. The signs of the heat reported are taken as related to the surroundings. Positive is exothermic in this summary and negative is

There are seven equations that represent the decomposition of ammonium nitrate:

1: The dissociation of solid or molten ammonium nitrate into nitric acid and ammonia as gases: $NH_4NO_3 \rightarrow HNO_3 + NH_3$. This reaction occurs at low temperature, about 100°C, the salt subliming in part. There is absorption of heat: -41,300 cal (endothermic or required) for the solid and -37,000 cal for the molten state.

2: Between 180 and 200°C the decomposition takes place more rapidly, mainly as indicated by the formula: $NH_4NO_2 \rightarrow N_2O + 2\ H_2O$. In the decomposition, nitrous oxide (laughing gas and water are formed, 10,200 cal being liberated for the solid and 14,000 cal for the molten state. At constant volume the temperature and pressure attained are about 555°C and 4500 kg/cm², respectively. With fairly good regulation of temperature, 98% of the decomposition products follow this reaction. Increasing the temperature causes an increase in production of other nitrogen oxides. Near 260°C ammonium nitrate burns with puffs of flame. If dissociation of the salt at lower temperatures has given nitric acid and ammonia, these may react to develop higher temperature as follows: HNO_3 (gas) + NH_3 (gas) $\rightarrow N_2O + 2\ H_2O$ + 51,500 cal (exothermic or released).

3: Explosive reactions of ammonium nitrate develop when the salt is heated rapidly, but explosion occurs only from a very strong initial impulse in an enclosed space or a restricted volume. One reaction produces nitrogen, oxygen, and steam as follows: 2 $NH_4NO_3 \rightarrow 2\ N_2 + O_2 + 4\ H_2O$ (gas) + 30,700 cal from solid and +35,000 cal from molten salts, with a possible temperature of 1500°C and pressures of 11,200 kg/cm^2 at constant volume.

4: The reaction may produce nitric oxide, nitrogen, and water according to the formula: 2 $NH_4NO_3 \rightarrow 2\ NO^* + N_2 + 4\ H_2O$ + 9200 cal for the solid and 13,000 cal for the molten salt; 518°C and 4860 kg/cm^2, respectively are the temperature and pressures obtained.

* The reference had this as N_2O_2

5. The reaction may give nitrogen peroxide, nitrogen, and water as follows: 4 $NH_4NO_3 \rightarrow N_2O_4 + 3\ N_2 + 8\ H_2O$ + 29,500 cal for the solid and +33,500 cal for the molten salt.

6. The reaction may give nitrogen trioxide, nitrogen, and water represented by the formula: 3 $NH_4NO_3 \rightarrow N_2O_3 + 2\ N_2 + 6\ H_2O$ + 23,300 cal for the solid and 27,000 for the molten salt. This reaction never takes place alone; however, as N_2O_3 exist in the dissociated state only as nitric oxide in nitrogen peroxide.

7. Finally, under certain conditions, such as the presence of a spongy platinum and gaseous nitric oxide, nitrogen and steam will be obtained following the formula: 5 $NH_4NO_3 \rightarrow 2\ HNO_3 + 4\ N_2 + 9\ H_2O$ +33,400 cal for the solid and +37,500 cal for the molten salt.

endothermic. The heat amounts are relative and do not represent the heat of reaction at 298.15 K, 1 atm. With the advent of modern instrumentation such as the differential scanning calorimeter, work has followed that includes adding a carbon source such as mineral oil to the ammonium nitrate [18]. For pure ammonium nitrate, the first step is the endothermic dissociation to ammonia and nitric acid. Subsequent reactions follow with nitrous oxide being the main gas formed, and nitrogen gas a by-product.

4 Simulation and Reactions used in this study

The work that follows involved using Aspen Plus, V9, as the simulation program. 4 reactions were modeled using a continuous stoichiometric reactor at 1 atm run isothermally (heat of reaction estimated) or adiabatically (adiabatic temperature reached). Each simulation used 1 mol/s of ammonium nitrate (solid) as the feed stream. No adjustment in the default parameters in the Aspen data base were made, and "ideal gas" was selected for the equation of state. Simulation runs were also done in a closed batch adiabatic reactor that was kept at 1 m^3 in volume. Progressive amounts of ammonium nitrate were added along with the necessary air to create a pressure of 1 atm within the reactor.

The 4 reactions investigated were:

Reaction 1:

Endothermic dissociation of ammonium nitrate to ammonia and nitrogen

NH_4NO_3 (solid) → HNO_3 (liquid) + NH_3 (gas)

ΔH_{rxn} at 25°C = +35,000 cal/mol or +438 cal/g; Adiabatic temperature prediction -263°C

Reaction 2:

Exothermic dissociation of ammonium nitrate to nitrous oxide and water

NH_4NO_3 (solid) → N_2O (gas) +2 H_2O

ΔH_{rxn} at 25°C = -28,518 cal/mol or -356 cal/g; Adiabatic temperature prediction 312°C

Reaction 3:

Exothermic dissociation of ammonium nitrate to nitrogen, oxygen, and water

NH_4NO_3 (solid) → N_2 (gas) +2 H_2O (vapor) + 0.5 O_2 (gas)

ΔH_{rxn} at 25°C = -47,876 cal/mol or -598 cal/g; Adiabatic temperature prediction +946°C

Reaction 4:

Exothermic dissociation of ammonium nitrate with carbon added to nitrogen, carbon dioxide, and water

NH_4NO_3 (solid) + ½ C (solid) → N_2 (gas)+ ½ CO_2 (gas)+ 2 H_2O (vapor)

ΔH_{rxn} at 25°C = -94865 cal/mol or - 1,185 cal/g; Adiabatic temperature prediction 2,100°C

Reaction 1 is a natural barrier that prevents the automatic initiation of ammonium nitrate to the molecules that actually form it. The author is aware from working with ammonia and nitrogen oxide, the mixture has the ability to form ammonium nitrate in downstream lines, especially when moisture is present. The salt formation occurs easily. As explained above, the salt is quite stable. In other words, ammonia plus nitric acid will easily form ammonium nitrate exothermically. Figure 4 show the heat released for each reaction shown above. Moving from the initial state to the first reaction, the reaction goes "uphill". External energy is required to reach the "peak" or "barrier" that the decomposition back to ammonium and nitric acid presents. This barrier is why the material can be stored in garden centers safely.

The barrier implies that an external heating source is required to initiate the full decomposition. An external fire, heat transfer by radiation, or other means is required to overcome this initial barrier. Thus the material can be stored in cool locations. The documented temperature needed to initiate the 1st step is around 200°C [18]. Adding a little contaminant decreases this temperature to as low as 160°C. If the ammonium nitrate, which naturally is adsorbs water, accumulates moisture, the temperature of the 1st step decomposition increases because water has to be evaporated before the 1st step will take place.

The next steps are more complicated and are summarized below. Several publications exist that have done extensive research in this area. For our purposes we'll presume that once the ammonia and nitric acid are liberated, they have the ability to recombine to form nitrogen gas, water as a vapor, and oxygen. The formation of oxygen in the full decomposition of soul ammonium nitrate is why it is considered an oxidizer each mole choose mole that decomposes liberates half a mole of oxygen.

As the reactions above show, each step is creating more non-condensable and condensable gases. In Reaction 3, one mole of solid has the potential to form 3.5 moles of gases at elevated temperatures. One mole is equivalent to 80 g of ammonium nitrate. With a specific gravity of 1.73 g/ml, this is 46 ml of material. At temperatures above the boiling point of water one mole of gas will occupy 30 liters times 3.5 moles gives an approximate 2000 times increase in volumetric expansion when the reaction is carried out. Include an adiabatic environment, and for pure ammonium nitrate, the temperature can reach over 940°C, and the final volume of gases increase another threefold due to expansion by being hotter. The final volumetric expansion from 46 mL to execs milliliters represents a 7500 fold increase in volume. Add a little carbon to this mixture. Then, carbon dioxide forms exothermically, so that the temperature increases approximately another thousand degrees Celsius, and the volume will increase further so that the initial 46 mL of material has become a volume at 1 atm of 680,000 mL, nearly a 14,000 fold increase in volume. This might explain why this material has been examined for use in automobile safety bags for example. Figure 5 show the progressive volume increase for each reaction run adiabatically

In Figure 6, one of the interesting findings of modeling the sequence of reactions shown above is the temperature change for the first reaction. Since the decomposition is endothermic in the first reaction, the temperature decreases, and because the energy is required to complete the reaction, the temperature of the system drops to an impossible -263°C. Then as the ammonia and nitric acid recombine to other products such as nitrogen dioxide, nitrogen, water, and oxygen, each step becomes more exothermic and the resultant temperatures rise rapidly. A runaway reaction so to speak; however, this one actually is an explosion, one of the fastest runaway reactions known. Note, that with the addition of carbon (Reaction 4), the combination of oxygen and carbon results in the formation of carbon dioxide with a significant amount of heat available to continue the heating the contents.

The next consideration in the modeling of these reactions is shown in Figure 7. Imagine a one m^3 volume, initially occupied by only air. Then, stepwise, add some ammonium nitrate, initiate it, and then measure the final pressure. The results show just how little mass (and volume) of ammonium nitrate builds pressure rapidly. 200 grams or 0.2 kg or 115 ml or 1.15×10^{-6} m^3 material in volume added to the closed box will create an overpressure of 1 atm within when well insulated (adiabatic decomposition).

5 Conclusion

Shown in Figure 8 is the wood frame building afire with the ammonium nitrate inside. We can see the fire, the soot formation, and the radiation all contributing energy to the ammonium nitrate storage pile. The soot, as it falls by gravity, into the pile adds carbon to the ammonium nitrate making it more lethal. The purpose of this paper is to assist process safety engineers in understanding the true nature of the hazards associated with ammonium nitrate. Certainly storage in wooden buildings, or wooden bins, must be considered carefully. Moisture mitigates the decomposition; but the trade-off is the ability to move the material from one point to another. Certainly, during an initial fire in a building holding ammonium nitrate, sprinkler systems should be installed and available to wet the storage piles automatically.

Figures

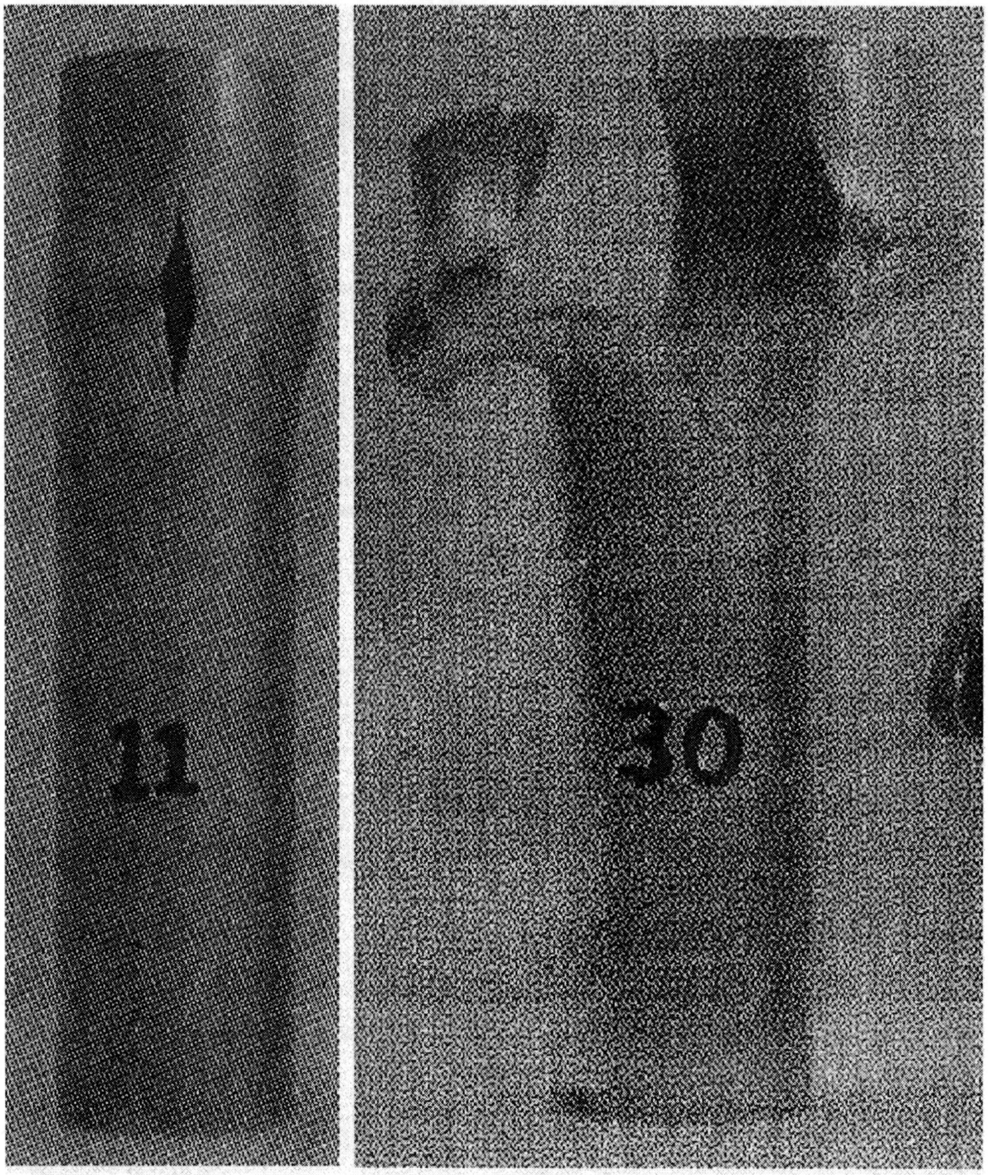

1a. Pure AN

1b. AN + 2.5% dodecane

Figure 1. Comparison of the detonation possibility between pure ammonium nitrate (a) and ammonium nitrate with a small amount of "fuel" source added (b) (from [15])

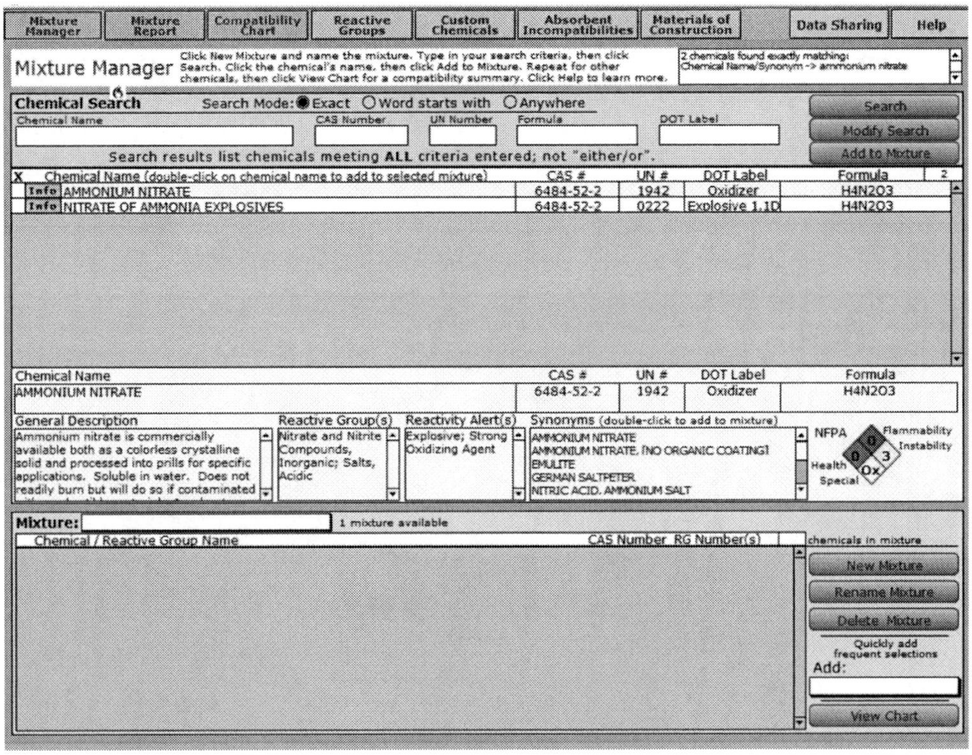

Figure 2. Ammonium Nitrate information from Chemical Reactivity Worksheet 4. [16].

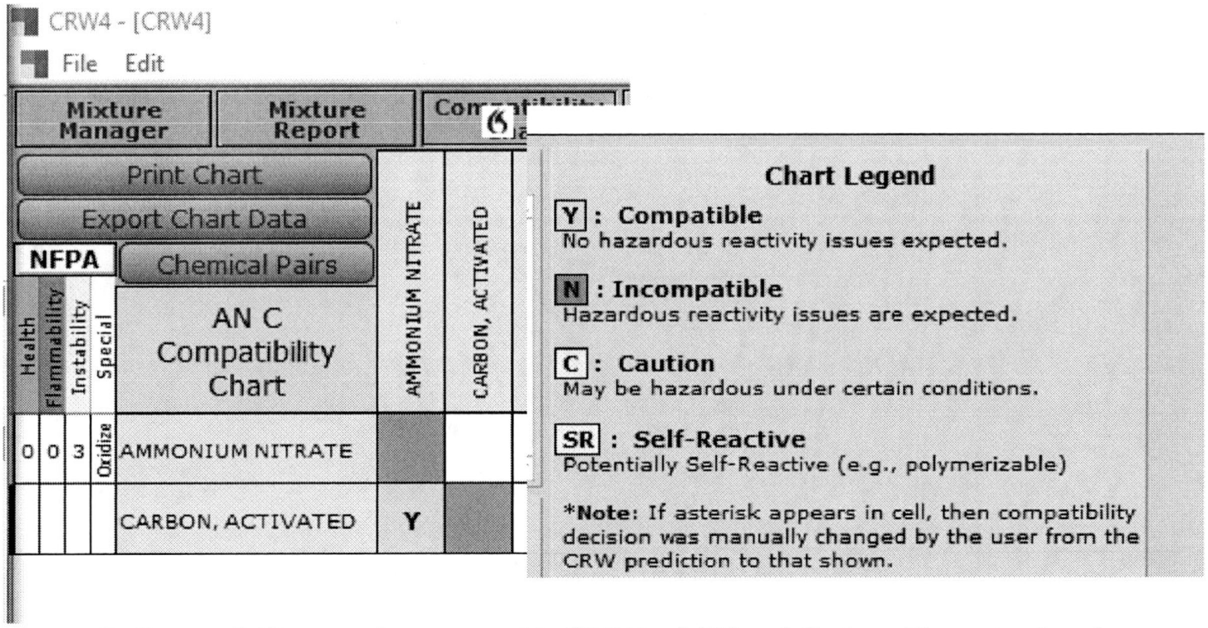

Figure 3. Compatibility matrix output with CRW4 of AN and Carbon. There needs to be a correction made.

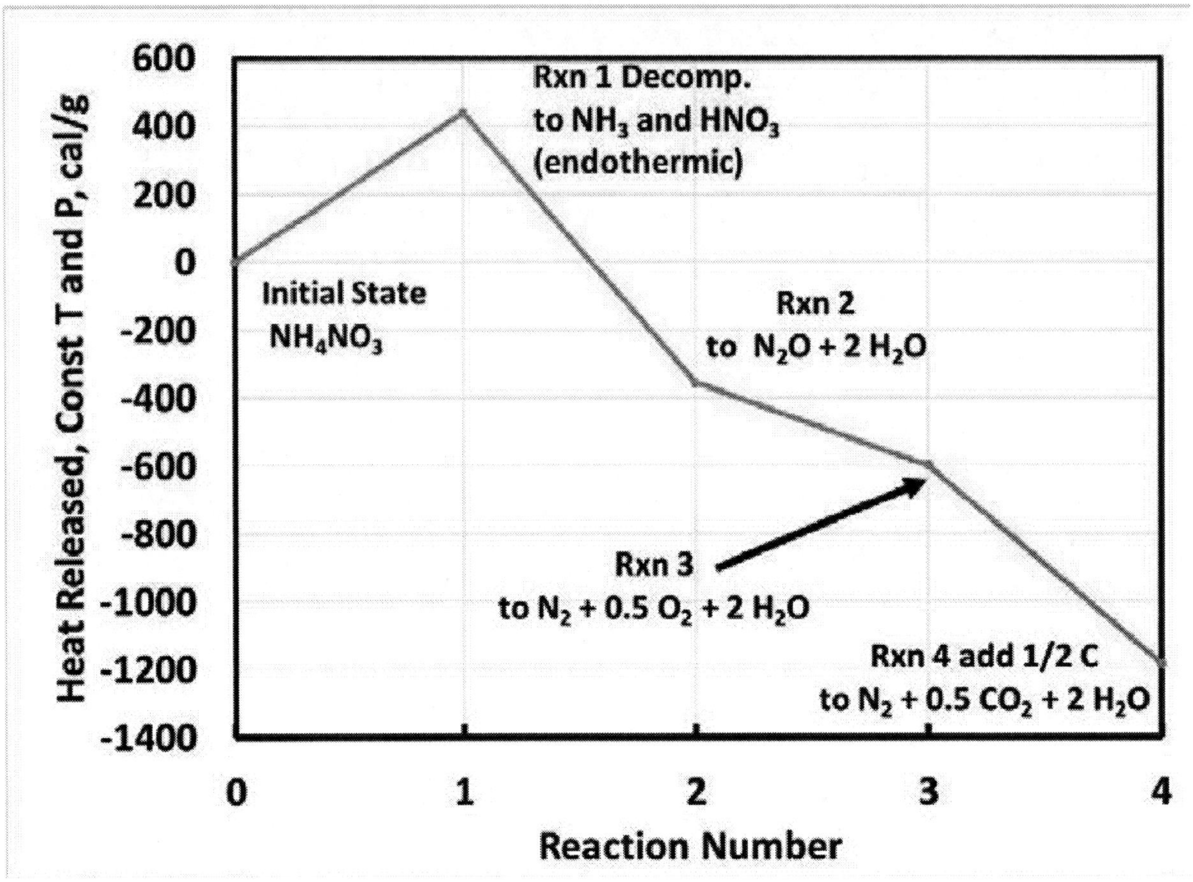

Figure 4. The heat released per gram of ammonium nitrate when decomposed at 25°C, 1 atm.

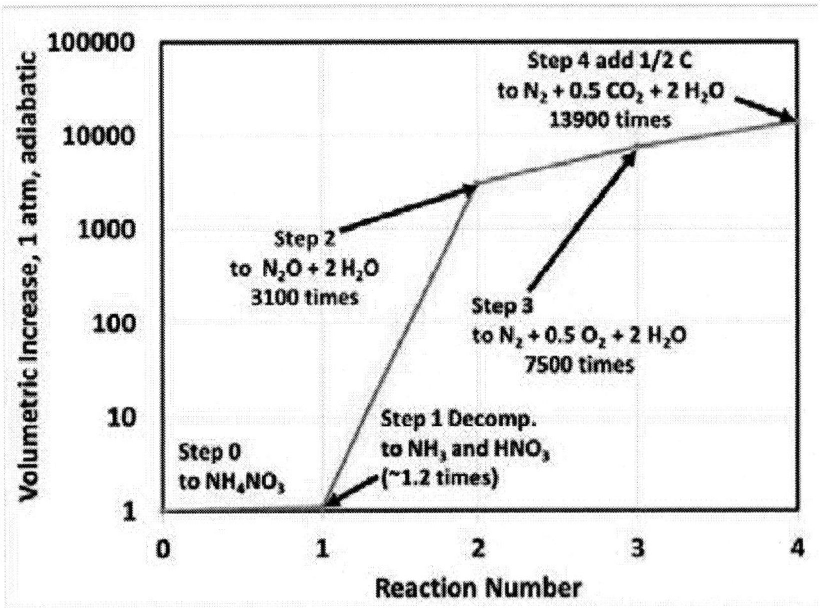

Figure 5. The volumetric increase created when ammonium nitrate reacts adiabatically in an open system with a surrounding pressure of 1 atm.

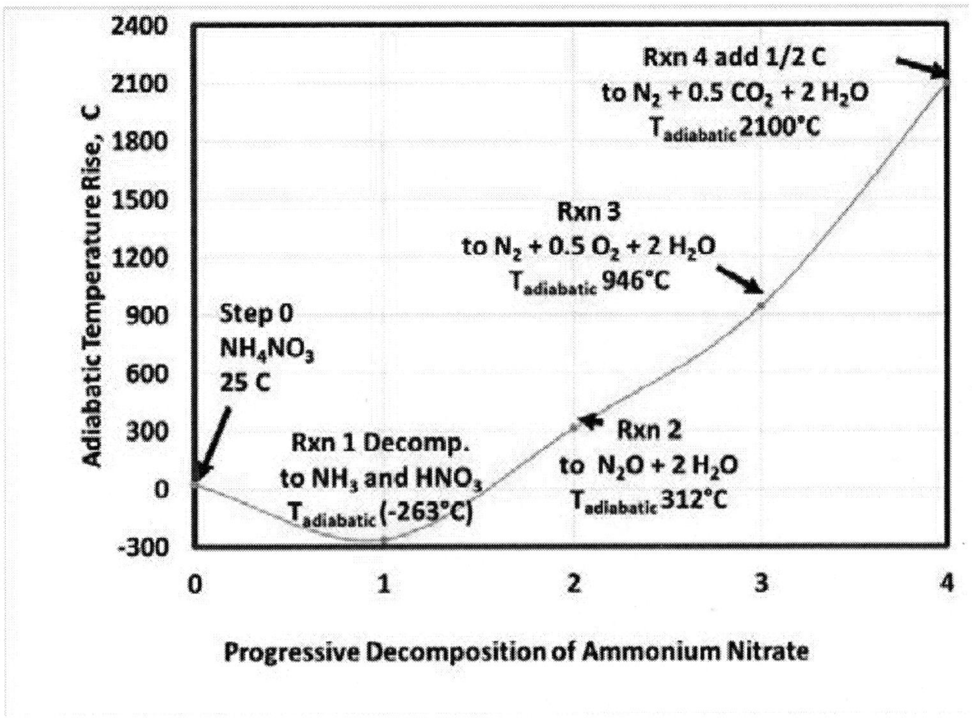

Figure 6. Adiabatic temperature predictions as the reactions progress from Reaction 1 to Reaction 4.

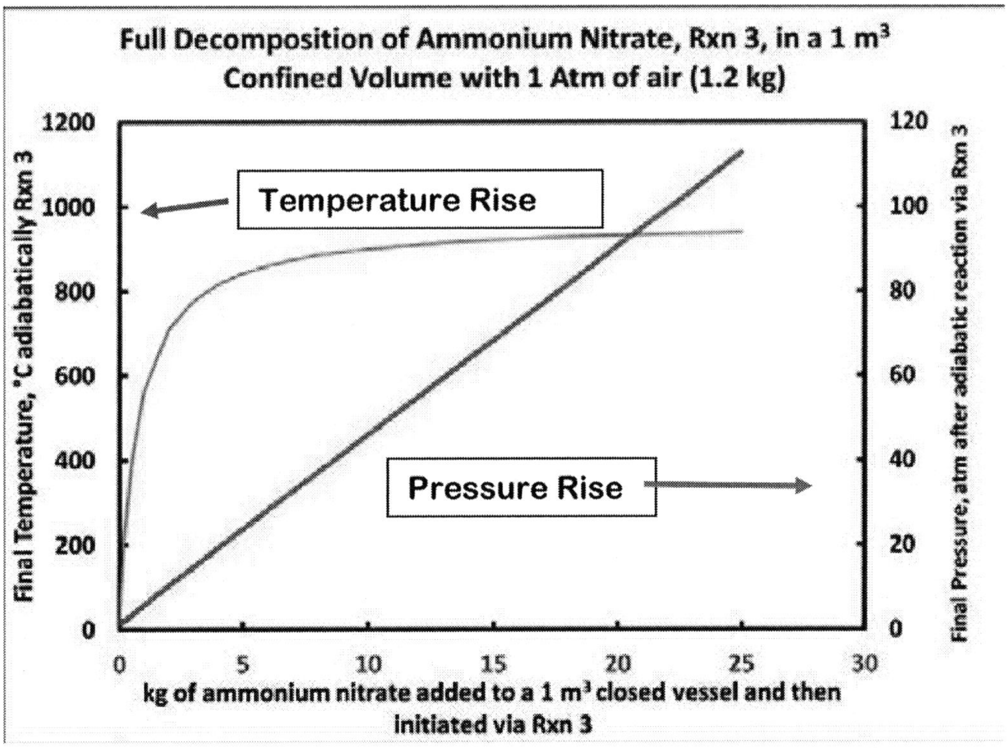

Figure 7. Constant volume (1 m^3) decomposition of ammonium nitrate by Reaction 3.

Figure 8. Ammonium nitrate storage area, afire, within a wooden structure that is seeding soot (carbon) above the ammonium nitrate pile (from [1]).

6 References

[1] U. S. Chemical Safety Board, "Final Report: West Fertilizer Final Investigation Report Report 2013-02-I-TX," (2016), 267, Retrieved from http://www.csb.gov/file.aspx?DocumentId=732, Access 2/9/2017

[2] J. Banks, "Dangerously close: The CSB's investigation into the fatal fire and explosion in West, Texas," *Process Safety Progress* 35 (4), (2016), 312-316, DOI: 10.1002/prs.11857.

[3] R. J. Willey, J. Murphy and A. Baulch, "The explosion in Tianjin, China, August 12, 2015," *Process Safety Progress* 34 (4), (2015), 312-312, DOI: 10.1002/prs.11790.

[4] B. Sun, "Tianjin port explosions," *Process Safety Progress* 34 (4), (2015), 315-315, DOI: 10.1002/prs.11788.

[5] P. Huang and J. Zhang, "Facts related to August 12, 2015 explosion accident in Tianjin, China," *Process Safety Progress* 34 (4), (2015), 313-314, DOI: 10.1002/prs.11789.

[6] R. J. Willey, "When it comes to ammonium nitrate, the initiator is free," *Process Safety Progress* 35 (3), (2016), 215-215, DOI: 10.1002/prs.11838.

[7] G. Fu, J. Wang and M. Yan, "Anatomy of Tianjin Port fire and explosion: Process and causes," *Process Safety Progress* 35 (3), (2016), 216-220, DOI: 10.1002/prs.11837.

[8] R.J.Willey, West Fertilizer Company Fire and Explosion: A Summary of the U.S. Chemical Safety and Hazard Investigation Board Report submitted ***Special Issue*** of ***Journal of Loss Prevention in the Process Industries***, Jan 2017.

[9] "Ammonium nitrate disasters," (2017), Retrieved 9 Feb 2017, from https://en.wikipedia.org/wiki/Ammonium_nitrate_disasters

[10] "Oppau explosion," (2017), Retrieved 9 Feb 2017, from https://en.wikipedia.org/wiki/Oppau_explosion

[11] "Texas City disaster," (2017), Retrieved 9 Feb 2017, from https://en.wikipedia.org/wiki/Texas_City_disaster.

[12] "Fertilizer Plant Explodes in France," *C&EN* 79 (40), (2001), 14.

[13] R.O.E.Davis, *Explosibility and Fire Hazard of Ammonium Nitrate Fertilizer, Circular No. 719,* United States Department of Agriculture, 1945

[14] R. W. V. Dolah, et al., *Explosion hazards of ammonium nitrate fire exposure*, U.S. Bureau of Mines. Report of Investigations 6773, 1966.

[15] S. Meyers, S. K. Singh and E. S. Shanley, "A Blasting Cap Test for Evaluating Self-Reactivity Hazards," *Process Safety Progress* 20 (1), (2001), 1-5.

[16] Center for Chemical Process Safety, AIChE, https://www.aiche.org/ccps/resources/downloadinstall, accessed 2/9/2017.

[17] M. Berthelot, *Sur la Force des Matières Explosives d'après la thermochimie,*, 3, Gauthier-Villars, Paris, 1883.

[18] J. C. Oxley, S. M. Kasushik and N. S. Gilson, "Thermal decomposition of ammonium nitrate based composites," *Thermochimica Acta* 153 (1), (1989), 139-147.

Calorimetry and Thermo-Kinetic Modeling to Determine SAPT and Safe Shipping Conditions for Self-Reactive Materials

Min Sheng
Reactive Chemical / the Dow Chemical Company
1897 Building F24, Midland, MI, 48667
Msheng@dow.com

Dow Co-Authors:
Steve Horsch (Freeport, TX)
Florin Dan (Midland, MI)
Robert Bellair (Freeport, TX)
Marabeth Holsinger (Midland, MI)
Steve Weinberg (Deer Park, TX)
Alan Sopchik (Deer Park, TX)

Prepared for Presentation at
American Institute of Chemical Engineers
2017 Spring Meeting and 13th Global Congress on Process Safety
San Antonio, TX
March 26 – 29, 2017

Calorimetry and Thermo-Kinetic Modeling to Determine SAPT and Safe Shipping Conditions for Self-Reactive Materials

Min Sheng
Reactive Chemical / the Dow Chemical Company
1897 Building F24, Midland, MI, 48667
Msheng@dow.com
Other Author(s)
Affiliation(s)

Dow Co-Authors:
Steve Horsch (Freeport, TX)
Florin Dan (Midland, MI)
Robert Bellair (Freeport, TX)
Marabeth Holsinger (Midland, MI)
Steve Weinberg (Deer Park, TX)
Alan Sopchik (Deer Park, TX)

Keywords: Runaway Reaction, SAPT, Kinetic modeling, Transportation, Monomer, Calorimetry

Abstract

Transportation of self-reactive materials such as monomers are potentially hazardous if the conditions of shipping (eg. temperature and inhibition level during transport) are favorable for polymerization to occur. The United Nations Recommendations on the Transport of Dangerous Goods, Model Regulations, Rev.19 (2015) has a new requirement for the determination of the SAPT (Self-Accelerating Polymerization Temperature) for polymerizing substances. Depending upon their SAPT and packaging dimensions, polymerizing substances may be subject to temperature monitoring or temperature control. The new regulation states that SAPT shall be determined in accordance with the test procedures established for the determination of SADT (Self Accelerating Decomposition Temperature) for self-reactive substances. There are multiple strategies for determination of SADT, each of which differ in their measurement techniques. Recent work published for Acrylic Acid has shown that SAPT results can differ significantly depending upon which SADT method was applied. To ensure consistent SAPT values for safe transportation, several industry consortia committees (Basic Acrylic Monomer Manufacturers, European Basic Acrylate Manufacturers and Methacrylate Producers Association) met in a joint session of which The Dow Chemical Company (Dow) is a member, and agreed to an inter-company effort to identify the best method on the basis of good science, readily available technology, and flexibility in packaging applications.

This paper highlights the technical approach developed and validated at Dow for inhibited methyl methacrylate (MMA) using isothermal microcalorimetry, modeling, simulations, and SADT H1 testing. In this study, the effect of inhibitor level, headspace volume, and mixing during sample preparation on polymer induction times (PIT) were carefully examined and managed. Several kinetic models were evaluated based on their fit to PIT data, full heat profiles, and accuracy of measured and predicted data. The H1 test results for MMA were consistent with the SAPT values predicted by simulation using kinetics obtained from microcalorimetry tests, and the measured heat transfer coefficient for the 1-gallon package. With the successful completion of this proof of concept study, Dow advocates for the determination of SAPT utilizing isothermal calorimetric data and modeling. This approach can be implemented at smaller companies using commercially available calorimeters, and can accommodate variations in size and shape of common transportation containers.

1 Introduction

The United Nations Recommendations on the Transport of Dangerous Goods, Model Regulations, Rev.19 (2015) [1] has a new requirement for the determination of the Self-Accelerating Polymerization Temperature (SAPT) for polymerizing substances. Accordingly, polymerizing substances are subject to temperature monitoring or temperature control depending upon their SAPT and type of packaging. The requirement also states that SAPT shall be determined in accordance with the test procedures established for the determination of Self Accelerating Decomposition Temperature (SADT) for self-reactive substances. According to "Recommendations on the Transport of Dangerous Goods, Manual of Tests and Criteria" [2], SADT is the lowest environmental temperature at which the temperature increase of a chemical substance in a specified commercial package is at least 6°C during a period of seven days, starting from the time when the temperature is 2°C lower than the environmental temperature. There are four recommended test methods for SADT determination which differ in their measurement techniques: H1, United States SADT test (isothermal oven test with an actual transport package); H2, Adiabatic storage test (adiabatic test with a 1.0 or 1.5 L glass Dewar and heat balance evaluation of the transport package); H3, Isothermal storage test (heat flow measurement of 20 g sample and heat balance evaluation of the transport package); and H4, Heat accumulation storage test (isothermal test with 400 mL sample in a Dewar vessel which has the heat loss per unit sample mass characteristics similar to the transport package). Obviously, performing an H1 test for large volume packages has a high cost and the challenge to mitigate potential hazards related to a runaway polymerization reaction. H4 method carries the difficulty to build a vessel with the same heat loss characteristics of packages [3]. Therefore, H2 or H3 methods are preferred in industry, especially with the wide application of commercially reliable calorimeters [4, 5]. Both H2 and H3 methods involve reaction kinetic measurements with a relatively small quantity of chemicals and then determining SADT by performing the heat balance calculation for a desired package. However, recent work [6] has shown that SADT results for materials with autocatalytic reaction or long induction times can differ significantly depending upon which SADT method was applied. Considering the cost to implement the temperature control as well as the potential hazards that could be present in a large transportation vessel, such as railcar or a ship, an incorrect SAPT result is unacceptable. To ensure consistent SAPT values for safe transportation, several industry consortia committees (Basic Acrylic Monomer Manufacturers, European Basic Acrylate Manufacturers and Methacrylate Producers Association), of which Dow is a member, met in a joint session and agreed to an inter-company effort to identify the best method on the basis of good science, readily available technology, and flexibility in packaging applications.

It is well known that spontaneous polymerization is a safety concern during transportation of polymerizing substances [7-9]. Large quantities of monomers are carried worldwide by railcars, ships and on the road, which are not designed to mitigate the hazards of a runaway polymerization. Consequently, inhibitors are widely used to prevent the polymerization from occurring [8, 10]. Inhibitors can stabilize monomers for some period of time (polymer induction time, PIT) until either oxygen or inhibitors have been depleted, after which the polymerization reaction proceeds at its normal rate. Typical inhibitors work by having the ability to stabilize free radicals and hence reducing the bulk polymerization rate. For instance, 4-methoxyphenol (MEHQ) can transfer a hydrogen atom to an active radical to form a less active, stabilized radical [11], while oxygen or quinone can be added to an active radical to reduce the activity [8]. However, the effectiveness of inhibitors depends on the temperature, the nature and the concentration of the added inhibitors, the availability of molecular oxygen (aerobic inhibitors), the byproducts spectrum and the concentration of impurities. The mutual interaction of these parameters makes the measurement and prediction of monomer stability more challenging [12, 13].

Since many effective inhibitors are used worldwide to extend the shelf-life of monomers for transportation purposes, a good method for monomer SAPT determination should include the PIT of monomers. In this paper, several calorimetric methods reported for SADT determination were evaluated to determine the PIT and full kinetics for a monomer, methyl methacrylate (MMA) with 4-methoxyphenol (MEHQ) as inhibitor. Further, theoretical and experimental work was done on the 1-gallon scale to determine an accurate approach to estimating the heat transfer properties of a package. The overall strategy is to use micro-calorimetry and kinetic modeling to determine the PIT for the appropriate transportation conditions. If the PIT is not exceeded for the desired transportation time no further work is required provided no heat is released during the induction period. Of course, a safety margin should be built into the transportation time because of potential delays. If the PIT is exceeded a full kinetic model can be built from micro-calorimetry data and combined with the aforementioned heat transfer properties of the vessel. The combination can be used to simulate the H1 tests for different monomers and environmental temperatures to determine the SAPT.

2 Experiment

2.1 Sample Preparation

Chemicals used in this study include MMA (Aldrich 99%) with less than 30 ppm MEHQ, MEHQ (Aldrich 99%) and Dimethyl phthalate (DMP, Aldrich 99%). Gas Chromatography (GC) analysis indicated the MEHQ level in the MMA sample was 25 ppm (Table 1).

A portion of the MMA sample was vacuum distilled to remove MEHQ and the freshly distilled MMA was portioned into appropriate amounts and varying amounts of MEHQ was added. The approach allowed for specific concentrations of MEHQ, which are not commercially available, to be obtained in order to study the effect of inhibitor level. The distillation was controlled at 62.4°C and 200 mmHg. The condenser and collector for uninhibited MMA was cooled with dry ice. The actual MEHQ levels of the samples was determined by GC analysis, as listed in Table 1..

Table 1. Actual MEHQ level in MMA samples by GC analysis

Sample	Actual MEHQ level by GC analysis (ppm)
Original MMA from Sigma-Aldrich	25
Fresh MMA from distillation	3
Fresh MMA with 5 ppm MEHQ	4.8
Fresh MMA with 10 ppm MEHQ	12
Fresh MMA with 20 ppm MEHQ	21.3
Fresh MMA with 25 ppm MEHQ	26
Fresh MMA with 50 ppm MEHQ	59

2.2 Isothermal calorimetry to determine kinetic parameters

Micro calorimeters offer a well-controlled temperature and a high sensitivity to heat signal. The micro calorimeters used in this study were Setaram C80 (temperature accuracy ±0.1°C, Accuracy 10 μW) and TAM III (temperature accuracy < ±0.1°C, Accuracy 0.2 μW). TAM III was used for a study with multiple samples at the same temperature, such as liquid filling level study and MEHQ level study. Glass cells with a total internal volume of 4 mL were used for isothermal calorimetric tests. Typically, 3.2 mL of liquid sample (corresponding to 80% filling level) was loaded in the glass cell, while in the reference side, there was another glass cell containing an equivalent mass of glass beads. Heat flow to/from a sample was recorded over the duration of the isothermal test. The moment the sample cell was placed into the calorimeter was taken as the 'zero' moment of the test.

Since an isothermal test records heat flow over the duration of the test, a sensitive calorimeter can measure both the small heat signal during the induction period and the continuous heat flow generated by the bulk polymerization reaction. The moment when the heat signal becomes significant is the point when the bulk polymerization reaction starts to take place. Polymer Induction Time (PIT) is defined by the time from the 'zero' moment to the start point of bulk polymerization (positive deviation from the baseline). As shown in Figure 1, a typical MMA thermogram has two onsets, one for kinetic control step (n^{th} order) and the other one for diffusion control step (autoacceleration) [14, 15]. But the first onset was always used to define the PIT for isothermal tests in this study (1390.92 min in Figure 1).

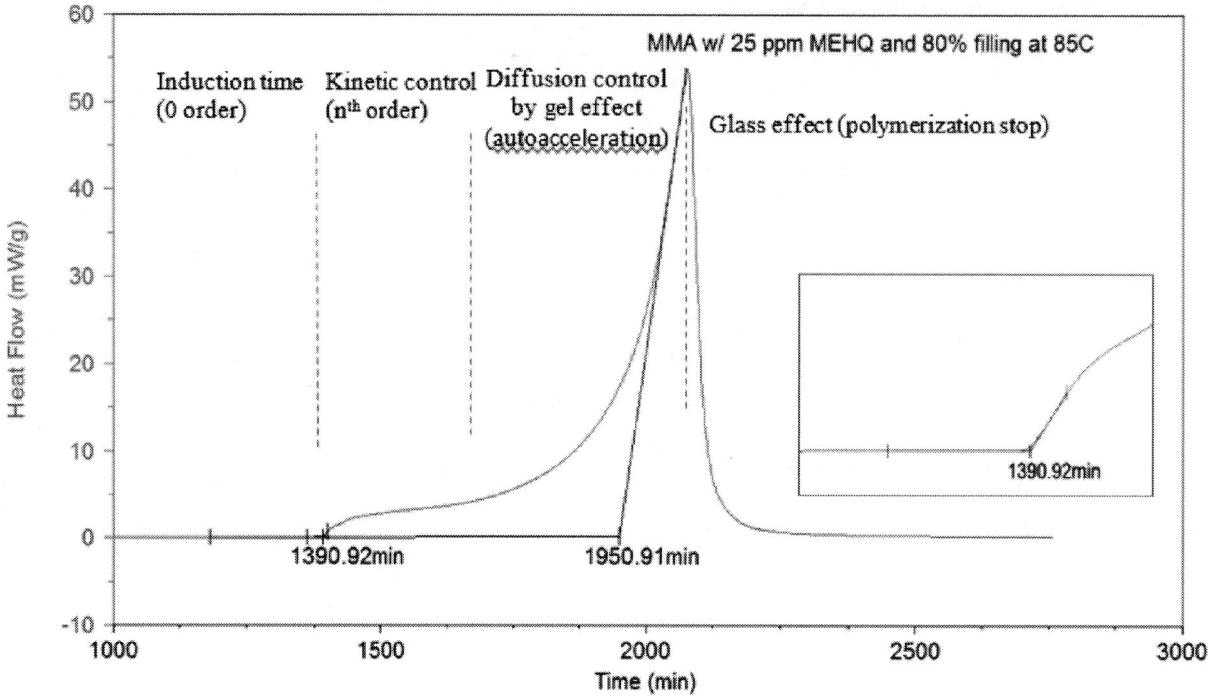

Figure 1. A typical thermogram of MMA sample by an isothermal microcalorimeter

2.3 Heat transfer coefficient measurement for 1 gallon drum

The 1 gallon steel drums used in this study have an UN rating code of 1A1/1.5/250/16/USA. They are 17 cm in diameter, 22 cm in height, 0.9 mm in wall thickness and 3.88 L of internal volume (Figure 2.a).

To measure the overall heat transfer coefficient, 3.49 L of chemical (City water, DMP or MMA) was loaded into a drum and then placed into an oven. For the test with MMA, the oven was located in a bunker in case the drum ruptured due to runaway polymerization reaction (no rupture actually occurred). Once the sample temperature reached the desired temperature, the drum was taken out from the oven and set on a wood block (for insulation purpose) at ambient conditions to cool. The temperature profiles of the sample and the ambient were monitored over the entire process. The thermocouples were placed at the inside wall of the drum, the middle of the drum and in the headspace of the drum to ensure adequate understanding of the temperature profile.

2.4 H1 test for MMA in 1 gallon Drum

In order to avoid forced convection, a conduction oven was used for H1 tests of MMA with 25 ppm MEHQ. For safety reason, this oven was located in a Bunker and monitored remotely. The oven was preheated to a desired temperature and then a drum with 3.49 L MMA sample (90% filling level) was transferred into the oven. Thermocouples were installed on the drum with

Swagelok fittings to monitor the sample temperature and an additional thermocouple for the oven temperature.

A venting line was also installed on the drum for pressure measurement. However, it was found that MMA condensed and polymerized in the line, so pressure data was not collected. The SADT tests resulted in a high internal pressure during the polymerization reaction of bulk MMA and bulging of the drum. Figure 2.b&c shows an example of the bulging top and bottom of a drum after being used for MMA H1 test.

Figure 2. One gallon drum used in this study: a), a new drum; b), a used drum after MMA H1 test; c), the bottom of a used drum after MMA H1 test

3 Results

3.1 Isothermal Calorimetric Testing to determine kinetic parameters

3.1.1 Isothermal method on micro calorimeter

An isothermal calorimetric method utilizing a TAM has been used for SADT evaluation of autocatalytic or auto-accelerated materials [16, 17]. The method allows for characterization of complex reaction mechanisms where autocatalysis or some physical phenomena takes place. Most importantly, a well-controlled temperature and high heat flux sensitivity which micro calorimeters can offer, are essential to reliable measurement of the monomer PIT.

Isothermal calorimetric tests were carried out for MMA samples with 25 ppm MEHQ at four different temperatures in micro calorimeters. The results are presented in Figure 3 and also listed in Table 2. The average total energy was -531.6 J/g with a standard deviation of 8.9 J/g. This corresponds to a conversion of 97.9±1.5%. Based on the shape of heat flow curves (Figure 3.b), isothermal calorimetric tests successfully demonstrated the different mechanisms of thermally initiated polymerization: induction time, kinetic control, diffusion control and glass effect (as shown in Figure 1). More importantly, a well-controlled temperature over the induction period ensures an accurate measurement of monomer PIT.

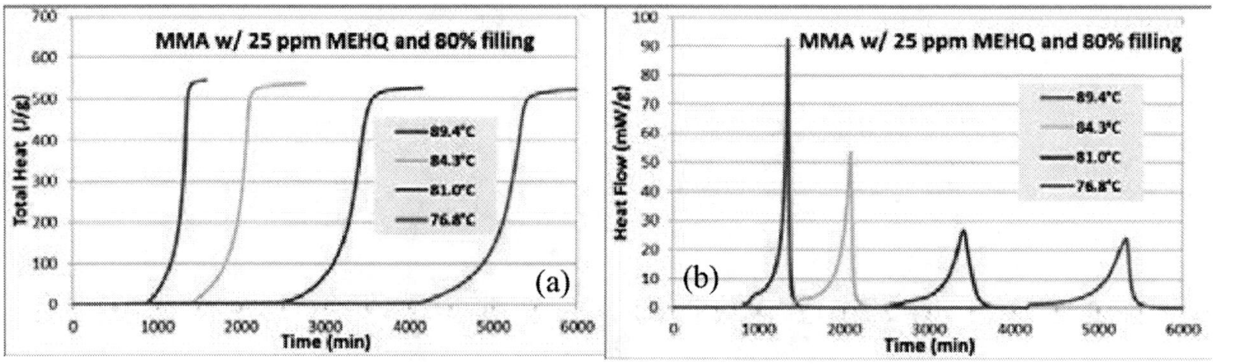

Figure 3. Heat curves of MMA with 25 ppm MEHQ by micro calorimeters at different temperature, (a): total released heat; (b): heat flow

Table 2. Isothermal calorimetric result of MMA with 25 ppm MEHQ and 80% filling level

Sample Temp. (°C)	Bath Temp. (°C)	Sample Size (g)	PIT (min)	Total Heat (J/g)
89.4	90.0	3.033	800.3	-544.4
84.3	85.0	3.058	1388.8	-536.5
81.0	81.6	3.051	2403.9	-526.8
76.8	77.5	3.063	4110.9	-528.9
69.9	70.5	3.041	10929.6	-521.5

The PIT is the period in which all inhibition related reactions take place. The reciprocal PIT is a measure of the overall reaction rate for those reactions. Thus, an Arrhenius-like relationship is found by plotting the logarithm of PIT against inverse temperature (1/Kelvin) [8-11, 18]. As shown in Figure 4a, a linear relationship between the logarithm of PIT from the isothermal calorimetric data (Figure 3) and inverse temperatures was obtained. The resultant linear equation was used to predict the PIT at a lower temperature (69.9°C). The prediction was confirmed by an additional Isothermal measurement at this temperature (7.4 days vs. 7.6 days) verifying the extrapolation or interpolation of the PIT from limited measurements can be used to estimate the monomers stability under the transportation conditions.

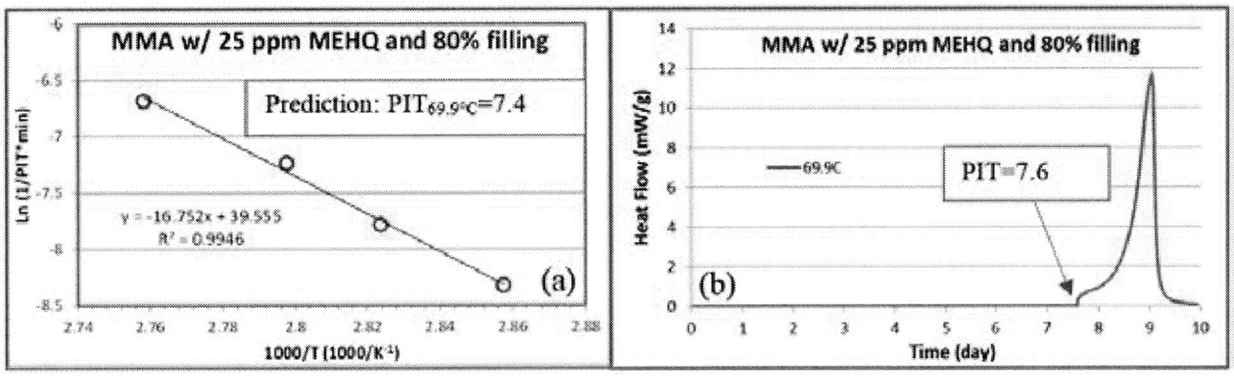

Figure 4. (a): PIT prediction at 69.9C based on four temperature fit, (b): Measurement of the PIT at 69.9C to validate prediction

In Figure 5, the shelf life of MMA with the same inhibitor level from Dow monomer reference (Rohm&Haas) [19] was plotted against the measured PIT in this study. It was noticed that the shelf life of Dow product was always longer than the measured PIT. There are multiple reasons for this deviation. First, the different sample sources might cause this difference. The Dow sample was freshly produced, while the sample in the measurements was shipped from Aldrich. In addition, Dow's monomer shelf life was determined using an open sample cell and didn't limit the oxygen to the monomer. Finally, historic stability tests were visually based and therefore required a significant amount of polymerization to occur in order to determine that the induction time had been exceeded. Conversely, almost no conversion has occurred when the micro calorimeter detects the exothermic polymerization reaction.

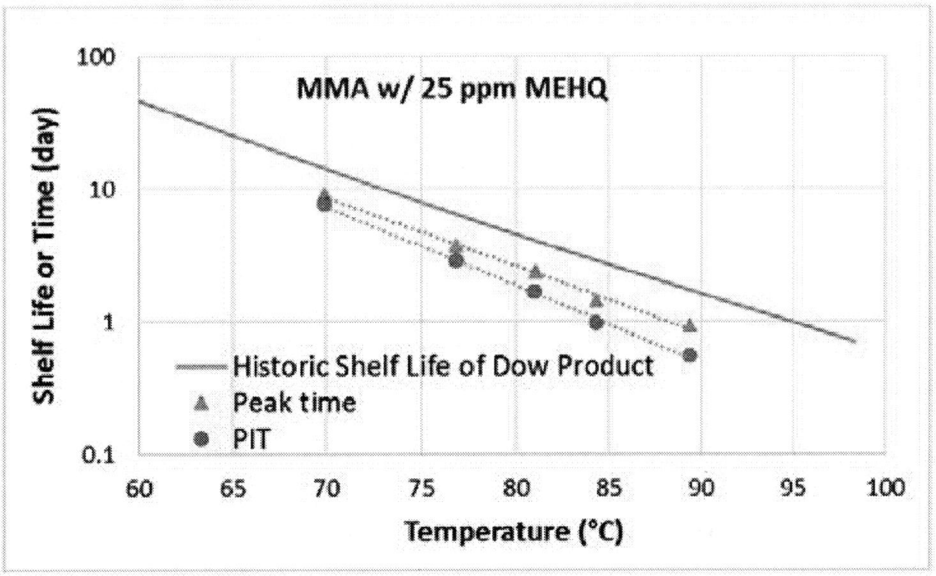

Figure 5. (a): PIT from tested MMA sample vs. the historic shelf life of Dow MMA product

For monomers inhibited with an aerobic inhibitor such as MEHQ, reducing the fill level of a vessel or sample container (and therefore increasing the headspace of the vessel) increases the amount of oxygen available and therefore extends the induction time. In order to confirm this effect of liquid filling level on monomer PIT, additional isothermal calorimetric tests were carried out by using the same MMA sample at the same temperature, but with varied fill levels. As shown in Figure 6.a, the monomer PIT decreased with an increase in the liquid fill level. In addition, getting more dissolving oxygen gas into monomer liquid increases the PIT. In this study, more dissolved oxygen was introduced into the monomer by shaking the MMA sample in a large vial with a 1:3 volume ratio of liquid to air prior to being loaded into the testing cell. As presented in Figure 6.b, the PITs of such shaken samples at different temperatures were 44% to 64% longer than that of the unshaken samples.

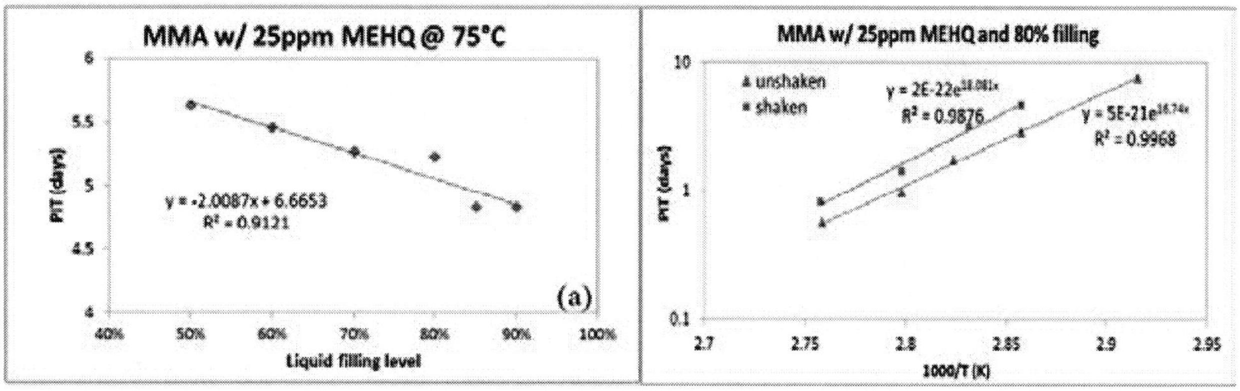

Figure 6. The Impact of oxygen availability on monomer PIT
(a): liquid filling level; (b): oxygen dissolution by shaking

Based on the inhibition mechanism proposed in literature [8-11, 18], MEHQ functions by reacting with peroxide radicals to form stable products. Without oxygen molecules, MEHQ is a poor inhibitor because it can't directly terminate the activity of the initial free radicals. However, even with the depletion of MEHQ, excess oxygen hinders the activity of the radicals resulting in some degree of inhibition.

In order to study the effect of inhibitor levels on PIT, a fresh MMA sample was prepared by distilling the MMA sample received from Aldrich and a series of MMA samples with different MEHQ levels was prepared by adding MEHQ to the distilled samples (Table 1). To reduce the variation caused by mixing, the samples were well shaken prior to being loaded into the testing cell. The PITs of the MMA samples at 84.3°C was plotted against the MEHQ level (Figure 7). A linear relationship between the PIT and the inhibitor level was found in the low range of MEHQ level, while PIT of the sample with a high MEHQ level (59 ppm) significantly deviated from this linear relationship. As discussed previously, the effectiveness of MEHQ requires the presence of oxygen. If there is not enough oxygen, MEHQ will fail to stabilize the monomer. For those samples with a low MEHQ level, there was enough oxygen present so that the addition of MEHQ extended the induction time of the monomer, while the oxygen becomes the limiting reagent for a high MEHQ level and the inhibitor failed to stabilize the monomer once the oxygen was depleted.

As shown in Figure 7, the as-received MMA sample (shaken) had a PIT of 34.1 hr., which was only 67.8% of the value calculated by the linear equation obtained from re-distilled and inhibited MMA samples. This indicated that the old sample had a shorter PIT than the distilled sample. It is believed that impurities introduced during the storage or transportation would decrease the effectiveness of the inhibitor. By distilling the MMA sample, most impurities were removed from the monomer. Without the impurities in the fresh sample, the effectiveness of inhibitor was regained. This was one of the reasons the measured PIT of as-received MMA sample in this study was shorter than the historic shelf life of Dow MMA product (Figure 5). Therefore, it is highly recommended to use the sample from the same source in the calorimetric tests as the sample shipped in the packages for SAPT evaluation.

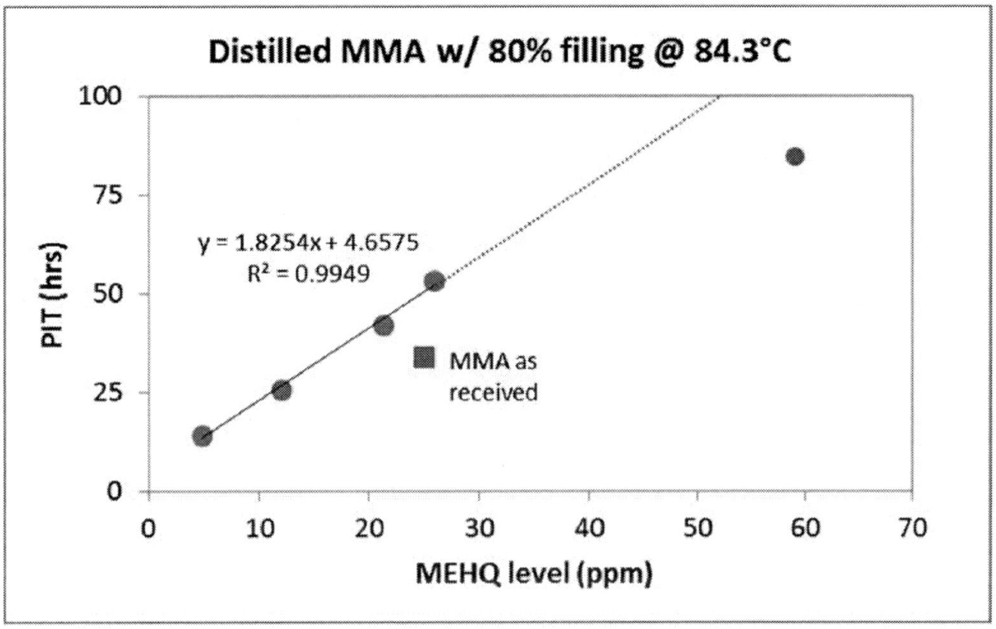

Figure 7. Impact of MEHQ level on the monomer PIT
(●): distilled MMA; (■): as-received MMA

Since the available oxygen is one of the critical variables for accurate PIT measurement, the calorimetric test to simulate a transportation condition should control the ratio of headspace to tested monomer liquid. Isothermal DSC test is widely used to compare the effectiveness of various inhibitors for monomers. However, due to small sample size used in a DSC test, the challenge to precisely control the ratio of headspace to monomer makes it difficult to measure a consistent PIT and accurately simulate the transportation condition. A calorimetric test with a relatively large sample size (normally > 1 ml) in a sealed vial is highly recommended.

In addition, it was observed that the escape of monomer sample from the testing vial during an isothermal calorimetric test resulted in a significantly longer PIT. This was caused by the loss of the monomer resulting in a higher concentration of the inhibitor in the remaining monomer. Therefore, it is also highly recommended to check the sample mass before and after the calorimetric test to verify no leaking occurred during the test.

3.1.2 Kinetic model for MMA

As shown in Figure 1, the thermally initiated polymerization reaction of MMA involves an induction step, a kinetic control step and a diffusion control step. The kinetic model for a full heat profile should be able to predict the reaction in each step. A simplified mechanism scheme for the chemistry occurring during these steps is presented in Figure 8 [8-11, 18].

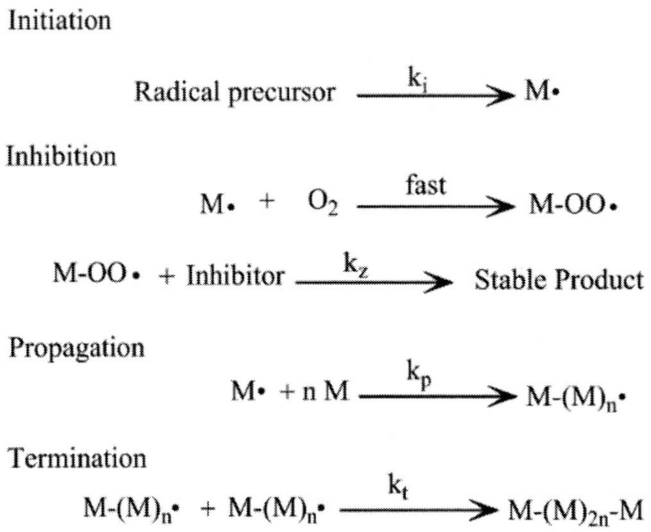

Figure 8. Simplified mechanism scheme for thermal initiated polymerization reactions

Due to thermal initiation, free radicals are generated from monomer molecule or other precursors. In the induction period, enough inhibitors are able to stabilize the monomer by combining with the free radicals to form stable products. Normally, an effective inhibitor has a high inhibitor constant (k_z/k_i). For instance, the inhibitor constant of MEHQ in acrylic acid is in the range of 2,400 to 3,500 at 55°C and that of oxygen in MMA is 33,000 at 50°C [11]. Such high value ensures the free radicals generated from the precursors get deactivated by inhibitors, as long as there are enough inhibitors, so that the concentration of active free radicals in the monomer is maintained at a significantly low level. With the steady state assumption for radical concentrations, the rate of inhibitor consumption is proportional to the rate of free radical formation from the precursors (eq.1).

$$\frac{d[Inhibitor]}{dt} = k_i \cdot f([Precursor]) \tag{1}$$

If the concentration of the monomer or other precursor is assumed to be constant during the induction period, the inhibitor consumption follows a pseudo-zero order reaction. The rate of inhibitor consumption can be expressed as in eq. 2:

$$\frac{d[Inhibitor]}{dt} = k_i C = CA_i \cdot \exp\left(-\frac{E_a}{RT}\right) = A_z \cdot \exp\left(-\frac{E_a}{RT}\right) \tag{2}$$

Since the inhibitor consumption rate is constant during the induction period in isothermal tests, this rate can be obtained as the total inhibitor concentration (X_{in}) divided by time (PIT). Therefore, there is a relationship between the PIT and temperature for the samples with the same inhibitor level, as in eq. 3:

$$\ln\left(\frac{1}{PIT}\right) = Ln\left(\frac{A_z}{X_{in}}\right) - \frac{E_a}{R} \cdot \frac{1}{T} \tag{3}$$

As plotted in Figure 4a, the measured PIT of MMA with 25 ppm MEHQ at 80% filling level and at different temperatures by isothermal calorimetric tests gives the kinetic information of the pseudo-zero order inhibitor consumption reaction (Table 3).

Determining whether the MEHQ is depleted first or the oxygen is depleted first at the end of induction period is not possible solely base on calorimetric results. Therefore, modification of the inhibitor concentration term (X_{in}) in the equation is not suggested without full understanding of the relationship of MEHQ, oxygen and free radicals. Assuming the oxygen is depleted first, a constant consumption rate at a single temperature results in a shorter PIT for a test with a higher filling level, since a high filling level reduces the volume of the headspace and the total amount of oxygen available to the monomer. Therefore, in order to maintain the linear relationship in eq. 3, it is important to keep the same filling level in calorimetric tests. For the same reason, it is also important to apply the same filling level in calorimetric tests as that in a transportation vessel, in order to predict the PIT of the monomer in the transportation vessel. Otherwise, a correction term is needed when the kinetic parameters from calorimetric tests are applied to predict the PIT of monomer in a particular transportation vessel at a particular fill level. Such correction can be obtained by performing micro calorimetry at different fill levels and measuring the corresponding PIT (Figure 6.a).

By the end of induction period, either the depletion of oxygen or MEHQ results in a reduction of free radical scavenging. As a consequence, active free radicals in the monomer accumulate to a level where the propagation reaction rate (Figure 8) dominates and initiates the bulk polymerization reaction. On the isothermal calorimetric thermogram (Figure 1), this moment corresponds to the onset of the kinetic control reaction, before which the inhibition reaction dominates the kinetics and after which the bulk polymerization reaction starts. Because of this, the onset of kinetic control reaction is used to define the PIT.

It was observed that the reaction rate in the kinetic control step followed an nth order reaction rate. Nising [20] suggested this step was related to decomposition of the small amount of peroxide species which was formed during the induction time. For the diffusion control step, most researchers reported an increase in the viscosity due to long polymer chains formed at a high conversion resulting in mass transfer limitation which reduced the chance of active chain ends meeting each other and reduced the rate of the chain termination reaction ((Figure 8.). With the decline of the chain termination reaction rate, diffusion control step follows an auto-accelerating kinetic mechanism [21].

The kinetic model for the thermally initiated polymerization reaction is proposed below:

Induction period: $Inhibitor \overset{k_z}{\rightarrow} Product$ (pseudo-zero order)

Kinetic control: $M \overset{k_1}{\rightarrow} polyM$ (n^{th} order)

Diffusion control: $M + polyM \overset{k_2}{\rightarrow} 2\, polyM$ (auto-acceleration)

The inhibitor consumption rate is determined by eq.4.

$$-\frac{d\alpha_{in}}{dt} = k_z \tag{4}$$

The monomer consumption rate is calculated by eq.5.

$$-\frac{d\alpha_M}{dt} = \delta \cdot [k_1(1-\alpha_M)^{n1} + k_2\alpha_M^{m2}(1-\alpha_M)^{n2}] \tag{5}$$

δ is a conditional factor corresponding to the inhibitor effectiveness. When there is still effective inhibitor, it turns off the bulk polymerization reaction. When the inhibitor is depleted, it turns on the bulk polymerization reaction. It can be defined as:

$$\begin{cases} \delta = 0, & \alpha_{in} < 1 \\ \delta = 1, & \alpha_{in} = 1 \end{cases}$$

The heat flux can be determined by eq.6.

$$q = \Delta Hr_z \cdot \frac{d\alpha_{in}}{dt} + \Delta Hr_M \cdot \frac{d\alpha_M}{dt} \tag{6}$$

Runge-Kutta 4th order method [22] was used to solve ordinary differential equations (ODEs). The kinetic parameters were determined by fitting the isothermal calorimetric data. An example of the fitting is presented in Figure 9 and the kinetics parameters for the MMA with 25 ppm MEHQ and 80% filling level are listed in Table 3.

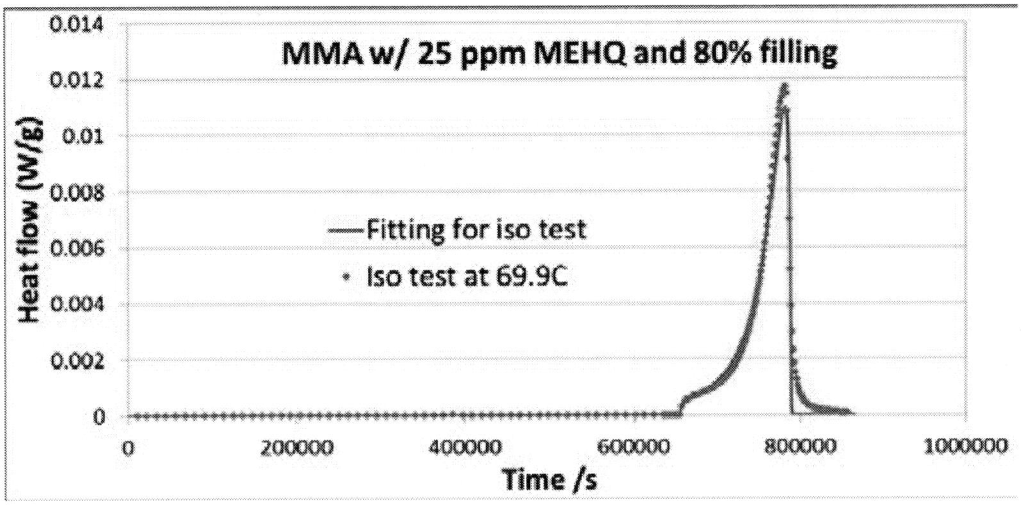

Figure 9. Kinetic fitting for isothermal calorimetric testing at 69.9°C

Table 3. Kinetic parameters for MMA with 25 ppm MEHQ and 80% filling level

	Induction period	Kinetic control	Diffusion control
ΔHr (J/g)	0	-543.3	-543.3
ΔEa (kJ/mol)	139.34	107.0	105.64
Ln(A)	38.69	23.9	26.92
RXN order (n)	0	1	0.155
RXN order (m)	-	-	1.282

3.2 Heat Transfer Coefficient Measurements and Estimation

3.2.1 Heat Transfer Coefficient Measurements

The overall heat transfer coefficient (U) can be estimated directly from measurement of the temperature-time profile during cooling of a heated vessel. Ideally, the U of a given system should be measured to give the most accurate predictions of SAPT. However, this is not always feasible for large vessels/packages, and it can be time and resource intensive. When feasible, the UN "Manual of Tests and Criteria"[2] recommends use of the half-time of cooling method, where a linear relationship can be found as eq. 7.

$$ln(T - T_a) = ln(T_0 - T_a) - \frac{UA}{mC_p}t \qquad (7)$$

Figure 10. shows the fit for the linear regression of the above equation, whose slope was used to estimate the heat transfer coefficient for multiple chemicals in this container.

Another method to determine the U value from test data is via nonlinear regression of the heat transfer equation (eq. 8), wherein the temperature profile can be fitted by applying the 4th order Runge-Kutta method [22] and minimizing the mean square error to solve for U. Figure 10.b gives an example of this approach.

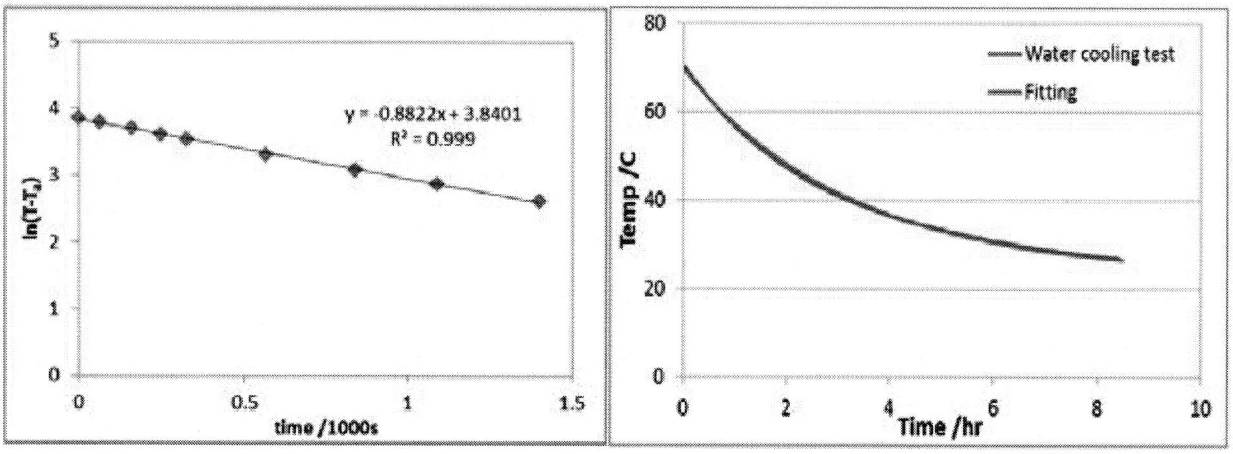

Figure 10. Data fitting for water cooling test
a) using the half-time of cooling method; b) nonlinear regression

$$mC_p\frac{dT}{dt} = UA(T - T_a) \qquad (8)$$

One advantage of the nonlinear regression approach over the half-time of cooling method is that it also can be applied to the heating portion of the temperature profile, as shown in Figure 11. Normally, use of the heating profile is not recommended for estimation of U because it is challenging to maintain the temperature of the oven atmosphere at a constant value. However, the heating profile of the unreacted monomer in a desired drum is available during a H1 test if a cold drum is transferred into a hot oven. Therefore, this heating profile and nonlinear regression approach is helpful to confirm the U value of each H1 test with that from cooling test.

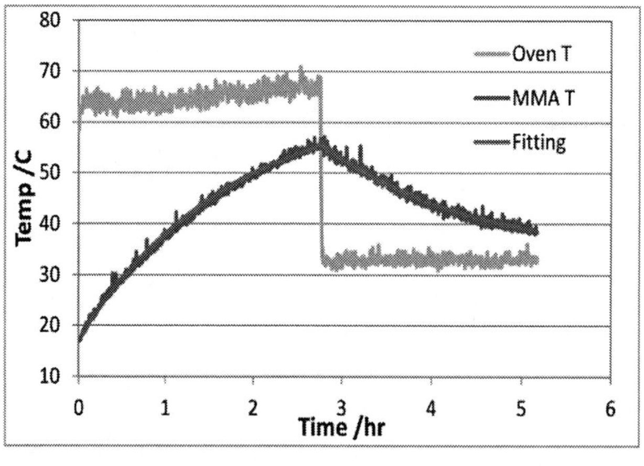

Figure 11. Nonlinear regression for both heating temperature profile and cooling temperature profile

Table 4. Overall heat transfer coefficient (U) of a 1 gallon drum with different chemicals

Chemicals in the drum	U for ambient cooling (W/m2/K)	Thermal conductivity (W/m/K, 25°C)	Viscosity (Pa-s, 25°C)
Literature [23-25]	11.3	-	-
Water	11.8[a]	0.609	8.88E-04
DMP	7.4[a]	0.149	0.00955
MMA	9.0[a]	0.149	6.32E-04

[a] *uncertainty of measurement result is approximately 10%*

The measured U values for 1 gallon drums loaded with water, DMP, and MMA, respectively are listed in Table 4. Previous studies on SADT of organic materials in drums have used a U value of 11.3 W/(m^2K) [23-25]. The measured U value for water is close to the literature value. However, the measured U values of MMA and DMP are smaller than the literature value, which means the application of the literature value for the SADT estimation of those chemicals is not conservative.

3.2.2 Heat Transfer Coefficient Estimations

For the three tested chemicals, it can be assumed that the heat transfer characteristics of the drum wall and outside the drum are identical, but the heat transfer characteristics inside the drum are different due to different physical properties. For such a storage vessel, the heat transfer inside the drum is primarily due to either heat conduction or natural convection. Heat conduction depends on thermal conductivity, and natural convection mainly relates to viscosity of the fluid. As can be seen from the results in Table 4, thermal conductivity and viscosity of the chemical in the drum can have a significant effect on the overall heat transfer coefficient. A good estimation method should include these properties in the calculations.

3.2.2.1 Engineering Estimation

The film coefficient is widely used to estimate the overall heat transfer coefficient of a fluid due to natural convection. The following isothermal surface model is used to estimate the U value:

Film coefficients:

$$\frac{hL}{k} = 0.59(Gr * Pr)^{0.25} \quad (9)$$

Overall heat transfer coefficient:

$$U_o = \frac{1}{\frac{1}{h_o} + \frac{(r_o - r_i)A_o}{kA_{lm}} + \frac{A_o}{A_i h_i}} \quad (10)$$

Where $A_{lm} = \frac{A_o - A_i}{ln\left(A_o/A_i\right)}$, $Pr = \frac{C_p \mu}{k}$ and $Gr = \frac{g\rho^2 \beta (T_w - T_\infty) L^3}{\mu^2}$

Table 5. Predicted vs measured heat transfer coefficient (U) of a 1 gallon drum with different chemicals

	Measurement	Film heat transfer coefficient	CFD steady state simulation
Water	11.8	4.8	12.5
DMP	7.4	2.9	6.4
MMA	9.0	4.2	9.8

The results of this method for estimating U for water, DMP and MMA is tabulated alongside the measured heat transfer coefficients and values obtained by CFD modeling (discussed in the next section) in Table 5. Compared with the measured value, this engineering estimation method significantly underestimates the U value in all cases.

3.2.2.2 CFD Simulation for Heat Transfer Coefficient

Another approach to estimate the U value is the simulation of the heat transfer phenomena via CFD modeling (Fluent 6.3). As shown in Figure 12, a 3-D finite element model was built using the physical dimensions of the 1 gallon drum. The simulated drum was assigned a constant temperature boundary condition on the vertical walls (25°C) and a non-heat flux boundary condition on the top and the bottom of the vessel. Two fluid domains were inside the drum: air headspace and liquid for the tested chemicals. The liquid domain inside the drum had an artificial constant volumetric heat generation of 4,600W/m^3.

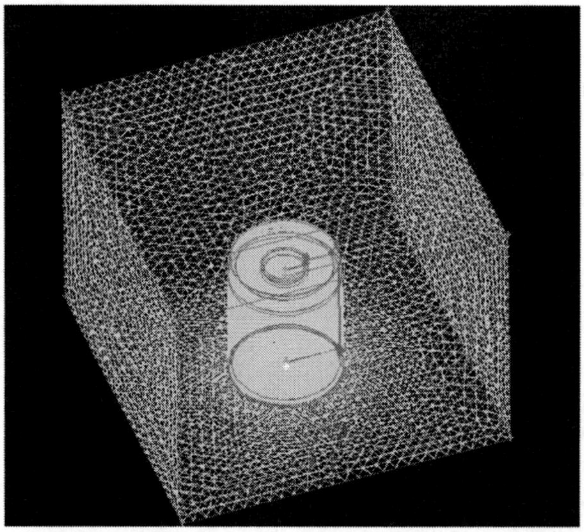

Figure 12. 3D finite element model for drum

A steady state simulation was performed to get the heat transfer due to temperature gradient caused by the heat generated from the chemicals and the mass transfer due to natural convection. With the average temperature of the liquid domain, the U value is calculated with eq. 11.

$$U = \frac{Q}{A(T_{average} - T_a)} \tag{11}$$

An example of temperature distribution from such simulation is presented in Figure 13. The results of this approach for estimating the U value of MMA is listed in Table 5. Compared to engineering estimation method, CFD simulation is a recommended method to estimate the heat transfer coefficient.

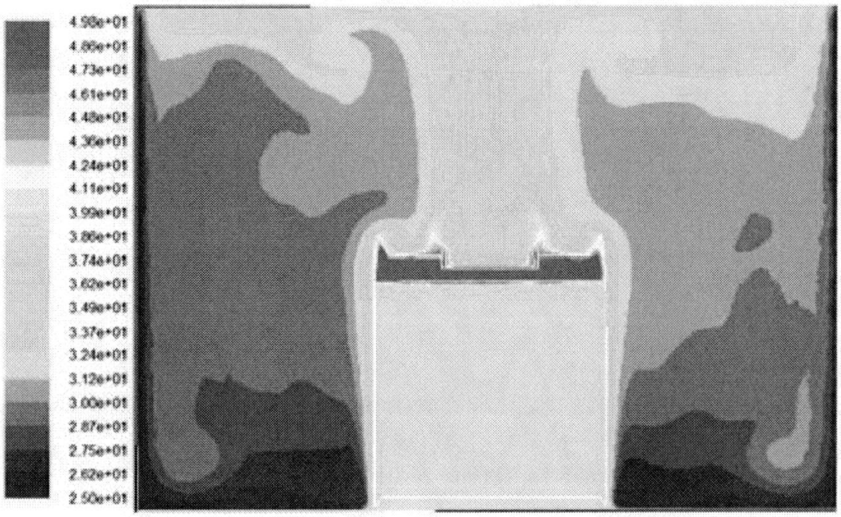

Figure 13. Temperature contours of the drum with MMA sample with a steady state simulation

3.3 H1 Test Validation

Two H1 tests were conducted using MMA with 25 ppm MEHQ, one at 76°C and another one at 72.2°C. The temperature-time profiles from both tests are presented in Figure 14. Although the oven temperature was not constant over the entire test, the first 4 hours of heat-up data from the test was sufficiently constant to confirm the previously measured U value with the nonlinear regression method. Calculated U values from both tests agreed with the U value previously determined from the cooling profile within the uncertainty range. The induction time of H1 test at 76°C was lower than that of 72.2°C, which is consistent with the result from isothermal micro calorimeter tests. The temperature profiles of the polymerization reaction were similar for both H1 tests. Due to the heat from the reaction, the oven temperature also increased with a max overheat of 20°C occurring at the same time as the peak sample temperature. The time required for the sample temperature to increase from 2°C lower to 6°C higher than the oven temperature was 3.4 days for the 76°C test and 6.1 days for the 72.2°C H1 test.

The kinetic parameters from isothermal micro calorimetry (Table 3 combined with filling level correction) and measured U value for MMA in a 1 gallon drum were used to simulate the H1 tests and the results are plotted in Figure 14. The energy balance equation used in the simulation is as below.

$$-mC_p\frac{dT}{dt} = m_m(\Delta Hr_z \cdot \frac{d\alpha_{in}}{dt} + \Delta Hr_M \cdot \frac{d\alpha_M}{dt}) + UA(T - T_a) \qquad (12)$$

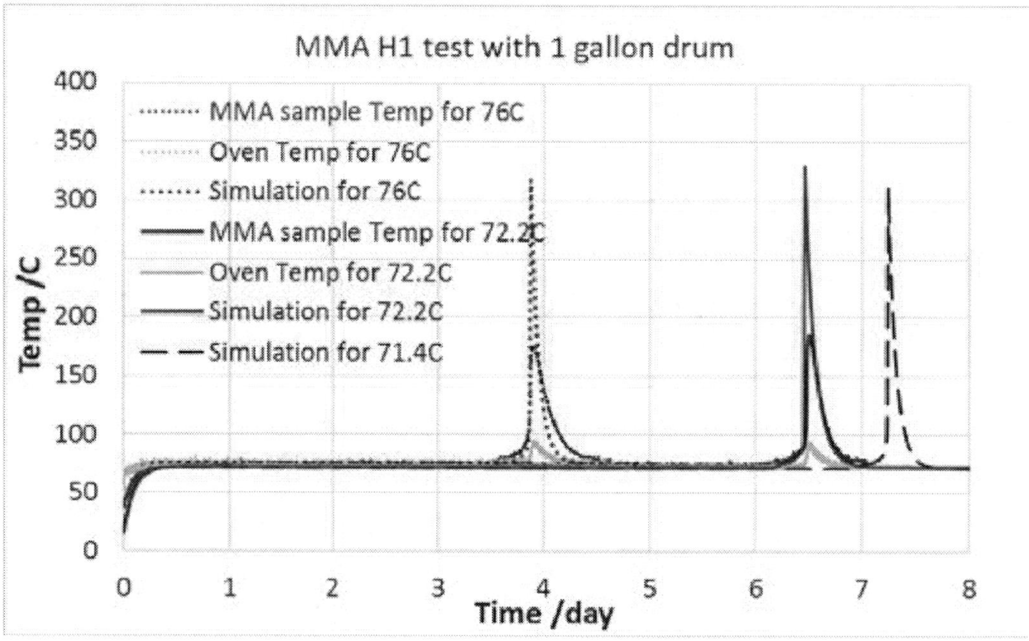

Figure 14. H1 test result with simulation

The simulation predicts the early portion of the polymerization exotherms of both H1 tests quite well, however it over predicts the peak sample temperature during the exotherm. This is not a concern for this application because the time for a 6C temperature rise occurs in the early part of polymerization with poor heat transfer. The reason for the deviation is not clear, however it is possibly a result of the monomer reaching the so called "ceiling temperature". As discussed by Dainton [26], the polymerization conversion of MMA is thermodynamically limited by the presence of a ceiling temperature (167~207°C), which is not integrated into this simulation. Additionally, it was observed after the tests that a small amount of monomer had leaked out around the thermocouple fittings meaning that some amount of evaporative cooling could have occurred during the test.

Utilizing the criteria of SADT, this simulation can be used to determine the monomer SAPT, which is the lowest environmental temperature at which the temperature increase of a monomer in a specified package is at least 6°C during a period of seven days, starting from the time when the temperature is 2°C lower than the environmental temperature. Based on the simulation, the SAPT of this MMA monomer with 25 ppm MEHQ in 1 gallon drum with 10% air headspace is 71.4°C. The simulated temperature profile for an H1 test at 71.4°C is presented in Figure 14 for illustrative purposes.

It is important to remember that just like SADT, the SAPT is not constant for a given monomer but rather it is dependent on the monomer properties as well as the packaging properties and size. If different sizes of package are used to ship the same monomer, the SAPT will vary because of the change in heat loss per unit monomer mass. The magnitude of the sensitivity of SAPT to changes in heat transfer will be a function of the rate of energy release once polymerization begins. By varying the U value in our simulation of MMA in the 1 gallon drum, the sensitivity of SAPT to package size in this system can be seen. It was found that for MMA, the SAPT only decreases from 71.6°C to 71.1°C when the U value changes from 12 $W/m^2/K$ to 0 $W/m^2/K$ (adiabatic) (Table 6). Out of the required 7 day of the H1 test at 71.4°C, the induction time lasts 6.7 days (96%). Since there is almost no heat generation from the sample over the induction time, the decrease of U value will not impact the sample temperature. Thus, only 4% out of the required 7 day can be impacted by the heat loss. Therefore, it can be concluded that the SAPT of this monomer is not extremely sensitive to heat loss, as well as package size.

If only PIT kinetics is used to determine the SAPT, the result is more conservative because the heat generation over PIT is insignificant so that the ending temperature will be less than 6°C higher. As listed in Table 6, SAPT from PIT equation is lower than the SAPT from simulation with full kinetics. Also, the SAPT from PIT is not dependent on the U value or package size at all. The advantage of this PIT approach for SAPT is that it can avoid the kinetic fitting, measurement of U value and complex simulation.

Table 6. Sensitivity of SAPT to the heat transfer coefficient

Proposed U value for 1 gallon drum (W/m2/K)	SAPT estimation /°C	
	Simulation with full kinetics	PIT Equation[a]
0	71.1	71.0
3	71.2	71.0
6	71.3	71.0
9	71.4	71.0
12	71.6	71.0

[a] *the method shown in Figure 4.*

3.4 Proposed methodology for SAPT evaluation

As discussed above, the kinetic parameters for the complete heat profile of a monomer can be determined from appropriate isothermal calorimetric tests. Combining this information with the heat transfer capacity of the transport container allows for the determination of the SAPT. This is accomplished by simulating the heat balance and finding the environmental temperature at which the monomer temperature in a package increases from 2°C lower to 6°C higher than the environmental temperature in a period of seven days (H1 method).

The determination of the PIT is less time consuming than determining the SAPT. Since the monomer is stabilized by inhibitors during the PIT, the heat generation is normally insignificant in this period of time. By simulating the transportation condition, such as the same sample source, the same inhibitor level, the same filling level and so on, measured PIT represents the induction time when there is no temperature rise in the package. Therefore, if the PIT of the monomer is greater than seven days at a temperature, the SAPT of this monomer in a commercial package should be greater than this temperature. With that, the following SAPT methodology was proposed, as presented in Figure 15. For the monomer passing the methodology (a), only one PIT measurement is needed and reported SAPT as >75°C. Otherwise, the heat transfer characteristics and heat balance of the monomer package are needed to determine the SAPT value. An alternative conservative approach to determine the SAPT value is only using PIT kinetics to determine the SAPT.

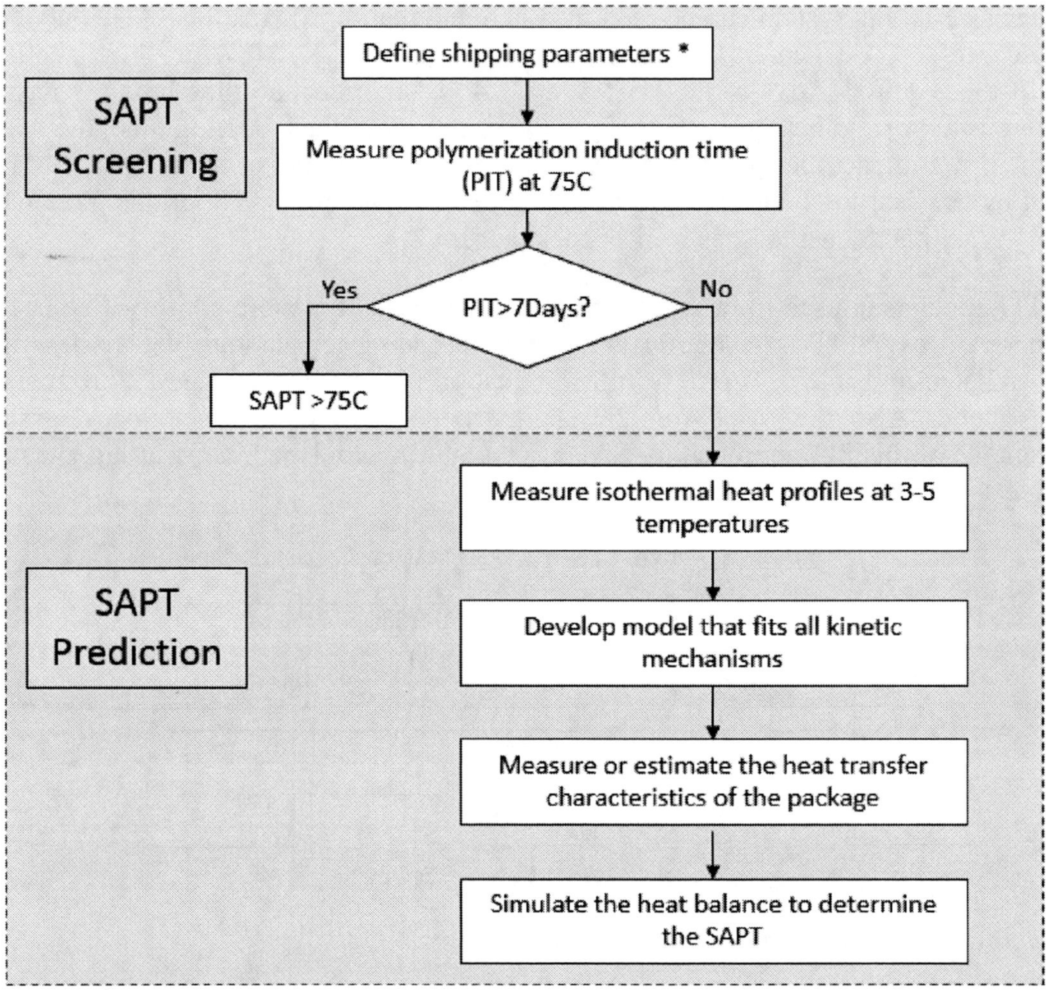

*The term "shipping parameters" here means the conditions under which the monomer will be shipped including inhibitor type/concentration, vessel headspace composition, headspace to liquid volume ratio etc.

Figure 15. Proposed monomer SAPT evaluation methodology

4 Conclusions

The United Nations Recommendations on the Transport of Dangerous Goods, Model Regulations, Rev.19 (2015) has a new requirement for the determination of the Self-Accelerating Polymerization Temperature (SAPT) for polymerizing substances. Recent work has shown that SADT results for materials with autocatalytic or auto-accelerated reactions can differ significantly depending upon which SADT method was applied. This research aims to identify the best method on the basis of good science, readily available technology, and flexibility in monomer packaging applications.

The heat flow curves of an isothermal calorimetric tests successfully demonstrated the different mechanisms of thermal initiation polymerization. Most importantly, a well-controlled temperature over the induction period results in an accurate measurement of PIT. Therefore, the isothermal method on a sensitive micro calorimeter was found to be an accurate method to determine the PIT, as well as the full kinetic parameters for monomers. The effect of inhibitor level, headspace volume, and mixing during sample preparation on PIT were carefully examined with this method. A simplified kinetic model was proposed to fit the isothermal calorimetric testing result. Pseudo-zero order, nth order and auto-acceleration reactions were used for induction period, kinetic control and diffusion control steps respectively.

Overall heat transfer coefficients of a 1 gallon drum with different chemicals were experimentally determined by measuring the cooling temperature profile and applying both half-time of cooling method and nonlinear regression method. For estimation of overall heat transfer coefficient, Engineering Estimation with film coefficient underestimates the value and CFD simulation agrees well with the experimentally obtained value. Two H1 tests were conducted for MMA sample, one at 76°C and another one at 72.2°C. The simulation with kinetic parameters from isothermal calorimetric tests and measured U value fitted the temperature profiles of H1 tests reasonably well. So this simulation was used to calculate the SAPT. The SAPT for MMA with 25 ppm MEHQ in a 1 gallon drum is 71.4°C.

Finally, a methodology for SAPT evaluation was proposed. According to this methodology, if PIT of the monomer is greater than seven days at the lowest allowed SAPT not requiring temperature monitoring (75°C), the SAPT of a commercial package should be greater than this temperature and no additional test should be required. Otherwise, full kinetics of the monomer and heat transfer capacity of the package are needed to evaluate the SAPT. An alternative conservative approach to determine the SAPT value is only using PIT kinetics to determine the SAPT.

5 Acknowledgements

The authors thank Jacob Crosthwaite for helping to distill the commercial MMA to remove the inhibitor and Bob Harfmann for performing GC analysis to determine the concentration of inhibitor in MMA.

6 References

[1] UN Recommendations on the Transport of Dangerous Goods, Model Regulations, in: U. Nations (Ed.), New York and Geneva, 2015.

[2] UN Recommendations on the Transport of Dangerous Goods, Manual of Tests and Criteria, in: U. Nations (Ed.), New York and Geneva, 2009.

[3] H. Fierz, Influence of heat transport mechanisms on transport classification by SADT-measurement as measured by the Dewar-method, Journal of Hazardous Materials, 96 (2003) 121-126.

[4] H.G. Fisher, D.D. Goetz, Determination of self-accelerating decomposition temperatures using the Accelerating Rate Calorimeter, Journal of Loss Prevention in the Process Industries, 4 (1991) 305-316.

[5] T. Kotoyori, The self-accelerating decomposition temperature (SADT) of solids of the quasi-autocatalytic decomposition type1, Journal of Hazardous Materials, 64 (1999) 1-19.

[6] A.A. Kossoy, I.Y. Sheinman, Comparative analysis of the methods for SADT determination, Journal of Hazardous Materials, 142 (2007) 626-638.

[7] G. Krause, K.-D. Wehrstedt, M. Malow, K. Budde, J. Mosler, Safe Transport of Acrylic Acid in Railroad Tank Cars. Part 1: Determination of the Self-Accelerating Decomposition Temperature, Chemical Engineering & Technology, 37 (2014) 1460-1467.

[8] A. Hartwig, R.H. Brand, C. Pfeifer, N. Dürr, A. Drochner, H. Vogel, Safety and Quality Aspects of Acrylic Monomers, Macromolecular Symposia, 302 (2011) 280-288.

[9] L.S. Kirch, J.A. Kargol, J.W. Magee, W.S. Stuper, Stability of acrylic monomers, Plant/Operations Progress, 7 (1988) 270-274.

[10] L.B. Levy, Inhibition of acrylic acid polymerization by phenothiazine and p-methoxyphenol. II. Catalytic inhibition by phenothiazine, Journal of Polymer Science Part A: Polymer Chemistry, 30 (1992) 569-576.

[11] S.S. Cutié, D.E. Henton, C. Powell, R.E. Reim, P.B. Smith, T.L. Staples, The effects of MEHQ on the polymerization of acrylic acid in the preparation of superabsorbent gels, Journal of Applied Polymer Science, 64 (1997) 577-589.

[12] R. Li, F.J. Schork, Modeling of the Inhibition Mechanism of Acrylic Acid Polymerization, Industrial & Engineering Chemistry Research, 45 (2006) 3001-3008.

[13] F. Tüdos, T. Földes-Berezsnich, Free-radical polymerization: Inhibition and retardation, Progress in Polymer Science, 14 (1989) 717-761.

[14] V. Casson, T. Snee, G. Maschio, Investigation of an accident in a resins manufacturing site: The role of accelerator on polymerisation of methyl methacrylate, Journal of Hazardous Materials, 270 (2014) 45-52.

[15] A. Faldi, M. Tirrell, T.P. Lodge, E. von Meerwall, Monomer Diffusion and the Kinetics of Methyl Methacrylate Radical Polymerization at Intermediate to High Conversion, Macromolecules, 27 (1994) 4184-4192.

[16] X.-R. Li, H. Koseki, SADT prediction of autocatalytic material using isothermal calorimetry analysis, Thermochimica Acta, 431 (2005) 113-116.

[17] X.-R. Li, H. Koseki, Thermal decomposition kinetic of liquid organic peroxides, Journal of Loss Prevention in the Process Industries, 18 (2005) 460-464.

[18] L.B. Levy, The inhibition of butyl acrylate by p-methoxyphenol, Journal of Applied Polymer Science, 60 (1996) 2481-2487.

[19] J.A.K. Larry S. Kirch, Joseph W. Magee, Wanee S. Stuper, 19 Stability of acrylic monomers, AICHE, 7 (1988) 270-274.

[20] P. Nising, T. Meyer, R. Carloff, M. Wicker, Thermal Initiation of MMA in High Temperature Radical Polymerizations, Macromolecular Materials and Engineering, 290 (2005) 311-318.

[21] J.-M. Dien, H. Fierz, F. Stoessel, G. Killé, The Thermal Risk of Autocatalytic Decompositions: A Kinetic Study, CHIMIA International Journal for Chemistry, 48 (1994) 542-550.

[22] M. Hazewinkel, Runge-Kutta Method, Springer, Berlin, 2001.

[23] S.H.W. W.H. Lin, G.Y. Shiu, S.S. Shieh, C.M. Shu, Self-Accelerating Decomposition Temperature (Sadt) Calculation Of Methyl Ethyl Ketone Peroxide Using An Adiabatic Calorimeter And Model, Journal of Thermal Analysis and Calorimetry, 95 (2009) 645-651.

[24] Y.-Y.C. Jo-Ming Tsenga, Teh-Sheng Suc, Chi-Min Shu, Study of thermal decomposition of methyl ethyl ketone peroxide using DSC and simulation, Journal of Hazardous Materials, 142 (2007) 765-770.

[25] S.H.W. C. C. Liao, T. S. Su, M. L. Shyu, C. M. Shu, Thermokinetics evaluation and simulations for the polymerization of styrene in the presence of various inhibitor concentrations, Journal of Thermal Analysis and Calorimetry, 85 (2006) 65-71.

[26] F.S. Dainton, K.J. Ivin, Some thermodynamic and kinetic aspects of addition polymerisation, Quarterly Reviews, Chemical Society, 12 (1958) 61-92.

478550 Thermal Decomposition of Mono-Nitrated Toluene (MNT) with Additives

Abstract:

Abstract: On September 21st 1992, an explosion happened in Castleford, UK, at the distillation unit, owned by Hickson & Welch Ltd. The incident killed 5 people and caused nearly 200 injuries. The main reason is due to the decomposition of mono-nitrated toluene (MNT). Nitrated aromatic compounds present a potential run-away reaction threat to the chemical process industry, due to its unstable structure and extensive heat release during upset situations. This study employs the Advanced Reactive System Screening Tool (ARSST) to study the effects of various common contaminants on MNT's thermal decomposition, including sodium nitrate, sodium hydroxide, and sodium carbonate. Experimental parameters and results such as 'onset' temperature, pressure increase rate, temperature increase rate and maximum temperature are present. Sodium hydroxide and sodium nitrate can lower the decomposition onset temperature of MNT and make the consequence more severe. Thermokinetics of the decomposition reaction is shown and SADT prediction of UN25 kg package storage is calculated.

479379 Addressing Challenges of Controlling Change By Implementing Best MOC Practices

Abstract:

Effectively controlling change in a plant environment results in a cumbersome process with an inefficient paper-based Management of Change process or use of a traditional and complex electronic Management of Change (MOC) system. Recent research indicates that the majority of chemical incidents are the outcome of poor implementation of Process Safety Management (PSM) and the lack of efficient MOC processes. As a result, MOC workflow steps are often skipped or prolonged resulting in an ineffective process that does not capture the change fully or safely; approvals are delayed, Process Safety Information (PSI) is not being updated, Pre-Startup Safety Review (PSSR's) are not being performed, and MOCs are not being closed in a timely manner.

The goal of this paper is to address industry challenges with controlling change, and how Siemens implements best practices for an effective MOC process.

479243 Process Change Impact on Relief System Design

Abstract:

Process change impact on relief system design

(Understanding equipment impact and the evaluation warranted)

Various scenarios can cause the pressure within equipment to exceed design pressure. Appropriately sized relief devices are one of the most common methods of protecting equipment and personnel operating this equipment. A key challenge to being in compliance is the dynamic nature of processing plants. Numerous changes can be expected to happen over the lifecycle of a processing plant. Management of Change, one of the Process Safety Management (PSM) elements, requires equipment impact evaluation when these changes occur to ensure adequate pressure relief. This presentation talks about numerous types of changes which will impact equipment and warrant overpressure protection re-evaluation.

479152 Identification and Local Impact Analysis of Projectile Hazard in the LNG Industry

Abstract:

With a strong growth in natural gas production in the United States, a number of liquefied natural gas (LNG) facilities have been proposed and authorized. These LNG facilities have sufficient safety, security, risk assessment and mitigation plans developed in close coordination with local, state, and federal authorities. However, to continue to ensure safe operation, it is vital to continue to improve the scope of risk assessment and management strategies. Recent incidents have highlighted the need to consider the potential hazard and risk from projectiles. This paper studies the sequence of phenomena that occur as part of the local impact of a projectile. The main objectives of this study were to: 1) list projectile (missile/fragment) parameters such as source, type, and characteristics (mass, diameter, velocity) and 2) conduct a local impact analysis for identified potential projectile hazard scenarios at an LNG facility. The paper concludes that the five different calculated thicknesses for concrete penetration, scabbing, and perforation, and steel penetration and perforation, depend directly on the kinetic energy and diameter of the projectile. Furthermore, the authors recommend that operators conduct a projectile hazard analyses for LNG facility projects, which would be useful in making recommendations for the thickness and type of storage tanks and other equipment and for other appropriate mitigation measures.

478563 Thermal Hazard Analysis and Chemical Incompatibility Tests with Custom High-Pressure Crucibles Made from Commonly Used Metals and Alloys

Abstract:

The assessment of reactive hazards and chemical incompatibilities is necessary for safe operations in the chemical and pharmaceutical industries. Commercial ARC (Accelerating Rate Calorimeter), RSST (Reactive System Screening Tools) and high pressure crucibles for Differential Scanning Calorimetry (DSC) can be used for this purpose [1-2]. However, ARC and RSST require significant capital investment and require relatively large amounts of sample (at least 1 g).The alternative, commercial high pressure crucibles for DSC, can also be expensive (e.g., several hundred U.S. dollars for each pair of gold-plated high-pressure crucibles) when multiple samples need to be investigated. These requirements make thermal hazard analysis prohibitively expensive for research laboratories that do not focus on chemical process safety, and where materials under development may be available in limited quantities (e.g., milligrams).

To address these short-comings of current hazard screening methods, inexpensive custom DSC high-pressure crucibles made from different commonly used metals and alloys (stainless steel 304, stainless steel 316, titanium, brass and tantalum) have been constructed in-house using a micro-welding system containing a jewelry spot welder and a microscope. Chemical compounds are tested in different crucibles made from different metal/alloys, which simulates different real-world storage, transportation, and process operations in the chemical and pharmaceutical industry. The cost for each pair of crucibles is less than two U.S. dollars. In a typical experiment, only 1-5 mg of sample is required. Kinetic parameters can be extracted from only a few DSC runs with these crucibles, allowing for prediction of the chemical's thermal behavior at larger scales. This method provides an inexpensive and efficient way to perform thermal hazard analysis and incompatibility tests. The application of these crucibles is demonstrated with case studies of ammonium nitrate (AN) and ionic liquids. Predictions of thermal behavior at larger scales have been verified with ARC tests.

References

1. Oxley, Jimmie C. 2003. "Handbook of Thermal Analysis and Calorimetry. Vol. 2: Applications to Inorganic and Miscellaneous Materials M.E. Brown and P.K. Gallagher, Editors 9 2003 Elsevier B.V. All Right Reserved. 349" 2: 349–69.

2. Han, Zhe, Sonny Sachdeva, Maria I. Papadaki, and M. Sam Mannan. 2014. "Ammonium Nitrate Thermal Decomposition with Additives." Journal of Loss Prevention in the Process Industries. Elsevier Ltd, 1–9. doi:10.1016/j.jlp.2014.10.011.

13TH GLOBAL CONGRESS ON PROCESS SAFETY

Do Dust Hazards Analysis (DHA) Really Differ from a Process Hazards Analysis (PHA) or a Process Analysis (PA)?

Michael E. Robertson

4094 Roslindale Ln
Buford, GA 30519
processesmatter@gmail.com

Prepared for Presentation at
American Institute of Chemical Engineers
2017 Spring Meeting and 13th Global Congress on Process Safety
San Antonio, TX
March 26 – 29, 2017

Do Dust Hazards Analysis (DHA) Really Differ from a Process Hazards Analysis (PHA) or a Process Analysis (PA)?

Michael E. Robertson

4094 Roslindale Ln
Buford, GA 30519
processesmatter@gmail.com

Keywords: NFPA 652, Dust Hazards Analysis, DHA

Abstract

Do Dust Hazards Analysis (DHA) Really Differ From a Process Hazards Analysis (PHA) or a Process Analysis (PA)?

This author reviews the latest NFPA 652 requirement on Dust Hazards Analyses (DHA) and compares it to OSHA PSM's Process Hazards Analyses (PHA) requirement and other process analyses required by industry specific NFPA dust regulations. Though process analyses have been a part of NFPA regulations for several years and in the OSHA PSM regulation for 25 years, the 2016 version of NFPA 652 has clearly defined that a process analysis called Dust Hazards Analysis must be done where there is a combustible dust risk. The subject of this paper is the Dust Hazards Analysis.

The aim of the paper is to find commonality in dust hazards analysis, process hazards analysis and process analysis for tips that may limit the number of analyses a company may have to perform. These tips provided will assist a PHA leader in doing their DHA's. This will not be an exhaustive review of standards or regulations, but an overview and guide for those starting DHA's. It has become obvious that NFPA regulations are modeling a PSM framework and some discussion will be made on the subject in the area of risk assessment and management systems. Some comments will be made concerning the software that can be used, but no software will be sold or marketed. The author is not sharing examples of DHA's from his company due to confidentiality and timing of his DHA's.

Some of the Dust Hazards Analysis differences detailed are the emphasis on dust inside and outside of equipment, the use of NFPA dust thickness calculations, building reviews, and frequency of re-validating the analyses. How does a Dust Hazards Analyses affect the process safety information required, team make-up or report after the review? What are the real requirements for a DHA leader?

1 Introduction

There have been several dust related incidents that resulted in loss of life and severe damage to industrial properties. Imperial Sugar in Savannah, GA (2008) and West Pharmaceutical in North Carolina (2003) are two that top the list. Since the early 2000's, the Chemical Safety Board (CSB) has been recommending the Occupational Safety and Health Administration (OSHA) develop standards surrounding combustible dust. The National Fire Protection Association (NFPA) has had guidelines in several of its standards concerning dust for years.

Table 1. Industry Specific NFPA Guidelines Pertaining to Dust

61 – Agriculture
484 - Metal
654 - General
664 - Wood

NFPA 654, 664, 61, 484, 86 and 499 have dealt with combustible dust for years.

It is 2017 and while OSHA still has not set forth comprehensive dust standards, NFPA has published a comprehensive guideline on the management of combustible dust as 2016 NFPA 652 Standards on the Fundamentals of Combustible Dust. This guideline requires dust hazards analysis (DHA) to be complete by September 7, 2018. long with renewed regulations, risk assessments are now becoming part of every safety performance system. NFPA has required process analysis and hazards analysis in the industry specific guidelines..

A DHA is similar to a Process Hazards Analysis (PHA), but there are some key differences that need to be understood by study leaders. PHA methodology is well- defined especially from the OSHA 1910 PSM regulation, the CCPS book *Guidelines for Hazards Evaluations* **[2]**,and other information from new and vintage PHA courses like Primatech's Process Hazards Analysis Leaders' Training and Process Improvement Institute's PHA Leader's Training.

The similarities between PHA's and DHA's are in the methodology. The methodologies used to conduct a PHA, such as HAZOP, Checklist, What-If, are applicable for identifying and assessing hazards in dust handling processes. Expertise with PHA's can assist with the completion of DHA's. The differences begin when the building has to be evaluated for combustible dust hazards. Process Analysis is defined in NFPA 664-12 – Standard for the Prevention of Fires and Explosions in Wood Processing and Woodworking Facilities. [3]. However, NFPA did not offer much instruction on how to do a Process Analysis (PA).

There are other tools that have to be employed for the building evaluation. Pictures, audits, root cause analysis and cause and effect diagrams can be used for the out of the process review.

Detailed building structure evaluations may require a code inspection or some other engineering tools.

This paper evaluates the elements of dust hazards analysis. Comparisons are made between dust hazards analysis, a process hazards analysis methodology called HAZOP (HAZard and OPerability Study), and a process analysis as described in older NFPA guidelines. These comparisons are made to assist PSM companies in aligning PHA's and DHA's and to assist non-PSM companies on resources to assist them in conducting DHA's.

2 What Is a Dust Hazards Analysis (DHA)

NFPA 652 requires that all facilities handling or creating combustible dust must do a Dust Hazards Analysis. A dust hazards analysis is a systematic review of processes and buildings where dust is created or handled and could create a deflagration.

Per the CCPS process safety glossary: A combustion that propagates by heat and mass transfer through the un-reacted medium at a velocity less than the speed of sound.

This requirement has many businesses scrambling in a similar way to the release of the PSM standard in 1992, which left many oil and gas and chemical companies reeling. The dust hazards analysis is not a PSM requirement nor is it a type of process hazards analysis. While dust hazards analyses are new, analyzing your process for dust hazards has been a part of industrial specific NFPA guidelines for a few years now. Now with NPFA 652, OSHA through Recognized and Generally Accepted Good Engineering Practice(RAGAGEP) has the standards it needs for inspection and enforcement to address the potential for a dust deflagration in a facility.

The dust hazards analysis must be complete by September 2018. The DHA requirement is set forth in Element 7.0 of the 2016 version of NFPA 652. There is a good example of the DHA in Appendix B of NFPA 652. Operating areas are reviewed for the probability of deflagration. Classifications are 'Not A Hazard', 'Maybe a Hazard' and 'Deflagration Hazard'. If the probability exists, then the owner must document how they manage or mitigate that risk. If the owner does not have current controls to prevent the deflagration, steps must be taken to document a hazards management plan with steps to reduce the risk.

Within a process system or building, when there is a sufficient quantity of combustible dust, the following conditions must be considered and documented:

1) Oxidizing atmosphere
2) Credible ignition source
3) Credible suspension mechanism

These are items from the fire triangle used in most fire prevention training. These are key in the dust hazards analysis and must be treated with high importance. Housekeeping also must be properly reviewed and documented to prevent the third item mentioned above – credible suspension mechanism.

3 How Do Dust Hazards Analysis (DHA) Compared to Process Hazards Analysis (PHA)

A major difference between the DHA and a PHA is the consideration of dust risk in a building. PHA's do evaluate the risk to occupied buildings in relation to the potential for releases of highly hazardous chemicals covered by the PSM regulation. This risk is evaluated through facility siting studies and more detailed reviews like a Quantitative Risk Analysis (QRA) or Occupied Building Risk Assessment (OBRA).

Table 2. PHA and DHA Methodology Comparison

Process Hazards Analysis (HAZOP)	Dust Hazards Analysis	Process Analysis
Source: OSHA 1910.119 (e)	Source: NFPA 652 7.0 and Appendix B – Dust Hazards Analysis	Source: NFPA 664 4.0 and Explanatory Material A4.3
Covers the process system	Covers the process system and the building construction	A methodical review of the facility, each operation housed within, and the identification of where a hazard exists.
Gather Data – Drawings and Process Information	Gather Data – Lab Data, Drawings and Process Information	
Have a Qualified PHA Leader a Team with an Operator and a Process Expert, along with other Subject Matter Experts as needed	Have a Qualified DHA Review Leader and at least one other for a team.	
Set Objectives and Boundaries		
Determine the Study Nodes by marking up the Piping and Instrument Diagrams	Determine the nodes on the drawings and sketches	
Review HAZOP Guidewords	Review DHA questions from Appendix B of NFPA 652	
Describe Process Node	Describe Process Node	
Use all Guidewords in all Nodes	Use all questions in all Nodes	
Review Global Node, Human Factors, Management of Change and Incidents		

With the PSM regulation, PHA's are required in facilities with highly hazardous chemicals. These chemicals are listed in the PSM requirement. PHA's are a group of tools that are used to review processes handling highly hazardous chemicals. PHA's include HAZOP, Failure Mode and Effect Analysis (FMEA), What If Studies, and Checklists.

The DHA requires buildings to be reviewed for the possibility of a fire from dust build-up. Building construction must minimize the effects of a deflagration or explosion from a combustible dust.

From the table above, you can see some differences in PHA's and DHA's. DHA's do not require a review of management of change and incident review history. The team composition is not set in a DHA. P&ID's are the basis, but DHA's do not have a required standard drawing type although P&ID's are applicable for some aspects of dust handling.

4 DHA Toolbox

Similar to PHA's, multiple tools are needed to complete a DHA.A PHA method could be used to evaluate the risk associated with combustible dust in a process. However, beyond typical PHA methods, field audits, pictures and video recordings of the areas may be need of areas where dust can build up.Alternatively, a root cause analysis method like 5 Why method might also be applied in areas where a dust build-up to determine sources of dust leakage.. The key is to locate where dust gets out of a process or accumulates beyond its minimum explosive concentration (MEC) in a process, why the dust gets outside the confines of the process and what can be done to prevent the leakage.

Whereas chemical analysis and laboratory data are helpful in a PHA, parameters like Maximum Explosion Concentration (MEC), Minimum Ignition Temperature (MIT) and Kst are very key in determining dust combustibility and risk for a process and building. NFPA 652 has information in the use of these parameters. There is also good information in NFPA's Guide to Combustible Dust and the AIChE Book on Solids Handling Equipment. This paper will not cover in detail any laboratory or chemical analysis of dust. [4], [5]

5 Documentation for Review

Piping and instrument diagrams (P&ID's) have long been the basis of the PHA techniques, including HAZOP, Checklist, and What-If. Supporting documentation may include SDS's, process descriptions, management of change documents, incident history and standard operating procedures. Plant layouts and general arrangement drawings will help orient the team and provide information for facility siting and items in the global node.

P&ID's will help for the process analysis in the DHA. You may also need general arrangement drawings and assembly drawings where machines are involved. Plant layouts and 3D renderings may help with the building analysis portion of the DHA. Pictures and video of areas where dust buildup can occur will also be helpful.

Lastly, plant conveying, processing and finishing OEM information such as screen sizes, motor power, pneumatic flow rates, and duct sizing are key in understanding out the equipment affects the creating or lack of creation of combustible dust. Information on seals and design pressures must be available for comparison with operating parameters.

The bottom line is DHA documentation is as important to the process as Process Safety Information is for PHA. However, it might not be as standard or available as PSI is for a PHA. Some of the information may also be very old since plant layouts and renderings might come from the original start of the plant.

6 The Dust Hazards Analysis Process

The DHA process can start like a PHA with an overview, process description and agenda. The DHA must cover the fugitive combustible dust that gets out into the building and the building's ability to withstand deflagrations and explosions.

Documentation must be available to follow the flow of the materials that contribute to creation of combustible dust. Equipment like fans and screens are especially sensitive since those components can create dust suspensions and impact the Minimum Explosion Concentration of the system. Critical laboratory must be available for discussion since it will provide the data to determine combustibility.
As with a PHA, a tour of the area would be very critical to see the areas were combustible dust accumulates the most. A start to the process is chosen and each part is reviewed by asking a series of question. Below are the questions from Appendix B of NFPA 652 used in a DHA.

Table 3. DHA Process Questions ([1] – B.4.5.1.1 – B.4.5.1.5)

I. Is the particulate deflagrable (explosible)?
II. Is the particulate suspended in air?
III. Is there sufficient concentration to propagate a flame front?
IV. Are there competent igniters available?
V. What hazard management is in place?
VI. Is there metal detection, spark detection, bonding and grounding, or other hazard management means in place?

These questions are asked for each piece of equipment, piping and ducting. The answers will should lead the team or competent person to the hazard management system measures in place and the gaps that will need attention from site management. The process must continue until the entire process is analyzed and all the outcomes have been reached.

As mentioned previously, Appendix B of NFPA 652 (2016) provides a good example to do a DHA. For those in the wood industry, the example is actually of a wood processing plant.

7 Outcomes

NFPA 652 requires that results from the Dust Hazards Analysis be documented. This includes necessary action items requiring change to the process materials, process itself, operations or facilities. NFPA 652 has requirements for managing the hazards under the heading of Hazards Management Program.

The results from a Process Hazards Analysis are the hazard scenarios, a list of safeguards currently in place, and any action items required to ensure that risk is effectively managed within the process handling highly hazardous materials. This usually involves a risk ranking system as well.

The Dust Hazards Analysis does not require a risk ranking system. Where a hazard exists, identification and evaluation of the fire scenario shall include safe operating ranges, safeguard identification, recommendations for additional safeguards as needed to properly manage the fire event, and lastly a plan for implementation.

8 Leaders

A meeting has a leader and teams have leaders. DHA's and PHA's are no different. Each review must have a leader. Her qualifications are listed in Table 2. There are no certifications or extensive training requirements. Many believe that certification is coming for PHA leaders because the competency of those who lead PSM is being discussed in government and industry especially as we enter the next 25 years of PSM.

Many experts agree that an experienced PHA leader can also make an effective DHA leader. This is a good start, but the DHA leader must also be familiar with the hazards of combustible dust. Additionally, it would be prudent to have a leader who is experienced in the process being reviewed or ensure that a process expert is present as a team member. I would review the qualifications listed in CCPS's book on Hazards Evaluation and information.

Table 4. Comparison of Leader Qualifications

Dust Hazards Analysis Leader	Process Hazards Analysis Leader
Must be able to lead DHA's Must be familiar with the hazards of combustible dust	Must be trained in PHA techniques Must be re-qualified every 5 years

9 Frequency

PHA's are done initially to establish a base evaluation and then every five years after the initial PHA to address changes or revisions to covered processes. . DHA's are done once for the original process and for any major upgrades. There is no renewal period, but the initial reviews have to be done by September 7, 2018.

10 Management Systems

What is a dust hazards management system? How does a dust hazards management system compare to the standard elements of PSM? Chapter 9 of NFPA 652 mentions its elements as listed in Table 6. Both PSM and NFPA 652 have the requirement of risk assessment that has been the subject of this paper.

PSM is embedded in many of the NFPA guidelines. Though the thought of PSM to many firms is frightening, PSM offers a good framework for managing risk whether it is combustible dust or other system or process risks.

Table 5. Management Systems Comparison

Dust Hazards Analysis Management System (Ch. 9)	Elements of Process Safety Management
1. Operating procedures and practices	1. Standard Operating Procedures
2. Inspection testing and maintenance	2. Mechanical Integrity
3. Training and Hazard Awareness	3. Training
4. Emergency Planning and Response	4. Emergency Planning and Response
5. Incident investigations	5. Incident investigations
6. Management of change	6. Management of change
7. Management system review,	7. Compliance Auditing
8. Contractor training	8. Contractor Safety
9. Employee participation	9. Employee Participation
10. Document retention	10. Process Safety Information
11. Ignition Control (Covered in Chapter 8)	11. Hot Work Permits
12. DHA	12. Trade Secrets
	13. Pre-Start Up Safety Review
	14. PHA

11 Conclusions

DHA's must be completed. It is similar but different than a PHA. Expertise with PHA's can assist with the completion of DHA's. The similarity between PHA's and DHA's is in the methodology of reviewing processes for dust. The differences start when the building has to be evaluated for combustible dust. Here other tools have to be employed. These tools as previously discussed may vary from audits and inspections to root cause analysis tools. Detailed building structure evaluations may require a code inspection or some other engineering tools.

12 References

[1] NFPA 652-15 – Standards on the Fundamentals of Combustible Dust, National Fire Protection Association, September 7, 2015.

[2] CCPS. *Guidelines for Hazards Evaluations.* Center for Chemical Process Safety, American Institute of Chemical Engineers, New York, NY, 2007.

[3] NFPA 664-12 – Standards on the Fundamentals of Combustible Dust, National Fire Protection Association, 2011.

[4] NFPA– Guide to Combustible Dusts, National Fire Protection Association, 2012.

[5] CCPS. *Guidelines for Safe Handling of Powders and Bulk Solids.* Center for Chemical Process Safety, American Institute of Chemical Engineers, New York, NY, 2005.

478758 A Leading Indicators Based Decision Support Tool to Predict Blowout Events

Abstract:

Predicting a potential blowout scenario timely and efficiently is often a challenging task due to the complexity of drilling or other well intervention activities. Blowouts are usually preceded by gas kicks and early prediction of gas kick events can provide some precious time window to take control of the well in concern. A leading risk indicators based approach for predicting gas kick or blowout events has been introduced in this work. A causal factors based step-by-step approach for identifying comprehensive sets of leading indicators are presented and discussed. Leading indicators were divided into two broad sections - real-time indicators and long-term organizational safety performance indicators. With the real-time indicators, different decision support algorithms are developed which can be used as training tools. A real-time indicator may not always successfully lead to a well control event but at the same time the situation needs to be assessed carefully considering other factors and indicators. Such uncertainties are taken into account while developing the algorithms. For evaluating the relative importance of different leading indicators Bayesian Network models are developed. The key causal factors for well control barriers failure are identified by conducting fault tree analysis and appropriate indicators are linked with relevant causal factors in a Bayesian model. At the end, a list of efficient leading risk indicators are presented for predicting primary well control barrier failure which can initiate gas kicks.

Consequence Modeling at the Eastman Kingsport Site

David Skelton
Principal Engineer
Eastman Chemical Company
P.O. Box 511, Building 18
Kingsport, TN 37662
davidskelton@eastman.com

Michael James
Senior Chemical Engineer
Eastman Chemical Company
P.O. Box 511, Building 18
Kingsport, TN 37662
mjames@eastman.com

Prepared for Presentation at
13th Global Congress on Process Safety
San Antonio, TX
March 26-29, 2017

UNPUBLISHED

Consequence Modeling at the Eastman Kingsport Site

Michael James
Senior Chemical Engineer
Eastman Chemical Company
P.O. Box 511, Building18
Kingsport, TN 37662
mjames@eastman.com

David Skelton
Principal Engineer
Eastman Chemical Company
P.O. Box 511, Building 18
Kingsport, TN 37662
davidskelton@eastman.com

Keywords: Dispersion modeling, consequence analysis, risk assessment, dosage, shelter-in-place, temporary haven, evacuation, toxicity, toxic chemical release, toxic load, downwind concentration, blast overpressure, vapor cloud explosion, pressure vessel burst

Abstract

The intent of this paper will be to review the methods utilized at the Eastman Kingsport site in determining magnitude of consequences of toxic, thermal and blast overpressure scenarios. This will include discussion on dispersion modeling, dose-response methods for toxic and thermal probits as well as assessing occupant vulnerability through the use of BakerRisk's Explosion Risk Cooperative tools.

1. Potential Consequences of Loss of Containment

Figure 1 is a typical event tree for potential losses of containment [1].

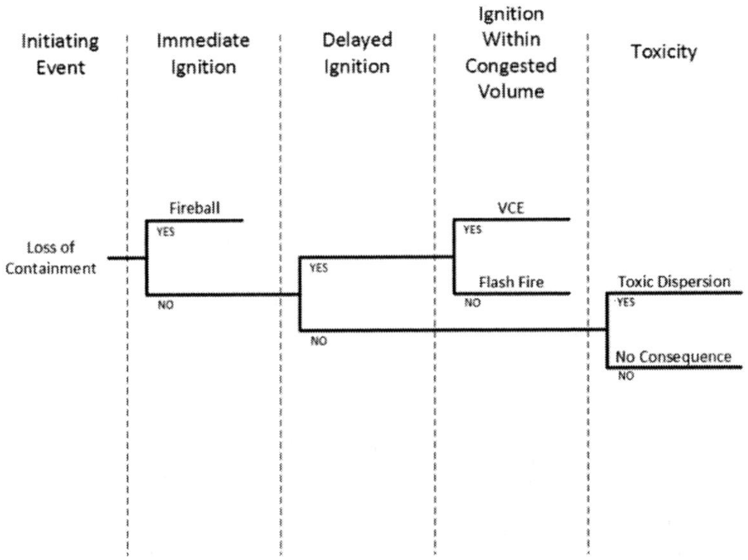

Figure 1. Potential Losses of Containment

The consequences of loss of containment in Figure 1 detail the potential release scenarios that are chiefly of concern. The method for consequence modeling for each type of scenario is described in this paper.

2. Source Terms

Figure 2 details the potential source terms that will need to be developed for each type of scenario.

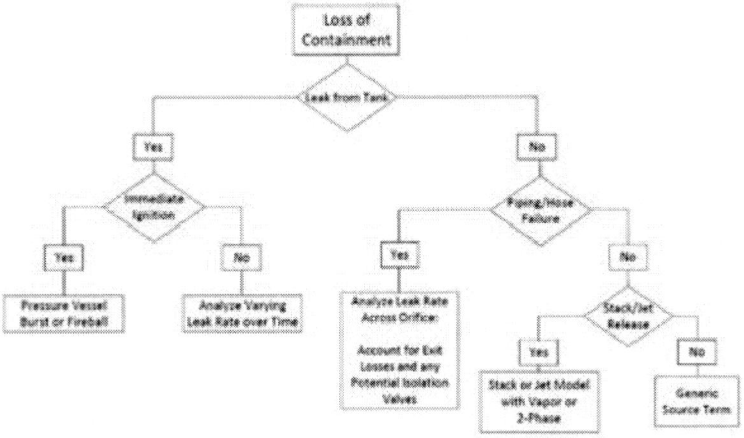

Figure 2. Potential Source Terms.

The source term may have any of the following characteristics:

- Initial momentum
  - Based on the linear velocity across the orifice and other physical properties of the release, the source term may need to account for significant initial momentum of the cloud.
  - In addition, if the linear velocity is sufficiently high, the source of the release may ingest air, resulting in a jetting condition. As a rule of thumb, 10 m/s is a lower limit for potentially jetting releases [2]. There are specific dispersion models to account for jetting conditions during a release.
- Pool Formation
  - Most dispersion models have an integrated pool evaporation model to determine the dynamic evolution of vapor from a liquid spill.
  - If the preferred evaporation model is outside of the dispersion model, the evaporation rate results will need to be analyzed for integration into the dispersion modeling tool.
- Flashing
  - Leaks across an orifice may partially or completely flash from liquid to vapor based on the pressure differential.
  - Modeling tools may report choke conditions at the outlet.
  - Further flash analysis may be required to determine the temperature and liquid/vapor composition from choke to atmospheric conditions.

- Piping Layout between Pressure Source and Leak Point
    - Ensure that all pressure losses are accounted for. This includes entrances, exits, elevation changes, elbows, pipe diameter changes, and other fittings.

- Indoor Release
    - If the release occurs indoors, analyze the concentration changes over time inside the structure based on the source term calculations of the release.
    - Concentration gradients throughout the structure may need to be ignored and perfect mixing assumed.
    - If the vapor portion of the leak will cause overpressuring of the indoor area, determine the amount of vapor required to be vented in order to balance pressure in the room.
    - A flash block can be employed to assess the amount of vapor generated from a leak that will need to be relieved from an indoor area.
    - Flash blocks can also be used to account for vapor suppression systems inside of buildings.
    - This level of modeling is beyond the scope of this paper. However, methods for assessing indoor releases can be referenced from "Consequence Modeling of Dynamic Source Terms" [3].

Presence of the above factors may result in models that require expert judgment and the development of conservative estimates. Most dispersion modeling tools do not have the levels of sophistication required to accurately model these types of source terms. However, the results of a dispersion model employed with conservative estimates may be suitable for site risk assessment.

The following section will discuss the potential leak scenarios and how source terms can be developed, accounting for each of the above factors.

3. Blast Modeling

Simple blast models can be used to predict potential consequences of explosions in lieu of more accurate but complicated methods such as computational fluid dynamic (CFD) models. The TNT equivalency model developed by the U. S. Navy and Air Force was largely used for years, even in industrial settings. Over time it became clear that the TNT model did not predict the behavior of all explosions accurately enough, particularly in regard to vapor cloud explosions more commonly occurring in refining and chemical producing industries. Today several simplified are available to estimate the potential of consequences of explosions, each method with its own strengths and weaknesses.

3.1 Vapor Cloud Explosion (VCE) Modeling

Baker Strehlow Tang (BST) is the simplified method predominantly used at Eastman Kingsport site to model vapor cloud explosions [4][5][7][8]. The BST model is based on the premise that only the portion of a flammable vapor cloud that is confined or contained within a congested area will significantly contribute energy to an explosion. Unconfined or uncongested gas clouds are more likely to result in flash fires or deflagrations. The first step to perform a BST analysis is to identify the confined/congested volume of the plant that has the potential to harbor a flammable cloud of vapor. Examples might include piping and equipment, closely spaced railcars or cars, solid walls, ceilings and floors. A flammable cloud is most often generated from a release of material as a result of an equipment failure, a leak, or a spill. Flammable clouds can drift a considerable distance under the right conditions, so it's important to account for all the equipment within a reasonable radius of the congestion. Once the confined/congested volume is calculated, two parameters contribute the greatest influence on explosion strength using BST methodology; the energy contained within the flammable vapor cloud and flame speed as it burns through the available fuel in the congested region. The first parameter is calculated using the spreadsheet shown in Figure 3. It is a simple tool developed by Eastman to calculate volume based on dimensions of the confined cloud, and then estimate the energy contained with the calculated volume.

ESTIMATION OF VAPOR CLOUD EXPLOSION ENERGY

The user enters dimensions (in ft.) of vapor cloud subject to confinement & obstructions, ambient temperature (F), and the chemical involved in the designated boxes. Thermodynamic data listed in the lookup table to the right is used, or if not available, an approximation is made based on typical hydrocarbons.

The spreadsheet will output the participating energy to be used in V-CLOUD for prediction of pressure and impulse by the Baker-Strehlow method. The methodology is based on the paper "Vapor Cloud Explosion Analysis" by Baker, Tang, and Scheier; Prepared for the AIChE Loss Prevention Symposium, April 1994.

Width (ft)	200	
Height (ft)	50	
Depth (ft)	100	
Volume (ft3)	1000000	
Volume (m3)	2.83E+04	
Ambient Temp (F)	90	
Ambient Temp (C)	32	
Chemical	Acetic Acid	
MW	60.05	
Hc(kJ/mole)	919.73	(heat of combustion in KJ/mole)
Hc(J/m3)	3.67E+07	(heat of combustion in J/m3)
Hc (BTU/lb)	6.59E+03	(heat of combustion BTU/lb for ref.)
Hcs(J/M3)	3.48E+06	(heat of combustion of a stoichiometric mixture with air)
Energy (J)	9.86E+10	
Energy (in-lb.)	8.72E+11	(total available energy in confined / obstructed volume)
% Participation	100	
Equiv Energy in-lb.	8.72E+11	(net energy of explosion in-lb.)

Figure 3. Spreadsheet used to determine energy content of vapor cloud

For a single congested area or multiple areas with little separation distance so they can be effectively considered a single area of congestion, the results from the spreadsheet are used as an input in VCloud®. VCloud® is a simple software tool developed by the BakerRisk Engineering Consultant to produce Pressure/Impulse (P-I) diagrams (this is not to be confused with PI Diagrams representing the response curve for structures and might better be referred to as the side-on blast pressure and impulse as a function of standoff distance). Notice the check box denoted "Plot Using Co-op Energy Factor." Membership in the BakerRisk Engineering Consultant Explosion Research Cooperative (ERC) includes access to experimentally verified energy factors that can reduce conservatism and provide a more accurate prediction of explosion effects. The first page of inputs for VCloud® is illustrated in Figure 4.

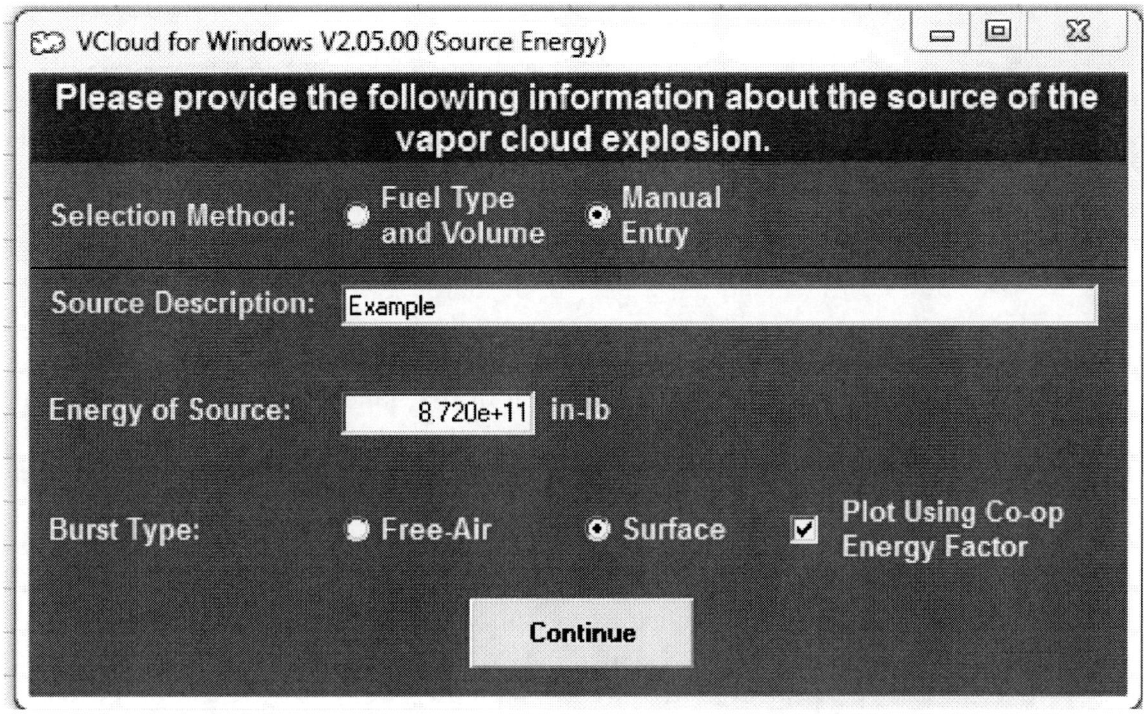

Figure 4. VCloud® Input to Determine Energy Content of Vapor Cloud. Image Courtesy of BakerRisk Engineering Consultant Explosion Research Cooperative.

The second parameter greatly influencing explosive strength when using BST methodology is flame speed, measured as a ratio to the speed of sound or Mach number. Four criterion most influence how fast a flame front burns through fuel: congestion, confinement, fuel reactivity, and the distance to free venting.

- Congestion
  - Burning rate is greatly dependent on the surface area of the flame as it travels through the fuel source. Turbulence causes the flame front to furrow and can greatly increase the surface area which contributes to increasing the speed of the advancing flame. In an industrial setting, turbulence can be generated by piping, equipment, gravel surfaces, vehicles, or any obstacle(s) that inhibits simple linear progression of the burning flame front.
  - The greater the induced turbulence, the faster a flame front will progress increasing the likelihood of detonation. BST method calculations qualify congestion into three categories; high, medium or low congestion. Congestion categorization is the most subjective parameter in BST methodology. Experience is probably the most important tool used to determine degree of congestion.
  - A good rule of thumb is if process equipment occupies an area but can be easily walked through, it is low congestion. If you can manage to slip through an area with difficulty, it's medium congestion. If it is impossible to physically pass through an area, it's high congestion.

- Confinement
  - Any solid surface that prevents the propagation of the flame in a given direction is considered confinement, with the exception of the ground. The BST method models a spherical, free-air vapor cloud explosion, but multiplies the energy of the explosion by a factor of 2 to account for ground reflection of the blast wave from its origin surface.
  - BST calculations categorize confinement in four ways; 3D, 2D, and 2½ D, and 1D.
    - 3D Confinement: A flame front burning through fuel is capable of expanding in all three dimensions with the exception of the ground. It applies to all free-field processing areas not bound by solid walls or ceilings.
    - *2D Confinement:* The propagation of the flame front is free to expand in only two axes because one dimension is blocked by a solid obstacle. A common example is a processing unit located between a solid concrete ceiling with open walls. It also applies if a processing unit is located between two solid walls with no roof. It is important to consider the aspect ratio of 2D confinement, which is the ratio of total width to height of the confined volume. Higher aspect ratios constrict venting capacity and allow additional pressure accumulation over those with a shorter distance to free venting. Typical values are available in a drop down menu in VCloud®, ranging as high as a 25:1 aspect ratio. Anything higher than 25:1 will not provide significantly more pressure accumulation, nor are aspect ratios greater than this commonly found in the field. Any aspect ratio greater than 4:1 should be represented in VCloud as 25:1 to maintain a conservative estimate.
    - 2½ D Confinement: This is a hybrid condition where neither 3D nor 2D accurately describe confinement. For instance, processing equipment contained within a pre-engineered building with lightweight metal walls and roof that will confine the flame front for a time but release at low pressures, before the impending explosion is fully realized. The roof and walls provide some containment, but in short order the blast wave is capable of 3D expansion. Tightly packed equipment on the boundaries of the congested area generating an explosion can also be considered 2½ D confinement, since it provides a small amount of venting, but not enough to consider it 3D expansion.
    - 1 D Confinement: Flame burn in a linear direction, such as inside a pipe.

- Fuel Reactivity
  - The BST method categorizes fuel reactivity in only three ways; low, medium, and high reactivity. Fuel reactivity is based on the laminar burning velocity at a specific concentration, determined by a standard testing procedure in stagnant conditions.
  - Fuels with a laminar burning velocity less than 45 cm/s are considered to have a low reactivity. Fuels with laminar burning velocities above 75 cm/s are considered highly reactive. Everything between is considered to be of medium reactivity.
  - Table 1 illustrates reactivity of a few common chemicals for reference here. NFPA 68 is a good reference for finding laminar burning velocities of many more chemicals so the reactivity category can be determined.

Table 1. Relative Reactivity

Reactivity	Fuel	Group	Relative Reactivity	Formula
Low	Methane	Alkanes	1	CH4
Low	Formic Acid	Acids	1	HCOOH
Low	Acetic Acid	Acids	1	CH3COOH
Medium	Ethane	Alkanes	3	C2H6
Medium	Propane	Alkanes	3	C3H8
Medium	Butane	Alkanes	3	C4H10
High	Ethylene	Alkenes	6	C2H4
High	Acetylene	Alkynes	7	C2H2
High	Hydrogen	Inorganics	7	H2

(Courtesy Eastman Chemical Company)

- Distance to Free Venting
  - Distance to free venting is simply the shortest distance a flame must travel to exit the area of congestion/confinement to find free space.
  - After free venting, turbulence is less likely to accelerate the flame front and under most circumstances it will begin to slow. This reduces the possibility of a transition from deflagration to detonation (DDT).

These four criterion are now input into VCloud® in the format shown in Figure 5. From the provided data, a Mach number (M_f) is assigned to the flame front.

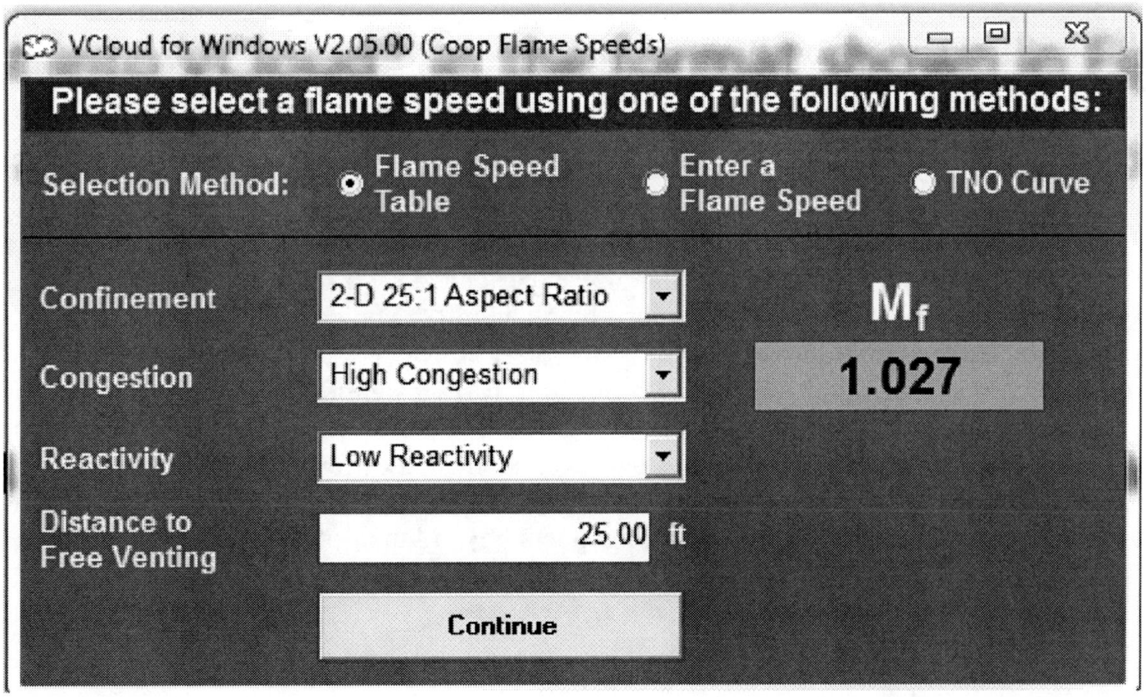

Figure 5. Vcloud® Input to Determine Mach Number. Image Courtesy of BakerRisk Engineering Consultant Explosion Research Cooperative.

A Mach number greater than one indicates the flame front accelerated faster than the speed of sound and beyond, meaning a detonation is more likely to occur. Some typical flame speeds in Mach numbers are provided in Table 2 for reference, with respect to fuel reactivity and degree of congestion.

Table 2. Flame Speeds for Soft Ignition Sources [9]. Provided as a reference but not relevant to the example calculation.

Flame Speed in Mach Numbers (M_w) for Soft Ignition Sources

Flame Expansion	Fuel Reactivity	Obstacle Density		
		High	Medium	Low
ID	High	5.2	5.2	5.2
	Medium	2.27	1.77	1.03
	Low	2.27	1.03	0.294
2D	High	1.77	1.03	0.588
	Medium	1.24	0.662	0.118
	Low	0.662	0.471	0.079
3D	High	0.588	0.153	0.071
	Medium	0.206	0.100	0.037
	Low	0.147	0.100	0.037

VCloud® uses the provided inputs to produce a Pressure/Impulse (PI) diagram. The PI diagram is used to determine the overpressure and impulse experienced by a target at a given distance. Figure 6 is an example of a PI diagram produced by VCloud®. The distance between the target and the explosion is manually input to determine the blast effects at that specified distance. The target at 150 feet from the explosion will experience a 10.12 psig side-on overpressure with a 316.05 psi-ms impulse. This is more than enough to completely demolish an unreinforced masonry building. A key attribute of vapor cloud explosions that differs from high explosives is a much higher impulse. Impulse is more threatening to rigid construction than overpressure, because it tends to distort supporting structural members beyond their plastic limits.

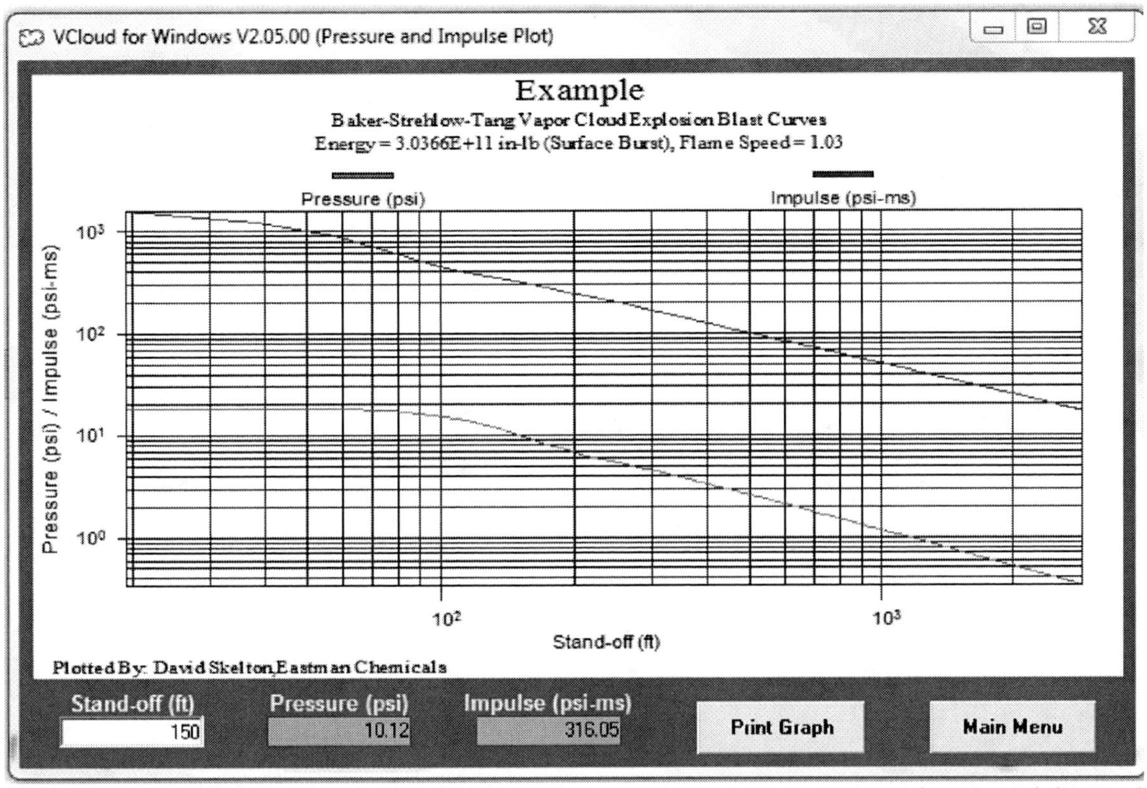

Figure 6. P-I Diagram Output From VCloud®. Image Courtesy of BakerRisk Engineering Consultant Explosion Research Cooperative.

Once a PI diagram is prepared, it can be used to predict the potential damage to conventional industrial buildings from the modeled blast event. Another software tool developed by the ERC can be used for this purpose, the Blast Evaluation and Screening Tool, affectionately known as BEAST®. BEAST® uses empirically based correlations from accidental explosions, shock tube tests, and government sponsored high explosive tests to predict the level of damage imparted by an explosion to a given building or structural component. Originally developed in 1995, BEAST® has been updated over the years to include more building and component damage curves, improve the user interface, and add more building types. The most recent version of BEAST® can evaluate 19 unique building types and many types of individual structural components, such as windows and doors. The first step is to define the building for evaluation. An example using the current BEAST® interface is included in Figures 7a and 7b. Required inputs are type of construction (drop down menu), building dimensions, blast overpressure and impulse (from PI Diagram), and whether wall and roof loading are side-on or reflected.

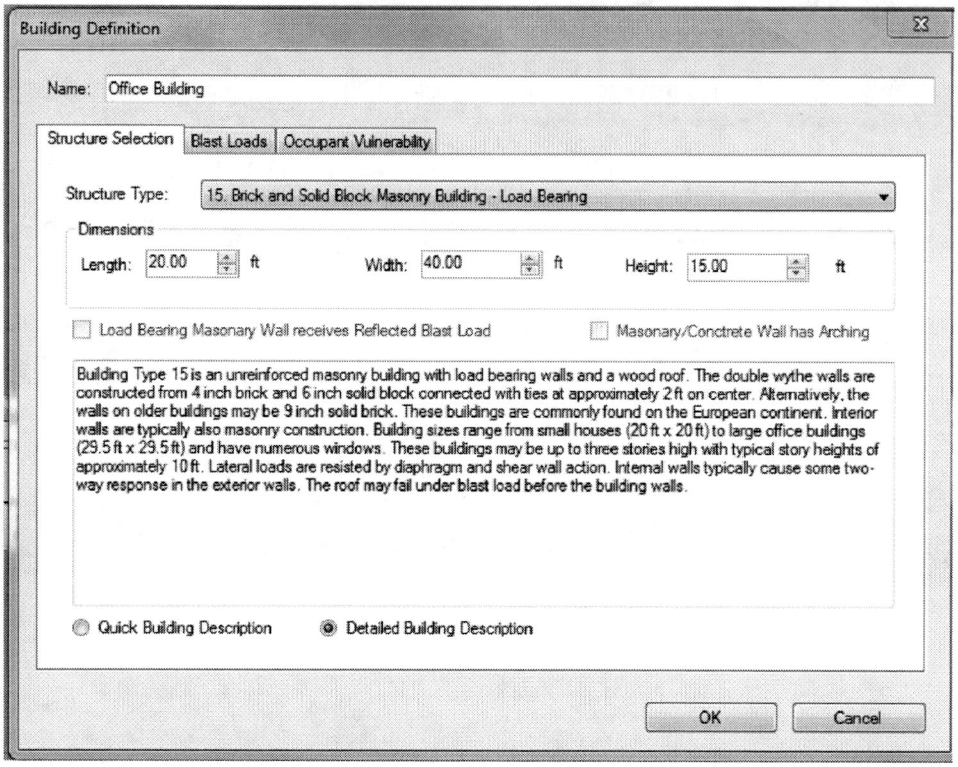

Figure 7a. Beast® Building Selection.

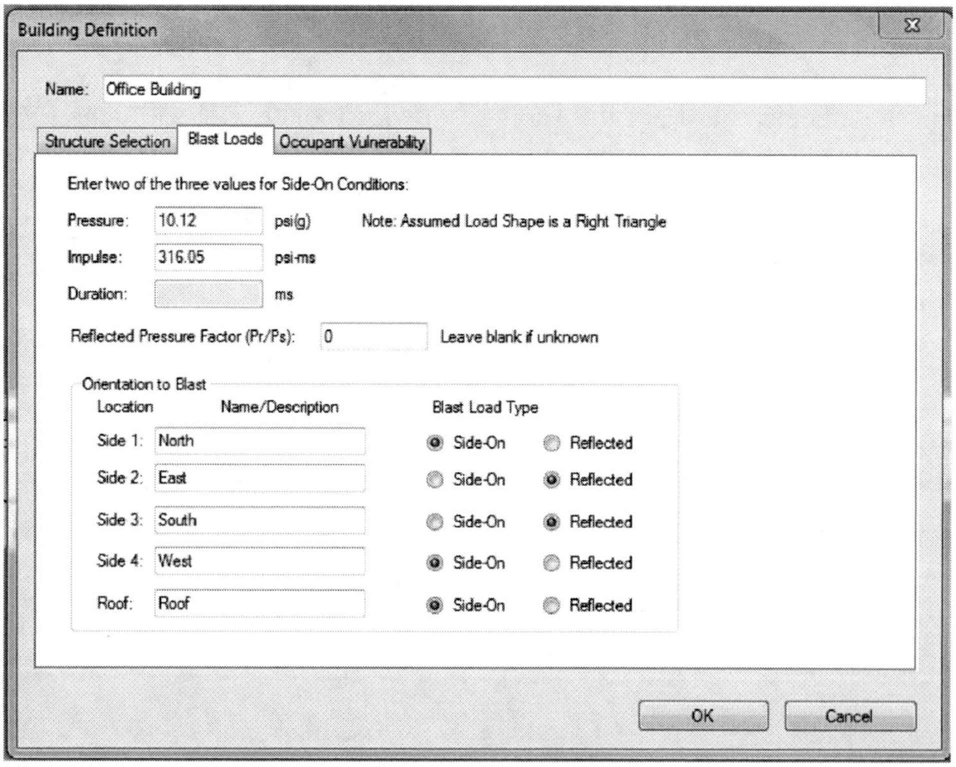

Figure 7b. Beast® Building Selection (cont'd).

Images Courtesy of BakerRisk Engineering Consultant Explosion Research Cooperative.

Given the appropriate inputs, BEAST® predicts a damage level to the building and an occupant vulnerability. Figure 8 illustrates the output provided from BEAST®.

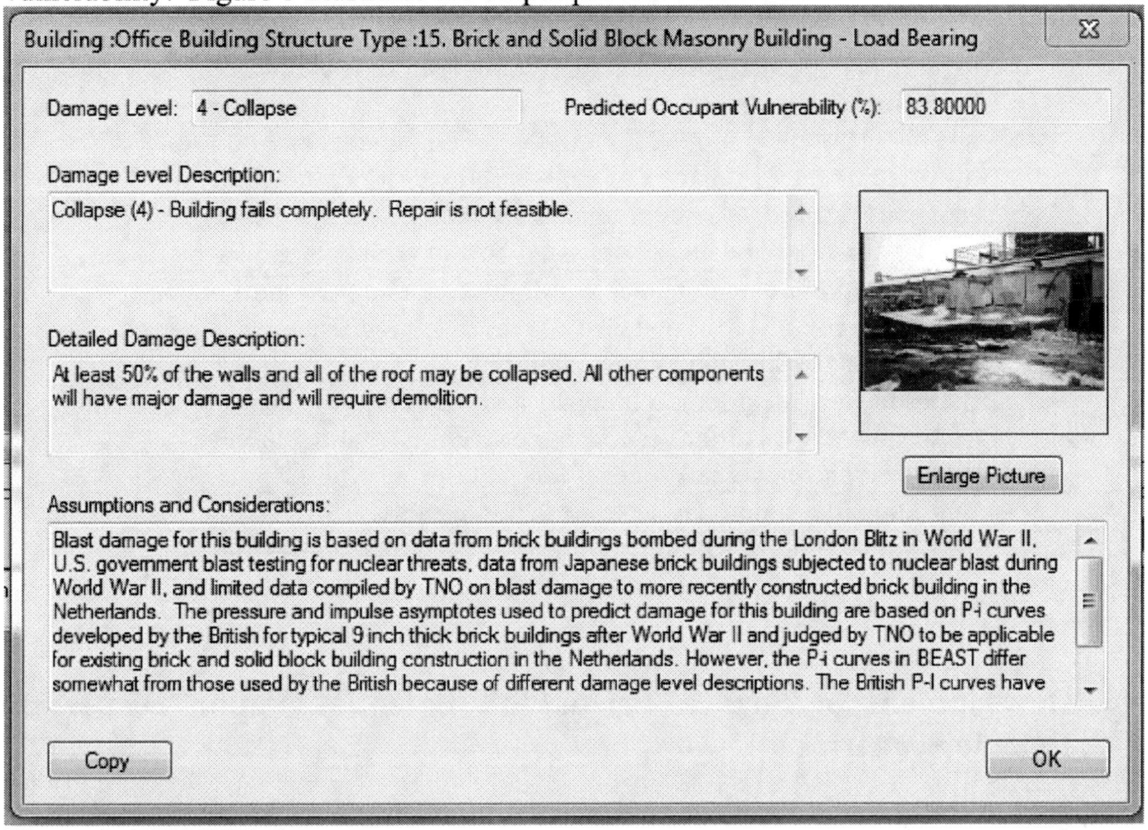

Figure 8. BEAST® Output

Of course, not all explosions result in the complete destruction of a building like the extreme example provided. BEAST® categorizes building damage using the criteria found in Table 3 [6]. A Building Damage Level up to 2B is still survivable by most occupants.

Table 3. BEAST® Building Damage Levels

Building Damage Level	Damage Description
1	Onset of visible damage to building walls facing explosive source. Repairs only needed for cosmetic reasons. Building is reusable following an explosion.
2A	Localized building damage, primarily to components on building walls facing the explosive source. Building performs function and can be used; however, major repairs are required to restore integrity of structural envelope. Total cost of repairs is moderate.
2B	Widespread building damage. Building cannot be used until major repairs are completed. Total cost of repairs is significant, approaching replacement cost of building.
3	Building has lost structural integrity and may collapse due to environmental conditions (i.e. snow or wind). Total repair cost exceeds replacement cost of building.
4	Building fails completely. Repair is not feasible.

Courtesy of BakerRisk Engineering Consultant Explosion Research Cooperative

4. Development of Source Terms for Potential Vapor Dispersion Release Scenarios

Methods for determining the source term for releases from tanks, liquid pools and piping systems are discussed here. All examples provided here do not reflect actual potential release conditions. For all examples, any relation to true conditions is incidental.

Prior to the modeling methods discussion, it is important to review a tool for determining losses due to piping layout.

IoMosaic SuperChems® v 8.21 (2016) allows for development of piping layouts to detail the piping diameters, elbows, and other potential connections. In simple models, as will be discussed in this paper, this can be used as the outlet piping for a pressure source (e.g., tanks) to a release point.

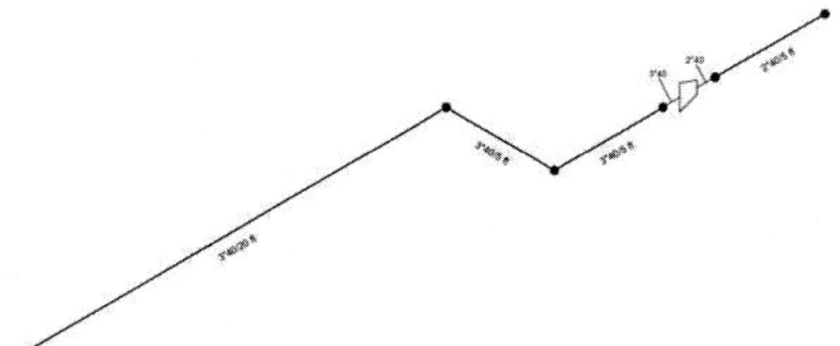

Figure 9. Example Piping Layout.

Figure 3 details a layout from the source at 3 inches with 2 90-degree elbows and a reduction to the 2-inch leak point. The release will be from a guillotine break on the 2-inch line.

4.1 Release from Tank

A reliable model for release from vessels is contained in ioMosaic SuperChems® v8.21 (2016).

This example will model the release of a liquid blanketed by nitrogen. The "Vessels Containing 2-Phases Model" allows for estimation of the vapor and liquid contents of a vessel, accounting for nitrogen blanketing or air in the head space. This same model can be utilized for relief from the vapor phase. There are alternative models for gas-filled vessels.

The overall process for using SuperChems® to model this is as follows:

1. Development of objects for mixture composition, vessel dimensions, piping layouts and other objects. Ensure piping layouts are properly applied.

   The following steps are within the "Vessels Containing 2-Phases" Model.

2. Input the vessel pressure and an initial estimate for overall composition inside of the vessel. The overall composition includes both the vapor and liquid phases. In this example, the modeled vessel is filled to 80% with methyl isocyanate and has a nitrogen blanket.

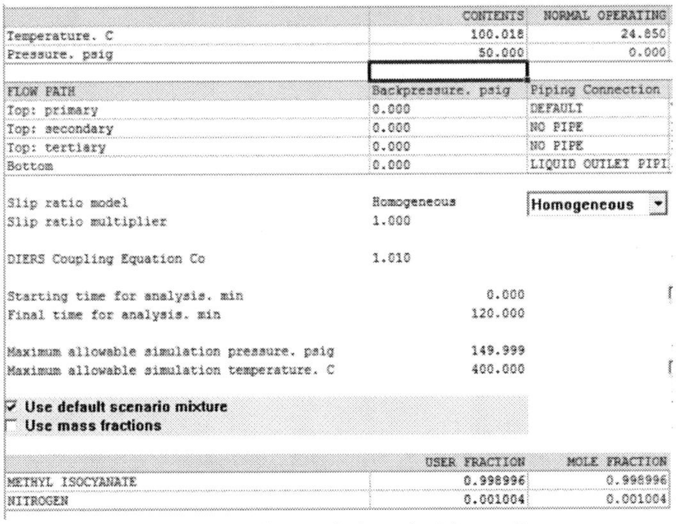

Figure 10. Vessels Containing 2-Phase Inputs

   Uncheck the "Use default scenario mixture" as defining this model is an iterative process.

3. Use "Specify Fill Fraction" to input the approximate vessel fill level. This example uses an estimate 80% fill level.

   The model it calculates the vessel temperature at the specified overall composition.

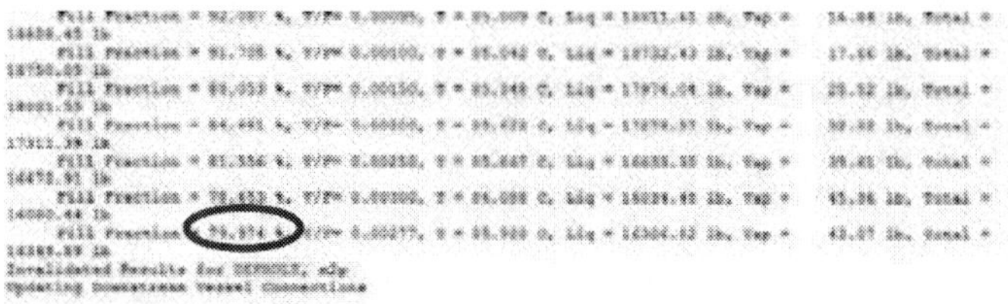

Figure 11. Model Iterations to determine vessel temperature at 80% liquid fill level.

4. Iterate over steps 2 and 3, varying the composition and re-entering the fill fraction until the modeled vessel temperature matches the true conditions. Increasing content of lower boilers will decrease the modeled vessel temperature.

   In the example, the composition was iterated until the storage temperature was approximately 25 deg C.

	CONTENTS	NORMAL OPERATING
Temperature, C	24.918	24.850
Pressure, psig	50.000	0.000

FLOW PATH	Backpressure, psig	Piping Connection
Top: primary	0.000	DEFAULT
Top: secondary	0.000	NO PIPE
Top: tertiary	0.000	NO PIPE
Bottom	0.000	LIQUID OUTLET PIPI

Slip ratio model	Fauske	Fauske
Slip ratio multiplier	1.000	
DIERS Coupling Equation Co	1.010	
Starting time for analysis, min	0.000	
Final time for analysis, min	120.000	
Maximum allowable simulation pressure, psig	149.999	
Maximum allowable simulation temperature, C	400.000	

☐ Use default scenario mixture
☐ Use mass fractions

	USER FRACTION	MOLE FRACTION
METHYL ISOCYANATE	0.991100	0.991100
NITROGEN	0.008900	0.008900

Figure 12. Final composition.

5. Running the model will determine the flow rate through the piping layout and guillotine break.

6. When the run is complete, a report will appear. The "Profiles-Bottom" tab is the most useful for this source term model.

TIME, min	VESSEL CONTENTS TEMPERATURE, C	VESSEL CONTENTS PRESSURE, psig	PRESSURE AT VESSEL BOTTOM, psig	BOTTOM PIPING LIQUID MASS FLOW RATE, lb/min	BOTTOM PIPING GAS/VAPOR MASS FLOW RATE, lb/min	bottom flow reduction factor (Actual flux with pipe/Ideal Flux)	BOTTOM FLOW NOZZLE PRESSURE, psig		BOTTOM FLOW DISCHARGE PRESSURE, psig	BOTTOM FLOW DISCHARGE TEMPERATURE, C
0.0000	24.92	50.000	56.394	2.005E+03	0.000	0.405		56.392	-0.001	25.094
0.0246	24.88	47.142	53.517	1.966E+03	0.000	0.394		53.516	0.000	25.051
0.0506	24.85	44.305	50.662	1.871E+03	0.000	0.383		50.660	-0.001	25.006
0.0782	24.81	41.472	47.809	1.775E+03	0.000	0.372		47.808	-0.001	24.960
0.1077	24.77	38.642	44.961	1.679E+03	0.000	0.360		44.959	-0.001	24.913
0.1394	24.73	35.817	42.116	1.583E+03	0.000	0.347		42.115	-0.001	24.864
0.1735	24.69	32.996	39.275	1.485E+03	0.000	0.334		39.274	-0.001	24.814
0.2104	24.64	30.181	36.440	1.387E+03	0.000	0.321		36.439	0.000	24.761
0.2506	24.60	27.371	33.610	1.289E+03	0.000	0.306		33.609	0.000	24.706
0.2946	24.55	24.568	30.787	1.190E+03	0.000	0.291		30.786	0.000	24.649
0.3433	24.50	21.773	27.971	1.090E+03	0.000	0.276		27.970	0.000	24.585

Figure 13. Vessels Containing 2-Phases Output.

The liquid and vapor release rates out of the piping as well as the discharge temperature and pressure are reported. The discharge pressure will be above atmospheric if there is choked flow inside the piping.

If the choke pressure is significantly above atmospheric, running some select points through the "2-Phase Expansion and Aerosol Model" to determine the final vapor/liquid composition and temperature is recommended.

7. As the release is liquid in this case, evaporation from a liquid forming pool will next be modeled. As stated previously, several dispersion models have integrated pool models. The SuperChems® pool evaporation will be employed in this case, and the output from the model will be input to an external dispersion model.

4.2 Liquid Pools

As with tank leaks, SuperChems® will be utilized for determining the source term of a liquid pool formation.

An important consideration is the variety of weather conditions over which a release could take place. As solar incidence is a major factor in pool evaporation, ensure that the most probable conditions for both daytime and nighttime are employed. Also, to account for solar incidence, ensure that the appropriate time of day and site coordinates are input for the scenario in SuperChems®. In addition, diked area dimensions and substrate composition can be specified.

SuperChems® pool evaporation model allows for a static release rate over a set period of time. As most releases will be dynamic in nature, this may involve expert judgment.

Once a characteristic release rate, duration and temperature have been determined, this information can be entered into the pool evaporation model.

Liquid discharge temperature. C	24.209
Liquid discharge rate. lb/min	583.903
Liquid discharge duration. min	1.949

Use default scenario mixture
Use mass fractions

COMPONENT	USER FRACTION
METHYL ISOCYANATE	1.0000
NITROGEN	0.0000
Average molecular weight	57.0520

Figure 14. Pool Evaporation Model Inputs

The model can perform evaporation on multi-component mixtures. In this case, the evolution of any dissolved nitrogen is not of concern. The liquid rate has been imported from the Vessels Containing 2-Phases model, and the source concentration has been set to pure methyl isocyanate.

Based on the inputs, the model will provide a conservative estimate for the evaporation of methyl isocyanate from the formed pool.

Time. min	OVERALL POOL EVAPORATION RATE. lb/min	Pool diameter. ft	Pool surface temperature. C
0.017	0.000	3.146	4.209
0.017	0.258	3.146	-5.182
0.017	0.257	3.146	-5.235
0.017	0.257	3.147	-5.287
0.017	0.256	3.147	-5.339
0.018	0.257	3.147	-5.389
0.028	0.249	3.171	-5.439
0.038	0.242	3.241	-7.525
0.049	0.252	3.371	-8.514
0.059	0.273	3.553	-8.996
0.070	0.302	3.765	-9.216
0.080	0.335	3.988	-9.306
0.090	0.371	4.212	-9.334
0.101	0.409	4.431	-9.335
0.111	0.447	4.644	-9.322
0.122	0.485	4.849	-9.305

Figure 15. Pool Evaporation Model Output.

Figure 15 details the most critical data points in the pool evaporation model. Understanding these will allow for straightforward input to a dispersion model.

While SuperChems® does have an integrated dispersion model, the author's preference is SAFER Trace® v10.2 (2012). Implementation of pool models into dispersion models is discussed later in this paper.

4.3 Leaks from Piping

For lengths of piping with isolation, it may be appropriate to treat the isolated volume as a vessel and analyze the varying leak rate over time. This section considers the situation where the pressure differential across the orifice may be assumed constant. In addition to the previous source terms, SuperChems® is utilized in determining the source term.

The "Source Term Wizard" can be employed to specify an upstream pressure and temperature to determine the flow rate across a specified piping layout or idealized nozzle. The example here will determine the release rate of methyl isocyanate pressurized to 50 psig and released through a guillotine break in the 2" segment of piping from Figure 9.

OPTIONS	
Liquid Density	Databank
PVT Behavior	Equation of state
VLE Search Direction	Dew point

☐ Use mass basis data to compute fractions
☐ Mixture is soluble in water
☐ Use an existing mixture

COMPOUND	FORMULA
METHYL ISOCYANATE	C2H3NO

Inlet temperature. C	25.00000
Inlet pressure. psig	50.00000
Back pressure. psig	0.00000

☑ Initial state is subcooled liquid

Figure 16. Liquid Source Term Wizard Inputs

The release is treated as a subcooled liquid in order to allow the model to estimate if any portion of the release will flash across the opening. It is not expected to have any flashing due to the release conditions in comparison to the chemical's vapor pressure at 50 psig. However, this model is employed for consistency.

** INLET CONDITIONS	
Inlet temperature. C	25.0
Inlet pressure. psig	50.0
Surroundings/back pressure. psig	-3.974E-06
Slip ratio exponent eta	0.001
Vapor quality	1.E-04
**** EXIT CONDITIONS**	
Exit temperature. C	25.160
Exit pressure. psig	-1.7331E-04
Barometric pressure at site elevation. psig	-3.974E-06
Mixture exit velocity. ft/s	21.213
Exit Mach Number	0.006
Maximum Mach Number	0.006
Exit rho*u*u in kg/m/s2 or in Pascals	38695.90
Maximum rho*u*u in kg/m/s2 or in Pascals	38695.90
Exit steady state impulse. lbf	68.146
Maximum steady state impulse difference. lbf	418.675
Pipe length travelled / total piping length	0.998
Flow rate. lb/min	1714.327

COMPONENT	MW	TOTAL FLOW. lb/min	LIQUID FLOW IN. lb/min	LIQUID FLOW OUT. lb/min	VAPOR FLOW IN. lb/min	VAPOR FLOW OUT. lb/min
METHYL ISOCYANATE	57.052	1713.817	1713.817	1713.817	0	0

Figure 17. Source Term Wizard Outputs.

The results from this model include liquid and vapor release rates as well as exit pressure and temperature. Had flow been choked in the piping, the exit pressure would be above atmospheric. As with previous models, it is recommended to utilize a flash block to determine conditions at atmospheric pressure for releases from systems with choked flow.

For releases from holes or cracks in piping, the piping layout can be developed with an orifice at the end of the layout. The orifice area would be equal to the area of the hole, and the orifice length would be equal to the width of the piping. In addition, due to constraints with the software, a short pipe segment of with the same diameter as the leak should be placed at the exit of the segment.

The results for the above cases can directly be input to a dispersion model, as described in the next section.

5. Application of Source Terms to Dispersion Models

This section will describe the inputs to dispersion models based on the results from the two-phase leak from the sour stream.

Section 5 will review how to utilize the data from the dispersion models in order to assess the hazard posed to a hypothetical shelter 500 ft downwind from the release.

5.1 Test for Jetting

In two-phase or vapor release cases, the linear velocity across an opening may be sufficient to cause jetting. As described previously, there are specific dispersion models for jetting conditions. The rule of thumb for jetting regime is 10 m/s [1]. A simple ideal gas calculation can provide an estimate for linear velocity, given a known release rate, hole size and temperature. This can be expanded to a real gas calculation for highly non-ideal cases. The temperature used in this calculation should be the flashed temperature to atmospheric pressure. The rate employed should be that of the vapor phase for two-phase releases as this will be the largest volumetric component once flashed to atmosphere.

In this example, a 200 psig stream of sour hydrocarbons at 150 deg C is released across a 1-inch hole in 6-inch piping. While several VLE models are available for hydrocarbon mixtures, ideality is assumed in this example.

Piping Isometric / Layout	6-INCH LINE WITH 1-INCH CRACK
Site Name	DEFAULT
Liquid Density Method	Databank
VLE/PVT Method	Equation of state
** INLET CONDITIONS	
Inlet temperature. C	150.691
Inlet pressure. psig	200.0
Surroundings/back pressure. psig	-3.974E-06
Slip ratio exponent eta	0.001
** EXIT CONDITIONS	
Exit temperature. C	133.870
Exit pressure. psig	138.432
Barometric pressure at site elevation. psig	-3.974E-06
Mixture exit velocity. ft/s	190.056
Exit Mach Number	1.0
Maximum Mach Number	1.0
Exit rho*u*u in kg/m/s2 or in Pascals	393239.785
Maximum rho*u*u in kg/m/s2 or in Pascals	393239.785
Exit steady state impulse. lbf	165.061
Maximum steady state impulse difference. lbf	6037.864
Pipe length travelled / total piping length	0.998
Flow rate. lb/min	455.119

COMPONENT	MW	TOTAL FLOW. lb/min	LIQUID FLOW IN. lb/min	LIQUID FLOW OUT. lb/min	VAPOR FLOW IN. lb/min	VAPOR FLOW OUT. lb/min
HYDROGEN SULFIDE	34.082	2.052	2.052	0.866	0.001	1.186
n-PENTANE	72.150	347.584	347.547	268.991	0.037	78.593
HEXANE	86.175	77.840	77.835	66.892	0.005	10.949
n-OCTANE	114.231	27.515	27.515	26.078	0.001	1.438

Figure 18. Flow of stream across 1-inch crack.

The exit conditions show that the flow has choked. As previously discussed, the "2-Phase Expansion and Aerosol Model" will be employed to determine the composition and temperature at atmospheric pressure. In addition, this model will report the linear velocity and cloud diameter when flashed to atmospheric pressure. The linear velocity can be used to determine if jetting is possible.

**TWO-PHASE EXPANSION / DEFAULT

	INITIAL	FINAL
Flow diameter. ft	8.3333E-02	2.3754E-01
Velocity. ft/s	190.42	651.59
Vapor density. lb/gal	0.24	0.02
Liquid density. lb/gal	4.20	5.19
Mixture density. lb/gal	0.98	0.04
Temperature. C	133.87	45.56
Pressure. psig	138.43	0.00
Vapor/Feed (V/F) mole fraction	0.21	0.71
Mass flow rate. lb/min	455.12	455.12
Available energy. BTU/lb	26.45	
Modified Weber number for thermal breakup	392.30	
Critical Weber number for mechanical breakup	12.01	
Viscosity group number	0.01	
Stable droplet diameter. Microns	39.50	
Thermal breakup droplet diameter. Microns	39.50	
Mechanical breakup droplet diameter. Microns	44.18	
Number of droplets	4.9421E+10	
Droplet mass. lb	4.4271E-11	

COMPONENT	TOTAL FLOW	INITIAL LIQ	FINAL LIQ	INITIAL VAP	FINAL VAP
HYDROGEN SULFIDE. lb/min	2.053	0.866	0.038	1.186	2.015
n-PENTANE. lb/min	347.681	269.076	80.158	78.605	267.523
HEXANE. lb/min	77.862	66.912	35.926	10.950	41.936
n-OCTANE. lb/min	27.523	26.085	23.812	1.438	3.711

Figure 19. Sour Stream Contents after Expansion.

Per Figure 19, the temperature of the release cloud will be 46 deg C at atmospheric pressure and will have a linear velocity of 652 ft/s (199 m/s). This is an order of magnitude above the 10 m/s guidance, and should result in significant jetting.

5.2 Input to Jet Model

The dispersion modeling tool utilized for these examples is SAFER Trace® 10.2 (2012).

Figure 20 details the inputs to the jet model based on our example. This is focusing on the vapor portion of the model, ignoring liquid entrainment.

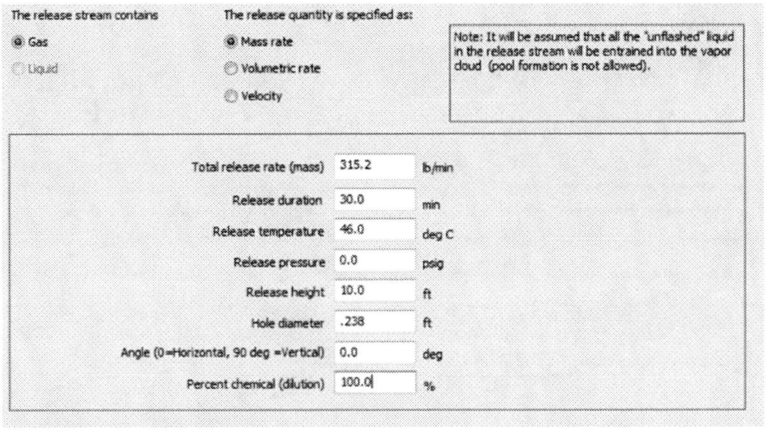

Figure 20. Jet Model Inputs.

The dispersion outputs will report concentrations and plots related to concentration of H2S in air.

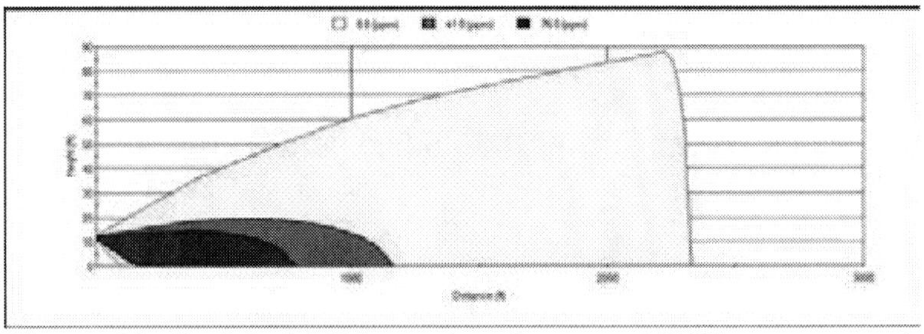

Figure 21. Jet Model Dispersion Output. AEGL concentration limits based on 10-minute exposure to H2S [10].

The H2S concentration profile at the outside of a shelter 500 ft downwind from the release point is shown in Figure 22.

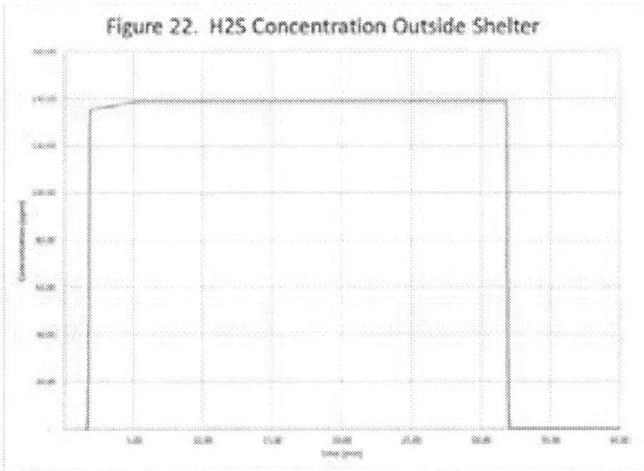

Consequence to personnel in this shelter will be analyzed using a probit function, detailed in Section 5.

5.3 Modeling Pool Evaporation

The liquid portion of the example two-phase release will be treated as the source of a liquid pool.

```
Spill Surface                                    DEFAULT

Liquid discharge temperature. C                   45.560 <<<
Liquid discharge rate. lb/min                    139.934
Liquid discharge duration. min                    15.000

[x] Use default scenario mixture
[ ] Use mass fractions

COMPONENT                                  USER FRACTION
HYDROGEN SULFIDE                                  0.0006
n-PENTANE                                         0.6394
HEXANE                                            0.2399
n-OCTANE                                          0.1200

Average molecular weight                         80.5399

RUN OPTIONS
```

Figure 23. Liquid Pool Model Inputs

The concentration profile of evaporate is highly dynamic over the studied period of the release.

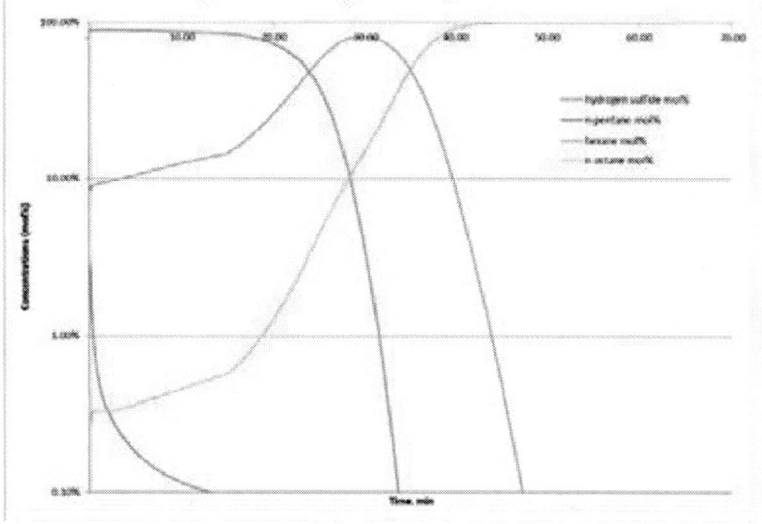

Figure 24. Pool Evaporate Concentration Profile

While the concentration is highly dynamic, the main toxicity concern is with H2S. The majority of H2S appears to evaporate in the first 15 minutes. An average rate will be developed based on the total evaporated mass of each component over that time. The effective concentrations and dispersion parameters are shown in Table 4.

Table 4. Molar Concentrations and Dispersion Parameters

Chemical	Mol%
Hydrogen Sulfide	0.1%
Pentane	87.0%
Hexane	12.4%
Octane	0.5%
Average Evaporation Rate	57.0 lb/min
Temperature	2°C
Duration	15 min

The data from this model is input to the SAFER Trace® "Single Phase" steady state model, per Figure 25.

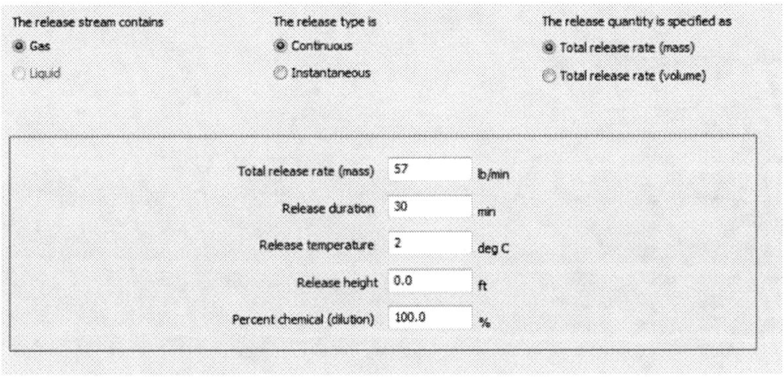

Figure 25. Steady State Model Inputs.

The downwind dispersion of the cloud, focusing on H2S concentrations in air is shown in Figure 26. The concentration is negligible at the 500-ft shelter, so the contribution from evaporation to toxic impact can be ignored.

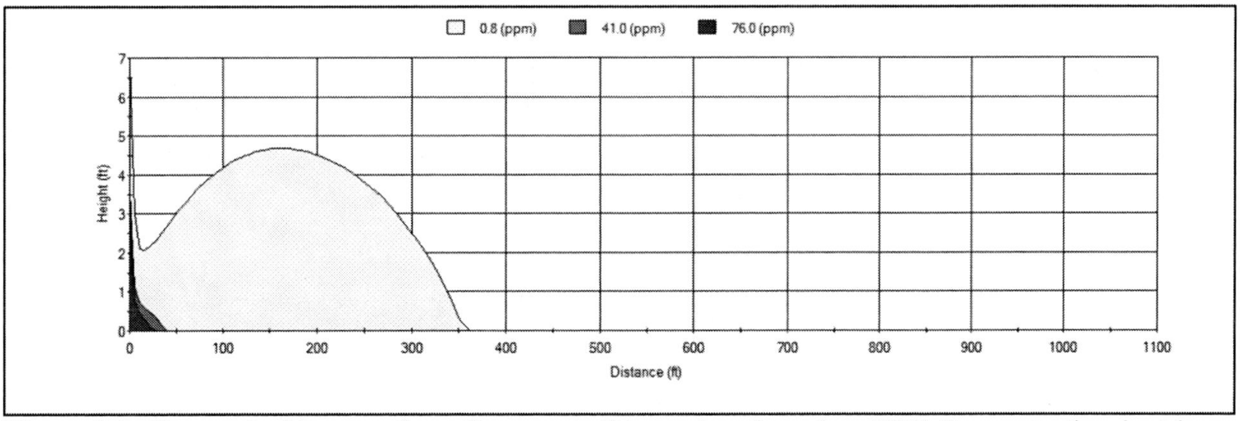

Figure 26. Downwind Impact from Evaporate Dispersion, based on H2S Concentration in Air.

6. Magnitude of Consequence Determination Using Probits

The methods utilized here are described in detail in "Simplified Methods of Using Probit Analysis in Consequence Analysis" [1].

To summarize, the impact to individuals may calculated using dose [5].

$$Dose = \int C^n dt \qquad \text{[Eq. 1]}$$

Dose may be transformed to probability of fatality using a probit transformation [5].

$$Y = A + B\log(Dose) \qquad \text{[Eq. 2]}$$

The constants "A", "B", and "n" are specific to the chemical. The Y value, or probit score is referenced on a probit table to determine probability of studied consequence.

Table 5. Probit table.

%	0	1	2	3	4	5	6	7	8	9
0	—	2.67	2.95	3.12	3.25	3.36	3.45	3.52	3.59	3.66
10	3.72	3.77	3.82	3.87	3.92	3.96	4.01	4.05	4.08	4.12
20	4.16	4.19	4.23	4.26	4.29	4.33	4.36	4.39	4.42	4.45
30	4.48	4.50	4.53	4.56	4.59	4.61	4.64	4.67	4.69	4.72
40	4.75	4.77	4.80	4.82	4.85	4.87	4.90	4.92	4.95	4.97
50	5.00	5.03	5.05	5.08	5.10	5.13	5.15	5.18	5.20	5.23
60	5.25	5.28	5.31	5.33	5.36	5.39	5.41	5.44	5.47	5.50
70	5.52	5.55	5.58	5.61	5.64	5.67	5.71	5.74	5.77	5.81
80	5.84	5.88	5.92	5.95	5.99	6.04	6.08	6.13	6.18	6.23
90	6.28	6.34	6.41	6.48	6.55	6.64	6.75	6.88	7.05	7.33
—	0.0	0.1	0.2	0.3	0.4	0.5	0.6	0.7	0.8	0.9
99	7.33	7.37	7.41	7.46	7.51	7.58	7.65	7.75	7.88	8.09

Source: Finney, 1952 [11]

The Microsoft Excel (2016) function "NORMSDIST" will convert the probit score to a probability.

The Equation 1 is combined with the "NORMSDIST" function to determine impact to occupants of the shelter, per Table 6. This is assuming that the air handling provides 10 air changes per hour to the shelter.

Table 6. Probability of Consequence to Sheltered Occupants.

Time (min)	Dosage ($ppm^{1.43}$-min)	% Probability
1.45		0%
1.46	-	0%
1.47	-	0%
1.48	-	0%
1.49	-	0%
1.50	-	0%
...	...	...
57.00	44,794	0%
58.00	45,354	0%
59.00	45,902	0%
60.00	46,436	0%
61.00	46,957	0%
62.00	47,466	0%
63.00	47,963	0%
64.00	48,448	0%
65.00	48,922	0%

In this example, the occupants can shelter for over an hour with 0% probability of consequence. Probit transformations can be utilized to study probability of fatal exposure from toxicity, thermal radiation, and blast overpressure. In addition, they can analyze probability of burn severity.

7. Conclusion

This paper has summarized the methods utilized in determining impact to personnel from toxic exposure and vapor cloud explosions. The covered models include pipe breach, tank release, and jetting models. These are highly utilized models for consequence analysis. In addition, the paper summarized methods to utilize dispersion modeling data to fully quantify impact to occupants of shelters downwind from releases.

For blast modeling, this paper provided an overview of one simplified method to predict explosion overpressure, impulse, and damage to buildings within the radius affected by the explosion. The method utilized is Baker-Strehlow-Tang (BST), one of several methods engaged by industry for this purpose. Most of the software tools used to perform the analysis were developed by the BakerRisk Engineering Consultant Explosion Research Cooperative, including VCloud® and BEAST®.

For all consequence models, more rigorous methods such as computational fluid dynamic (CFD) modeling will produce more accurate results. However, the methods discussed in this paper are excellent screening tools whose value lies in their ease of use and are suitable for risk analysis.

References

[1] Baybutt, Paul, "Using Layers of Protection Analysis to Evaluate Fire and Gas Systems", Process Safety Progress, Volume 31, Issue 3, 2011

[2] "SAFER Trace® Online Help", SAFER Systems, 2003

[3] James, Michael D., "Consequence Modeling of Dynamic Source Terms", Process Safety Progress, Volume 35, Issue 4, 2016

[4] Baker, Q. A., Tang, M. J., E. A. Scheier, and G. J. Silva, "Vapor Cloud Explosion Analysis," Process Safety Progress, Vol. 15, No. 2, 1996, pp. 106-109

[5] A. Sari, "Comparison of TNO Multienergy and Baker-Strehlow-Tang Models," Process Safety Progress 30 (2011), pp. 23-26

[6] BEAST® Help File, 2000 BEAST® Report, 2000

[7] "VCE Evaluation Method Comparison," The Quest Quarterly, Volume 4, Issue 1, Spring 1999

[8] Baker-Strehlow-Tang Methodology, Guidelines for Evaluating the Characteristics of Vapor Cloud Explosions, Flash Fires, and BLEVEs, CCPS, New York, 1994

[9] Pierorazio, A.J., J.K. Thomas, Q.A. Baker and D.E. Ketchum (2005) "An Update to the Baker-Strehlow-Tang Vapor Cloud Explosion Prediction Methodology Flame Speed Table," Process Safety Progress, 24(1): 59-65

[10] "About Acute Exposure Guideline Levels (AEGLs)", 2016, https://www.epa.gov/aegl/about-acute-exposure-guideline-levels-aegls (Accessed Mar 2, 2017)

[11] D. J. Finney, *Probit Analysis*, Cambridge University Press, New York 1952

Relief Systems Design: The Importance of Accounting for Static Head

Nicholas N. Cristea
Smith & Burgess, LLC
7600 W Tidwell Rd, Ste 600, Houston, TX 77040
Nicholas.cristea@smithburgess.com

Zubin Kumana
Smith & Burgess, LLC
7600 W Tidwell Rd, Ste 600, Houston, TX 77040
Zubin.kumana@smithburgess.com

Guomin Wang
Smith & Burgess, LLC
7600 W Tidwell Rd, Ste 600, Houston, TX 77040
Guomin.wang@smithburgess.com

Prepared for Presentation at
American Institute of Chemical Engineers
2017 Spring Meeting
13th Global Congress on Process Safety
San Antonio, Texas
March 26-29, 2017

Relief Systems Design: The Importance of Accounting for Static Head

Nicholas N. Cristea
Smith & Burgess, LLC
7600 W Tidwell Rd, Ste 600, Houston, TX 77040
Nicholas.cristea@smithburgess.com

Zubin Kumana
Smith & Burgess, LLC

Guomin Wang
Smith & Burgess, LLC

Keywords: Relief Systems Design, Static Head, Upstream, Downstream

Abstract

The process of evaluating an entire facility's relief and flare system design can be time consuming and costly. During the process, there will be systems to be evaluated that have liquid scenarios that may be protecting other assets within the facility. If the static head for the liquid scenarios is not accounted for, then the analysis can become even more costly due to the analysis not being performed correctly the first time. The result could be not meeting code by not adequately protecting system. This paper will look at two different cases where the static head was not accounted for, and how the issue was mitigated in the most cost-effective way. First, when dealing with the liquid heading on an offshore platform, there are many different levels on the platform and ensuring that the static head is accounted for is important to protect the different assets in the liquid handling train. Second, when evaluating a column system, it is important to ensure that the relief device provides protection for every ASME Section VIII Div 1 stamped equipment within the system. These two case studies highlight the necessity of accounting for static head in relief systems design.

1 Introduction

In the process safety industry, relief devices are set such that at least one relief device set pressure is equal to the lowest system MAWP within the protected system, to meet the requirement of ASME Section VIII Div 1 to provide adequate protection against overpressure. ASME Section VIII Div 1 UG-125 states "All pressure vessels within the scope of this Division, irrespective of size or pressure, shall be provided with overpressure protection in accordance with the requirements of UG-125 through UG-138, or with overpressure protection by system design in accordance with requirements of UG-140, or a combination of the two". ASME Section VIII Division VIII UG-125(c) and (1) also state that "when a pressure relief device is provided, it shall prevent the pressure from rising more than 10% or 3 psi (20 kpa), whichever is greater, above maximum allowable working pressure" and "when multiple pressure relief devices are provided and set in accordance with UG-134(a), they shall prevent the pressure from rising more than 16% or 4 psi (30 kpa), whichever is greater, above the maximum allowable working pressure". The vessel is normally hydrotested without accounting for the potential impact of inlet piping to the relief devices. Therefore, when the elevation difference between the top of the vessel and the relief device becomes significant [2], the vessel may experience a pressure higher than its allowable accumulation due to the static head, especially during liquid relief scenarios. The equation below is a conversion from the liquid height to the equivalent liquid head (psig).

$$P = 0.433\ x\ SG\ x\ H$$

P: liquid head in psi

SG: Specific Gravity

H: liquid head in feet

Equation 1: Liquid height to pressure conversion

This paper will look at two different cases where the static head was not accounted for and how the issue was mitigated for each case. The first case will look at a blocked system liquid outlet scenario on a platform where the vessel of interest has the same MAWP as the protecting vessel on an upper level. The second case will look at the liquid relief case (i.e. overfilling) and the mitigation for a column system that contains several vessels. These two cases will show the importance of accounting for static head and the potential mitigation options for these types of concerns.

2 Case 1 – Produced Water Desanders and Accumulators on an Offshore Platform

2.1 System overview

Figure 1 below is an illustration of case study 1. A produced water desander on the upper deck of a platform is fed by a booster pump. The desander then feeds the desander accumulator. The desander and the accumulator both have the same MAWP of 350 psig. The maximum discharge pressure of the booster pump is 415 psig. In the event of a blocked outlet, the relief device PSV-1 (4" P 6") on the desander is intended to provide adequate protection for the desander and the downstream accumulator. The relief device PSV-2 (1" E 2") on the accumulator is intended only for the fire scenario, based on the assumption that the upstream relief device (PSV-1) would limit the maximum deliverable pressure to the accumulator to 350 psig for the blocked outlet scenario. However, the vessels are located on different levels of the platform. The elevation difference was never accounted for in the choice of relief device set pressure. When PSV-1 is relieving at 385 psig, the pressure inside the accumulator would be the relief pressure (385 psig) plus the static head due to the difference in elevation. If the difference in elevation is large, the downstream accumulator may no longer be adequately protected by PSV-1, because the vessel would experience a pressure higher than its allowable accumulation. The elevation difference between the two relief devices illustrated in Figure 1 is roughly 27 feet. The static head from the elevation difference is about 12 psig. This means that when the desander is at the relief pressure, the pressure in the accumulator would be 397 psig, which exceeds the 10% allowable accumulation [2] (385 psig) of the accumulator.

2.2 Case 1 mitigation

The common mitigation option to account for the static head is to lower the relief device set pressure on the desander to account for the static head, so that the downstream accumulator would be adequately protected during the liquid relief event. However, for this particular case study, a more detailed engineering analysis revealed that the installation was acceptable as-is.

However, it was determined that there are technically two relief devices protecting the system in the event a blocked outlet on the accumulator. The allowable accumulation for the accumulator can go up to 16% of the system (406 psig) because of the multiple relief devices. Since the manual valve between the two systems will remain open (the initiating event is the closure of the downstream block valve), the accumulator is still protected. If the manual valve between the two systems were to be closed, then the blocked outlet case on the accumulator would no longer be applicable as the source of pressure would be lost as well.

If the static head is greater than 21 psig (~49 feet), the accumulator may experience overpressure in excess of the allowable (a pressure of 406 psig). The potential mitigation would be to lower the set pressure of the desander relief device to account for the static head such that the maximum pressure did not exceed this limit. For this case study, PSV-1 set pressure is not required to be lowered because the accumulator is still within the 16% allowable accumulation after accounting for the static head. Additionally, if the relief device set pressure were lowered, the potential exists that the relief device may be undersized, and additional mitigations may be required to resolve the concerns.

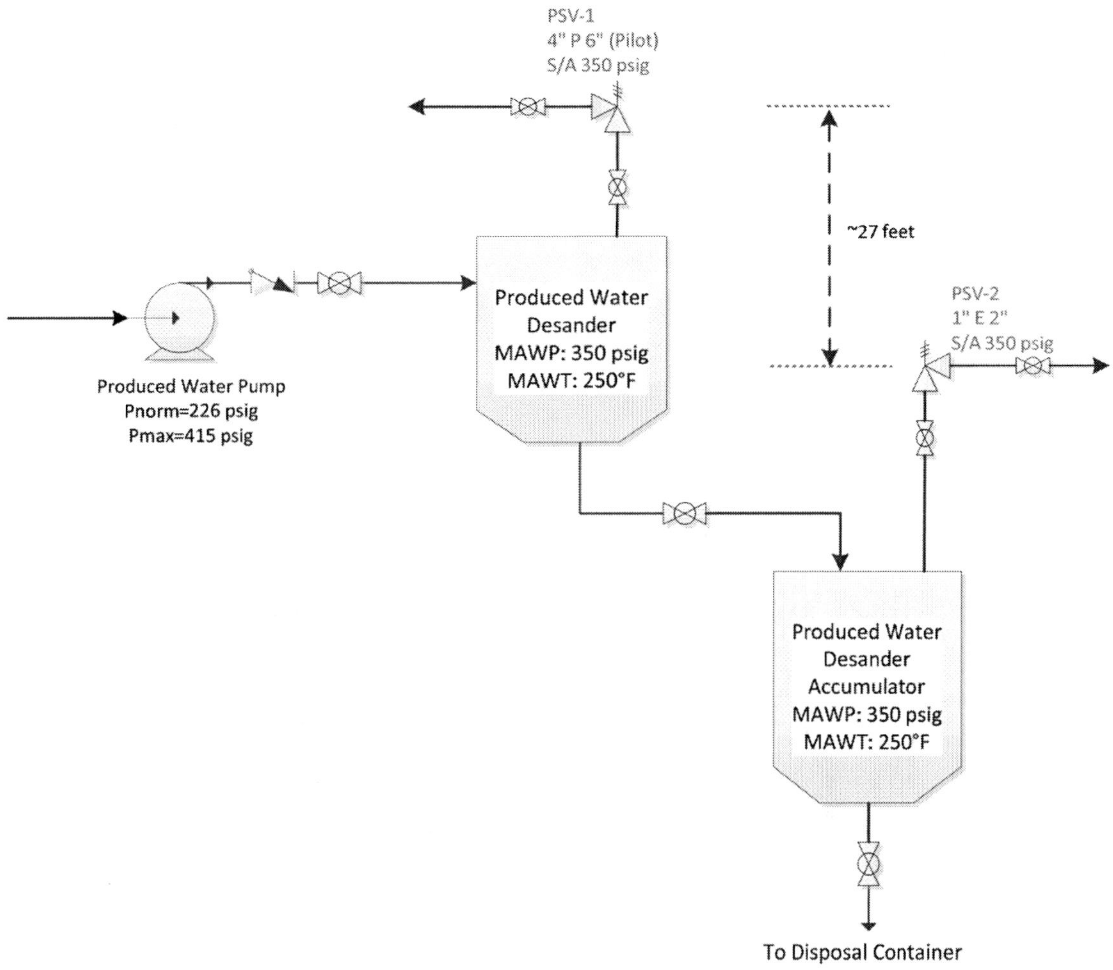

Figure 1, Case 1 – Produced Water System Overview

3 Case 2 – Common Column System

3.1 System Overview

Figure 2 below is a simplified column system that contains several pieces of equipment. The top of column is 50 feet from grade. The reboiler is located at grade. The relief device (PSV-3) is 5 feet above the top of the column and set at 330 psig. The lowest system MAWP is 330 psig based on the MAWPs of the column and the reboiler (shell side). This case study only looks at the "overfilling" scenario for the column. As the lowest system MAWP and the set pressure are the same, it may appear at first glance that the relief device is set correctly to protect the system. However, during the overfilling scenario, there are 55 feet of static head on top of the reboiler (assuming the reboiler is flooded). Therefore, the pressure that the reboiler (shell side) experiences during the liquid relieving event is higher than the allowable accumulation. For this case study, the specific gravity of the fluid is assumed to be 0.7, which would result in roughly 17 psig of static head. This means that during the relief event, the pressure in the reboiler (shell side) is 380 psig (~15% accumulation), which exceeds the 10% allowable accumulation. Therefore, mitigations may be required as the reboiler is not protected due to the static head.

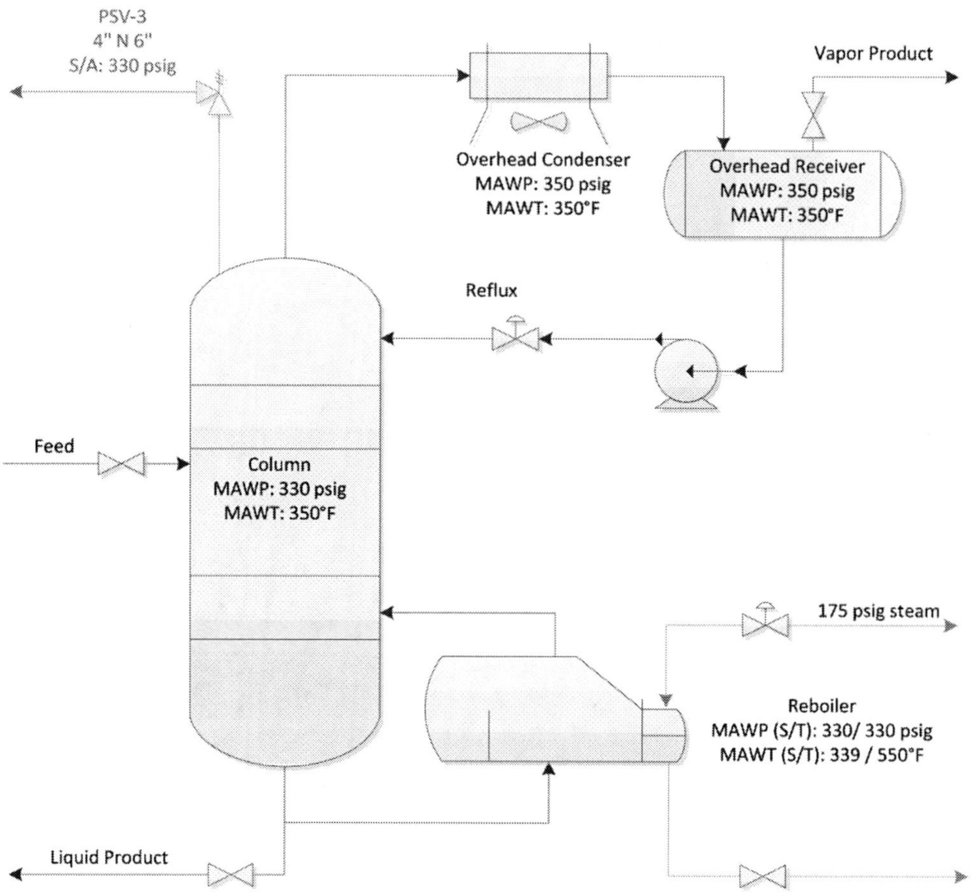

Figure 2, Case 2 - Common Column System Overview

3.2 Case 2 mitigation

There are multiple mitigation options for this type of concern. One mitigation option is to re-rate the reboiler to a higher MAWP to account for the static head. Another mitigation is to lower the set pressure of the relief device to account for the static head. Therefore, the new set pressure for this case study can be 314.5 psig or lower if the reboiler MAWP plus allowable accumulation is not exceeded by the relief pressure plus static head. When the relief device set pressure is lowered to account for the static head during the liquid relief events, engineers must confirm the relief device can still provide adequate capacity at the reduced pressure for all applicable overpressure scenarios. Also, the operating pressure of the column needs to be considered to ensure that the lower set pressure would not result in simmering of the relief device during normal operation. In practice, many plants would also have high independent liquid level alarms and fill time criteria to remove the potential liquid relief scenarios by taking credit for operator responses. Another option is to look at risk assessing the concern as the MAWP is not exceeded by 16% accumulation.

There are many ways to mitigate this concern; this case shows one of the most common mitigation option.

4 Conclusion

The two case studies discussed in this paper show the need to consider the static head when designing the relief system. The issue of static head should be considered during the design phase to ensure the system is code compliant. The cases studies show that the system may not be protected by code, but may be able to be risk accepted based on what the resulting pressure at the vessel is question becomes. It is important during the design phase to be cognizant on the elevation of the systems and the protecting relief devices to ensure the static head has been considered. This can be especially true when designing an offshore platform with multiple levels. Also, when commissioning columns or other vessels, it is important to consider the MAWP of all the vessels in the system to ensure all aspects of the system are protected by the relief device. The case studies give a brief overview of some of the possible mitigation options for static head concerns. It is the responsibility of the engineer reviewing the information to select the option that best fits the facility's needs.

5 References

[1] ASME Section VIII, "Boiler and Pressure Vessel Code," American Society of Mechanical Engineers, 2013.

[2] ANSI/API Standard 520 Part 2, "Sizing, Selection, and Installation of Pressure-relieving Devices," American Petroleum Institute, 6th ed., 2015.

[3] ANSI/API Standard 521, "Pressure-relieving and Depressuring Systems," American Petroleum Institute, 6th ed., 2014.

478548 Study of Factors That Affect Dust Explosion Characteristics of Fibrous Dust Materials

Abstract:

Abstract for 13th Global Congress on Process Safety

Track: Annual Loss Prevention Symposium (LPS)

Study of factors that affect dust explosion characteristics of fibrous dust materials

Bharatvaaj Ravi, Yan-Ru Lin, Chad Mashuga, M. Sam Mannan

Mary Kay O'Connor Process Safety Center

Artie McFerrin Department of Chemical Engineering

Texas A&M University

College Station, TX-77843-3122, U.S.A

+ 1 (774)-262-5215, vaaj26@tamu.edu

Combustible dust hazard is a serious threat to the process industries. Out of the 281 combustible dust incidents reported by the U.S. Chemical Safety and Health Investigation Board (CSB), 66 incidents have been caused by wood dust. In spite of the National Fire Protection Agency's (NFPA) efforts in specifically catering to the wood working and wood-based products manufacturing industries and Occupational Safety and Health Administration's (OSHA) National Emphasis Program on combustible dust hazards, to create awareness and minimize wood dust hazards, the wood dust incidents continue to occur. The fundamental difference between traditional dusts and wood dusts is that the latter are flocculent and fibrous. They are characterized effectively by an aspect ratio (length to diameter ratio) rather than a single dimension, such as diameter, as in the case of traditional - spherical or near spherical - dusts. Various studies in the literature have revealed that the standard dust explosion testing procedures and the categorization of these non-traditional dust based on particle diameter (mesh sizes) might not be valid. This research is concerned with similar line of thought and exploring the effect of aspect ratio on the time of suspension of these fibrous wood dusts. The likelihood of ignition increases with the time of suspension given other conditions to form a combustible dust hazard. The study also deals with characterizing the participating and non-participating portions when a dust explosion occurs through analysis of samples before and after explosion testing. Does the dust concentration give the entire picture of explosion characteristics such as maximum explosion pressure and explosion pressure rise rate? Or is there a better indicator of these characteristics when it comes to flocculent dust? These are the questions that the current research is trying to answer.

Key words: Combustible dust, Fibrous dust, Dust explosion testing, Dust concentration, Dust explosion characteristics, Non-traditional dust.

High Integrity Pressure Protection System (HIPPS) and Implementation Challenges

Janardhanan Kallambettu P.Eng.
Principal Engineer-Control Systems
Bechtel Corporation
3000 Post Oak Blvd, Houston, TX 77459
jkallamb@bechtel.com

Prepared for Presentation at
American Institute of Chemical Engineers
2017 Spring Meeting and 13th Global Congress on Process Safety
San Antonio, TX
March 26 – 29, 2017

High Integrity Pressure Protection System (HIPPS) and Implementation Challenges

Janardhanan Kallambettu P.Eng.
Principal Engineer-Control Systems
Bechtel Corporation
3000 Post Oak Blvd, Houston, TX 77459
jkallamb@bechtel.com

Keywords: PHA, HIPPS, SIS, SIF, SIL, PLC

Abstract

Over the years, the process industries have heavily utilized pressure relief devices (PRDs) to protect the mechanical integrity of system components. While PRDs offer a straightforward and practical way of controlling overpressure in process plants, they do have some inherent limitations such as impacts to flare system design, environmental concerns, and some physical limitations in PRD sizing. In these cases, where the use of PRDs is limited, High Integrity Pressure Protection System (HIPPS) can be utilized as a substitute for PRDs.

While HIPPS offers a solution to address the PRD limitations, there are some challenges associated with a successful HIPPS implementation. The HIPPS are classified as Safety Instrumented Systems (SIS) and require compliance with the functional safety standard IEC 61511[**1**]. The aim of this paper is to guide the reader with the necessary tools to successfully perform all the safety lifecycle activities of HIPPS, per functional safety standards and codes. Challenges in the various phases of the safety lifecycle, from process hazard analysis (PHA) and Safety Integrity Level (SIL) selection to operation and maintenance of HIPPS, are discussed in this paper. Additionally, this article describes many types of HIPPS, their pros and cons, to choose the right system for a given application. The paper concludes with an example HIPPS to achieve high-level risk reductions and demonstrates the appropriate use of quantitative tools, such as fault tree analysis (FTA), during the HIPPS design process.

1 Introduction

HIPPS is an acceptable solution instead of PRDs as a means of overpressure protection for vessels and pipelines. Codes such as ASME Section VIII, Division 1[**2**] and recommended practices and standards such as API Standard 521[**3**], API 17O [**4**] and HSE, U.K [**5**] provide criteria for the usage of HIPPS. The reader is advised to refer to the cited references to understand in detail the selection criteria for HIPPS.

The HIPPS is a safety instrumented function (SIF) implemented in a SIS and, therefore, the codes and the standards referenced above require compliance with the functional safety standards IEC 61508[**6**] and IEC 61511.

The functional safety standards require adherence to a safety lifecycle model. The various safety lifecycle activities are:

- process hazard analysis;
- SIL selection;
- safety requirements specification (SRS);
- SIS design per SRS;
- design verification meeting the target SIL including systematic capability, and architectural constraints (redundancy);
- installation, commissioning, validation, operation, and maintenance per SRS; and
- change management.

The intent of this paper is to expose particularly process engineers and other non-control systems engineers, about specific aspects of the HIPPS in a straightforward language.

2 PHA and SIL Selection

Whenever a HIPPS is indicated instead of PRDs, process and process safety engineers assign an SIL 3 target to the HIPPS as supported by Lawley and Kletz's [7] requirement that the reliability is ten times better than that of a relief device. For clean services, the probability of failure on demand (PFD) of a relief device is typically two orders of magnitude, hence the SIL 3 target for a HIPPS. API Standard 521, while emphasizing the need for an appropriate SIL selection process per established risk tolerance criteria, also suggests SIL 3 target for HIPPS in the absence of a documented risk tolerance criteria.

An assumption, however, that the HIPPS's SIL target is always SIL 3 may not be correct in some circumstances. Applicable codes and standards mandate a requirement for a documented PHA to identify and examine all potential overpressure scenarios. The severity of consequences typically ranges from moderate to catastrophic and, based on the owner's risk tolerance criteria, the HIPPS target can vary from SIL 2 to SIL 4. For example, if the likelihood of an initiating event causing the overpressure is 1 in 10 years and the owner's risk criteria for catastrophic severity is 1in 100,000 years, and in the absence of any other independent protection layers or conditional modifiers, the required risk reduction is 10,000, a SIL 4 target.

Designing a SIL 4 HIPPS is a significant challenge and not recommended. The functional safety standards discourage such high-level risk reduction measures by a single SIF. Every effort should be made to remove or reduce the hazard at the source, introduce other risk reduction measures which are not based on the SIS or minimize the likelihood of the causes of the hazardous event. If the above steps do not help, then use of multiple independent HIPPS, each with a lower integrity, to meet the overall risk reduction requirements should be considered. See section 6 for a hybrid HIPPS design solution meeting an overall risk reduction of 10,000 or more.

3 Types of HIPPS

Many in the process industry believe that HIPPS always includes 2oo3-voted pressure transmitters (sensors), Triple Modular Redundant (TMR) safety PLC (logic solver), and 1oo2-voted ball valves (final elements). In reality, there are many types of HIPPS available to choose from based on the specific application.

HIPPS can be broadly classified as **Electronic**, **Mechanical** or **Hybrid** based on the technology used for the logic solver as shown in **Figure 1**.

A brief description, pros and cons, and applications for each of the above types are given in the following subsections.

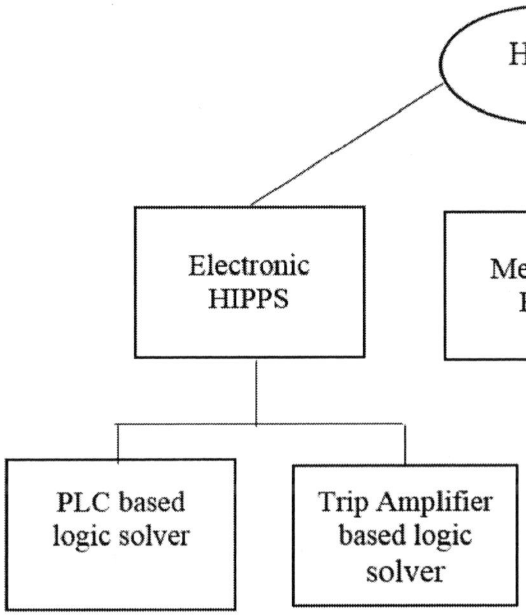

Figure1. Types of HIPPS

3.1 Electronic-HIPPS

An electronic-HIPPS uses either a programmable logic controller (PLC) or a trip amplifier system as a logic solver.

3.1.1 PLC based HIPPS

There are multiple varieties of PLCs equipped with microprocessor or solid state technology in a multitude of architectures such 1oo2, 1oo2D, 2oo3, 2oo4, and 2oo4D. Typically, all these PLCs will meet SIL 3 reliability. These PLCs are further categorized as complex devices and use application programming languages such as ladder logic or functional block diagrams. Refer to **Figure 2**.

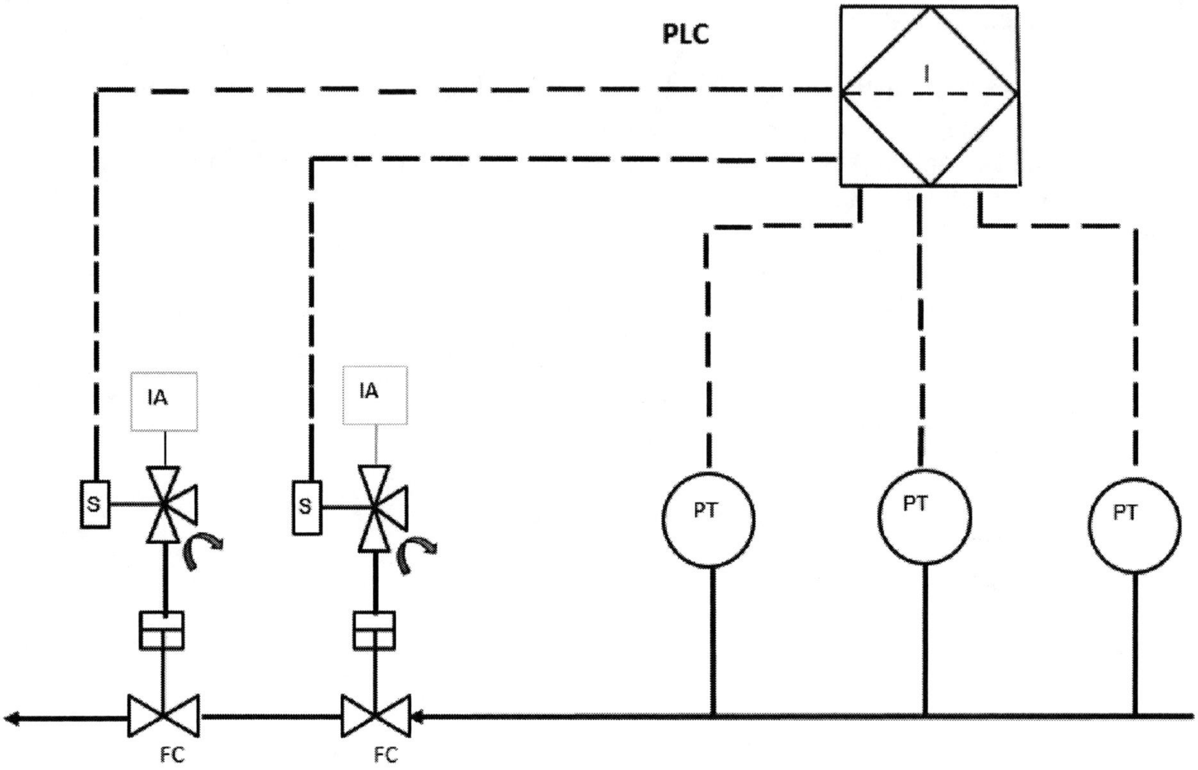

Figure 2. PLC based HIPPS

Pros:

- Flexible in developing the logic program and accommodates higher number of field inputs and outputs (I/Os);
- Equipped with special features for application programming such as variable valve closure speeds while closing the valve to mitigate consequences associated with slamming shut.
- Accommodate multiple HIPPS in a single logic solver within a plant facility;
- Facilitates partial stroke testing to meet target SILs or extend the testing interval, when practicable;
- Easily interfaces with the plant operator stations (human machine interfaces, HMIs).

Cons:

- Often needs specialized application programming expertise, in addition to vendor-specific training for selection of hardware and complex control strategies;
- Validation of the vendor claimed SIL against the safety manual is not a trivial exercise;
- Requires protection against cyber security threats;
- Potential for introducing systematic errors in application program;
- Application program using a low variability language (LVL) calls for special considerations per IEC 61508/IEC 61511 standards including application program safety requirements specification;
- Requires rigorous Factory Acceptance Test (FAT);
- Large footprint is needed -typically 800mmX600mmX1000mm panel;

- Extensive installation requirements such as controlled environment and special grounding;
- Requires complex, and redundant large capacity power supply;
- Voluminous documentation is involved;
- Requires a high level of competence in commissioning, validation, maintenance and operation; and
- Expensive for simple HIPPS applications ($50,000 for a logic solver alone).

Applications:

In a large facility, where many HIPPS are to be implemented for different hazardous event scenarios requiring complex functions or features, a PLC based HIPPS is recommended. For example, over pressure protection of multiple vessels, various piping segments, and process equipment within a facility to avoid loading of the flare system.

3.1.2 Trip-Amplifier based HIPPS

Trip-amplifiers are discrete logic solvers of low complexity. A trip amplifier accepts analog 4-20mA input from a field device such as a pressure transmitter and de-energizes contact output(s) when the process value exceeds the trip set point. Each sensor needs a dedicated trip amplifier. Refer to **Figure 3**.

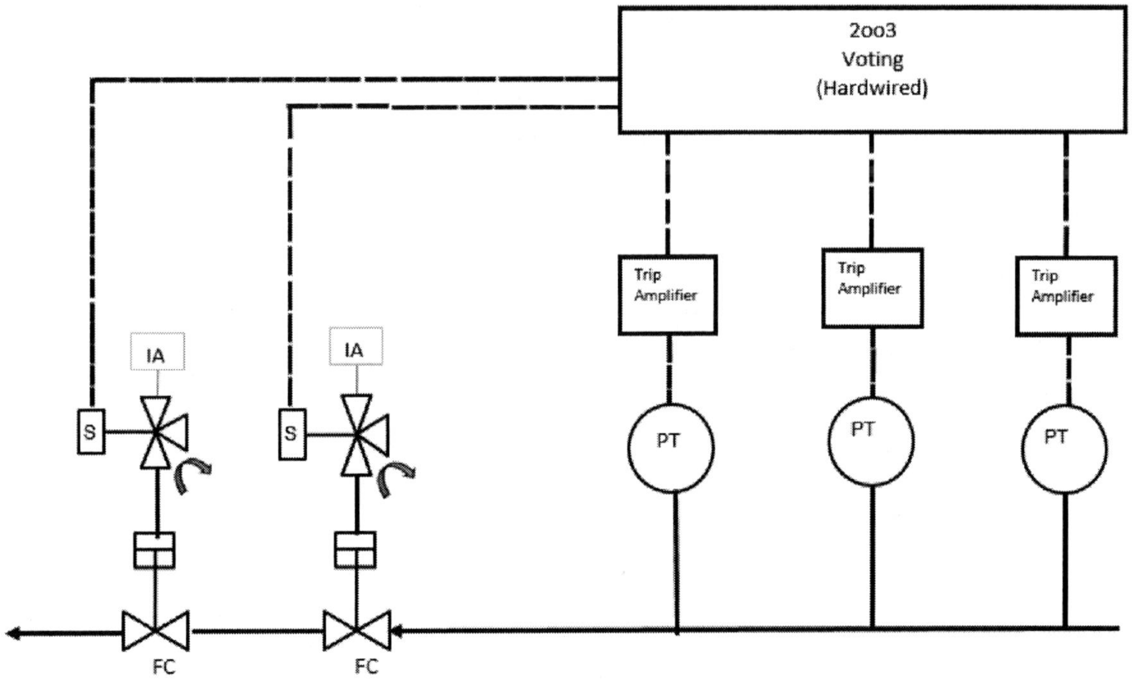

Figure 3. Trip Amplifier based HIPPS

Pros:

- Simple, reliable, with low design complexity;
- Meets SIL 2 reliability without redundancy for Trip Amplifiers and SIL 3 reliability with redundant architecture;
- Does not require application programming; classified as a device with fixed programming language (FPL) per IEC 61508/IEC 61511and, therefore, eliminates the need for application program safety requirements specification;
- No cybersecurity related issues;
- Less potential for systematic errors;
- Well suited for simple logic such as HIPPS logic;
- Low space requirement-typically DIN rail mounted;
- Low power device with no special requirements;
- No need for a controlled environment;
- Being a simple device FAT is not required for the logic solver;
- Easy to install and maintain; and
- Low cost; (logic solver range less than $10000)

Cons:

- Requires external devices for complex logic such as variable speed closing times for valves to avoid slam shut scenarios;
- Interfacing with operator stations requires additional I/Os in the plant DCS;
- Difficult to interface and facilitate partial stroke testing;
- Voting among the trip-amplifiers are external to the trip-amplifiers using the trip contacts and physical wiring which is less straightforward; and
- Implementation of maintenance bypasses is not practical;

Applications:

Well suited for applications involving straightforward, simple logic and for facilities which do not use PLCs. Also, appropriate for applications requiring a small footprint, low lifecycle cost and facilities not having multiple HIPPS. For example, remote pipeline compressor stations or storage tanks.

3.2 Mechanical-HIPPS

A mechanical HIPPS uses mechanical initiators. The mechanical initiator not only senses the over pressure but also serves as a logic solver to detect high pressure and actuate the valve. Mechanical HIPPS do not use any electrical or electronic components and, therefore, are inherently safer. Standalone versions of mechanical HIPPS are available which do not require any energy sources, such as instrument air supplies. Refer to **Figure 4**.

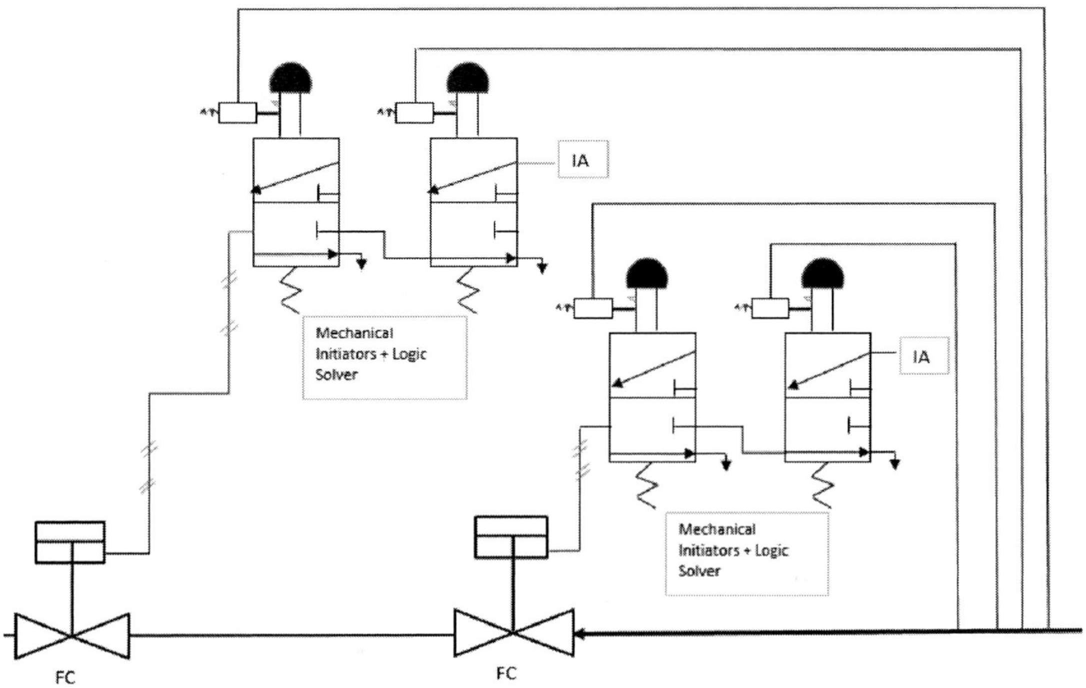

Figure 4. Mechanical HIPPS

Pro:

- Simple, reliable and minimally complex;
- Does not involve wiring and programming;
- Less susceptible to systematic errors;
- Cables, cable trays, conduits, etc. are not required;
- Can be made standalone without external energy sources;
- Single valve system can meet SIL 2 targets, redundant valve configuration can even meet SIL 3;
- Third party agencies certify an entire system in contrast to individual elements/subsystems of electronic HIPPS;
- Fewer components to fail;
- Easy to install, commission, validate, maintain and operate;
- Minimal documentation: no wiring diagrams or, installation drawings for logic solvers and initiators, application program SRS required;
- Competence in programming or maintaining a complex electronic system is not required;
- There are no EMC interference issues;
- There are no impacts due to cyber security threats;
- Suitable for harsh environments-no environmental climate control is needed;
- Easy to comprehend by operators and maintenance mechanics;
- Lowest life cycle cost among all the types of HIPPS;

Cons:

- Limited distance between the pressure sensing points, mechanical initiators and the valve- all components must be in the close proximities to avoid transportation time lags impacting the response times;
- Mechanical initiators need to be carefully selected as they are sensitive to setpoints and their accuracies;
- Interfacing with operator stations requires additional transmitters in the field and Inputs/outputs (I/Os) in the plant DCS;
- Requires external devices to facilitate partial stroke testing;
- Implementing maintenance bypasses consistent with the overall facility philosophy is not practical;

Applications:

Mechanical HIPPS are attractive for standalone applications in remote plant locations, for areas where electrical cabling is impractical and in facilities, like pipeline pump/compressor stations, where PLCs are not commonly used.

3.3 Hybrid-HIPPS

Hybrid-HIPPS are a combination of electronic and mechanical HIPPS.

Pros:

- High level of risk reduction- 10,000 (SIL 4) or higher can be obtained;
- Significant reduction of common cause and common mode failures due to diversity (mechanical and electronic);
- Low-level of systematic errors;

Cons:

- Complex system;
- Requires high level of competence in the installation, commissioning, validation, maintenance and operation;
- Requires a sizeable spare parts inventory;
- Many of the Cons of electronic and mechanical HIPPSs described above are applicable for the hybrid HIPPS as well.

Application:

High-pressure feed gas to a facility typically requires a high-risk reduction in the absence of non-instrumented IPLs. When a high level of risk reduction (greater than 10,000) is required, hybrid systems are well-suited as they are diverse systems with minimum common cause and common mode failures.

4 Safety requirements specification (SRS)

Functional safety standard IEC 61511 requires the development of a SIS safety requirements specification (SRS), including application program safety requirements, when applicable. HIPPS as a functional SIS, requires an SRS. There are a minimum of 29 SRS requirements specified by the standard. While every element is important, a few present unique challenges that can significantly impact the selection, design, and quantification of random hardware failures of the HIPPS.

4.1 Process safety time (PST)

Process safety time, per clause 3.2.52.1, IEC 61511, is the period between a failure occurring in the process or the basic process control system and the occurrence of a hazardous event, if the SIF is not performed. The SIF has to detect the failure and complete its action soon enough to prevent the hazardous event taking into account any lag in the process.

HIPPS that implements the overpressure protection function must complete the action to bring the process to a safe state within the process safety time. Quite often, the process safety times for HIPPS are less than few seconds. The valve subsystems may need quick exhaust valves, larger size air supply tubing, or solenoid valves to meet the response times. Therefore, the PST determination is an important specification which in turn determines various components within the HIPPS to achieve the response time.

4.2 HIPPS output action success criteria

Output actions of a SIF will prevent a particular consequence of concern only when the action is completed with certain performance criteria. For example, stating an action as to close a valve is inadequate. The criteria for the successful operation should be specified. Many HIPPS applications do not require tight shut-off requirements for valves. However, there are situations where larger leakage rates may impact the process safe states. In such scenarios, the process engineer must specify an acceptable leakage rate for the valves, and ensure the design shall have means to achieve such low leakage rates. The design may require unique valve types and trims. Also, additional pressure transmitters and isolation/bypass valves are required for measuring the extent of leakages.

Low leakage rate specification may impact the proof test coverages for the final elements in the quantification of random hardware failures (PFD_{avg}). Achieving a SIL 3 target with tight shut-off requirements is a significant challenge.

4.3 Proof test interval and overhaul interval (mission time)

IEC 61511 defines the proof test as a periodic test performed to detect dangerous (hidden) failures in a SIS so that, if necessary, a repair can restore the system to an "as new" condition or as close as practical to this condition.

The frequency of testing plays a significant role in determining the HIPPS's availability to perform intended safety function on demand. However, frequent testing may affect the production, introduce spurious shutdowns due to human error or affect the life of a device. If only a full stroke test is performed for a valve, then the test is imperfect since it will not detect all the covert failures. The undetected, hidden failures are detected and repaired during overhaul periods (mission time). The longer the mission time, the higher the PFDavg; therefore, it will be difficult to achieve SIL 3 targets with longer mission times.

Bypass valves and isolation valves are required for proof testing intervals shorter than the overhaul periods to avoid process interruption. Provision of bypass and isolation valves increases the complexity of the design, footprint, capital and the life cycle costs of the facilities.

4.4 Survivability

As per IEC 61511, the SRS shall specify any requirements for a SIF to survive a major accident event, e.g. the time required for the HIPPS valve to remain in operation in the event of fire shall be specified. Survivability specification impacts the selection of valve trims and additional fireproofing requirements.

4.5 Common cause failures

The SRS must specify the measures to identify and minimize common cause failures. Common cause failures will significantly impact PFD_{avg} for high SILs, such as SIL 3 and SIL 4. Common cause failures can be due to process conditions or similar redundant equipment configurations or systematic failures.

Physical and electrical separation and use of diverse devices which respond differently to the common stresses will minimize common cause failures. An effective functional safety management plan and verification processes will reduce systematic failures.

5 Quantification of random failure of HIPPS

The HIPPS design must be evaluated to ensure that the calculated failure measure (PFDavg) is less than the target related to the SIL as specified in the SRS.

The functional safety standard mandates assessment of the following as part of the SIL verification:

- Systematic capability of all the elements associated with a HIPPS meets the specified SIL target
- Calculated PFD_{avg} is less than the target PFD_{avg} of the selected SIL
- Hardware fault tolerance (HFT) complies with the required SIL target

5.1 Systematic capability

IEC 61511 defines the "systematic capability" as a measure (expressed on a scale of SC 1 to SC 4) of the confidence that the systematic safety integrity of a device meets the requirements of the specified SIL, in respect of the specified safety function, when the device is applied in accordance with the instructions specified in the device safety manual.

Systematic capability assessment is for addressing systematic failures which are not quantified and hence not included in the SIL verification calculations. IEC 61508 and IEC 61511explicitly identify a systematic capability parameter (SC). A systematic capability of "SC N" for a device means that the systematic safety integrity of "SC N" has been met when the device is applied following the instructions specified in the instrument safety manual. If a device is to be used in an SIL 3 application, then it must be assessed for the systematic capability of SC 3.

The systematic capability of a device depends on the nature of the device. If a device has no software, then only hardware systematic failure mechanisms are considered. If a device has both hardware and software, then both hardware and software failure mechanisms shall be considered.

There are agencies, such as TUV, Exida, and SIRA that perform systematic capability assessment of a product as per IEC 61508 and provide a certificate with a report specifying an SC parameter. Following a well-established functional safety management plan during detailed engineering, installation, operation and maintenance phases ensures that the systematic safety integrity of a device meets the requirements of the specified SIL.

5.2 PFD_{avg}

Many factors influence the PFD_{avg} of HIPPS function. A few of the factors that pose unique challenges are considered in this section.

5.2.1 Failure rates

There is always an uncertainty in the quantification of random failures. The failure rate is, probably, the least precise parameter used in performing SIL verifications [**8**]. Therefore, one must acknowledge that the results of the quantification are not real values, but rather a basis for comparing different design options and for monitoring the reliability performance during the operational phase of the safety lifecycle. The analysis is only a prediction. Historical performance is not the same as future performance.

If the failure rates are conservative, then achieving target SIL will be demanding. However, optimistic failure rates will result in inadequate protection integrity. Therefore, the failure rates of components used in the analysis should be sufficiently realistic.

There are many sources of failure rate data, such as application-specific, site-specific, company-specific, manufacturer-specific, industry-specific, and generic data. These failure rates can vary significantly by several orders of magnitude. While application specific or site specific data is the best option, but it can often be difficult to obtain. A detailed discussion on the selection of failure rates to use in the analysis is beyond the scope of this paper, and the reader is advised to refer to other sources [**9**].

5.2.2 Proof test interval, test coverage and mission time

As explained in section 4.3, a proof test may be imperfect based on the following:

- The extent of tests that could be performed to reveal the hidden failures. For example, if the valve is only full stroke tested without examining the internals then the proof test is imperfect;
- The conditions under which the proof test is carried out may differ from a real demand;
- Lack of competence of testing crew;
- Inadequate procedures.

Perfect testing is not a realistic assumption. IEC 61511/IEC 61508 requires consideration of the impact of imperfect testing in the analysis. In the SIL verification calculations, a proof test coverage (PTC) is considered for incomplete testing. The incomplete testing will increase the PFDavg. The lower the proof test coverage, the higher the PFD_{avg}. Undetected dangerous failures due to imperfect proof testing will be detected either during the complete overhaul period (when repaired or refurbished) or when the component is replaced or during an actual demand. Mission time is the terminology used by many reliability specialists for the above overhaul period.

For example, if an overhaul period or demand interval for HIPPS is ten years and the imperfect proof test interval for a valve is one year with a proof test coverage of 90%, then achieving an SIL 3 target for HIPPS is an arduous task. If tight shut-off requirement is specified, then the proof test coverage could be even less than 30%.

From the above, it is evident, that the significance of proof test interval, the effectiveness of proof test, and the mission time cannot be ignored and must be taken into account in the analysis.

5.2.3 Partial stroke testing

Partial stroke test will help to improve the PFD_{avg}, extend the proof test interval, or increase the safe failure fraction. A single valve with partial stroke test may achieve an SIL 2 integrity. However, for many HIPPS with short process safety times, partial stroke test may not be useful. Moreover, partial stroke tests may result in spurious trips and cause potential damage/wear to the valve seat due to the frequent testing.

A perusal of literature reveals that the partial stroke testing coverage is in the range of 40-80% [10]. However, the above-stated coverage is widely debated.

If the partial stroke testing is not automated, then it is not considered as the diagnostic testing per functional safety standards. Also, the analyst shall take into consideration the partial stroke testing interval with the repair time in the PFD calculations.

5.2.4 Common cause failures

Common cause failures impact the redundant components such as valves and sensors. The common cause failures occur simultaneously or, at least, within a short interval as to be perceived as simultaneous.

The following are the defenses against the common cause failures:

- Diversity (different types of valves and systems);
- Separation (physically and electrically);
- Staggered testing;
- Use of low complexity devices- simpler equipment less prone to common cause failures;
- Well administered and complied procedures;
- Environmental controls;

When diverse components are used, the common cause factor (Beta factor) is minimal. Using IEC 61508, Part 6 [5], Annex D, Tables, D.1, D.2 and D.3, the Beta factor can be estimated to be as low as 2% with diverse components. However, the diverse components must be suitable for the application with comparable performance without any degradation.

Also, staggered testing reduces the overall PFD_{avg} by a factor of 30 to 50% depending upon the voting configurations [**10**].

5.2.5 Modeling of components

Modeling of HIPPS elements in the analysis must include all the components that are involved in bringing the process to a safe state within the process safety time.

Many HIPPS valves are equipped with quick exhaust valves (QEVs) to reduce the valve closing time to meet the process safety time. Sometimes, redundant QEVs are provided in various voting configurations such 1oo2 or 3oo4, etc. The contribution of QEV failures to the PFD_{avg} is significant and cannot be ignored. Therefore, the reliability analyst must consider the QEVs with their appropriate voting architecture in the PFD_{avg} calculations. Similarly, the likelihood of plugging of sensing lines must be taken into account in freezing applications.

Any interfacing components such as barriers, isolators, interposing relays, and poppet valves, which are part of the safety instrumented function to achieve or maintain the process to a safe state must be modeled and analyzed.

5.3 Hardware fault tolerance (HFT)

Irrespective of the hardware reliability calculated (PFD_{avg}) for the design, the functional safety standard IEC 61511 specifies minimum levels of redundancy for each safety integrity level.

There are two approaches:

- Route 1_H: Determination of HFT based on the calculated safe failure fraction (SFF);
- Route 2_H: Determination of HFT based on field experience and field data (proven-in-use)

5.3.1 Route1$_H$

The Route1$_H$ approach uses SFF and the SIL target to determine the hardware fault tolerance. The SFF is defined as the sum of the potentially dangerous failures revealed by auto-test and those failures which result in a safe state, as a fraction of a total number of failures.

$$SFF=\frac{[dangerous\ detected\ failures(self\ test)+safe\ failures]}{Total\ failures}$$

The above is calculated for each element (sensor, complete valve assembly, logic solver). Two tables, Table 2 and Table 3, are provided in IEC 61508- Part 2. Table 2 is for "Type A" devices which are simple devices such as valves (no microprocessors and programming are involved), and Table 3 for "Type B" devices which are complex devices such as electronic transmitters, trip amplifiers, and PLCs. Based on the SFF, type of the device, and the SIL target, the hardware fault tolerance is determined using Table 2 and Table 3.

5.3.2 Route2$_H$

In Route 2$_H$, the hardware fault tolerance is specified by the functional safety standard for each SIL target if the device is selected based on well documented and verified field based failure rate data with 90% statistical confidence. For low demand application with SIL 1 or SIL 2 target, there is no need for any redundancies (HFT=0), and SIL 3 requires one hardware fault tolerance.

Unless the field failure data with 90% confidence levels are available, the Route 2$_H$ cannot be justified. There are third party agencies who provide Route 2$_H$ certifications and should be evaluated carefully before adopting the Route 2$_H$ approach based on those certifications.

6 High level risk reductions (RRF> 10,000)

As discussed in section 2, sometimes, PHA/SIL selection may indicate a required risk reduction greater than 10,000 in the absence of any non-instrumented protection layer or, human intervention is impractical due to process safety time considerations or safety reasons.

If there are no options other than providing HIPPS with a risk reduction greater than 10,000, then the functional safety standard IEC 61511 mandates certain requirements. The following are a few of the clauses and their requirements:

Clause 9.2.6: *If further consideration of application and confirmation that a risk reduction requirement >10,000....is still required, then* consideration *should be given to achieving the safety integrity requirement using a* <u>*number of protection layers with lower risk reduction requirements*</u>. *If the risk reduction is allocated to multiple protection layers, then such protection layers shall be independent from each other or the lack of independence shall be assessed and shown to be sufficiently low compared to the risk reduction requirements.*

Clause 9.2.7: *If a risk reduction greater than 10,000....is to be implemented, whether allocated to a single SIS or multiple SIS or SIS in conjunction with a BPCS protection layer, then a further risk assessment shall be carried out using a quantitative methodology to confirm that the safety integrity requirements are achieved.*

Clause 9.2.8: *If the risk reduction required for a hazardous event is allocated to multiple SIFs in a single SIS, then the SIS shall meet the overall risk reduction requirement.*

A hybrid HIPPS using the combination of electronic and mechanical HIPPS is a solution to achieve risk reductions greater than 10,000. Each HIPPS can be designed to be better than SIL 2, and the overall risk reduction will be better than 10,000. The performance of electronic and mechanical HIPPS are comparable.

High level of risk reduction is achieved due to the diversity and very low common cause failures between the electronic and mechanical HIPPS. All the subsystems of the HIPPSs are diverse with different technologies. Also, systematic failures are minimal. The integrity of such multiple HIPPS is best analyzed using a fault tree analysis (FTA).

An example fault tree analysis for a hybrid HIPPS using ISOGRAPH's Reliability Workbench™ is shown in **Figure 5**.

The example hybrid HIPPS achieved a risk reduction of 12,815(PFD_{avg} = 7.80E-05).

The FTA shows an approach for modeling of various components including the common cause failures. The failure rates assumed are generic for the purpose of demonstration. The actual failure rates and other factors are application specific and cannot be generalized.

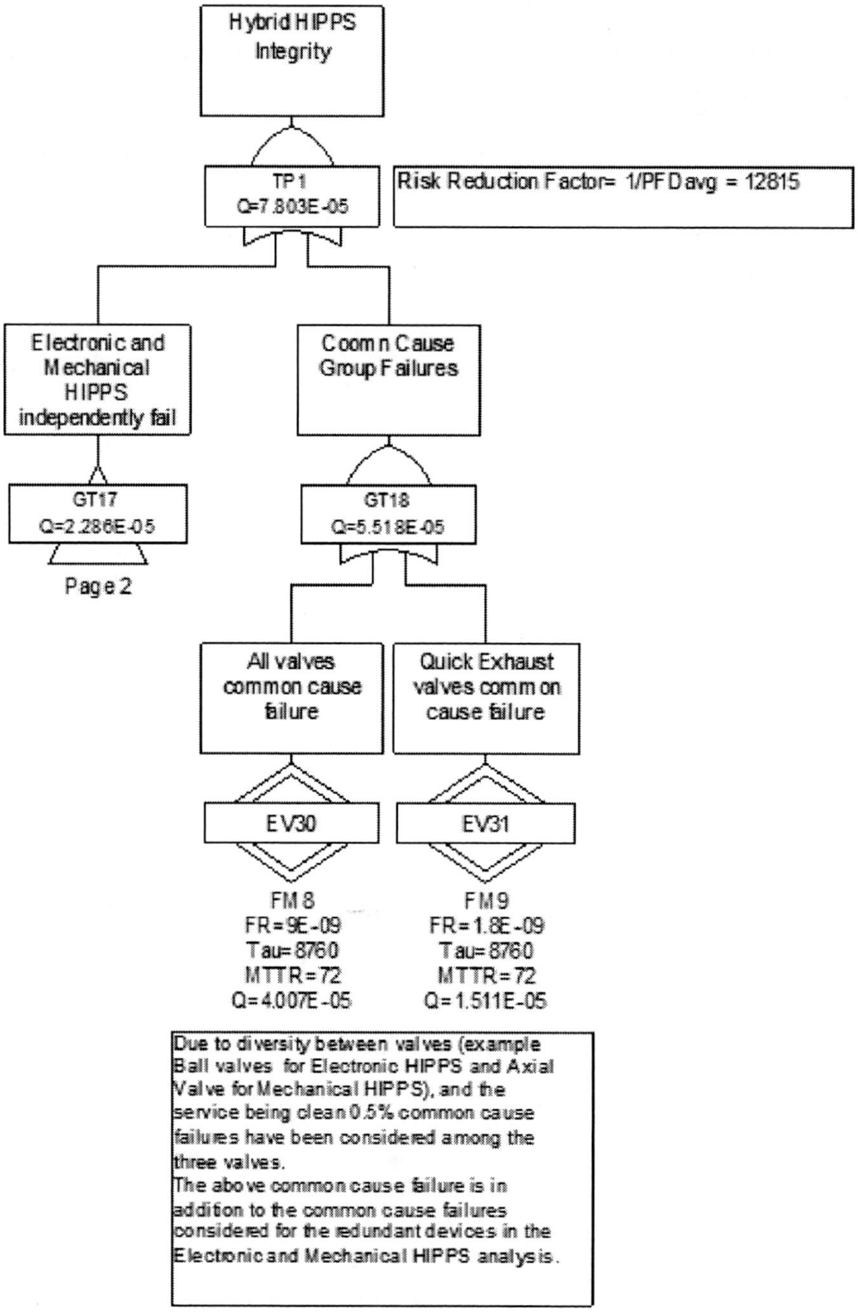

Figure 5. FTA for multiple HIPPS to achieve high risk reduction

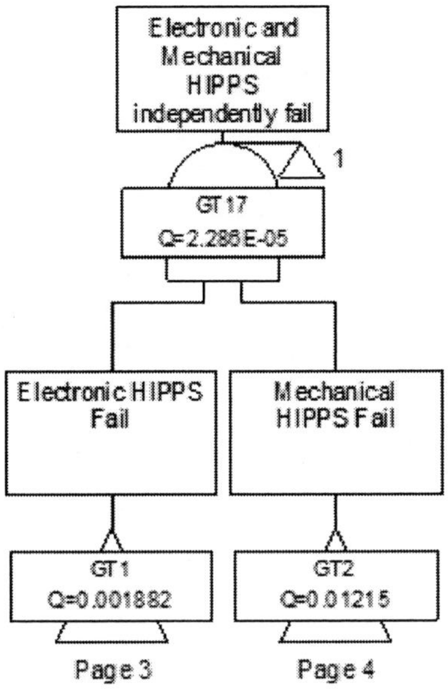

Figure 5. FTA for multiple HIPPS to achieve high risk reduction (contd..)

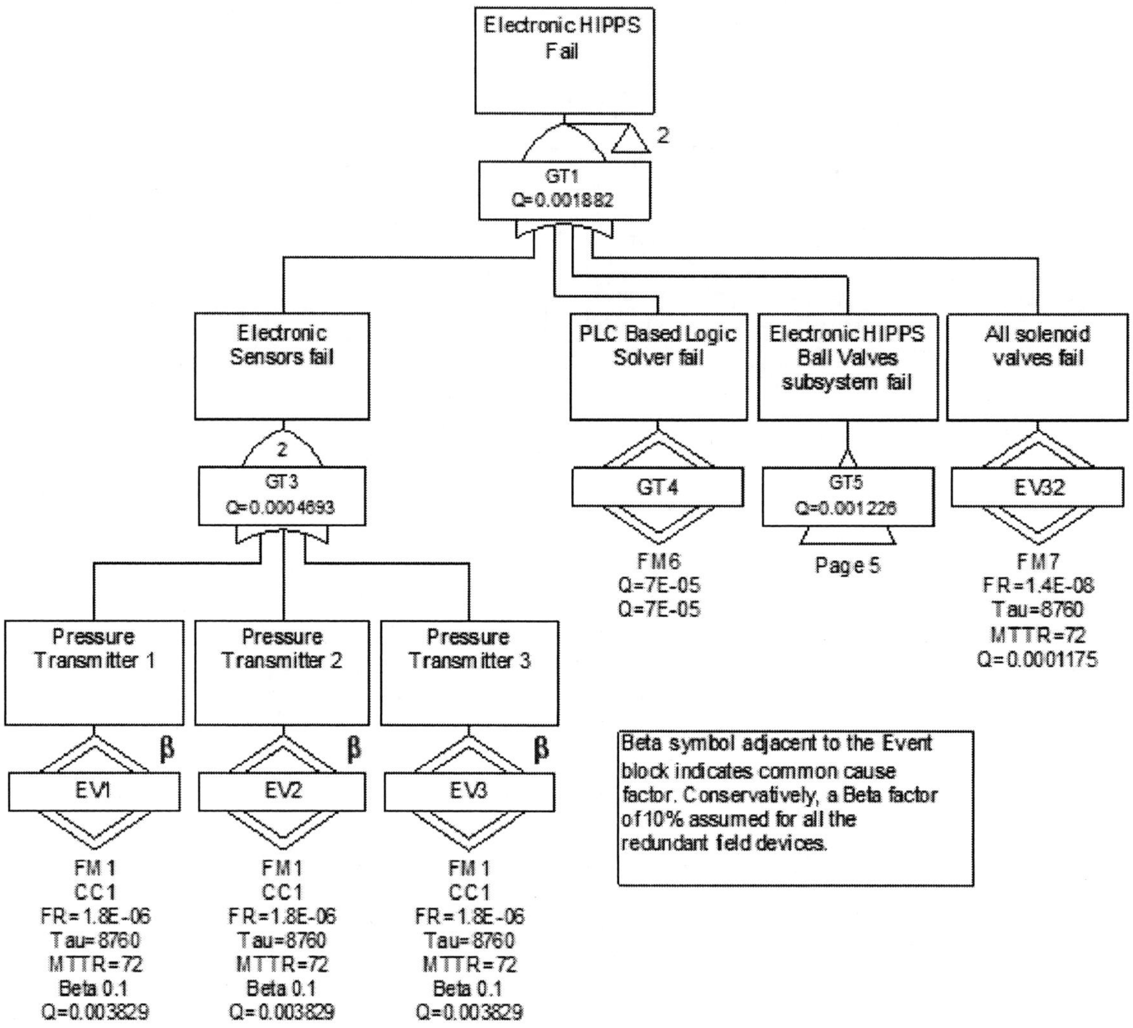

Figure 5. FTA for multiple HIPPS to achieve high risk reduction (contd..)

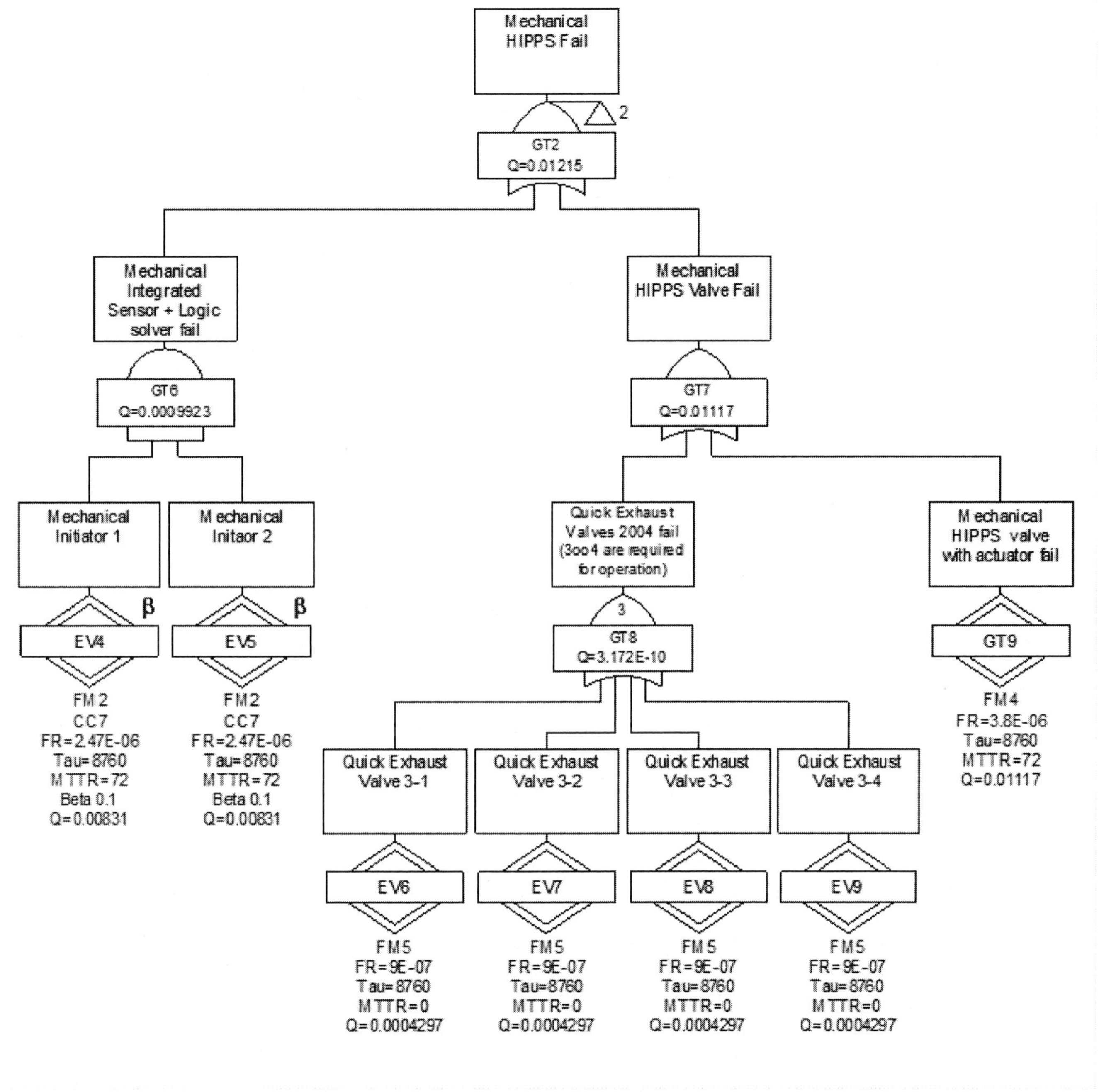

Figure 5. FTA for multiple HIPPS to achieve high risk reduction (contd..)

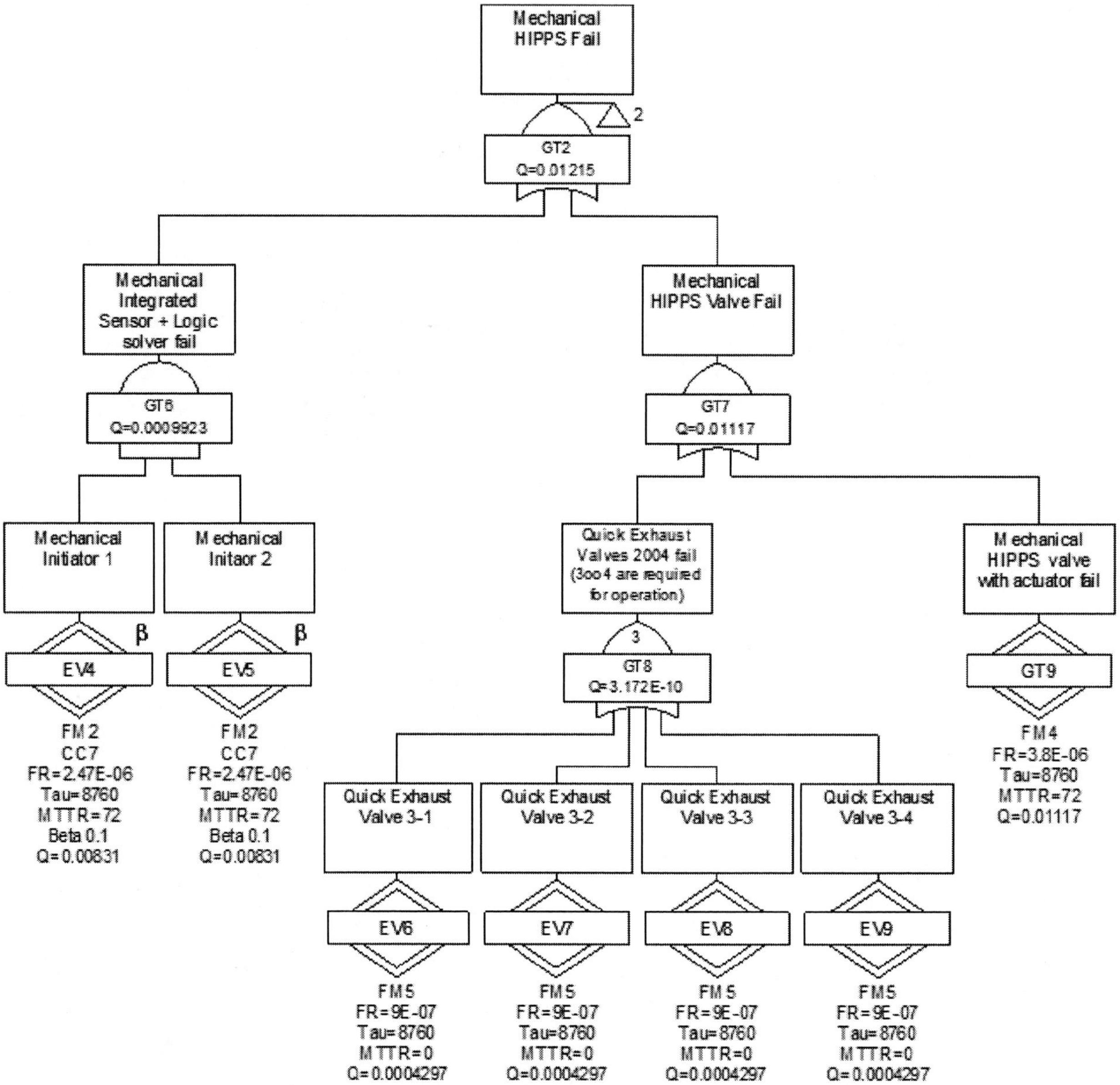

Figure 5. FTA for multiple HIPPS to achieve high risk reduction (contd..)

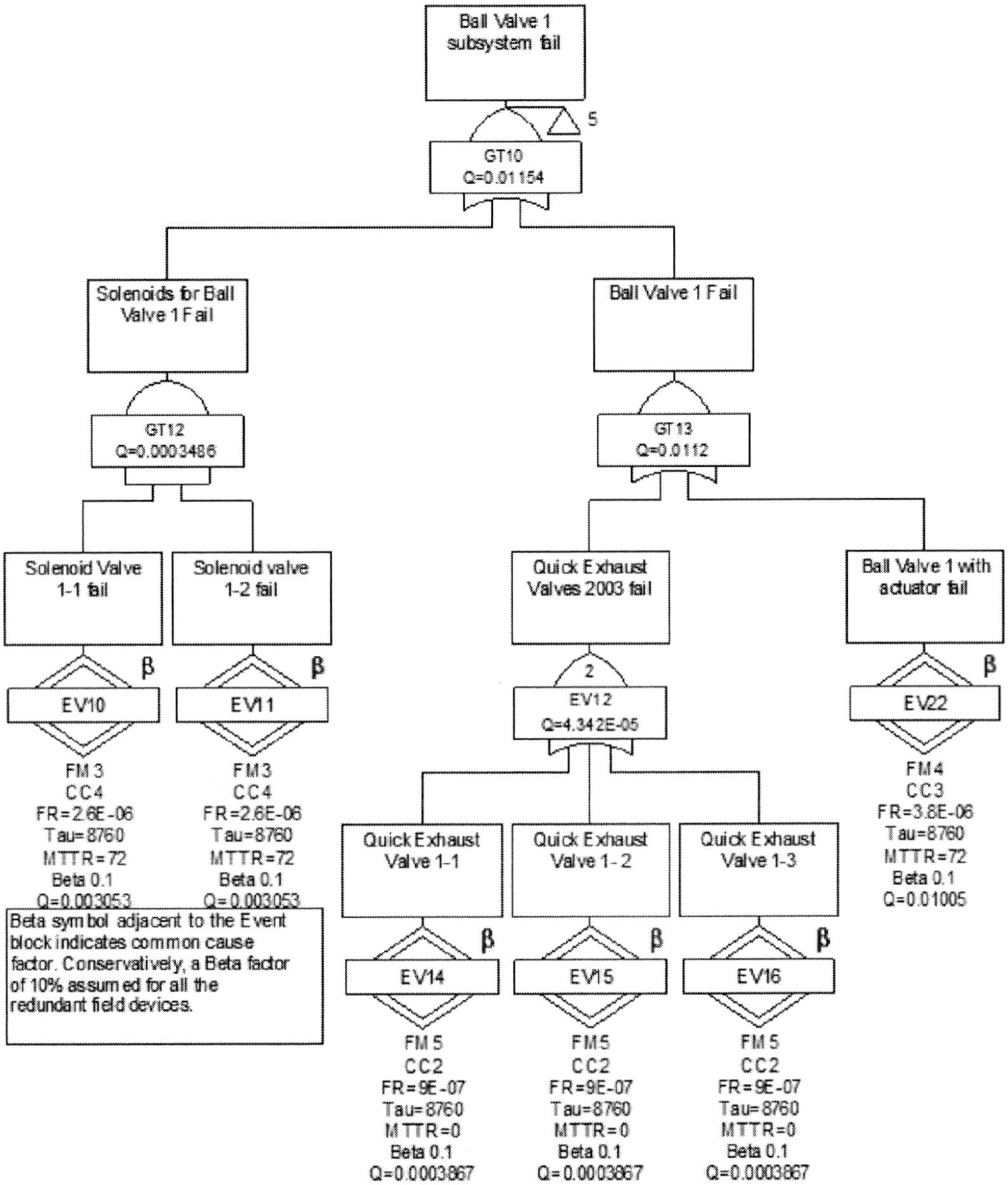

Figure 5. FTA for multiple HIPPS to achieve high risk reduction (contd..)

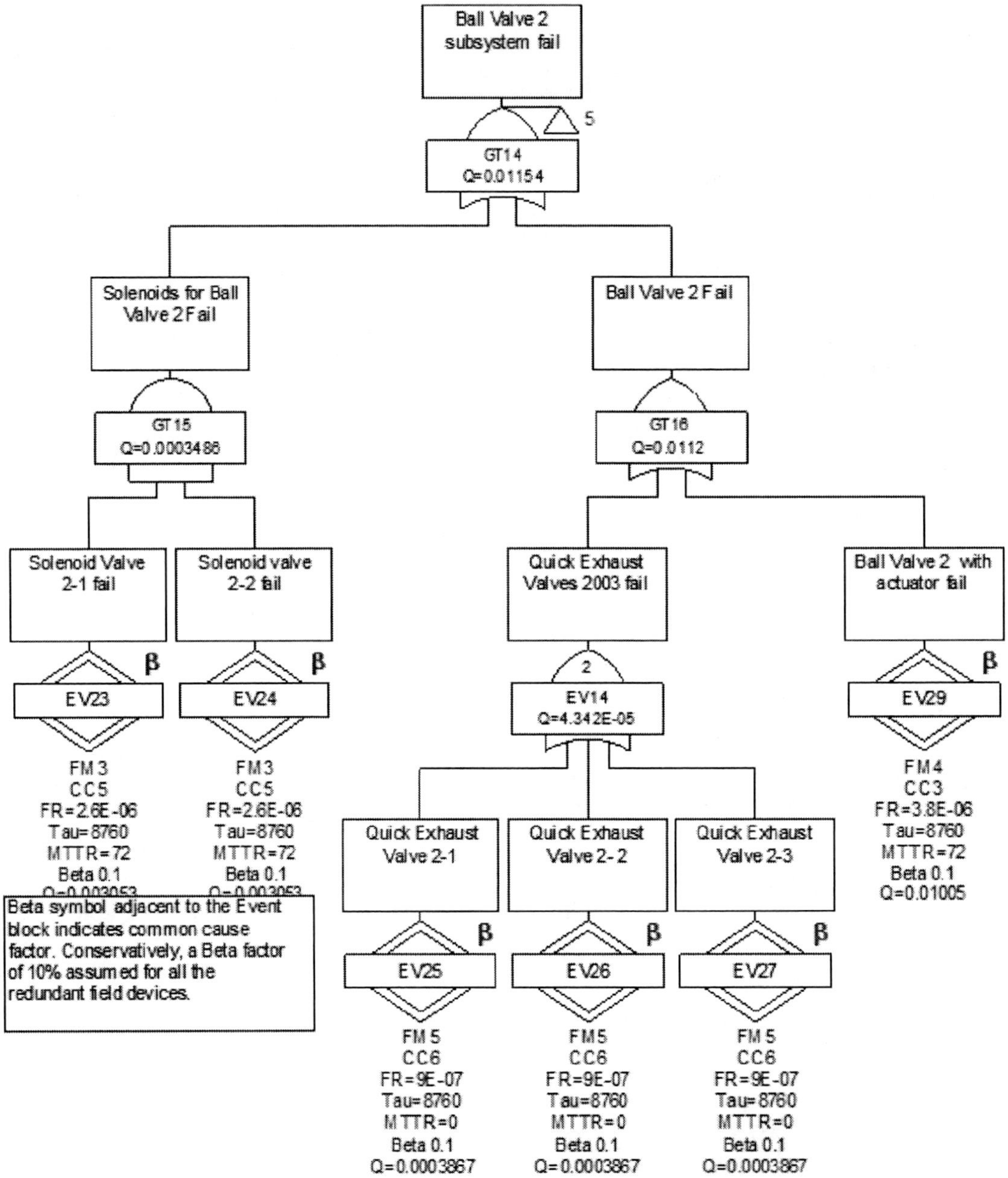

Figure 5. FTA for multiple HIPPS to achieve high risk reduction (contd..)

7 Installation, commissioning, operation and maintenance

Installation and commissioning of the HIPPS must be completed in accordance with the installation drawings, safety manuals, and the SRS.

Operation and maintenance is the longest period in the safety life cycle. The functional safety standards mandate continuous performance monitoring, including demand rates, failure modes and failure rates of various components. The actual performance must be compared with the predicted performance to identify any gaps so that remedial measures can be provided with appropriate safety review.

A robust change management process must be in place to ensure HIPPS integrity throughout the operation and maintenance phase.

8 Conclusion

Assignment of an SIL target and choosing an appropriate technology for a HIPPS is not trivial. In this paper, the significance of PHA and SIL selection for HIPPS has been re-emphasized and a brief account of available options for HIPPS to choose from, their pros and cons, and the selection criteria with examples are provided.

The design, installation, operation, and maintenance of HIPPS shall comply with the requirements of functional safety standards IEC 61508 and IEC 61511. In this article, the challenges in each of the above activities are discussed and appropriate solutions to mitigate the above challenges are provided.

In applications where a high risk reduction greater than 10,000 is required a hybrid HIPPS design approach is proposed. Using FTA demonstrated the modeling of complex hybrid HIPPS to achieve high level risk reduction.

9 References

[1] IEC. IEC 61511 Functional Safety: Safety Instrumented Systems for the Process Industry Sector- Part 1-3, Edition 2, Geneva, 2016.

[2] American Society of Mechanical Engineers (ASME), Boiler and Pressure Vessel Code, Section VIII-Pressure Vessels, New York, 2010.

[3] API, Standard 521, Pressure-relieving and Depressuring Systems, 6th Edition, Washington, DC, 2014.

[4] API, RP17O, Recommended Practice for Subsea High Integrity Pressure Protection Systems (HIPPS), 1 Edition, Washington, DC 2009.

[5] HSE, UK. SPC/TECH/OSD/31, Safety instrumented systems for the overpressure protection of pipeline risers, Available at

http://www.hse.gov.uk/foi/internalops/hid_circs/technical_osd/spc_tech_osd_31.htm, Accessed on Jan 29, 2017.

[6] IEC. IEC 61508 Functional safety of electrical/electronic/programmable electronic safety related systems- Part 1-7, Edition 2, Geneva, 2010.

[7] H.G. Lawley and T.A. Kletz. High-pressure-trip systems for vessel protection, Chemical Engineering, Vol 82, (1975).

[8] Dr. D. J. Smith, Reliability, Maintainability and Risk, 8th Edition, Butterworth-Heinemann, Waltham, MA, 2011.

[9] S. Hauge, S. Habrekke, M.A. Lundeteigen. Reliability Prediction Method for Safety Instrumented Systems- PDS Example Collection, SINTEF Technology and Society, Safety Research, 2010.

[10] Dr. D.J. Smith, and K.G.L. Simpson, The Safety Critical Systems Handbook, 4th Edition, Butterworth-Heinemann, Cambridge, MA, 2016.

475693 Managing Risk Associated w/ Electrical Energy Ignition Sources

Abstract:

In facilities that handle flammable hydrocarbons it is extremely important that all forms of potential ignition sources be carefully managed. This presentation will focus on key points associated with process electrical safety, open flame process heaters, and mechanical forms of ignition sources. OSHA 1910.119 has for many years required that hazardous area classification documentation remain in an evergreen state that requires re-certification at least once every five years. For most process facilities this OSHA mandate presents many problems. In most cases documentation has not been maintained in a current state, the facility perhaps has changed owners if not once many times over the years. The presentation attendee will be exposed to a common sense methodology that focuses primarily on minimizing electrical energy from becoming a source of ignition. A process for updating hazardous area classification documentation will be presented with tips on providing the end-user with an audit proof set of documentation. Once the hazardous area map has been developed then methods of properly documenting and conducting a site compliance assessment will be explained. This compliance assessment is a gap analysis that compares the installed electrical systems against the requirements of the National Electrical Code (NFPA 70 Article 500). Methods of presenting this documentation will be explained so the end-user has a clear understanding of the proper steps for remediation.

The presentation attendee will be exposed to the following:

- Managing risk associated with roadways that are impinged by hazardous classified boundaries
- Open Flame Process heaters encroachment by hazardous classified locations
- Methodology for area classification assessment *(it's determined by the process conditions)*
- Hazard's compliance for instrumentation and electrical system installations in hazardous process locations
- How to interpret documentation in support of the area classification assessments which include:
  - Plan Drawings (maps), Elevations, Site Specific Details. Legends, and notes
  - Assessment report documentation that addresses General Considerations, Identification of Point Sources of Emissions, Physical properties of Materials being Handled, Processed, or Stored, and Gas Dispersion Modeling Graphs and plots
- Methods for mathematical determination of hazard radius thru the concept of commercially available computer assisted gas/vapor dispersion modeling software.
- Application of common standards, codes and installation practices i.e. (NFPA, ISA, API and OSHA)

Detailed handouts as well as a complete audio-visual presentation are used to supplement the presentation.

479320 Review and Analysis of Recent Cyber Security Attacks

Abstract:

Process plants need to ensure that adequate safeguards are in place for a variety of upset conditions, including cyber security attacks, which can result in loss of product, damage to equipment, downtime, and potential safety and environmental disasters. Select cyber security incidents across a variety of industries will be reviewed to determine the modes of cyber security attack and their subsequent impacts. Specific areas of vulnerability will be identified with corresponding mitigation strategies. Possible solutions will be presented for incorporating cyber security into an existing process safety program to ensure a tolerable level of protection for high-consequence events.

478316 Resilience-Based Survival Analysis in the Event of a Process Upset: A Case Study of Utility Plant

Abstract:

Abstract: In a process plant system, safe and reliable operations are highly sensitive to the power or steam failure, loss of cooling water or nitrogen or instrument air. Most of these products and services from utilities plant play an important role as safety barriers. The main focus of this paper is to propose a model for survival evaluation of a process system under an upset condition. This model is based on the holistic resilience-based method that integrates both technical (process parameters variations) and social (policy/regulations, human and organizational) factors. The resilience methodology emphasizes non-linear dynamics, new types of threats, uncertainty, survival through upset conditions and recovery from catastrophic situations [1], [2].

The purpose of this paper is to maximize the performance of safety barriers that will be utilized in case of a process upset condition to prevent the escalation of the event into a catastrophic incident. In this work, a methodology is presented to select the best maintenance strategy [3] while considering the statutory safety regulations for maintenance of equipment like boilers, generators and more. Three different types of maintenance alternatives considered as part of this analysis are: predictive, preventative and condition-based. The key ideas of survival evaluation are presented using a utility plant as the case study.

References

1. Dinh, L. T., Pasman, H., Gao, X., & Mannan, M. S. (2012). Resilience engineering of industrial processes: principles and contributing factors. Journal of Loss Prevention in the Process Industries, 25(2), 233-241.
2. Jackson, S. (2009). Architecting resilient systems: Accident avoidance and survival and recovery from disruptions (Vol. 66). John Wiley & Sons.
3. Vassiliadis, C. G., & Pistikopoulos, E. N. (2001). Maintenance scheduling and process optimization under uncertainty. *Computers & Chemical Engineering*, *25*(2), 217-236.

478816 Model-Based Pressure Relief Design for Reactive Systems

Abstract:

Abstract

Reactive systems behave unpredictably under overpressure conditions. Varying composition impacts the physical properties and response of the mixture to temperature and time. Pressure generation in a reactive system is often a function of both vapor pressure and gas generation rate. Standard methodologies applied in non-reactive pressure relief system design cannot be relied upon to give appropriate sizing for reactive systems. Consistent with DIERS methodologies, model-based design provides accurate, reliable vent sizing. A comprehensive approach to modeling is presented which incorporates vapor-liquid equilibrium, reaction stoichiometry, and reaction rate.

Dynamic Modeling of Inbreathing Requirements for Low-Pressure Storage Tanks

Dona Abou-Chakra
ioMosaic Corporation
93 Stiles Road | Salem, NH 03079
chakra.d.nh@iomosaic.com

Georges A. Melhem, PhD, FAIChE
ioMosaic Corporation
93 Stiles Road | Salem, NH 03079
melhem@iomosaic.com

Ronald J. Willey, PhD
Northeastern University
360 Huntington Avenue | Boston, MA 02115
r.willey@neu.edu

Prepared for Presentation at
American Institute of Chemical Engineers
2017 Spring Meeting and 13th Global Congress on Process Safety
San Antonio, TX
March 26 – 29, 2017

Dynamic Modeling of Inbreathing Requirements for Low-Pressure Storage Tanks

Dona Abou-Chakra
ioMosaic Corporation
93 Stiles Road | Salem, NH 03079
chakra.d.nh@iomosaic.com

Georges A. Melhem, PhD, FAIChE
ioMosaic Corporation
93 Stiles Road | Salem, NH 03079
melhem@iomosaic.com

Ronald J. Willey, PhD
Northeastern University
360 Huntington Avenue | Boston, MA 02115
r.willey@neu.edu

Keywords: Thermal Inbreathing, API 2000, Storage Tanks, Venting.

Abstract

A Process Safety Office™ - SuperChems™ component based model is proposed to study the dynamic behavior of the thermal inbreathing of low-pressure storage tanks upon the occurrence of a sudden cooling rainstorm. The objective here is to establish a model that is capable of properly calculating the design relief capacity of the relief device to be installed on such tanks. The impact of phase-change, particularly vapor condensation, on the thermal inbreathing of the tank assessed. The model was validated against two large-scale experiment datasets and some other published theoretical models. The results were then discussed and compared to the relief requirements obtained using the API 2000 7th Edition method, which is the current standard for venting atmospheric tanks.

1 Introduction

Low-pressure storage tanks are protected against extreme weather changes following the API 2000 (7th edition) standard. The threat of creating a vacuum occurs when the contents of the tank vapor space are suddenly cooled by an intense rainfall causing the vapor space to contract and condense, hence, rapidly decreasing the internal pressure. This pressure drop will pull outside air into the tank's vapor space and acts as the driving force for thermal inbreathing.

The current limitation with the API standard is that it only addresses tanks storing non-condensable gases and have no liquid residue, where in fact most storage tanks contain condensable vapors such as water vapor which will have a big impact on the thermal inbreathing of the tank. This paper assesses the API 2000 relief requirement calculations for low-pressure storage tanks by comparing them to (1) the results of the dynamic simulations from a proposed SuperChems™ based model and (2) to several methods found in the literature (PTB W-22, Salatino, etc.) for both condensable and non-condensable vapor services.

2 API 2000 Methods

The tank contents are affected by the ambient conditions especially when the tank is uninsulated, which is the case for most storage tanks. This method uses air at normal conditions (0°C and 101.325 kPa) as the inbreathed gas. The maximum inbreathing rate, $\dot{V}_{IT}$, is primarily a function of tank volume, insulation, and a C-factor (found in Table 1 below) as shown in Equation (1) below:

$$\dot{V}_{IT} = C\, V_{tk}^{0.7} R_i \tag{1}$$

where $\dot{V}_{IT}$ is the maximum inbreathing volumetric flow rate of air at normal conditions (Nm^3/h), C is the dimensionless factor that depends on vapor pressure, storage temperature and latitude, V_{tk} is the tank volume (m^3), and R_i is the reduction factor for insulation.

Table 1. API 2000 *C*-factor for Thermal Inbreathing

Latitude, φ	**C-factor**			
	Vapor pressure similar or below hexane		Vapor pressure higher than hexane or unknown	
	Average storage temperature (°C)			
	< 25	≥ 25	< 25	≥ 25
φ < 42°	4	6.5	6.5	6.5
42° ≤ φ ≤ 58°	3	5	5	5
φ > 58°	2.5	4	4	4

The assumptions used for the scenario of "tank cooling by rain" are found in Annex E of API 2000 [1] and are listed below:

- Tank is filled with air only.
- Initial vapor space and shell wall temperature are in thermal equilibrium at 55 °C.
- Rain temperature and intensity are constant at 15°C and 225 kg/m^2h (This rainstorm has a probability of 1/100 year) respectively; rain angle is at 30° from the horizontal.
- Inside free convection heat-transfer coefficient is 5 W/m^2K.
- Outside film cooling heat-transfer coefficient is 5,000 W/m^2K.

2.1 API 2000 Annex A Method

The method found in Annex A was used prior to 7th Edition (2008) and is quite different than the current sizing method for venting devices. In order to quantify the thermal inbreathing for storage tanks volumes smaller than 3180 m^3, Annex A [1] assumes that the tank starts cooling from an initial temperature of 48.9°C and at a constant temperature rate of -56 K/h. The thermal inbreathing rate, $\dot{V}_{IT}$ (Nm^3/h) is as follows:

$$\dot{V}_{IT} = 0.169\ V_{tk} \quad (2)$$

where V_{tk} is the tank volume (m^3). For tank volumes equal or greater than 3180 m^3, the thermal inbreathing is assumed to be limited by a constant heat flux, $h\Delta T$, equal to 63 W/m^2. The thermal inbreathing rate in this case becomes:

$$\dot{V}_{IT} = 0.577\ A_{exp} \quad (3)$$

where A_{exp} is the external tank area exposed to rain (m^2). Table 2 below summarizes the thermal normal venting requirements as found in Annex A of API 2000. Volumes greater than 30,000 m^3 are outside the scope of this method. However, interpolation is allowed for Table 2 as long as the tank volumes lie between the specified volume limits of 10 m^3 to 30,000 m^3. [1]

Table 2. API 2000 Annex-A Normal Venting Requirements Due to Thermal Effects [1]

Tank Capacity	Inbreathing	Out-breathing	
Column 1 [a]	Column 2 [b]	Column 3 [c]	Column 4 [d]
		Flash Point ≥ 37.8 °C or Normal Boiling Point ≥ 149 °C	Flash Point < 37.8 °C or Normal Boiling Point < 149 °C
m^3	Nm^3/h of air	Nm^3/h of air	Nm^3/h of air
10	1.69	1.01	1.69
20	3.38	2.02	3.38
100	16.9	10.1	16.9
200	33.8	20.3	33.8
300	50.4	30.4	50.4
500	84.5	50.7	84.5
700	118	71.0	118
1,000	169	101	169
1,500	254	152	254
2,000	338	203	338
3,000	507	304	507
3,180	537	322	537
4,000	647	388	647
5,000	787	472	787
6,000	896	538	896
7,000	1,003	602	1,003
8,000	1,077	646	1,077
9,000	1,136	682	1,136
10,000	1,210	726	1,210
12,000	1,345	807	1,345
14,000	1,480	888	1,480
16,000	1,615	969	1,615
18,000	1,750	1,047	1,750
20,000	1,877	1,126	1,877
25,000	2,179	1,307	2,179
30,000	2,495	1,497	2,495

3 Thermal Inbreathing of Non-Condensable Vapors

One of the major assumptions of API 2000 is that the tank experiencing inbreathing is empty (i.e. only filled with dry air) and the vapor space is considered non-condensable. This section will review recent work on inbreathing dynamics for non-condensable dry air services only.

3.1 Salatino's Model

Salatino et al. [2] compare their tank inbreathing model to the older version of API 2000 (1992) that utilized the method of Annex A: a method based on the maximum estimated heat loss rate of 63 W/m^2 upon sudden cooling by a rainstorm of 225 kg/m^{2}h rain density. They recognized that all existing thermal inbreathing models assume a uniform vapor-space temperature, thus his team's contribution was accounting for the temperature non-uniformity inside the tank by dividing the tank into two individual volumes: the shell and the roof.

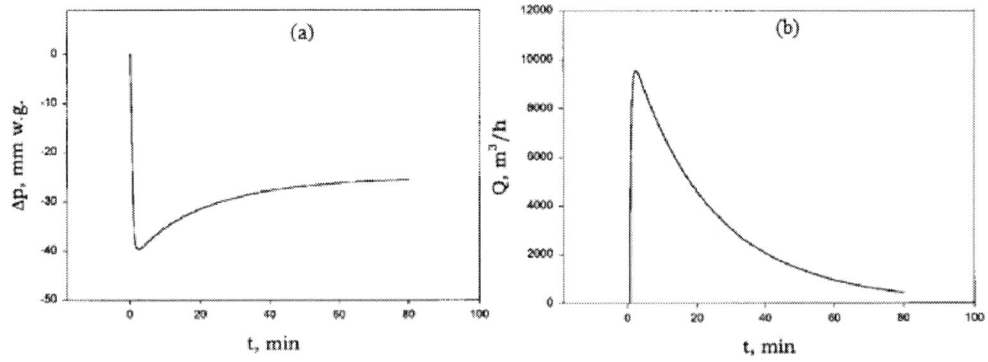

Figure 1. Pressure drop (a) and thermal inbreathing (b) profiles of Salatino's Model [2]

Salatino derived an equation for the maximum thermal inbreathing (Equation 4), stating that the maximum occurs at time t=0 (Figure 1) when the sudden weather change causes (1) vanishing of the radiative flux absorbed by the tank, (2) abrupt drop in ambient temperature and finally (3) rainfall.

$$\dot{Q}_{max} = -\frac{V}{T_G}\frac{dT_G}{dt}\bigg|_{t=0} = \frac{V}{T_{G,1}}\frac{1}{\tau}\left(T_{G,1} - T_{G,2}\right) \tag{4}$$

$$\tau = \frac{pV}{RT_G}\frac{c}{\sum A_i h_{iG}} \tag{4a}$$

where $\dot{Q}_{max}$ is the maximum thermal inbreathing rate (m^3/h), V is the tank volume (m^3), $T_{G,1}$ and $T_{G,2}$ are the gas temperatures (K) before and after the rainstorm, respectively. τ is the time constant that sums the effects of pressure, volume, heat transfer parameters (Equation 4a).

Salatino et al. further simplified the model by assuming an average value for the heat transfer coefficient of 4 W/m^{2}K and a gas temperature drop (after rainfall) of T_{G1}-$T_{G2} \cong 25$°C claiming that these values will result in a conservative approximation of the inbreathing rate. Replacing these values in Equation (4) yields the simplified maximum inbreathing rate, $\dot{Q}_{max}$, which is a function of the tank volume, *V,* and the height-to-diameter ratio, *f:*

$$\dot{Q}_{max} = 2.6\, V^{2/3} \left(\frac{1+2f}{\sqrt[3]{f}} \right) \tag{5}$$

For a 63,000 m^3 cone-roof tank, Salatino predicted a maximum thermal inbreathing value of 9,900 m^3/h, using Equation 5. However, using the API 2000 Annex A method, the maximum inbreathing for the same tank was found to be 4,100 m^3/h. Salatino proved that his method is more conservative than the API 2000 equations and advised that an oversized thermal relief device is considered a safer design in this case.

3.2 Lloyd's Register Energy Americas, Inc. (LR) Model

Lloyd's Register (LR) Energy Americas, Inc. analytically solved the heat balances of a storage tank that is suddenly cooled by rain [3] and compared their results against ISO 28300, which is an equivalent of API 2000 6th edition (2009). The model results are summarized in Figure 2 below.

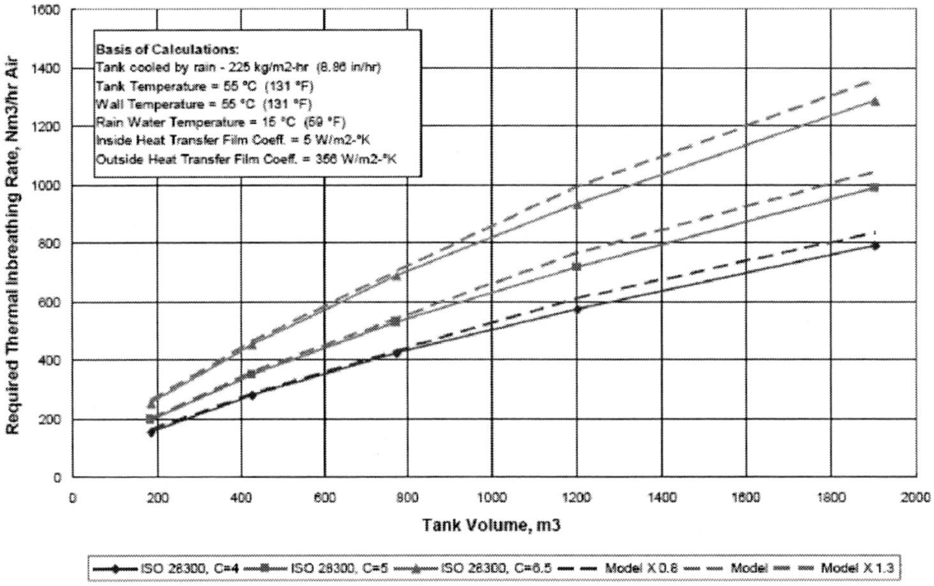

Figure 2. LR Model versus API 2000 6th ed. (ISO 28300)

The LR model was slightly more conservative than the API 2000 thermal inbreathing predictions by factors ranging between 0.5 and 1.3. Note that the C value in Figure 2 refers to the C-factor in the API 2000 thermal inbreathing in Equation (1) of this paper.

The model was then used to study the impact of several variables, some of which had minimal impact on the predicted thermal inbreathing (less than 5% error) such as the tank weight and initial gas temperature. Other variables, like the outside heat-transfer coefficient (Table 3) and rain intensity, had substantial impact on the calculated thermal inbreathing rates.

Table 3. LR Model - Impact of Outside Heat-Transfer Coefficient, h_o [3]

Case No.	Tank Volume m3	Predicted Thermal Inbreathing Rate, Nm3/hr		
		ho = 200 W/m2- K	ho = 300 W/m2- K	ho = 400 W/m2- K
1	186	169.6	191.2	205.1
2	427	308.3	341.4	362.7
3	775	475.1	520.7	548.6
4	1202	678.9	740.5	778.6
5	1903	935.2	1011.6	1058.5

The LR model was used to derive a relationship between the rain intensity (mm/h) and the heat transfer coefficient between the outer tank wall and rain (W/m^{2}K) as shown in Figure 3 below. This linear dependence relationship was implemented in SuperChems^{TM} for this study.

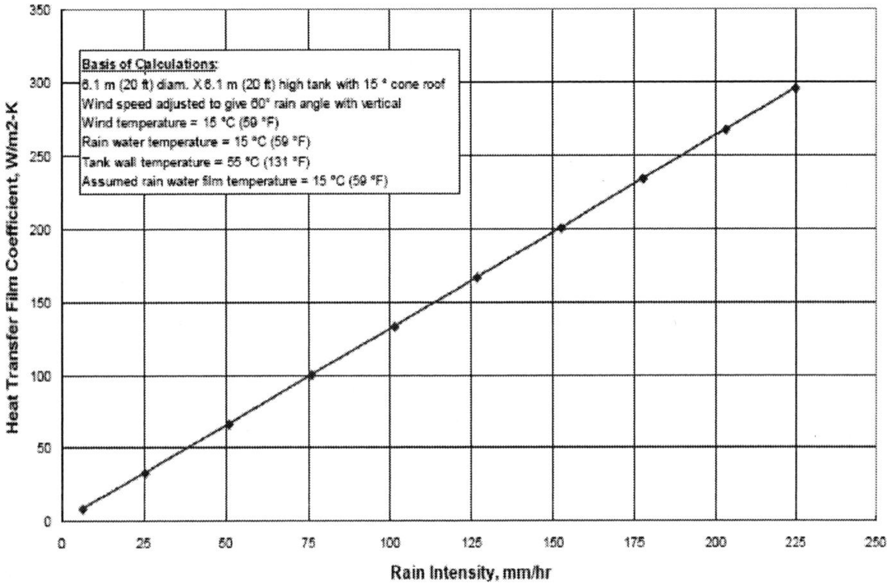

Figure 3. Rain/wall heat transfer coefficient versus of rain intensity

3.3 PROTEGO© Model

Scholz and Weber [4] from PROTEGO© derived the API 2000 7th Edition thermal inbreathing formula (Equation 1 of this paper) and determined that the maximum thermal inbreathing was a function of primarily tank volume with a coefficient that is less than or equal to 9.99:

$$\dot{V}_{max} = Cofficient \cdot V^{2/3} \tag{6}$$

The above equation is applicable to both flat-roof and conical-roof storage tanks. The derivation of Equation (6) was obtained by performing energy balances on the stored gas, the tank walls, and on the rainwater film. The differential equation and its solution are not included in this document, but can be found in Scholz and Weber's publication [4].

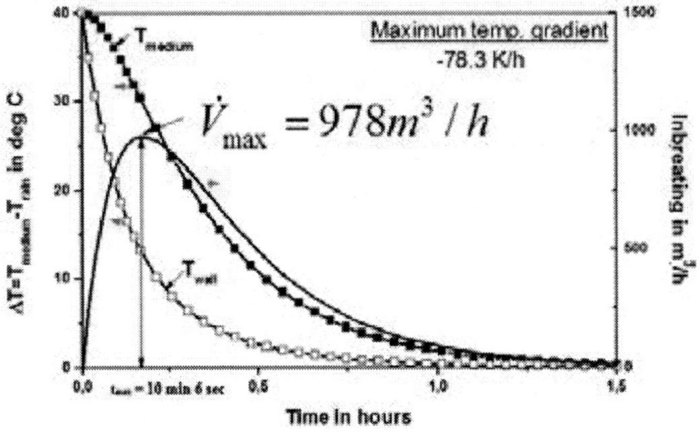

Figure 4. PROTEGO© Model Results

PROTEGO© then compared their results with the Foerster's model [5], whose main assumption was a fully immersed tank, which proved to be in good agreement with the published data. Figure 4 contains the results of a simulation using a conical-roof storage tank of 3,978 m^3 volume and an aspect ratio f = 0.6. Light rain was assumed (75 kg/m^2h) at 15°C. The inside and outside convective heat-transfer coefficients were set to 5 and 37 W/m^2K respectively.

PROTEGO© used their model to study 5 different types of condensable vapors and their effect on the maximum thermal inbreathing rate, however, the results will be discussed in Section 4 of this paper.

4 Thermal Inbreathing of Condensable Vapors

API 2000 neglects the possibility of vapor condensation for the thermal inbreathing/out-breathing relief calculation since only dry air is present in the tank i.e. the tank is empty. One reason to consider the effects of having condensable vapors is the steam cleaning that storage tanks undergo. Hot steam will occupy the tank after the cleaning process is complete and if a rainstorm occurs, the water vapor will condense/contract and pull a vacuum inside the tank.

4.1 Fullarton Model – Water Vapor

Fullarton [6] introduced two dimensionless factors in his model: (1) The reduction factor "η" that accounts for the tank wall thickness by factoring in the metal heat capacity of the walls and (2) the condensation factor "χ" that accounts for the stored-vapor condensation i.e. the additional condensation mass flux and the additional heat of condensation. The formulas of each "η" and "χ" can be found in Fullarton's paper [6].

It is expected that the tank wall temperature drops much faster than the temperature of the vapor stored inside the storage tank because of the metal's higher heat capacity. However, a thicker metal wall acts as a resistance to heat transfer. This means that the tank walls cool at a slower rate than when the wall thickness is neglected. The maximum inbreathing in this case will be lower, hence the reduction factor "η" which "reduces" the maximum thermal inbreathing of a storage tank with thicker walls.

Table 4. Range of χ-factor for Different Condensable Vapors

Condensable Vapor	χ-factor
Water vapor Heptane Hexane	1.5-2
Ethanol	2-2.5
Methanol	3-4

The condensation factor "χ", on the other hand, is primarily a function of the volatility of the condensable vapor being studied. For that reason, Fullarton established a range for the "χ" factor for different vapors as seen in Table 4 above [6].

Therefore, the maximum thermal inbreathing for a condensable vapor space per Fullarton can be obtained by using each of the Naumann approximation, $\dot{V}_N$, the reduction factor, η, and the condensation factor, χ.

$$\dot{V}_{K,max} = \dot{V}_N \eta \chi \tag{7}$$

One must keep in mind that properties of air at 288K were used to derive this equation in addition to the assumption of the initial temperature gradient ($T_{G,0} - T_A$) equal to 40°C. If other values are to be used, the above equation must be re-derived to account for the appropriate conditions.

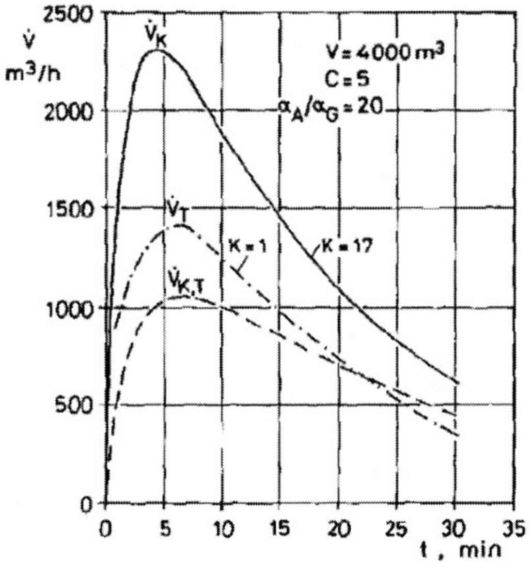

Figure 5. $\dot{V}_{inbreathing}$ **profile – Fullarton Model** [6]

Fullarton performed a dynamic simulation on a 4,000 m^3 empty tank intended for the storage of ethanol. The initial gas temperature was 55°C and was suddenly cooled by rain at 15°C. The rain fall was assumed to be an extremely heavy downpour (probability of 1 every 100 years) with $^{\alpha_A}/_{\alpha_G} = 20$ and $\alpha_G = 5$ W/m^{2}K. Figure 5 above shows the resultant inbreathing profile for each of dry air, K=1, and saturated water vapor, K=17. The maximum thermal inbreathing occurs at time t = 6 min where the inbreathing of the condensable vapor space ($\dot{V}_{K=17}$) is *1.6 times larger* than that of dry air ($\dot{V}_{K=1}$).

4.2 PROTEGO© Model – Condensable Vapors

PROTEGO© used their model [4] to obtain the inbreathing curves of five different vapors: methanol, ethanol, saturated water vapor, heptane and decane (Figure 6) and comparing them to the inbreathing curve of dry air in the same tank.

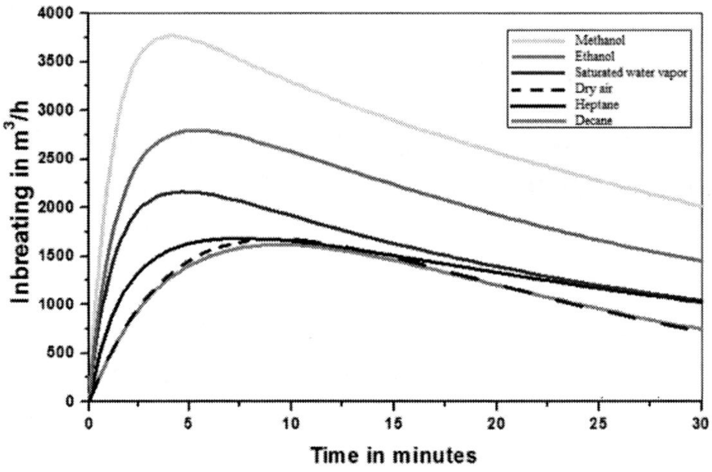

Figure 6. $\dot{V}_{inbreathing}$ profile of different vapors – PROTEGO© Model [4]

It can be seen that the condensation of volatile vapors has a significant impact on the inbreathing rate causing it to have a twofold increase. On the other hand, non-volatiles such as decane do not affect the inbreathing rate and yields the same inbreathing profile as with dry air.

Table 5. Condensation Factor "χ" Results

Vapor	$\chi_{PROTEGO}$	$\chi_{Fullarton}$
Methanol	4.76	3-4
Ethanol	2.23	2-2.5
Heptane	1.61	1.5-2
Water	1.29	1.5-2
Decane	1.01	N/A

PROTEGO© calculated the condensation factor "χ" for each of the simulated vapors and compared their values to that of Fullarton's (Table 4) and summarized the results in Table 5 above. The condensation factor values from this model are all in agreement with the published values [6] except the $\chi_{methanol}$ which is greater than the range specified by Fullarton (1986). A possible explanation for the latter is that PROTEGO© may have been more conservative (higher safety margin) when dealing with extremely volatile vapors like methanol.

5 Large-Scale Experimental Data

A thorough literature review yielded two sets of large scale data from actual experiments to study the thermal inbreathing of low-pressure storage tanks. Large scale datasets are preferred for model validation. The first dataset is for non-condensable gases (air) by *Sigel et al.* [7] and the second dataset is for condensable gases by *Holtkoetter et al.* [8].

5.1 Sigel's Experiment

Sigel performed actual data collection for the thermal inbreathing rate from a large tank of 617.5 m^3 capacity exposed to high solar radiation. The tank was empty without any trace of product vapor or liquid residue.

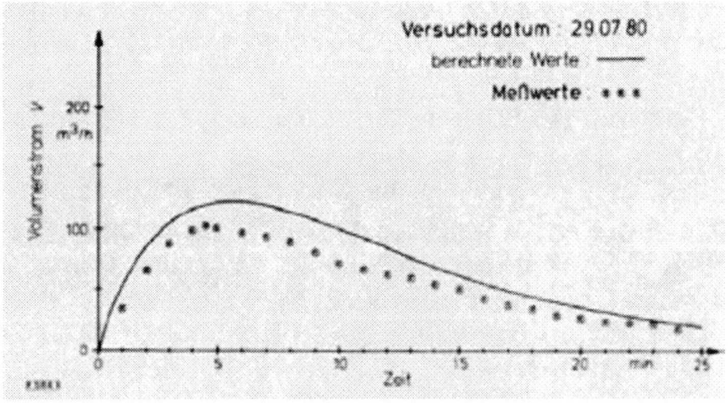

Figure 7. Sigel Experimental Inbreathing Profile

Sigel et al. [7] developed their own thermal inbreathing model to compare it with their experimental results. Figure 7 above displays the measured values of inbreathing (markers on the graph) versus the model (solid line). The inbreathing rates predicted by his model proved to be more conservative than the measured rates.

5.2 Holkoetter's Experiment

Holtkoetter performed an experiment on a 1.18 m^3 tank to measure the actual inbreathing rates and compare the results to his proposed model. This experimental investigation aimed at studying the effects of condensation on thermal inbreathing for three different vapors: water, isopropanol, and methanol.

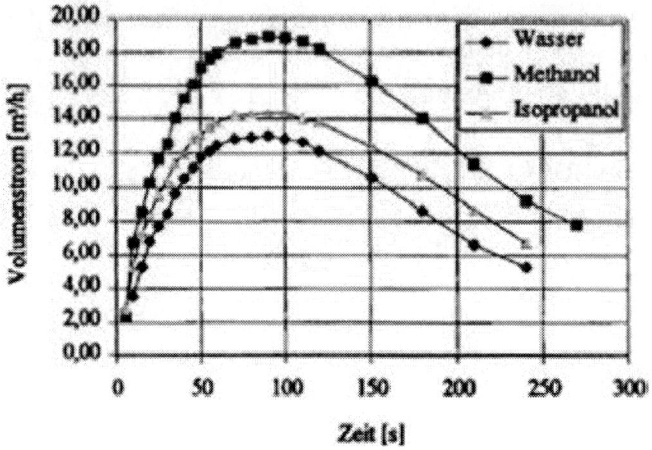

Figure 8. Holtkoetter Experimental Inbreathing Profile

The maximum thermal inbreathing is recorded after almost 95 seconds for all three gases. The maximum thermal inbreathing of methanol $\dot{V}_{max} \approx 19$ m^3/h is significantly higher than that for both water ($\dot{V}_{max} \approx 13$ m^3/h) and isopropanol ($\dot{V}_{max} \approx 14$ m^3/h) due to the fact that methanol is the most volatile vapor.

6 SuperChems™ proposed Model

The inbreathing profiles were generated using SuperChems™ dynamic simulation of storage tanks experiencing a sudden weather change. Dynamic models continually generate results that are used as the initial conditions for the next time interval calculation which perpetually drives the model to completion.

The proposed model first starts by defining a variety of objects to represent components of the system such as the storage tank vessel, the mixture that is stored, the piping segment and relief device, and the water sprays that are used to simulate a rainstorm. A built-in thermodynamics package is selected (modified Peng-Robinson EOS [9]) and scenarios are then created to collect the objects. Once the scenarios are created, single and multi-phase flow dynamics models are associated with the individual scenarios and the dynamic solutions for the inbreathing profiles are produced.

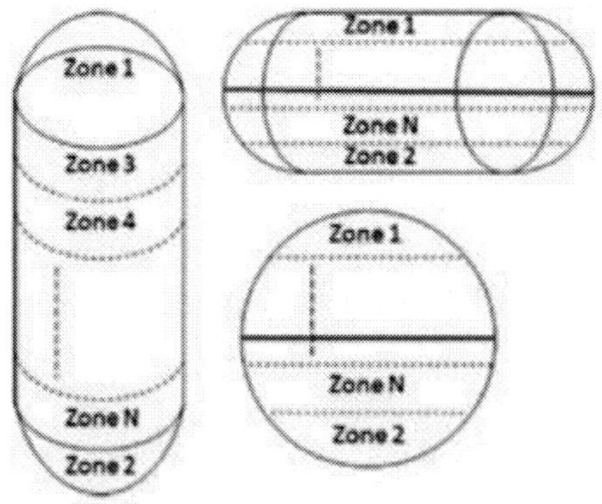

Figure 9. SuperChems™ Segmentation Scheme

An important feature of the software is wall dynamics which segments the vessel into multiple zones (Figure 9) allowing the user to account for heat conduction across the wall and between the zones using the effective conduction length method.

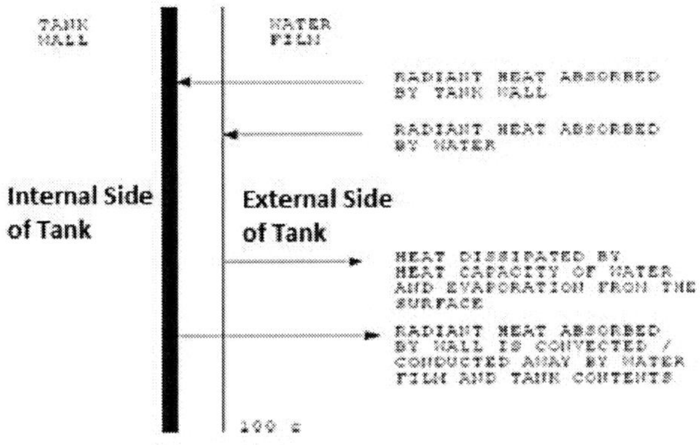

Figure 10: Heat Transfer Mechanisms for a Tank during a Rainstorm

Once segmentation is defined, heat transfer analysis is applied per wall segment (Figure 10) where the impact of rain, solar radiation, wind, and insulation is considered. Inside the tank, the wall-to-fluid heat transfer includes radiation and natural convection for the vapor phase; and nucleate, film and pool boiling for the liquid phase.

It should be noted that many parameters from the existing models and large-scale datasets were missing (vessel, dimensions, wall thickness, mixture composition, heat transfer coefficients, rain intensity, etc.) and had to be scientifically estimated in our dynamic simulations.

6.1 SuperChems™ Model Prediction for Non-Condensable Vapors (Large-scale Data)

The dimensions of the 617.5 m^3 empty tank used in Sigel's experiment [7] were introduced as input parameters to the SuperChems™ vessel object. The rainstorm used in this simulation had an intensity of 0.38 inch/h and a duration of 30 minutes with a temperature of 15°C. The inside heat transfer coefficient value was set to 5 W/m^2K as specified by API 2000 [1].

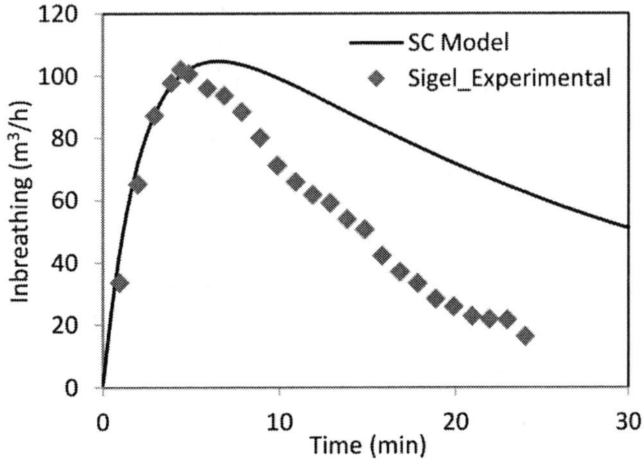

Figure 11. Simulation of Sigel's Tank using SuperChems™

The maximum value for inbreathing predicted by SuperChems was 104 m^3/h at time t = 6.6 min, as seen in Figure 11 above. This value is at a 2% deviation from the actual recorded maximum. The model's slope matches the first part of the inbreathing curve up until the maximum. Although the slopes of the decay of the inbreathing are different, they do not contribute to the design of the relief device because the most important value extracted from such studies is the maximum inbreathing that will act as the design relief capacity of the venting device.

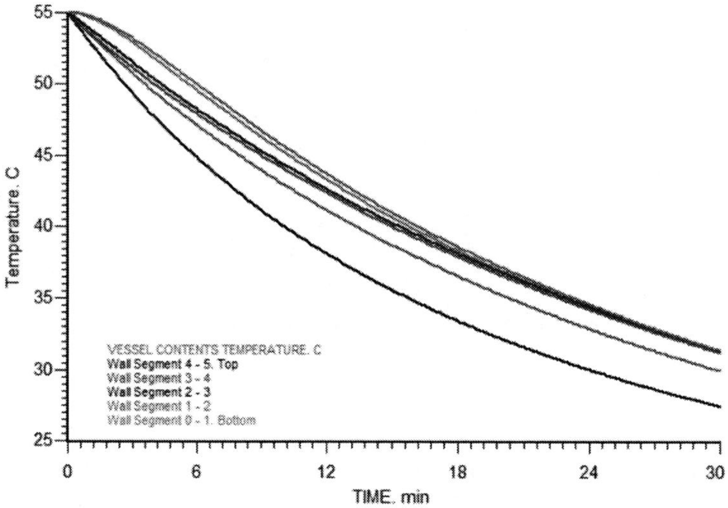

Figure 12: Temperature Profiles of Sigel's Tank using SuperChems™

Figure 12 above shows the temperature profiles of the wall segments and air inside the tank from the SuperChems™ simulation. Since Sigel [7] did not disclose the actual temperature profiles, therefore, we could not compare the computed temperature profile from SuperChems™ to the actual readings from Sigel's experiment.

6.2 SuperChems™ Model Prediction for Condensable Vapors (Large-scale Data)

Holtkoetter's experimental setup [8] consisted of a 1.18m^3 tank with walls that are 1.2 mm wall thick. The three vapors Holtkoetter studied were: methanol, isopropanol and water. For each type of vapor, the tank was filled with a small amount of liquid which was then vaporized using three infrared radiators (2.5-4.2 kW). After vaporization step was complete, water sprays were activated to cool the tank and the thermal inbreathing was recorded (Figure 8).

Using SuperChems™, the vessel was defined, then the rain-wall heat transfer coefficient, h_{rain}, was fixed at 600W/m^2K that corresponds to a rain intensity of 0.1262 kg/m^2s as specified by Holtkoetter. The duration of the rain for this simulation was 300 seconds. The h_{gas} was initially set to the API 2000 value of 5 W/m^2K, but was then adjusted to compensate for the vapor condensation inside the tank.

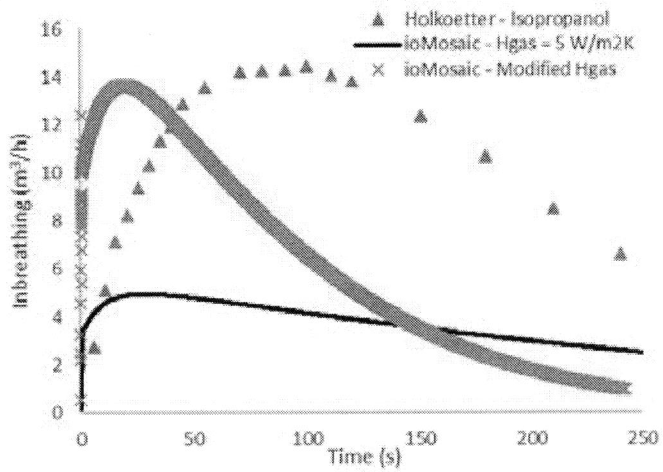

Figure 13. SuperChems™ Simulation of Holtkoetter's Tank with Isopropanol

Figure 13 illustrates the inbreathing profile for two different values of the free-convection heat transfer coefficient of the isopropanol vapor, h_{gas}, inside the tank. It can be clearly seen that using the API 2000 value $h_{gas} = 5 W/m^2K$ underpredicts the thermal inbreathing by almost 62%. After numerous iterations, a modified value for h_{gas} was found to be 30 W/m^2K. This higher value makes physical sense since the condensation of isopropanol occurring on the inner walls of the tank enhances the heat transfer between the vapor and the wall.

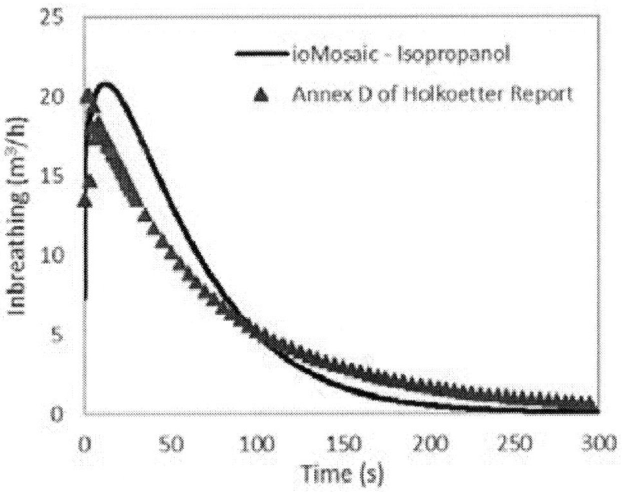

Figure 14. SuperChems™ Reproduction of Holtkoetter's Inbreathing Model (Isopropanol)

Although *Holtkoetter et al.* [8] performed experimental work to measure the maximum inbreathing, they did not, however, compare their experimental results to their proposed model. Hence, the SuperChems™ results were compared to Holtkoetter's model found in "Annex D" of his publication [8].

Figure 14 above proves that the SuperChems™ based model shows good agreement with the Holtkoetter model that considers spontaneous condensation. The predicted value for the $\dot{V}_{max}$ using SuperChems™ was 20.7 m^3/h and is at a +3% error from the value reported by Holtkoetter [8].

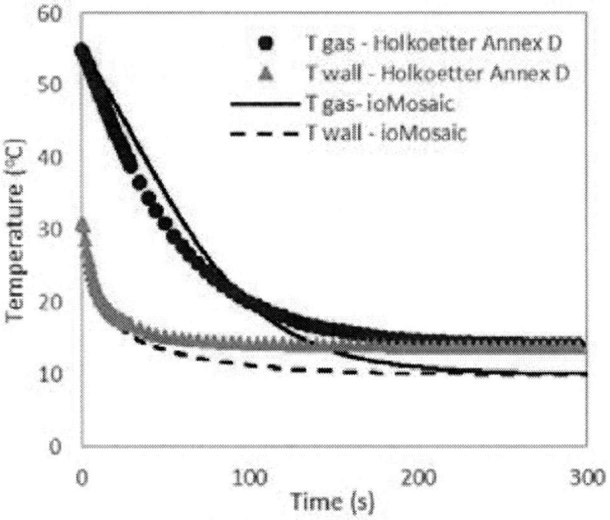

Figure 15. SuperChems™ Reproduction of Holtkoetter's Temperature Model (Isopropanol)

The gas and wall temperature profiles were almost accurately reproduced using the SuperChems™ model and are shown in Figure 15. Holtkoetter attributes the discrepancy between his model and his experimental result to the fact that the experiment was done in the month of December where the available water had a temperature of 8°C instead of 15°C. Colder rain will eventually lead to higher inbreathing rate values, therefore, the inbreathing should be inspected as a function of temperature to account for this 7°C difference in the water temperature. For more information about this method, please refer to the actual publication [8].

6.3 SuperChems™ Model vs. Published Models

In addition to validating our proposed model against large-scale experimental data, the SuperChems™ based model was also compared to the published models listed in Section 3 and 4 of this paper [10].

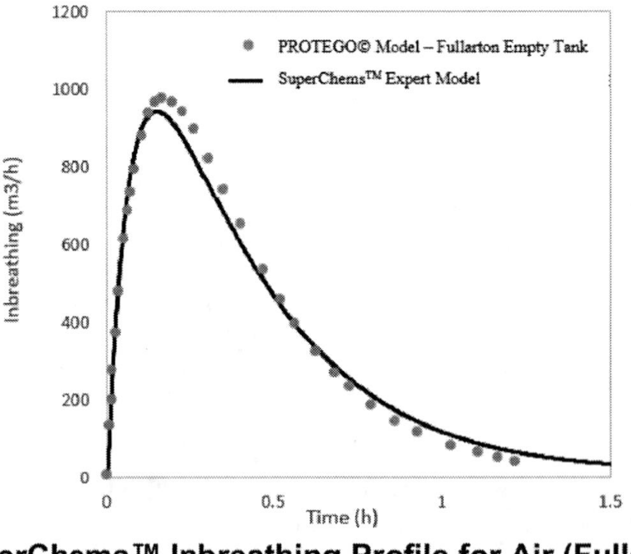

Figure 16. SuperChems™ Inbreathing Profile for Air (Fullarton's tank)

Figure 16 shows the predicted inbreathing profile of dry air using the SuperChems™ model and compared against the PROTEGO© model. The maximum recorded inbreathing from the SuperChems™ Expert simulation is around 940 Nm^3/h which has a -4% deviation from $\dot{V}_{max}$ predicted by PROTEGO©. The temperature and temperature-rate profiles were also calculated and can be found in the actual publication along with the fits for the different models and vapors using SuperChems™ Expert [10].

6.4 Modified Heat Transfer Coefficient

The heat transfer coefficient between the stored vapor and the inner-tank walls, h_{gas}, was modified for each type of volatile condensable vapor [10]. Table 6 below summarizes the values of h_{gas} used in the two-phase dynamic model for a tank-wall thickness of 5 mm. These results show that condensation "enhances" heat transfer thus providing evidence that use of the API 2000 value heat transfer coefficients h_{gas} will underpredict the maximum thermal inbreathing (Figure 13). Also, note that h_{gas} is directly proportional with the vapor's volatility.

Table 6. Modified h_{gas} Values for SuperChems™ Two-Phase Model

h_{gas} (W/m^2K)	Wall Thickness = 5 mm
Methanol	40
Ethanol	30
Water	20
Heptane	5
Decane	7

7 API 2000 Conservativeness

The maximum thermal inbreathing rates, V_{max}, obtained from the SuperChems™ model were compared to those obtained using the API 2000 thermal inbreathing equation, Equation (1), to assess the performance of the API 2000 inbreathing method.

In order to determine the appropriate value for the C-factor in Equation (1), the vapor pressure and storage temperature of the stored gas are needed.

- Since all the investigated vapors (methanol, ethanol, water vapor, etc.) do not have vapor pressures higher than hexane, the vapor pressure was classified as "Similar to Hexane".
- The storage temperature of the gas is 55°C, which is greater than 25°C, so according to Table 1, the C-factor has a *value of 5.*

The storage tanks in this study were all uninsulated, hence, the insulation reduction factor R_i has a value of one. The API 2000 inbreathing equation used for this case now becomes:

$$\dot{V}_{IT} = 5\, V_{tk}^{0.7} \tag{8}$$

To determine whether the API 2000 inbreathing relief requirement is adequate, a deviation percentage of the simulated (or experimental) results the API 2000 $\dot{V}_{max}$ method is computed as follows:

$$\% \, Deviation = \frac{\dot{V}_{max,SC} - \dot{V}_{max,\ API\,2000}}{\dot{V}_{max,\ API\,2000}} \times 100 \tag{9}$$

where $\dot{V}_{max,\ API\,2000}$ is the maximum thermal inbreathing obtained from Equation (8) (m^3/h) and $\dot{V}_{max,SC}$ is the maximum inbreathing from SuperChems™ model simulation (m^3/h).

7.1 API 2000 performance – Large-scale Experiments

Sigel et al. [7] used a tank volume of 600 m^3, which when replaced in Equation (8), results in a maximum inbreathing value of 440 m^3/h. Figure 17 compares $\dot{V}_{max}$ using the API 2000 method to the actual recorded $\dot{V}_{max}$ values during Sigel's experiment. The API 2000 inbreathing prediction is conservative for Sigel's empty tank where the relief device is designed to handle an inbreathing rate 4.2 times larger than the recorded $\dot{V}_{max}$.

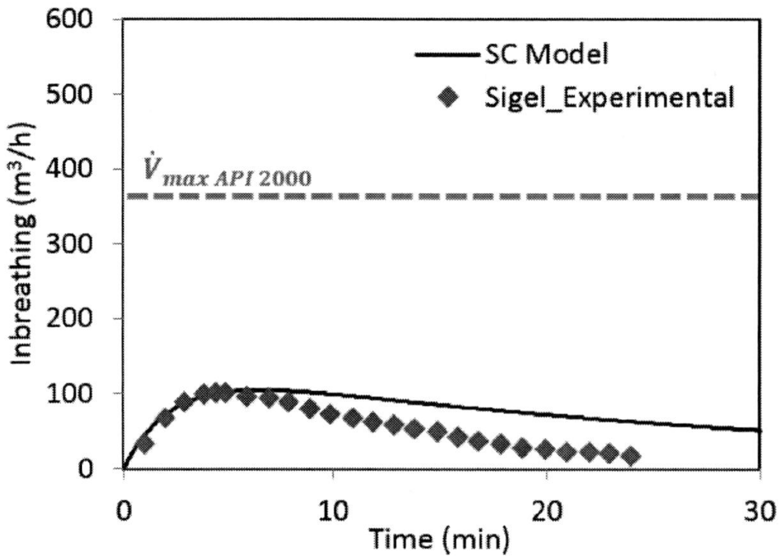

Figure 17. API 2000 Inbreathing Requirements – Sigel's Tank with Dry Air

On the other hand, *Holtkoetter et al.* [8] used a smaller tank with a capacity equal to 1.18 m^3 which results in a maximum inbreathing rate of 5.6 m^3/h according to the API 2000 method – Equation (8). Figure 18 shows that the API 2000 inbreathing relief requirement is insufficient for water, isopropanol, and methanol vapors in Holtkoetter's tank.

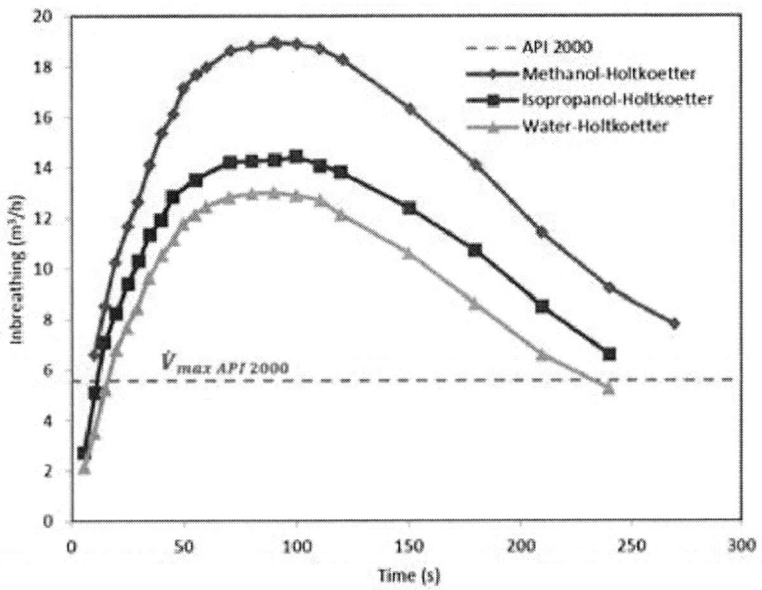

Figure 18. API 2000 Inbreathing Requirements – Holtkoetter's Tank

The recorded inbreathing rates are 2.3 to 3.7 times larger than the rate calculated by API 2000; this leads to an undersized relief device that puts the storage tank at a high risk of failure if a sudden change of weather occurs.

7.2 API 2000 performance – Computer Model

The tank size used in Scholz and Weber's model [4] was 6,283 m^3 to study the condensable vapors. Replacing the tank volumes in Equation (8), results in a maximum inbreathing, $\dot{V}_{max}$ = 2,278 m^3/h. Figure 19 graphically shows the API 2000 $\dot{V}_{max}$ in comparison to the dynamically simulated inbreathing profiles for the different types of vapors, including dry air.

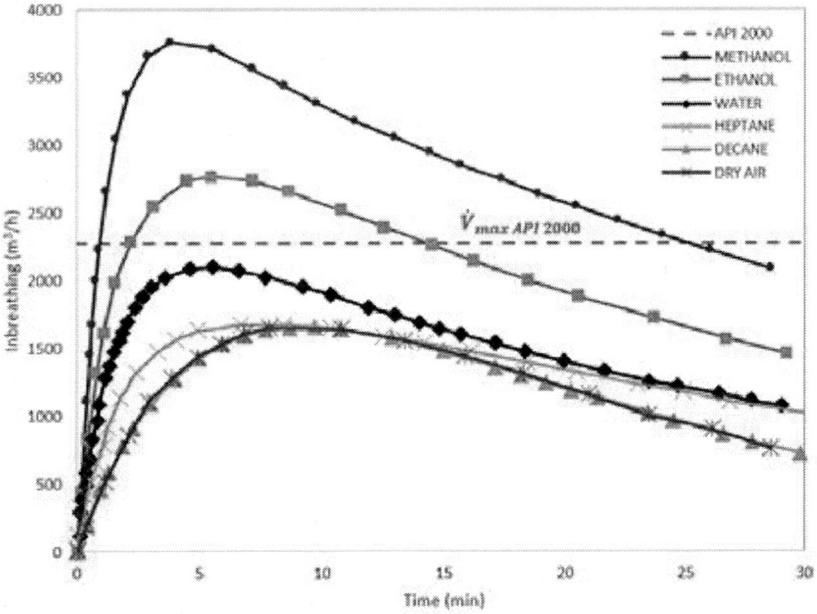

Figure 19. API 2000 Inbreathing Requirements – Different Types of Vapors

It can be seen that API 2000 method has a conservative prediction for the maximum thermal inbreathing of: (1) dry air, (2) heptane, (3) decane, and (4) water services. However, the API 2000 method underpredicts the relief requirement for the thermal inbreathing of a tank saturated with *methanol* and *ethanol*. The relieving capacity, for this case, has to be 1.7 times higher for methanol and 1.2 times higher for ethanol.

8 Conclusion

A dynamic simulation of the thermal inbreathing of low pressure storage tanks experiencing sudden change of weather conditions was performed using a SuperChems™ model. This model was validated by successfully reproducing the thermal inbreathing and temperature profiles of large scale experimental datasets and other available inbreathing models.

The thermodynamic variables specified in the API 2000 7th Edition standard have been tested for adequacy. It was found that the gas/wall free-convection heat transfer coefficient, $h_{gas} = 5W/m^2K$, specified by API 2000 was adequate when the tank was filled with non-condensable vapors like air. However, higher values were suggested for condensing vapors with a range of 10 W/m^2K to 45 W/m^2K depending on the type and volatility of the vapor. As for the rain/wall heat transfer coefficient, the API 2000 value of 5,000 W/m^2K proved to be too conservative, and even unrealistic, therefore leading to an oversized relief device. Suggested values for the rain/wall heat transfer coefficient were found to be in the range of 12 to 600 W/m^2K depending on the intensity and angle of the rainfall.

Finally, the API 2000 7th Edition method for calculating thermal inbreathing was compared with several tanks of different vapor services. The method proved to calculate an adequate relief requirement for tanks storing air, heptane, decane, and even water vapor in one case. However, the application of the API 2000 thermal inbreathing method was not conservative enough when the vapor services became more volatile, like in the case of isopropanol, ethanol, and methanol, that resulted in high risk of failure for these types of storage tanks.

9 References

[1] API Standard 2000, Venting Atmospheric and Low-Pressure Storage Tanks, 7th Edition, March 2014

[2] P. Salatino, G. Volpicelli and P. Volpe. "On the design of thermal breathing devices for liquid storage tanks", Process Safety and Environmental Protection, 77.6, pp. 354-359, 1999.

[3] D. Brooks, "Thermal Inbreathing Requirements of Low Pressure Storage Tanks at Elevated Temperatures". Lloyd's Register Energy Americas, Inc. Spring DIERS Users Group Meeting, Presentation, May 2012.

[4] R. Scholz and R. Weber, "Cooling Down Industrial Tanks". Prepared for PROTEGO Braunschweiger Flammenfilter at API 2000 Meeting Presentation, Fall 2010.

[5] H. Foerster, K. Schampel, and H. Steen, "Atmungsvorgänge infolge von Witterungseinflüssen an Lagertanks für brennbare Flüssigkeiten", Physikalisch Technische Bundesanstalt (PTB) W-22. Braunschweig Univ. Phys.-Tech. Bundesanst, 1984.

[6] D. Fullarton, J. Evripidis and E.U. Shclünder, "*Influence of Product Vapour Condensation on Venting of Storage Tanks*", Institut für Thermische Verfahrenstechnik, Universität Karlsruhe (TH), 1986.

[7] R. Sigel, B. Kuxdorf, R. Meiss and H. Schwarz, "Praktische und theoretische Untersuchung der Beatmung von Festdachtanks". Chemie Ingenieur Technik, 55.5, pp. 377-379, 1983.

[8] T. Holtkötter, J. Shang and H.G. Scheckerm "Behälteratmung - Experimentelle Untersuchungen und Entwicklung eines Vorhersagemodells" (Master Thesis), 1995.

[9] G.A. Melhem, R. Saini, and B. M. Goodwin, "A Modified Peng-Robinson Equation of State," *Fluid Phase Equilibria,* 47, pp. 189-237, 1991.

[10] D.D. Abou-Chakra, G.A. Melhem, and R.J. Willey, "Dynamic Modeling of Inbreathing Requirements for Low-Pressure Storage Tanks" (Master's thesis), 2016. Retrieved from **http://hdl.handle.net/2047/D20209806**

478223 Studying the Relationship Between Inherently Safer Design and Equipment Reliability

Abstract:

Risk associated with a process depends on both likelihood and consequence of an incident. Inherently safer design is based on the use of technologies and chemicals with intrinsic properties that reduce or eliminate hazards. This philosophy is mainly focused on reducing the consequences associated with an incident. However, recent research has revealed that applying principles of inherent safety can have an effect on the likelihood element of risk as well. One of the major drawbacks of implementation of an inherently safer design (ISD) is that it can lead to risk migration, Thus, attempting risk reduction through implementation of ISD by mitigating the consequence element of risk can undesirably lead to an increase in risk by affecting the likelihood element of risk. This likelihood element of risk is the mathematical complement of system reliability. This study mainly focuses on assessing the effect of implementation of ISD principles on the reliability of a chemical process as it progresses along different stages of process design and manipulating this relationship to achieve maximum economic benefits while simultaneously minimizing risk associated with the process.

In this study, safety indices are used for quantification of inherent safety of a process. The choice of safety index depends on the stage of chemical project under consideration. Reliability of an equipment involved in a process is determined through the use of offshore and onshore reliability data (OREDA) and center for chemical process safety (CCPS) reliability databases for process equipment. The reliability of process system is then determined from the reliability of the individual equipment involved in the process. Lastly knowing values of safety indices and reliability of systems for competing designs during different stages of a chemical project (such as process selection stage, conceptual stage etc.), the required relationship between ISD and reliability is established for that particular stage.

Applying this methodology to process selection stage reveals that selecting a process that's inherently safer in terms of the involved reactions, chemicals and equipment might lead to risk migration, since the selected process might have lesser system reliability as compared to other competing process. When this methodology is applied to conceptual stage, where-in the competing processes are similar in terms of reactions and chemicals and vary only in terms of the equipment involved, it is observed that a process that has lesser number of equipment might be more reliable and inherently safer as compared to other competing processes. Lastly, when this methodology is applied for a specific equipment such as a reactor where the reliability is modeled as a function of design parameters of the reactor, it has been observed that within a certain range of design parameters, the reactor becomes sufficiently reliable and inherently safer.

Thus, this study provides a framework for designing of process systems and equipment that are reliable, economically profitable and inherently safer during different stages of project.

479377 Pressure Relief Analysis - It's about More Than Just Compliance

Abstract:

Pressure relief valves and disposal systems are the last layer of defense in petrochemical plants and refineries. Therefore, there are many codes and standards that must be followed to design those systems. Failure to ensure safe plant operations can result in fatalities, injuries, and environmental releases which can result in significant cost and poor public relations.

OSHA's CFR 1910.119 - Process Safety Management of Highly Hazardous Chemicals is designed to manage the hazards associated with facilities which handle highly hazardous chemicals. Many of the requirements in PSM are also valuable for facilities which are not governed by the requirement and application of these principals may improve safety for all facilities.

This paper will present the proposal that Process Hazard Analysis (PHA) be the heart of process for ensuring safe operations to prevent negative impact on companies' image and goodwill with the community.

480249 How to Leverage Limted Design Data Along with Historical Incident Data to Assess Inherent Process Safety During Early Stage of Design?

Abstract:

Inherent safety level quantification remains as one of the challenges for the inherent safety concept to gain industrial acceptance comprehensively. In current practice, most of the time safety assessment is conducted once the process design has been completed. At this stage of design implemented strategies look to control the hazard level. Any design modification necessary as a result of unfavorable safety performance at this stage of design, would be costly when compared to alteration during conceptual design stage [1]. It is beneficial to design industrial processes that are user friendly, inherently safer and have a level of tolerance for the deviations from ideal performance, which will eventually reduce the number of hazardous accidents. Therefore, both industrial experts and researchers are interested in safety indices that can be applied during the early design stage of a process, where the extent of knowledge regarding the process is very limited. Again, the available indices to quantify inherent safety still leave room for enhancement to represent process stream condition more accurately as a mixture rather than pure component(s) and to account for the interaction of all process parameters in determining the level of inherent safety [2].

In this work, a property based methodology to quantify inherent safety level has been proposed by assessing chemical and process unit risk at earlier stage of design. This will assist process designers in producing safer designs by the application of inherent safety principles in a more efficient and cost effective manner. It would offer process/design engineers an alternative to include safety without waiting until detailed design stage, where safety costs increase. Degrees of freedom at this stage would include alternative route selection, equipment sizing, and selection of operating conditions. The additional benefit of safety consideration in the early design stages is that it decreases the level of dependence on safety systems. The proposed safety assessment method here considers the perspective of five main safety factors associated with chemicals such as flammability, explosiveness, corrosiveness, toxicity & reactivity. These factors were linked to the process operating conditions (temperature and pressure). The process unit risk are calculated by classifying them into six main categories (i.e., pump, fired heat exchanger, pressure vessels, etc.) based on API 14C standard and identifying the undesirable events associated with each category. Major incidents database and the probability of unwanted incidents were used to link the undesirable events to operating conditions; this allows for the calculation of unit safety factors using the limited available information. Lastly, an enhancement technique to reflect the contribution of the individual components in the mixture is introduced based on the mixture properties, where the severity score was calculated for the mixture of chemicals in each unit at the actual operating conditions (whenever applicable) and multiplied by the flow rate of the mixture.

This proposed methodology of safety assessment gives a more intuitive, easy-to-understand way to represent the chemical and equipment safety factors posed by each individual unit. The steps in the methodology calculate, compare and prioritize the level of inherent safety of process streams during simulation work that influences the abovementioned safety factors in order to offer the user an insight into the risk associated with alternative design decisions. By prioritizing the streams based on the potential for chemical and equipment safety factors, the design engineers can easily identify the critical streams/units to be considered for improvement in order to avoid or minimize the risk. This methodology of quantifying inherent safety level will provide users another dimension to consider when selecting optimal design with multi objectives (e.g. techno, economic, and environmental).

The assessment of inherent safety level using the proposed method is demonstrated through a case study. HYSYS simulation is used to design two alternative designs for ammonia production.

1. Thiruvenkataswamy, P., et al., *Safety and techno-economic analysis of ethylene technologies.* Journal of Loss Prevention in the Process Industries, 2016. **39**: p. 74-84.

2. Leong, C.T. and A.M. Shariff, *Process route index (PRI) to assess level of explosiveness for inherent safety quantification.* Journal of Loss Prevention in the Process Industries, 2009. **22**(2): p. 216-221.

Inherently Safer Design from Overpressure Protection Viewpoint

Chris Ng
Sachin Kanade
Nancy Faulk
Siemens Energy, Inc.
4615 Southwest Freeway, Suite 900
Houston, TX 77027
chris.ng@siemens.com

Prepared for Presentation at
American Institute of Chemical Engineers
2017 Spring Meeting and 13th Global Congress on Process Safety
San Antonio, TX
March 26 – 29, 2017

Inherently Safer Design from Overpressure Protection Viewpoint

Chris Ng, Sachin Kanade, and Nancy Faulk
Siemens Energy, Inc.
4615 Southwest Freeway, Suite 900
Houston, TX 77027
chris.ng@siemens.com

Keywords: pressure relief, process safety, inherent safety

Abstract

Pressure relief and flare system design is mostly evolving or iterative throughout the various engineering process design stages. This iterative nature stems from the need to capture the impact of engineering design changes on this critical layer of protection, which often is designed as the last line of defense for overpressure events.

The systematic review of pressure relief and flare system design basis at various engineering design and project execution phases will not only help to ensure adequate overpressure protection, but can also highlight opportunities for inherently safer processes and project cost savings. Often, evaluating trade-offs between process safety design improvements against project economics may be challenging.

In this paper, several case studies are presented to demonstrate how inherently safer design options and cost-effective solutions could be implemented across the various engineering design and project execution phases to minimize gaps in terms of overpressure protection between early and as-built designs.

1 Introduction

Figure 1 depicts the stages of a typical engineering project lifecycle, excluding decommissioning. Each stage has associated activities in terms of selection and refinement of design details which contribute to the overall integrity of the facility and economic feasibility of the project. The general approach involving timely review of pressure relief and flare system design could be applied to both large projects involving major Capital Expenditures (CAPEX) as well as smaller projects with limited scope.

Figure 1: Typical CAPEX Project Life Cycle.

During the early phases of design such as conceptual design and Front End Engineering Design (FEED) phases, the scope, objectives, and requirements are first defined, followed by development of a process basis (i.e., heat and material balance), Process Flow Diagrams (PFDs), and preliminary Piping and Instrumentation Diagrams (P&IDs). Equipment, line and instrument lists are then prepared, and preliminary equipment sizing and specifications for process and utility systems are generated. Design options are selected, and design drawings such as equipment layout plans are completed. A budget is developed and costs are estimated.

During later phases such as Detailed Engineering and Procurement, the remaining design information is added, such as vendor data, drains and vents, and piping details. The actual piping layout is verified and updates are incorporated. In addition, electrical, instrumentation, structural and building requirements to support the project are specified. Often Hazard and Operability (HAZOP) reviews are carried out. Typically procurement of the main equipment and auxiliaries takes place followed by construction, installation, commissioning, and Pre-Startup Safety Reviews (PSSRs), in preparation for start-up, operation and maintenance.

However, what is often overlooked during the early phases is input from pressure relief and effluent handling system design engineers. This paper will describe how early involvement by these engineers could help validate relief system design and reduce overall costs and delays later. The case studies described in Section 3 below offer examples of the potential savings.

Relief systems engineering analysis should be performed during various process design and project implemented phases, which allows identification of potential gaps as well as opportunities for improvement. Typical activities include:

- Overpressure scenario analysis
- Determination of relieving loads
- Selection of relief technology
- Sizing of relief systems
- Evaluation of relief device inlet loss and backpressure effects
- Depressuring analysis
- Hydraulic evaluation of common relief headers
- Flare radiation calculations
- Dispersion analysis, including suitability of atmospheric release
- Acoustic induced vibration (AIV) analysis
- Presenting findings and developing mitigation
- Developing comprehensive documentation
- Reviewing impacts of design changes and updating documentation to capture changes are they are implemented.

It is recommended that the reviewer(s) be well-versed in the appropriate industry standards, e.g., American Petroleum Institute (API) Standards 520 Parts I and II, 521, and 2000 [1][2] [3][4]; American Society of Mechanical Engineers Boiler and Pressure Vessel Code (ASME BPVC) Sections I and VIII [5][6]; as well as corporate guidelines and site-specific requirements. In addition, it is recommended that the reviewer establish and utilize a consistent approach for review of all designs, irrespective of the Engineering, Procurement, and Construction (EPC) companies working on the project.

2 Benefits of Early Review of Relief Systems Requirements and Design

2.1 Early Involvement Allows Critical Input

Early involvement of relief system engineers can allow critical input to be incorporated which influences selected design options from an overpressure protection point of view, thus promoting an inherently safer design. By careful consideration of a number of factors, the design team has the potential to reduce overpressure relief and flaring requirements which in turn can reduce sizing of relief devices and associated piping, as well as flare components such as headers, knockout drums, flare stack, and flare tip. Potential reduction of environmental emissions is also realized. The examples below demonstrate some ways inherently safer design options and cost-effective solutions were implemented to minimize gaps in terms of overpressure protection between early and as-built design. In addition, early review can help to reduce or minimize the need for later design changes or mitigating measures, which could add significant costs and delay completion schedule for the project.

2.2 Identify Opportunities to Increase Inherent Safety

2.2.1 Centrifugal Pump Design

Consider a centrifugal pump as a pressure source for downstream equipment, as in Figure 2. As it stands, P-1 is capable of overpressuring the downstream vessel, V-1, in the event of a blocked outlet or an inlet control valve failure.

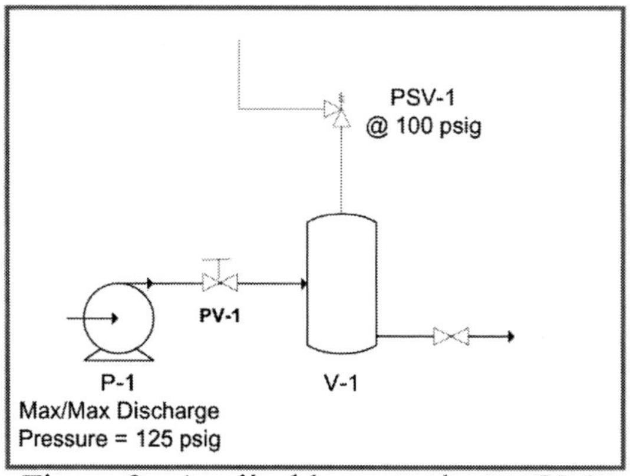

Figure 2: Applicable scenarios on V-1.

By properly sizing this pump during the design phase, however, it may be possible to minimize the relief rates for blocked outlet and control valve failure-open overpressure scenarios of the downstream equipment or even to eliminate the applicability of such scenarios altogether. Working with the process and mechanical engineers as well as rotating equipment designers, the maximum pump impeller size or speed and/ or maximum pump suction pressure could be carefully specified within process operation limits, which in turn limits the maximum pump discharge pressure. By reducing the maximum discharge pressure of the pump to a value less than the downstream equipment Maximum Allowable Working Pressure (MAWP), the above-noted scenarios may not be applicable, as in Figure 3a. An alternative to this, as shown in Figure 3b, is to ensure that the downstream equipment MAWP would be higher than the maximum discharge pressure of the upstream pump. These engineering design choices, which would result in minimizing relief loads or eliminating applicable scenarios, helps make the process inherently safer.

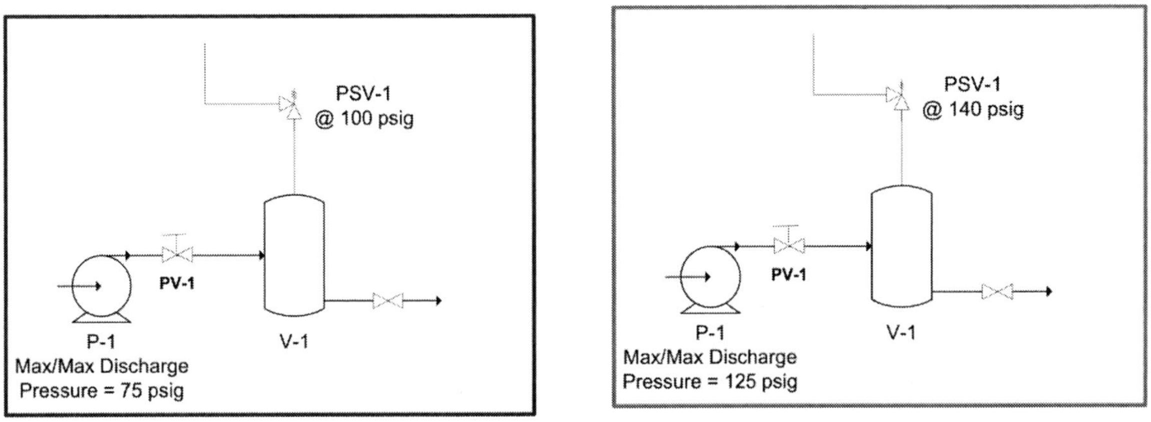

Figure 3: a) Limiting pump discharge pressure. b) Increasing vessel design pressure.

2.2.2 Pump and Compressor Discharge Check Valve Design

A related example involving pumps (and compressors, for that matter) is to ensure appropriate discharge check valve(s) are allocated to mitigate potential for reverse flow, impacting upstream equipment. In some instances, such as in Figure 4, feed could continue into an upstream vessel such as a feed surge drum if the liquid outlet pump is stopped with a discharge check valve stuck open. If this relief scenario is not identified by the relief designer, the relief valve on the feed surge drum may not provide adequate relief capacity.

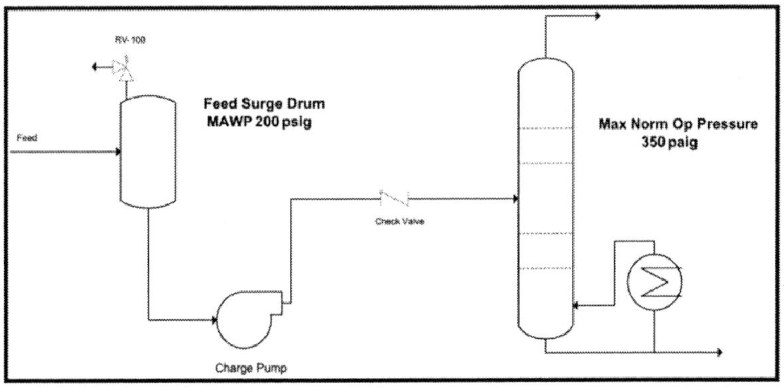

Figure 4: Applicable scenario on Feed Surge Drum

Allocating dual check valves in series of similar (or diverse) design may be required to minimize the backflow rate in the event of a check valve failure; if they meet the requirements set forth in API STD 521, the backflow rate may be limited to check valve leakage instead of check valve failure [7]. Careful specification of the MAWP of the upstream vessel may also help mitigate the overall relief requirement. Consideration of such issues early on is critical to avoid potentially costly design changes later as well as to increase the safety of the process; however, this is commonly missed due to factors such as improper management of change or poor communication between different groups designing upstream and downstream systems.

In a similar fashion, an early design review of several capital projects was performed, focusing on inherently safer design. One recommendation was proposed and accepted to raise the design pressure of the feed surge drum to mitigate the scenario in which a shutdown of a downstream charge pump results in reverse flow to the surge drum, while normal feed to the surge drum continues. By doing so, the feed and reverse flow sources would no longer be capable to overpressure the drum. Early review made the design inherently safer as the scenario was no longer applicable. Without this recommendation, installation of either a safety-instrumented system (SIS) or three 8T10 relief devices would have been necessary.

2.2.3 Control and Bypass Valve Design

Early input is critical in the selection and sizing of a control valve and its bypass valve, since the relief designer must account for the scenario in which *both* the control valve and its bypass valve fail open, as discussed in API STD 521 on "Pressure-relieving and Depressuring Systems". Early engagement of a pressure relief systems expert will help optimize the control valve/bypass and pressure relief requirements to provide adequate overpressure protection within the process operation envelope and requirements, while maintaining industry design and safety best practices. In one instance, a 2" gate valve was specified as the bypass for a 1" control valve, as shown in Figure 5. During a later design stage, it was discovered that this installation resulted in a very undersized relief device for the scenario described above, and thus the need to replace the bypass valve with a different and smaller-capacity type of valve, such as a globe valve. As in this instance, if the nominal size of the installed bypass valve is bigger than the control valve size, it necessitates careful review and possible re-design.

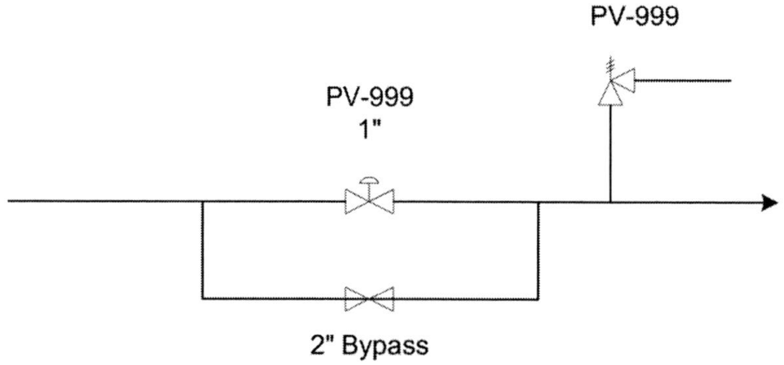

Figure 5: Bypass valve initially specified results in undersized relief device

2.2.4 Static Head

Static head is often overlooked in the preliminary design phases as equipment layouts may not yet be available. Relief designers should encourage vessel designers to specify vessels to be designed to be liquid full at the requested MAWP. Alternatively, relief device set pressures can be reduced to compensate for static head effects provided that the vessels can be filled above the top weld seam (into the top head). In another instance, the effects of static head on the set pressure of multiple relief valves had not been accounted for by an EPC company during detailed engineering. The potential for the internal pressure of affected equipment to exceed MAWP plus allowable accumulation per code was identified, allowing the relief valve set pressures to be adjusted prior to field installation.

2.2.5 Detailed Design Reviews

An independent pressure relief design audit was conducted for a large operating company. Relief design analysis and calculations had been performed by an EPC company as part of an expansion project. A review was conducted during the detailed engineering phase, during which it was found that most of the relief valve and associated piping upgrades which were being proposed by the EPC company were not required. This review included one application for which the EPC company had recommended upgrades to 6 relief valves (including recommendation to install two additional 8T10 relief valves), based on a recirculation flow rate which had been incorrectly included in a blocked outlet scenario relief rate estimate. Upon review, it was concluded that the existing relief valves were adequate and did not need to be upgraded. This reduced the potential for oversized relief capacity.

2.2.6 Design Gaps

In another example of early design review, engineers were able to identify a proposed design gap which would have resulted in costly and unnecessary modifications to a flare system. The proposed design project involved tie-ins of multiple large atmospheric relief valves to the associated flare for a process unit, resulting in inadequate flare capacity and the necessity of making expensive upgrades to this flare system. By performing detailed flare hydraulic modeling, relief engineers were able to show that re-routing the additional relief loads to an alternate adjacent flare having some spare capacity would be more practical.

2.2.7 Design of Feed Surge Drum

During a design review of a feed surge drum, additional examples of recommendations were developed to further increase the inherent safety, such as:

- By limiting the maximum discharge pressure of a feed pump to the proposed design pressure of the downstream equipment, a blocked outlet overpressure scenario was no longer credible for that equipment.
- Rerating the feed/bottom exchangers to account for static head in the event of overfilling in the associated distillation column.
- Car-sealing open (CSO) appropriate valves to ensure that various overpressure scenarios would no longer be considered as design contingencies and relief paths would be available for impacted equipment.

2.2.8 Car-Sealing Strategic Valves

In a final example, several recommendations were developed and adopted prior to start-up for a Coker Vapor Recovery Unit (VRU) revamp project. These recommendations include the following:

- Re-rating equipment
- Car-sealing closed (CSC) the control valve bypass valves
- Reducing the control valve flow coefficient (C_v) to limit maximum flow through the valve in the event of failure open
- Increasing the vessel design pressure to account for static head
- Adding dual check valves to mitigate reverse flow

These changes helped make the overall design inherently safer, so fewer safeguards were required to reduce the risks associated with overpressure scenarios for multiple systems.

2.3 Relief Systems Engineer's Role

Numerous other examples of inadequate design issues as well as missed opportunities to increase the inherent safety of the design are routinely identified by the relief systems engineers after the design has already progressed to a late stage or even after field implementation. Design review can begin early in the process, when preliminary P&IDs are issued for HAZOP or other types of Process Hazard Analysis (PHA). The relief systems engineer should also validate the design once piping isometrics are issued for approval and construction or after other design changes are proposed later. Taking these steps can help to ensure that design shortfalls can be resolved at the earliest opportunity. Making design changes early on is easier and less expensive than during later design stages and even more so than after commissioning and startup, thereby preventing subsequent shutdown, delay or major re-design work.

3 Critical Factors for Success

All pressure relief and effluent handling design documentation and supporting data should be stored in a central repository, for which access can easily be provided to all who have a need for such information. This access can be controlled, so that most viewers will have "read-only" access, and only authorized personnel can make required changes associated with Management of Change (MOC) items, thus keeping overall content up-to-date (evergreen) throughout the life of each facility as MOC activities, major expansion projects, or process modifications are implemented. These activities should trigger design review by relief systems engineers to ensure that relief design continues to be designed to provide adequate relief for all credible overpressure scenarios and that associated documentation is kept up-to-date.

In order for the review process to achieve the desired results, the three parties which are commonly involved (relief systems team, operating company, and EPC company) should work together to understand the goals of each party and how the design inputs from each party should be integrated in the various stages of engineering design and project execution. The following factors have been identified as critical for success:

- Conducting independent relief system design review at the right time in the project scope

- Communication among all parties
- Sharing of critical design data
- Effective management of change
- Discussing and resolving identified issues promptly
- Integrating relief system design review team with project design and execution teams

4 Conclusion

As shown in several examples in this paper, the systematic review of pressure relief and flare system calculations at various process design phases will not only help to ensure adequate overpressure protection, but can also highlight opportunities for inherently safer processes and promote overall project lifecycle cost savings. In addition, such a review process can easily be extended to resolve similar design issues commonly addressed during HAZOP reviews and integrated with the MOC workflow.

5 References

[1] American Petroleum Institute, *API Standard* 520, *Sizing, Selection, and Installation of Pressure-Relieving Devices, Part I-Sizing and Selection*, 9th Edition, 2014.

[2] American Petroleum Institute, *API Standard* 520, *Sizing, Selection, and Installation of Pressure-Relieving Devices, Part II-Installation*, 6th Edition, 2015.

[3] American Petroleum Institute, *API Standard 521, Pressure-relieving and Depressuring Systems,* 6th Edition, 2014.

[4] American Petroleum Institute, *API Standard 2000, Venting Atmospheric and Low-pressure Tanks*, 7th Edition, 2014.

[5] American Society of Mechanical Engineers, *ASME BPVC Sec I, Rules for Construction of Power Boilers*, 2015 Edition.

[6] American Society of Mechanical Engineers, *ASME BPVC Sec VIII-1, Rules for Construction of Pressure Vessels*, 2015 Edition.

[7] API STD 521, §4.4.9.3

Disclaimer:

exida®

Three Reasons Major Event Risk Management Fails on Capital Projects

Denise Chastain-Knight

P.E., CFSE

exida®

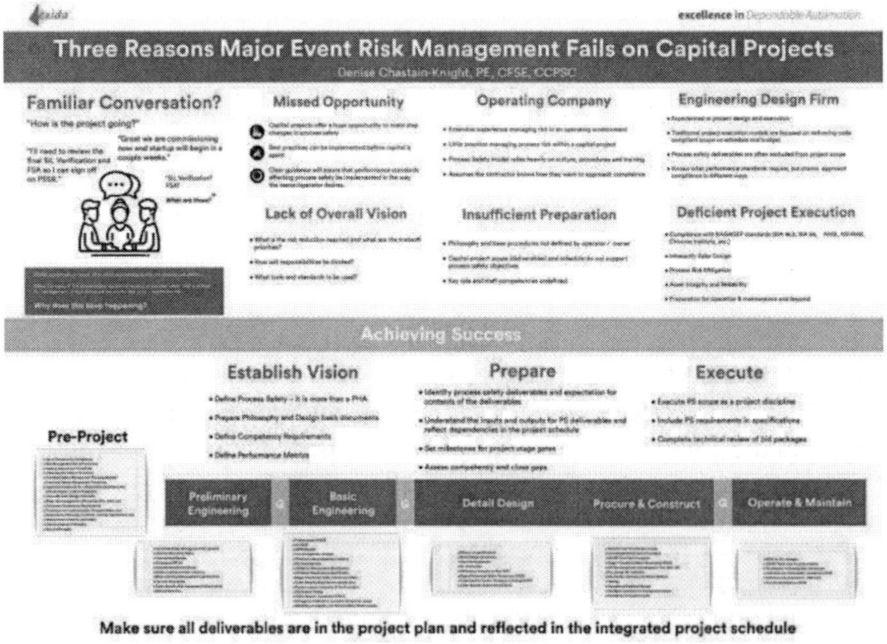

 2

exida

About the Presenter

- **Denise Chastain-Knight, PE, CFSE, CCPSP**
  - BS ChE, Georgia Institute of Technology
  - Senior Functional Safety Engineer
  - More than 31 years experience in engineering design, process improvements, process risk analysis and safety instrumented system implementation
  - Experienced in the chemicals, resins, coatings, monomers, polymers, pharmaceutical, silicate, bio-fuels, petroleum, power…
- **Affiliations**
  - CCPS Technical Steering Committee
  - Past Chair NCEES PE (ChE) exam development committee
  - Fellow AIChE

 3

exida

Familiar Conversation?

How is the project going?

Great we are commissioning now and startup will begin in a couple weeks

I'll need to review the final SIL Verification and FSA so I can sign off on PSSR

SIL Verification? FSA? What are those?

 4

exida

Familiar Conversation?

Similar conversations occur all to often as capital projects are nearing completion.

Often the topic of the conversation involves an aspect of process safety that is critical for management of high consequence events that pose a significant risk.

Why does this keep happening?

 5

exida

Missed Opportunity

- Capital projects offer a huge opportunity to make step changes in process safety
- Best practices can be implemented before capital is spent
- Clear guidance will assure that performance standards effecting process safety be implemented in the way the owner/operator desires.

 6

Operating Company

- Extensive experience managing risk in an operating environment
- Little practice managing process risk within a capital project
- Process Safety model relies heavily on culture, procedures and training
- Assumes the contractor knows how they want to approach complaince

Engineering Design Firm

- Experienced at project design and execution
- Traditional project execution models are focused on delivering code compliant scope on schedule and budget
- Process safety deliverables are often excluded from project scope
- Knows what performance standards require, but clients approach compliance in different ways

exida

PS Activities that Usually Happen

- Hazards Identification and Risk Analysis
- Relief system design
- Compliance with code mandated standards (ASME VIII, NFPA 30, NFPA 70)

 9

exida

PS Activities that Often Fail

- Compliance with RAGAGEP standards (ISA 18.2, ISA 84, ANSI, ASHRAE, Chlorine Institute)
- Inherently Safer Design
- Process Risk Mitigation
- Asset Integrity and Reliability
- Preparation for operation & maintenance and beyond

 10

Why We Fail

- No or inadequate overall vision
- Insufficient preparation
- Deficient project execution

Communication

 11

Lack of Overall Vision

- What is the risk reduction required and what are the tradeoff priorities?
- How will responsibilities be divided?
- What tools and standards to be used?

 12

Insufficient Preparation

- Philosophy and base procedures not defined by operator owner
- Capital project scope (deliverables) and schedule do not support process safety objectives
- Key role and staff competencies undefined

 13

Deficient Project Execution

- Underestimate of the workload required and dependencies with other scope objectives
- Process Safety /Risk Management not treated as a discipline
- Reactive not proactive

 14

How Project work

- Stage gate process
- Scope identified, schedule and budget developed to support scope
- Detail built in layers. Early definition important to minimize rework
- Often involve many contractor teams working on a subset of the scope

Establish Vision

- Define Process Safety – it is more than a PHA
- Prepare Philosophy and Design basis documents
- Define Competency Requirements
- Define Performance Metrics

exida

Prepare

- Identify process safety deliverables and expectation for contents of the deliverables
- Set milestones for project stage gates
- Understand the inputs and outputs for PS deliverables and reflect dependencies in the project schedule
- Assess competency and close gaps

exida

Execute

- Execute PS scope as a project discipline
- Include PS requirements in specifications
- Complete technical review of bid packages

Pre-Project Documents

- List of Standards for Compliance
- Risk Management Plan & Procedures
- Hazards Assessment Procedures
- Cybersecurity Policy & Procedure
- Functional Safety Management Plan (organization)
- Functional Safety Management Procedures
- Approved vendors list for critical/safety equipment and instrumentation
- Alarm Philosophy
- Inherently Safer Design Philosophy
- Major risk management philosophies (fire, relief, etc)
- Contractor Competency Requirements
- Procedures for required studies (dropped object, etc)
- Maintenance Philosophy (isolation, clearing requirements etc)
- Measurement & Metrics philosophy
- Communications Philosophy
- Spares philosophy

Preliminary Engineering

- Functional Safety Management Plan (project)
- Chemical Reactivity Matrix
- Hazards identification
- Conceptual SIF list
- Hazards Identification Study
- Define credible accident scenarios
- Major and long lead equipment specifications
- Controls Description
- Cyber Security Risk Assessment & Gap Analysis
- Maintenance Plan

exida

Basic Engineering

- Project phase HAZOP
- CHAZOP
- HIPPS Studies
- Fire and Explosion Analysis
- Chemical release dispersion modeling
- SIL Determination
- SIS Safety Requirements Specification
- SIF Safety Requirements Specification
- Stage 1 Functional Safety Assessment (FSA1)
- Cyber Security Requirements Specification
- Countermeasure Inspection & Test Procedure
- Penetration Testing
- Cyber Security Assessment (CSA-1)
- Emergency notification, evacuation & response review
- Reliability, Availability, and Maintainability (RAM) analysis

 21

exida

Detailed Design

- Balance of specifications
- Detail Design documents
- Alarm Rationalization
- SIL Verification
- SIS Factory Acceptance Test (FAT)
- Stage 2 Functional Safety Assessment (FSA2)
- Cybersecurity Factory Acceptance Testing (CFAT)
- Cyber Security Assessment (CSA-1)

 22

Procure and Construct

- Technical and financial bid reviews
- Operating & Maintenance Procedures
- SIS/SIF Proof Test Procedures
- Stage 3 Functional Safety Assessment (FSA3)
- SIS Site Acceptance and Integration Test (SAT, SIT)
- Pre-startup SIF Validation
- Pre-startup Cybersecurity Review (CSA 3)
- Training
- Operational Readiness Review
- Configure maintenance management system
- Emergency management plan

 23

Operate & Maintain

- MOC for ALL changes
- SIS/SIF Proof tests & results analysis
- Revalidation of Detailed Risk Assessment
- Cybersecurity Vulnerability Assessment (CVA)
- Cybersecurity Assessment - CSA 4 & 5
- Periodic Revalidation of PHA

 24

25

A Comparative Analysis of Safety Rating System

Linlin Wang[a]

[a]Mechanical and electrical engineering college, China University of Petroleum (East China)

Abstract: With the development of petrochemical enterprises, the scale and petrochemical enterprises frequent accidents, the high risk of petrochemical enterprises has been more and more extensive attention. It's particularly important to establish a sound safety rating system for the timely detection and elimination of risk factors, reduce the probability of accidents appears. However, in China there is less research and application on safety rating system. This paper summarizes the similarities and differences between the various systems and puts forward the development of China's safety rating system.

Keyword: Safety Rating System; Petrochemical enterprises

1 Introduction

Petrochemical enterprises have a lot of flammable and explosive, toxic substances, and the installation of large-scale and continuous production, can easily cause major accidents. In 2013 Sinopec Huangdao Oil pipeline explosion, 62 people were killed, 136 people were injured. With the occurrence of petrochemical enterprises accidents, the safety production of petrochemical enterprises has become one of the issues that government and national focus on. Evaluation and rating of the status quo of enterprise security can promote the enterprises to find their own problems, develop improvement goals and implement continuous improvement. At the same time, it can also improve the government safety supervision departments to find out the status quo of enterprises, determine the key supervision enterprises, and optimize the allocation of resources, enhance the safe production level of enterprises in China, it is also an important means to improve business efficiency. The Chinese government's security supervision department attaches great importance to the assessment of enterprise safety, since 2006, has issued a "Basic norms of enterprise safety production standards", " General specification for safety standardization of hazardous chemicals", "Guidelines for oil industry safety production standardization "etc., The government promotes the implementation of the safety production standardization assessment throughout the country, take the "enterprise self-evaluation, government review" approach to the enterprise rating, and achieved important work results. But in the implementation a lot of problems were also found. Based on the status quo of safety rating of petrochemical enterprises in China, this paper analyzes and compares the implementation process of enterprise safety rating system such as ISRS (International Safety Rating System) ,NOSA five-star comprehensive evaluation system and VPP project implemented by OSHA. It is expected to provide support for the selection and development of our country's petrochemical enterprise safety rating tool.

2 Status Quo of Safety Rating of Petrochemical Enterprises in China

Safety standardization rating is developed, implemented and reviewed by China's relevant laws and regulations, industry standards. It is led by the State Administration of Safety Supervision and implemented on a national scale. It is the most extensive and highest commitment safety rating system in China. In addition, some petrochemical enterprises received the support of government, and independently introduced tools such as ISRS and NOSA. For example, 2014 China Petroleum West Pipeline Company reached the rate 7 of ISRS, and became first enterprise which reached this level in China.

Based on the safety standardization assessment of petrochemical enterprises in China, it is found that there are three main problems in the process of implementing safety standardization assessment in our country:

(1) Safety standardization system and some other implementation systems, such as HSE system, have incompatible problems, resulting in enterprises to repeat engaging in several sets of systems, repeat records, repeat systems. Companies not only feel cumbersome, but also waste human and material resources;

(2) Enterprises have low motivation and a bad attitude to practice safety standardization. Standardization of safety production is a measure imposed by the government to improve the safety level of the enterprise, and enterprises do not really realize its importance. So that the work is easy to become formalize and surface, and did not achieve the purpose of improving the level of safety, but only to let the enterprises to meet the rating and inspection;

(3) China's safety standardization assessment consists of three grades. The grade of the score range is quite large. And the flexibility of the assessment is not sufficient. Besides, in the absence of experimental circumstances, the assessment has a low recognition for some enterprises which take a new process or a new management approach.

3 System Introduction

3.1 ISRS（International Safety Rating System）

ISRS（International Safety Rating System）is a rating system of international security measure, and improving and showing the level of health, safety, environmental and business performance and other aspects of the organization, designed to help organizations manage risk and promote continuous improvement of the evaluation system. Det Norske Veritas (DNV) international safety rating system (ISRS) eighth edition published in 2009 after decades of practicing and improving.

ISRS [7] international safety rating system has 15 elements, each element is divided into several sub-elements of a total of 119 sub-elements, and list the number of points for each sub-element of the assessment, the total number is 661. 15 elements are: leadership, planning and administration, risk assessment, human resources, compliance assurance, project management, training and skills, communication and marketing, risk control, asset management, procurement and contractor management, emergency preparedness , the event learning, risk monitoring, and review the results, in which risk management (including 3 risk assessment,9 risk control, 14 risk monitoring) system for the trunk. During the assessment process, the customer selects the subroutine according to the grade to be evaluated, as follows::

Grade	1	2	3	4	5	6	7	8	9	10
Level range	1-4				5-6		7-8		9-10	
Required number of subroutines	30				60		90		119	
The number of customer chooses	20				15		10		0	
The total number of subroutines	50				75		100		119	
The lowest score of each program%	10	15	20	25	30	35	40	50	60	70
Average minimum score%	20	30	40	60	60	70	70	80	80	90
Live tour lowest score %	65	65	65	65	70	75	80	85	90	95

3.2 NOSA integrated five-star management system

In 1951, the International Occupational Safety Association (NOSA) created a five-star safety management system by conducting a comprehensive hazardous and harmful factors identification and risk assessment to identify all possible hazard identification and develop appropriate management for reduce hazard and risk. NOSA management system regards hazard identification and risk management as the core, Heinrich's "iceberg theory" as the basis and PDCA manner as its operation mode. System consists of five main elements, 72 elements, more than 300 minimum standards, the implementation of 1200 rules and standard instructions (against NOSA Integrated Five Star system revision and reorganization, the system consists of 75 elements). System evaluation content mainly in the following two aspects:

(1) Performance evaluation

Evaluation content of NOSA five-star rating system mainly consists of five areas, particularly in the following table:

NOSA integrated five-star rating system

project	Unit content	The number of elements	score	Share of the share
1	Building and housekeeping (factory) management	11	450	15
2	Mechanical electrical and personal protection	17	600	20
3	Fire and other emergency risk	8	400	13.3

	management			
4	HSE accident recording and investigation	5	400	13.3
5	HSE management organization	31	1150	38.4
total		72	3000	100

Through five major categories of 72 elements of qualitative evaluation, the project almost involved all activities of production and operation. It evaluates each element from the "four levels, three aspects, five grades" to carry out a comprehensive calculation to measure whether the risk of the element has been effectively controlled. "Four levels" means: risk awareness, the file system, compliance, implementation effect. "Three areas" means: safety, health, environment. "Five grades" means: 0, 25, 50, 75, 100 five scoring levels。

（2） Experience assessment

Introducing DIIR (Disabling Injury Incidence Rate) rate of accidents concept of enterprise, and performance assessment combined score, drawn 1 to 5-star level. DIIR shows annual percentage of the total number of employees were injured, according to NOSA claims, statistics should include employees of the enterprise, the tertiary industry and outsourcing construction team personnel. DIIR calculated as follows：

$$\text{DIIR} = \frac{\textit{The number of accidents caused by work} \times 200000}{\textit{Number of employees working hours per year}}$$

Where：DIIR——Disabling injury incidence rate；

The number of accidents caused by work - refers to the work caused by or caused by the injury, so that the injured cannot or cannot fully perform the daily designated job work up to one or more working days；

200000-- base by 100 people work for a year (50 weeks), 40 hours a week calculated；

Number of employees working hours per year - Total number of employees × Actual hours per person per year

The final combination of the above two aspects of the results given the final star rating results, the corresponding assessment of the star method as shown in the following table：

NOSA integrated five - star rating system assessment standard table

Star rating	**score（%）**	**DIIR（%）**
NOSCAR	≥95	≤0.8
Five star	≥91	≤1.0
Four star	≥75	≤2.0
Three star	≥61	≤3.0
Two star	≥51	≤4.0
One star	≥40	≤5.0

NOTE: DIIR = the number of industrial accidents × 200000 (base) ÷ annual working hours of employees

3.3 VPP（The Voluntary Protection Programs）

The Voluntary Protection Programs (VPP) is a safety and health management system to prevent workplace accidents and death, by allowing enterprises to implement additional higher than the US Occupational Safety and Health Administration (OSHA) Voluntary standards measures to achieve excellent safety and health management goals. OSHA is through a detailed site assessment and comprehensive site accident rate / illness rate to assess whether the corporate site has a good protection system。

The major elements of the program: management leadership and employee involvement; worksite analysis; hazard prevention and control and safety and health training. To assess something like this, there are three main forms of evaluation methods, including the inspection of documents, on-site inspection and interviews。

Leadership and employee participation	Written safety and health management system, management commitment and leadership, Authority and Line Accountability (authority and responsibility), contractor employees, employees' participation
Workplace analysis	Baseline Hazard Analysis baseline risk analysis, hazard analysis regular activities, regular inspection, hazard reporting systems, accident / incident investigation, trend analysis
Hazard Prevention and Control	Whether there are effective measures to combat control and control, as well as written contingency plans safety and health training

In addition, team members also need to The Total Case Incidence Rate (TCIR)以及The Days Away from Work, Restricted Activity and Job Transfer (DART) Case Incidence Rate evaluated to confirm the company level must be below the national average. After a series of reviews, OSHA decides that the applicant's qualifications are Star, Merit, Demonstration program, if the applicant fails to meet the requirements, return the application, and VPP level is relatively flexible, for those who do not meet the Star designing 1-Year Conditional Star Participants and Star Demonstration Programs.

3.4 Safety standardization assessment

Safety production standardization refers to the establishment of safety production responsibility system, the development of safety management systems and operating procedures, troubleshooting and monitoring of major hazards, the establishment of preventive mechanisms to regulate production behavior, so that the production links in line with relevant safety laws and regulations and standards Standard requirements, people, machines, objects, ring in a good state of production, and continuous improvement, and constantly strengthen the standardization of enterprise safety construction.

Safety standardization evaluation mainly taken independent assessment, external review of the way, self-assessed score of 60 points or more companies can apply for formal standardization rating, the final assessment is divided into one, two, three, one is the highest, from the security Production supervision and management departments to review the classification of supervision and management. During the review process, the following aspects are mainly reviewed: objectives, organization and responsibilities, safety production, laws and regulations and safety management system, education and training, production equipment and facilities, operation safety, hidden trouble investigation and management, major hazard monitoring , Occupational health, emergency rescue, accident learning, performance assessment and continuous improvement of the thirteen major aspects of compliance assessment, the final score to score how much. The enterprises that need to be graded need to carry out self-evaluation on a yearly basis to determine the safety standardization status of the enterprises and to re-order them every three years to urge enterprises to continuously improve the standardization work level.

4 System comparison results

Through the understanding of the information, compared the above systems, found the following the same and different:

1. Purpose

In essence, the above mentioned rating methods are all aimed at checking and enhancing the enterprise to achieve a certain level of security. However, the basic purpose of the rating system is different because of the different backgrounds of the development period. For example, the NOSA five-star rating system was developed in the 20th century, the rapid development of industrial and mining in South Africa and the quality of personnel and other issues caused the high rate of injury accidents, so its basic purpose is to reduce the high incidence of work accidents. In the final analysis, the rating is only a means and security is the purpose.

2. Applicable scope

ISRS is available in 11 languages and has been implemented on thousands of sites worldwide. The industries involved include marine, oil and gas, energy, nuclear, etc., and ISRS is introduced into the oil industry in China, such as China Petroleum West Pipeline Company reached the international security rating of 7.

NOSA five-star rating also can evaluate several kinds of industries including mining, forestry, railways, health care and natural gas, but its vast application in China is applied to the mining industry.

The VPP assessment program is applicable to all companies that use OSHA standards. VPP's participation companies include construction, agriculture, marine, oil and gas, health care and other industries.

Since the implementation of the Safety Production Standardization, under the guidance of the government, Safety Production Standardization has been applied in most enterprises in China, including industry and trade, power industry, transportation, hazardous industries, non-coal mines, construction industry, fireworks, coal industry, Its application covers most of the production enterprises.

3. Evaluators

Safety Production Standardization rating system take the way of "enterprise independent assessment then external assessment " to carry out the rating, Safety Supervision and Management Departments supervises the classification. The external review is conducted by the relevant experts. Similarly, the NOSA five-star rating system is evaluated by the company's team of experts. In the composition of the assessment team, VPP projects and ISRS systems have made innovations.

The VPP site assessment team is composed of experts from all areas of OSHA and SGE (The Special Government Employee), which has performed on-site evaluation. The implementation of the SGE project is intended for enterprises employees to participate in the process of VPP's on-site assessment. The implementation of this innovative project is more conducive to government and corporate information and ideas sharing.

ISRS needs to form an evaluation team consisting of two or more persons prior to evaluation, and to ensure the depth of the HSEQ field in evaluation, the team members contain relevant experts and do not restrict internal / external members, and All evaluation members will receive DNV training courses.

4. Rating method

Whether it is ISRS system, NOSA, VPP or safety standardization rating system are all required to compare the relevant standards and scoring rules with the actual situation of enterprises, through the document inspection, on-site visits and consultation to decide the company's compliance to standards.

However, NOSA and VPP projects need to combine other indicators before the final results. VPP not only requires companies to meet or exceed OSHA standards, but also requires companies to maintain below the average rate of disease released by National Bureau of Labor Statistics. And NOSA combines the evaluation rate of qualified with DIIR ratio to determine the stars should be given.

5. Follow up later

The "PDCA" model has been adequately applied in several rating systems, "plan, implement, check, action", adopt the infinite loop method of " find problems then solve problems, and then find the problem, and then solve the problem" to promote the continuous improvement of enterprise safety management system and continuously improve the level of enterprise security.

But for the rating implementation unit, the follow-up degree is different from others. Among them, the VPP project's follow-up work is very good. VPP participants need to submit an annual self-assessment report, OSHA's site assessment team will confirm whether the enterprise has made a corresponding improvement, and re-examination time interval varies by the enterprise level, such as star business, every 3-5 years conduct a reassessment. For companies with a one-year correction, if they do not meet the requirements after one year, they will withdraw the VPP application.

System comparison table

	ISRS	NOSA	VPP	Safety Standardization
Basic Purpose	get management	Reduce the	prevent	Promote

	system to work more effectively	incidence of industrial accidents	workplace accidents and death	enterprises to achieve standardization and standardization
Applicable scope	ISRS is available in 11 languages. The industries involved include marine, oil and gas, energy, nuclear, etc.,	Including mining, forestry, railways, health care and natural gas, but its vast application in China is applied to the mining industry	all companies that use OSHA standards	Covers most of the production enterprises, such as industry and trade, power industry, transportation, hazardous industries, non-coal mines, construction industry, fireworks, coal industry
Evaluators	Two or more people (DNV training)	Company experts team	OSHA experts in various fields and SGEs who have conducted on-site evaluations	Independent assessment, external review The external review is conducted by the relevant experts
Rating method	Performance evaluation	Performance evaluation and DIIR	companies to meet or exceed OSHA standards, low disease rate	Performance evaluation
Assessment Frequency		Self-evaluati on Twice a year Re-rating depend on clients	Self-evaluation once a year re-examination time interval varies by the enterprise level	Self-evaluation once a year Every three years have re-examination
Grades	10	6	Have Merit、Star、Demonstratio n program etc., grade flexibility is high	3
Voluntary	voluntary	voluntary	voluntary	Forced

5 Conclusion

Combine comparison and thinking above, for the Chinese petrochemical enterprise safety rating system selection, provide the following suggestions:

1. For small private enterprises, Safety Standardization has been fully compatible with security requirements, for other state-owned enterprises, large-scale Sino-foreign joint ventures can choose to introduce foreign security rating system.
2. ISRS system covers the ISO standard and OSHA standard requirements, and also developed a special version of ISRS-Pipe, the rating system also covers the majority of indicators in China oil safety management system, ISRS has been applied in a lot of petrochemical enterprises in China But the ISRS system uses the lowest score of the total score in the first-level elements as the criterion for determining the final grade, so it is not easy to reflect the advantages of the enterprise in the final level judgment.
3. NOSA five-star rating system considers the performance evaluation and DIIR value together in the final evaluation to get the overall level, but when the enterprise finally decided to use the NOSA system, NOSA was originally used for mining assessment should be considered; For the company implementing OSHA standards can apply for the VPP program because VPP determines the enterprise level by checking whether the company is performing a higher-than-OSHA standard.

Acknowledgement

This work is supported by the National Natural Science Foundation of China (21406271) and the China Scholarship Council "Innovative Talents International Cooperation Program" (Liu Jin Mei [2016] 7988). Thank you very much.

Reference

1. https://www.dnvgl.com/services/isrs-for-the-health-of-your-business-2458
2. https://www.nosa.co.za/

478942 The Difficulty and Prospect of the Development of Quantitative Assessment Technology in the Field of Chemical Safety in China

Abstract:

Abstract: With the increasingly high demand for safety of chemical industry quantitative assessment technology has been widely recognized in chemical industry safety field because of its more accurate assessment results. Nowadays many existing quantitative assessment techniques,(such as quantitative risk assessment (QRA), reliability assessment, safety integrity level assessment (SIL) .etc)have been gradually introduced into the filed of China chemical safety to provide more assistance for the risk management of chinese chemical enterprises. However, some basic data of the quantitative assessment on chemical industry in China are lost due to the lack of systematic statistical analysis of data(such as equipment failure data, accident event data .etc), which leads to the situation that our country often adopt foreign related data when conducting a quantitative assessment. In result, results of quantitative assessment are out of their authenticity. In addition, because there is no more specific and more operational assessment guidelines for the part of quantitative assessment techniques issued in our country, it also affects the accuracy and scientific of the results. According to analyzing some misunderstandings and difficulties in chinese chemical safety field when carrying out the quantitative assessment technology, the paper puts forward some suggestions on the change and improvement of related industries to make quantitative assessment technology to provide greater help for the development of chemical safety field in China.

Recipe for a *Sizzling* HAZOP

Poster Presentation to 13th GCPS

Angus Keddie, Process Safety Matters

Benefits of Running an Effective HAZOP

- Commercial: Meeting schedule; Risk optimisation of facility
- Safety: Pick up all significant potential hazards
- Psychological: Limiting stress on participants
- Chairman: High probability that the chairman will be invited back

Recipe Success? – Key Components

- Recipe Disasters……
- Provenance
- Preparation
- Supporting Team
- Kitchen Knives
- Ingredients
- Spice and Flavour
- Oven Conditions
- ….and How to Avoid Them
- Ta Dah!

Recipe Disasters

Recipe Disasters

- Behaviour to avoid if Effective HAZOP to be achieved:
  - Falling asleep
  - Fighting with team members
  - Allowing fighting between team members
  - Silence of participants
  - Permitting interminable discussions on minor issues
  - Scribing yourself
  - Working well in excess of 7 hours per day

HAZOP Chair Provenance

To paraphrase the Bard "Some or born to HAZOPness, some achieve HAZOPness and some have HAZOPness thrust upon them"

There are around 5000 HAZOP chairmen worldwide all of them being equal, but some are more equal than others.

In order to optimise process safety for existing and new facilities worldwide, excellent PHSERs need to be carried out. A key ingredient of this is the performance of the facilitator. What personal characteristics combine to create this performance?

PROCESS SAFETY MATTERS

HAZOP Chair Provenance

- Undergraduate degree in Chemical Engineering
- Professional exposure to relevant aspects of industry:
  - Process Design
  - Operations/Commissioning
  - Team Leadership/Management
  - Process Safety
- Top up with Masters in Process Safety
- Modal age range: 40 to 60
- CPD of IChemE Training Course
- Observation of Other respected HAZOP chairs to evolve best practice
- But like all great recipes, a little variety doesn't do any harm and may enhance the final product.

PROCESS SAFETY MATTERS

HAZOP Chair Characteristics

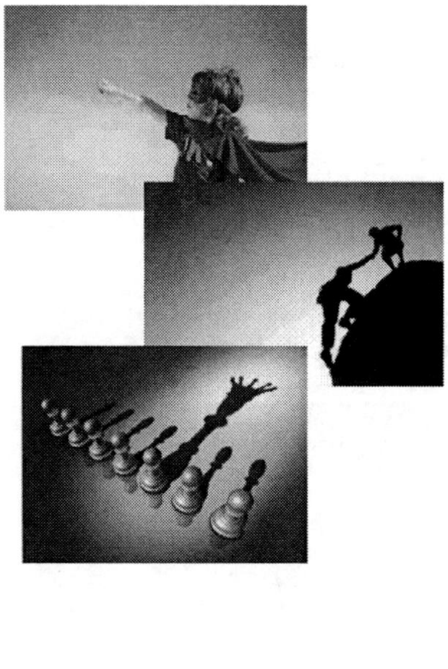

Desirable attributes for the leader include:

- wide experience of all the stages of process hazard studies, including quantitative risk assessment;
- extensive experience as a HAZOP study member and, preferably, some as a HAZOP study scribe;
- training in the leadership of HAZOP studies;
- technical competence and the ability to quickly understand the system and its operation;
- meticulous attention to relevant detail;
- good analytical thinking;
- motivational skills including the encouragement of creativity and open speaking;
- independence from the project itself with no other direct responsibility to the project manager other than completing the scope of the HAZOP study.

Fail to Prepare……

PROCESS SAFETY MATTERS

- Hold a Pre-meeting with key project stakeholders to establish significant draft tenets:
    - Project Background and Scope
    - Stakeholder Drivers
    - Optimal Methodology
    - Key Hazards
    - Putative Assumptions
    - Principal Personalities
- Outcomes and Terms of Reference inform Intro presentation and initial approach

The Scribe - Sous chef or trusty side kick?

PROCESS SAFETY MATTERS

PROCESS SAFETY MATTERS

LinkedIn HAZOP Discussion:

- **Does a PHA scribe need to have a technical background? – Ron Brachten, Nov 2014**
- **AK contribution: scribal attributes -**
  - Fast and accurate typing skills
  - Native or excellent English language skills
  - Sufficient *confidence* to capture the consensus and to overlook the inevitable spelling and grammar gremlins
  - ***Stamina***
  - Good humour
  - Interest in the process
  - Threshold technical knowledge

  Technical knowledge is handy but....

PROCESS SAFETY MATTERS

HAZOP Tools – Sharpen those knives

- Climate controlled room, good natural light
- A1 PIDs on walls or second screen
- Availability of Flip charts or White boards
- Wifi/LAN access to upload project tech info and wider references
- Good food during breaks and lunch
- Effective HAZOP recording software

PROCESS SAFETY MATTERS

Ingredients

- Meat and 2 Veg:
  - Process Description
  - Noded P&IDs
  - Intro Presentatiion
- Side Dishes
  - PFDs
  - Cause and Effect Diagram
  - Report from any previous HAZOP(s), HAZID
  - HAZOP Terms of Reference
  - Facility Design Standards
  - Plot Plan
  - Line List
  - Piping Classification Overview
  - Toxicity/flammability data for materials being handled
  - Relevant National/International Codes/Laws/Standards

PROCESS SAFETY MATTERS

The texture and flavour – the local element!

The experience of HAZOP can be vary significantly depending on technical, cultural and stakeholder perspectives:

Stakeholders

- Operators focus on operability and consistency of design. (gooey and sweet)
- The EPC contractor's focus is on schedule and no changes (hard and dry)
- The operator's safety team focus on strict code compliance (tart and bitter)

Technical

- Upstream – no steady state; little standardisation
- Downstream – Large through puts/inventories; replication

Cultural

- "Many nations separated by a common language – HAZOP!"

PROCESS SAFETY MATTERS

Oven Conditions – HAZOP Ambience

The team works better if relaxed and motivated – try to make this happen early on:

- Introduce yourself to team members
- Get everyone to introduce themselves round the table, including what contribution they expect to make
- Directly engage all team members in the first 30 mins
- Allocate responsibilities based on member skills and attributes
- If team is flagging, call a break
- It two team members become adversarial, take a time out
- Smile

PROCESS SAFETY MATTERS

Recipe Disasters – and their remedies

- Behaviour to avoid if Effective HAZOP to be achieved:
  - Falling asleep – stand up and walk around when facilitating
  - Fighting with team members – develop excellent rapport with team from the off;
  - Allowing fighting between team members – call time out; verbal warnings; firing culprit(s)
  - Silence of participants – engage in meetings (and in breaks).
  - Permitting interminable discussions on minor issues – set time limits for discussion; "is there anything we need to record at this point?"
  - Self Scribing – refuse!
  - Working well in excess of 7 hours per day – planning; negotiate schedule; identify key risks up front; prioritise nodes (Shell approach – 6 working hours per day, 4 days per week, 4 weeks out of 5)

Ta Dah!

PROCESS SAFETY MATTERS

Angus Keddie
Director
Process Safety Matters
+44 (0) 7515897869
+44 (0) 1344 455800
info@processsafetymatters.com

477894 Bridging the Divide - OHS and Process Safety

Abstract:

The occupational health and safety field is well established, with a number of professional bodies, undergraduate programs and a comprehensive body of knowledge. Conversely, the process safety field has evolved largely over the past 40 years as an engineering discipline, with a focus on engineering solutions. Both of these fields utilise a risk management basis for decisions and actions. But both fields still largely operate independently of each other. There is much to be learned from sharing knowledge and responsible collaboration to deliver safer plants and operations from an OHS and a process safety perspective. This paper will explore how we can bridge the divide between OHS and process safety to achieve better safety outcomes. This includes better education and information flow between both professions as well as leveraging off existing systems to advance process safety.

20 Years Of Personal Process Safety Lessons

Steve Streblow
ACS Engineering
10914 Milano Ct.
Richmond, TX 77406
streblowsr@yahoo.com

Prepared for Presentation at
American Institute of Chemical Engineers
2017 Spring Meeting and 13th Global Congress on Process Safety
San Antonio, TX
March 26 – 29, 2017

20 Years of Personal Process Safety Lessons

Steve Streblow
ACS Engineering
10914 Milano Ct.
Richmond, TX 77406
streblowsr@yahoo.com

Keywords:

Amine tower; carbon dioxide; communication and restriction; competency; compliance; culture; documentation; failure analysis; intimidation; investigation; lessons learned; material of construction; mechanical integrity; pressure; stress; risk assessment; vessel repair.

Abstract

This paper presents in detail one of the many process safety issues encountered by the author in his 20 years of industrial experience. This issue deals with an amine tower, but touches upon the following topics: carbon dioxide; communication and restriction; competency; compliance; culture; documentation; failure analysis; intimidation; investigation; lessons learned; material of construction; mechanical integrity; pressure; stress, risk assessment and vessel repair.

It describes better ways to address the situation and applicable RAGAGEP. It is hoped that sharing this information will enable others to learn from the past and prevent similar issues in the future.

1 Introduction

This paper presents in detail one of the many process safety issues encountered by the author in his 20 years of industrial experience. This issue deals with an amine tower, but touches upon the following topics: carbon dioxide; communication and restriction; competency; compliance; culture; documentation; drawings; failure analysis; intimidation; investigation; lessons learned; material of construction; mechanical integrity; pressure; stress, risk assessment and vessel repair.

It describes better ways to address the situation and applicable RAGAGEP. It is hoped that sharing this information will enable others to learn from the past and prevent similar issues in the future.

2 Amine Tower Leak

In one of the operating companies I worked for, I was supposed to be the technical support for many gas processing facilities, which included amine systems for removal of carbon dioxide. I was verbally informed of a small leak in one of the facilities amine regenerator towers which was stopped by placing a screw in the side of the tower, as shown in the photo (taken later) immediately below.

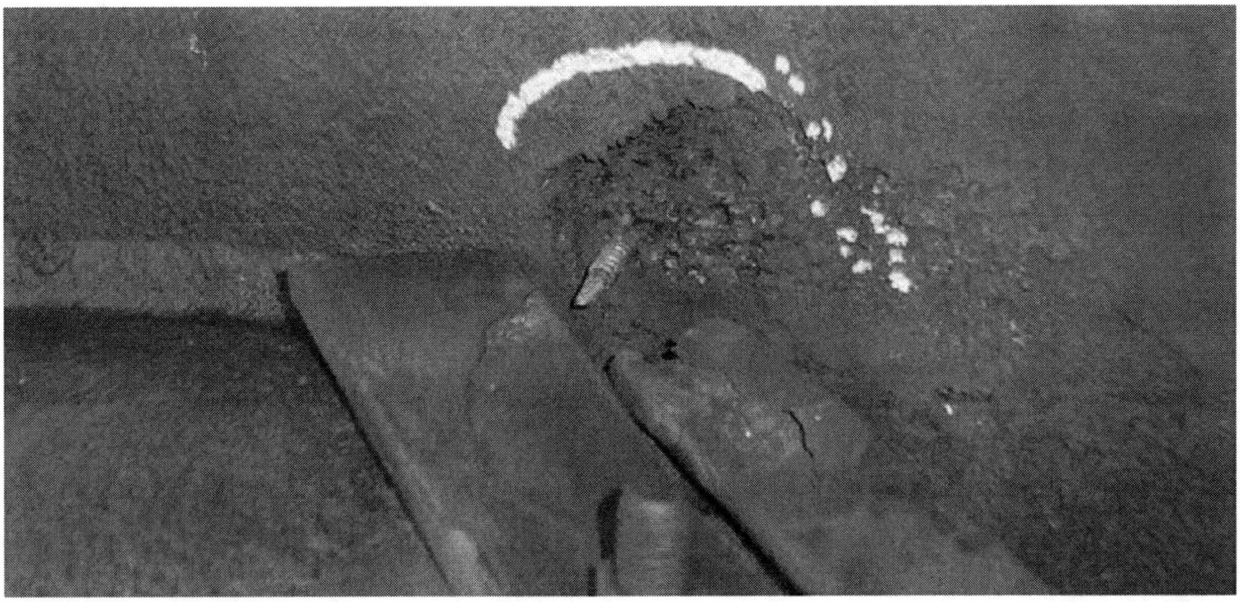

I was shocked for several reasons. First, I was not consulted for my technical opinion on how to proceed. Second, because after I communicated with the plant, they stated they were told by a service provider that they normally use that this type of repair was proper.

So much of this situation seemed wrong to me, so I provided the plant suggestions to restrict access to the area nearby the tower, while I determined how to proceed. The plant did put this area restriction in place.

I started to research what to do. First, I verified the pressure vessel law, as one can do by checking NB-370 at:
https://www.nationalboard.org/PrintAllSynopsis.aspx?Jurisdiction=Select
The location was in the state of Texas, which did not have a pressure vessel law. I next tried to find RAGAGEP for this. I called some pressure vessel manufacturers and had some brief conversations, which helped me focus my search. I found out that to keep the ASME code status, the repair should have been performed through a company which has an R stamp and filled out the appropriate paperwork. The R stamp status of a company can be confirmed at:
http://www.nationalboard.org/ManufacturerDirectory.aspx
The RAGAGEP to be used to maintain ASME code compliance is NBBI NB 23 (NBIC), National Board Inspection Code, Parts 1-3, reference 1.

If the vessel is a non-code vessel or if ASME code status does not need to be maintained, then one can refer to RAGAGEP API 510, Pressure Vessel Inspection Code: In-service Inspection, Rating, Repair, and Alteration, Tenth Edition, May 2014, reference 2.

After seeing that there was much technical knowledge needed for this activity that I did not have, I checked with management to see if there was someone better qualified than I inside the company, and I was disappointed to learn that there was not. Management did not seem concerned about the lack of expertise or the fact that I was being asked to work outside my area of expertise. It would be good to have a dedicated mechanical integrity expert in the company, since there is a great breadth and depth of technical knowledge needed.

I then tried to get the results of the previous mechanical integrity inspections for the tower, which I was told existed, but I was not provided. This withholding of information without any explanation seemed suspicious to me. This is a weakness in the culture. So, I tried to determine what the inspections should be for this item.

I tried to get the paperwork on the tower screw "repair", but none was provided, again showing a cultural problem.

I informed the plant that the company they were using did not have an R stamp and the screw action was not a proper repair.

I suggested contacting an inspection company to perform some external vessel wall thickness inspections, to see if additional thin spots were present. The external inspection was performed, which showed some additional thin areas.

I was verbally informed of a second small leak in the subject amine regenerator tower which was stopped by placing a screw in the side of the tower, as shown in the photo (taken later) immediately below.

It was disappointing that the facility took the same actions as last time, when they had been advised otherwise. I notified my management, and they did not seem concerned. I again reminded the plant that the company they were using did not have an R stamp and the screw action was not a proper repair.

I was verbally informed of a third small leak in the subject amine regenerator tower which was stopped by placing a screw in the side of the tower, as shown in the photo (taken later) immediately below.

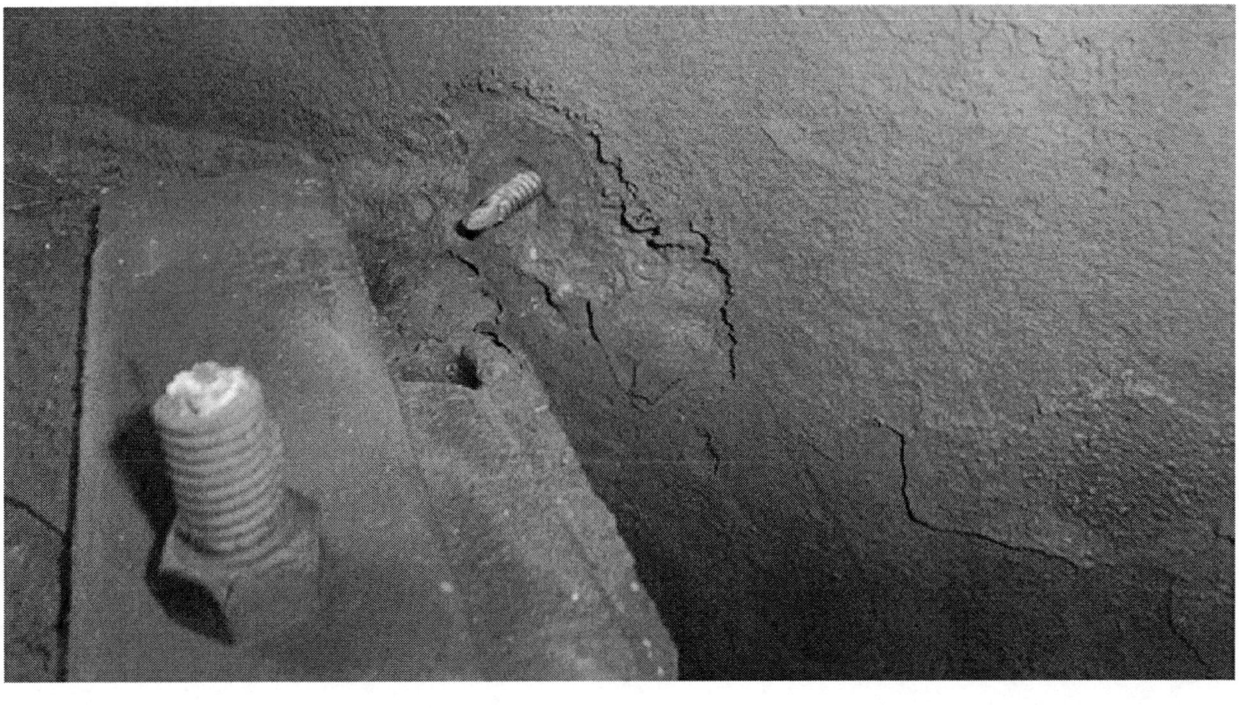

It was disappointing that the facility took the same actions as the last two times, when they had been advised otherwise. I again reminded the plant that the company they were using did not have an R stamp and the screw action was not a proper repair.

This time when I notified my management, they may have over reacted, since they now changed their direction drastically in stating that the right thing to do was shut down the facility, again without trying to get input from me or the plant manager. I was concerned about this drastic change in direction. When I spoke with management about this, they said this was a great success, since the facility was shut down in response to the latest leak. I informed management that I did not view it as a success to take this action in response to the third instance, but would have viewed it more positively if more action was taken on the first issue. A consequence analysis could have been conducted to determine the extent of the consequences. Then the likelihood could have been determined. Combining the consequence and the likelihood results in the risk, which could be used to help make a quantitative risk based determination, rather than a subjective decision.

I then asked management if we could share the learnings from this one site with the other 30+ sites in the company, and they said no, they would not do so until the incident happened at 3 or more facilities. It greatly disturbed me that management would wait for more issues to happen rather than sharing knowledge to prevent future incidents. This shows another cultural weakness.

Since the facility was now shutdown, it presented an opportunity to do some internal inspection, which was done. This effort resulted in many pictures, including the ones in this paper, as well as many wall thickness readings. The wall thickness readings were as thin as 0.110".

The original vessel wall thickness was supposed to be 0.375", with 0.125" corrosion allowance, which shows that the end of life wall thickness should be 0.250", well above the 0.110" measured. The vessel had a 50 psig design pressure, was 78" ID, with an 18 Ft skirt, 64 feet shell height and was made from SA-516-70.

I put together a spreadsheet, based on ASME Boiler and Pressure Vessel Code, Section VIII, Division I, 1980, reference 9, to evaluate what wall thickness was needed to handle different scenarios. Joint efficiencies were taken from reference 9, UW-12. Allowable stresses and overstress for combined loads were from reference 9, UG-23.

Seismic and wind speed were considered by calculating the moment at the bottom head to shell weld, and then converting to metal thickness.

I got the original calculations for the tower, and found that the original seismic calculations were in error. Seismic tension resulted in a required wall thickness of 0.048", which is less than the 0.110" measured wall thickness, so the wall thickness was acceptable. Seismic compression resulted in a required wall thickness of 0.113", which is just above the 0.110" measured, meaning the wall is not thick enough. A useful reference for seismic design is
Guidelines for Seismic Evaluation and Design of Petrochemical Facilities, Second Edition, ASCE, Reston, Virginia, 2011, reference 10.

For the wind loading, the tower was divided into 4 different zones. Wind tension resulted in a required wall thickness of 0.072", which is less than the 0.110" measured wall thickness, so the wall thickness is acceptable. Seismic compression resulted in a required wall thickness of 0.169", which is above the 0.110 measured, meaning the wall is not thick enough. A useful reference for wind loading is ASCE 7-05, Chapter 6, Figure 1A Basic Wind Speed – Western Gulf of Mexico Hurricane Coastline, reference 11. A second useful reference is Wind Loads for Petrochemical Facilities and Other Industrial Facilities, ASCE, Reston, Virginia, 2011, reference 12.

Reference 9, UG-27 was used to determine the minimum thickness for internal pressure and liquid height. The thickness needed for circumferential stress is 0.176", which is above the 0.110 measured, meaning the wall is not thick enough. The thickness needed for longitudinal stress is 0.052", which is less than the 0.110" measured wall thickness, so the wall thickness is acceptable. The thickness needed for compressive stress is 0.085", which is less than the 0.110" measured wall thickness, so the wall thickness is acceptable.

For the combined load temperature, pressure, level, wind and seismic, the thickness needed for longitudinal stress is 0.144", which is above the 0.110 measured, meaning the wall is not thick enough. The thickness needed for compressive stress is 0.306", which is above the 0.110 measured, meaning the wall is not thick enough.

Management tried to direct me to just solve the issue by reducing design pressure until the wall was thick enough. However, this was not a reasonable path forward, for several reasons, such as that if the same corrosion allowance was used, the remaining life (0.110" measurement minus 0.125" corrosion allowance) was 0. Also, there was no mention of the other aspects that would need to be considered, such as increased relief valve sizing at lower pressure. Also, the wall was not thick enough for several other cases listed above, such as seismic and wind, which are independent of the design pressure. Reducing the design pressure from 50 to 15 psig, meant that the wall thickness was still not adequate for the combined loading.

I also did some process simulation to determine if there could be changes made to the process and/or equipment to reduce the corrosion, such as:

- Changing type of amine
- Changing flowrate of amine
- Changing tower diameter
- Changing tower materials of construction
- Changing heat input to the reboiler

I was told to stop on this effort, which meant no better design would be in place when the next tower has problems. Production again won out over process safety when I was told to get a new tower the same as the last one.

A fitness for service assessment could have been conducted, however, production again won out over process safety when I was told to get a new tower the same as the last one.

A risk based mechanical integrity program could have been developed and put into place, but it wasn't. The emphasis was on production over process safety.

A root cause analysis could have been conducted, but the emphasis was on production and appearance over process safety. This is demonstrated by the operations direction to rapidly take the tower offsite to keep the plant appearance attractive. There was not an effort to engage in dialogue to discuss the issue. A better approach is to engage in dialogue with employees and seek their input and comments on suggestions. When management eventually realized the value of the investigation process, the tower was already destroyed. This was unfortunate, because the same type of failures starting appearing at other sites.

I told management that they needed to institute a mechanical integrity group. They eventually did, however, when I was informed of the head of the mechanical integrity group and the recent hire for our area, I could see that they were not qualified. I informed management of this, and I was told that was not my concern. This again shows a poor process safety culture. No attempt was made to resolve my valid concern, just to dismiss it.

I finally put many items in writing to my management, and was provided an individual visit by HR from headquarters. In this closed-door meeting, I was told many unacceptable things, including that I should not write about process safety, but management is glad to discuss orally. This shows a poor process safety culture. I resorted to writing when it was clear that oral was not effective. I did not agree with HR. I was then sent to an individual training session with the head of compliance, which was titled "Write Now, Pay Later." It is clear that the company tried to restrict internal written communication on safety and health, when it is an employee right to write written complaints to external agency, OSHA. This does not show a safety culture.

3 Additional RAGAGEP

One good resource is Guidelines for Mechanical Integrity Systems, AIChE CCPS, 2006, Wiley, reference 3. For pressure vessels, a useful item is "Table 9-1 RAGAGEPs for Pressure Vessels".

Another useful item in reference 3 is "Table 9-13 Mechanical Integrity Activities for Pressure Vessels". It has example activities and typical frequencies for:

- New Equipment Design, Fabrication, and Installation
- Inspection and Testing
- Preventive Maintenance
- Repair

Reference 3 also has Technical Basis for Activity and Frequency, Sources of Acceptance Criteria, Typical Failures of Interest, Personnel Qualifications, Procedure Requirements, Documentation Requirements for the following categories:

- New Equipment Design, Fabrication, and Installation
- Inspection and Testing
- Preventive Maintenance
- Repair

Reference 3, Chapter 11 contains an introduction to a risk based approach for mechanical integrity.

Reference 3, Section "12.4 Equipment Failure and Root Cause Analyses" provides useful information such as:

- "Step 1 - Evaluate Conditions at the Site."
- "Step 2 - Perform Preliminary Component Assessment."
- "Step 3 - Preserve "Fragile" Data Sources."
- "Step 4 - Perform Macroscopic Examination."
- "Step 5 - Perform Microscopic Examination (if warranted)."
- "Step 6 - Perform More Detailed Testing (if warranted)."
- "Step 7 - Test Under Simulated Conditions (if warranted)."
- "Step 8 - Make the Failure Mechanism Determination and Develop Corrective Actions."

Listed below are several additional API documents that contain excellent information related to mechanical integrity, with specific applicable sections listed.

API Recommended Practice 571, Damage Mechanisms Affecting Fixed Equipment in the Refining Industry, First Edition, December 2003, is reference 4. See sections:

- "4.2.14 Erosion / Erosion-Corrosion"
- "4.3.6 CO2 Corrosion"
- "5.1.1.1 Amine Corrosion"
- "5.1.2.2 Amine Stress Corrosion Cracking"

API Recommended Practice 572, Inspection of Pressure Vessels (Towers, Drums, Reactors, Heat Exchangers, and Condensers), Second Edition, February 2001, is reference 5. See sections:

- 7 for reasons for inspection
- 8 for causes of deterioration
- 9 for frequency and time of inspection
- 10 for inspection methods and limitations
- 11 for methods of repair

API Recommended Practice 579, Fitness for Service, First Edition, March 2000 is reference 6. Reference 6 states "The ASME and API design codes and standards for pressurized equipment provide rules for the design, fabrication, inspection and testing of new pressure vessels, piping systems, and storage tanks. These codes do not address the fact that equipment degrades while in-service and that deficiencies due to degradation or from original fabrication may be found during subsequent inspections. Fitness-For-Service (FFS) assessments are quantitative engineering evaluations which are performed to demonstrate the structural integrity of an in-service component containing a flaw or damage. This Recommended Practice provides guidance for conducting FFS assessments using methodologies specifically prepared for equipment in the refining and petrochemical industry. The guidelines provided in this recommended practice can be used to make run-repair-replace decisions to help ensure that pressurized equipment containing flaws which have been identified by inspection can continue to operate safely."

API Recommended Practice 580, Risk-based Inspection, First Edition, May 2002 is reference 7. Reference 7 states "1.1 PURPOSE -The purpose of this document is to provide users with the basic elements for developing and implementing a risk-based inspection (RBI) program. The methodology is presented in a step-by-step manner to the maximum extent practicable."

API Recommended Practice 581, Risk Based Inspection Technology, Second Edition, September 2008 is reference 8. In reference 8, "PART 1- INSPECTION PLANNING USING API RBI TECHNOLOGY", see sections:

- 4 for RBI Concepts
- 5 for Pressure Vessels

In reference 8, "PART 2 - DETERMINATION OF PROBABILITY OF FAILURE IN AN API RBI ASSESSMENT", see section 8 for "Damage Factor – Amine Cracking".

Reference 8, "PART 3 - CONSEQUENCE ANALYSIS IN AN API RBI ASSESSMENT" covers hole size, release rate calculation, release inventory, impact of detection systems, flammable and explosive consequences, toxic consequences, non-flammable non-toxic consequences, component damage and personnel injury consequence areas, financial consequences.

4 Conclusions

This paper presented in detail one of the many process safety issues encountered by the author in his 20 years of industrial experience. This issue deals with an amine tower, but touches upon the following topics: carbon dioxide; communication and restriction; competency; compliance; culture; documentation; failure analysis; intimidation; investigation; lessons learned; material of construction; mechanical integrity; pressure; stress, risk assessment and vessel repair.

It describes better ways to address the situation and applicable RAGAGEP. It is hoped that sharing this information will enable others to learn from the past and prevent similar issues in the future.

5 References

[1] NBBI, NB 23 (NBIC), National Board Inspection Code, Parts 1-3.

[2] API 510, Pressure Vessel Inspection Code: In-service Inspection, Rating, Repair, and Alteration, Tenth Edition, May 2014.

[3] Guidelines for Mechanical Integrity Systems, AIChE CCPS, 2006, Wiley.

[4] API Recommended Practice 571, Damage Mechanisms Affecting Fixed Equipment in the Refining Industry, First Edition, December 2003.

[5] API Recommended Practice 572, Inspection of Pressure Vessels (Towers, Drums, Reactors, Heat Exchangers, and Condensers), Second Edition, February 2001.

[6] API Recommended Practice 579, Fitness for Service, First Edition, March 2000.

[7] API Recommended Practice 580, Risk-based Inspection, First Edition, May 2002.

[8] API Recommended Practice 581, Risk Based Inspection Technology, Second Edition, September 2008.

[9] ASME Boiler and Pressure Vessel Code, Section VIII, Division I, 1980.

[10] Guidelines for Seismic Evaluation and Design of Petrochemical Facilities, Second Edition, ASCE, Reston, Virginia, 2011.

[11] ASCE 7-05, Chapter 6, Figure 1A Basic Wind Speed – Western Gulf of Mexico Hurricane Coastline.

[12] Wind Loads for Petrochemical Facilities and Other Industrial Facilities, ASCE, Reston, Virginia, 2011.

Identifying Opportunities of Enhancing Safeguard Stewardship through IPL Rationalization

SreeRaj R Nair
Technical Safety, Chevron U.S.A Inc.
Houston, USA
snair@chevron.com

Bernadette Thornton
Safety Systems, Chevron
ETC, Houston, USA
bernadette.thornton@chevron.com

[Rev2]

Prepared for Presentation at
American Institute of Chemical Engineers
2017 Spring Meeting and 13th Global Congress on Process Safety
San Antonio, TX
March 26 – 29, 2017

Identifying Opportunities of Enhancing Safeguard Stewardship through IPL Rationalization

SreeRaj R Nair
Technical Safety, Chevron U.S.A Inc.
Houston, USA
snair@chevron.com

Bernadette Thornton
Safety Systems, Chevron
bernadette.thornton@chevron.com

Keywords: Process Hazard Analysis, Layer of Protection Analysis, Independent Protection Layers, Rationalization

Abstract

Ensuring effective safeguard stewardship is an integral part of the journey to address the challenge of moving from 'I think" to "I know" that a major accident hazard is managed. One of the steps in achieving effectiveness is to ensure that the process design is robust and that the safeguards provide the necessary protection to reduce the process risk.

Process Hazard Analysis and Layer of Protection Analysis are carried out to identify hazards, determine the potential consequence and evaluate the adequacy of the safeguards. Opportunities for further risk reduction are identified and risks are eliminated or reduced through established risk management programs. Though the study methodologies have been used for decades, there are opportunities for improvement.

This paper specifically addresses the opportunities in the evaluation of safeguards during the design of major capital projects. It explores some of the key concerns including lack of independence and variations in modifier application that could lead to over or under engineering; this could in turn may result in increased risk exposure. Additionally, this paper explains how the use of a defined methodology for the rationalization of Independent Protection Layers can assist in overcoming these challenges. The significance of performing IPL rationalization at appropriate stage of engineering progress is key in ensuring effective safeguards through design. A comparison of the advantages of rationalization exercise performed at various project phases is also discussed in the paper.

1 Introduction

A critical starting point in process safety management lies with the identification of facilities with potential for high impact (Safety and Environment) events and an understanding of the high consequence scenarios. For the high impact facilities, a detailed Process Hazard Analysis is conducted to determine the scenarios of concern. The potential impact from realizing those scenarios and the likelihood considering the safeguards (prevention and mitigation) is evaluated. A process safety management system is established with a system of "controls" (including critical safeguards verifications, and validation) to prevent the occurrence of these incidents or reduce their severity.

Layers of Protection Analysis (LOPA) was introduced as a detailed frequency evaluation method compared to Hazard and Operability Studies and as a simpler version compared to Quantitative Risk Assessment. This method is used to assess the adequacy of independent protection layers (IPL) and to ensure that required risk reduction is achievable (i.e., there will be no gaps post implementation of the risk reduction plan). LOPA is a systematic approach for applying IPL which follows guided principles and rulesets as prescribed in the industrial standards (AIChE 2001). CCPS publication (AIChE, 2007) provides guidance on choosing and evaluating initiating event frequency and independent layers of protection.

An independent protection layer (IPL) is a device, system, or action that is capable of preventing a scenario from proceeding to its undesired consequence independent of the initiating event or the action of any other layer of protection associated with the scenario (AIChE, 2015).

An effective IPL should be independent of the initiating causes, survive the initial event, and available to perform at the expected level for the intended duration. In order to determine the effectiveness, each safeguard which has been identified as a potential IPL needs to have the entire core attributes; the attributes are independence, specificity, integrity, reliability, auditability, access security, management of change. IPL can be either 'Active' or 'Passive' with Active IPL being those with the ability to detect (sense a condition or problem), decide (whether an action or intervention is necessary or not) and deflect (action capable of preventing consequence in a timely manner) and Passive IPL being those which primarily deflect to reduce the impact.

The following sections review some of the challenges in determining the effectiveness of active IPL and how to overcome the challenges through a structured approach known as IPL rationalization. It is followed by discussion on the inherently safer approach and its significance in improving the performance of the LOPA results. The potential economic benefits of IPL rationalization (Capital and Operational) are also evaluated with relevant project case studies.

2 LOPA Process Concerns

After evaluation of multiple projects, we have determined that the major LOPA concerns tend to fall into three categories: lack of independence, variations in modifier application, and appropriate documentation. The challenges in IPL related documentation can be addressed effectively by implementation of Process Safety Information management plan.

2.1 *Lack of Independence*

The first issue, lack of independence, relatively an easy one to eliminate, yet it continues to appear in reviews. Most of this is due to a failure to understand packaged equipment controls and a failure to identify the critical actions required by a safety function. Packaged equipment controls typically are not well defined in early Process Hazard Analysis (PHA) / Hazard and Operability (HAZOP) studies. Assumptions concerning the independence of systems and poorly defined functionality tend to lead to IPL credit being given to these systems in error. For example, vibration monitoring systems tend to send requests to other systems, such as the Safety Instrumented System (SIS), to complete critical actions such as the closure of valves or motor stops. If a Safety Instrumented Function (SIF) exists with the same critical actions, then credit cannot be taken for both. The critical actions of each would have to be defined to remove the conflict and the system common cause would need to be reviewed to ensure that the combination will provide the appropriate layers of protection. Failure to validate packaged equipment independence as the project progresses leads to discovery late in the project that the required number of IPL is unavailable.

2.2 *Modifier Application*

Differences in the team make-up and the facilitation of the LOPA studies leading to differences in the application of enabling events and conditional modifiers and therefore poor application of IPLs. Enabling events and conditional modifiers are applied to change the likelihood of a consequence or severity of a consequence, so these take into account factors such as the likelihood of personnel being in the area and the phase of operation. Failing to apply enabling events and conditional modifiers properly can lead to major operational issues, including alarm floods and excessive instrumentation. For example, if one facilitator applies a conditional modifier to a scenario and another facilitator does not, the required risk reduction is greater for the second facilitator and will require the application of additional IPL. This inconsistency in the application of enabling events and conditional modifiers leads to under-design (with gaps) and over-design (most-often) of instrumented systems and, potentially requiring additional maintenance and testing requirements.

Effective IPL rationalization can reduce the concerns of lack of independence and modifier application by providing an opportunity to look at the application of IPL on a facility basis rather than a scenario by scenario basis.

3 IPL Implementation Challenges

Poor application of IPL in the processes could result in facilities operated at elevated risk levels. The key challenges in the implementation of LOPA findings include the following:

- Risk reduction during design primarily relies on engineered systems, leaning to the route of 'Add-on' safety;
- Lack of credits for non-engineered / soft barriers as additional layers of protection;
- Operational inefficiency due to overly engineered protection system (multiple interlocks, trips for dedicated systems designed without considering overall facility reliability requirements);

- Regulatory requirements – prescriptive versus performance based approach;

Kletz and Amyotte [Ref. 3] have said, "Intensification, substitution, attenuation, and limitation of effects produce inherently safer design because they avoid hazards instead of controlling them by adding protective equipment. The term 'inherently safer' implies that the process is safer because of its very nature and not because equipment has been added to make it safer". However, this very point is often overlooked during design and often adding additional layers (mainly instrumented and hardware) are considered as improvement to safety. Industries' traditional solution to increased levels of hazard has been through more controls. For example, an offshore facility designed to be operated as 'normally unmanned' could become 'normally manned' just because of the manning distribution required to ensure the effectiveness (through inspection and maintenance) of the 'added on' engineered safeguards.

Projects have become more complex and personnel involved in projects rarely see the effect of their designs, so design complexity is less of a consideration and the more cost effective solution, typically an instrumented one, is implemented to ensure enough safeguards are present to manage the risk. Unfortunately, this leads to the massive problem of facilities having to manage large numbers of safeguards and IPL.

While industry is moving towards performance based approaches to define risk reduction, prescriptive regulatory environments still exist. Prescriptive environments ensure a minimal level of risk reduction is implemented, but they also can add challenges to the implementation of a performance based approach. Prescribed testing and testing procedures may provide some reduction in the overall risk of a facility, the tests rarely are sufficient to provide assurance of the effectiveness of the prescribed functions as IPL; therefore, additional testing must be performed on top of the regulatory required testing potentially requiring more personnel to complete ensure the functions meet their risk reduction target.

4 How to Overcome the Challenge

Eliminating hazards using the inherently safer design route is the most effective way to overcome the challenge of safeguard stewardship simply because there will be only fewer safeguards as IPL (Kletz and Amyotte, 2010). Time and effort can be spent on managing the asset integrity of pipe, vessels, and major equipment rather than managing hundreds of safeguards. In such cases, the projects compare options for managing risk. If elimination is not a viable option considering safety and operability factors, decisions are made to proceed with acceptable levels of risk using adequate safeguards/IPL. The following words of wisdom shall be considered as part of ISD approach (Nair 2009, Nair 2017, Stuart, et. al 2003, Kletz and Anyotte, 2010):

- More is not always safer
- Identify all hazards (you can't eliminate or manage a hazard which have not been identified)
- Design against accidental loads
- Design plants for easy and safe maintenance (*reliable equipment which requires less maintenance*)
- Simplify the process and minimizing the requirement to rely on human intervention

While companies acknowledge that management of IPL is necessary, the number of IPL being identified prioritizes finding ways to reduce the load on maintenance and testing personnel. One way to reduce the load is by using IPL rationalization. IPL rationalization can be defined as a process of defining the most effective set of IPL to provide the appropriate risk reduction while also reducing the number of management activities and operator tasks and the maintenance and testing load a facility must endure. So the question of how to define the most effective set of IPL must be answered.

IPL rationalization uses the results of a LOPA to identify IPL which are unnecessary to provide the facility with the risk reduction required. The process requires a functional safety Subject Matter Expert (SME), a process engineer and an operations representative to review the application of enabling events and conditional modifiers, identify SIF with multiple Safety Integrity Level (SIL) and ensure that SIF, critical alarms and basic process control system (BPCS) loops identified as IPL are, in fact, effective. During a LOPA, many of these items may be addressed, but as projects require more and more facilitators to make large HAZOP more manageable, more and more issues creep into the design. The smaller group allows for a more in-depth review of independence, consistency in identification of the SIF and its actions, and a way to compare the results of one study with another study to ensure IPL chosen between similar units and similar scenarios provide the appropriate risk reduction while applying the same enabling events and conditional modifiers. Most importantly, IPL rationalization provides a way for the project to review the IPL from a bird's eye view as well as scenario by scenario to identify whether IPL are appropriate to mitigate the risk for the overall facility as well as the individual scenarios. The most surprising element of the process is that it requires very little time to complete, especially with the correct pre-work completed.

The process of IPL rationalization is simple. First, the team must first identify the order of preference for implementation of IPL. This preference will take into account the facility's maintenance staff size, the competency of the staff and the operating basis of the facility. For example, some owners may prefer to take more SIFs as they have a good management program for them and no management program for other engineering IPL. Other owner may prefer other instrumented IPL over SIF as they do not expect to have the competency they would prefer to manage the IPL.

Secondly, the team reviews the application of enabling events and conditional modifiers to guarantee their effectiveness or applicability. Next, the IPL are reviewed in order of preference and where different IPL have been provided for scenarios with the same consequence, the use of different IPL is either rationalized or the IPL are revised to provide the most effective set. Unused IPL are either removed or removed from the IPL list.

Finally, the team reviews any recommendations to identify if there are LOPA gaps that may be addressed using existing IPL or whether new IPL must be created.

Experience has demonstrated that the process can be effectively used for brownfield facilities as well. However, it need to be noted that there could be differences in the IPL preferences and that unused IPL are less commonly found as facilities choose to use existing IPL rather than add new ones, even if the final result may be less efficient.

5 A Representation of the Benefits

The benefits of completing IPL rationalization depends on three things:

1. The objective of the project
2. The timing of the study
3. The regulatory environment of the facility

The objective of the project may affect IPL rationalization results as looking for ways to eliminate or reduce the numbers of a particular IPL type or may require the addition of IPL in other categories. This activity may not be a reduction of IPL, but rather a reallocation of IPL. Alternatively, the project may be looking to reduce the number of IPL overall by ensuring the application of IPL is streamlined. Illustrations of IPL rationalization benefits in terms of IPL and potential economic saving are given in Figure 1 and Figure 2. A recent project looking to reduce the number of alarms was able to achieve a 90% reduction in alarms with little effect to the number in other IPL categories due to the proper application of conditional modifiers. If one assumes that maintenance of these alarms requires approximately $100 per year and that the number of alarms was reduced from 1000 to 100, then it can be found that the facility saves approximately $90000 per year in alarm maintenance alone (*representative figures only*).

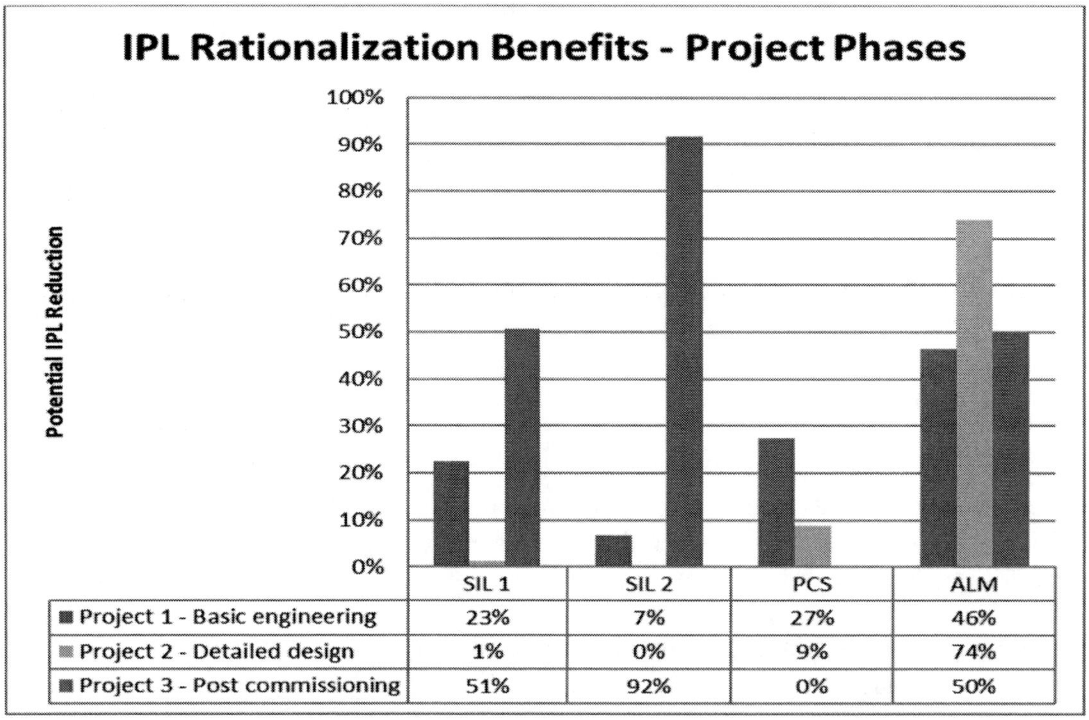

	SIL 1	SIL 2	PCS	ALM
Project 1 - Basic engineering	23%	7%	27%	46%
Project 2 - Detailed design	1%	0%	9%	74%
Project 3 - Post commissioning	51%	92%	0%	50%

Figure 1: IPL Rationalization Benefits – At Different Phases of Project

The timing of the study has a significant effect on the outcome of the IPL rationalization. Performing rationalization before the design is complete (i.e., before construction) provides opportunity to eliminate IPL, such as SIF, that are found to be unnecessary (Figure 1). A project may choose to keep the functions or alarms for operational or reasonable risk reduction justification/demonstration; but take the opportunity to reduce the testing or maintenance associated with these. In addition, the team may be able to identify designs which could be modified to provide a more inherently safer facility. A study performed late in the project will not achieve the same benefits. They may be able to make reductions in some areas, but if the design or installation is almost complete, there is only limited opportunity to remove unnecessary IPL (Figure 1). Alarm priorities may be reduced and maintenance and testing requirements may be relaxed, but the facility will likely be forced to retain most IPL. Performance of IPL rationalization post-commissioning provides even less opportunity for savings. IPL have not only been designed and installed, but they have already been commissioned and, in many cases, changing maintenance and testing requirements will require changes to maintenance managements systems that likely have already been completed, potentially adding cost short-term. A review of projects showed potential IPL reduction of 25-30% for projects in Basic Design / Front End Engineering Design (FEED) and 10% reduction for those in Detailed Design or later.

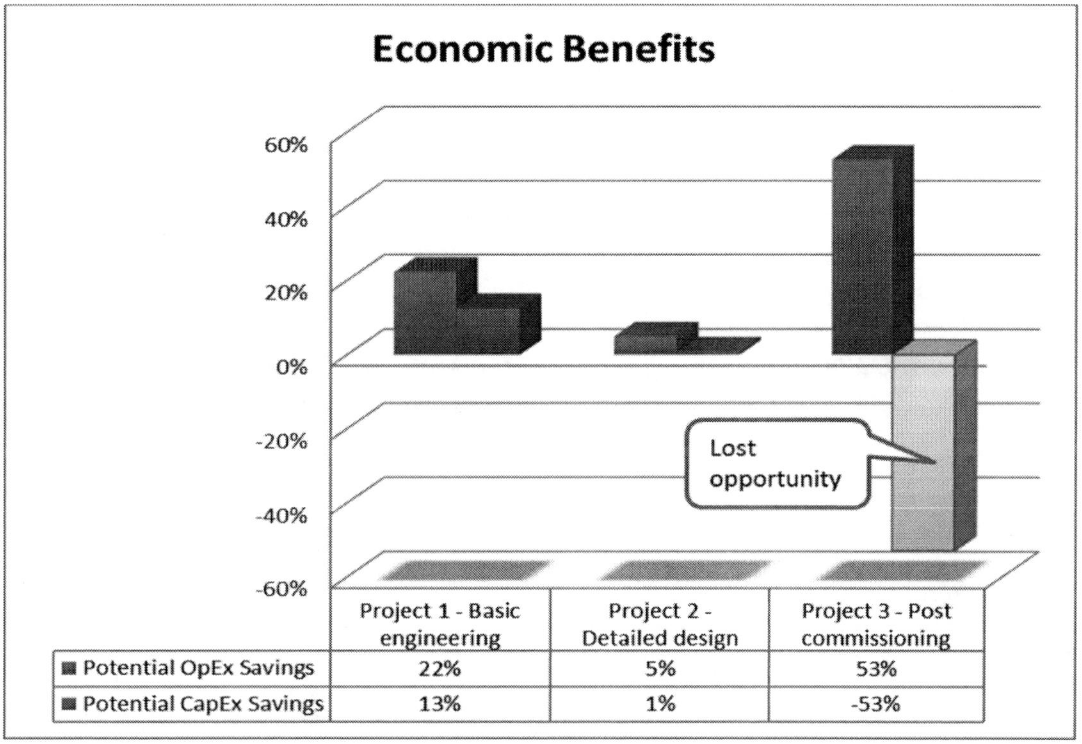

Figure 2: IPL Rationalization and Potential Economic Savings

When taking into account both Capital Expenditure (engineering design, construction, installation) and Operational Expenditure (inspection, testing and maintenance), the savings to a facility can be significant if the appropriate evaluation is carried out in early project stages; even in later stages of the project, the suggested approach can reduce IPL requirement and remove some of the maintenance and testing cost of the so-called "gold plating" that occurs on projects (Figure 2).

Finally, the regulatory environment can have an impact on the study. Where regulators follow a prescriptive regimen for safety instrumented systems (e.g., U.S. Gulf of Mexico), then the benefits of IPL rationalization are realized through the removal of extensive testing and maintenance related to IPL, but not through removal of IPL or through the removal of regulatory required testing. The potential to gain savings still exists though, so the exercise provides benefit by removing the "gold plating" of the facility.

6 Conclusion

Critical safeguards identified, performance requirements established and an evergreen process to ensure safeguards remain in place are key to keep high consequence scenarios from occurring. Through safety studies and by following industrial codes and standards, multiple layers of protection are factored in to prevent or mitigate potential events. However, it is experienced that often the designs and facility become complicated due to 'add-on' safety. Such facilities could experience operational inefficiency, high cost of maintenance and potentially higher risk exposure to personnel.

IPL rationalization should be part of the design engineering process to overcome the challenges and a systematic approach will help in:

- establishing 'effective' safeguard
- optimizing IPLs
- lifecycle benefits

The paper reviewed some of the challenges in determining the effectiveness of IPL and how to overcome the challenges through a IPL rationalization exercise. The paper discussed the inherently safer approach and its significance in improving the performance of the LOPA results. The timing of the exercise has a significant effect on the outcome of the IPL rationalization. The potential economic benefits (Capital and Operational) are also evaluated with relevant project case studies.

7 References

1. CCPS. "Guidelines for Independent Protection Layers and Initiating Events", American Institute of Chemical Engineers, New York, NY, 2007
2. CCPS, "Guidelines for Initiating Events and Independent Protection Layers in Layer of Protection Analysis", American Institute of Chemical Engineers, New York, NY, 2015
3. CCPS, "Layer of Protection Analysis – Simplified Process Risk Assessment", American Institute of Chemical Engineers, New York, NY, 2001
4. Nair SreeRaj, "Determining the criteria for evaluation of toxic hazards, Journal of HSE & Fire Engineering", ASFE, Issue 2 March 2009
5. Nair SreeRaj, "'Inherently Safer' – the Concept and its Application in Oil & Gas Sector," International Conference on Safety and Fire Engineering, SAFE-'17, 2017
6. Stuart Chia, Kieran Walshe and Ed Corpuz, "Application of inherent safety challenge to an offshore platform design for a new gas field development – approaches and experiences," Symposium Series No. 149, IChemE, 2003
7. Trevor Kletz and Paul Amyotte, "Process Plants: A Handbook for Inherently Safer Design," Second Edition, 2010

Process Safety Site Assessment Program (PSSAP) – 5 years of Assessments

Andrew P. Broadbent
American Petroleum Institute (API)
1800 West Loop South, Suite 475
Houston, TX 77027
broadbenta@api.org

Prepared for Presentation at
American Institute of Chemical Engineers
2017 Spring Meeting and 13th Global Congress on Process Safety
San Antonio, TX
March 26 – 29, 2017

Process Safety Site Assessment Program (PSSAP) – 5 years of Assessments

Andrew P. Broadbent
American Petroleum Institute (API)
1800 West Loop South, Suite 475
Houston, TX 77027
broadbenta@api.org

Keywords: Process safety, assessments, benchmarking, protocols, scoring, rating, weighting

Abstract

The API Process Safety Site Assessment Program was developed in 2011 as part of the AFPM and API Advancing Process Safety Programs. PSSAP began conducting assessments in 2012. By the end of 2016, the API Site Assessment program will have conducted 52 General (7 protocols) and 27 HF Alkylation/RP 751 assessments. These assessments have been conducted 59 different refineries and petrochemical facilities. The areas assessed are: Process Safety Leadership, Management of Change, Mechanical Integrity, Safe Work Practices, Operating Practices, Facility Siting, Process Hazards Analysis, and HF Alkylation/RP 751.

This presentation will provide a brief overview of the program and present some the learnings, trends and benchmarking developed from the assessments based on the five-year history of the program.

In 2010, the American Petroleum Institute (API), American Fuel & Petrochemicals Manufacturers (AFPM) and industry joined together to address improving process safety performance with the goal of preventing process safety events at refineries and petrochemical plants. API, AFPM and the industry wanted this effort, known as Advancing Process Safety, to be:

- Focused on higher risk process safety activities;
- Refining & petrochemical focused;
- Voluntary; and
- An avenue to communicate and share experiences and knowledge.

Advancing Process Safety (APS) consists of six different programs which are:

- Process Safety Metrics and Analysis (RP 754)
- Process Safety Hazard Identification
- Event Sharing
- Focused Improvement
- Regional Meetings
- Process Safety Site Assessments Program (PSSAP).

The intent of all the programs is to raise the level of process safety performance at the sites to prevent process safety events.

The management of these programs is split between API and AFPM. API manages the Process Safety Site Assessment Program, Process Safety Metrics and Analysis, and the Focused Improvements effort. AFPM manages Process Safety Hazard Identification, Event Sharing and Regional Networks.

Benefits/Opportunities

The Process Safety Site Assessment Program is perhaps the most hands-on of the six programs since it involves a team of assessors being onsite at a refinery or petrochemical plant for an entire week. The program uses credible, third-party assessors who have specific expertise in one or more of the eight areas being assessed. The team utilizes a set of protocols that are applicable to both refineries and chemical plants. These protocols were developed through a collaborative effort of subject matter experts from a wide range of companies.

Through these site assessments, the program provides the sites with feedback regarding what they are doing well and what improvements they can make to their process safety programs. Additionally, the program provides the sites with benchmarking data so the sites can compare how they are doing versus their industry peers.

The assessments are not process safety management compliance audits, and so we refer to them as assessments and not audits. We speak to process safety rather than process safety management so people will understand that these assessments are not the three-year process safety management (i.e., 29 CFR 1910.119) compliance audits, rather they are assessments of how sites are performing relative to 'what good looks like" based on the experience of industry and the expert assessors. As you will see, the assessments only cover seven areas of process safety (compared with the 14 areas in OSHA 1910.119) as well as HF alkylation/API RP 751.

Protocols

The eight protocols address:

- Process Safety Leadership
- Operating Practices
    - Procedures
    - Shift Handover
    - Operator Rounds
    - Alarm Management, and
    - Workplace Conditions
- Mechanical Integrity – Fixed Equipment
- Safe Work Practices
    - Procedures
    - Work Permit Procedures
    - Utilities, etc.
- Management Of Change
- Process Hazards Analysis
- Facility Siting
    - Permanent Buildings
    - Portable Buildings, and
    - Tents
- HF Alkylation/API Recommended Practice (RP) 751

The first seven elements are areas that could prevent a process safety event from occurring or reduce the severity of an event (i.e., Facility Siting). The eighth area – HF alkylation – deals with process safety in this specialized operation. Assessments involving the first 7 protocols are referred to as "General" Assessments and the HF Alkylation/RP 751 ones are known as "HF In-Depth RP 751" Assessments.

Note: Since the HF Alkylation protocol specifically follows API RP 751, the HF-In-Depth RP 751 Assessment does meet the RP 751 requirement to conduct an audit once every three years.

Sites may also choose to do just some of the protocols (for example, MOC and Facility Siting) rather than all seven at a time. These are known as "Partial" assessments.

These protocols were developed by industry for industry utilizing 54 subject matter experts from a wide range of companies. The protocols were developed, balloted using the API standards development consensus process, and then piloted at three sites. The protocols were then reviewed and modified based on learnings from the pilot assessments, including input from the assessors.

Keys to the Program

There are three very important components the PSSAP program, which are:

- Benchmarking
- Mentorship
- Expertise of the Assessors

Benchmarking

Refineries and petrochemical plants always like to see how they compare to their peers in the industry. In the PSSAP program this is done by providing a score for each protocol and each protocol question.

The 7 General Assessment protocols have a total of 577 questions and the HF In-Depth protocol has 394 questions.

When developed, each protocol question was given a weighting indicating the question's importance to preventing a process safety event from occurring. The question's weighting is a fixed number between 1 and 4 in half increments (i.e., 0, 0.5, 1, 1.5, etc.) with 4 indicating the most importance. During the assessment the assessor will rate each question based on the information and evidence the site provides. This is done on a scale of 0 to 4, also in half increments. The score is a simple calculation of the weighting times the rating.

Keeping the scoring objective is done through ensuring the protocols are well written so that different assessors will understand and interpret the question the same way. Consistency is also addressed through the training provided and calibration sessions with the assessors at the start of the assessments. Additionally, the PSSAP Program Manager provides oversight so that a consistent approach is used by the assessors.

Mentorship

One of the most important pieces of the program is the information exchange the sites get with the assessors. As mentioned, the assessors have considerable experience and expertise based on their years of service in the industry and through performing PSSAP assessments.

We inform the sites (we typically visit the site six months prior to the assessment to set up all the logistics), to get the most value out of the assessment, that their employees should be interviewing the assessors as much as the assessor is interviewing the employee.

As an example, on one assessment the assessor sat down with three site employees for an hour and discussed various ways to manage car seals.

We encourage sites to take advantage of the time with the assessors throughout the week, including the daily debriefs, lunch time discussions, and the closing conference of Friday.

Assessors

As mentioned, the assessors are major piece of the PSSAP program. They are handpicked based on their expertise and industry background/knowledge. Typically they are retired from industry companies, though some work for consulting firms (or have their own firm) but worked in the industry at some point in time.

An assessor candidate with appropriate qualifications and references is selected to attend a 1-1/2 day training course which explains the API Process Safety Site Assessment Program to them. The candidate must undergo an interview with a three-person panel – the PSSAP Manager and two industry representatives before being selected to be an assessor. Not all candidates that attend the training get selected. Also, assessors are picked for specific protocols (i.e., we do not look for Mechanical Integrity assessors to be Operating Practices experts and vice versa). Currently, over 70 people have inquired about being an assessor and we have around 20 in our assessor pool.

To date, a typical assessment team is averaging 39 years of industry experience. Typically the team has 7 assessors.

An in-depth HF alkylation unit assessment follows the API RP 751 Recommended Practice in detail. This assessment can be done at the same time as a "basic" assessment or completely separate and consists of 5 assessors. As noted, an In-Depth RP 751 assessment can be used by the site as their 3-year RP 751 audit.

Assessments

An assessment gets started when the site contacts the PSSAP Program Manager to schedule a week for the assessment. It is important that the site assessment date, be selected well in advance – at least 6 months – so appropriate time is available for planning, including scheduling the assessors.

A pre-assessment visit is then scheduled and conducted at the site by the PSSAP Program Manager. This meeting serves two purposes. First, it allows the site management team to learn more about the program and ask any questions. The second purpose is to meet with the site contact (the person who will work with the PSSAP Program Manager to ensure all arrangements are made prior to the start of the assessment) to go over the logistical needs for the assessment - including conference rooms, scheduling interviews, documents needed prior to the assessment, etc.

Conducting the Assessment

On the week of the assessment the team arrives on Sunday afternoon/evening ready to go. The assessors will have already spent about 8 hours reviewing site documents, provided by the site through an FTP file share site, so they will be well versed on the site's programs relative to the protocols they are assessing.

Monday begins with an opening conference to review the ground rules and usually a quick tour of the site. After the tour, each assessor interviews key people to learn more about the site's management systems, programs and procedures. These interviews have been pre-scheduled by the site using a template provided to them by the PSSAP Program Manager.

The assessors also work to spend as much time in the field as possible. The intent is to not only review whether the site has good management systems, but to provide them with insight on how well their systems are being implemented in the field.

On Tuesday, Wednesday, and Thursday, the assessment team meets with site staff, spends time in the field, and holds a debrief with the site personnel, typically around 4:00 pm. These daily debriefs have been extremely beneficial as they are excellent opportunities for site personnel to continue to interact with the assessors. Often when the debrief has been completed, there are follow-up conversations with the assessors so site staff can gain more insights.

Thursday night and Friday morning is when the assessors compile all the information and the protocol ratings. The assessors enter observations and ratings for each line item in each protocol, thus the facility is provided with a considerable amount of information.

On Friday, an hour to an hour and a half closing conference is held with the site personnel, including the site management team. The site receives a PowerPoint overview, the completed protocol spreadsheets, and some preliminary benchmarking data. The draft report is provided to the site the following week. Distribution of the reports is as designated by the site.

When the assessment team leaves the site on Friday, the assessors return all hard copy documents and delete any electronic documents that were provided to them by the site.

Handling Report & Data

The site will typically be given two to three weeks to comment on the draft final report. This review period is not a discussion of the scores, but whether the wording of the Observations and Good Practices is clear. A final report is then issued by the program manager to the site.

Once the final report is issued, the data and information including the good practices and observations, with the permission of the site, are blinded and entered into the database for trending and benchmarking purposes. API does not ask for corrective action plans or updates.

Program Status, Experience & Keys

As of year-end 2016, we have conducted 53 General site assessments, 27 HF alkylation/API RP 751 in-depth assessments and 5 Partial assessments. The program has been to 28 different companies at 60 different locations (note: several companies have had a site assessment at more than one site.)

Item	Total	Average per Assessment
Protocol Questions	> 41,200	577 Basic 394 HF
Interviews	> 8,000	>100
Observations	> 2,800	~ 35
Good Practices	> 2,400	~ 30

Below are a couple of comments received from sites that have been assessed:

- "You cannot fool these Assessors — they have

done *your* job for 30 years"

- *"In one week the assessment team found issues that our internal teams could not find in 3+ weeks, and it was stuff we needed to know and address. Well focused, deep dive process."*

Benchmarking Data

The benchmarking data is provided in several formats. For example, the data is analyzed and presented by:

- Protocol
- Sub-protocol (some protocols have multiple areas of review – Shift Handover, Operator Rounds, etc.)
- Specific protocol sections (groupings of questions for specific areas)
- Management Systems vs. Implementation (having the procedure vs. how is it being done in the field)
- Size of site
- Size of company
- Quartile

Benchmarking data updates are provided generally every six months to sites that have had an assessment.

Data Review – Good Practices and Observations

Below are some examples of the types of Good Practices and Observations that the assessors have been identifying through the site assessments. If you note the last bullet on both slides this is an example of exactly what we are after. At some sites it's an item that needs to be addressed while at other sites we have seen them doing it well.

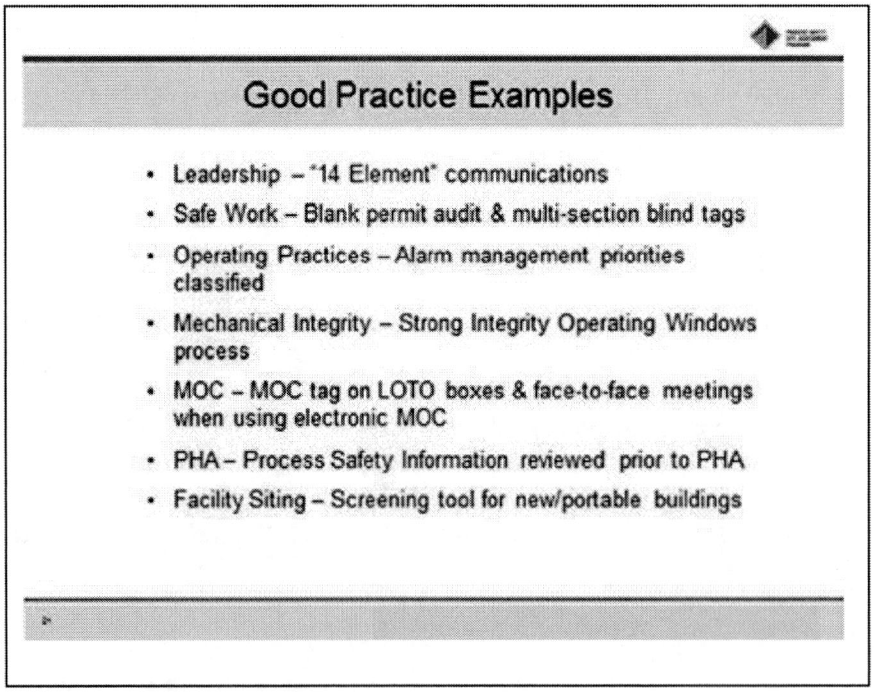

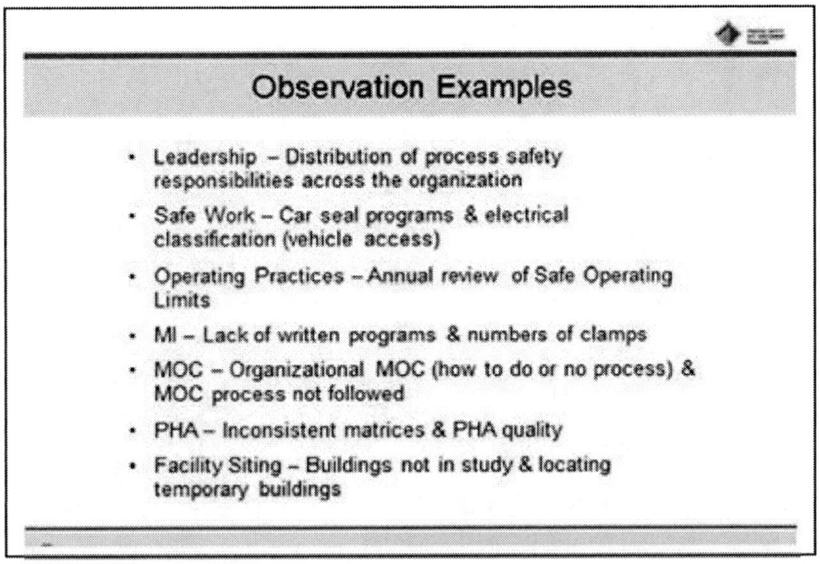

This points to a key element of the site assessment program - sharing the information from sites doing it well with others that are working to put systems/programs/procedures in place at their site. A variety of methods are used for sharing the practices learned through the site assessment program including:

- The Advancing Process Safety Program
  - Regional networks
  - Hazard Identification program
- Industry Conferences and Seminars
- API standards

The data, information, and learnings from the site assessments have been used to focus resources to develop informational and guidance documents regarding areas identified as improvement opportunities. For example, an informational document for mechanical integrity regarding small bore piping was developed based on observations during PSSAP site assessments and data from the 754 metrics and Event Sharing database..

<u>Conclusion</u>

In conclusion, the "boots on the ground" performance of site assessments began April of 2012. Feedback from these assessments has told us that the process of preparing for an assessment (going through the protocols prior to the assessment) and the mentorship/interaction with the assessors has been very beneficial to the sites. And now that we have done enough sites to allow the sharing of benchmarking data the sites are getting even more value from the program.

Alarm Management Meets SIS

SIS Alarm Requirements

How many alarms must be configured per the SIS Standard (ISA-84 Part 1) for an SIF?
Zero

How many alarms must be configured per the AM Standard (ISA-18.2) for an SIF?
Zero

How many alarms are necessary for an SIF to correctly perform its function?
Zero

Pre-alarm – What makes sense?

Pre-alarm before an SIS impending trip?
True – This gives the operator an opportunity to prevent the trip, keeping the plant running and reducing the demand frequency on the SIF.

Pre-alarm should always be on the BPCS control or indicator tag rather than the SIS transmitters?
False – Choice of where to put the pre-alarm is up to the facility, and it's a good idea to document this in the Alarm Philosophy.

Each transmitter in a voting arrangement should alarm individually?
False – Are we trying to keep our plants operating safely or do we just throw multiple alarms at the operator?

SIS and BPCS transmitter disagreements among redundant SIS transmitters should be alarmed?
True – With this alarm in place, there is no need for a pre-alarm on more than one transmitter.

Introduction

Most plants focus on meeting the alarm management requirements laid out in ISA 18.2. Even if the proper approach is taken in configuring the BPCS, there is one area that is sorely neglected, the SIS system. SIS is addressed based on the requirements laid out in ISA 84. Unfortunately, this standard does not address alarming related to SIS events. ProSys contends that a successfully implemented alarm management philosophy is only effective when taken as a holistic view of the BPCS and SIS systems.

Unified Alarming Scheme

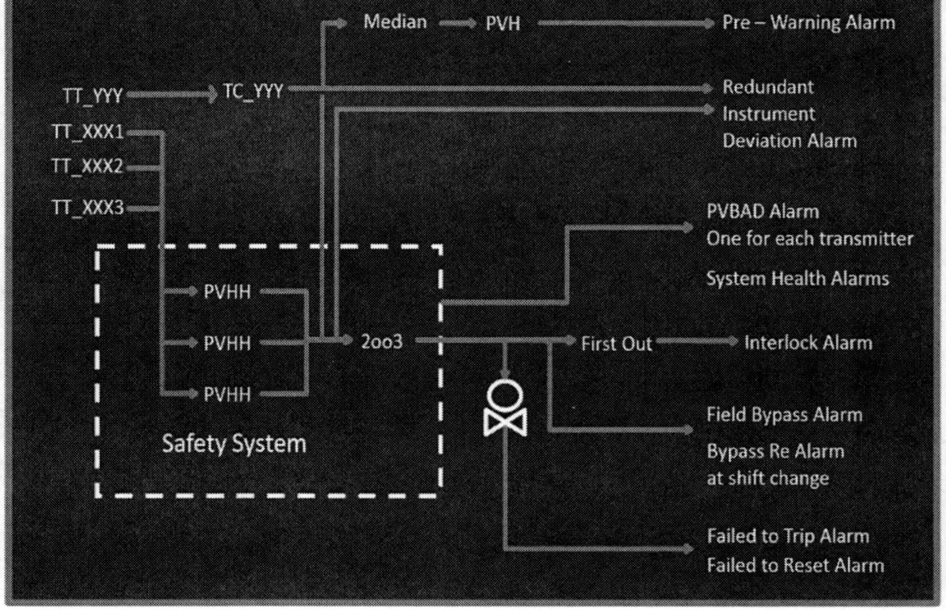

Trip – What makes sense?

First out alarm indicating the reason for a trip?
True – There is no better way to announce to the operator that a trip has been initiated.

Alarm SIS valve moving to trip position?
False – Going to the trip position is the expected response to an SIS trip – better, provide a failed to trip alarm, which makes noise if the valve doesn't work as expected.

Bypass – What makes sense?

Alarm board operator initiated SIS bypass?
False – Although a common practice, this alarm adds little or nothing to the integrity of the system: "Hey Joe/Mary/Sam, you just pressed a button!"

Alarm field operator initiated SIS bypass?
True – This may come as a surprise to the console operator, so it's not a bad thing to annunciate.

If a bypass is in place, trip point alarm on the SIS transmitter?
False – Why give the operator an alarm that isn't reliable or won't be believed? Find an alternative monitoring plan.

Re-alarm on SIS Bypass?
Neutral – Bypass and shift changeover procedures exist, but do we follow them 100%? This is a facility decision. If used, re-alarm shortly after shift change.

PROSYS® Maximize Your Plant™ www.ProSys.com

ENHANCING PROCESS SAFETY MANAGEMENT INTEGRATION TO IMPROVE PROCESS SAFETY AND PROFITABILITY IN THE FUELS AND PETROCHEMICALS INDUSTRY

Thomas D. Watson, P.E.
TDWatson Consulting LLC
Pittsburgh, PA USA
twatson@tdwatson.com

Ronald D. Fuchs, P.E.
RD Fuchs Engineering LLC
Lake Jackson, TX USA
rdfuchs13@aol.com

Harold Phillippi, P.E.
Independent Process Safety Consultant
Charleston, SC USA
butch.phillippi@gmail.com

Keywords: PSM, Process Safety Management, Relief Systems, Incident Investigation, Mechanical Integrity.

Abstract

Process Safety Management (PSM) in the Fuels and Petrochemicals industry has the paradox of being well established and yet constantly challenged to improve its effectiveness and impact on process safety and profitability. Such a paradox is a natural consequence of an industry exhibiting a positive process safety culture that maintains a sense of vulnerability and sees the imperative to improve process safety. This paper addresses the frequently asked question, "How do we continue to improve our organization's process safety culture?" and presents opportunities and examples for organizations to enhance their existing PSM systems and culture by further integrating PSM elements both within an organization's established PSM systems and its other systems, processes, and procedures. The paper illustrates integration and enhancement opportunities using Relief System Design Basis (RSDB) documentation, a component of the PSM element - Process Safety Information (PSI). Illustrations show RSDB integration with the PSM elements of Mechanical Integrity (MI), Process Hazard Analysis (PHA), and Incident Investigation (II). The paper then illustrates opportunities to increase profitability by highlighting RSDB integration with partial downtime throughput opportunities and improved project definition. The illustrations challenge organizations that truly "own" a process safety culture to look for additional opportunities to further enhance and integrate established PSM systems into their organization to further improve safety and profitability.

1. Introduction

The Fuels and Petrochemicals Industry in partnership with governing regulatory agencies have established effective Process Safety Management (PSM) systems to handle hazardous materials and energies over the past several decades. Examples include the U.S. Occupational Health and Safety Administration (OSHA) PSM Standard and the European Commission Seveso II Directive. The Center for Chemical Process Safety (CCPS) offers an extensive reference list of regulatory/industry/system requirements.[1] The catalyst for these industry systems have often been the tragic circumstances around industry incidents and catastrophes. Today, PSM in the Fuels and Petrochemicals Industry has the paradox of being well established and yet constantly challenged to improve its effectiveness and impact on process safety and profitability. Such a paradox is a natural consequence of an industry exhibiting a positive process safety culture that maintains a sense of vulnerability and sees the imperative to improve process safety.[2] The vulnerability and sense of urgency are well founded as the industry continues to experience preventable incidents. The Marsh Report states the 100 largest property damage losses in the hydrocarbon industry from 1974 to 2013 had a combined property damage cost of approximately US$34 billion.[3] Organizations demonstrating a committed process safety culture continually challenge and address the question, "How do we improve our organization's process safety culture?"

Organizations that have successfully managed the PSM implementation and sustainment phases have already experienced and moved past the common misconception that process safety and profitability have conflicting objectives. These organizations further realize they must move beyond a "separate but equal" PSM system established among their current business management systems, and fully integrate PSM into their systems to further improve their organization's impact on process safety and profitability. Recent efforts in the Fuels and Petrochemicals Industry to improve process safety through management system integration include industry efforts such as:

- The American Fuel and Petrochemical Manufacturers (AFPM) "Advancing Process Safety Programs," which promotes sharing PSM learnings across member organizations of AFPM
- The CCPS publication containing guidelines for integrating management systems and metrics to improve process safety performance.[4]

This paper identifies and presents specific examples and opportunities for organizations to continue to enhance their existing PSM systems and process safety culture by further integrating PSM elements both within an organization's established PSM systems and its other business systems, processes, and procedures. The paper illustrates integration and enhancement opportunities using Relief System Design Basis (RSDB) documentation, a component of the PSM Process Safety Information (PSI) element. Specifically, the examples illustrate RSDB integration with the PSM elements of:

- Mechanical Integrity (MI)
- Process Hazard Analysis (PHA)
- Incident Investigation (II)

RSDB integration with other business processes also present opportunities to directly improve profitability. Two examples presented here include RSDB integration with:

- Production planning (partial downtime throughput opportunities)
- Improved project definition

The following sections provide a general introduction to RSDB and illustrate the five integration opportunities in more detail.

2. An Introduction to RSDB and its Relationship to PSM

Within the Fuels and Petrochemicals Industry, many processes involve highly hazardous chemicals. To adequately assess and address a process' hazards an organization must adequately define the Process Safety Information of the process. Process Safety Information (PSI), includes information regarding the process' hazards, technology, and equipment.[5] Examples of hazard information include toxicity and permissible exposure limits. Examples of technology information include a process' chemistry and operating limits. Finally, examples of equipment information include materials of construction and the Relief Systems Design Basis (RSDB). There are many references for what constitutes RSDB and how to effectively manage RSDB documentation.[6] The American Petroleum Institute states RSDB documentation should demonstrate that all pressure-containing equipment has been identified and the relief requirements adequately analyzed. Documentation should include specific engineering guidance, methodology, analysis, and reference information that collectively reflects Recognized and Generally Accepted Good Engineering Practices (RAGAGEP). [7]

RSDB is a key component of PSI, but RSDB is also an important component in maintaining other Process Safety Management (PSM) elements such as Mechanical Integrity (MI), Process Hazard Analysis (PHA) and an Incident Investigation (II). Furthermore, RSDB can be a significant aspect of process production constraints and project opportunities associated with processes. The next sections discuss integrating RSDB into the aforementioned PSM elements, production planning, and project development processes.

3. RSDB Integration with Mechanical Integrity (MI)

Integrating Relief System Design Basis (RSDB) analysis with Mechanical Integrity's (MI's) Risk Based Inspection (RBI) through a combined risk assessment can identify acute process equipment vulnerabilities. Mechanical Integrity (MI) activities ensure the proper design, installation, operation, and maintenance of process equipment. Risk Based Inspection (RBI) is one component of an effective MI program that links risk assessment and mitigation activities with appropriate inspection and process control activities. RSDB analysis, on the other hand, identifies the applicable relieving scenarios and required relieving rates affecting pressure-containing process equipment to ensure the relieving scenarios do not expose the equipment to operating conditions outside of design limits. The American Petroleum Institute's Recommended Practice (RP) 581 identifies the interdependency of the Pressure Relief Device (PRD) with the fixed equipment the PRD is protecting.[8] RP 581 evaluates PRD criticality to set mechanical integrity activities such as inspection frequency. The RBI methodology combines the PRD's anticipated demand rate with the probability of failure on demand of the PRD to determine an overall risk for the PRD and the process equipment the PRD protects. While the API RP 581 methodology is sound, when the RSDB documentation is not available or the RSDB analysis has identified deficiencies in the PRD design, the RBI methodology develops shortcomings that should be addressed.

As an example, consider a process vessel containing a flammable hydrocarbon liquid. An RBI of the vessel defines a risk associated with the vessel, the vessel's PRD and the anticipated overpressure scenarios. The RBI analysis is predicated in part, however, on the PRD adequately protecting the vessel in the anticipated overpressure scenarios should the PRD open. The RSDB documentation identifies fire as the only applicable overpressure scenario but concludes the PRD is undersized. The RSDB can provide additional insight into the level of overpressure that would occur (and the anticipated consequence). The additional insights would significantly impact the RBI risk assessment and potentially warrant a reassessment of inspection and mitigation actions for the PRD and the protected vessel.

The example illustrates how integrating RSDB with MI's RBI can identify acute equipment vulnerabilities and aid in further prioritizing MI efforts and RSDB deficiency resolution, including risk mitigation efforts. Consequently, integrating both RSDB and MI can further enhance PSM integration and improve process safety and profitability.

4. RSDB Integration with Process Hazard Analysis (PHA)

Integrating RSDB with PHA can reduce costs and improve the effectiveness of both the PHA and the RSDB. As most readers are aware, a Process Hazard Analysis (PHA) in the United States and many other countries, is a requirement of a Process Safety Management (PSM) program, (for example, 29 CFR 1910.119 requires PHAs and periodic revalidations). No matter which methodology a PHA employs [Hazard and Operability study (HAZOP), "what-if," Failure Modes and Effects Analysis (FMEA), etc.], capable teams must perform the analysis. The teams typically range from 4 to 8 (or more) persons, consisting of owner's personnel and often outside consultants/specialists. No matter who is present, the participants represent a necessary and significant staff resourcing cost. Furthermore, each PHA finding should be well documented, especially those requiring additional evaluation and stewardship before the findings are declared either a confirmed issue or not. The PHA finding documentation and corresponding follow-up/stewardship also represent a significant staff resourcing cost. Finally, actions required to address the confirmed issues, whether facilities changes, work process changes, or administrative changes represent significant costs.

With the PHA costs sufficiently identified, now consider how integrating RSDB with PHA reduces costs and improves PHA and RSDB effectiveness. PHAs always identify questions and concerns about the adequacy of the over (or under) pressure protection systems as well as items such as Maximum Allowable Working Pressure (MAWP), hydrotest pressure, materials of construction, etc. Furthermore, PHAs require the most recent drawings, heat and material balances, and other Process Safety Information (PSI). Similarly, the most recent RSDB should have many (if not all) of these items assembled and documented, with any changes since the most recent RSDB tracked and provided as a part of the facility's Management of Change (MOC) program. The overlap of PSI and resources between RSDB and PHA is substantial. Consequently, completing RSDB analysis prior to and integrating it with the PHA process and follow-up activities introduces cost savings. Common savings occur relating to over (or under) pressure findings. If properly completed and documented, the RSDB can provide the PHA participants with immediate feedback regarding the adequacy of pressure protection devices, which in most cases are code items and must be addressed. A benefit of having a quality RSDB done prior to the PHA is that the over or under pressure concerns can be quickly resolved. On many occasions, no action item follow-up work is necessary. This reduces costs both during the PHA and the PHA follow-up. Conversely, the PHA serves as an audit check of the RSDB as a capable team reviews the RSDB basis as needed.

As a quick example, a PHA team questioned the adequacy of an exchanger shell's design pressure during a blocked-in condition. The maximum pressure generated from a blocked-in condition from a manual valve closure was greater than the design pressure by approximately 10 percent. The PHA team referenced the RSDB. The RSDB previously identified the same issue and investigated the situation in more detail. The RSDB identified the exchanger's MAWP to be approximately 20 percent more than the noted design pressure, and the hydrotest pressure was based on 1.5 times the MAWP. The U-1 form that was part of the RSDB package clearly indicated this. The RSDB package also noted the exchanger shell was adequate based on ASME VIII, Appendix M, M-5.7, in this situation, because the pressure was below 116% of MAWP and it would be immediately noted by operators, the valve closure was not identified as an issue and the PHA follow-up item was closed. Because the PHA utilized the RSDB, the PHA saved costs on discussion and documentation of an action item and the necessary follow-up and tracking of a code related item. Conversely, the PHA team validated the RSDB, enhancing PSM integration and improving process safety and profitability.

5. RSDB Integration with Incident Investigation (II)

Incident Investigation (II) is also a mainstay of any effective personnel or Process Safety Management (PSM) program. II is an opportunity to learn from mistakes, system failures, equipment failures, and/or overlooked vulnerabilities. II is also an opportunity to evaluate the effectiveness of preventive and control measures installed to prevent or mitigate an incident. In the CCPS Vision 20/20 Characteristics of Companies with Great Process Safety Performance the authors write: "To reduce incidents, employers and employees must enthusiastically support a culture that is driven to learn from many sources including benchmarking, near misses and incidents, and jobs done well."[9] For incidents involving release from a relief device and/or a flaring event, II intersects Relief System Design Basis (RSDB) and together they provide a learning opportunity that enhances PSM and improves process safety.

Yogi Berra said; "When you come to a fork in the road, take it." When an incident occurs investigate it. Definitions may vary for what constitutes an incident. Indeed, there are legal requirements for investigation of safety and environmental incidents. These form the baseline for what organizations must do but do not limit organizations from doing more. For instance, some organizations with very effective PSM programs investigate all flaring events. The organization calling the event an incident, a near miss, or flaring event is least important, while the organization's learning opportunity from the event is most important.

By integrating RSDB with II organizations can do real case study evaluation of the RSDB's assumptions and technical predictions. Integration can help answer such questions as:

- Did the scenarios of the RSDB represent the incident that occurred?
- Did the relief valve effectively handle a release rate lower than the controlling case without issues such as chattering?
- Were the release rate, pressure limitation, and system temperature change what the calculations predicted they would be?
- Did the flare handle the relieving event without an environmental release or safety issue? For example: flare smoking, toxic release, or high temperature exposure of personnel and adjacent equipment.

RSDB integration with II may facilitate getting to the root cause of the incident. The RSDB contains valuable information about the relief system response and may help identify the original process design intent. If the relief system response was not as intended, the relief system may be the cause of the incident. Additionally, if there are layers of protection designed to make flaring or release valve opening an infrequent event but flaring is occurring often, the preventing layers of protection may need further evaluation.

The RSDB and II integration may also help identify additional inspections and corrective actions. In this case II also integrates Mechanical Integrity (MI). For example, damage to the relief system may have occurred from too-high or too-low temperature or pressure. Mechanical damage resulting from corrosive chemicals or relief valve chattering may have occurred. Additionally, the RSDB may shed light on damage that the incident caused to other process equipment. Finally, if the RSBD does not accurately predict what will occur during such an incident there should be a corrective action to change the RSDB and possibly modify the relief system. Consequently, the II serves to further validate the RSDB and the RSDB enhances the II. An important point to note: incorporating RSDB into II requires someone knowledgeable in RSDB be on the II team.

In conclusion, a relief device release or flaring event II that does not consider the RSDB leaves a valuable tool unused. Integrating RSDB with II enhances PSM integration and improves process safety and profitability by gaining additional benefit from both RSDB and II.

6. RSDB Integration with Production Planning

Opportunities often present themselves in complex facilities to ensure process safety and increase profitability by integrating Relief System Design Basis (RSDB) with production planning. Many facilities integrate processing units in a complex manner. Such facilities often have multiple production constraints preventing individual processing units from running maximum capacities simultaneously. Production constraints may include downstream processing capacity, feed qualities, processing unit downtimes, permitting limits, or even facility RSDB constraints. When a facility RSDB constraint limits production, understanding the relationship between other production constraints and the facility RSDB constraint becomes critical to ensure process safety and optimal profitability. Hence the need and opportunity to integrate RSDB with production planning.

Before presenting RSDB and production planning integration examples, there are a few points to note. First, RSDB analysis for an individual processing unit typically establishes the unit's safe operating window at a maximum capacity. A facility RSDB, however, reflects multiple processing units operating simultaneously. Second, due to the myriad of production constraints a complex facility may have, the facility RSDB often reflects a limited set of facility production constraints (often with a very conservative basis). Third, in some instances, the facility RSDB may constrain various processing units to a fraction of their maximum production capacity. Finally, when production constraints setting the facility RSDB change, production planning needs to evaluate alternative production options that the facility RSDB does not adequately reflect.

In one example, a complex facility relief system constraint of excessive relief system backpressure for low set point relief devices exists. The excessive backpressures result from a potential utility failure scenario impacting multiple processing units. This facility RSDB constraint limits the facility's production. When the processing unit with the low set point relief devices is down for maintenance, however, the facility RSDB constraint relaxes. Thus, opportunities present themselves for other processing units to increase production. In another example, a complex facility schedules a portion of the facility relief system maintenance downtime with processing units having a substantial impact on the facility relief system constraint. During the downtime, however, the startup timing for the relief system is delayed due to discovery work and opportunities arise to increase profitability by minimizing production cuts within a newly defined facility RSDB constraint.

7. RSDB Integration with Project Development

Integrating Relief System Design Basis (RSDB) with projects presents opportunities to improve project definition, increase project profitability (through project cost savings), and enhance Process Safety Management (PSM). Having a preliminary well documented RSDB at the on-set of a project — often called the Front End Engineering Design (FEED) stage — will identify the various scenarios that should be a part of the sizing of not only the individual relief devices, but also the relief device collection system functional requirements. The potential costs of the project's impact on the relief devices and systems can then be reviewed and optimized in coordination with the project's functional equipment. The following examples best illustrate these points.

In a larger project a liquid feed drum was required. The feed drum's initial design was 50 psig, and the feed sources ranged from 35 psig to 100 psig. Because the feeds to the drum were greater than the drum's proposed design pressure, the overpressure device and corresponding system had to be designed for liquid overfill. All the sources that were greater than 50 psig represented an overfill potential of around 70,000 barrels a day, and consequently the relief device and the corresponding collection system size and costs were unattractive. Upon further investigation and review of the options, the project team increased the design pressure of the feed drum to 110 psig, which had minimal cost differential when compared to the necessary relief system design. Thus, the 110 psig design was chosen and the relief device was then sized for fire and some potential back-flow on stoppage of the feed pumps. Cost savings were in the millions of dollars.

A second project involved design and installation of equipment that had to be tied into the flare system. A preliminary RSDB was done early in the project. Upon analysis of some typical global scenarios such as loss of power, loss of cooling water, etc., the new project's primary relief device(s) would require increasing the collection system's piping and flare size. Furthermore, the sizing of the relief device and piping alone had a large estimated cost.

The relief device's size was driven by the main tower's selected design pressure and steam reboiler pressure available. The project team, in coordination with the RSDB personnel, were tasked with optimizing the reboiler and the relief system. During the optimization phase, the team identified 2 key items: 1) the reboiler steam had only one pressure available, but it could be let down to a lower pressure, and 2) at a slightly lower pressure the steam could easily provide the available heat necessary for the process.

Additional investigation by the team revealed that the tower material and thickness had not initially taken advantage of the potential MAWP. The material thickness needed was ~0.82 inches, and 0.875 inches was to be selected. This allowed for a MAWP higher than design. However, because the steam pressure was reduced, good engineering practice would require that the design consider failure of the steam pressure control valve.

Failure of the steam pressure control valve would provide sufficient heat to overpressure the tower based on the new design pressure and the relief device to the flare was still needed. To address this failure, the steam system (downstream of the pressure let-down valve) was provided with its own pressure relief valve which could be vented to atmosphere.

With the team's recommended changes noted above, the only scenario for the main tower relief device was fire, which had a lower cost of installation and did not require an update of the flare system. Having the RSDB done early in the project allowed the designers and project team to make some easy (paper) cost effective changes, and did not require any modifications to the existing collection and flare systems.

The examples above demonstrate organizations integrating RSDB during project development to improve project definition, reduce project cost, and even make processes inherently safer. Organizations realize further PSM enhancements by integrating a process unit's existing RSDB documentation with project development activities. In addition to the benefits illustrated above, organizations incorporate project impacts on a process unit's RSDB with minimal duplication of effort (minimal cost), and enhance the process unit's RSDB integrity. From an overall PSM perspective, the project development integration enhances both the PSM and project development system.

8. Conclusion

This paper presents opportunities for organizations to enhance their PSM systems and process safety culture by integrating Relief System Design Basis (RSDB) with the following PSM elements and business processes:

- Mechanical Integrity (MI)
- Process Hazard Analysis (PHA)
- Incident Investigation (II)
- Production planning
- Project development.

Organizations that truly "own" a process safety culture should use each successful opportunity implementation as a spring board to further enhance PSM integration and improve process safety and profitability.

9. References

1. CCPS, "Guidelines for Integrating Management Systems and Metrics to Improve Process Safety Performance", Appendix A, Center for Chemical Process Safety, American Institute of Chemical Engineers, New York, NY 2016.

2. CCPS, "Vision 20/20 Booklet – Committed Culture", Center for Chemical Process Safety, American Institute of Chemical Engineers, New York, NY 2016.

3. Marsh and McLennan Companies, "The 100 Largest Losses", 23rd edition (2014).

4. CCPS, "Guidelines for Integrating Management Systems and Metrics to Improve Process Safety Performance", Center for Chemical Process Safety, American Institute of Chemical Engineers, New York, NY 2016.

5. Occupational Health and Safety (OSHA) "Process Safety management of Highly Hazards Chemicals" 29CFR1910.119(d).

6. Thomas Watson, "AM-13-24 Considerations for Effectively Managing Emergency Relief Systems Design," American Fuels & Petrochemicals Manufacturers Annual Meeting, 2013.

7. American Petroleum Institute, "API Standard 521 – Pressure-relieving and Depressuring Systems", section 4.7.3, sixth edition (2014).

8. American Petroleum Institute, "API RP 581 – Risk-Based Inspection Technology, Part 1", Section 7, second edition (2008).

9. Center for Chemical Process Safety's Vision 20/20: "Characteristics of Companies with Great Process Safety Performance" (2013, McCavit, Berger, and Nara).

479098 Reactive Hazards Management: Lessons Learnt from Major Incidents

Abstract:

Chemical process industry has long suffered from major incidents due to lack of proper understanding of reactive hazards. The methyl-isocyanate leak in Bhopal, India, has been by far the worst industrial incident in history which led to the development of many process safety regulations. However, the formation and implementation of regulations has not prevented the occurrence of these incidents. Many incidents such as Concept Sciences Inc. where a hydroxylamine manufacturing unit was destroyed, the explosion at the Sterigenics International Inc., ethylene oxide sterilization facility, Bayer CropScience Pesticide Chemical Runaway Reaction Explosion and West Fertilizer Explosion have highlighted the importance of having a better understanding of the reactive hazards. The focus of this presentation is to identify different types of reactive hazards through examples of chemicals incidents and analyze how a better understanding of these hazards could have avoided these incidents.

479245 The Study of Safety Evaluation Technique for High Temperature Process

Abstract:

There are many accidents of a firing furnace and kneading machine by overheating and the abnormal injection of the product. When the gas which suddenly expanded by thermal decomposition destroys plant equipment and enters the explosion concentration range, a gas explosion is caused and damage spreads. The material which it is easy to decompose is invested by mistake by a firing furnace and has the trouble that the pressure of the furnace abnormally rises to in our factory. Also, troubles such as an explosion of the plumbing by the thermal decomposition of the polymer by the overheat, vent explosion by the thermal decomposition by the false injection of an additive having low heat stability has broken out with kneading machine

These troubles are caused by the thermal decomposition by a sudden temperature rise. The temperature program of the analytical instrument which is generally used by a thermal analysis is 10-20 /min. The conventional thermal analysis equipment cannot evaluate the temperature rise rate at the time of the trouble exactly to become than 20 /min. It becomes the problem that the slow temperature rise may be different from rapid heating up in a decomposing reaction.

We produced a device capable of quickly raising temperature to solve these problems.

This device is comprised of the vertical plumbing which it sealed and heating furnace, and it is very simple structure. The sample holder is fixed to the upper part in the plumbing by a magnet. And the holder drops in the heating zone by keeping a magnet away from the plumbing and is heated rapidly. The materials of the plumbing are SUS316L, and, as for the use upper limit temperature, 500 , the use upper limit pressure are 1MPa. Because the deterioration of SUS materials by the heat is intense when it is raised the temperature of the device to 750 , the plumbing is disposable.

The inside of the plumbing is replaced with inert gases such as nitrogen, argon. The maximum temperature rising rate is 150 /min (furnace temperature: 750).

We introduce an example of the trouble analysis of a firing furnace and kneading machine using this device.

[Case example 1] Study of firing furnace trouble

After burning with a new additive at 720 , abnormal pressure rise happened in a furnace, and a plant stopped. After measuring quantity of outbreak gas using the rapid heating device, it was recognized that gas of 7 times was generated and a gasification rate tripled in comparison with a conventional additive.

[Case example 2] Study of kneading machine trouble by overheating

The additive which had low heat stability was spent mistakenly in large quantities to a kneading machine, and a vent board exploded. After performing a rapid temperature rise examination under a kneading temperature condition, as for the pressure raising rate when an additive disintegrated, 1.8MPa/min, the quantity of gasification became 190L/kg. The additive concentration became (mixing ratio with the resin) more than at least 75%, and it was recognized that a large amount of additives were invested mistakenly.

Fortunately, this example does not become the big accident with resulting in injury or death. In this explosion example, a vent board penetrated a checker plate. It might become the casualty when there was a person near a kneading machine.

In comparison with the plant of the petrochemical and fine chemistry that heat risk analytical technique establishes, the firing furnace and kneading machine is often run without detailed risk assessment. The rapid heating device becomes the powerful tool to do the heat risk evaluation of these high temperature processes.

In future, it is going to be improved the rapid heating device changes the plumbing materials, and to use it to the temperature ranges more than 1,000 .

Handling uncertainty in LOPA and SIF design

Akshat Khirwal
IRESC
Unit 2801, Enterprise Square 3
39 Wang Chiu Road
Kowloon Bay
Hong Kong
Akshat.khirwal@irescglobal.com

Arjan Abeynaike, Venkatesh Sourirajan
IRESC
Arjan.abeynaike@irescglobal.com, Venkatesh.s@irescglobal.com

Keywords: Uncertainties, LOPA, SIL, SIF design, reliability

Abstract

Layer of protection analysis (LOPA) is a widely used semi-quantitative method for determining the reliability requirement for safety instrumented functions (SIFs). An advantage of LOPA is that it specifies a numerical reliability target - the average probability of failure on demand (PFD_{avg}). However, uncertainties in the input parameters and simplifications within the LOPA method lead to a margin of error in the target PFD_{avg} that is often not accounted for in the design of the SIF. Further uncertainties are introduced when carrying out a reliability analysis of the SIF design to ensure that the target PFD_{avg} is met. The result of overlooking such uncertainties is that the target PFD_{avg} may be set low and the calculation of the actual PFD_{avg} achieved by the design may be optimistic, leading to a SIF that is under-designed. The need to account for uncertainty is emphasized in the latest version of IEC 61511, the international standard for SIF design.

This paper investigates and highlights the sources of such uncertainties in LOPA and SIF reliability analysis. In particular, the quality of the data used as input parameters to LOPA and reliability models has a large effect on the accuracy of the results. These parameters include failure rates, conditional modifiers, beta factors for common cause failures, and proof test coverage factors. Available data for these parameters can be sparse and may not be representative of the conditions in which the SIF is applied. Where the confidence intervals of the data are available, these can be incorporated into the calculations to reach a more conservative result. Alternatively, a safety factor can be applied to try and capture all the uncertainties. In this approach, the target PFD_{avg} obtained from LOPA is reduced by the safety factor to give a more stringent target PFD_{avg} for the SIF design to meet. The methods for handling uncertainty are reviewed in this paper.

1 Introduction

SIL Determination is the process of assigning a Safety Integrity Level (SIL) to Safety Instrumented Functions (SIFs) following any of the methods i.e. Risk Graph, Risk Matrix or Layer of Protection Analysis (LOPA); as laid down in IEC 61511 and IEC 61508 [1] [2]. Of these methods, the semi-quantitative LOPA approach is increasingly being adopted for SIL Determination as compared to other qualitative approaches such as Risk Graph or Risk Matrix. An advantage of LOPA is that it specifies a numerical reliability target for a safety instrumented functions (SIF), which is expressed in terms of the target average probability of failure on demand (target PFD_{avg}), or target risk reduction factor (RRF, where RRF = 1/ PFD_{avg}). This target PFD_{avg} then forms the basis for design of the SIF wherein a reliability analysis is carried out to ensure that the target PFD_{avg} is met. However, uncertainties in the input parameters and simplifications within the LOPA method lead to a margin of error in the target PFD_{avg} that is often not accounted for in the design of the SIF. Further uncertainties are introduced when carrying out a reliability analysis of the SIF design. Without accounting for these uncertainties in the assessment, the SIF may be under-designed since LOPA may result in under prediction of the target PFD_{avg} or the reliability analysis may show that the target PFD_{avg} has been achieved although this may not be the case. This paper has reviewed the sources of uncertainties in the LOPA method and design of the SIF (SIF reliability analysis) and provides a way forward for handling such uncertainties.

2 Uncertainties in LOPA

Initiating event frequency, consequence severity, and likelihood of failure of Independent Protection Layers (IPLs) are the key parameters used in LOPA to approximate the risk of a scenario in case the SIF fails to act when the demand arises. In addition to the above, conditional modifiers [3] may also be used to factor in specific parameters pertaining to the demand on the SIF. The simplicity of LOPA allows it to be easily understood by engineers of all disciplines, leading to its widespread use. The drawback of this simplicity is the greater uncertainty in the results, as described in the following sub-sections. The governing equations and the LOPA worksheet used for assessing the scenarios discussed in this paper are referenced from IEC 61511-3 [1] and CCPS, Simplified Process Risk Assessment [4].

2.1 Initiating event frequency

The demand on any SIF may arise due to one or more events, each with a frequency in the range of once per year to once in plant life (for example, 30 years). There is some approximation involved in the selected initiating event frequency values especially when using generic databases. For instance, the source data in the generic database may not be appropriate for application to the facility under review or there may not be enough confidence in the dataset. Power failure or failure of other utilities such as cooling water or steam is a good example. The frequency of these failures from generic databases is typically 1×10^{-1} per year (once in 10 years) [4]. However, in the author's experience, based on feedback from many operators, these values are likely to be somewhere between once per year to once in ten years. Similarly, the frequency of pump trip based on generic databases is 1×10^{-1} per year, however operational experience suggests that a value in the range of once in 2 years to once in 3 years may be more appropriate. An analysis of the OREDA [5] data also provides a value close to once in 3 years. Additionally, when there are multiple demand scenarios for a particular SIF, failure to consider the cumulative frequency of all the demand scenarios can also lead to under-estimation of the overall initiating frequency. A simple approach is to consider a suitable approximation for the cumulative demand frequency [6] [7]. In addition, common cause events specific to the facility is also a contributor which needs to be identified. Without such considerations, the required target PFD_{avg} may be under-predicted. All of the above introduce uncertainties in the LOPA assessment noting that the initiating event frequency is the starting point of a LOPA assessment. Therefore, care must be taken in selecting an appropriate value for this parameter.

To better understand the dependence on the initiating event frequency, consider a typical distillation column as shown in Figure 1. There are many scenarios such as overhead condenser failure, reflux pump failure, reboiler control failure, etc. which may lead to a demand on the high high pressure SIF provided on the column. For the sake of brevity, only the case of Reflux Pump trip is used further in this example.

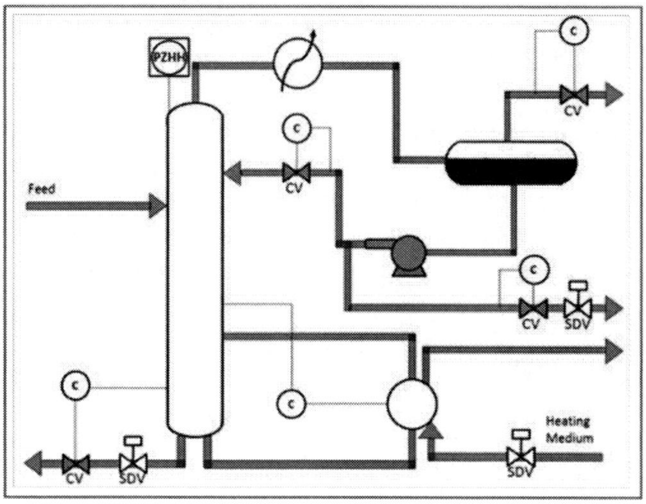

Figure 1. Typical Distillation Column

Assuming an initiating event frequency of 1×10^{-1} per year for pump trip and considering a Target Mitigated Event Likelihood (TMEL) of 1×10^{-5} per year for multiple fatalities along with an ignition probability of 0.1 and taking credit for pressure safety valve (PSV) on the column (PFD_{avg} of 1×10^{-2}), this would result in a SIL 1 with Risk Reduction Factor (RRF) 10 (target PFD_{avg} of 1×10^{-1}). However, if the initiating event frequency for pump trip is changed to 3.33×10^{-1} per year (once in 3 years), this translates to SIL 1 with RRF of 33 (target PFD_{avg} of 3×10^{-2}). This example demonstrates the dependence of initiating event frequency selection on the target PFD_{avg}.

2.2 Consequence Severity and Conditional modifiers

The TMEL assigned to the SIF is directly associated with the degree or the severity of the consequence assumed (e.g. single fatality: 1×10^{-4} per year, multiple fatalities: 1×10^{-5} per year). This in turn has a direct impact on the SIL assigned to the SIF as well as its target PFD_{avg}. Therefore, careful consideration must be given to the selection of the consequence severity during a LOPA review. For instance, the impact area or hazard zone due to a loss of containment event can be determined using consequence modelling [7], the result from which can be used to assign the severity (e.g. single fatality vs multiple fatalities). Additionally, this can be used to determine the relevant conditional modifiers such as occupancy factor within the impact zone. This approach is more reliable than making coarse assumptions which may introduce uncertainties in the analysis. Such refinement can also be applied to other modifiers such as ignition probability where using a more specific value representative of the hazardous fluid and/or the facility features may be appropriate as compared to use of generic values.

If we revisit the example of typically distillation column presented in Section 2.1; considering an initiating event frequency of 3.33×10^{-1} per year and assuming a TMEL of 1×10^{-4} per year (single fatality), an ignition probability of 0.1 and taking credit for PSV on the column (PFD_{avg}

of 1 x 10^{-2}), this would result in a RRF of 3 (target PFD_{avg} of 3 x 10^{-1}), which is below SIL 1 requirement. However, if a higher severity (TMEL: 1 x 10^{-5} per year corresponding to multiple fatalities) is selected based on consequence modelling results and an ignition probability of 0.3 is selected based on refinement, the assigned SIL will increase significantly to SIL 2 with RRF 100 (target PFD_{avg} of 1 x 10^{-2}). Although this example shows an increase in the assigned SIL, the application of detailed assessment for determining severity could also show a reduction.

2.3 Independent Protection Layers

Independent Protection Layers (IPLs) are devices, systems or actions that prevent the realization of an undesired consequence (e.g. the consequence the SIF is safeguarding against). Each IPL has an associated PFD_{avg} which is assigned and used in the LOPA methodology for SIL Determination. For example, process deviation alarm with sufficient time for operator intervention has a PFD_{avg} of 1 x 10^{-1}, an independent BPCS control has a PFD_{avg} of 1 x 10^{-1} while a PSV has a PFD_{avg} of 1 x 10^{-2}.

2.3.1 Effect of Common Cause Failures

One of the fundamental assumptions in the LOPA approach is that these IPLs are completely independent and do not share any common cause failures (CCF). In reality, multiple protection layers may be impacted or their performance may be affected due to similar environmental conditions or deviations from desired process parameters. Corrosion or deposition leading to simultaneous degradation of a pressure safety valve and a shutdown valve is one such example. Other examples of CCF include scenarios where multiple PSVs are required to meet total relief requirements, or where an alarm and control, although independent safeguards, both reside on the BPCS. Additionally, dual check valves for preventing backflow could also be impacted by CCF. Although such scenarios may take a period of time to develop and the likelihood may be low, such CCF can have a direct impact on the PFD_{avg}. For example, consider a case where the safety system comprises a SIF and an IPL and the target PFD_{avg} for this system is 1 x 10^{-3}. The PFD_{avg} of the IPL is 5 x 10^{-2}. If the SIF and IPL fail independently only, then the target PFD_{avg} required of the SIF is (1 x 10^{-3}) / (5 x 10^{-2}) = 2 x 10^{-2}. However on inclusion of CCF, even if a low probability of CCF is assumed (1% of the PFD_{avg} of the IPL), the target PFD_{avg} for the SIF then becomes 1 x 10^{-2}.

The authors do acknowledge that for the purpose of time-effective LOPA review, it may not be practical to dwell deeper into CCF for IPLs as it may require an in-depth analysis (for which sufficient details may also not be available at the time of LOPA review). However, it is important to be aware of the inherent uncertainties introduced in the analysis by not accounting for the CCF associated with IPLs.

2.3.2 Effect of Multiple Protection Layers

There is another inherent uncertainty associated with IPLs apart from CCF. The PFD_{avg} for any protection layer (whether mechanical protection, alarm with operator response, BPCS or SIF) is not static but increases over time until a required test or maintenance activity is conducted to check for faults [4]. This relationship between PFD and time is shown graphically in Figure 2. In LOPA, the PFD_{avg} values of the IPLs are multiplied together to estimate the PFD_{avg} of the overall safety system [4]. This calculation, while simple to perform, may tend to underestimate the overall PFD_{avg}. In a more quantitative analysis of reliability, the time-dependent PFDs of each protection layer are multiplied together to yield a result similar to Figure 3, which shows the PFD of the safety system. The PFD_{avg} for the system can then be calculated.

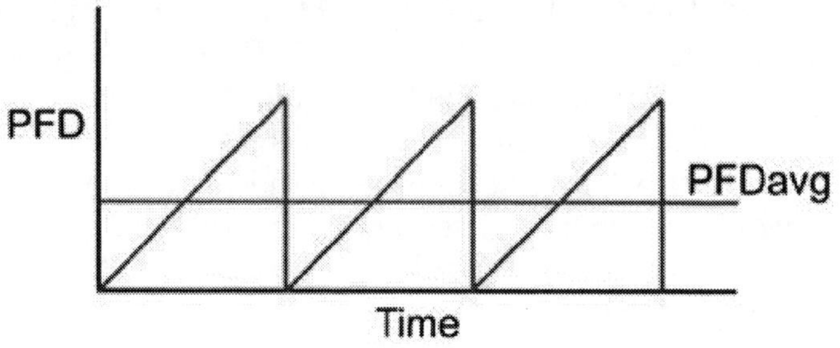

Figure 2. PFD as a function of time for a single protection layer. Periodic testing / maintenance was assumed to detect all faults in this example.

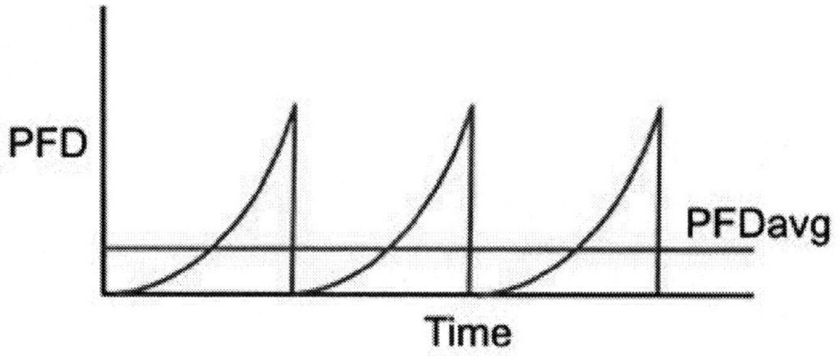

Figure 3. PFD as a function of time for a safety system with two protection layers (one SIF and one other IPL). Proof testing was assumed to detect all faults in this example.

To illustrate the difference in the results provided by LOPA and a more rigorous analysis, the same example from Section 2.3.1 is used. LOPA estimates the PFD_{avg} of the SIF to be 2×10^{-2} (excluding CCF), considering one IPL with a PFD_{avg} of 5×10^{-2} per year and a target overall PFD for the system of 1×10^{-3} (see calculation in Section 2.3.1). Using the PDS method of reliability analysis [8] gives a SIF PFD_{avg} of 1.5×10^{-2}, i.e. LOPA overestimates the target PFD_{avg} by 33%. As the number of protection layers increases, this error becomes more significant, as shown in Table 1. For two or more IPLs with a SIF, the error in the target PFD_{avg} for the SIF is at least 100% using the PDS method; i.e. the LOPA value for target PFD_{avg} is double that given by the PDS method. This gross overestimation in the target PFD_{avg} for the SIF suggests that the LOPA method may not be sufficiently accurate where two or more IPLs are considered. In such cases a more rigorous analysis may be conducted outside of the LOPA session (refer Section 3). Such rigorous approach, if adopted, may also help reduce the error for cases where only one IPL is applicable.

Table 1: Effect of the number of IPLs on the error in PFD_{avg} for a safety system as calculated by LOPA

Number of IPLs	Error in target PFD_{avg} for SIF
0	0%
1	33%
2	100%
3	220%
n	$\frac{2^{n+1}}{n+2} - 1$

2.4 Assessing the cumulative uncertainty for multiple parameters in LOPA

The above sections have discussed how the uncertainty of each individual parameter affects the target PFD_{avg} calculation. In practice, however, the cumulative effect of the uncertainties for multiple parameters also needs to be considered. As a simple illustration and keeping in line with the discussion above, the same example of the distillation column high high pressure SIF is used. As shown in the table below, the changes in various parameters results in a significant difference in the target PFD_{avg} requirements. One additional parameter presented in the table below is the use of PFD_{avg} of 3×10^{-2} for IPLs considering multiple PSVs are required to meet relief requirements. As expected, the required RRF is much more stringent when the uncertainties are modelled. However, as highlighted in earlier sections, depending on the SIF under review, in some cases, the requirements may also be less stringent upon further refinement.

Table 2: Effect of cumulative uncertainty of multiple parameters in LOPA

Description*	Initiating Event Frequency	IPL	Ign. Prob.	Total Mitigated Event Frequency	TMEL	Target PFD_{avg}	RRF
Case 1	1.00E-01	1.00E-02	1.00E-01	1.00E-04	1.00E-04	1.00E+00	1
Case 2	3.33E-01	1.00E-02	3.00E-01	1.00E-03	1.00E-04	1.00E-01	10
Case 3	1.00E-01	1.00E-02	1.00E-01	1.00E-04	1.00E-05	1.00E-01	10
Case 4	3.33E-01	1.00E-02	3.00E-01	1.00E-03	1.00E-05	1.00E-02	100
Case 5	1.00E-01	3.00E-02	1.00E-01	3.00E-04	1.00E-04	3.33E-01	3
Case 6	3.33E-01	3.00E-02	3.00E-01	9.00E-04	1.00E-04	1.11E-01	9

3 Approaches for handling uncertainty in LOPA

Based on the various modes of introduction of uncertainties in the LOPA analysis as discussed in the preceding sections, the question that arises is whether the direct use of target PFD_{avg} from LOPA as input to SIF design is optimal or is it desirable to investigate possible ways to refine the target PFD_{avg} from LOPA to ensure the SIF is not under-designed. Some ways to refine the target PFD_{avg} are discussed in the following paragraphs.

One proposed approach for dealing with uncertainty without compromising the simplicity that makes LOPA an attractive method is to conduct a more rigorous risk assessment outside of the LOPA session for high consequence / high risk scenarios to confirm that the SIF reliability target PFD_{avg} obtained from LOPA is sufficient. Such an assessment may include a quantitative reliability analysis such as a Fault Tree Analysis or Markov Model to obtain a more accurate estimate of the required target PFD_{avg}. These methods can also take into account common cause failures and the time dependency of PFD. In fact the most recent version of IEC 61511 [1] requires that a quantitative methodology be used for cases where the required risk reduction is greater than 10,000 (target PFD_{avg} from LOPA is equal to or lower than 1×10^{-4} per year). By using LOPA as a screening tool, this strategy can ensure that high consequence / high risk scenarios receive the more rigorous assessment they deserve, while for lower risk scenarios the more time-efficient LOPA can be considered sufficient.

Alternatively, a number of different cases can be considered for each LOPA scenario, such as in Table 2. Each case considers different values for the various parameters used in the LOPA calculation, resulting in a range of RRFs, from the more conservative to the more liberal. Taking Table 2 as an example, the results range from an RRF of 1 to 100. A final RRF at some point in between these two extreme cases may be selected for the SIF and carried forward into the SIF design. In this way, a level of uncertainty can be accounted for without being overly conservative.

Another approach is to introduce a "Safety Factor" to the reliability target set during LOPA. In this approach, it is recommended that the target Risk Reduction Factor obtained from LOPA should be increased by a safety factor to give a more stringent target for the SIF design to meet. For instance, if the target based on results from LOPA is RRF 100, the SIF design can conservatively consider a target PFD_{avg} associated with RRF 300. Similarly, if the target is RRF 500, then SIF design may consider RRF 800. However, when the RRF is at or close to borderline values, the SIF design may consider the next interval, e.g. for RRF 800 and above, the SIF design may consider RRF 1300 or alternately the RRF used can be justified by conducting a more rigorous quantitative analysis as mentioned above. Some engineering judgement should be applied when selecting this safety factor depending on the level of conservatism adopted in the initial LOPA assessment. As can be inferred from Table 2 above, when uncertainties are accounted for, the target PFD_{avg} can actually vary by one order of magnitude. Applying a safety factor to the results will therefore account for the uncertainties.

4 What PFD_{avg} to assume when Risk Graph Method is adopted for SIL Determination?

Risk Graph Method [1] is the alternate method commonly used for SIL Determination. Unlike the semi-quantitative LOPA approach, which specifies a numerical reliability target (PFD_{avg}) for the SIF in addition to the assigned SIL, the Risk Graph approach relies on the qualitative definitions for the parameters involved to deduce the assigned SIL without setting an actual numerical target (PFD_{avg}). Furthermore, this target PFD_{avg} value from Risk Graph is provided as a range. Hence, where Risk Graph method is used, it is not an uncommon practice to use the upper bound of the PFD_{avg} associated with the assigned SIL. For example, the LOPA method may provide an RRF of 40 (PFD_{avg} of 2.5×10^{-2} per year) for a SIL 1 function, while the same SIL 1 function following the Risk Graph method may adopt the upper bound RRF of 10 (PFD_{avg} of 1×10^{-1} per year).

Consider the same example of a distillation column as discussed in Section 2. If the column high high pressure SIF is assessed using the Risk Graph approach assuming that a trip of Reflux Pump could be expected once a year, taking credit for the pressure safety valve on the column as an IPL (credit of 100) and taking no credit for occupancy and avoidance; the resulting analysis show that Risk Graph method also determines a SIL 2 which is similar to the results from LOPA are accounting for uncertainties. The key difference being the overall PFD_{avg} of 3.33×10^{-3} per year (RRF 300) obtained from the LOPA which will be more stringent than the PFD_{avg} of 1×10^{-2} per year (RRF 100) in case the upper bound value of SIL 2 is used.

The above example shows that using the output from the Risk Graph directly may lead to under-prediction of the target PFD_{avg} and thereby leading to under design of the SIF. The concept of "Safety Factor" as discussed in Section 3 may therefore also be applied where Risk Graph is used. Based on this, it is recommended that the SIF design can use the mid-point of the upper bound and lower bound values based on PFD_{avg} range applicable to the SIF corresponding to the assigned SIL. The safety factor may also be selected on a case by case basis depending on the severity of the associated consequences (for example, using a value above the mid-point of upper and lower bound may be justified for the most severe consequences).

5 Uncertainties in SIL reliability analysis

The uncertainties in terms of obtaining an appropriate target PFD_{avg} for a SIF using LOPA methodology are discussed in Section 2. It is however, worth noting that potential for introduction of uncertainties into the analysis is not limited to this step alone. Moving on in the safety life cycle, the next step after LOPA involves a reliability analysis to ensure that the target PFD_{avg} is met for each SIF. This activity is commonly termed as "SIL Verification" in the industry, but strictly with reference to IEC 61511, it is considered within the domain of "SIS Design and Engineering". The target PFD_{avg} that is considered for the reliability analysis should be the same as the target PFD_{avg} that was calculated in LOPA (a safety factor may be applied as described in Section 3). Where Risk Graph methodology is applied, the target PFD_{avg} can be a mid-point value in the assigned SIL band as suggested in Section 4.

This reliability analysis also involves a number of input parameters, some of which may introduce uncertainties into the analysis. It is important to understand and evaluate the nature of these uncertainties associated with input parameters such as failure rate data for SIF components, beta factor for common causes and proof test coverage factors in SIF reliability analysis as discussed in the subsequent sub-sections. In this section, PFD_{avg} results that were calculated using exSILentia software are presented to illustrate the effects of the input parameters on the PFD_{avg}. The numerical PFD_{avg} results are only applicable for the specific systems and assumptions used in the calculations.

5.1 Failure Rate Data

Failure rate data for the individual SIF components is the basis of the reliability analysis. It is therefore of utmost importance that the data used is based on sufficient sample size and is representative of the conditions in which the SIF is applied. The quality of the data used as input parameters to reliability models has a large effect on the accuracy of the results.

Ideally, an appropriate method for determining the failure rate is to collect field data from a large, randomized sample of units. The operational time until failure should be recorded for each unit. Such data mining however; is a long term project, requiring the co-operation of multiple operators, ideally from a wide range of different facilities and operating environments. For this reason, such studies are rare and normally "generic data", i.e. data applicable for a broad type of component (such as pressure transmitter or shut-down valve) is used rather than data for a specific model or a specific environment. The OREDA database [5] is one example, which provides such generic data. Alternately, some vendors may commission a field study to collect failure data specific to their components or build their own failure data based on reported failures. However, given the smaller sample size as well as the gaps in ensuring that each failure is reported appropriately by the operating facilities, there will be uncertainties in the reported data and these values may tend to be on the optimistic side. For example, while the failure rate for a ball valve from a generic data source is 9.8×10^{-7} per hour [9], several vendors of ball valves report failure rates of less than 1.0×10^{-7} per hour. Another issue is that such data may only be sourced from a limited number of facilities operating in a specific environment and hence, this data may not be applied to a range of facility types or operating environments.

One approach to iron out these uncertainties is to conduct a Failure Mode and Effect Analysis which is a workshop activity bringing together a group of experienced personnel (operations, maintenance, instrumentation, design, etc.) for brainstorming the possible failure modes associated with each component and its associated failure rate using operational experience and field data. The quality of such an exercise is highly dependent on the experience of the participants and their engineering judgement. Hence, even though conducting such an exercise will be fruitful, some margin of error is still expected.

5.2 Beta Factor

Typically, the overall reliability of a SIF is estimated based on both independent and dependent failures. For functions with redundant components, multiple components may fail simultaneously due to a common failure mode (e.g. stress, maintenance errors, etc.). These dependent failures are the percentage of the failures attributed to the common causes and are denoted as Beta Factor in the reliability analysis.

Taking note of the uncertainties in failure data for components as highlighted in earlier sections, it is worth highlighting that the chances of availability of detailed field data for beta factors covering a large sample size is even lower, thereby leading to a greater degree of uncertainty. IEC 61508 [2] advocates a method for estimation of the beta factor using a questionnaire and a scoring system. Based on the scoring system, typically the beta factor value ranges from 1 - 10%. This is a somewhat subjective approach. In reality, beta factors higher than 10% have also been reported in the industry [10]. This may however represent an extreme operating environment pertaining to a particular facility. Therefore, some engineering judgement is required.

It is to be noted that beta factor has a direct impact on the PFD_{avg} achieved for systems with redundant components. Since common cause failures are often the dominating contributor to PFD_{avg}, an increase in beta factor leads to a significant increase in PFD_{avg}. This relationship can be seen in the PFD_{avg} results given in Figure 4, for SIFs with various configurations (1oo2, 2oo3, etc.). Hence, selection of a low beta factor may lead to underestimating the PFD_{avg} and potentially also under design of a SIF. For configurations without redundancy (e.g. 2oo2), the beta factor has little impact on the PFD_{avg}.

In the author's experience, some references and guidelines from operating companies provide recommended beta factor values. However, taking note of the findings presented in this section, it should be ensured that such recommended values have a sound basis and are as accurate as possible. Over-estimating beta factor value may lead to over-design of the SIF which in turn also has an associated cost impact. Conversely, under-estimating the beta factor may lead to SIF being under-designed.

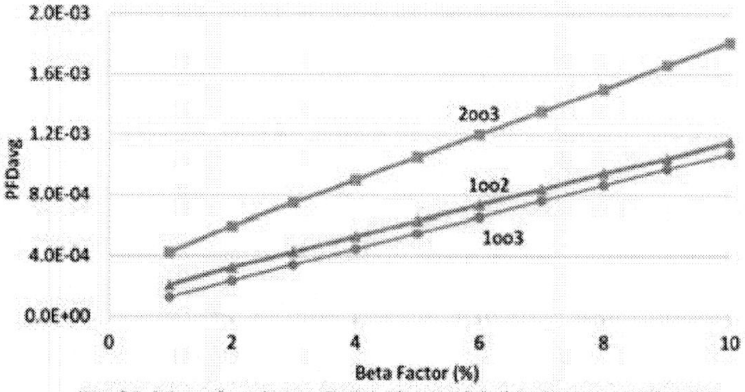

Note: Calculation performed using exSILentia software with the following parameters: λ = 2675 FIT; MTTR = 8 hrs; PTI = 1 yr; PTC = 100%; Mission Time = 10 yrs.

Figure 4. Impact of Beta Factor on PFD_{avg} calculation

5.3 Proof Test Interval

A proof test is performed periodically to detect failures and restore the system / component to an "as good as new" condition. The actual proof test interval may be governed by factors associated with SIF design and operating philosophy specific to a facility. However, the selected interval forms the basis of SIF design since the reliability of the SIF can be compromised with regard to the desired target PFD_{avg} in case the proof test interval used for SIF design is not followed. The uncertainty in this regard lies in whether the recommended proof test interval is rigidly followed or is only conducted on as-need basis in the actual operational phase. Operators may tend to postpone a shutdown for testing purposes if it may result in a loss of production, thereby compromising the SIF reliability. The emphasis to rigidly adopt the recommended proof test intervals therefore rests with the Operating Company and the kinds of safety culture they advocate or follow.

5.4 Proof Test Coverage

One of the important parameters associated with proof tests is the coverage factor which represents the fraction of the potential failures that will be identified during a full proof test. A perfect proof test (i.e. coverage factor of 100%) is unlikely. As far as the authors are aware, there is also no field data available to determine the effectiveness of proof tests. Vendors may recommend a coverage factor based on a detailed FMEA exercise for the associated components. Such an FMEA can identify which failure modes will be successfully captured by a proof test, and thus estimate the coverage factor. Proof test coverage factors can sometimes be lower for components with good diagnostic coverage because the remaining failure modes that are undetected by diagnostics are more difficult for a proof test to detect. In the author's experience, typical coverage factors range from 80 - 99% for sensors and logic solvers and 70 - 80% for final elements (e.g. shutdown valves). While this is currently the best approach for determining the coverage factor, the uncertainty in the result is usually not estimated.

The proof test coverage factor has a significant impact on the calculated PFD_{avg} of a SIF element, as can be seen in Figure 5. For example, for an element with a 1oo1 configuration, decreasing the coverage factor from 90% to 80% resulted in a 46% increase in the PFD_{avg}. Given this effect, caution should be exercised when selecting coverage factors for use in SIF design since they will impact the risk reduction that can be achieved. The actual test performance should ideally be verified to match the assumptions (this becomes especially critical when higher factors e.g. 90% or above are used).

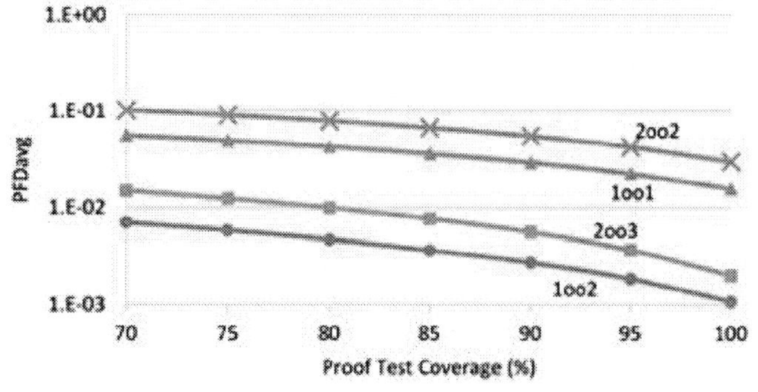

Note: Calculation performed using exSILentia software with the following parameters: λ = 1810 FIT; MTTR = 8 hrs; PTI = 2 yrs; Mission Time = 20 yrs; beta factor = 5%

Figure 5. Effect of Proof Test Coverage Factor on PFD_{avg} for various voting configurations

An alternative method for considering failures that are not detected by a proof test is to include a "probability of test independent failures", or P_{TIF}. This approach, which is advocated in the

PDS method [8], differs from the proof test coverage approach in that P_{TIF} is a constant contributor to the PFD_{avg}, whereas the effect of proof test coverage on the PFD_{avg} increases over time. Even when using P_{TIF}, however, there remains uncertainty in the value used for this parameter, with implications on the accuracy of the calculated PFD_{avg}.

5.5 SIF Operational Life

Operational life of a SIF refers to the life time of the various components of the SIF, sometimes also denoted as mission time. This also serves as a key input in SIF design. In the author's experience, typically a mission time in the range of 20 to 30 years is assumed. Lower mission times such as 5 or 10 years may be impractical in terms of actual implementation on site since this implies that the corresponding components will be replaced or completely overhauled after a 5 or 10 year period, which also has a significant cost impact. This assumption on operational life of the SIF has a direct impact on the risk reduction achieved by a SIF, provided that all other parameters are kept constant. Figures 6 and 7 show that a higher mission time is likely to result in a higher PFD_{avg} (or in other terms, a lower risk reduction). Figure 7 shows that the relationship between mission time and PFD_{avg} is almost linear, with a steeper gradient at a lower proof test coverage. This shows that the mission time has a greater impact on PFD_{avg} when the proof test coverage is lower. If the proof test coverage is taken as 100%, the mission time has no effect on PFD_{avg}. However, as discussed in Section 5.4, 100% proof test coverage is not likely to be realistic.

While operators may typically consider the cost of replacing equipment when deciding upon mission time, consideration should also be given to the effect that mission time will have on the reliability of SIFs. A shorter mission time can help SIFs to meet the target PFD_{avg} and may even allow a lower level of redundancy to be used for the SIF components. This is particularly relevant for SIL 3 rated SIFs, which may require a high level of redundancy as well as high reliability of components, resulting in a high capital cost. Reducing the mission time may help optimize the capital cost of the SIF. A cost benefit analysis may be conducted to justify optimization of the mission time. However, some engineering judgement must be exercised in order to keep the mission time within practical limits. In this regard, 15 years may be considered as a practical lower limit for mission time but this may vary on a case by case basis also depending on the owner's operating philosophy. In such cases, it is important to be aware that in case the capital replacement is not performed after 15 years, this will potentially leave the system exposed to higher probability of failures.

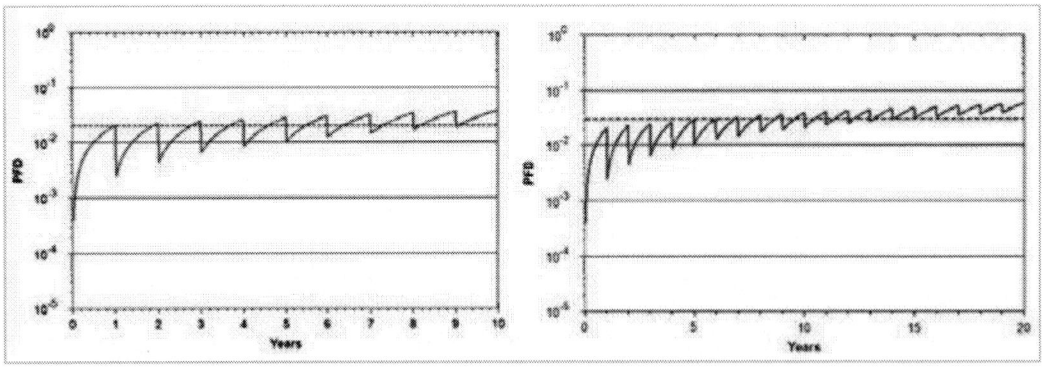

Figure 6. PFD_{avg} vs SIF Operational Life / Mission Time (Years)

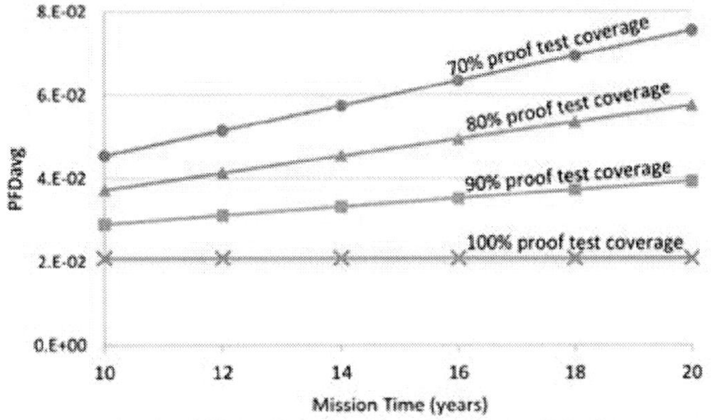

Note: Calculation performed using exSILentia software with the following parameters: λ = 2675 FIT; MTTR = 8 hrs; PTI = 2 yrs; beta factor = 5%

Figure 7. Effect of Mission Time on PFD_{avg} for a SIF with 1oo1 configuration

5.6 Assessing the cumulative uncertainty for multiple parameters in SIF Reliability Analysis

The above sections discuss how the uncertainty of each individual parameter affects the PFD_{avg} calculation. In practice, however, the cumulative effect of the uncertainties for multiple parameters also needs to be considered. As a simple illustration, Table 3 presents PFD_{avg} values calculated for a high pressure SIF with 1oo2 component configuration, using three different combinations of beta factor and proof test coverage, representing a best case, base case and worst case scenario. The beta factor was varied between 1% and 10%, while the proof test coverage was varied between 70% and 90%, which are ranges that can be reasonably expected for these parameters. The PFD_{avg} for the worst case scenario was more than double that of the base case and more than five times that of the best case. In these scenarios, beta factor was found to be the main contributor to the variation in PFD_{avg}, as the uncertainty in the beta factor may be +100% / -60%, whereas the proof test coverage may typically have an uncertainty of only ±13%.

Table 3: PFD_{avg} for a high pressure SIF in three scenarios

Scenario	Beta Factor	Proof Test Coverage	PFD_{avg}	Variation from base case
Best Case	1%	90%	2.93E-04	-70%
Base Case	5%	80%	9.88E-04	0
Worst Case	10%	70%	2.10E-03	+113%

Note: Calculation performed using exSILentia software with the following parameters: λ = 600 FIT; MTTR = 8 hrs; PTI =3 yrs; Mission Time = 15 yrs.

6 Conclusion

In this paper, the authors have highlighted multiple modes through which uncertainties are introduced into the LOPA review and subsequently in the reliability analysis for SIF design. Through working examples, this paper demonstrated that without accounting for such uncertainties, the SIF can easily be under-designed or over-engineered. The emphasis on accounting for such uncertainties is also highlighted in the most recent edition of IEC 61511. This paper investigated further into the nature of uncertainties associated with various input parameters associated with LOPA and SIF reliability analysis including approaches using which such uncertainties can be accounted appropriately.

7 Acknowledgement

The authors wish to acknowledge the support received from IRESC including some of the information included in this paper, as well as the contribution from Ms. Nicole Yeung.

8 References

[1] The International Electrotechnical Commission (IEC), Functional Safety – Safety Instrumented Systems for the Process Industry Sector, IEC-61511, Second Edition

[2] The International Electrotechnical Commission (IEC), Functional Safety of Electrical/ Electronic/ Programmable Electronic Safety Related Systems, IEC-61508, First Edition

[3] Guidelines for Enabling Conditions and Conditional Modifiers in Layer of Protection Analysis, CCPS

[4] Layer of Protection Analysis, Simplified Process Risk Assessment, CCPS

[5] Offshore and Onshore Reliability Data (OREDA), Volume 1 – Topside Equipment, 6th Edition, 2015

[6] Akshat Khirwal, Venkatesh S., Consideration of Cumulative Frequency of Demand Scenarios in SIL/LOPA Review, Hazards Asia Pacific, Kuala Lumpur, Malaysia, 2015

[7] Akshat Khirwal, Venkatesh S., Inherently Safer Design & Robust Protection Layers for High Consequence Hazards, Mary Kay O'Connor Process Safety Center, 19th Annual International Symposium, College Station, Texas, USA, 2016

[8] Reliability Prediction Method for Safety Instrumented Systems, PDS Method Handbook, SINTEF, 2010 edition

[9] Safety Equipment Reliability Handbook, Exida, Third Edition, 2007

[10] S. Hauge et al., Common cause failures in safety-instrumented systems: Using field experience from the petroleum industry. Reliab. Eng. Sys. Saf. 151 (2016), 34-45.

479264 Fundamental Research at Purdue Process Safety Research Center

Abstract:

Fundamental Research at Purdue Process Safety Research Center

Significant industrial incidents continue to occur such as fires, explosions and gas releases resulting in serious injuries, loss of life, and damage to the environment and facilities. Such incidents are not solely germane to the oil & gas industry, but occur in a wide variety of industries including agriculture, pharmaceuticals, and manufacturing. Two years ago the 'Purdue Process Safety and Assurance Center' (P2SAC) was established within the School of Chemical Engineering to conduct fundamental research to prevent such incidents. The presentation will address the four research projects the Center is currently pursuing with funding from its initial Sponsors: BP, Chevron, ExxonMobil, Honeywell, Eli Lilly and Shell.

The initial projects include:

- **optimal placement of gas detectors to prevent fires and explosions**: the research uses CFD-based dispersion simulations and rigorous mathematics to perform optimal placement over hundreds of thousands of leak scenarios and potential detector locations.

- **design & synthesis of catalysts to prevent undesired hazardous byproducts**: zeolite catalysts are widely used in the refining industry and certain structures can promote undesired byproducts leading to potential over-pressures and explosions. Zeolite synthesis and characterization methods are being developed.

- **real time process optimization to reduce risk**: Novel continuous manufacturing processes can reduce the risk related to powder handling. Continuous manufacturing, as with pharmaceuticals, requires efficient process monitoring and control.

- **application of inherently safer separation technology** to significantly reduce processing volume, solvent use, and extent of unit operations. Advanced mixed solvent extraction & simulated moving bed techniques are used to recover polycarbonates and flame retardants from plastic waste.

Indirect benefits from the Center are the depth of the core Chemical Process Safety course required of all undergraduates, as well as undergraduate process safety research initiatives in such as areas as dusts, reactive chemicals and nano-materials.

Current focus is on growing the Center and encouraging participation by a wide variety of industries such as pharmaceuticals, agriculture and manufacturing, in addition to oil & gas, and chemicals.

10-13-16

479277 Bringing New Life to Process Safety

Abstract:

<>Abstract:

Whilst risk cannot be eliminated, it can be better managed. The world of Process Safety Management continues to evolve as organizations find better ways of managing risks. In the simplest form, managing the risk is to reducing inherent process risk to the tolerable level.

First step in managing risks is to identify source of Hazards and quantifying risk associated to Hazards. Traditionally organizations heavily rely on expert judgment and brainstorming methodologies like HAZOP for finding operational and design risks. In addition, quality of such analysis are directly related to the precision of assumptions and information from operation and maintenance of the equipment under control.

In this session we will explore how organizations are using new techniques such as dynamic risk visualization, cumulative risk techniques, operational risk assessments and cyber Hazop's to advance the boundaries of process safety management.

This paper will explore how existing systems and data sources, risk visualization and intelligence techniques are being used to bring static process safety data to life. Flat "1 dimensional" information such as lists, forms and databases are now being transformed into "2 dimensional" dynamic information such as plot plans, barrier models, risk matrix, bow ties etc. that provide greater context of relationships between information.

This information will resonate with engineers and technician who rather working with meaningful information of operation and maintenance systems instead of some abstract numbers and concepts that need extensive expertise in the reliability engineering.

We also will look at how advances in Operational Risk Assessments and Cumulative Risk allow operators to determine in real time if the overall cumulative risk is tolerable before actions are taken, devices are removed from service, maintenance activities are deferred etc. In other words, leading indicators and metrics will be in disposal of operators to proactively facilitate decision making for managing risk. This could avoid incidents that developing in the operation due to degradation of control and safety systems or deviations from initial operation and maintenance procedures and may lead to severe incidents.

And finally, since new technologies and architectures for instrumentation, control and safety systems introduced new risks of cyber intrusions and attacks that need to be considered in the risk management and the world of security standards (ISA 62443-3-2 Security Risk Assessment and System Design) and safety standards (ISA-TR84.00.09-2013 Security countermeasures related to Safety Instrumented Systems and Associated IACS) advanced in the last few years, we will look at how Cyber Process Hazard Analysis (PHA) techniques are being applied to provide the same systematic approach to security as traditionally applied to safety.

"Good process safety management does not happen by chance and require constant active engagement", in this paper we suggested novel approaches that can revive and improve the process safety.

479367 Developing an Effective System for Action Item Resolution

Abstract:

PSM action items are are regularly identified through many different means in a typical chemical processing facility. Knowing what to do with these action items and how best to proceed is no easy task. Not all action items are created equally, or even with the same goal in mind. An action item identified in a PHA may present a direct hazard to an employee, while an action item generated from a pressure relief systems study may present a documentation concern, a deviation from current practices, a direct hazard to an employee, or any combination thereof. Many facilities have a system in place for shepherding action items to completion, but these systems typically assume all action items are created equally. In this paper, we will demonstrate the importance of having a system that handles each action item in the most effective way; from evaluating the nature of the concern, considering the risks associated with the action item, developing the action plan, to completing the action plan. We will also cover whose input should be considered at each stage in the process to help ensure the best solutions possible are developed.

13TH GLOBAL CONGRESS ON PROCESS SAFETY

Where Does Management of Change (MOC) Begin?

Juliana Schmitz
Praxair, Inc.
Av. Pastor Martin Luther King Jr., 126, bloco 10/ala A,
Corporate Shopping Nova América, Rio de Janeiro, RJ, Brazil
juliana_schmitz@praxair.com

Patti Jones
Praxair, Inc.
175 E Park Dr, Tonawanda, NY, 14150, USA
patti_jones@praxair.com

Gabriel Drummond
Praxair, Inc.
Av. Pastor Martin Luther King Jr., 126, bloco 10/ala A,
Corporate Shopping Nova América, Rio de Janeiro, RJ, Brazil
gabriel_drummond@praxair.com

Prepared for Presentation at
American Institute of Chemical Engineers
2017 Spring Meeting and 13th Global Congress on Process Safety
San Antonio, TX
March 26 – 29, 2017

Where Does Management of Change (MOC) Begin?

Juliana Schmitz
Praxair, Inc.
Av. Pastor Martin Luther King Jr., 126, bloco 10/ala A, Corporate Shopping Nova América, Rio de Janeiro, RJ, Brazil
juliana_schmitz@praxair.com

Patti Jones
Praxair, Inc.
175 E Park Dr, Tonawanda, NY, 14150, USA
patti_jones@praxair.com

Gabriel Drummond
Praxair, Inc.
Av. Pastor Martin Luther King Jr., 126, bloco 10/ala A, Corporate Shopping Nova América, Rio de Janeiro, RJ, Brazil
gabriel_drummond@praxair.com

Keywords: MOC, engineering, process safety

Abstract

Management of Change is often affiliated with an operating facility. However, several projects have the common challenge of being developed within the constraints of an aggressive schedule, considering the time from project kick-off until unit start-up. In order to comply with the deadlines, early releases of engineering design documentation become subject to changes. During project development, these changes may come as scope changes, supplier updates, etc. that may affect more than one engineering discipline and any on-going construction/assembly activities. Another change source may be non-conformances identified during the construction/assembly phase. In this case, considering project milestones and costs, an option taken into account is to leave the system or portion the way it is instead of re-working to match the original design. For all of the given cases, a robust Engineering Management of Change (EMOC) process must be in place throughout the project life cycle to guarantee modifications are properly identified, communicated, and evaluated by technical experts for possible effects on process and design safety prior to implementation, or at a minimum, prior to startup of the system/subsystem. In addition, an EMOC process gives the opportunity to analyze the reasons for the modifications in order to avoid increased costs and design errors on future projects, leading to improved quality control. This paper presents some previous accidents related to management of changes, and addresses EMOC implementation, work process and elements of a successful EMOC program.

1 Introduction

Preliminary design activities begin early in the process to define scope, cost and product delivery requirements. Changes are a natural part of these efforts.

As the project progresses through the engineering detail design phase, a process hazard analysis (PHA) is conducted. At this moment, a multidisciplinary team evaluates if the process is safe and if any modifications are required. The project team is informed of the revised documentation and proceeds with each skill activity.

Subsequent changes may occur as, for example: updated supplier documentation, the team that will effectively operate the plant in the future may raise a new need for the project, some may be identified during detail design requiring changes and change requests during the construction phase. Any modification after the PHA may not be properly communicated to the adequate professionals and analyzed, resulting in a possible unsafe condition.

This paper focuses on project execution phases until plant startup and turnover to operations where the Management of Change (MOC) work process begins. It also explores the challenges identified when developing an Engineering Management of Change (EMOC) work process and shares an approach taken within Praxair, Inc.

2 Why is the Engineering Management of Change Important?

Process Safety is a relatively new engineering area, being developed from the early 60's and mostly based on catastrophic incidents, which negatively affected the industry. Kerin [5] confirms this point of view and lists a series of landmark events, their key learnings and influences in process safety. These events had different causes, mostly related to poor management of various safety areas, such as hazard identification, training, maintenance, management of change (MOC) and others.

The numerous and notable MOC related accidents that happened in the industry over the last century showed the world the need of a well-developed and maintained MOC procedure.

2.1 Titanic, 1912

The sinking of the Titanic, in 1912, is a well-known catastrophic incident. "The design of the ship included 44 water tight compartments giving rise to the perception that the ship was unsinkable and in the event of an accident would be its own lifeboat." [2] An accelerated construction schedule was the catalyst in making changes that proved to be fatal. A robust management of change process and a cultivated safety culture may have prevented the incident.

2.2 Flixborough, 1974

Venart [3] describes the Flixborough accident, which occurred on June 1st, 1974, causing 28 fatalities and several property damages near the Nypro plant in North Lincolnshire, UK. In this plant, there was a process area responsible to produce caprolactam, an intermediary for the Nylon production. Caprolactam production was conducted through the oxidation of the cyclohexane, a highly flammable chemical. There were six reactors in series where this oxidation reaction was performed.

Two months before the explosion, a crack was observed in the 5th reactor of the chain. With the objective of reducing the impact in production, the responsible plant personnel decided to connect directly the 4th and 6th reactors while the 5th was removed for maintenance. To proceed with the change, a "dog-leg" pipe was used to connect the reactors 4 and 6, while the original installation considered a straight pipe between these reactors. This is a modification that clearly changed the characteristics of fluid flow between reactors 4 and 6. The dog-leg line support was a temporary installation, made of scaffolding bars. These changes were never evaluated in detail and during start-up. On June 1st, the bellows connected to the reactors flanges failed due to an overstress and tons of cyclohexane was released into the atmosphere. With the vapor dispersion, an ignition source was found in the plant, and a vapor cloud explosion occurred resulting in fatalities, as well as significant property damage. [4]

2.3 CONDEA Vista Company, 1998

U.S. Chemical Safety and Hazard Investigation Board [5] promotes a discussion about the reaction vessel explosion in CONDEA Vista Company, on October 13th, 1998. In this unit, linear alkyl benzene was produced by mixing aluminum chloride and hydrocarbons. Around 3 months before the incident the process was modified, involving a mixing of powdered aluminum with hydrogen chloride to form the aluminum chloride. After this change was implemented, the bottom nozzle of the reactor got plugged due to the formation of a sludge residue, which the operators were not able to reach and remove during maintenance. Considering this scenario, the chemist of the site and the process engineer decided to use water to dissolve the sludge and to absorb the heat generated due to an exothermic reaction between the water and the sludge. This process was started and once it was apparently under control, the process engineer and chemist left the plant.

The shift supervisor decided to make a new change and inject steam at the nozzle. The process supervisor who was not in the plant agreed with the new procedure. At the shift turnover, the day shift supervisor talked with the night shift supervisor about the steam injection. The night shift supervisor took responsibility to inject steam and oriented an operator to execute this procedure.

After a few minutes of continuous steam injection, the reactor exploded, leaving three workers injured and property damage estimated in $13 million USD. Among other shortcomings related to the catalyst change that resulted in sludge formation, U.S. Chemical Safety and Hazard Investigation Board **[5]** affirms that this accident highlights the value of having an authorization or approval step as part of an MOC system for abnormal situations. If such a procedure had been in place, a technical manager would have reviewed the proposed procedure involving steam and may have detected its deficiencies.

2.4 Prevention

Among others, these accidents demonstrate the importance of Management of Change in process safety. This was emphasized when MOC was included as one of the main pillars of the Process Safety Management (PSM) standard, issued by the U.S. Occupational Safety and Health

Administration and of the Risk Management Plan of the Environmental Protection Agency (EPA).

However, not only changes performed after the plant startup will result in safety accidents. Engineering Management of Change (EMOC) is also a key area to prevent accidents and should be addressed in a proper manner by industrial companies.

One example of a poor EMOC related event can be an updated drawing sent by an equipment supplier during the project execution phase indicating a decrease of a vessel Maximum Allowable Working Pressure (MAWP). If this modification is not communicated to the proper professionals, the pressure relief device set point designed for the equipment may not be adjusted as required, which can lead to lack of vessel overpressure protection and possible vessel rupture.

Another example can be a change of a pipe routing during project phase, without revision of the flexibility study. This could result in pipe over stress, leading to a line rupture and energy release.

Several other examples of poor EMOC related accidents could be discussed showing proper management of changes during engineering is indispensable for the safety of the plant, workers and environment. Figure 1 indicates some sources of modifications that could impact the safe startup/operation of the system.

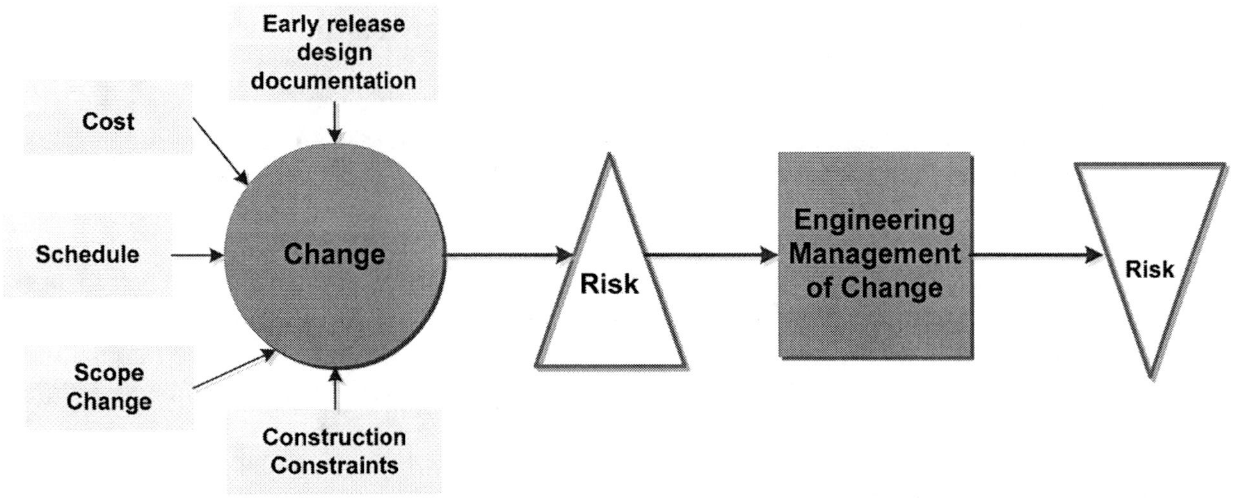

Figure 1 Integrated management of change system

Unfortunately, EMOC use is not as widespread as MOC, which still gives space for industrial accidents to happen. The main purpose of defining a work process for managing changes, even during the project development phases, is to guarantee all changes are adequately communicated, reviewed, new hazards are identified and analyzed and a mitigation plan is defined to ensure that a safe installation and operation is delivered to the customer.

3 EMOC Implementation

A change is known as a deviation from the original design. Modifications may come as scope changes, supplier updates, or other sources, and they may affect more than one engineering discipline and any on-going construction/assembly activities.

"Changes in construction projects are very common and likely to occur from different sources, by various causes, at any stage of a project, and may have considerable negative impacts on item such as costs and schedule delays." [6] If there is no formalized change management process, impacts not only on cost and schedule may be observed, but also process safety events, as mentioned on the previous section of this paper.

According to CCPS [7], the level of detail covered by the management of change system should be based on

- The hazards or risks of the process
- The existing process safety culture at the location

The company should develop a system for the control of modifications implemented on the project under design, construction, commissioning or startup phases. This work process should include approval of the change by a qualified professional or team, depending on the complexity of the hazards associated.

Based on the hazards involved in the specific process, the type of changes that need evaluation by an engineering expert from the appropriate discipline, as well as changes that do not, shall be defined when designing the specific EMOC work process for a certain company. For example; changes that can affect safety systems, critical operating parameters, shutdown set points, relief valve settings, materials of construction, etc. should undergo EMOC approval. Minor corrections to documentation or clarifications may not.

The management of change work process may not be the same in every industry and this does not mean one is better than another. The culture of the company and the location may require different applications of the MOC activities. [7]

3.1 Critical Phase

Changes during construction inherently introduce unexpected risk, especially under schedule and cost pressure. Some changes that can affect process safety may not be obvious to construction personnel. A simple yet important example is relocating a support on the discharge of a relief valve due to interference. Another may be substituting flange bolt material. Regardless of the origin of the change, evaluation and approval should take place prior to starting up the system/subsystem. Identification of all sources which can generate an engineering change must be considered as a natural part of the overall work process.

4 Proposed EMOC Work Process

Typically, during the engineering phase, several design safety meetings are conducted, such as P&ID Reviews, Layout Reviews, PHAs and FSSs. These reviews shall be conducted with the participation of all engineering disciplines related to the plant design. After these project landmark meetings, several changes may occur and their evaluation is a critical part of risk management throughout the project life cycle. At this moment, after the main safety reviews are performed, the EMOC work process is expected to begin.

As shown on Figure 2, all changes related to the project design are identified and communicated by the professional who initiated the modification, to a single point of contact within the project team.

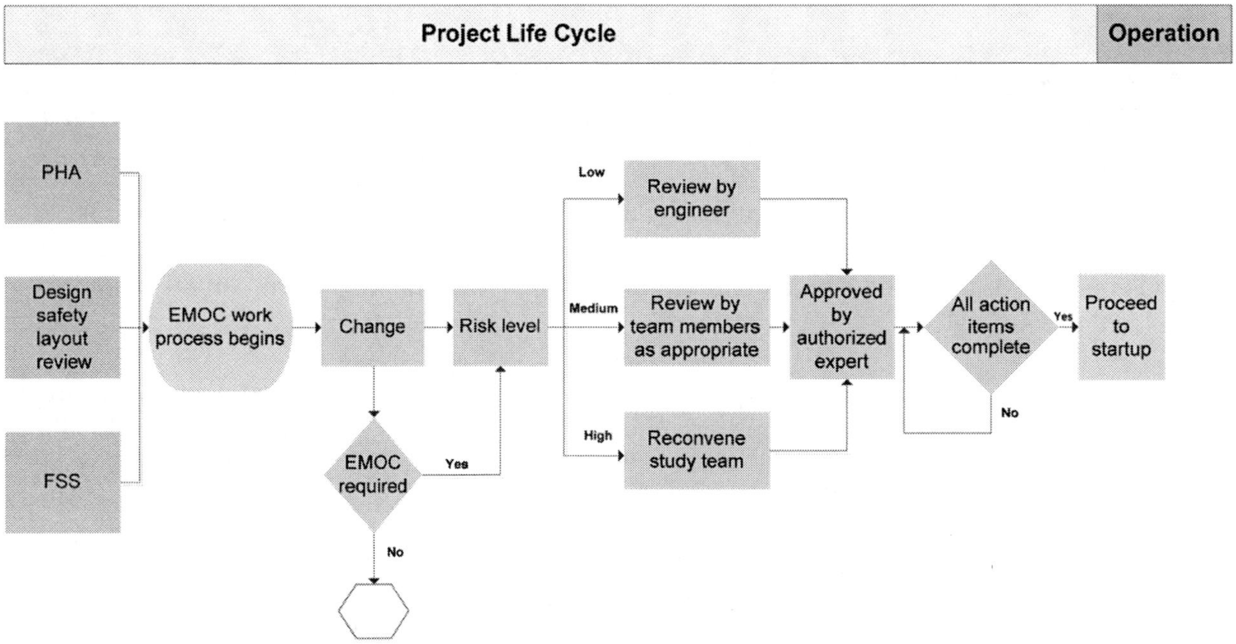

Figure 2 High-level EMOC Workflow Diagram

In cases where a modification is only a correction, clarification or replacement in a kind, there is no need to go through the full EMOC work process. The change can be recorded in a simple spreadsheet containing the type of change, a brief explanation and a mitigation plan to make sure nothing has been missed.

The modification should then be classified as low, medium or high risk per the EMOC owner and the level of review is defined based on the associated risk. The risk ranking criteria shall be described on the company's EMOC program; e.g.:

- Low: addition of a local pressure indicator on the design for maintenance purposes only.
- Medium: manual isolation valve removal, while valve not indicated at PHA as part of a hazard scenario.

- High: customer addition of a tie in point near the pipeline battery limit, which could lead to misdirected or reverse flow, with potential for process safety impact.

After reviewing the change, an approval should be required by a Subject Matter Expert (SME) before proceeding to change implementation and mitigation plan. The plant will be allowed to start-up only after all EMOC items were reviewed and approved.

It will be responsibility of the company implementing EMOC to determine which kind of changes will require a complete review, the risk categories that will be available to be assigned, both qualitatively and/or quantitatively and the required level of review according to these risks. In the example presented in Figure 2, a low risk results in a review by an engineer, a medium risk results in a review by the team members and a high-risk change will require the engineering team involved in the HazOp, Design Safety Layout review or the FSS evaluations to reconvene.

Several different approaches are possible, as long as they are simple, easy to understand by those responsible for implementation and provide a robust EMOC work process.

5 Elements of a Successful Engineering Management of Change Program

Successful EMOC programs define the key elements applicable to their industry. Some will be common such as commitment from leadership, roles and responsibilities, etc. Included below are some elements to take into consideration.

- Commitment from Leadership;
- Roles and responsibilities as they apply to specific skill sets (i.e. process engineering, controls system engineering, manufacturing, construction management, etc.);
- Clear definition of changes that must undergo engineering evaluation, including examples;
- Definition of changes that are not considered EMOC related;
- Risk level associated with the change;
- Approvers limited to subject matter experts;
- A streamlined work process focused on function over form, aligned with the various phases of the project life cycle;
- Identification of change sources;
- In process verification at predetermined milestones;
- Robust training that includes internal and external examples accompanied by a test with feasible scenarios;
- Associated PSI updated accordingly;

- Auditable;
- Evergreen
  - Periodic calibration of the work process;
  - Input from those responsible to implement the process;
  - Captures internal data to enhance understanding of what constitutes a change that requires EMOC review.
- Lessons learned captured and implemented on future projects;
- Leverage key learnings from external events in where MOC was identified as the cause or a contributing factor.

6 Conclusion

An aggressive schedule may lead to early release of design documentation that is still subject to change, which could result in the execution of PHA studies earlier than the optimal period. Although the early study may provide an opportunity to identify key risks earlier in the process, a significant number of action items and changes will probably occur, which may affect the integrity of the PHA.

A well-defined and implemented EMOC program is an essential part of risk management. The methodology applied allows systematic control of project changes which helps to ensure process, personnel and environmental safety. An additional benefit is facilitating a proactive means to analyze the underlying cause for unplanned changes. These lessons learned can assist on future projects to manage cost and improved quality control.

The main purpose of the management of change is to avoid loss of primary containment (LOPC), especially of hazardous chemicals, which could lead to catastrophic consequence. Once the EMOC process is successfully implemented, the reduction of quality issues will become a natural part of the process, resulting in a safer and higher quality facility/assembly/operation. An effective Engineering Management of Change work process is critical to the safe commissioning and startup of a chemical plant.

7 References

[1] T. Kerin. The evolution of process safety standards and legislation following landmark events – What have we Learnt?, Process Safety Progress 35 (2016), 165-170.

[2] CCPS. *Incidents that Define Process Safety*, Center for Chemical Process Safety, American Institute of Chemical Engineers, New Jersey, 2008.

[3] J.E.S Venart, Flixborough: The explosion and its aftermath. Process Safety and Environmental Protection 82 (B2): 105 – 127 (2004).

[4] C. Sadee, D.E. Samuels and T.P. O'Brien, The Characteristics of the explosion of cyclohexane at the Nypro (UK) Flixborough plant on June 1974, Journal of Occupational Accidents, 1 (1976/1977) 203 – 205.

[5] U.S. Chemical Safety and Hazard Investigation Board, Management of Change (No. 2001-04-SB) (2001).

[6] Q. Hao et all, Change Management in Construction Process, International Conference on Information Technology in Construction, Santiago, Chile, 2008

[7] CCPS. *Guidelines for Management of Change*, Center for Chemical Process Safety, American Institute of Chemical Engineers, New Jersey, 2008.

479400 Effectively Utilizing Calarp/RMP/PSM Contractors to Assist in Optimizing Process Safety Culture, Compliance with Standards, and Workforce Involvement

Abstract:

Since the inception of Occupational Safety and Health Administration (OSHA) Process Safety Management (PSM) in 1992, facilities have frequently contracted process safety consultants to satisfy regulatory needs, as well as assist in complying with industry best practices.

The primary objective of this paper is to provide tangible suggestions for effectively utilizing process safety contractors for assisting in optimizing process safety culture, compliance with regulatory requirements and industry best practices, as well as tips to improve workforce involvement and coordination. The core of this paper will detail various management styles of the prevention program and an evaluation of influences on process safety culture based on varying degrees of involvement from the facility and usage of contractors. This paper aims to provide an in-depth analysis of the pros and cons of contractor involvement and its efficacy relative to the scale of the facility and intricacy of corporate structure based on anecdotal evidence and industry experience.

Key topics include:

- Overview of different types of process safety culture in relation to prevention program management and contractor usage,
- The Challenge - Striking an optimal balance between contractor use and facility involvement whilst maintaining economic viability,
- The Challenge for Multi-Facility Corporations - Ensuring changes in corporate standards are successfully communicated to all children facilities ranging from documentation updates and organizational changes,
- Suggestions and Improvements - Implementation of effective prevention program management systems to ensure workforce involvement and accountability, as well as practical allocation of contractor resources to accomplish critical tasks whilst balancing facility resource commitment.

A Streamlined Approach for Full Compliance with SIF Implementation Standards

William G. Bridges, President
PROCESS IMPROVEMENT INSTITUTE, INC. (PII)
1321 Waterside Lane, Knoxville, TN 37922
Phone: (865) 675-3458
Fax: (865) 622-6800
e-mail: wbridges@piii.com

Art Dowell III, Principal Engineer
PROCESS IMPROVEMENT INSTITUTE, INC. (PII)
2437 Bay Area Blvd PMB 260
Houston, TX 77058
Phone: 713-865-6135
e-mail: adowell@piii.com

Prepared for Presentation at
13th Global Congress on Process Safety
San Antonio, TX
March 27-29, 2017

UNPUBLISHED

13TH GLOBAL CONGRESS ON PROCESS SAFETY

A Streamlined Approach for Full Compliance with SIL Implementation Standards

William G. Bridges, President
PROCESS IMPROVEMENT INSTITUTE, INC. (PII)
e-mail: wbridges@piii.com

Art Dowell III, Principal Engineer
PROCESS IMPROVEMENT INSTITUTE, INC. (PII)
e-mail: adowell@piii.com

Keywords: Safety Instrumented System, SIS, Safety Integrity Level, SIL, LOPA, Layer of Protection Analysis, Independent Protection Layers, IPLs, human factors

Abstract

Many companies put FAR too much redundant effort into determining what SIL (safety integrity level) is needed and then verifying the SIF (safety instrumented function) design will give the SIL targeted. This paper shows how to apply the qualitative definition of independent protection layers (IPLs) within the setting of a process hazard analysis (PHA) to get most of the gain from Layer of Protection Analysis (*LOPA)* without doing a LOPA (without using numerical values). We show how we use a PHA team to identify when a SIF is needed and to select the proper target SIL. This portion of the SIL evaluation and the identification and labeling of the IPLs during the PHA/HAZOP does not take any longer than a normal PHA/HAZOP, once the right habits are established. Note that this approach eliminates the need for a separate SIL Evaluation Study to identify the SIFs and select the target SIL. Then, this paper describes how to perform the SIL Verification and Safety Requirements Specification (SRS) remotely, again without the need for a redundant team meeting. This approach has been used at many sites and for thousands of SIFs.

Background

Typically, hazardous consequences, their causes, and safeguards are identified in a PHA/HAZOP (process hazard analysis / hazard and operability analysis)[1]. The PHA may determine if additional protection layers are required or if additional analysis is required.[2]

In many organizations, a separate LOPA (layer of protection analysis)[3] meeting is held to determine which HAZOP safeguards are independent protection layers (IPLs) and to determine if additional IPLs are needed to meet the risk tolerance criteria. If SIFs are needed, the required risk reduction (PFD, probability of failure on demand) is determined in LOPA, which then sets the SIL.

Some organizations may even have a separate SIL evaluation study, which is certainly redundant to a well-run PHA and is wasteful if the PHA is properly staffed and managed and/or if any subsequent LOPA are performed well.

Fundamentals of SIL Assessment

The lifecycle requirements for safety instrumented functions are specified in industry standards IEC 61511 (ANSI/ISA 84.00.01-2004, Part 1)[4,5]. The required steps to specify the SIS (safety instrumented system) are:

1. Safety Instrumented Function (SIF) identification.
2. Determining the Safety Integrity Level (SIL) for each SIF.
3. Designing the SIF to meet the required SIL.
   SIL Verification Calculation (actually, this calculation is iterative with step 3, but the end calculation is a deliverable that proves 3 is correct). The SIL standards allow the following methods to use to determine the required SIL (ANSI/ISA 84.00.01-2004, Part 3, Section 3.8)[4]:
   - Fault Tree Analysis (FTA), Full Quantitative Risk Analysis (QRA)
   - LOPA – simplified quantitative analysis
   - Qualitatively within PHA (HAZOP, etc.)

   "A qualitative method may be used as a first pass to determine the required SIL of all SIFs. Those which are assigned a SIL 3 or 4 by this method should then be considered in greater detail using a quantitative method to gain a more rigorous understanding of their required safety integrity." – *ANSI/ISA 84.00.01-2004, Part 3, Section 3.8*[4]

Table 1 compares risk analysis approaches for Steps 1 and 2.

FTA and QRA are very detailed and labor-intensive and would be required for complicated systems, particularly where there is not complete independence between safeguards.

LOPA is a simplified analysis tool that requires fewer resources, but is heavily dependent on the requirement that all the safeguards be completely independent.

However, for many processes, there are a number of scenarios (cause leading to a consequence) where the PHA team can readily determine sufficient independent protection layers are already in place.

Table 1: Comparison of Risk Analysis Approaches

Approach	Assessment Methods	Risk Judgment Method	Risk Judgment Method	Estimated Range of the Results
Qualitative Only	HAZOP, FEMA	Expert Voting, focusing on site data	Capable >95% of time	Plus or minus 1/2 order of magnitude
Simplified Quantitative	LOPA, Risk Graph	Multiplication of statistical averages of general failure rate data; with broad assumptions on management systems	Needed on about 5% of the scenarios	Plus or minus 1 order of magnitude
Full Quantitative	FTA, ETA, QRA, HRA		Needed for less than 0.01% of the scenarios	Plus or minus 1 order of magnitude

Figure 1 contrasts the typical approach to SIL assessments with the optimized streamlined approach.

Typical SIL Determination versus Streamlined Approach

In the typical approach,

- **Step 1**: The PHA/HAZOP is performed using a multidisciplinary team of engineers, operators, instrumentation staff, maintenance staff, and perhaps vendors. Then in a separate SIL study, a multidisciplinary team of engineers, instrumentation staff, maintenance staff, and perhaps vendors brainstorm the hazard scenarios (again) and decide which are candidates for SIFs. A semi quantitative approach, such as RiskGraph, or layer of protection analysis, determines if an SIF is required.
- **Step 2:** The RiskGraph dictates the required SIL, or the LOPA compares the mitigated risk to the risk tolerance criteria and determines the SIL from the order-of-magnitude risk reduction needed.

In contrast, in the streamlined approach,

- **Step 1:** The PHA/HAZOP is performed using a multidisciplinary team of engineers, operators, instrumentation staff, maintenance staff, and perhaps vendors (to this point the same as the typical approach). Within the PHA/HAZOP, the team estimates the risk of each scenario using qualitative judgment (based on experience, knowledge, and memory of site-specific data). The PHA/HAZOP team then qualitatively determines if the best risk reduction method is an SIF.
- **Step 2:** The PHA/HAZOP team qualitatively determines the SIL needed (typically by judging the level of redundancy needed based on the value they place on each IPL and the new or modified SIF needed)

Figure 1: Comparison of the Typical and the Optimized Streamlined Approach from PII for SIL Assessments (Steps 1 and 2)

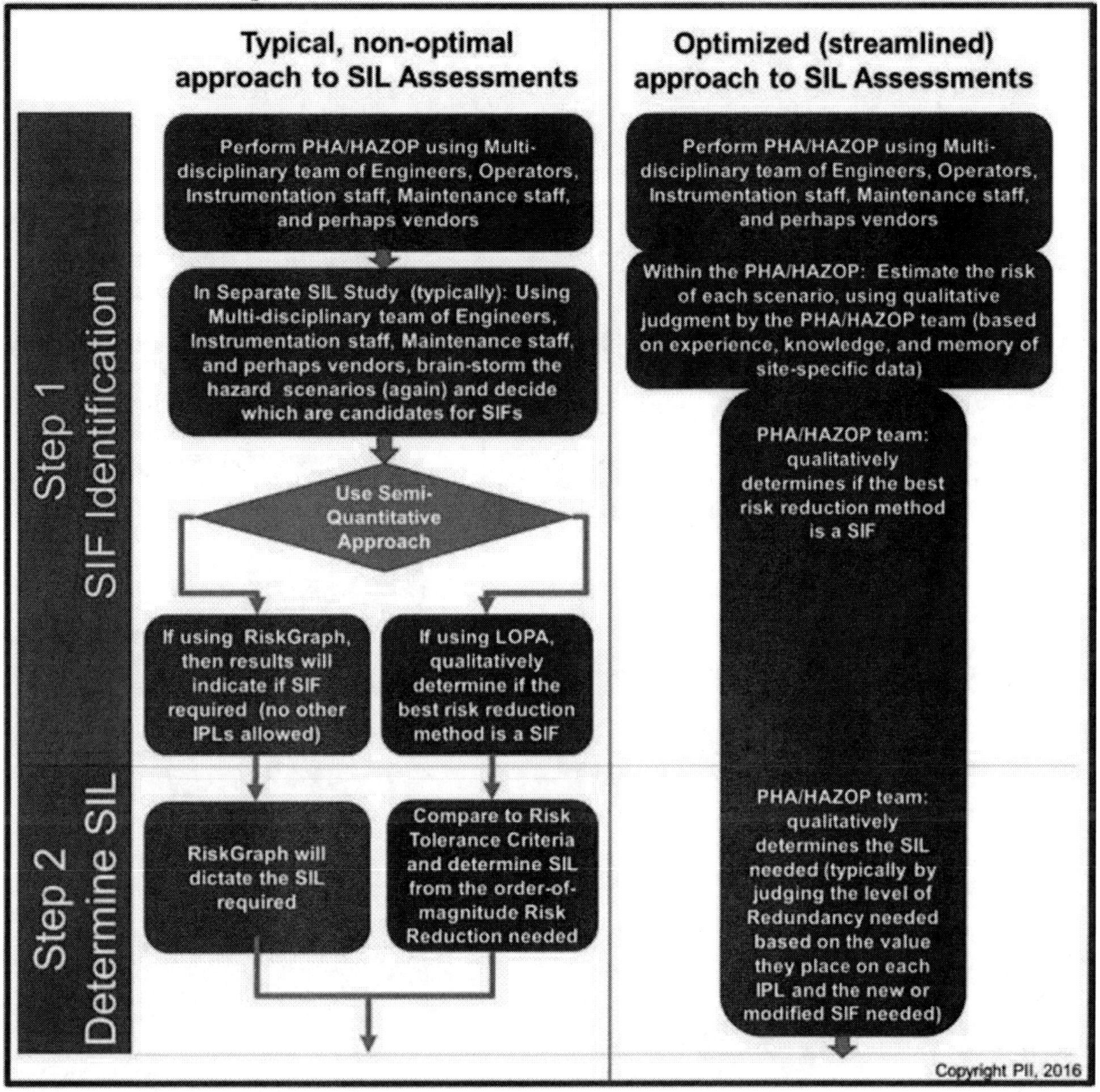

Figure 2 also contrasts the typical approach to SIF design and verification with the streamlined approach. Steps 3 and 4 are iterative.

In the typical approach,

- **Steps 3 and 4**:
  - o Decide on basic architecture of SIF, from experience with other SIFs.
  - o Decide on vendors of components
  - o Provide full design of SIFs from experience

- Perform trial SIL verification calculations using software such as exSILentia™ using actual component data
- If the target SIL is achieved, issue the SIL verification calculation report and the safety requirement specification (SRS)
- If the target SIL is not achieved, revise the design and recalculate.

Figure 2: Comparison of the Typical and the Optimized Streamlined Approach from PII for SIL Assessments (Steps 3 and 4)

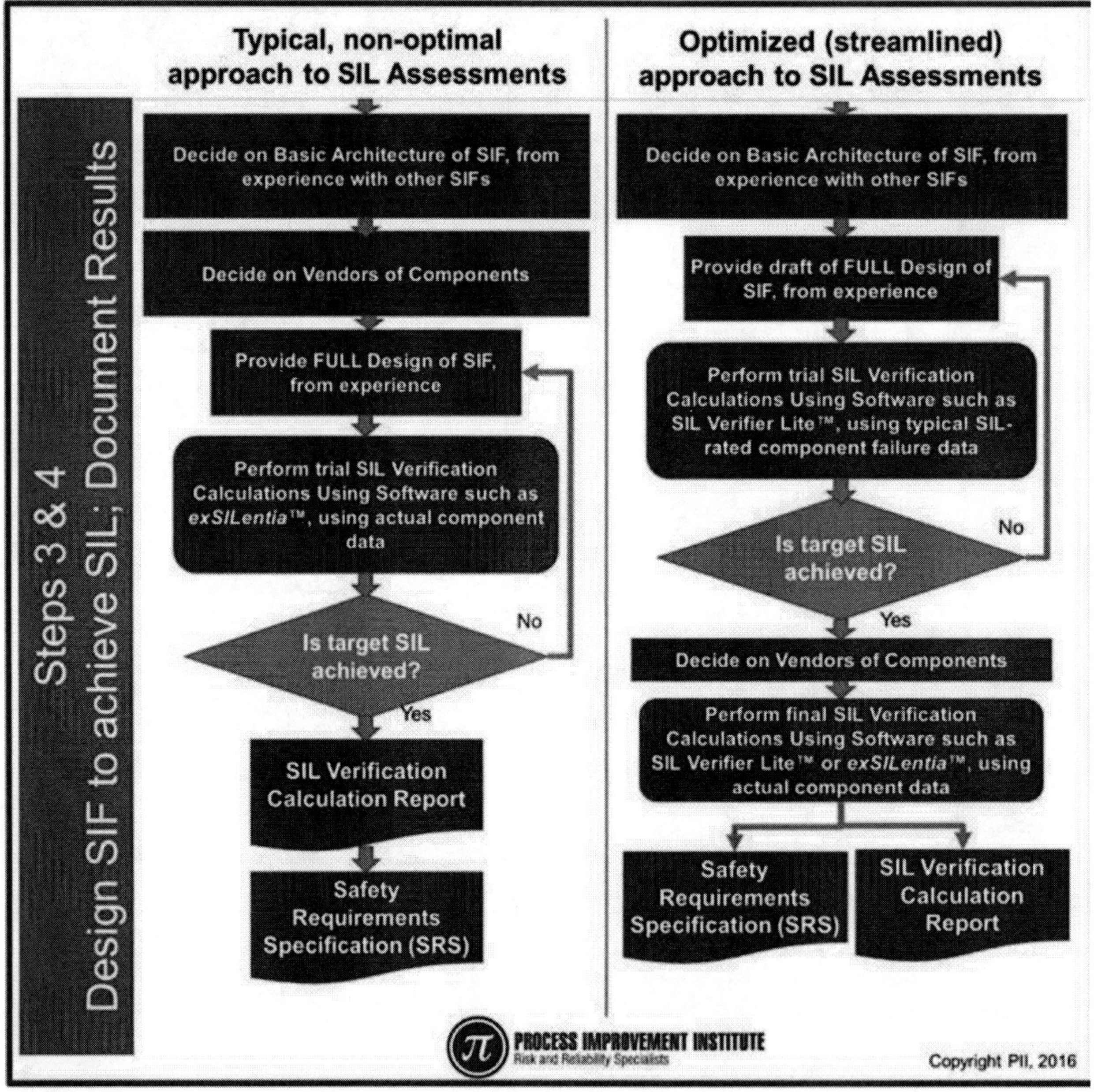

In the streamlined approach,

- **Steps 3 and 4**:
  - Decide on basic architecture of SIF, from experience with other SIFs.

- Provide draft of full design of SIF, from experience.
- Perform trial SIL verification calculations using software such as SIL Verifier Lite™ using typical SIL-rated component failure data.
- If the target SIL is not achieved, revise the design and recalculate.
- If the target SIL is achieved,
    - Issue the SIL verification calculation report and the safety requirement specification (SRS).
    - Decide on vendors of components.
    - Perform final SIL Verification Calculations Using Software such as *SIL Verifier Lite*™ (from PII) or *exSILentia*™ (from exida) using actual component data.
    - Issue the SIL verification calculation report and the safety requirement specification (SRS).

The advantage of the streamlined approach is determination of the need for an SIF and of its SIL quicker and with fewer resources than the typical approach.

Requirements for the Qualitative Streamlined Approach for SIL Determination

For the qualitative streamlined approach to be used effectively, there are important requirements for the PHA/HAZOP team.

- PHA/HAZOP team leader is knowledgeable and proficient in LOPA.
- PHA/HAZOP team members have the following information in memory:
    - Consequence severity.
    - Initiating cause frequency.
    - Independent protection layer (IPL capability).
    - Risk tolerance criteria.

An experienced PHA team, who is knowledgeable in the process technology, can effectively and efficiently analyze a scenario and mentally perform a layer of protection analysis. The team determines the consequence severity and the initiating cause frequency. A highly competent team understands which safeguards are truly independent protection layers and can judge the risk reduction provided by each. Then the team can determine if additional IPLs are required to meet the risk tolerance criteria. For some scenarios, an SIF may be required and the team can determine the required probability of failure on demand for the SIF and can thus determine the required SIL.

A best practice is that the PHA team recommends the functional description and the SIL for the proposed SIF. The functional description does not specify the details of the SIF design, rather it specifies what process variables should be considered and what action should be taken to detect the upset and to prevent the consequence. For example, the functional description may say "monitor the level of a vessel and stop the flow into the vessel to prevent overflow".

If the PHA team is uncertain, or if the scenario is complex, then the PHA can recommend a detailed LOPA be done for the scenario.

These conclusions are based on more than 8000 PHA/HAZOPs, 3000 LOPAs, and about 100 QRAs and HRAs. We have observed PHA participants becoming knowledgeable about LOPA and IPL requirements. The participants then began to apply the principles of LOPA spontaneously in the PHA. With good leadership, the PHA team could then easily move to the full streamlined approach.

The experience and competency of the PHA leader and the PHA team is extremely important. If the PHA team is not comfortable with the principles of LOPA, and especially the principles of IPLs, the streamlined approach is not appropriate. Likewise, the PHA team must have the required information and understanding in its memory. Otherwise, the qualitative approach can degenerate into the "arm wrestling" arguments that PHA teams used qualitatively to determine SIL requirements before the introduction of LOPA.

Suggestions for the Streamlined Approach for SIF Design

Likewise, there is a streamlined approach for the SIF design. Organizations have developed a recipe book of typical SIF designs for SIL 1, SIL 2, and SIL 3. Depending on the required SIL, the designer can choose a design from the recipe book (as described in Figure 2, "from experience"). The recipe book provides a starting point for the detailed SIF design. Figure 3 shows a small sample of SIF designs that could be included in a recipe book.

Figure 3: Typical SIF Architectures for Information

Typical SIL 1

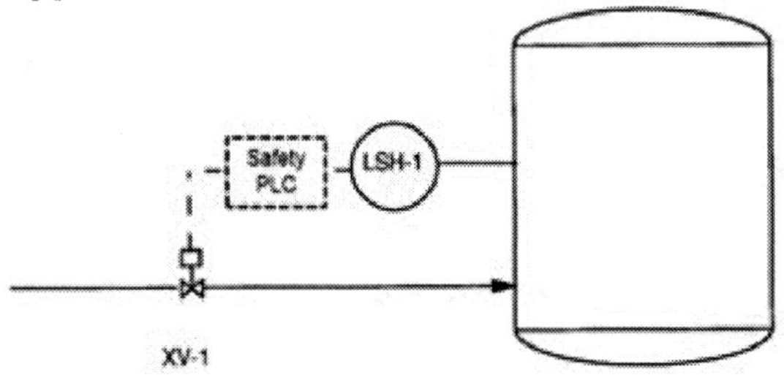

Typical SIL 2

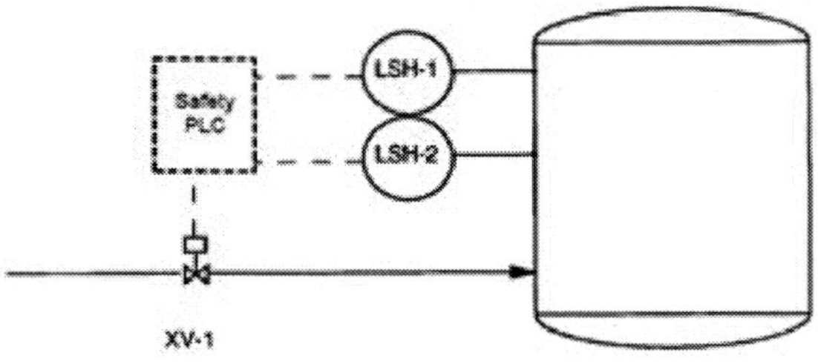

Typical SIL 3

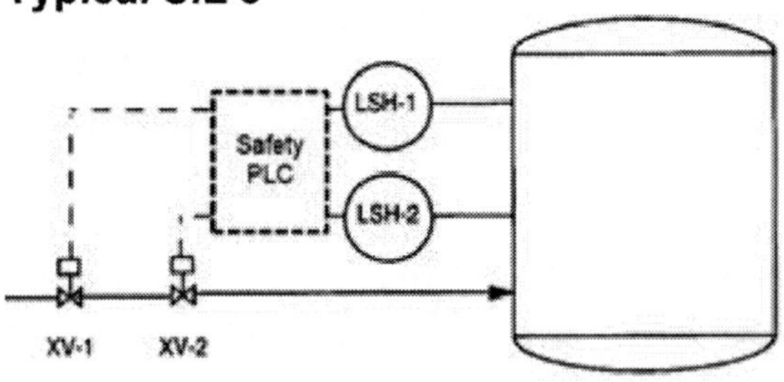

Case Study 1

An example of the streamlined approach is summarized in Table 2 for a facility with storage spheres (in China; the PHA and other analyses were performed in 2014). The table shows two deviations from the PHA. IPLs are identified and existing SIFs are assigned an SIL, based on the architecture. (Note that the achieved SIL for each of the SIFs (existing or recommended) must be confirmed by a SIL verification calculation that includes the failure rate of the components, the voting architecture, and the proof test interval.)

For the high-level deviation, the high-level SIF was assigned SIL 1. The overflow through the pressure equalization line to other spheres was determined to be an IPL.

For the low level deviation, the low level indication and low level alarm were determined not to be an IPL because they are part of the initiating cause. The PHA team concluded that feeding from two spheres at all times is an IPL for this facility because of the unlikelihood for both spheres to have low level at the same time. The PHA team also thought that an additional SIF level alarm with operator response to switch tanks within 60 minutes could be a possible IPL, if the action of the operator is quick enough. The team wrote a recommendation to ensure that the operator response could be an IPL.

In this case study, the PHA team was able to determine in the PHA meeting which safeguards were IPLs. They also determined the nominal SIL for existing SIFs (to be confirmed in a later SIL verification). And the team determined that operator response could be an IPL if there was a documented procedure and an annual drill. For these two deviations, the PHA team did the work that would be done in a later LOPA meeting in the typical approach.

Next, the SIL Verification was performed for the SIL 1 SIF (and all other SIFs in the unit). This SIL Verification was accomplished using exSILentia™ software and the analysis and documentation took less than 2 hours.

Thus, the entire requirements for a SIL identification, SIL target level determination, and SIL Verification were accomplished very quickly compared to typical approaches.

Table 2: Storage Sphere PHA

No.: 2 XXXX storage spheres xxx-T-XX A/B/C/D/E/F/G/H/I/J/K/L (1 of 12)					
#	**Dev.**	**Causes**	**Consequences**	**Safeguards**	**Recommendations**
2.1	High level	Too much flow to one sphere from XX Plant (through their pump; about 40 bar MDH)	High pressure (see 2.5)	High level SIF with level sensors voted 2oo2, to close inlet valve - **SIL 1** Overflow thru pressure equalization line to other spheres (through normally open [NO] valve) - **IPL**	
2.2	Low level	Failing to switch from the sphere with low level in time (based on level indication)	Low/no flow - Liquid from spheres through high pressure product pumps to the vaporizer (see 4.2) Low/no flow - Unqualified liquid from spheres back to Plant (see 6.2)	Level indication and low level alarm, inspected each year, per government regulation **(not IPL; part of the cause)** 9 other spheres with possibly enough level to switch to Feeding from two spheres at all times, so unlikely for BOTH spheres to have low level at the same time - **IPL** Two level indication from SIS level transmitter, with low level alarm, with more than 60 min available to switch tanks (SIF driven alarm and response) - **possible IPL, if action of the operator is quick enough**	Rec 4. Make sure the Human IPL of response to low level in all spheres and tanks is described in a trouble-shooting guide (like an SOP) and practiced once per year per unit operator. This will make this response a valid IPL.

Case Study 2

In 2014 and 2015, an identical approach to that of Case Study 1 (the streamlined approach of this paper) was performed for a large gas producer in South America. In that work, the PHAs for more than 500 nodes plus all procedures of all operating modes was completed in about 10 weeks of PHA meetings. The PHA team identified more than 800 SIFs, along with their target SILs, using the approach in Steps 1 and 2.

Next, staff within PII but separate from the PHA team performed about 60 LOPA to clarify certain issues, some related to SIFs. But less than 5% of the SIFs were established with LOPA; 95% were established with the PHA alone.

Finally, staff separate from the PHA performed SIL Verification of the 800+ SIFs, using exSILentia™. The verification was for a combination of SIL 1 and SIL 2 SIFs; a couple of SIL 3 SIFs were also analyzed and debated. The relatively few LOPAs took about 2-3 hours each (due to the complexity of the 5% of scenarios that went to LOPA) and each SIL Verification took about 2-3 hours to perform and document.

Caution – the issue of human factors for SIL 2 and SIL 3 SIFs[6]

There is an important aspect of important aspect of SIFs and the possible achieved SIL that is not addressed in this paper. Relying on a high integrity SIF, without accounting for human error during interventions, is also a waste of resources. You may believe that you have reduced the risk by a factor of 100 for a SIL 2 SIF, or a factor of 1000 for a SIL 3 SIF, when in fact the human factors during commissioning and proof testing may degrade the overall risk reduction factor to about 10. If the human factor is accounted for in the SIF design phase and the SIL verification phase, then the human error can also be prevented to allow the higher reduction in risk that is targeted.

Conclusions

Based on experience in a multitude of PHAs, LOPAs, QRAs, and HRAs, we have seen that experienced, competent PHA teams with an experienced, competent leader can effectively apply the principles of LOPA to many scenarios during the PHA meeting. The team can determine the approximate SIL of existing SIFs, the team can determine which safeguards are truly IPLs, and the team can recommend additional IPLs, including SIFs with specific SILs. If the scenario is too complex, or the PHA team is uncertain, the team can recommend a detailed LOPA. The streamlined qualitative makes better use of resources and time than the typical approach with separate meetings for PHA and LOPA.

It is critical that the PHA leader is experienced and competent in LOPA. It is critical that the PHA team has in its memory the data for consequence severity, initiating cause frequency, IPL requirements, and the risk tolerance criteria.

Acronyms Used

AIChE– American Institute of Chemical Engineers
CCPS – Center for Chemical Process Safety (an AIChE technology alliance)
HAZOP – Hazard and Operability Analysis
HRA – Human Reliability Analysis
IEC – International Electrotechnical Commission
IEF – Initiating Event Frequency
ETA – Event Tree Analysis
FMEA – Failure Mode Effect Analysis
FTA – Fault Tree Analysis
HRA – Human Reliability Analysis
IPL - Independent Protection Layer
LOPA – Layer of Protection Analysis
PHA – Process Hazard Analysis
PII – Process Improvement Institute, Inc.
QRA – Quantitative Risk Analysis
SIF – Safety Instrumented Function
SIL – Safety Integrity Level
SIS – Safety Instrumented System
SRS – Safety Requirements Specification

References

1. *Guidelines for Hazard Evaluation Procedures*, 2008, CCPS/AIChE.
2. Bridges, W., & Dowell, A., "Identify SIF and Specify Necessary SIL, and other IPLs, as part of PHA/HAZOP," *12th Global Congress on Process Safety*, AIChE, 2015.
3. *Layer of Protection Analysis: Simplified Process Risk Assessment,* 2001, CCPS/AIChE.
4. *Functional Safety: Safety Instrumented Systems for the Process Industry Sector - Part 3: Guidance for the Determination of the Required Safety Integrity Levels*; ANSI/ISA 84.00.01 Part 3, 2004.
5. IEC 61511, *Functional Safety: Safety Instrumented Systems for the Process Industry Sector - Part 1: Framework, Definitions, System, Hardware and Software Requirements*, International Electrotechnical Commission
6. Bridges, W., and Thomas, H., "Accounting for Human Error Probability in SIL Verification Calculations," *8th Global Congress on Process Safety,* Houston, AIChE, April 2012.

479509 The Art of the Audit: Being Audit Ready, Everyday

Abstract:

Have you ever made an appointment for an annual physical and suddenly join a gym? We all know we should be exercising every day, but it becomes more important before going in for that checkup. In reality, not exercising has consequences to your health and not your doctors, so why do we do this? To impress them, instead of doing it to take care of ourselves? This phenomena of doing things you do not normally do in preparation for an event is unfortunately how too many of us treat audits. We hear an auditor is coming and end up running around trying to organize paperwork, convince the operators to clean up the plant, and everyone in management becomes much more attentive. Does this behavior really set the proper tone to develop a safety culture? Auditors are there to check up on our compliance of rules that benefit only us to follow, particularly in the field of safety.

A successful operation will be 'Audit Ready' every day. 'Audit ready' means that operations are in compliance with rules, personnel are safe and exhibit safe behaviors, and all of the team members understand why they perform the functions of their job. Compliance tools, personnel awareness and routine self-audits are three key topics in becoming ready for an audit every day.

Personnel awareness is among the most vital areas to ensuring safe and effective operations, as well as successful audits. Audits involve people from across almost all departments, and as such mirror the requirements of multifunctional teams in everyday operations.

Compliance tools are important to track work that is being completed, and create convenient locations to store data which proves compliance. This can include tracking the progress of MOCs through weekly notifications, using databases to track when procedures need to be reviewed, and more. Visibility of what needs to be accomplished is important.

Routine self-audits help us to find gaps in compliance, and reduce complacency. These can include a simple spot check of one area of compliance, or be more comprehensive, and there are pros and cons to both. A spot check keeps everyone on their toes and takes a short enough time such that they can be done more frequently.

Ensuring sustainable records and checking everyday habits will go a long way to creating a safe operating environment, and with that a workforce that is audit ready, every day.

479546 Economic Justification for Better Maintenance in a Safety System

Abstract:

It is not difficult to think of numerous ways in which human factors might impact functional safety. For example, imperfect repair, improper calibration, faulty installation, etc. can all contribute to decreases in functional safety performance. However, if actions are to be taken to mitigate these impacts then it is essential to have some quantitative measure of how this impacts the PFDavg, so that the actions' effectiveness can be assessed. We will show how a simple measure system called the Site Safety Index^{TM} (SSI) is used to adjust safety metrics, including the PFDavg and the MTTFS, on a site specific basis. The lifecycle economic impact of these metrics is then used to show how improved operations and maintenance can be justified. Economic justification for better maintenance can be easily verified through this approach being further mindful of the safety culture.

479547 How Important Is Realistic Failure Rates for Safety Instrumented Functions?

Abstract:

A given Safety Instrumented Function design must be verified to meet the required SIL level by calculating probability of failure on demand. However, this does require realistic failure rate data for each device. Fortunately, many device manufacturers are getting their products certified per IEC 61508 and many of the certificates now have failure rate data. Other manufacturers supply analysis reports with failure rate data. But how good is this data supplied by the manufacturers? All such data must be checked to make certain it is applicable for the process industries as some data may have been derived based on machine safety or automotive applications.

This paper will show some examples of manufacturer provided failure rate data and a methodology created to validate device type failure rate data for the process industries.

479917 Enabling Sustainable Relief System Management Programs through Independent Reviews

Abstract:

Overpressure protection and flare systems usually represent the last line of protection for personnel and the environment in industrial facilities operating under high temperatures and pressures. During the design phase of either Greenfield or Brownfield projects, assessing the basis for major relief systems and making clear recommendations to reduce the risks associated with each system has helped companies realize substantial safety and economic benefits. Independent and safety focused reviews, with an emphasis on value added solutions has enabled organizations to set the right foundation for a sustainable relief system management program. This, in turn, has helped enable the facilities in meeting the ever-changing regulatory needs in a timely manner.

This paper will give a typical breakdown of the workflow process for each of the entities involved in the project - the client, the contractor, and the independent reviewer. Four distinct work activities are typically identified for such a review - development of preliminary overpressure protection, engineering design of relief systems, validation of the design, and validation of as-built documentation. After each stage, the output from the independent analysis is compared with the contractor's work and updates are performed before continuing to the next stage. This paper will also give an overview of these stages and discuss the various steps and value added for each stage of the project.

Jack Rabbit II: Experimental Large-Scale Chlorine Release Trials

Mark Whitmire[1], Shannon Fox[2], Thomas Spicer[3], Leo Stockham[4], Steven Hanna[5], Joseph Chang[6], Michael D. Sohn[7], Thomas Mazzola[8], Ronald Meris[9], Damon Nicholson[10], Andy Byrnes[11]

[1]Noblis, Falls Church, VA, USA; [2]DHS S&T Chemical Security and Analysis Center, Aberdeen Proving Ground, MD, USA; [3]University of Arkansas, Fayetteville, AR, USA; [4] Leidos, Albuquerque, NM, USA; [5]Hanna Consultants, Kennebunkport, ME, USA; [6]ANSER, Falls Church VA, USA; [7]Lawrence Berkeley National Laboratory, Berkeley CA, USA; [8] Engility, Lorton, VA, USA; [9]DTRA, Fort Belvoir, VA, USA; [10]Dugway Proving Ground, UT, USA; [11]Utah Valley University, Provo, UT USA

INTRODUCTION

Chlorine is a potent toxic inhalation hazard (TIH) chemical, and each year millions of tons of it are transported in bulk via road, water, and rail. Consequently, an accidental or deliberate chlorine release near a populated area may cause a considerable number of fatalities, and would create other health, safety, and environmental problems. Hazard prediction models are inconsistent with the evidence and observations from chlorine disasters, and the understanding of such releases is otherwise insufficient to guide and inform an effective emergency response. So beginning in 2010, DHS and its Health, Safety and Environment (HSE) partners and stakeholders planned the Jack Rabbit (JR) Program. The program was led by the DHS Chemical Security Analysis Center (CSAC) located at Aberdeen Proving Grounds-Edgewood Area MD.

OVERVIEW

The Jack Rabbit (JR) program is comprised of a series of projects that includes JRI (2010), JRII Phase 1 (2015), and JRII Phase 2 (2016). The JRI (Fox and Storwold, 2011) project was completed to develop, test, and evaluate a mechanism for the controlled, rapid release of liquefied, pressurized gases (specifically, ammonia and chlorine), to observe the transport and dispersion of a disseminated vapor and aerosol cloud, and to investigate chemical deposition and reactions with water and soil. The project also evaluated instruments, test methods, and strategies that may be used to study more industrial scale tests. Using either chlorine or ammonia, two 'pilot' tests and eight 'record' tests of either 1 or 2 tons were completed. For chlorine, among other findings, the results demonstrated that downwind transport and turbulent mixing are initially resisted by a dense persistent gas and aerosol cloud, that source phenomena were non-linear with increasing release volumes, and that the reactivity with soil containing water and organic matter is an important removal mechanism.

JACK RABBIT II

JRII leveraged and extended the work of JRI with controlled chlorine field experiments on a scale corresponding to railcars and tank-trucks; that is, 5 to 20 tons each. The goals were to collect data on the release source, measure chemical reactions with the environment, observe and evaluate exposure effects on equipment and infrastructure, and assess urban impacts using a Mock Urban Environment (MUE) or an unobstructed dissemination and downwind dispersion environment, both located on the Urban Test Grid (UTG) at Dugway Proving Grounds UT. Data was collected in real time via ground based and unmanned aerial vehicle systems, video cameras, and spectroscopic instruments, in situ point detectors, path integrated detectors, and concentration and dosage determination instruments. The data was thoroughly and qualitatively reviewed and shared with all project participants in prescribed and common formats. The generated data and findings will improve the understanding of and operational hazard prediction modelling for TIH releases, and lead to more effective emergency response and training, national preparedness, and mitigation strategies.

URBAN TEST GRID (UTG)

The UTG (Nicholson, Hedrick, and Lian, N; 2017) was the designation for the JRII chlorine release site surveyed and built specifically for the JRII tests; it was at Dugway Proving Grounds west of Granite Mountain on Goodyear Road. It included the playa and adjacent support areas, concrete pad, gravel pad, and surveyed grid extending 11 km downrange and aligned roughly south to north.

The UTG included a circular, 0.5 ft. deep, 82 ft. diameter, rebar reinforced concrete disseminator vessel pad with 3 embedded temperature sensor stations, and a 1 inch steel lip to contain chlorine liquid. The center of the circular pad was the center of the UTG site, Northing 4445633.945, Easting 288109.182 (Latitude, 40.134347; Longitude, -113.487076). The concrete pad rested on a 24 inch deep, 400 ft. × 600 ft. gravel pad level with the concrete pad and symmetric about it.

Figure 1. Chlorine release from the disseminator on the concrete pad, with CONEX container array on the gravel pad of MUE, looking upwind (south) from atop the 2 x 3 CONEX container stack. The dense cool cloud remains low as it moves downwind and laterally around and through the Mock Urban Environment made from CONEX containers.

Winds predominate from approximately 165° from June to September, so the UTG was oriented about a centerline from that direction. The UTG included the playa within a 90° angle bisected by the UTG centerline with vertex at the disseminator vessel location, and extending to the 11 km arc. The "Near Field" was the UTG region from the disseminator up to and including the 2 km arc and inside the 90°

angle; the "Extended Field" was the UTG region past the 2 km arc to the 11 km arc and inside the 90° angle.

The Mock Urban Environment (MUE) was constructed on the gravel pad and surrounding the concrete disseminator pad using a staggered row array of 86 CONEX containers (CONtainer EXpress shipping containers, either 18 or 40 ft. long) to observe and study the transport and dispersion of dense gas around and above urban structures. The array included a two wide by three high (2 x 3) stack of six 20 ft.-CONEX containers located directly downwind of the source, and two 2-axle trailers nearby with insulation, dry wall, and trailer skirting for infiltration studies. Aside from the CONEX containers and trailers, up to seven vehicles were included in the array. Three fire trucks, one ambulance, and three cars were located on the north edge of the gravel pad during Trials 1-3. During Trial 4, four vehicles were placed on the playa past the 250 m arc; during trial 5, three vehicles were moved between the rows of CONEX containers but adjacent to the concrete disseminator pad, and three vehicles remained on the playa past the 250 m arc.

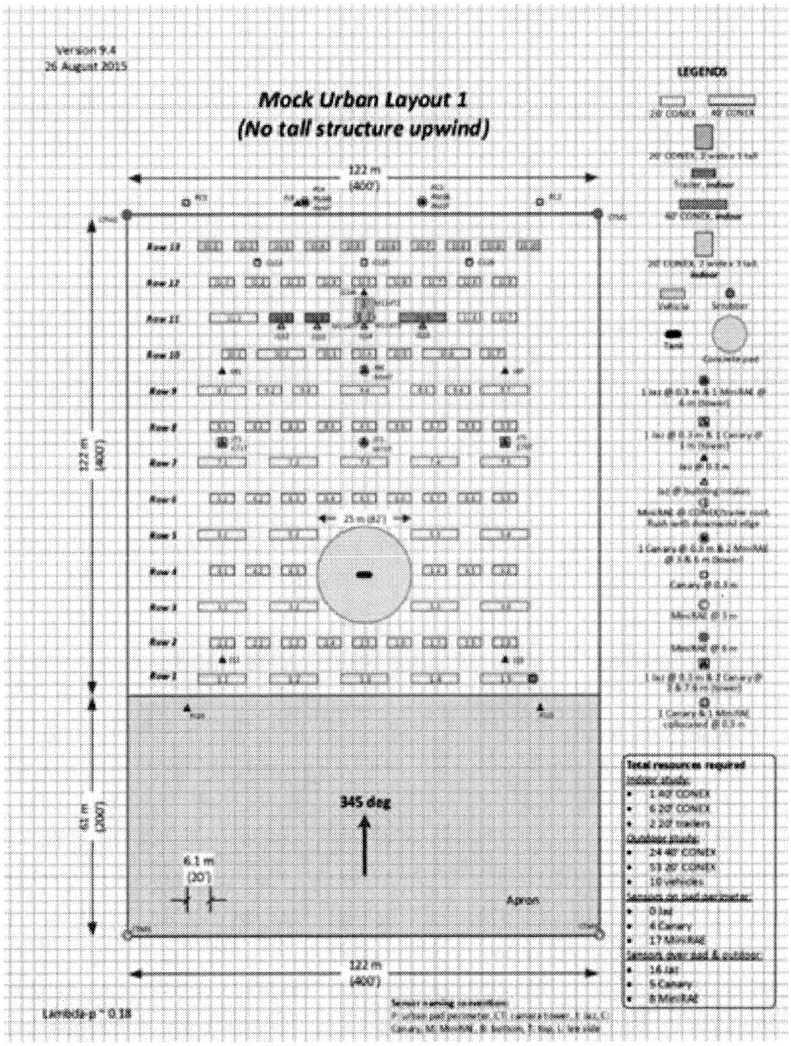

Figure 2 Mock Urban Environment (MUE) schematic showing gravel pad around circular pad for Chlorine disseminator; CONEX containers and instrument locations are also shown.

Instrument locations for the UTG "Near Field" and "Extended Field" were surveyed, recorded, and reported. Point sensors included about 170 MiniRAE 3000s equipped with volatile organic compound

(VOC) photoionization detectors; about 60 ToxiRAE Pro units; a Gasmet™ DX-4000 (FTIR) gas analyzer; 28 Ultraviolet (UV) Canary units for UTG monitoring and indoor studies; and 16 Ocean Optics Jaz instruments for UTG monitoring and indoor studies. Standoff Detectors included two Cerex Monitoring Solutions UV Sentry chlorine-specific integrated path detector, three Ultraviolet (UV) Light Detection and Ranging (LiDAR) systems for monitoring the moving chlorine gas cloud, and elastic backscatter LiDAR systems for monitoring the moving aerosol cloud. A GRIMM Aerosol portable laser aerosol spectrometer and dust monitor was also used for particle sizing model using laser scattering techniques. Other equipment included remotely operated high-definition (HD), high-speed (HS), and FLIR® IR Camera SC660 micro-bolometer cameras positioned throughout the test grid to visualize and record the chlorine release and resulting plume.

Meteorology equipment included 34 Portable Weather Information Display Systems (PWIDS) collecting data at 2 m; three 32 m towers each equipped with sonic anemometers collecting data at heights between 2 and 32 m; and 31 3D ultrasonic anemometer to gather wind and turbulence statistics, and fluxes of heat and momentum. Energy Balance Station (EBS) provided in situ measurements of the surface vertical fluxes of sensible heat, latent heat, net radiation, and soil surface heat flux. The MiniSODAR™ high-frequency Doppler acoustic sounding system designed to measure the atmospheric vertical wind profile from 15 to 200 m. Numerical gridded forecast meteorology using embedded observations and real-time meteorological data from the test grid were used for HPAC modelling leading up to each trial and to determine "GO/NO GO" test status.

DOWNWIND DISPERSION AND TRANSPORT ENVIRONMENT

For tests in 2016, and other than two vehicles, a CONEX container, and a modular office trailer used for follow-up infiltration experiments, the UTG was cleared of the simulated buildings and obstructions used in 2015. The project goals were to monitor and collect the cloud concentration data from the initial phase of a very dense, two-phase cloud near the source to dispersion of the vapor cloud farther downwind and up to 11 km away using point detectors and stand-off spectroscopic instruments.

DISSEMINATION

The disseminator vessel for use as the chlorine release source during the tests was located at the center of the concrete pad with its major axis perpendicular to the UTG centerline. The 2000 gal. tank was mounted in a 3-dimensional dynamic mass measurement load cell system. Other vessel mounted instrumentation included a redundant system for determining the remaining mass of chlorine in the vessel, differential pressure sensors on the vertical axis at various depths in and below the vapor space, guided wave radar liquid level measurement, and instrumentation for absolute pressure and direct contact temperature profiles inside the vessel. At 80-90% operating capacity, the vessel could contain from 8.5 to 10 tons chlorine.

Each release was accomplished using a blind flange with exploding bolts rated to safely hold the pressure of a vessel loaded with liquid chlorine at ambient temperature. To blow the flange and release the chlorine, the bolts were remotely detonated; when the bolts broke, the blind flange was blown off with the aid of the chlorine weight and head pressure, and chlorine flowed through the vessel's circular opening. A quick release valve was not used to simulate the release because of the likelihood of internal flashing flow in the piping and valves; that is, the mass release rate would be less than what would be observed from a breach of containment. In 2015, every release was directly down (180°) through a six inch hole. In 2016, each release was either directly down (Test 6); at 135° and downwind (Test 7), or directly up (Test 8). After Tests 7 and 8 any liquid chlorine remaining in the vessel was discharged via a remotely operated valve in the bottom of the tank. Before inert gas purging and chlorine refill operations, the ejected blind flange was replaced using standard bolts. The last release was from a tanker truck that had been removed from service, and containing up 19 t chlorine. The tanker truck's skin was breached with a remotely activated explosive shape charge.

Figure 3. Test Release 6; 10 Tons chlorine on August 31, 2016. A moment after the release the liquid chlorine may be seen splashing onto the concrete pad. As the chlorine aerosolizes and vaporizes it cools the surroundings and condenses water vapor.

OBSERVATIONS AND FINDINGS

From the JRII trials the program observed that:

- Upon breaching the tank, liquid chlorine exits the tank and rapidly flashes to a two phase aerosol and vapor cloud with a liquid layer deposited on the pad, ultimately all vaporizing and transporting downwind.
- The escaping chlorine was found to auto-refrigerate to its -34 degree Celsius boiling point and lower due to further evaporative cooling outside the tank.
- The force of the escaping chlorine jet generated strong initial turbulence and momentum to drive the expanding cloud over nearby Conex structures.
- Evacuation of the tank during 5-ton releases took approximately 35 seconds.
- Evacuation of the tank during 9-ton releases took approximately 50 seconds.
- Structures were observed to channel the escaping and expanding cloud laterally.
- Significant upwind movement of the chlorine clouds was observed due to the momentum of the escaping gas, expansion of the gas, and gravity-slumping.
- Visible clouds were observed to persist in UTG for approximately 20-30 minutes depending upon release amount and winds.
- Long visible clouds observed to stretch from UTG to approximately 2 km as downwind dispersion progressed.

- Data sets from Jack Rabbit II have been quality-assured, finalized, and recently released, enabling in-depth studies to begin on cloud concentrations, movement, and behavior to be quantified.

EMERGENCY RESPONSE EXPERIMENTS

For each test, researchers from Utah Valley University placed two "Witness Boards" on surveyed arcs adjacent to the disseminator in the urban environment to observe the effect of liquid chlorine and high concentrations of chlorine gas (approaching 25%) on common building components, urban surfaces, and fire equipment. Go-Pro® and HD color surveillance cameras captured the exposure on video. The selection of the test materials emphasized high organic content (creosote treated telephone pole and railroad tie, road asphalt, shingles, rubber, and bare wood), but copper, aluminum, galvanized steel, and carbon steel coupons were included. The copper and carbon steel surfaces began to be discolored 1-2 hours after the exposure, and the chlorine sometimes dissolved the petroleum based materials.

INDOOR INFILTRATION STUDIES

The Defense Threat Reduction Agency (DTRA) is evaluating various hazardous airborne contaminant applications for transition to DOD CBRN consequence models. Their current indoor application models presume that any contaminant is inert and neutrally buoyant, thus the models may not be useful for predicting the indoor effects of a dense, reactive gas like chlorine. Lawrence Berkeley National Laboratories (LBNL) proposed a series of experiments to DTRA and the JR program to assess the applicability of the models for predicting the indoor concentration of chlorine in the event of an outdoor chlorine release.

LBNL successfully completed a series of indoor infiltration experiments in 2015 in support of the DTRA effort. The LBNL experiments were completed to understand and characterize the dominant infiltration pathways for reactive dense gases into buildings, the uptake of such gases into structures, and the re-emission of them from the structures after the peak of the plume had passed. Three single-story CONEX structures, a single three-story CONEX structure, and 2 modular office trailers placed on the gravel pad were used to represent indoor office, school, or home environments. Follow-on studies were completed in 2016 using a single-story CONEX structure at the same, but unobstructed site. The CONEX structures were configured for controlled experiments of airflow and transport in single- or multi-story structures, and modified to include insulation and temperature-control. Chlorine penetration into and through the structure, and transfer of it across the building envelope was monitored via controlled artificial "leaks" and exhaust fans. The model offices were furnished with realistic insulation, surface materials, windows, and HVAC system characteristics, thus the simulated modular offices helped to relate the data from the controlled CONEX structures to real buildings. The data may be used for calibration, verification and/or validation of the indoor model, and to support DOD concepts of operation for sheltering in the event of an outdoor CBRN release.

PROGRAM STATUS

The program is entering a new phase after completing field testing. In 2017, the program will be processing and reporting data from 2015 and 2016, and transitioning it for permanent data storage, and to stakeholders. CSAC will also be presenting a JR overview, narrative, and plans in presentations to interested groups. The program will be proposing, designing, managing, and performing small and large scale experiments during chamber testing, and reducing and analyzing data to demonstrate or elucidate the reaction of chlorine with various substrates to suggest large-scale toxic inhalation hazard (TIH) chemical release mitigation strategies, inform emergency response strategies that may reduce the impact to affected populations and infrastructure, calculate the value of modeling parameters, and otherwise support stakeholders interests.

ACKNOWLEDGEMENTS
This research has been supported by the US Department of Homeland Security (DHS) and the US Defence Threat Reduction Agency (DTRA). Additional support (funding, personnel and instruments) is coming from numerous partners, with the list growing with time. We appreciate guidance from many scientists who are members of the planning groups and/or have provided comments on the scientific approach.

REFERENCES

1. Nicholson, D., Hedrick, A., and Lian, N., Final Data Report for Jack Rabbit (JRII), 2017

2. Fox, S. and D. Storwold, 2011: Project Jack Rabbit: Field Tests. Chemical Security Analysis Center, Science and Technology Directorate, U.S. Department of Homeland Security. CSAC 11-006, 162 pp.

479394 Functional Safety - Pitfalls and Good Practices

Abstract:

The implementation of effective functional safety is valuable in ensuring that safety critical control and protective systems are appropriately specified, designed, installed and maintained. Through the International Electro-technical Commission (IEC) 61508 and 61511 suite of functional safety standards, the concepts of the Safety Lifecycle and Safety Integrity Levels (SIL) are now well known. However, DNV GL's experience from numerous projects across different industry sectors suggests that there remain challenges to be overcome.

DNV GL has conducted a number of Functional Safety Assessments (FSAs) on various electrical/ electronic and programmable electronic safety systems in the process sector. The FSAs are a mandatory element of IEC 61511 which provide an independent judgment on the integrity of the safety system and should be performed by competent independent Functional Safety Practitioners. At least one FSA must be performed on a Safety Instrumented System and should be completed prior to the introduction of hazards. Through DNV GL's experiences in functional safety assessments this paper will discuss;

- How to avoid common mistakes in the management and implementation of functional safety;
- Examples of good practice in the application of functional safety;
- How good practice can help optimize test regimes, increase plant availability and reduce maintenance workload whilst ensuring the system continues to provide the required risk reduction.

General Method for Uncertainty Evaluation of Safety Integrity Level (SIL) Calculations – Part 2 Analytical Methods

Raymond "Randy" Freeman
S&PP Consulting

For Presentation at the American Institute of Chemical Engineers
2017 Spring Meeting in San Antonio, Texas -- March 26-30, 2017

Abstract

The IEC 61511 standard requires a verification calculation that a proposed design for a safety instrumented function (SIF) achieves the desired safety integrity level (SIL). The evaluation of the safety integrity level of a new or existing safety instrumented system requires detailed calculations based on the failure rates of the device and the planned maintenance/testing cycle for the system. The failure rates of the devices are often taken from standard failure rate tabulations of equipment. The maintenance and testing plans are developed based on plant experience. All of the data used in the SIL calculations are uncertain. This paper develops a general method for uncertainty analysis of the SIL calculations. The general method is based on the application of probability theory - variance contribution analysis (VCA) – to the equations presented in ISA TR 84.00.02-2115. An example is worked to demonstrate the methodology.

Background

The calculation of the probability of failure on demand (PFD) is a common engineering task when designing an interlock or safety system that is to be in compliance with IEC 61511 [Ref. 1]. The calculation of the PFD is often done using approximate equations defined in the ISA TR84.00.02 technical report [Ref. 2]. The simplified equation method in the ISA report is commonly used and is based on the use of reliability block diagrams where the field sensors, Safety Instrumented System (SIS) logic solver and final control elements are considered independent of each other in the sense of not sharing common devices or systems. The PFD is then calculated using the failure rates of the devices, planned test intervals, vendor supplied estimates on diagnostic coverage of the devices and an allowance for the potential for common cause failures. Almost all of these parameters are uncertain. The failure rate data is often taken from generic data sources which show wide ranges in the observed values.

Because of the uncertainty in the parameters, the design engineer makes allowances in the design by the use of safety factors or rules of thumb to improve the chances that the final interlock installation will work as intended. Since each engineer has a different set of safety factors and rules of thumb, two designs may differ significantly in the way a hazard is controlled.

A more formal method for handling the underlying uncertainty in the calculation of the PFD of an interlock is needed. Previously, Freeman and Summers [Ref. 3] published an uncertainty analysis of the PFD of an interlock. Two different methods were used in this analysis:

- Monte Carlo Simulation
- Variance Contribution Analysis (VCA)

Monte Carlo Simulation requires that the engineer build a model of the interlock using specialized computer software. The use of VCA requires that the sensitivity of the interlock model be determined either by numerical methods or by direct analytical calculations. The Freeman paper demonstrates that VCA can be used for the uncertainty analysis. However, the paper does not present a complete analysis method that can be applied to any system defined in the ISA technical report TR84.00.02. The goal of this paper is to develop a general set of analytical equations that will allow VCA to be used in uncertainty analysis of any interlock developed per the IEC 61511 standard.

Review of Interlock Design

The IEC standard looks at an interlock as a series of three major components (see Figure 1). First is the sensor set which sends an indication of an abnormal event to a logic solver. The logic solver determines is the signal from the sensor meets the conditions required to activate (trip) the interlock. If the interlock is to be activated, a signal is sent to the final control elements to take action (stop flow, close valve, etc) to prevent a process safety event from occurring.

Based on the devices selected the design engineer then evaluates whether the proposed interlock design meets the design criterion for probability of failure on demand. Typically, the design engineer is given the target PFD as a statement such as "provide a SIL-2 interlock to stop flow on high level in a vessel." A SIL-2 interlock requires a PFD of at most 0.01 or a risk reduction factor of at least 100. The needed risk reduction is often determined in a Layer of Protection Analysis (LOPA) [Ref. 4] and represents the managements decision on how risk is to be managed in a system. The resulting SIL-2 interlock design must have a high likelihood of achieving the target risk reduction as the LOPA team may be relying on it to be part of an overall risk reduction plan to prevent a process safety incident.

The ISA technical report (TR 84.00.02) [Ref. 2] provides shortcut methods for the evaluation of the PFD of an interlock design. Tables 1 and 2 are taken from the ISA technical report [Ref. 2]. Typical parameters need to compute the PFD include:

- DC is the diagnostic coverage;
- DI is the diagnostic interval;
- TI is the proof test interval,
- λ^D is the dangerous failure rate;
- MTTR is the mean time to restore the system to operation
- β is the common cause failure parameter that is always is between 0 and 1

In practice most if not all of these parameters are uncertain. Using parameter values obtained from plant records, vendor data or generic data bases the design engineer proceeds to develop the model for the interlock.

What is needed is a means to rapidly evaluate the impact of parameter uncertainty on the interlock PFD. The remainder of this paper applies the methods of variance contribution analysis (VCA) to determine the mean (expected value) and the standard deviation of the interlock PFD for a given design.

Table 1. Simplified PFD_{avg} formulas for Non-Repairable System without considering CCF, Diagnostics or MTTR.			
Configuration	**Function**	**PFD based on "Average before" failure rate**	**Appendix A Equation Number**
1oo1	$F_1 =$	$\left[\lambda^D \times \frac{TI}{2}\right]$	A-1
1oo2	$F_2 =$	$\left[(\lambda^D)^2 \times \frac{TI^2}{4}\right]$	A- 2
1oo3	$F_3 =$	$\left[(\lambda^D)^3 \times \frac{TI^3}{8}\right]$	A-3
2oo2	$F_4 =$	$\left[\lambda^D \times TI\right]$	A-4
2oo3	F5 =	$\frac{3}{4} \times \left[(\lambda^D)^2 \times TI^2\right]$	A-5
3oo3	F6 =	$3 \times \left[\lambda^D \times \frac{TI}{2}\right]$	A-6

Where TI is the proof test interval, λ^D is the dangerous failure rate.

Table 2. Simplified PFD_{avg} Formulas for Repairable System considering CCF, Diagnostics and MTTR

Configuration	Function	PFD based on "Average before" failure rate	Appendix A Equation Number
1oo1	F_7 =	$\left[\frac{(1-DC)\times\lambda^D\times TI}{2}+\frac{DC\times\lambda^D\times DI}{2}+\lambda^D\times MTTR\right]$	A-24
1oo2	F_8 =	$\left[\frac{(1-DC)\times(1-\beta)\times\lambda^D\times TI}{2}+\frac{DC\times(1-\beta)\times\lambda^D\times DI}{2}+(1-\beta)\times\lambda^D\times MTTR\right]^2+\left[\frac{(1-DC)\times\beta\times\lambda^D\times TI}{2}+\frac{DC\times\beta\times\lambda^D\times DI}{2}+\beta\times\lambda^D\times MTTR\right]$	A- 25
1oo3	F_9 =	$\left[\frac{(1-DC)\times(1-\beta)\times\lambda^D\times TI}{2}+\frac{DC\times(1-\beta)\times\lambda^D\times DI}{2}+(1-\beta)\times\lambda^D\times MTTR\right]^3+\left[\frac{(1-DC)\times\beta\times\lambda^D\times TI}{2}+\frac{DC\times\beta\times\lambda^D\times DI}{2}+\beta\times\lambda^D\times MTTR\right]$	A-26
2oo2	F_{10} =	$2\times\left[\frac{(1-DC)\times\lambda^D\times TI}{2}+\frac{DC\times\lambda^D\times DI}{2}+\lambda^D\times MTTR\right]$	A-27

Table 2. Simplified PFD_{avg} Formulas for Repairable System considering CCF, Diagnostics and MTTR			
Configuration	**Function**	**PFD based on "Average before" failure rate**	**Appendix A Equation Number**
2oo3	$F_{11} =$	$3\times\left[\frac{(1-DC)\times(1-\beta)\times\lambda^D\times TI}{2}+\frac{DC\times(1-\beta)\times\lambda^D\times DI}{2}+(1-\beta)\times\lambda^D\times MTTR\right]^2+\left[\frac{(1-DC)\times\beta\times\lambda^D\times TI}{2}+\frac{DC\times\beta\times\lambda^D\times DI}{2}+\beta\times\lambda^D\times MTTR\right]$	A-28
3oo3	$F_{12} =$	$3\times\left[\frac{(1-DC)\times\lambda^D\times TI}{2}+\frac{DC\times\lambda^D\times DI}{2}+\lambda^D\times MTTR\right]$	A-29

Review of Variance Contribution Analysis (VCA) Methodology

The mean and variance of a function of random variables can be approximated using the method described by Haugen [Ref. 5] and applied by Freeman [Ref. 3, 6, 7]. Define an arbitrary function of a set of random variables, x_i, as:

Let

$$Y = F(x_i) \qquad \text{(Eq 1)}$$

The mean of Y can be estimated using the following approximation:

$$E(Y) = F[\,E(x_i)] \qquad \text{(Eq 2)}$$

Where:
E(Y) = expected value of random variable Y = mean of Y
$E(x_i)$ = expected value of random variable x_i = mean of x_i

The variance of Y can likewise be estimated as:

$$V(Y) = \sum_{i=1}^{n} \left[\frac{\partial Y}{\partial x_i}\right]^2 V(x_i) \qquad \text{(Eq 3)}$$

Where:

V(Y) = variance of random variable Y as defined above in Equation 1

$V(x_i)$ = variance of random variable x_i as defined above in Equation 1

Note that the variance is simply the square of the standard deviation. Using the variance will simplify the mathematics that is described below. The contribution of each independent random variable to the overall variance in the function is:

$$V(Y \text{ from } x_i) = \left[\frac{\partial Y}{\partial x_i}\right]^2 V(x_i) \qquad \text{(Eq 4)}$$

The relative contribution of each term to the overall variance V(Y) is a measure of the importance in the uncertainty in the particular random variable, x_i. In effect, this is a sensitivity analysis combined with a uncertainty evaluation. The variance contribution combines the sensitivity in the answer to changes in the uncertain random variable, x_i, with a measure of the uncertainty in the random variable, x_i. The overall variance in Y is found by summing the sensitivity weighted variances from each random variable.

Recommended Interlock PFD Uncertainty Analysis Method

Previously, Freeman and Summers [Ref. 3] suggested a framework for the inclusion of uncertainty analysis in the calculations completed for a new interlock per the ISA TR [Ref. 2]. The following uncertainty analysis method has been expanded to incorporate critical decisions that the process management must make in the design process.

1. Complete the interlock design using the methods outlined in IEC 61511 [Ref. 1].

2. Review with the process system management and determine if the proposed interlock should be considered repairable or non-repairable. A simple flow chart for this decision making is presented in Figure 2. The basic question is: "can the interlock be repaired safely while the process operates." This is a management question and management should be the one that decides the answer to this important question.

3. Create interlock performance equation as the mathematical model for the combination of sensor, logic solver and final control elements using the methods outlined in ISA technical report (TR84.00.02) [Ref. 2]. Equations needed to evaluate the uncertainty of the models specified in the ISS report are developed in Appendix A of this paper. For non-repairable systems, Table 3 can be used. For more complex systems such as non-repairable redundant systems with the potential for common cause failures, use the repairable equations of Table 4 and set DI, DC, and MTTR all equal to zero. For systems where common cause failures (CCF), Diagnostics and repair are to be considered, use the recommendations of Table 4. There are design situation where the equations developed in Appendix A will not exactly apply. Specifically, designs where the system contains redundancy subject to common cause failures (CCF) that should be considered non-repairable during operation. The failure models and the equations needed to evaluate the uncertainty in these systems are developed in Appendix B to this paper.

Table 3. Roadmap for Mean and Variance of Non-Repairable System Without Considering CCF, Diagnostics and MTTR			
Configuration	**Function**	**Mean PFD Using Appendix A Equation**	**PFD Variance Using Appendix A Equation**
1oo1	F_1 =	A-9	A-18
1oo2	F_2 =	A- 10	A- 19
1oo3	F_3 =	A-11	A-20
2oo2	F_4 =	A-12	A-21
2oo3	F_5 =	A-13	A-22
3oo3	F_6 =	A-14	A-23

Table 4. Roadmap for Mean and Variance of Repairable System Considering CCF, Diagnostics and MTTR			
Configuration	**Function**	**Mean PFD Using Appendix A Equation**	**PFD Variance Using Appendix A Equation**
1oo1	F_7 =	A-30	A-44
1oo2	F_8 =	A- 31	A- 86
1oo3	F_9 =	A-32	A-94
2oo2	F_{10} =	A-33	A-48
2oo3	F_{11} =	A-34	A-102
3oo3	F_{12} =	A-35	A-50

Note that in all of the calculations indicated by the equations referenced in Table 4, the expected value of any the random variables is used to complete the calculations.

4. Define the uncertainty in the parameters and variables of the interlock model specified in step 2. The uncertainty can be given as the upper and lower range of the possible values (uniform probability distribution), as the upper, lower and recommended values (triangular distribution), or as a mean and standard deviation (normal distribution). See the example above for guidance in the evaluation of safety instrumented system interlocks.

5. Compute the expected value of each variable in the interlock performance equation. The equations for the mean and variance for the uniform, triangular and normal probability distributions are presented for various probability distributions in Vose [Ref. 8]. The example interlock calculations used a triangular distribution to represent the uncertainty in the parameters.

6. Compute the expected value or mean of the interlock PFD using the mean value of each of the variables in Step 5. See Tables 4 and 5 for recommended equations.

7. Compute the sensitivity of the result from the interlock performance equation by use of the partial derivative of the basic interlock performance equation with respect to each of the variables as presented in Appendix A.

8. Compute the variance of the interlock performance equation PFD by use of the variance contribution using equations from Tables 3 or 4. This entails multiplying the variance of each of the uncertain variables in the basic interlock performance equation by the square of its sensitivity (obtained in step 7), as evaluated at the variable mean. Sum the resulting terms to obtain the overall variance of the PFD in the interlock performance equation. See Tables 4 and 5 for recommended equations.

9. Determine the level of risk that the owner/operator wishes to take that the final interlock will not work. Note that this is a management decision not an instrument engineer decision! In this paper, the 95% level of risk reduction has been used:
   - 5% chance of failure or a 95% chance of the interlock achieving the desired risk reduction

10. Assuming that the interlock owner operator wishes to take a low risk (5%) of the interlock failing to achieve its design target PFD, compute the 95% upper confidence limit on the computed PFD by use of the standard normal factor, Z, [Ref. 9] as:

$$Z = \left[\frac{xi - E(x)}{\sigma}\right] \qquad \text{(Eq 5)}$$

Where:

σ = standard deviation of the PFD of the interlock of interest from the interlock performance equation obtained from step 7. Note that the variance of a random variable is the square of the standard deviation of the random variable.

E(x) = the expected value of the PFD of the interlock of interest from the interlock performance equation obtained from step 5

For the 95% upper limit, Z = 1.645. Rearranging Eq. 5 allows for the direct calculation of the corresponding value of the 95% upper confidence limit on the PFD as:

$$X95\% = 1.645\,\sigma + E(x) \qquad \text{(Eq 6)}$$

Where:

X95% = the upper 95% limit on the computed PFD of the interlock of interest from the interlock performance equation.

Compare the 95% upper confidence limit on the PFD of the interlock of concern with that established as the desired PFD for risk reduction. If the 95% confidence of the RRF is greater than the desired RRF, the design is complete. If not, revise the design or change inspection test intervals to achieve the desired RRF. If it is not possible to achieve the desired target RRF economically, revisit the LOPA study accordingly to incorporate better information obtained in the uncertainty analysis. Improve the integrity of the LOPA IPLs or identify additional IPLs to drive the risk to a tolerable level. Continue this process until the computed RRFs are greater than the desired RRFs for risk reduction and risk management.

Example Interlock Problem

Previously, Freeman and Summers [Ref. 3] published an example in the use of VCA to determine the likelihood that the interlock would provide the risk reduction desired by the management and the LOPA team. The process system is shown in Figure 3. The process uses a compressor to increase the pressure of a process stream prior to additional processing. The gas being compressed is toxic and flammable. Of concern is a slug of liquid being sent to the compressor. If a liquid slug is sent to the compressor, significant damage to the compressor is expected with probable seal damage and a subsequent release of flammable and toxic material into the work area. A large fire and/or explosion could result if the release were to be ignited. If the release is not ignited, the nearby workers could be exposed to the toxic gas resulting in death or injury. A Layer of Protection Analysis (LOPA) review of this system has been completed. Among several recommendations, the LOPA team recommended the installation of a SIL-2 high level interlock in the Compressor Knock Out Drum to stop the compressor prior to liquids entering the system.

Figure 1 shows a simplified diagram for the interlock. Three level sensors are provided in the Compressor Knock Out Drum. The SIS logic solver will use 2oo3 voting to detect high level in the Compressor Knock Out Drum. The SIS logic solver will also monitor the difference in level signal from each of the three level sensors and will activate an alarm if the deviation is excessive. Two independent methods of stopping the compressor are provided. The SIS will directly signal the motor controller on the compressor to stop. In addition, the SIS logic solver will also signal two additional relays to open the power supply to the compressor motor. These two different shutdown will both be activated in the event of high level in the Compressor Knock Out Drum. Either one of the two shutdowns is capable of stopping the compressor by turning off the electric power supply to the motor.

Models

The first step in the calculation of the "goodness" of an interlock is to establish the model to be used in the calculations. Note that the sensors are 2oo3 voting and the final control elements are each 1oo1. Appendix A presents the equations for various models that can be used for describing this system.

The overall probability of failure on demand (PFD) of the interlock is given as:

PFD = PFDs + PFDsis + PFDfce (Eq 7)

Where:

PFD = Probability of failure on demand of the interlock as a whole

PFDs = Probability of failure on demand of the sensors (voting as 2oo3)

PFDsis = Probability of failure on demand of the SIS logic solver

PFDfce = Probability of failure on demand of the final control elements.

Since there are two final control elements arranged in series, the PFDfce = sum of the PFDs of the final control elements (Relay PFD and MCC PFD).

PFDce = PFDr + PFDmcc (Eq 8)

Where:

PFDr = Probability of failure on demand of the two relays voting as 1oo2 to shutdown gas compressor.

PFDmcc = Probability of failure on demand of the MCC to shutdown the gas compressor

For this example, the following selections are made to model the performance of the interlock.

Model for Sensors (2oo3)

The sensors are considered repairable as one sensor can be replaced while the other two function and provide continuous functionality of the interlock. There are three sensors that will be used in a 2oo3 voting system. From Appendix A, using Equation A-7, the model for the sensors becomes:

$$\text{PFDsavg} = 3 \times \left[\frac{(1-DCs)\times(1-\beta s)\times \lambda^{Ds} \times TIs}{2} + \frac{DCs\times(1-\beta s)\times \lambda^{Ds} \times DIs}{2} + (1-\beta s)\times \lambda^{Ds} \times MTTRs \right]^2 +$$

$$\left[\frac{(1-DCs)\times \beta s \times \lambda^{Ds} \times TIs}{2} + \frac{DCs\times \beta s \times \lambda^{D} \times DIs}{2} + \beta s \times \lambda^{Ds} \times MTTRs \right] \qquad \text{(Eq 9)}$$

Where:

PFD_{savg} is the average probability of failure on demand of the sensors

DC_s is the diagnostic coverage for sensor failure

DI_s is the diagnostic interval for the sensors

$MTTR_s$ is the mean time to restore the sensors to functionality given a sensor failure

TI_s is the test interval for the sensors

β_s is the common cause failure parameter

λ^{Ds} is the failure rate to a dangerous condition for the sensors

MTTR is the mean time to restore the system from the time that failure occurs

Model for SIS Logic Solver

Use a fixed probability of failure on demand as specified by the vendor. Unless detailed design information is provided by the SIS logic solver vendor, this will be the normal default condition for most studies. For this problem a typical PFD of 1.30×10^{-4} was selected to represent the SIS Logic Solver.

Model for Final Control Elements

There are two separate paths to shutdown the gas compressor. First is by the SIS logic solver commanding the MCC to shutdown power to the the gas compressor motor. The second is for the SIS logic solver to issue a shutdown command to two interposing relays (R1 and R2) which will cause the power to the gas compressor motor to stop. Two different models are needed to the final control elements. These two subsystems are considered non-repairable in this analysis,

Model for Interposing Relays (1oo2)

There are two interposing relays (R1 and R2 in the interlock). From Appendix A, use Equation A-25 setting parameters DC=DI=MTTR=0 for the non-repairable interposing relays, the model becomes:

$$PFDravg = \left[\frac{(1-\beta r)\times\lambda^{Dr}\times TIr}{2}+\right]^2+\left[\frac{\beta r\times\lambda^{Dr}\times TIr}{2}+\right] \quad \text{(Eq 10)}$$

Where:

PFD_{ravg} is the average probability of failure on demand of the relays voting 1oo2 to shutoff the gas compressor.

TI_r is the test interval for the relays

β_r is the common cause failure parameter

λ^{Dr} is the failure rate to a dangerous condition for the relays

Model for MCC (1oo1)

From Appendix A and using equation A-1 for the non-repairable MCC, the model becomes:

$$\boldsymbol{PFDmccavg} = \frac{\lambda^{Dmcc} * \mathbf{TImcc}}{2} \quad \text{(Eq 11)}$$

Where:

PFD_{mccavg} is the average probability of failure on demand of the MCC to shutoff the gas compressor.

TI_{mcc} is the test interval for the MCC

λ^{Dmcc} is the failure rate to a dangerous condition for the MCC

Data

The calculation of the PFD of the interlock requires a set of data to be used to represent the system. Tables 5, 6, and 7 are taken from the Freeman-Summers paper [Ref. 3] and presents the data used to represent the interlock system. Note that these data were originally taken from generic data sources and do not represent any particular device or system.

Table 5. Uncertainty Data for Level Sensors Used in Example Interlock
(All variable probability distributions assumed to be triangularly distributed)

Variable	Min	Mode	Max	Mean	Variance
λ^{Ds}- Fail Dangerous Rate	$2.84\text{x}10^{-3}$ Yr^{-1}	$5\text{x}10^{-3}$ Yr^{-1}	$8.5\text{x}10^{-3}$ Yr^{-1}	$5.45\text{x}10^{-3}$ Yr^{-1}	$1.36\text{x}10^{-6}$ Yr^{-1}
DC_s – Diagnostic Coverage *	0.8	0.9	0.99	0.897	$1.51\text{x}10^{-3}$
DI_s – Diagnostic Interval **	$5.71\text{x}10^{-5}$	$5.71\text{x}10^{-5}$	$5.71\text{x}10^{-5}$	$5.71\text{x}10^{-5}$	0
TI_s – Test Interval	4 Yr	5 Yr	6 Yr	5 Yr	$5.56\text{x}10^{-2}$ Yr^2
β_s- Common Cause Failure Fraction *	0	0.02	0.1	0.04	$4.67\text{x}10^{-4}$
$MTTR_s$ - Mean Time to Restore ***	$1.37\text{x}10^{-3}$ Yr	$8.22\text{x}10^{-3}$ Yr	$1.92\text{x}10^{-2}$ Yr	$9.59\text{x}10^{-3}$ Yr	$1.34\text{x}10^{-5}$ Yr^2

* Unit less

** 0.5 hours, assumed to be deterministic

***Min of 12 hr, Mode of 72 hours, Max of 168 hrs

Table 6. Uncertainty Data for Relays Used in Example Interlock (All variable probability distributions assumed to be triangularly distributed)					
Variable	**Min**	**Mode**	**Max**	**Mean**	**Variance**
λ^{Dr}- Fail Dangerous Rate	$8.76x10^{-9}$ Yr^{-1}	$2.00x10^{-3}$ Yr^{-1}	$4.73x10^{-2}$ Yr^{-1}	$1.64x10^{-2}$ Yr^{-1}	$1.19x10^{-4}$ Yr^{-2}
DC_r – Diagnostic Coverage *	NA	NA	NA	NA	NA
DI_r – Diagnostic Interval*	NA	NA	NA	NA	NA
TI_r – Test Interval	1 Yr	1 Yr	2 Yr	1.33 Yr	0.056 Yr^2
β_r- Common Cause Failure Fraction	0	0.02	0.1	0.04	$4.67x10^{-4}$
$MTTR_r$ - Mean Time to Restore*	NA	NA	NA	NA	NA

* Not used in relay model

Table 7. Uncertainty Data for MCC Used in Example Interlock (All variable probability distributions assumed to be triangularly distributed)					
Variable	**Min**	**Mode**	**Max**	**Mean**	**Variance**
λ^{Dr}- Fail Dangerous Rate	$1.74x10^{-4}$ Yr^{-1}	$1.31x10^{-3}$ Yr^{-1}	$3.00x10^{-2}$ Yr^{-1}	$1.05x10^{-2}$ Yr^{-1}	$4.76x10^{-5}$ Yr^{-2}
DC_r – Diagnostic Coverage*	NA	NA	NA	NA	NA
DI_r – Diagnostic Interval*	NA	NA	NA	NA	NA
TI_r – Test Interval	1 Yr	1 Yr	2 Yr	1.33 Yr	$5.6x10^{-2}$ Yr^2
β_r- Common Cause Failure Fraction*	NA	NA	NA	NA	NA
$MTTR_r$ - Mean Time to Restore*	NA	NA	NA	NA	NA

* Not used in MCC model

Uncertainty Analysis Results

Using the VCA method to estimate the PFD of the interlock design yields the results shown in Table 8.

Table 8. Results of VCA on Example Interlock Design		
Subsystem	**Mean PFD**	**Variance of PFD**
Sensors	6.41E-5	1.7657E-9
Logic Solver	1.34E-4	0
Relays	5.46E-4	2.0333E-7
MCC	6.98E-3	4.52E-5
Totals	**7.73E-3**	**4.54E-5**

Note: Calculations are carried to an adequate number of places to allow for the resulting PFD to be determined. Rounding to significant number of figures is done at the end.

The mean PFD of 0.00773 implies a mean risk reduction factor of:

RRF = 1/PFD = 1/0.00773 = 129 (Eq 11)

The variance of 4.54E-5 implies a standard deviation in the PFD of

Std Dev = $(\text{Variance})^{1/2}$ = 0.006737 (Eq 12)

The corresponding 95% level of confidence in the PFD:

X95% = 1.645 std Dev + Mean PFD (Eq 13)

X95% = 1.645 (0.006737) + 0.00773= 0.01881 (Eq 14)

The RRF for 95% certain PFD is

RRF95% = 1/0.01881 => 53 (Eq 15)

This is essentially the same result previously reported by Freeman and Summers [Ref. 3] using either Monte Carlo Simulation or numerical approximation methods for the VCA sensitivities. The calculated PFD at the 95% level of 0.01881 indicates that there is only a 5% chance that the interlock will provide an RRF worse that 53. Since this level is only SIL-1 capable and not SIL-2 capable as desired by management, a revised design will be needed to achieve the desired risk reduction or a different protective measure will be needed to control risk.

Acknowledgement

This paper is an expanded version of a paper entitled ''General Method for Uncertainty Evaluation of Safety Integrity Level (SIL) Calculations'' presented at the Mary Kay O'Connor 2016 Process Safety Symposium, October 26, 2016, College Station, TX. The author wishes to thank those individuals who gave comments and suggestions for improving this article.

References

1. International Society of Automation, Functional Safety: Safety Instrumented Systems for the Process Sector—Parts 1, 2, and 3, ANSI/ISA 84.00.01, 2004 [USA implementation of IEC 61511].
2. International Society of Automation, *Safety Integrity Level (SIL) Verification of Safety Instrumented Functions*, ISA TR84.00.02, Research Triangle Park, NC. (2015).
3. Freeman, R. And Summers, A, "Evaluation of Uncertainty in Safety Integrity Level Calculations," *Process Safety Progress*, Vol 35, No 4, pp 341-348, December 2016. Published Wiley Online Library (wileyonlinelibrary.com). DOI 10.1002/prs.11805, 23 January 2016
4. Center for Chemical Process Safety, *Layer of Protection Analysis – Simplified Process Risk Assessment*, American Institute of Chemical Engineers, New York, 2001
5. Edward B. Haugen, *Probabilistic Approaches to Design*, John Wiley, New York, 1968
6. R.A. Freeman, "Quantifying LOPA Uncertainty," *Process Safety Progress*, Vol 31, No. 3, pp 240-247, 2012
7. R. A. Freeman, "Simplified Uncertainty Analysis of Layer of Protection Analysis Results," *Process Safety Progress*, Vol. 32, No 4, pp 351-360, December 2013, Published online in Wiley Online Library (wileyonlinelibrary.com) DOI 10.1002/ prs11585.pdf, 3 June 2013
8. David Vose, *Risk Analysis – A Quantitative Guide*, 3rd Edition, John Wiley, 2008
9. Paul L. Meyer, *Introductory Probability and Statistical Applications*, John Wiley, New York, 1972

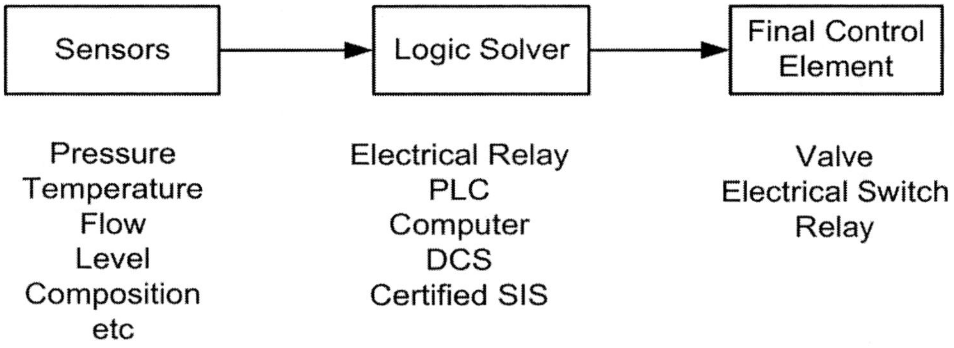

Figure 1. Typical Components of an SIS Interlock

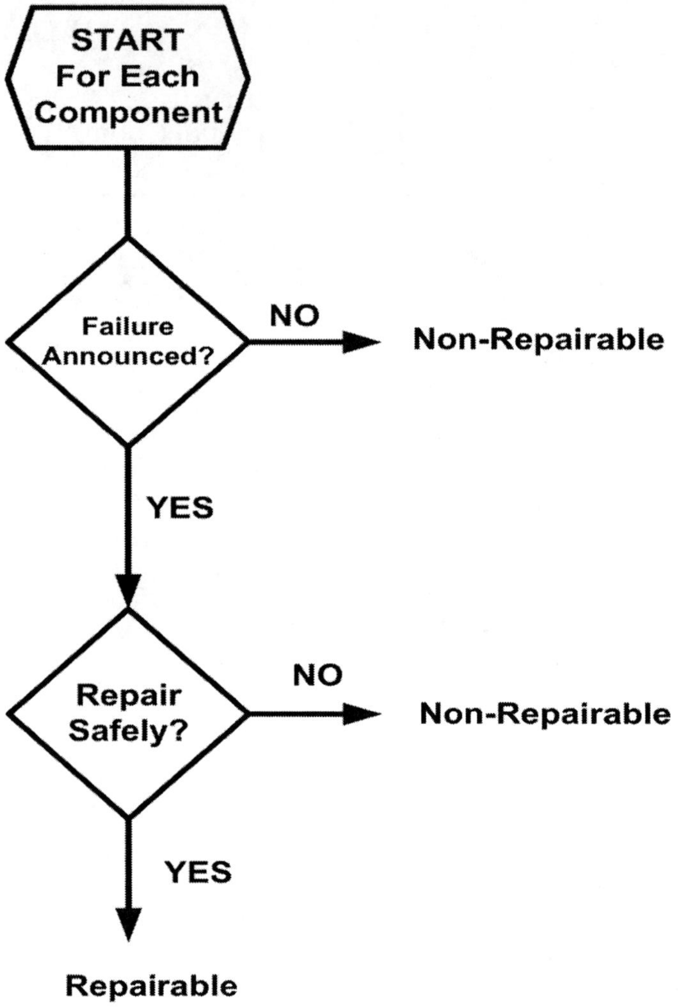

Figure 2. Decision Tree for Determination of Repairable or Non- Repairable

Figure 3. P&ID Sketch for Example Interlock

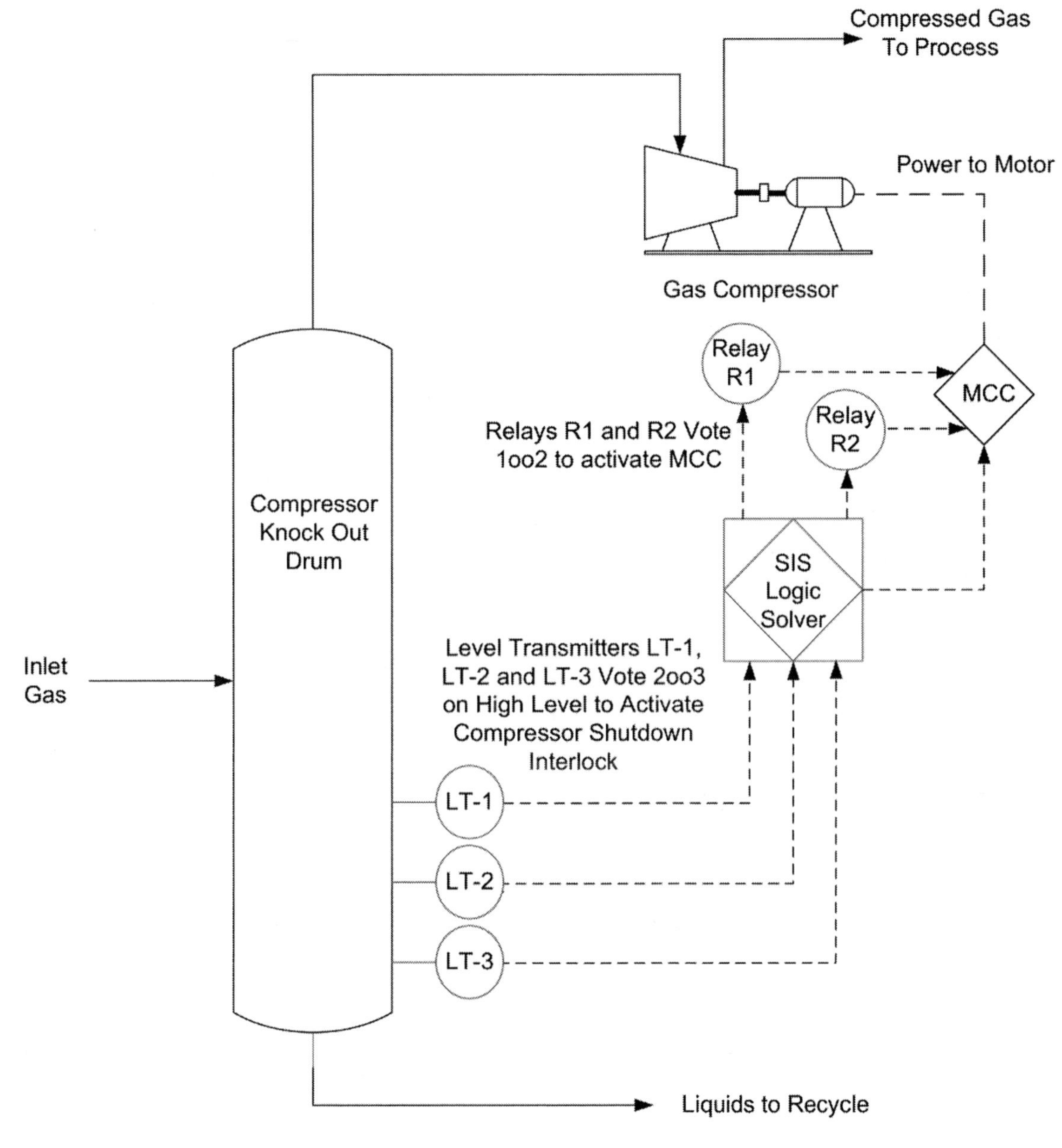

APPENDIX A -- DEVELOPMENT OF UNCERTAINTY EQUATIONS

This appendix develops the equations needed to evaluate the uncertainty in the SIL PFD as computed using the methods outlined in the ISA Technical Report [Ref. 2]. This appendix is divided into two sections:

- Non-Repairable Systems
- Repairable Systems

The uncertainty equations for other configurations not covered in the ISA report are developed in Appendix B.

NON-REPAIRABLE SYSTEMS

For systems that are considered non-repairable, the simplified equations of Table A.1 (taken from the ISA Technical Report [Ref. 2]) are used for the analysis.

Table A.1 Simplified PFD_{avg}formulas for Non-Repairable System without considering CCF, Diagnostics or MTTR.

Configuration	Function	PFD based on "Average before" failure rate	Equation Number
1oo1	F_1 =	$\left[\lambda^D \times \frac{TI}{2}\right]$	A-1
1oo2	F_2 =	$\left[(\lambda^D)^2 \times \frac{TI^2}{4}\right]$	A- 2
1oo3	F_3 =	$\left[(\lambda^D)^3 \times \frac{TI^3}{8}\right]$	A-3
2oo2	F_4 =	$\left[\lambda^D \times TI\right]$	A-4
2oo3	F5 =	$\frac{3}{4} \times \left[(\lambda^D)^2 \times TI^2\right]$	A-5
3oo3	F6 =	$3 \times \left[\lambda^D \times \frac{TI}{2}\right]$	A-6

Where TI is the proof test interval, λ^D is the dangerous failure rate.

The expected value of the probability of failure on demand is found by calculating the PFD using the expected value (mean) of the random variables.

Let y a function of some random variables, x_i, as:

$$y = F(x_i) \quad \text{(Eq A-7)}$$

The mean of y can be estimated using the following approximation:

$$E(y) = F[E(x_i)] \quad \text{(Eq A-8)}$$

Where:

E(y) = expected value of random variable y = mean of y

$E(x_i)$ = expected value of random variable x_i = mean of x_i

For the configurations defined in Table A.1, the corresponding expected PFD are presented in Table A.2.

Table A.2 Expected Value of PFD$_{avg}$formulas for Non-Repairable System without considering CCF, Diagnostics or MTTR.

Configuration	Function	Expected PFD based on "Average before" failure rate	Equation Number
1oo1	$E(F_1) =$	$\left[E(\lambda^D)\times\frac{E(\mathrm{TI})}{2}\right]$	A-9
1oo2	$E(F_2) =$	$\left[(E(\lambda^D))^2\times\frac{E(\mathrm{TI})^2}{4}\right]$	A- 10
1oo3	$E(F_3) =$	$\left[(E(\lambda^D))^3\times\frac{E(\mathrm{TI})^3}{8}\right]$	A-11
2oo2	$E(F_4) =$	$\left[E(\lambda^D)\times E(\mathrm{TI})\right]$	A-12
2oo3	$E(F_5) =$	$\frac{3}{4}\times\left[(E(\lambda^D))^2\times E(\mathrm{TI})^2\right]$	A-13
3oo3	$E(F_6) =$	$3\times\left[E(\lambda^D)\times\frac{E(\mathrm{TI})}{2}\right]$	A-14

Where E(x) is the expected value of random variable x.

The variance of the PFD is found by taking the weighted sum of the variance of the random variables. The weighting function is the sensitivity of the PFD with respect to random variable. In mathematical terms the variance of a function may be found in terms of the variance of the random variables as:

Once again let

$y = F(x_i)$ (Eq A-15)

The variance of Y can likewise be estimated as:

$$V(y) = \sum_{i=1}^{n} \left[\frac{\partial y}{\partial x_i}\right]^2 V(x_i)$$ (Eq A-16)

Where:

$V(y)$ = variance of random variable y as defined above

$V(x_i)$ = variance of random variable x_i as defined above

The sensitivity of y with respect to a random variable x_i is:

Sensitivity of y with respect to $x_i = \left[\frac{\partial y}{\partial x_i}\right]$ (Eq A-17)

For the configurations defined in Table A.1, the corresponding variance of the PFD are presented in Table A.3.

Table A.3 Variance of PFD_{avg} formulas for Non-Repairable System without considering CCF, Diagnostics or MTTR.			
Configuration	**Function**	**Variance in Function F_i based on "Average before" failure rate**	**Equation Number**
1oo1	$V(F_1) =$	$\left[\frac{E(TI)}{2}\right]^2 V(\lambda^D) + \left[\frac{E(\lambda^D)}{2}\right]^2 V(TI)$	A-18
1oo2	$V(F_2) =$	$\left[E(\lambda^D)^2 \frac{E(TI)^2}{2}\right]^2 V(\lambda^D) +$ $\left[E(\lambda^D)^2 \frac{E(TI)}{2}\right]^2 V(TI)$	A- 19
1oo3	$V(F_3) =$	$\left[3 \times E(\lambda^D)^2 \frac{E(TI)^3}{2}\right]^2 V(\lambda^D) +$ $\left[\frac{3}{8} E(\lambda^D)^3 E(TI)^2\right]^2 V(TI)$	A-20
2oo2	$V(F_4) =$	$E(TI)^2 V(\lambda^D) + E(\lambda^D)^2 \times V(TI)$	A-21
2oo3	$V(F_5) =$	$\left[\frac{3}{2} \times E(\lambda^D) \times E(TI)^2\right]^2 V(\lambda^D) +$ $\left[\frac{3}{2} \times E(\lambda^D)^2 \times E(TI)\right]^2 V(TI)$	A-22
3oo3	$V(F_6) =$	$\left[\frac{3}{2} \times E(TI)\right]^2 V(\lambda^D) +$ $\left[\frac{3}{2} \times E(\lambda^D)\right]^2 V(TI)$	A-23

Note that for the calculation of the variance functions, V(Fi), the random variables are evaluated at the expected value (mean).

REPAIRABLE SYSTEMS

For systems that are considered repairable, the simplified equations of Table A.4 are used for the analysis. The expected value of the probability of failure on demand (PFD) is found by substituting the expected value of each of the random variables into the corresponding equation. The expected value of the PFD for each of the system configurations is presented in Table A.5

Table A.4 Simplified PFD_{avg} Formulas for Repairable System considering CCF, Diagnostics and MTTR

Configuration	Function	PFD based on "Average before" failure rate	Equation Number
1oo1	$F_7 =$	$\left[\frac{(1-DC)\times\lambda^D\times TI}{2}+\frac{DC\times\lambda^D\times DI}{2}+\lambda^D\times MTTR\right]$	A-24
1oo2	$F_8 =$	$\left[\frac{(1-DC)\times(1-\beta)\times\lambda^D\times TI}{2}+\frac{DC\times(1-\beta)\times\lambda^D\times DI}{2}+(1-\beta)\times\lambda^D\times MTTR\right]^2+\left[\frac{(1-DC)\times\beta\times\lambda^D\times TI}{2}+\frac{DC\times\beta\times\lambda^D\times DI}{2}+\beta\times\lambda^D\times MTTR\right]$	A- 25
1oo3	$F_9 =$	$\left[\frac{(1-DC)\times(1-\beta)\times\lambda^D\times TI}{2}+\frac{DC\times(1-\beta)\times\lambda^D\times DI}{2}+(1-\beta)\times\lambda^D\times MTTR\right]^3+\left[\frac{(1-DC)\times\beta\times\lambda^D\times TI}{2}+\frac{DC\times\beta\times\lambda^D\times DI}{2}+\beta\times\lambda^D\times MTTR\right]$	A-26
2oo2	$F_{10} =$	$2\times\left[\frac{(1-DC)\times\lambda^D\times TI}{2}+\frac{DC\times\lambda^D\times DI}{2}+\lambda^D\times MTTR\right]$	A-27

Table A.4 Simplified PFD_{avg} Formulas for Repairable System considering CCF, Diagnostics and MTTR			
Configuration	**Function**	**PFD based on "Average before" failure rate**	**Equation Number**
2oo3	F_{11} =	$3\times\left[\frac{(1-DC)\times(1-\beta)\times\lambda^D\times TI}{2}+\frac{DC\times(1-\beta)\times\lambda^D\times DI}{2}+(1-\beta)\times\lambda^D\times MTTR\right]^2+\left[\frac{(1-DC)\times\beta\times\lambda^D\times TI}{2}+\frac{DC\times\beta\times\lambda^D\times DI}{2}+\beta\times\lambda^D\times MTTR\right]$	A-28
3oo3	F_{12} =	$3\times\left[\frac{(1-DC)\times\lambda^D\times TI}{2}+\frac{DC\times\lambda^D\times DI}{2}+\lambda^D\times MTTR\right]$	A-29

Where:

DC is the diagnostic coverage;

DI is the diagnostic interval;

TI is the proof test interval,

λ^D is the dangerous failure rate;

MTTR is the mean time to restore the system to operation

β is the common cause failure parameter that is always is between 0 and 1

Table A.5 Expected PFD Formulas for Repairable System considering CCF, Diagnostics and MTTR			
Configuration	**Function**	**PFD based on "Average before" failure rate**	**Equation Number**
1oo1	$E(F_7) =$	$\left[\frac{(1-E(DC))\times E(\lambda^D)\times E(TI))}{2}+\frac{E(DC)\times E(\lambda^D)\times E(DI)}{2} + E(\lambda^D)\times E(MTTR)\right]$	A-30
1oo2	$E(F_8) =$	$\left[\frac{(1-E(DC))\times(1-E(\beta))\times E(\lambda^D)\times E(TI)}{2}+\frac{E(DC)\times(1-E(\beta))\times E(\lambda^D)\times E(DI)}{2}+(1-E(\beta))\times E(\lambda^D)\times E(MTTR)\right]^2+\left[\frac{(1-E(DC))\times E(\beta)\times E(\lambda^D)\times E(TI)}{2}+\frac{E(DC)\times E(\beta)\times E(\lambda^D)\times E(DI}{2})+E(\beta)\times E(\lambda^D)\times E(MTTR)\right]$	A- 31
1oo3	$E(F_9) =$	$\left[\frac{(1-E(DC))\times(1-E(\beta))\times E(\lambda^D)\times E(TI)}{2}+\frac{E(DC)\times(1-E(\beta))\times E(\lambda^D)\times E(DI)}{2}+(1-E(\beta))\times E(\lambda^D)\times E(MTTR)\right]^3+\left[\frac{(1-E(DC))\times E(\beta)\times E(\lambda^D)\times E(TI)}{2}+\frac{E(DC)\times E(\beta)\times E(\lambda^D)\times E(DI)}{2}+E(\beta)\times E(\lambda^D)\times E(MTTR)\right]$	A-32

Table A.5 Expected PFD Formulas for Repairable System considering CCF, Diagnostics and MTTR			
Configuration	**Function**	**PFD based on "Average before" failure rate**	**Equation Number**
2oo2	$E(F_{10}) =$	$2\times\left[\frac{(1-E(DC))\times E(\lambda^D)\times E(TI))}{2}+\frac{E(DC)\times E(\lambda^D)\times E(DI)}{2}+E(\lambda^D)\times E(MTTR)\right]$	A-33
2oo3	$E(F_{11}) =$	$3\times\left[\frac{(1-E(DC))\times(1-E(\beta))\times E(\lambda^D)\times E(TI)}{2}+\frac{E(DC)\times(1-E(\beta))\times E(\lambda^D)\times E(DI)}{2}+(1-E(\beta))\times E(\lambda^D)\times E(MTTR)\right]^2+\left[\frac{(1-E(DC))\times E(\beta)\times E(\lambda^D)\times E(TI)}{2}+\frac{E(DC)\times E(\beta)\times E(\lambda^D)\times E(DI)}{2}+E(\beta)\times E(\lambda^D)\times E(MTTR)\right]$	A-34
3oo3	$E(F_{12}) =$	$3\times\left[\frac{(1-E(DC))\times E(\lambda^D)\times E(TI))}{2}+\frac{E(DC)\times E(\lambda^D)\times E(DI)}{2}+E(\lambda^D)\times E(MTTR)\right]$	A-35

The variance of the PFD of the repairable systems is found in the same manner as that in the non-repairable cases. Once again let:

$$y = F(x_i) \qquad \text{(Eq A-36)}$$

The variance of Y can likewise be estimated as:

$$V(y) = \sum_{i=1}^{n} \left[\frac{\partial y}{\partial x_i}\right]^2 V(x_i) \qquad \text{(Eq A-37)}$$

Where:

V(y) = variance of random variable y as defined above

$V(x_i)$ = variance of random variable x_i as defined above

The sensitivity of y with respect to a random variable x_i is:

In the case of the repairable systems, there are several more potentially uncertain or random variables:

- DC is the diagnostic coverage;
- DI is the diagnostic interval;
- TI is the proof test interval,
- λ^D is the dangerous failure rate;
- MTTR is the mean time to restore the system to operation
- β is the common cause failure parameter that is always is between 0 and 1

We must evaluate the partial derivative of the PFD function with respect to each and then combine them using equation A-37. The partial derivatives of function F7 with respect to each random variable are found as follows

$$\frac{\partial F_7}{\partial DC} = \frac{-\lambda^D \times TI}{2} + \frac{\lambda^D \times DI}{2} \qquad \text{(Eq A-38)}$$

$$\frac{\partial F_7}{\partial \lambda^D} = \frac{(1-DC)\times TI}{2} + \frac{DC \times DI}{2} + MTTR \qquad \text{(Eq A-39)}$$

$$\frac{\partial F_7}{\partial TI} = \frac{(1-DC)\times \lambda^D}{2} \qquad \text{(Eq A-40)}$$

$$\frac{\partial F_7}{\partial DI} = \frac{DC \times \lambda^D}{2} \qquad \text{(Eq A-41)}$$

$$\frac{\partial F_7}{\partial MTTR} = \lambda^D \quad \text{(Eq A-42)}$$

We can now compute the variance of function F7 as:

$$V\left(F_7\right) = \sum_{i=1}^{n} \left[\frac{\partial F_7}{\partial x_i}\right]^2 V(x_i) \quad \text{(Eq A-43)}$$

$$V\left(F_7\right) = \left[\frac{-\lambda^D \times TI}{2} + \frac{\lambda^D \times DI}{2}\right]^2 \text{V(DC)} + \left[\frac{(1-DC)\times TI}{2} + \frac{DC\times DI}{2} + MTTR\right]^2 \text{V}(\lambda^D) +$$

$$\left[\frac{(1-DC)\times\lambda^D}{2}\right]^2 \text{V(TI)} + \left[\frac{DC\times\lambda^D}{2}\right]^2 \text{(DI)} + \left[\lambda^D\right]^2 \text{V(MTTR)} \quad \text{(Eq A-44)}$$

Equation 44 represents the variance of the probability of failure on demand of the 1001 configuration when the system is repairable. We note the following relationships between the 1001, 2oo2 and 3oo3 configurations:

$F_{10} = 2\ F_7$ (Eq A-45)

$F_{12} = 3\ F_7$ (Eq A-46)

We can directly determine the variance of functions F_{10} and F_{12} from the properties of the variance operator, V(X).

$V(F_{10}) = V(2\ F_7)$ (Eq A-47)

$V(F_{10}) = 4\ V(F_7)$ (Eq A-48)

$V(F_{12}) = V(3\ F_7)$ (Eq A-49)

$V(F_{12}) = 9\ V(F_7)$ (Eq A-50)

The determination of the variance of functions F_8, F_{10}, and F_{11} is done in a similar manner. To simplify the presentation of the derivation of variance of functions F_8, F_{10}, and F_{11}, we introduce two functions H and Q as:

$$\text{H} = \left[\begin{array}{l} \dfrac{(1-DC)\times(1-\beta)\times\lambda^D\times TI}{2} + \\ \dfrac{DC\times(1-\beta)\times\lambda^D\times DI}{2} + (1-\beta)\times\lambda^D\times MTTR \end{array}\right] \quad \text{(Eq A-51)}$$

$$\text{Q} = \left[\frac{(1-DC)\times\beta\times\lambda^D\times TI}{2} + \frac{DC\times\beta\times\lambda^D\times DI}{2} + \beta\times\lambda^D\times MTTR\right] \quad \text{(Eq A-52)}$$

We note the following relationships:

$F_8 = H^2 + Q$ (Eq A-53)

$F_9 = H^3 + Q$ (Eq A-54)

$F_{11} = 3\,H^2 + Q$ (Eq A-55)

The variance of F_8, F_9 and F_{11} are found as:

$V(F_8) = V(H^2 + Q)$ (Eq A-56)

$V(F_9) = V(H^3 + Q)$ (Eq A-57)

$V(F_{11}) = V(3\,H^2 + Q)$ (Eq A-58)

We will need the sensitivity of F_8, F_9 and F_{11} to each of the potential random variables.

$$\frac{\partial F_8}{\partial x} = \frac{\partial(H^2 + Q)}{\partial x} = \frac{\partial(H^2)}{\partial x} + \frac{\partial(Q)}{\partial x} \qquad \text{(Eq A-59)}$$

Likewise

$$\frac{\partial F_9}{\partial x} = \frac{\partial(H^3 + Q)}{\partial x} = \frac{\partial(H^3)}{\partial x} + \frac{\partial(Q)}{\partial x} \qquad \text{(Eq A-60)}$$

$$\frac{\partial F_{11}}{\partial x} = \frac{\partial 3(H^2 + Q)}{\partial x} = 3\frac{\partial(H^2)}{\partial x} + 3\frac{\partial(Q)}{\partial x} \qquad \text{(Eq A-61)}$$

Using the chain rule for partial derivatives, we now write:

$$\frac{\partial F_8}{\partial x} = 2H\frac{\partial(H)}{\partial x} + \frac{\partial(Q)}{\partial x} \qquad \text{(Eq A-62)}$$

$$\frac{\partial F_9}{\partial x} = 3H^2\frac{\partial(H)}{\partial x} + \frac{\partial(Q)}{\partial x} \qquad \text{(Eq A-63)}$$

$$\frac{\partial F_{11}}{\partial x} = 6H\frac{\partial(H)}{\partial x} + \frac{\partial(Q)}{\partial x} \qquad \text{(Eq A-64)}$$

We now evaluate the derivatives of function H with respect to each potential random variables:

- DC is the diagnostic coverage;
- DI is the diagnostic interval;

- TI is the proof test interval,
- λ^D is the dangerous failure rate;
- MTTR is the mean time to restore the system to operation
- β is the common cause failure parameter that is always is between 0 and 1

$$\frac{\partial(\mathrm{H})}{\partial DC} = \frac{-(1-\beta)\times\lambda^D\times TI}{2} + \frac{(1-\beta)\times\lambda^D\times DI}{2} \quad \text{(Eq A-65)}$$

$$\frac{\partial(\mathrm{H})}{\partial TI} = \frac{(1-DC)\times(1-\beta)\times\lambda^D}{2} \quad \text{(Eq A-66)}$$

$$\frac{\partial(\mathrm{H})}{\partial\lambda^D} = \left[\begin{array}{l}\dfrac{(1-DC)\times(1-\beta)\times TI}{2}+\\ \dfrac{DC\times(1-\beta)\times DI}{2}+(1-\beta)\times MTTR\end{array}\right] \quad \text{(Eq A-67)}$$

$$\frac{\partial(\mathrm{H})}{\partial MTTR} = (1-\beta)\times\lambda^D \quad \text{(Eq A-68)}$$

$$\frac{\partial(\mathrm{H})}{\partial\beta} = \left[\begin{array}{l}\dfrac{(1-DC)\times(-1)\times\lambda^D\times TI}{2}+\\ \dfrac{DC\times(-1)\times\lambda^D\times DI}{2}+(-1)\times\lambda^D\times MTTR\end{array}\right] \quad \text{(Eq A-69)}$$

$$\frac{\partial(\mathrm{H})}{\partial DI} = \frac{DC\times(1-\beta)\times\lambda^D}{2} \quad \text{(Eq A-70)}$$

In a similar manner we evaluate the partial derivatives of function Q with respect to each potential random variables.

$$Q = \left[\frac{(1-DC)\times\beta\times\lambda^D\times TI}{2} + \frac{DC\times\beta\times\lambda^D\times DI}{2} + \beta\times\lambda^D\times MTTR \right] \quad \text{(Eq A-71)}$$

- DC is the diagnostic coverage;
- DI is the diagnostic interval;
- TI is the proof test interval,
- λ^D is the dangerous failure rate;
- MTTR is the mean time to restore the system to operation
- β is the common cause failure parameter that is always is between 0 and 1

$$\frac{\partial(Q)}{\partial DC} = \frac{(-1)\times\beta\times\lambda^D\times TI}{2} + \frac{\beta\times\lambda^D\times DI}{2} \quad \text{(Eq A-72)}$$

$$\frac{\partial(Q)}{\partial TI} = \frac{(1-DC)\times\beta\times\lambda^D}{2} \quad \text{(Eq A-73)}$$

$$\frac{\partial(Q)}{\partial\lambda^D} = \left[\frac{(1-DC)\times\beta\times TI}{2} + \frac{DC\times\beta\times DI}{2} + \beta\times MTTR \right] \quad \text{(Eq A-74)}$$

$$\frac{\partial(Q)}{\partial MTTR} = \beta\times\lambda^D\times MTTR \quad \text{(Eq A-75)}$$

$$\frac{\partial(Q)}{\partial\beta} = \left[\frac{(1-DC)\times\lambda^D\times TI}{2} + \frac{DC\times\lambda^D\times DI}{2} + \lambda^D\times MTTR \right] \quad \text{(Eq A-76)}$$

$$\frac{\partial(Q)}{\partial DI} = \frac{DC\times\beta\times\lambda^D}{2} \quad \text{(Eq A-77)}$$

We can now determine the sensitivity of the functions F8, F9 , F11 to the random variables of interest. Restating equation A-62 for sensitivity of F8 with respect to the random variables.

$$\frac{\partial F_8}{\partial x} = 2H\frac{\partial(H)}{\partial x} + \frac{\partial(Q)}{\partial x} \quad \text{(Eq A-78)}$$

For F_8 we find

$$\frac{\partial F_8}{\partial x} = 2\mathrm{H}\frac{\partial(\mathrm{H})}{\partial x} + \frac{\partial(\mathrm{Q})}{\partial x} \quad \text{(Eq A-80)}$$

$$\frac{\partial F_8}{\partial DC} = 2\left[\frac{(1-DC)\times(1-\beta)\times\lambda^D\times TI}{2} + \frac{DC\times(1-\beta)\times\lambda^D\times DI}{2} + (1-\beta)\times\lambda^D\times MTTR\right]\times \left[\frac{-(1-\beta)\times\lambda^D\times TI}{2} + \frac{(1-\beta)\times\lambda^D\times DI}{2}\right] + \frac{(-1)\times\beta\times\lambda^D\times TI}{2} + \frac{\beta\times\lambda^D\times DI}{2} \quad \text{(Eq A-79)}$$

$$\frac{\partial F_8}{\partial TI} = 2\left[\frac{(1-DC)\times(1-\beta)\times\lambda^D\times TI}{2} + \frac{DC\times(1-\beta)\times\lambda^D\times DI}{2} + (1-\beta)\times\lambda^D\times MTTR\right]\times\left[\frac{(1-DC)\times(1-\beta)\times\lambda^D}{2}\right] + \frac{(1-DC)\times\beta\times\lambda^D}{2} \quad \text{(Eq A-81)}$$

$$\frac{\partial F_8}{\partial MTTR} = 2\left[\frac{(1-DC)\times(1-\beta)\times\lambda^D\times TI}{2} + \frac{DC\times(1-\beta)\times\lambda^D\times DI}{2} + (1-\beta)\times\lambda^D\times MTTR\right]\times \left[(1-\beta)\times\lambda^D\right] + \beta\times\lambda^D\times MTTR \quad \text{(Eq A-82)}$$

$$\frac{\partial F_8}{\partial \beta}=2\left[\frac{(1-DC)\times(1-\beta)\times\lambda^D\times TI}{2}+\frac{DC\times(1-\beta)\times\lambda^D\times DI}{2}+(1-\beta)\times\lambda^D\times MTTR\right]\times$$

$$\left[\left[\frac{(1-DC)\times(-1)\times\lambda^D\times TI}{2}+\frac{DC\times(-1)\times\lambda^D\times DI}{2}+(-1)\times\lambda^D\times MTTR\right]\right]+$$

$$\left[\frac{(1-DC)\times\lambda^D\times TI}{2}+\frac{DC\times\lambda^D\times DI}{2}+\lambda^D\times MTTR\right] \quad \text{(Eq A-83)}$$

$$\frac{\partial F_8}{\partial DI}=2\left[\frac{(1-DC)\times(1-\beta)\times\lambda^D\times TI}{2}+\frac{DC\times(1-\beta)\times\lambda^D\times DI}{2}+(1-\beta)\times\lambda^D\times MTTR\right]\times\left[\frac{DC\times(1-\beta)\times\lambda^D}{2}\right]+$$

$$\frac{DC\times\beta\times\lambda^D}{2} \quad \text{(Eq A-84)}$$

$$\left[\frac{\partial F_8}{\partial \lambda^D}\right]=2\left[\frac{(1-DC)\times(1-\beta)\times\lambda^D\times TI}{2}+\frac{DC\times(1-\beta)\times\lambda^D\times DI}{2}+(1-\beta)\times\lambda^D\times MTTR\right]\times\left[\frac{(1-DC)\times(1-\beta)\times TI}{2}+\frac{DC\times(1-\beta)\times DI}{2}+(1-\beta)\times MTTR\right]+$$

$$\left[\frac{(1-DC)\times\beta\times TI}{2}+\frac{DC\times\beta\times DI}{2}+\beta\times MTTR\right] \quad \text{(Eq A-85)}$$

Using the sensitivities for F8 calculated above, the variance in F8 due to uncertain or random variables is found as:

$$V(F_8)=\left[\frac{\partial F_8}{\partial DC}\right]^2 V(DC)+\left[\frac{\partial F_8}{\partial TI}\right]^2 V(TI)+\left[\frac{\partial F_8}{\partial \lambda^D}\right]^2 V(\lambda^D)+$$

$$\left[\frac{\partial F_8}{\partial MTTR}\right]^2 V(MTTR)+\left[\frac{\partial F_8}{\partial \beta}\right]^2 V(\beta)+\left[\frac{\partial F_8}{\partial DI}\right]^2 V(DI) \quad \text{(Eq A-86)}$$

Using the equation A-63, we now determine the variance of F_9.

$$\frac{\partial F_9}{\partial x} = 3\mathrm{H}^2 \frac{\partial(\mathrm{H})}{\partial x} + \frac{\partial(\mathrm{Q})}{\partial x} \qquad \text{(Eq A-87)}$$

$$\frac{\partial F_9}{\partial DC} = 3\left[\frac{(1-DC)\times(1-\beta)\times\lambda^D\times TI}{2} + \frac{DC\times(1-\beta)\times\lambda^D\times DI}{2} + (1-\beta)\times\lambda^D\times MTTR\right]^2 \times$$

$$\left[\frac{-(1-\beta)\times\lambda^D\times TI}{2} + \frac{(1-\beta)\times\lambda^D\times DI}{2}\right] +$$

$$\frac{(-1)\times\beta\times\lambda^D\times TI}{2} + \frac{\beta\times\lambda^D\times DI}{2} \qquad \text{(Eq A-88)}$$

$$\frac{\partial F_9}{\partial TI}=3\left[\begin{array}{l}\frac{(1-DC)\times(1-\beta)\times\lambda^D\times TI}{2}+\\ \frac{DC\times(1-\beta)\times\lambda^D\times DI}{2}+(1-\beta)\times\lambda^D\times MTTR\end{array}\right]^2\times\left[\frac{(1-DC)\times(1-\beta)\times\lambda^D}{2}\right]+$$

$$\frac{(1-DC)\times\beta\times\lambda^D}{2} \qquad \text{(Eq A-89)}$$

$$\frac{\partial F_9}{\partial MTTR}=3\left[\begin{array}{l}\frac{(1-DC)\times(1-\beta)\times\lambda^D\times TI}{2}+\\ \frac{DC\times(1-\beta)\times\lambda^D\times DI}{2}+(1-\beta)\times\lambda^D\times MTTR\end{array}\right]^2\times\left[(1-\beta)\times\lambda^D\right]+$$

$$\beta\times\lambda^D\times MTTR \qquad \text{(Eq A-90)}$$

$$\frac{\partial F_9}{\partial\beta}=3\left[\begin{array}{l}\frac{(1-DC)\times(1-\beta)\times\lambda^D\times TI}{2}+\\ \frac{DC\times(1-\beta)\times\lambda^D\times DI}{2}+(1-\beta)\times\lambda^D\times MTTR\end{array}\right]^2\times$$

$$\left[\left[\begin{array}{l}\frac{(1-DC)\times(-1)\times\lambda^D\times TI}{2}+\\ \frac{DC\times(-1)\times\lambda^D\times DI}{2}+(-1)\times\lambda^D\times MTTR\end{array}\right]\right]+$$

$$\left[\frac{(1-DC)\times\lambda^D\times TI}{2}+\frac{DC\times\lambda^D\times DI}{2}+\lambda^D\times MTTR\right] \qquad \text{(Eq A-91)}$$

$$\frac{\partial F_9}{\partial DI} = 3\left[\begin{array}{l}\frac{(1-DC)\times(1-\beta)\times\lambda^D\times TI}{2} + \\ \frac{DC\times(1-\beta)\times\lambda^D\times DI}{2} + (1-\beta)\times\lambda^D\times MTTR\end{array}\right]^2 \times \left[\frac{DC\times(1-\beta)\times\lambda^D}{2}\right] +$$

$$\frac{DC\times\beta\times\lambda^D}{2} \qquad \text{(Eq A-92)}$$

$$\left[\frac{\partial F_9}{\partial \lambda^D}\right] = 3 \times \left[\begin{array}{l}\frac{(1-DC)\times(1-\beta)\times\lambda^D\times TI}{2} + \\ \frac{DC\times(1-\beta)\times\lambda^D\times DI}{2} + (1-\beta)\times\lambda^D\times MTTR\end{array}\right]^2 \times$$

$$\left[\begin{array}{l}\frac{(1-DC)\times(1-\beta)\times TI}{2} + \\ \frac{DC\times(1-\beta)\times DI}{2} + (1-\beta)\times MTTR\end{array}\right] +$$

$$\left[\frac{(1-DC)\times\beta\times TI}{2} + \frac{DC\times\beta\times DI}{2} + \beta\times MTTR\right] \qquad \text{(Eq A-93)}$$

Using the sensitivities for F_9 calculated above, the variance in F_9 due to uncertain or random variables is found as:

$$V(F_9) = \left[\frac{\partial F_9}{\partial DC}\right]^2 V(DC) + \left[\frac{\partial F_9}{\partial TI}\right]^2 V(TI) + \left[\frac{\partial F_9}{\partial \lambda^D}\right]^2 V(\lambda^D) +$$

$$\left[\frac{\partial F_9}{\partial MTTR}\right]^2 V(MTTR) + \left[\frac{\partial F_9}{\partial \beta}\right]^2 V(\beta) + \left[\frac{\partial F_9}{\partial DI}\right]^2 V(DI) \qquad \text{(Eq A-94)}$$

Finally, in a similar manner the variance in F_{11} is computed as:

$$\frac{\partial F_{11}}{\partial x} = 6\text{H}\frac{\partial(\text{H})}{\partial x} + \frac{\partial(\text{Q})}{\partial x} \qquad \text{(Eq A-95)}$$

$$\frac{\partial F_{11}}{\partial DC} = 6\left[\begin{array}{l}\dfrac{(1-DC)\times(1-\beta)\times\lambda^D\times TI}{2}+\\ \dfrac{DC\times(1-\beta)\times\lambda^D\times DI}{2}+(1-\beta)\times\lambda^D\times MTTR\end{array}\right]\times$$

$$\left[\frac{-(1-\beta)\times\lambda^D\times TI}{2}+\frac{(1-\beta)\times\lambda^D\times DI}{2}\right]+$$

$$\frac{(-1)\times\beta\times\lambda^D\times TI}{2}+\frac{\beta\times\lambda^D\times DI}{2} \qquad \text{(Eq A-96)}$$

$$\frac{\partial F_{11}}{\partial TI} = 6\left[\begin{array}{l}\dfrac{(1-DC)\times(1-\beta)\times\lambda^D\times TI}{2}+\\ \dfrac{DC\times(1-\beta)\times\lambda^D\times DI}{2}+(1-\beta)\times\lambda^D\times MTTR\end{array}\right]\times\left[\frac{(1-DC)\times(1-\beta)\times\lambda^D}{2}\right]+$$

$$\frac{(1-DC)\times\beta\times\lambda^D}{2} \qquad \text{(Eq A-97)}$$

$$\frac{\partial F_{11}}{\partial MTTR} = 6\left[\begin{array}{l}\dfrac{(1-DC)\times(1-\beta)\times\lambda^D\times TI}{2}+\\ \dfrac{DC\times(1-\beta)\times\lambda^D\times DI}{2}+(1-\beta)\times\lambda^D\times MTTR\end{array}\right]\times\left[(1-\beta)\times\lambda^D\right]+$$

$$\beta\times\lambda^D\times MTTR \qquad \text{(Eq A-98)}$$

$$\frac{\partial F_{11}}{\partial \beta} = 6\left[\begin{array}{l}\frac{(1-DC)\times(1-\beta)\times\lambda^D\times TI}{2}+ \\ \frac{DC\times(1-\beta)\times\lambda^D\times DI}{2}+(1-\beta)\times\lambda^D\times MTTR\end{array}\right]\times$$

$$\left[\left[\begin{array}{l}\frac{(1-DC)\times(-1)\times\lambda^D\times TI}{2}+ \\ \frac{DC\times(-1)\times\lambda^D\times DI}{2}+(-1)\times\lambda^D\times MTTR\end{array}\right]\right]+$$

$$\left[\frac{(1-DC)\times\lambda^D\times TI}{2}+\frac{DC\times\lambda^D\times DI}{2}+\lambda^D\times MTTR\right] \qquad \text{(Eq A-99)}$$

$$\frac{\partial F_{11}}{\partial DI} = 6\left[\begin{array}{l}\frac{(1-DC)\times(1-\beta)\times\lambda^D\times TI}{2}+ \\ \frac{DC\times(1-\beta)\times\lambda^D\times DI}{2}+(1-\beta)\times\lambda^D\times MTTR\end{array}\right]\times\left[\frac{DC\times(1-\beta)\times\lambda^D}{2}\right]+$$

$$\frac{DC\times\beta\times\lambda^D}{2} \qquad \text{(Eq A-100)}$$

$$\left[\frac{\partial F_{11}}{\partial \lambda^D}\right] = 6\left[\begin{array}{l}\frac{(1-DC)\times(1-\beta)\times\lambda^D\times TI}{2}+ \\ \frac{DC\times(1-\beta)\times\lambda^D\times DI}{2}+(1-\beta)\times\lambda^D\times MTTR\end{array}\right]\times$$

$$\left[\begin{array}{l}\frac{(1-DC)\times(1-\beta)\times TI}{2}+ \\ \frac{DC\times(1-\beta)\times DI}{2}+(1-\beta)\times MTTR\end{array}\right]+$$

$$\left[\frac{(1-DC)\times\beta\times TI}{2}+\frac{DC\times\beta\times DI}{2}+\beta\times MTTR\right] \qquad \text{(Eq A-101)}$$

Using the sensitivities for F_{11} calculated above, the variance in F_{11} due to uncertain or random variables is found as:

$$V(F_{11}) = \left[\frac{\partial F_{11}}{\partial DC}\right]^2 V(DC) + \left[\frac{\partial F_{11}}{\partial TI}\right]^2 V(TI) + \left[\frac{\partial F_{11}}{\partial \lambda^D}\right]^2 V(\lambda^D) +$$

$$\left[\frac{\partial F_{11}}{\partial MTTR}\right]^2 V(MTTR) + \left[\frac{\partial F_{11}}{\partial \beta}\right]^2 V(\beta) + \left[\frac{\partial F_{11}}{\partial DI}\right]^2 V(DI) \qquad \text{(Eq A-102)}$$

PROCEDURE FOR DETERMINATION OF VARIANCE FOR REPAIRABLE 1oo2, 1oo3 AND 2oo3 SYSTEMS

Based on the above derivations we may now write the formulas for the variance of various configurations of repairable equipment.

Step 1. Determine the configuration of equipment to be studied.

Step 2. Evaluate Function H using the expected value of all random or uncertain variables, equation A-51

Step 3. Evaluate Function Q using the expected value of all random or uncertain variables, equation A-52

Step 4. Evaluate the sensitivity of function H with respect to the random variables using equations A-65 thru A-70.

Step 5. Evaluate the sensitivity of function Q with respect to the random variables using equations A-72 thru A-77.

Step 6. Evaluate the variance as:

a. For 1oo2 systems use equation A-86 for $V(F_8)$
b. For 1oo3 systems use equation A-94 for $V(F_9)$
c. For 2oo3 systems use equation A-102 for $V(F_{11})$

The end result of the above calculations is the determination of the mean and variance of the configuration used in the sensor, logic solver or final control element systems. Table 4 of the main body of this paper presents a cross reference roadmap for the determination of the mean and variance of various of the probability of failure on demand (PFD) of various hardware configurations. The mean and variance for the particular configuration is then returned to the overall calculations of the PFD of the interlock.

APPENDIX B -- UNCERTAINTY EQUATIONS FOR CONFIGURATIONS NOT COVERED IN ISA TR 84.00.02

The ISA Technical report [Ref. 2] does not define the calculation equations for the PFDavg for non-repairable systems where duplicated devices create the potential for common cause failures. This appendix derives the equations needed to evaluate the uncertainty for non-repairable systems with the potential for common cause failures.

To derive the needed functions for the non-repairable systems with common cause failure (CCF) potential, we start with the equations for the repairable system with common cause failure allowed (Table A.4). First we note that a system composed of only one set of components (1oo1) without duplication has no potential for a common cause failure. In addition, a review of the equations of Table 2 finds that systems 1oo1, 2oo2 and 3oo3 do not contain the common cause failure (CCF) parameter β. Therefore, systems 1oo1, 2oo2, and 3oo3 can be eliminated from consideration as common cause failures cannot lead to failure of the specified interlocks.

For the other situations [1oo2, 1oo3 and 2oo3] we note that no repair will be allowed during the mission of the non-repairable system. This observation implies that the diagnostic coverage (DC), the diagnostic interval (DI) and the mean time for restoration (MTTR) are all zero (0). We will set DC = DI = MTTR = 0

Substitution of the zero value for DC, DI and MTTR into the equations of Table A.4 yields the non-repairable system PFD equations where CCF is allowed. Table B.1 presents the results of these algebraic substitutions.

Table B.1 Simplified PFD_{avg} formulas for Non-Repairable System Allowing for CCF

Configuration	Function	PFD based on "Average before" failure rate	Equation Number
1oo2	F_{13} =	$\left[\frac{(1-\beta)\times\lambda^D\times TI}{2}\right]^2+\left[\frac{\beta\times\lambda^D\times TI}{2}\right]$	B-1
1oo3	F_{14} =	$\left[\frac{(1-\beta)\times\lambda^D\times TI}{2}\right]^3+\left[\frac{\beta\times\lambda^D\times TI}{2}\right]$	B-2
2oo3	F_{15} =	$3\times\left[\frac{(1-\beta)\times\lambda^D\times TI}{2}\right]^2+\left[\frac{\beta\times\lambda^D\times TI}{2}\right]$	B-3

Where TI is the proof test interval, λ^D is the dangerous failure rate.

The expected value of the probability of failure on demand is found by calculating the PFD using the expected value (mean) of the random variables.

Let y a function of some random variables, x_i, as:

$$y = F(x_i) \qquad \text{(Eq B-4)}$$

The mean of y can be estimated using the following approximation:

$$E(y) = F[E(x_i)] \qquad \text{(Eq B-5)}$$

Where:

$E(y)$ = expected value of random variable y = mean of y

$E(x_i)$ = expected value of random variable x_i = mean of x_i

For the configurations defined in Table A.1, the corresponding expected PFD are presented in Table B.2.

Table B.2 Expected Value of PFD_{avg} formulas for Non-Repairable System Allowing CCF			
Configuration	**Function**	**Expected PFD based on "Average before" failure rate**	**Equation Number**
1oo2	$E(F_{13}) =$	$\left[\frac{(1-E(\beta))\times E(\lambda^D)\times E(TI)}{2}+\right]^2 + \left[\frac{E(\beta)\times E(\lambda^D)\times E(TI)}{2}\right]$	B-6
1oo3	$E(F_{14}) =$	$\left[\frac{(1-E(\beta))\times E(\lambda^D)\times E(TI)}{2}+\right]^3 + \left[\frac{E(\beta)\times E(\lambda^D)\times E(TI)}{2}\right]$	B-7
2oo3	$E(F_{15}) =$	$3\times\left[\frac{(1-E(\beta))\times E(\lambda^D)\times E(TI)}{2}+\right]^2 + \left[\frac{E(\beta)\times E(\lambda^D)\times E(TI)}{2}\right]$	B-8

Where E(x) is the expected value of random variable x.

The variance of the PFD is found by taking the weighted sum of the variance of the random variables. The weighting function is the sensitivity of the PFD with respect to random variable. In mathematical terms the variance of a function may be found in terms of the variance of the random variables as:

Once again let

$$y = F(x_i) \qquad \text{(Eq B-9)}$$

The variance of Y can likewise be estimated as:

$$V(y) = \sum_{i=1}^{n} \left[\frac{\partial y}{\partial x_i}\right]^2 V(x_i)$$ (Eq B-10)

Where:

V(y) = variance of random variable y as defined above

$V(x_i)$ = variance of random variable x_i as defined above

The sensitivity of y with respect to a random variable x_i is:

$$\text{Sensitivity of y with respect to } x_i = \left[\frac{\partial y}{\partial x_i}\right]$$ (Eq B-11)

The variance of the PFD for the repairable cases 1oo2, 1oo3 and 2oo3 are presented in equations of Appendix A as:

- 1oo2, the variance is specified by equation A-86
- 1oo3, the variance is specified by equation A-94
- 2oo3, the variance is specified by equation A-102

These three equations are the basis upon which the equations for the non-repairable system with common cause failure (CCF) are developed.

Variance of PFD for Non-Repairable 1oo2 with CCF

To derive the variance for F_{13} we start with the general equation for the variance as:

$$V(F_{13}) = \left[\frac{\partial F_{13}}{\partial DC}\right]^2 V(DC) + \left[\frac{\partial F_{13}}{\partial TI}\right]^2 V(TI) + \left[\frac{\partial F_{13}}{\partial \lambda^D}\right]^2 V(\lambda^D) +$$

$$\left[\frac{\partial F_{13}}{\partial MTTR}\right]^2 V(MTTR) + \left[\frac{\partial F_{13}}{\partial \beta}\right]^2 V(\beta) + \left[\frac{\partial F_{13}}{\partial DI}\right]^2 V(DI)$$ (Eq B-12)

The variance of a constant is zero. Thus, setting DC=DI=MTTR =0 results in the variance of these variables as becoming zero.

$$V(DC) = V(DI) = V(MTTR) = 0$$ (Eq B-13)

Thus, the variance for the 1oo2 case becomes:

$$V(F_{13}) = \left[\frac{\partial F_{13}}{\partial TI}\right]^2 V(TI) + \left[\frac{\partial F_{13}}{\partial \lambda^D}\right]^2 V(\lambda^D) + \left[\frac{\partial F_{13}}{\partial \beta}\right]^2 V(\beta) \qquad \text{(Eq B-14)}$$

The evaluation of the sensitive's of the PFD may now be taken from appendix A equations as:

$\left[\frac{\partial F_{13}}{\partial TI}\right]$ from equation A-81

$\left[\frac{\partial F_{13}}{\partial \lambda^D}\right]$ from equation A-85

$\left[\frac{\partial F_{13}}{\partial \beta}\right]$ from equation A-83

Substituting DC=DI=MTTR=0 into these sensitivity equations yields:

$$\frac{\partial F_{13}}{\partial TI} = 2\left[\frac{(1-\beta)\times\lambda^D\times TI}{2} + \right]\times\left[\frac{(1-\beta)\times\lambda^D}{2}\right] + \frac{\beta\times\lambda^D}{2} \qquad \text{(Eq B-15)}$$

$$\frac{\partial F_{13}}{\partial \lambda^D} = 2\left[\frac{(1-\beta)\times\lambda^D\times TI}{2} + \right]\times\left[\frac{(1-\beta)\times TI}{2}\right] + \left[\frac{\beta\times TI}{2}\right] \qquad \text{(Eq B-16)}$$

$$\frac{\partial F_{13}}{\partial \beta} = 2\left[\frac{(1-\beta)\times\lambda^D\times TI}{2}\right]\times\left[\frac{(-1)\times\lambda^D\times TI}{2}\right] + \left[\frac{\lambda^D\times TI}{2}\right] \qquad \text{(Eq B-17)}$$

Variance of PFD for Non-Repairable 1oo3 with CCF

To derive the variance for F_{14} we start with the general equation for the variance as:

$$V(F_{14}) = \left[\frac{\partial F_{14}}{\partial DC}\right]^2 V(DC) + \left[\frac{\partial F_{14}}{\partial TI}\right]^2 V(TI) + \left[\frac{\partial F_{14}}{\partial \lambda^D}\right]^2 V(\lambda^D) +$$

$$\left[\frac{\partial F_{14}}{\partial MTTR}\right]^2 \text{V(MTTR)} + \left[\frac{\partial F_{14}}{\partial \beta}\right]^2 \text{V}(\beta) + \left[\frac{\partial F_{14}}{\partial DI}\right]^2 \text{V(DI)} \qquad \text{(Eq B-18)}$$

The variance of a constant is zero. Thus, setting DC=DI=MTTR =0 results in the variance of these variables as becoming zero.

$$\text{V(DC)} = \text{V(DI)} = \text{V(MTTR)} = 0 \qquad \text{(Eq B-19)}$$

Thus, the variance for the 1oo3 case becomes:

$$\text{V}(F_{14}) = \left[\frac{\partial F_{14}}{\partial TI}\right]^2 \text{V(TI)} + \left[\frac{\partial F_{14}}{\partial \lambda^D}\right]^2 \text{V}(\lambda^D) + \left[\frac{\partial F_{14}}{\partial \beta}\right]^2 \text{V}(\beta) \qquad \text{(Eq B-20)}$$

The evaluation of the sensitive's of the PFD may now be taken from appendix A equations as:

$\left[\frac{\partial F_{14}}{\partial TI}\right]$ from equation A-89

$\left[\frac{\partial F_{14}}{\partial \lambda^D}\right]$ from equation A-93

$\left[\frac{\partial F_{14}}{\partial \beta}\right]$ from equation A-91

Substituting DC=DI=MTTR=0 into these sensitivity equations yields:

$$\frac{\partial F_{14}}{\partial TI} = 3\left[\frac{(1-\beta)\times\lambda^D\times TI}{2} + \right]^2 \times \left[\frac{(1-\beta)\times\lambda^D}{2}\right] + \frac{\beta\times\lambda^D}{2} \qquad \text{(Eq B-21)}$$

$$\frac{\partial F_{14}}{\partial \lambda^D} = 3\left[\frac{(1-\beta)\times\lambda^D\times TI}{2}\right]^2 \times \left[\frac{(1-\beta)\times TI}{2}\right] + \left[\frac{\beta\times TI}{2}\right] \qquad \text{(Eq B-22)}$$

$$\frac{\partial F_{14}}{\partial \beta} = 3\left[\frac{(1-\beta)\times\lambda^{D}\times TI}{2}\right]^{2}\times\left[\frac{(-1)\times\lambda^{D}\times TI}{2}\right]+\left[\frac{\lambda^{D}\times TI}{2}\right] \quad \text{(Eq B-23)}$$

Variance of PFD for Non-Repairable 2oo3 with CCF

To derive the variance for F_{15} we start with the general equation for the variance as:

$$V(F_{15}) = \left[\frac{\partial F_{15}}{\partial DC}\right]^{2} V(DC) + \left[\frac{\partial F_{15}}{\partial TI}\right]^{2} V(TI) + \left[\frac{\partial F_{15}}{\partial \lambda^{D}}\right]^{2} V(\lambda^{D}) +$$

$$\left[\frac{\partial F_{15}}{\partial MTTR}\right]^{2} V(MTTR) + \left[\frac{\partial F_{15}}{\partial \beta}\right]^{2} V(\beta) + \left[\frac{\partial F_{15}}{\partial DI}\right]^{2} V(DI) \quad \text{(Eq B-24)}$$

The variance of a constant is zero. Thus, setting DC=DI=MTTR =0 results in the variance of these variables as becoming zero.

$$V(DC) = V(DI) = V(MTTR) = 0 \quad \text{(Eq B-25)}$$

Thus, the variance for the 2oo3 case becomes:

$$V(F_{15}) = \left[\frac{\partial F_{15}}{\partial TI}\right]^{2} V(TI) + \left[\frac{\partial F_{15}}{\partial \lambda^{D}}\right]^{2} V(\lambda^{D}) + \left[\frac{\partial F_{15}}{\partial \beta}\right]^{2} V(\beta) \quad \text{(Eq B-26)}$$

The evaluation of the sensitive's of the PFD may now be taken from appendix A equations as:

$\left[\frac{\partial F_{15}}{\partial TI}\right]$ from equation A-97

$\left[\frac{\partial F_{15}}{\partial \lambda^{D}}\right]$ from equation A-101

$\left[\frac{\partial F_{15}}{\partial \beta}\right]$ from equation A-99

Substituting DC=DI=MTTR=0 into these sensitivity equations yields:

$$\frac{\partial F_{15}}{\partial TI} = 6\left[\frac{(1-\beta)\times\lambda^{D}\times TI}{2}+\right]\times\left[\frac{(1-\beta)\times\lambda^{D}}{2}\right]+\frac{\beta\times\lambda^{D}}{2} \qquad \text{(Eq B-27)}$$

$$\frac{\partial F_{15}}{\partial \lambda^{D}} = 6\left[\frac{(1-\beta)\times\lambda^{D}\times TI}{2}\right]\times\left[\frac{(1-\beta)\times TI}{2}\right]+\left[\frac{\beta\times TI}{2}\right] \qquad \text{(Eq B-28)}$$

$$\frac{\partial F_{15}}{\partial \beta} = 6\left[\frac{(1-\beta)\times\lambda^{D}\times TI}{2}\right]\times\left[\frac{(-1)\times\lambda^{D}\times TI}{2}\right]+\left[\frac{\lambda^{D}\times TI}{2}\right] \qquad \text{(Eq B-29)}$$

The evaluation of the functions for the variance are always done using the expected value of the variable. For example when the variable TI appears in an equation, the numerical evaluation of the function would be using E(TI) where E(TI) is the expected value of the random variable TI.

Comparison of Batch versus Continuous Process in Pharmaceutical Industry Based on Safety Consideration

Shiqi Chen
Mary Kay O'Connor Process Safety Center
Artie McFerrin Department of Chemical Engineering
Texas A&M University
College Station, Texas 77843-3122, USA Email address
shiqi_chen@tamu.edu

Yogesh Koirala, M. Sam Mannan
Mary Kay O'Connor Process Safety Center
Artie McFerrin Department of Chemical Engineering
Texas A&M University
College Station, Texas 77843-3122, USA Email address
ykoirala@tamu.edu, mannan@tamu.edu

Prepared for Presentation at
American Institute of Chemical Engineers
2017 Spring Meeting and 13th Global Congress on Process Safety
San Antonio, TX
March 26 – 29, 2017

Comparison of Batch versus Continuous Process in Pharmaceutical Industry Based on Safety Consideration

Shiqi Chen
Mary Kay O'Connor Process Safety Center
Artie McFerrin Department of Chemical Engineering
Texas A&M University
College Station, Texas 77843-3122, USA Email address
shiqi_chen@tamu.edu

Yogesh Koirala, M. Sam Mannan
Mary Kay O'Connor Process Safety Center
Artie McFerrin Department of Chemical Engineering

Keywords: Continuous Process, Hazard Identification, Pharmaceutical Manufacturing

Abstract

As opposed to the petro-chemical and bulk chemical industry, where continuous processes are widely applied, the pharmaceutical industry still primarily relies on traditional batch process due to the complexity of product, multi-step operation and low-volume production. Considering these conditions, the versatile batch process is more appropriate. Nowadays, driven by the contradiction between increasing demand for drugs and inefficient batch production mode, there is a trend in pharmaceutical industry, that is the transformation from traditional batch process to novel continuous process. Related projects and research that is aimed at analyzing this transition are conducted in worldwide, and the scale of these studies ranges from lab-scale reactions to overall arrangement of the factories.

Although continuous pharmaceutical process is thriving, the safety issue in this field is not promoted at the same time. Since the continuous pharmaceutical process is a novel technology, little information can be provided to evaluate its safety level. Moreover, process conditions are usually intensified for continuous process comparing to batch process. It also may bring potential risks and make continuous manufacturing inappropriate for some of pharmaceutical productions.

This research provides a comprehensive comparison for batch versus continuous pharmaceutical process by application of Dow's Fire and Explosion Index. In addition to this conventional safety evaluation, influences from production efficiency and specialties in pharmaceutical production are integrated into the comparison. Production of 2-methyl benzimidazole and peracetic acid via batch and continuous processes are conducted in this research. Via these integrative and systematic studies, safety levels of batch and continuous processes is analyzed and compared. The ways in which process conditions, production efficiency, and other requirements influence safety level for different production modes are illustrated.

1 Introduction

Continuous process is widely applied in petro-chemical and bulk chemical industry for its high production rate, automated operation and saving in cost[1]. In contrast, the pharmaceutical industry still primarily relies on traditional batch process. This aspect stems from intrinsic limitations in active pharmaceutical ingredients (APIs) production. Complexity of medical substances means multistep synthesis. An average of 8.1 steps is needed for one API synthesis[2]. Another limitation is production volume. Demand for drugs are not comparable with most of the bulk chemicals. These limitations make batch process dominate pharmaceutical industry for a long time.

However, due to the gradual expansion of the pharmaceutical market, batch process is behind the time because of its low production rate caused by low yield and inefficient operation mode. Driven by this contradiction between increasing demand for drugs and inefficient production mode, many pharmaceutical manufacturers are trying to shift from a traditional batch process to a novel continuous process. Experimental and industrial trials that are aimed at this transitioning have been done and exhibit the practicability of this new production mode. Several companies such as Lonza, Phoenix Chemical, and Corning make significant progress in this area[3-5]. Novel continuous process exhibits great potential for the pharmaceutical industry.

"Which process should be applied?" and "Is there enough benefit to move forward?" are the questions that are raised when an innovative process is introduced. To answer these questions, comprehensive comparison between batch process and continuous process is required.

This comparison should be specific in the early stages. For example, for a given substance, based on its intrinsic properties and demand in market, reasonable comparison is required to make a selection between batch and continuous process. After composing a conclusive summary that is based on number of different specific cases, the result of these comparisons can be modified and applied to generate a practical process screening methodology.

For the questions mentioned above, economic benefits, safety and environmental impacts should all be considered to get a reliable answer. In the batch versus continuous process comparisons, related economic analysis and environmental analysis have been done by many researchers[6-8]. However, there remains gaps in research and understanding of safety considerations in comparison of batch versus continuous process. Therefore, the need arises for related comparison within safety consideration and systematic methodology for safety evaluation.

In the following sessions, the development in continuous pharmaceutical process, the previous studies in process screening, selecting and comparing, and concept of inherent safety assessment are discussed. The statement of research objective, the description of methodology and results of case studies are illuminated as well.

2 Literature Survey

To decide whether to head to novel continuous process or not, evaluations of batch and continuous process are in demand for pharmaceutical companies.

In preliminary stage of the evaluation, fundamental information about reaction kinetics, heat transmission and mass transfer is collected and analyzed. The objective of this investigation is looking into the technical feasibility of chosen reaction. After these basic assessments for intrinsic properties of reaction, far-reaching analyze in economic, environmental and safety aspect can be achieved. In this more comprehensive comparison based on overall picture, more factors such as expenditure, wastewater treatment and loss control measure should be covered. Among these factors, which factor should be selected as criterion, how these factors interact, and what's the correlation between these factors and comparison results are questions require to be answered.

To answer these questions, some researchers put efforts in individual comparisons of manufacturing for specific active pharmaceutical ingredient (API). Most of these studies focus on economic advantage of continuous process. Spencer D. Schaber et al.[12] estimated the capital, operating and the overall cost for production of a given API from synthesis to tablet formulation by integrated continuous and batch pharmaceutical manufacturing respectively. Based upon the equal annual yield, the result shows that the capital expenditures for continuous production could be markedly lower than those of batch mode. Differences in operating and overall cost are not such significant but potential saving by material handling and labor in continuous process is explicitly illustrated in the study. Hessel, V. et al. compared continuous vs. batch pharmaceutical manufacturing for production of ibuprofen and artemisinin on the standpoint of cost and environmental impact. Capital expenditure and operating expenditure savings are up to 57.0% and 51.6% respectively for ibuprofen, while capital expenditure and operating expenditure savings are up to 19.6% and 29.3% respectively for artemisinin[6, 8]. These study shows the potential economic benefits in application of continuous pharmaceutical processes.

In addition to process comparison, some researchers developed general criterions or created systematic methodologies for process selection. Dominique M. Roberge et al.[13] classified reactions used in the fine and pharmaceutical industry into 4 categories according to kinetic properties. Based on this classification, the very fast reactions (reactions time less than 1s) and the fast reactions (reactions occurring from 1s to10min) would benefit the most from continuous process. These two types of reactions occupy 44% of all reactions investigated. Ryan L. Harman and Jonathan P. McMullen[14] developed a decision roadmap of flow process for use in lab-scale. Considering objectives of the study, the intrinsic properties of the chemistry and requirements for mixing, heat transferring, related questions are asked at the various nodes in this flow map to get the final choice. Gary S. Calabrese and Sergio Pissavini[15] established an conceptual methodology to evaluate applicability of continuous reactor in the form of reaction properties test. They measured applicability of flow reactor via series of questions. These questions covered issues in reaction properties and some requirements on process. Three "zones" that tagged with different level of applicability are defined. Different questions are included for each "zone". For any given reaction, the more answers with "yes" in the "zone", the higher possibility of this reaction to be at the level of applicability that the "zone" stands for.

3 Problem Statement

Although advantages of continuous process in pharmaceutical manufacturing are illustrated by many studies, it is still too early to claim that continuous process is the right direction to go forward. Several gaps in the literature survey are required to be focused on.

Firstly, most of process selection methodologies are confined to analyzing separate properties of reaction taken in process, rather than evaluation in a systematic way. For example, small inventory is usually identified as one of significant advantage of continuous process, however, it is possible that to achieve similar yield in the corresponding batch process, the reaction in continuous process requires more intensified conditions like higher temperature, higher pressure, and higher concentration of hazardous chemicals. Selection and comparison by independent and individual parameters without integration is not enough.

Secondly, existing comparisons between batch and continuous processes mainly focus on economic aspect, while lacking combination of safety issue and other important aspects from real production. Although the driving force for the application of this novel process is the economic interests, safety issues directly determine its feasibility. A choice between batch and continuous process needs to be made based on business performance within acceptable safety levels. Those comparisons that merely consider the profits are not enough.

Thirdly, the existing studies are limited to fixed production system. Potential problems in a different scale are not considered. In the pharmaceutical industry, gradually increasing production scales is required which is different from the bulk chemical industry. Hence, safety of process conditions needs to be guaranteed for all scale production. Comparison only based on fixed scale is not comprehensive.

Therefore, the primary objective of this study is to provide a comprehensive and systematic comparison between batch and continuous process used in pharmaceutical industry. Then on the basis of this analysis, this study will evaluate how main factors considered in Dow F&EI impact the safety level for these two different processes. Finally, this study will provide support to further improve existing methodology for process screening and selection for pharmaceutical industry.

It is worth mentioning that there are other risks such as chemical exposure for pharmaceutical manufacturing, however, this study mainly focuses on potential fire, explosion and reactivity incident. In addition, comparison in this study focuses on intrinsic properties of the reaction and process which are the major hazards of the reaction system. Therefore, other factors like loss control measures and emergency equipment are supposed to be effective for both batch and continuous systems.

4 Methodology

To measure the realistic fire, explosion and reactivity potential of batch and continuous process, Dow's Fire and Explosion Index (F&EI) methodology is applied as hazard index in this study, since it is the most widely used hazard index that recognized by the chemical industry[18].

From the first edition of Dow's Fire & Explosion Index Hazard Classification Guide in 1964, this index has been continuously improving though the past 50 years. The last version is the seventh edition that published in 1994. The purposes of this index include forecasting quantitative damage of potential explosion, reactivity incidents and fire; determining hazardous equipment which is possible to give rise to incidents in the system; conveying this F&EI risk potential to related managements and organizations. Since the aim of this study is to identify major hazard that affect selection, application of Dow F&EI will be mainly focus on its function of hazard identification.

In the Dow's Fire & Explosion Index Hazard Classification Guide, economic analysis is confined to financial loss after an incident. Correlation between safety and economic gain from production is not built. Actually, this kind of connection could be established via yield of reactions and geometry and inventory of units, since these factors take important role in both safety assessment and process gain. In the current study, this correlation between safety and profit is constructed on the foundation of Dow's Fire & Explosion Index.

5 Case Study I: Benzimidazole Derivative Synthesis

5.1 overview

Benzimidazole derivative is an important type of intermediates in pharmaceutical industry. It is widely exhibited in a variety of pharmaceutical substance, such as proton pump inhibitors, antimicrobials, antivirals, anticancer, anti-inflammatory, hormone modulators as well as depressants, antidiabetics and so on[22]. Famous drug with benzimidazole nucleus includes Omeprazole and Albendazole. Omeprazole is used to treat gastroesophageal reflux disease and peptic ulcer disease[23], and Albendazole is an effective drug in treatment of various parasitic worm infestations[24].

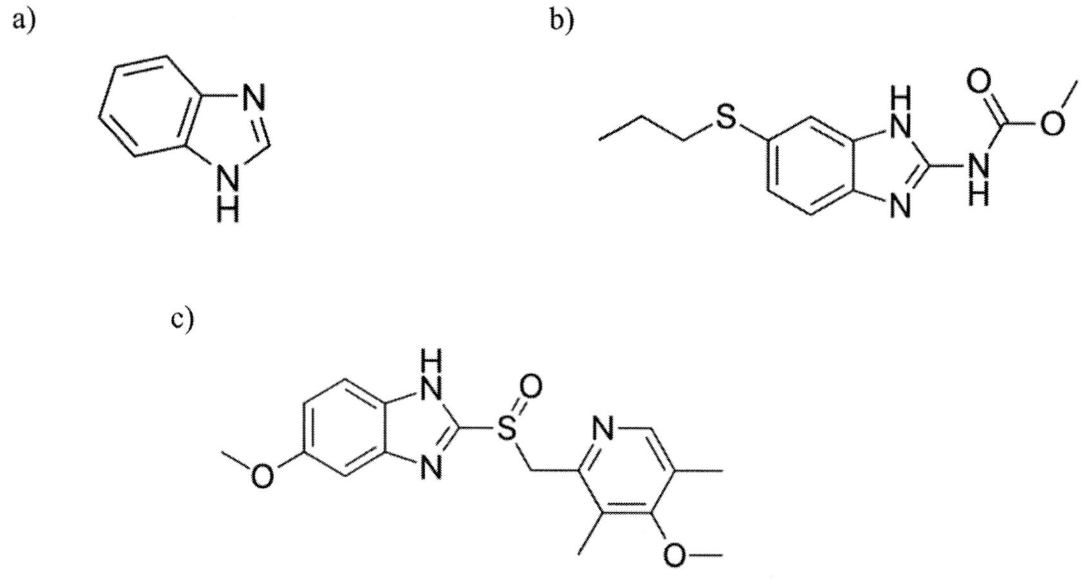

Figure 1. a) Molecular Structure of Benzimidazole, b) Molecular Structure of Albendazole, c) Molecular Structure of Omeprazole

2-methyl benzimidazole is a pharmaceutical-potential substance that belongs to this benzomidazole ring system. The condensation of o-phenylenediamines with carboxylic acids is one of the most common synthesis route for 2-methyl benzimidazole[25]. Conditions and production comparisons of o-phenylenediamines condensation by batch vs. continuous processes are showed in Table 1. These fundamental parameters are used for safety evaluation by Dow's F&EI in the following study.

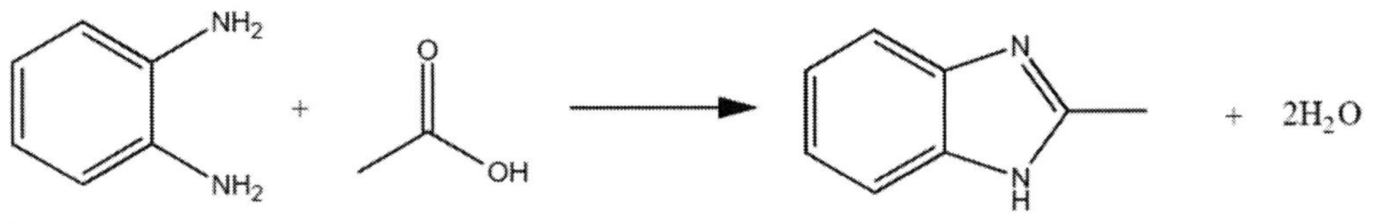

Figure 2. Synthesis of 2-methyl Benzimidazole by Condensation of o-phenylenediamine

Table 1. Production Conditions of Batch and Continuous Processes for Synthesis of 2-methyl benzimidazole

	Batch Process	Continuous Process
Temperature (℃)	200	270
Pressure (psi)	145	1885
Reaction Time (min)	5	0.5
Yield (%)	98	94
Substrate Concentration (M)	5	1
Reaction Volume (ml)	1000	4

In order to concentrate on study in intrinsic differences between two processes, and simplify the hazard identification in F&EI, several assumptions are adopted:

1. The reaction mass is perfect mixed in reactors;
2. Adjunct cooling systems in both processes are in perfect performances;
3. Emergency equipment is same and in perfect performances in both processes;
4. Scalabilities of both processes are good enough that reaction conditions remain same in all scales.

5.2. Safety Comparison Result

The comparison is conducted based on material properties and process features. Related explanation is stated below.

5.2.1. comparison on material

In this study, impacts from intrinsic properties of reactants are same for both processes. Material factors of all substances are listed in Table 2. According to Dow's Fire and Explosion Index Hazard Classification Guide, impact from the most hazardous material dominates the material factor of the whole process, therefore only impact from acetic acid is considered for both processes. In addition, although temperatures and composition are different in two processes and may lead to different chemical properties, these features are neglected in this part and contained in process comparison.

Table 2. Material Factor for Batch and Continuous Reactor

	Batch Reactor	Continuous Reactor
o-phenylenediamine	10	10
2-methylbenzimidazole	1	1
Acetic Acid	14	14

5.2.2. comparison on process

First of all, it's worth mentioning that albeit lots of distinctions exist between batch and continuous processes, comparisons in this study are confined to synthesis in reactor on account of its core role in the whole process: operation mode of reactor – continuously or intermittently – determines whether the process is continuous or not from the start. Therefore, the reactor is selected as the pertinent process unit without conventional unit screening procedure.

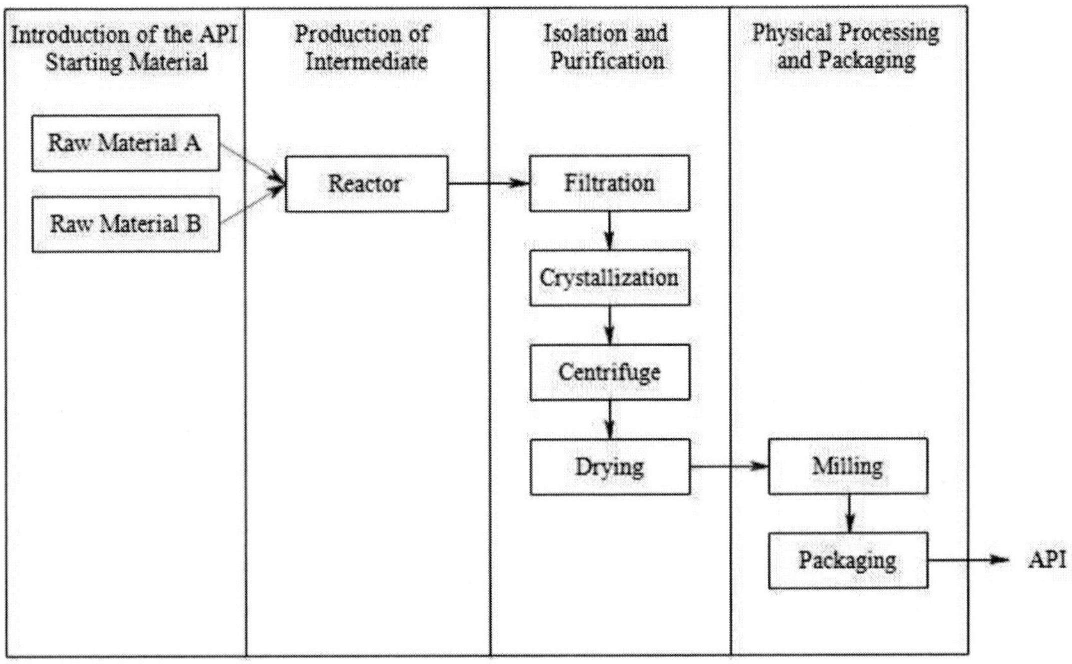

Figure 3. Procedure for Pharmaceutical Production

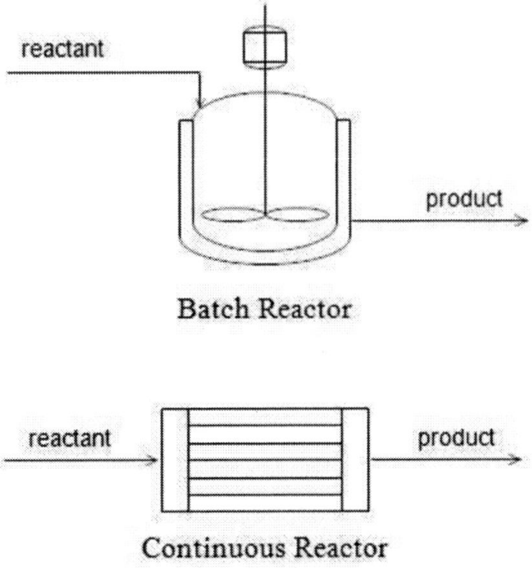

Figure 4. Differences in Structure between Batch and Continuous Reactors

On the foundation of safety evaluation by Dow's Fire & Explosion Index, quantitative judges between batch and continuous process are illustrated through comparison in sensitivity and productivity within safety consideration.

This sensitivity analysis focus on correlation between safety level and inventory of reactor. Inventory of unit plays an important role in the safety. In addition, this sensitivity analysis can also provide supportive information for scaling up procedure.

Inventory of unit plays an important part in the safety via quantity of flammable or unstable material. This potential is considered in both the General Process Hazards (F1) and the Special Process Hazards (F2) according to Dow's F&EI Hazard Classification Guide. In general, more the flammable or unstable material, higher the risk of fire and explosion within the process. For different unit (process unit or storage unit) and different phase (liquid, gas or solid), different correlation formula between inventory and penalty is given in the Special Process Hazards (F2) part of the Dow's F&EI Guide. Moreover, other issues such as whether these materials are processed indoor or outdoor, what's the temperature to handle these materials influences the safety level as well. These questions are included in the General Process Hazards (F1) part.

In this sensitivity analysis, F&EI that combines both the general and the special process hazards is expressed as a dependent variable that changes with the variation of total inventory of reactor. Other operation conditions remain the same for different inventory. Correlations between the F&EI and material inventory for batch and continuous process are shown in the Figure 5.

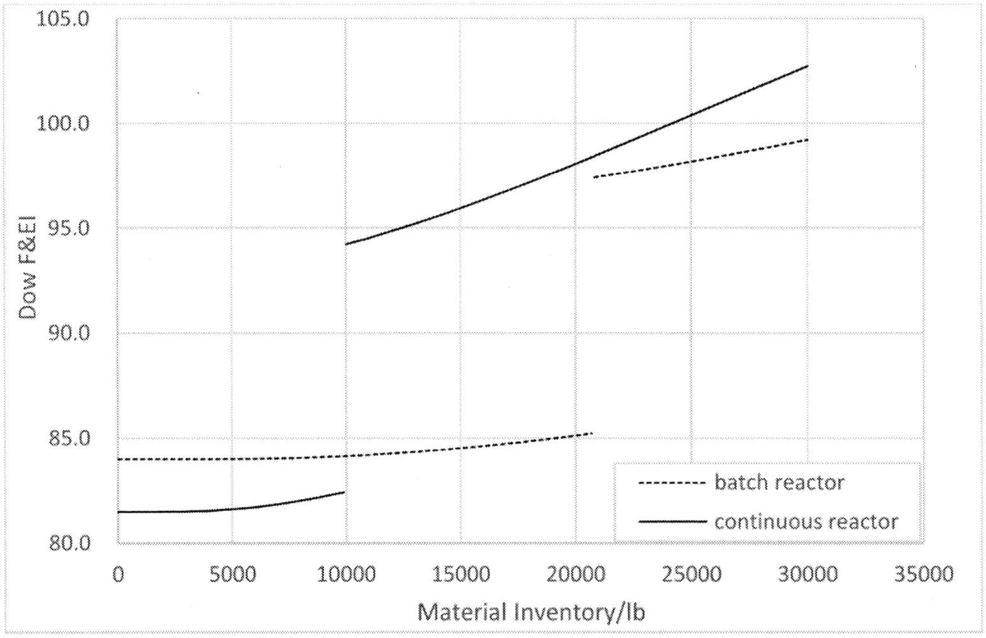

Figure 5. Analysis of Sensitivity to Inventory for Batch and Continuous Reactors Based on 2-methyl Benzimidazole Synthesis

for 0 lb < Inventory < 10000 lb:

F&EI of batch reactor is slightly greater than continuous reactor. This difference mainly comes from difference in usage of stirrer and reaction pressure. Due to requirement for homogeneous mixture, stirrer is necessary in batch reactor, while such a stirrer is not needed (or not allowed) in a continuous tubular reactor. Thus, stirrer, as a hazardous rotating equipment, adds extra penalty to batch process. As for difference in pressure, higher penalty is given to continuous process for its intensified condition. From the result, influence from rotating equipment is greater than that from pressure in this case.

Another observation is that, the magnitude of the difference in the F&EI between two reactors decreases with the growing inventory. This trend stems from the increasing impact from quantity of hazardous material. When inventory approaches zero, difference in F&EI is mainly caused by stirrer and distinct pressure aforementioned. As inventory increases, influence from unstable material extends. This influence depends on composition of mixture, or more specifically, concentration of acetic acid that defines the Material Factor (MF). Based on information of reactions, molar concentrations of acetic acid in batch and continuous reactor are 8.31M and 15.66M respectively. Thus, for similar total inventory, penalty from hazardous material in continuous reactor is greater than batch reactor.

when 10000 lb < Inventory < 20833 lb:

F&EI of continuous reactor is signally higher than batch reactor, and this difference grows along with the increasing inventory. In this range, part of influences come from rotating equipment, pressure and increasing inventory that mentioned above. More importantly, quantity of flammable and unstable liquids handled above boiling point exceeds threshold of 10,000lb for continuous reactor, therefore give a vertical rise to F&EI for continuous reactor, as shown in the Figure 5. This threshold is illustrated clearly in Dow's F&EI Hazard Classification Guide: when quantity of flammable and unstable liquids above its boiling point is less than 10,000 lb, a penalty of 0.60 is determined. A penalty of 0.90 is given when this liquids quantity above its boiling point is greater than 10,000 lb. In this case study, operation temperature of continuous reactor (270℃) is higher than boiling point of both o-phenylenediamine (257℃) and acetic acid (118℃), while operation temperature of batch reactor (200℃) merely higher than boiling point of acetic acid (118℃). Therefore, for inventory in the range of 10,000lb to 20833lb, total mass in continuous reactor is defined as unsafe, while merely acetic acid mass is defined as unsafe in the batch reactor. This divergence leads to different penalty in this range of inventory.

when 20833 lb < Inventory < 30000 lb:

F&EI of batch reactor drastically increases. However, F&EI of continuous reactor is still slightly greater than batch reactor, and the magnitude of this difference increases with the growing inventory.

This vertical rise in the F&EI of batch reactor is also caused by critical quantity of material above the boiling point. When total quantity in batch reactor goes beyond 20,833lb, quantity of flammable and unstable substance - acetic acid - also goes beyond threshold of 10,000 lb. Thus, penalty from material above boiling point is similar for both of processes.

The higher F&EI of continuous reactor originated from quantity of flammable and unstable material in this range of inventory. Since higher concentration of acetic acid is applied in continuous reactor, more proportion of substance is identified as unsafe for continuous reactor, thus higher penalty is given. This explanation is similar to that in range of low inventory (< 10,000 lb). However, unlike appearance in the low inventory range, influence from these flammable and unstable materials exceeds impact from pressure and stirrer. Therefore, continuous reactor becomes more dangerous for this range of total inventory.

It's worth mentioning that higher inventory is not considered here. Upper limit of inventory (30,000 lb) is set based on the general highest inventory of reactor used in pharmaceutical manufacturing.

Degree of hazard for batch versus continuous reactor used in this case study is shown in Table 3. In the range of inventory investigated, batch reactor is identified as moderate hazardous for a broader range of inventory, while continuous reactor is identified as intermediate hazardous for a broader range of inventory. Even so, It's worth mentioning that the F&EI values for both reactors are in a relative safe range (F&EI > 128 indicates heavy even severe hazard).

Table 3. Degree of Hazard for Batch and Continuous Reactors

Material Inventory X	Batch Reactor	Continuous Reactor
X < 15,152 lb	Moderate	Moderate
15,152 lb < X < 20,833 lb	Moderate	Intermediate
20,833 lb < X < 30,000 lb	Intermediate	Intermediate

In summary, F&EI for both batch and continuous reactors increase as material inventory increases. Moreover, F&EI of continuous reactor is more sensitive to change of inventory comparing to batch reactor. This higher sensitivity caused by comprehensive influence from higher concentration of unstable and flammable material, higher operation temperature and higher pressure it applied.

Evaluation in this part mainly focuses on safety of batch versus continuous process base on productivity considerations. Emphasis on productivity scale of this study primary comes from flexibility required for pharmaceutical production. Pharmaceutical industry is highly supply driven. Requirement on production rate is changed according to demand of market all the time. Therefore, better understanding on correlation between safety and productivity provides practicable support for research in process selection.

To approach this target, safety level is evaluated for the process at the same productivity capacity. In this part, F&EI is regarded as indicator for safety evaluation as above. Annual production is set as indicator for productivity. Related formula is shown below.

General calculations for productivity capacity are:

$$\text{Annual Production} = \text{Product Yield} \times \text{Annual Operation Time}$$

$$\text{Product Yield} = \text{Mass Flow of Reagent} \times \text{Conversion}$$

Mass flow is supposed to be proportional to reactor volume as set in the laboratory trial.

According to definition of fine chemical, annual production of fine chemical is limited to 1,000 tons per year[26]. Therefore, 1,000 tons per year is set as the upper limit in this evaluation.

As for product yield and annual operation time, the former is provided in Table 1, and expression for the latter is formed referring to the normal operation time[27].

Annual operation time for batch reactor is:

$$\text{Annual Operaton Time} = 300\ \text{day/yr} \times 24\ \text{hr/day}$$

Annual operation time for continuous reactor is:

$$\text{Annual Operation Time} = 8000\ \text{hr/yr}$$

Normally, in pharmaceutical industry, multi products are produced in one production line. Therefore, the same product cannot occupy the producing line for the whole year. In this study, 2-methyl benzimidazole is assumed to be produced in 2 months of one year.

In addition, circulating time in batch reactor need to be considered. According to the literature[11], the ratio of reaction time to total operation time for this reaction is shown below. Excepted reaction, other operations include filling, cleaning and cooling and other additional work.

$$\frac{\text{Reaction Time}}{\text{Total Operation Time}} = 0.185$$

Based on these formulas and assumptions, correlation between annual productions and F&EI is shown in the Figure 6.

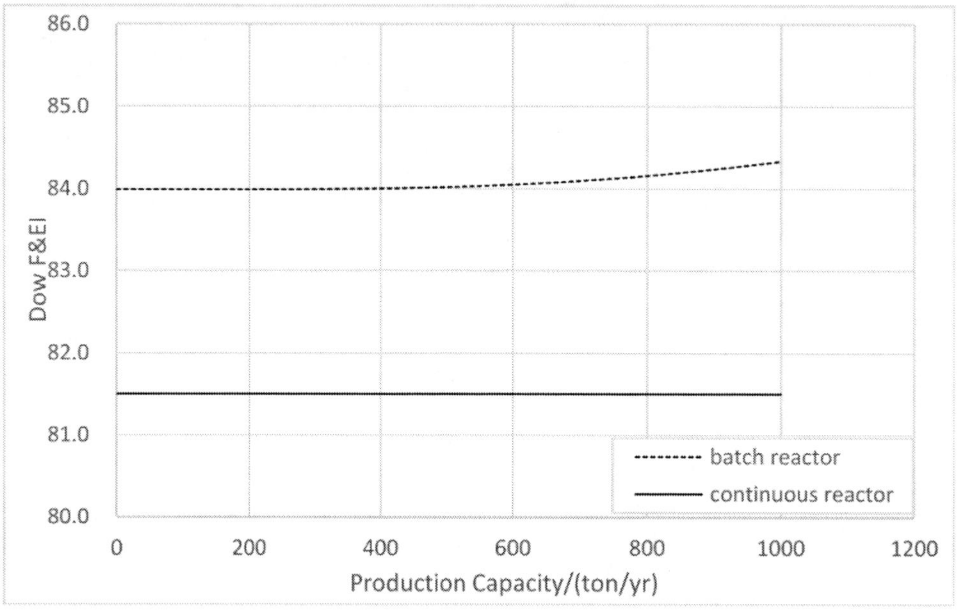

Figure 6. Analysis of Safety Based on Production for Batch and Continuous Reactors Based on 2-methyl Benzimidazole Synthesis

According to the Figure 6, F&EI value is constant in the range of practicable annual production (0 ~ 1000 tons per year). Plus, the F&EI for both reactors corresponds to a moderate level of hazard (61 < F&EI < 96).

This low F&EI resulted from relatively low demand for production. According to the sensitivity analysis, F&EI for both batch and continuous reactor increase as material inventory increases. Both processes could be intermediate hazardous once its inventory is higher than a specific value. However, these values are unapproachable in practical pharmaceutical manufacturing. Since demand for fine chemical, always less than 1,000 tons per year, confines scale of production.

Another observation is that, the F&EI of continuous reactor is always lower than that of batch reactor for same production. Plus, slight rise appears for batch reactor when production capacity exceeds 500 tons per year. These appearances predicate consistent superiority in safety for continuous reactor than batch reactor. Explanation is stated below:

Firstly, as stated in the sensitivity analysis, settled elements, such as operation pressure, temperature and use of rotating equipment, take major account for the difference at small-scale production. Advantages of continuous process in these aspects determine its superiority in small-scale pharmaceutical production.

Secondly, continuous reactor possesses higher space-time yield compared to batch reactor. This advantage stems from the intrinsic property of continuous operation mode - the low residence time and high flow throughput. That is to say, it costs fewer time and lower inventory for continuous reactor to produce same amount of product than batch reactor. These benefits in production efficiency influence safety level indirectly but significantly: Smaller inventory means fewer unstable materials in process, and shorter operation time means less chance to fail. This point is not considered in separate safety evaluation, but can be clearly illustrated by correlation between production capacity and F&EI. In this case, when demand on production growing, demand on inventory grows faster for batch reactor than continuous reactor. This faster rise in inventory lead to faster rise in amount of unstable material in process so as to the faster rise in F&EI value. It explains why slight rise appears for batch reactor when production capacity exceeds 500 tons per year.

In fact, the comparable batch and continuous reactor are rarely at the same level of volume. Referring to literature, for commercial production of fine chemical, the volumes of continuous reactor are usually in a range of 0.1 to 3 liter, while the volume of batch reactor is usually around the range of 2 to 10 cubic meters[6]. For this case, based on assumption of same annual production, ratio of batch reactor inventory and continuous reactor inventory is as much as one hundred.

6 Case Study II: Oxidation with Hydrogen Peroxide

6.1. overview

Hydrogen peroxide is one kind of widely used oxidation agent in pharmaceutical production. Typical reaction involving hydrogen peroxide includes Baeyer–Villiger oxidations for synthesis of lactone[9], N-oxidation of alkyl pyridine for synthesis of alkyl pyridine N-oxidant[28] and benzylic hydro-peroxide rearrangement for synthesis of Brivanib alaninate[29].

This common oxidation agent is also used in production of peracetic acid. Peracetic acid is produced via oxidation of acetic acid by hydrogen peroxide with sulfuric acid as the catalyst. Conditions for peracetic acid production by batch vs. continuous process are showed in Table 4[30].

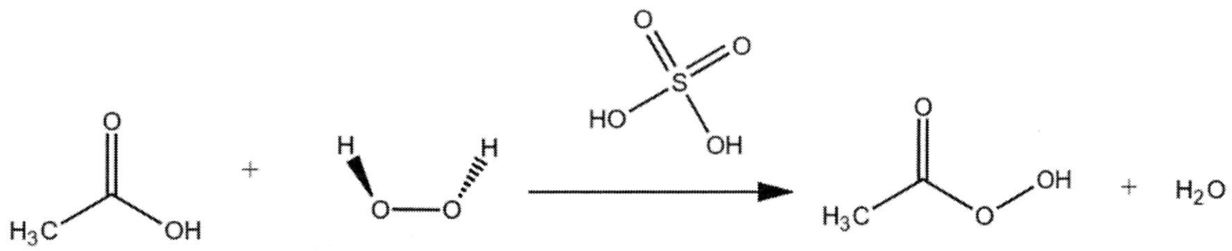

Figure 7. Synthesis of Peracetic Acid by Oxidation of Acetic Acid

Table 4. Production Conditions of Batch and Continuous Process

for Synthesis of Peracetic Acid

	Batch Process	Continuous Process
Temperature (℃)	60	80
Pressure (psi)	14.5	14.5
Reaction Time (min)	30	15
Yield (kg/h)	170	20
Acetic Acid Concentration (M)	7.3	7.3
Reaction Volume (L)	4000	10

Assumption for this case study is listed below:

1. The reaction mass is perfect mixed in reactors.
2. Since the temperature for decomposition of hydrogen peroxide in peracetic acid synthesis reaction is higher than 120 ℃ [31], which is larger than operation temperature for both processes in the case study, decomposition of hydrogen peroxide is assumed to be inexistent.
3. Sulfuric acid is dissociated completely to H^+ and HSO_4^-, in addition, concentration of H^+ is assumed to be constant of 0.67 mol/L during the reactions for both processes.
4. Influence from sulfuric acid is not considered in safety evaluation.
5. Emergency equipment and cooling system are perfectly performed for both of processes.
6. Scalabilities of both processes are good enough that reaction conditions remain same in all scales.

6.2. safety comparison result

The comparison is conducted based on material property and process feature. Related explanations are stated in the following paragraph.

6.2.1. comparison on material

In this case, impact from material within the reactor is same for both of processes. Material factors for reactants involved are listed in Table 5. According to Dow's Fire and Explosion Index Hazard Classification Guide, influence from the most hazardous material dominates the material factor. Although material factor values are same for hydrogen peroxide and peracetic acid, since concentration of hydrogen peroxide is higher than peracetic acid, hydrogen peroxide is chosen to be the dominant material. Therefore, only impact from hydrogen peroxide is considered for both processes. Change on substances' property from temperatures are neglected in this part and reconsidered in process comparison.

Table 5. Material Factor for Batch and Continuous Reactor

	Batch Reactor	Continuous Reactor
Acetic Acid	14	14
Hydrogen Peroxide	24	24
Peracetic Acid	24	24

6.2.2. comparison on process

The major difference between these two production processes come from distinctions in reaction conditions and reactor constructions. These differences are shown in Table 4. Based on safety evaluation by Dow's Fire & Explosion Index, quantitative judges between batch and continuous process are illustrated through comparison in sensitivity and productivity within safety consideration.

Procedure for sensitivity comparison is the same as the first case study. Correlations between the F&EI and material inventory for batch and continuous process are shown in the Figure 8.

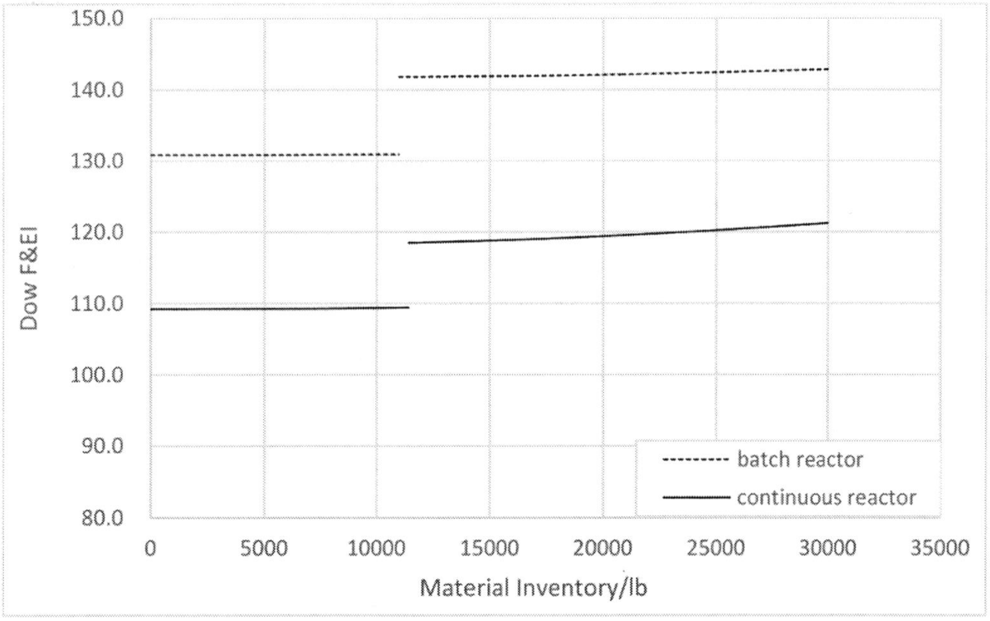

Figure 8. Analysis of Sensitivity to Inventory for Batch and Continuous Reactors Based on Peracetic Acid Synthesis

for 0 lb < Inventory < 11000 lb:

F&EI of batch reactor is larger than continuous reactor. This difference is almost caused by usage of stirrer in batch reactor. Application of such a rotating equipment increases penalty for batch process.

when 11000 lb < Inventory < 11430 lb:

F&EI of batch reactor is still higher than continuous reactor, but this difference grows drastically due to the sharp transition in F&EI of batch reactor. This transitions are caused by quantity of flammable liquids that handled above flashing point exceeds threshold of 10,000lb. This threshold is illustrated clearly in Dow's F&EI Hazard Classification Guide: when quantity of flammable liquids above their flashing point is less than 10,000 lb, a penalty of 0.30 is determined. A penalty of 0.45 will be given when this quantity of flammable liquids above their flashing point is larger than 10,000 lb. For peracetic acid production, operation temperature for batch reactor (60℃) and continuous reactor (80℃) is higher than flashing point of both acetic acid (39℃) and peracetic acid(40.5℃). Therefore, the value of inventory for higher penalty is determined by compositions of peracetic acid and acetic acid in reactors. Since concentrations of these two materials is slightly higher for batch comparing to continuous reactor, batch reactor approaches this transition in F&EI at a smaller total inventory.

when 11430 lb < Inventory < 30000 lb:

F&EI of continuous reactor exceed the transition caused by quantity of unsafe material above flashing point. However, it is still lower than F&EI of batch reactor.

Compared to the first case study, value of slope for both processes is similar low. It results in the consistent superiority of continuous reactor in safety level compared to batch reactor. This similarity in slopes comes from similarity in operation conditions. Differences only appear in operation temperature and hydrogen peroxide concentration. These differences are not such significant that lead to huge divergence between slopes. Thus, difference in F&EI value between two reactors does not change along with increasing inventory.

Degree of hazard of this case study is exhibited in Table 6. Although marked transitions appears for both reactor, safety levels for two reactors do not change based on scaling up of reactor size. The safety levels are majorly decided by settled parameters, such as operation pressure and usage of stirrer. These parameters do not change with inventory as variable. Hence, inventory does not play a significant role in judgement on degree of hazard.

Table 6. Degree of Hazard for Batch and Continuous Reactors

Reactor Type	Batch Reactor	Continuous Reactor
Degree of Hazard	Heavy	Intermediate

In sum, F&EI for both batch and continuous reactor increase as material inventory increases due to the increases of quantity above flashing point. However, since operation conditions are relatively mild and similar for these two reactors, both reactor do not exhibit high sensitivity when inventory is growing gradually.

For evaluation based upon productivity, assumption for practical production are set as following:

1. Confined by the conversion of this reaction route and volume of reactor, upper limit for annual production is settled to be 580 tons per year.
2. Same as the first case study, referring to normal operation time for this two operation mode[27], annual operation time for batch reactor is 8760 hour per day, and annual operation time for continuous reactor is 8000 hour per day.
3. Peracetic acid is assumed to be produced for 2 months in one year.
4. According to the literature[30, 32], the ratio of reaction time to total operation time for this reaction is 0.143

Based upon these formulas and assumptions, correlation between annual productions and F&EI is shown in the Figure 9.

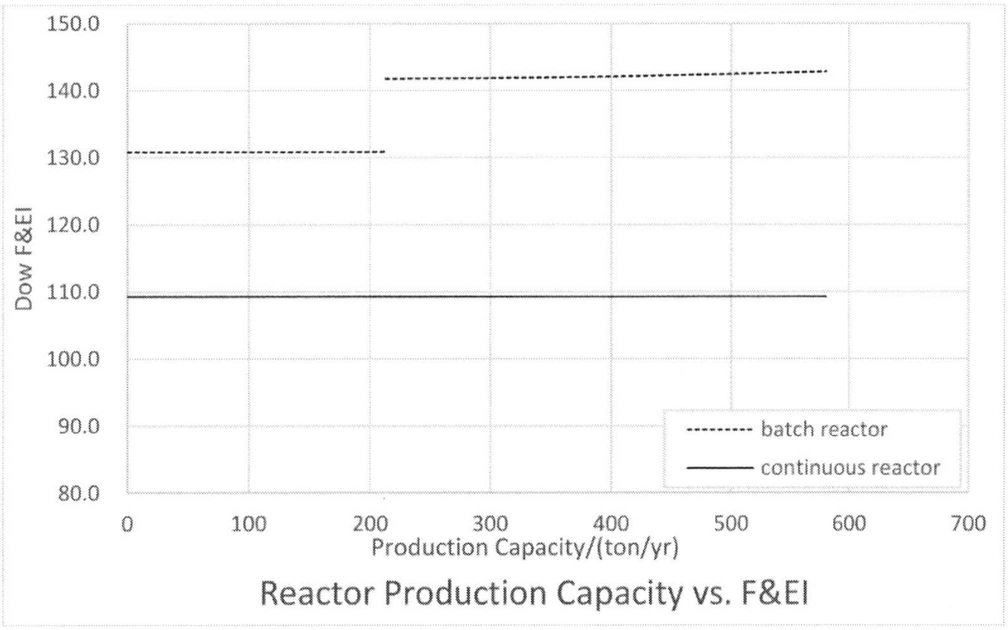

Figure 9. Analysis of Safety Based on Production for Batch and Continuous Reactors Based on Peracetic Acid Synthesis

According to the Figure 9, F&EI is stabilized at 109 in the range of practicable annual production for continuous reactor. For batch reactor, the F&EI value rise drastically when the annual production is higher than 220 tons per year.

For the batch production mode, relatively low production efficiency leads to substantial increase in reactor volume when demand on production grows. This increase in reactor volume is so significant that exceed the threshold for large quantity of flammable and unstable material above flashing point. Therefore, the F&EI value jump from 131 to 143 when the annual production is higher than 220 tons per year.

For continuous production mode, high production efficiency causes only a few change in volume hence only few change in F&EI in as well. This difference in F&EI between batch and continuous reactor means that, to produce the same amount of peracetic acid in the given period, the volume of continuous reactor is much smaller than batch reactor.

7 Result Summary

Figure 10 represents the comparisons of two case studies. The X-axis represents two different case studies, and the Y-axis represents the average Dow's F&EI value which can be used to define the degree of hazard.

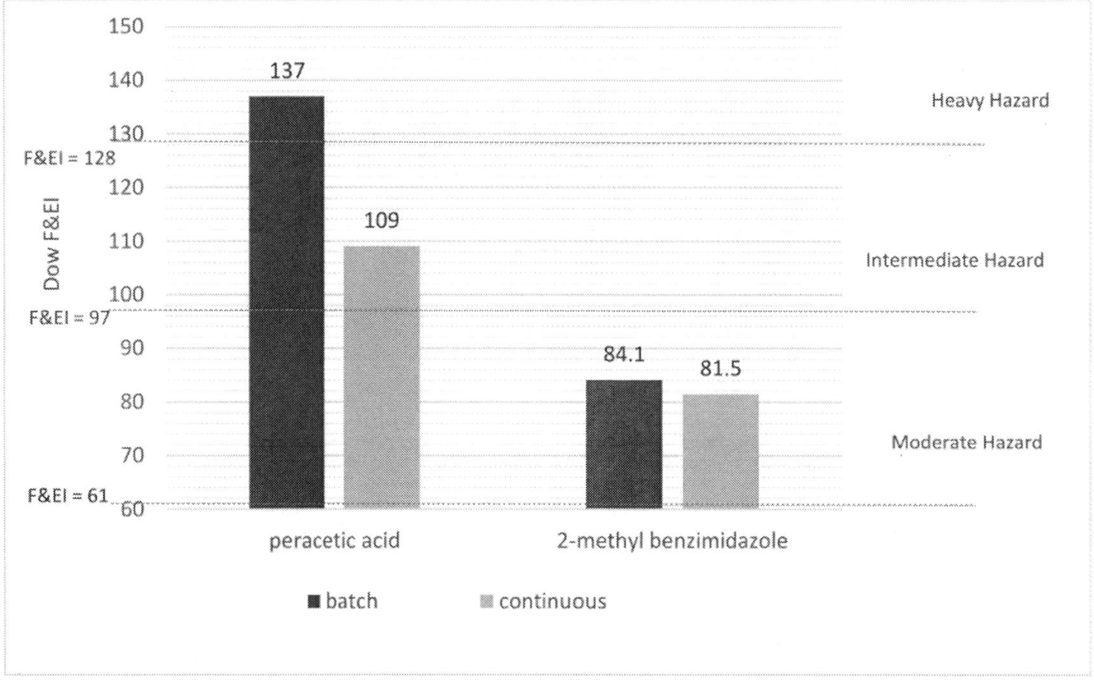

Figure 10. Comparison of Safety Level by Batch and Continuous Productions for Two Case Studies

Regardless of the reactor type, the safety level is predetermined by properties of the synthesis reaction. Although distinctions in F&EI appear when using different reactors for both cases, peracetic acid production is generally more hazardous than 2-methyl benzimidazole production, no matter which reactor it uses. This higher level of danger comes from the intrinsic properties, which consist of the flammability and reactivity of the chemicals involved in the synthesis of peracetic acid.

Taking batch and continuous reactors into consideration, distinction in F&EI comes from the differences in operation conditions and production efficiency. These varying factors such as temperature, pressure, and yield do not influence F&EI as individual factors. Instead, they are correlated and impact the F&EI as an ensemble. Difference in these ensemble decides the variance in F&EI value between reactors, then determines to which extent the selection of reactor can make a difference. For this study, the variance in F&EI value between batch and continuous reactor is greater for peracetic acid production (variance in F&EI = 28) than 2-methyl benzimidazole production (variance in F&EI = 2.6). Therefore, selecting a reactor is more important for peracetic acid comparing to 2-methyl benzimidazole production.

The importance of choosing a reactor is different for the two cases, the F&EI value for a continuous reactor is lower than a batch reactor for both of these cases. This consistency exhibits the superiority of a continuous reactor in safety.

8 Conclusion

A comprehensive comparison between the batch and continuous process used in the pharmaceutical industry based upon Dow's F&EI is made in this study. Two typical pharmaceutical productions, the production of 2-methyl benzimidazole and the production of peracetic acid are analyzed. Safety levels for batch and continuous reactors are compared based on both inventory and annual production being used as variables. The conclusions on the foundation of these two case studies are stated below:

1. The comparison of batch versus continuous processes in the pharmaceutical industry is complex. It's determined by reaction properties, operation conditions, structures of equipment, production efficiency, scale of production, and many other factors. To get a conclusive comparison between these two production modes, more systematic studies based on the integrated batch and continuous process need to be conducted for safety consideration.

2. Just focusing on the reactor in which part of API synthesis occurs, the continuous reactor is more sensitive to inventory in general. The magnitude of this sensitivity is determined by operation conditions such as operation temperature, operation pressure and concentration of hazardous material. This sensitivity to inventory might lead to problems for continuous production in a relatively large scale.

3. However, when considering practical pharmaceutical production, which is always in a small scale (no larger than 1000 tons per year), hazardous large-volume continuous reactor is not necessary. In addition, the space time yield of continuous production mode is usually higher than the batch production mode, application of a continuous reactor is relatively safer when compared to batch reactors.

9 References

1. Plumb, K., Continuous processing in the pharmaceutical industry: changing the mind set. *Chemical Engineering Research and Design* **2005,** *83* (6), 730-738.

2. Laird, T., What Type of Reactions Do Process Chemists Use on Scale? *Organic Process Research & Development* **2006,** *10* (5), 851-852.

3. Braune, S.; Poechlauer, P.; Reintjens, R.; Steinhofer, S.; Winter, M.; Lobet, O.; Guidat, R.; Woehl, P.; Guermeur, C., Selective nitration in a microreactor for pharmaceutical production under cGMP conditions. *Chimica oggi* **2009,** *27* (1), 26-29.

4. Proctor, L. D.; Warr, A. J., Development of a Continuous Process for the Industrial Generation of Diazomethane1. *Organic process research & development* **2002,** *6* (6), 884-892.

5. Roberge, D. M.; Gottsponer, M.; Eyholzer, M.; Kockmann, N., Industrial design, scale-up, and use of microreactors. *Chemistry Today* **2009,** *27* (4), 8-11.

6. Hessel, V.; Vural Gürsel, I.; Wang, Q.; Noel, T.; Lang, J., Potential analysis of smart flow processing and micro process technology for fastening process development: use of chemistry and process design as intensification fields. *Chemical Engineering & Technology* **2012,** *35* (7), 1184-1204.

7. Schaber, S. D.; Gerogiorgis, D. I.; Ramachandran, R.; Evans, J. M.; Barton, P. I.; Trout, B. L., Economic analysis of integrated continuous and batch pharmaceutical manufacturing: a case study. *Industrial & Engineering Chemistry Research* **2011,** *50* (17), 10083-10092.

8. Jolliffe, H. G.; Gerogiorgis, D. I., Plantwide design and economic evaluation of two continuous pharmaceutical manufacturing (CPM) cases: ibuprofen and artemisinin. *Computers & Chemical Engineering* **2016,** *91*, 269-288.

9. Gutmann, B.; Cantillo, D.; Kappe, C. O., Continuous - Flow Technology—A Tool for the Safe Manufacturing of Active Pharmaceutical Ingredients. *Angewandte Chemie International Edition* **2015,** *54* (23), 6688-6728.

10. Wakami, H.; Yoshida, J.-i., Grignard exchange reaction using a microflow system: from bench to pilot plant. *Organic process research & development* **2005,** *9* (6), 787-791.

11. Damm, M.; Glasnov, T. N.; Kappe, C. O., Translating high-temperature microwave chemistry to scalable continuous flow processes. *Organic Process Research & Development* **2009,** *14* (1), 215-224.

12. Schaber, S. D.; Gerogiorgis, D. I.; Ramachandran, R.; Evans, J. M. B.; Barton, P. I.; Trout, B. L., Economic Analysis of Integrated Continuous and Batch Pharmaceutical Manufacturing: A Case Study. *Ind Eng Chem Res* **2011,** *50* (17), 10083-10092.

13. Roberge, D. M.; Ducry, L.; Bieler, N.; Cretton, P.; Zimmermann, B., Microreactor technology: A revolution for the fine chemical and pharmaceutical industries? *Chem Eng Technol* **2005,** *28* (3), 318-323.

14. Hartman, R. L.; McMullen, J. P.; Jensen, K. F., Deciding Whether To Go with the Flow: Evaluating the Merits of Flow Reactors for Synthesis. *Angew Chem Int Edit* **2011,** *50* (33), 7502-7519.

15. Calabrese, G. S.; Pissavini, S., From Batch to Continuous Flow Processing in Chemicals Manufacturing. *Aiche J* **2011,** *57* (4), 828-834.

16. Kletz, T. A., What you don't have, can't leak. *Chemistry and Industry* **1978,** *6*, 287-292.

17. Kletz, T. A., *Plant design for safety: a user-friendly approach.* Taylor & Francis: 1991.

18. Fire, D., Explosion Index Hazard Classification Guide. *AICHE New York* **1994**.

19. Lawrence, D. Quantifying inherent safety of chemical process routes. © Duncan Lawrence, 1996.

20. Heikkilä, A.-M., *Inherent safety in process plant design: an index-based approach.* VTT Technical Research Centre of Finland: 1999.

21. Gentile, M.; Rogers, W.; Mannan, M., Development of a fuzzy logic-based inherent safety index. *Process Safety and Environmental Protection* **2003,** *81* (6), 444-456.

22. Bansal, Y.; Silakari, O., The therapeutic journey of benzimidazoles: a review. *Bioorganic & medicinal chemistry* **2012,** *20* (21), 6208-6236.

23. omeprazole. https://www.drugs.com/monograph/omeprazole.html.

24. albendazole. https://www.drugs.com/monograph/albendazole.html.

25. Patil, A.; Ganguly, S.; Surana, S., A systematic review of benzimidazole derivatives as an antiulcer agent. *Rasayan J Chem* **2008,** *1* (3), 447-460.

26. Patt, R.; Kordsachia, O.; Süttinger, R.; Ohtani, Y.; Hoesch, J.; Ehrler, P.; Eichinger, R.; Holik, H.; Hamm, U.; Rohmann, M., Ullmann's encyclopedia of industrial chemistry. Wiley: 2002.

27. Annual Production. http://www.eng-tips.com/viewthread.cfm?qid=389883.

28. Pineda-Solano, A.; Saenz-Noval, L.; Nayak, S.; Waldram, S.; Papadaki, M.; Mannan, M. S., Inherently safer reactors: Improved efficiency of 3-picoline N-oxidation in the temperature range 110–125° C. *Process Safety and Environmental Protection* **2012,** *90* (5), 404-410.

29. LaPorte, T. L.; Spangler, L.; Hamedi, M.; Lobben, P.; Chan, S. H.; Muslehiddinoglu, J.; Wang, S. S., Development of a Continuous Plug Flow Process for Preparation of a Key Intermediate for Brivanib Alaninate. *Organic Process Research & Development* **2014,** *18* (11), 1492-1502.

30. Ebrahimi, F.; Kolehmainen, E.; Turunen, I., Safety advantages of on-site microprocesses. *Organic Process Research & Development* **2009,** *13* (5), 965-969.

31. Kadla, J. F.; Chang, H., The reactions of peroxides with lignin and lignin model compounds. ACS Publications: 2001.

32. Ebrahimi, F.; Kolehmainen, E.; Oinas, P.; Hietapelto, V.; Turunen, I., Production of unstable percarboxylic acids in a microstructured reactor. *Chemical engineering journal* **2011,** *167* (2), 713-717.

478238 Risk Based Decision Making Methods for Evaluating Complex Technologies, a Comparative Analysis on Pipelines and Railroad for Crude Oil Transport

Abstract:

Decision making for long-term capital intensive projects involve large number of uncertainties. Traditional risk assessments are based on probabilities which are estimated from limited data sets and provide limited flexibility for complex technologies. Further, the environment and specific conditions under which the data is obtained may be unknown and yield to erroneous extrapolations. The method proposed in this work utilizes a novel method for estimation of the risk based on the net present value of technologies and provides the basis for a decision support tool based on the expected value theorem. This is demonstrated using the case of the safety and economic crisis in onshore crude oil transport in the continental United States. The exponential growth of crude oil production in the continental United States has led to crisis in the mid-stream industry with existing pipeline infrastructure under pressure to service the increased production. This fact is reflected in the 4000% increase in the carloads in a span of five years from 2009 to 2013. This has been compounded by the volatile fluctuation in crude oil price from $100 to $30 over the past few years. This has put enormous constraints on capital decision making. Since railroad infrastructure albeit aged, already exists in many regions, loading and unloading crude oil is relatively easy and their connection to the refining centers along the Gulf of Mexico apart from sites in the east and west coast make it an option to many producers. In this paper a novel decision making model is proposed using the concept of 'Net present value' of a technology to evaluate the option between railroad and as suitable mid-stream alternatives for the transportation of crude oil. The model is built on using a prescriptive decision making technique for the quantification of factors affecting the decision, incorporating also the utility of wealth of the decision maker. Monte-Carlo simulation is further used to obtain results and trends based on variation of initial parameters. This analysis can provide a basis for making decisions given the complex environment of crude oil transport operations and can be extended to a variety of geographies given that they satisfy the requirements of the model. The method can also be further extended to other technology evaluation given appropriate factors can be identified to build the model.

479492 Key Modeling Parameters and Process Safety Information for Effective Atmospheric Dispersion Modeling

Abstract:

Dispersion modeling is an important part of Quantitative risk assessment (QRA) documentation. Dispersion modeling can be performed using commercially available software. Variety of modeling parameters and Process Safety Information (PSI) data are needed for atmospheric dispersion modeling. Some of the parameters include weather, facility geographical location, flammability and toxicity data of the service fluid, radiation intensity, software tool, and model. Some PSI data needs include process and instrumentation diagrams (PIDs), Process Flow Diagrams (PFDs), relief system evaluation documentation, Heat and Material Balance (H&MB), and equipment data sheets. This paper describes the key parameters and their selection for effective dispersion modeling.

479373 Implementation of Hazard Mitigation Solutions: Consequence Vs. Risk-Based Building Specifications

Abstract:

Implementation of Hazard Mitigation Solutions: Consequence vs. Risk-Based Building Specifications

Travis Holland, Joshua Bruce-Black

Baker Engineering and Risk Consultants

3330 Oakwell Court, Suite 100

San Antonio, TX 78218

tholland@bakerrisk.com, jbblack@bakerrisk.com

Abstract

There are various strategies to mitigate occupant vulnerability and risk after completion of a consequence based facility siting study (FSS) or quantitative risk assessment (QRA). One potentially cost-effective mitigation strategy is to focus on reducing the occupant vulnerability and risk to personnel by upgrading existing buildings or designing new buildings. A building specification is commonly used to relay a small set of building design parameters based on the results of a consequence based FSS to engineers or contractors. However, with the industry shift towards risk based studies (QRAs), there has been a growing perspective that the information required to develop an appropriate building specification is more complicated given the increased number of variables involved in a QRA. Using the development of a new control room specification as an example, this paper will compare and contrast the steps and benefits/drawbacks involved in developing an appropriate building design basis employing a consequence and risk based approach. This paper will highlight how establishing tolerance criteria, emergency response strategy, personal protective equipment (PPE), and building occupancy early in the project definition stage promotes effective streamlined communication in the specification and reduces the differences between consequence and risk-based building specifications.

Generalization of Modeling Along-Wind Dispersion

Jessica M. Morris
Tom Spicer, Ph.D., P.E.
Ralph E. Martin Department of Chemical Engineering
University of Arkansas, Fayetteville, AR
Jmm065@uark.edu

Prepared for Presentation at
American Institute of Chemical Engineers
2017 Spring Meeting and 13th Global Congress on Process Safety
San Antonio, TX
March 26 – 29, 2017

Generalization of Modeling Along-Wind Dispersion

Jessica M. Morris
Tom Spicer, Ph.D., P.E.
Ralph E. Martin Department of Chemical Engineering
University of Arkansas, Fayetteville, AR
Jmm065@uark.edu

Keywords: along-wind, dispersion, finite duration, time-varying, atmosphere, releases, modeling

Abstract

The method to model along-wind dispersion is an important aspect of currently used commercial atmospheric dispersion models when considering time-limited or time-varying releases of toxic or flammable materials. Past wind tunnel experimental data has shown the validity of this approach. Inclusion of the effects of along-wind dispersion has proven to be critically important to modeling available field-scale data with current models for such time-varying releases. Many investigators have considered modeling along-wind dispersion from a theoretical point of view as well as basic parameterization of available field test data. This paper reviews past work and proposes a non-dimensionalized parameterization which shows improved agreement with available field-scale data.

1 Introduction

Continued world-wide industrial development has brought the wide spread use of industrial chemicals, many of which are hazardous. Whether released accidentally (or intentionally), air borne hazardous chemicals effect large populations quickly as evidenced by several industrial chemical accidents (e.g., Bhopal[1], Flixborough[2], Seveso[3]). Understanding and modeling hazardous chemical releases is vital to protecting the public. Many researchers have worked to assess these hazards using physical and computational models. Predicting the potential impact of hazardous gas releases in the atmosphere is imperative to properly determine where transport, storage, and use of such chemicals should be located. In the case of an actual loss of containment, understanding where air borne chemicals can travel is important to determine the proper response to an emergency.

While models exist to address these questions, there are opportunities to improve modeling methods particularly for (realistic) finite duration and time varying hazardous gas releases, which depend strongly on parameterizing the effects of along-wind dispersion (the elongation of the cloud in the along-wind direction). Modeling methods for along-wind dispersion are frequently based on established theoretical approaches which provide guidance for analyzing currently available field scale data. This paper considers experimental data and established theoretical methods to consider how the effects of along-wind dispersion can be improved.

2 Theoretical Description

2.1 Theoretical Models

Along-wind dispersion can be calculated under many different conditions and assumptions. There have been many established theories accounting for different conditions and assumptions. Scientists choose the along-wind dispersion coefficient as either a function of time or distance. Majority of scientists choose the along-wind dispersion coefficient as a function of the friction velocity. Some of the most prominent theories about along-wind dispersion are discussed below.

2.2 Batchelor

Batchelor (1964)[4] determined the Lagrangian Similarity Hypothesis. This hypothesis relates Eulerian and Lagrangian quantities when deriving theoretical solutions to problems. Applying known properties of fluid flow to the atmosphere, Batchelor was able to establish a generalized equations under neutral stability conditions, with no restraints on maximum height, for center line concentration at ground level and wind speed dependence with height. Batchelor derived many equations that are the foundation for modern theoretical equations commonly utilized. Batchelor's relationship between turbulence in a constant stress region on a boundary layer is a function of the friction velocity and time. This relationship and the Lagrangian Similarity Hypothesis serve as a template for along-wind dispersion equations to come.

Release height can be interpreted differently on its effect on dispersion. Restating Batchelor's similarity theory, "the statistical properties of the velocity of a marked fluid particle at time, t, after release at the ground depend only on friction velocity, u_* and t." This theory is then extended to the case of an elevated release: "The statistical properties of the velocity of a marked particle at time, t, after release at height, h, are the same as those of a particle released at the ground (z=0) at the instance $-t_1$, provided $t \gg t_1$ where t_1 is expected to be of the order of magnitude of the time-scale of the turbulence at height h, that is, of order h/u_*". Under Batchelor's similarity theory, after a sufficient amount of time, the height of release can be ignored.

2.3 Chatwin

Chatwin (1968)[5] was the first mathematician to theoretically derive a full equation with numerical values for the along-wind dispersion coefficient. He also derived the shear stress component of the along-wind dispersion equation. Chatwin used Batchelor's similarity theory along with the assumptions of the ability of turbulence to be described as an eddy diffusivity to solve for along-wind dispersion, σ_x. He makes further assumptions of neutral stability, ground release, and an unbounded boundary layer. The generalized equations for both overall along-wind dispersion, σ_x , and wind shear, σ_{xs} , by Chatwin are as followed

$$\sigma_x = 0.803\frac{u_* t}{\kappa} \quad (1)$$

$$\sigma_{xs} = 0.596\frac{u_* t}{\kappa} \quad (2)$$

where u_* is the friction velocity and κ is the van Karman's constant, approximately equal to 0.41.

2.4 Van Ulden

Van Ulden (1992)[6] derived an equation under slightly different dispersion definitions. Van Ulden made the distinction between absolute and relative dispersion. Wheatley and Chatwin were able to authenticate their equations with the assumption of a hypothetical atmosphere: where no large horizontal eddies exist. Taking into consideration this distinction of absolute and relative dispersion, van Ulden derived a new equation for σ_x, the total variance of the concentration distribution in the wind direction

$$\sigma_x^2 = \sigma_c^2 + \sigma_d^2 \quad (3)$$

where σ_c is the deterministic part (or centroid variance), and σ_d is the random part (or diffusive variance). The total longitudinal variance, σ_x, is dependent on three separate quantities: centroid variance, diffusive variance due to shear effects, and direct dispersion. These three variables were able to be solved for shear, centroid, and diffusive effects respectively as a function of friction velocity and time

$$\sigma_s = 0.5\frac{u_* t}{\kappa} \quad (4)$$

$$\sigma_c \simeq 0.54\frac{u_* t}{\kappa} \quad (5)$$

$$\sigma_d = 0.67\frac{u_* t}{\kappa}. \quad (6)$$

Under the assumption that $\sigma_c \simeq \sigma_s = 0.5\frac{u_* t}{\kappa}$, these variables can be substituted into equation (3) to get the total longitudinal variance. An exact equation for σ_x is obtained:

$$\sigma_x = 0.84\frac{u_* t}{\kappa} \quad (7)$$

Van Ulden not only derived an along-wind dispersion equation under neutral conditions from a ground release in an unbounded boundary layer, but also for an along-wind dispersion equation for very stable conditions. Under very stable conditions, the wind speed profile and eddy diffusivities change. When applying the new constraints for wind speed and eddy diffusivities in a different stability class, the total variance of concentration distribution in the wind direction is shown to be:

$$\sigma_x = \{(\frac{3}{4} - \frac{16}{9\pi})\frac{5u_*^3 t^3}{kL}\}^{1/2} \simeq 0.43(\frac{5u_*^3 t^3}{kL})^{1/2}. \quad (8)$$

where L is the Monin-Obukov length.

2.5 Summary

The relationship between advection cloud speed, time, and distance to determine the along-wind diffusion coefficient is complex. Changing conditions such as stability class, surface roughness, wind speed, function of distance or time, release height, and other parameters all have an effect on determining the along-wind diffusion coefficient. With the varying conditions, not all the equations are able to be integrated together into an overarching equation meeting all conditions. Different assumptions or criteria result in diverse equations for the same diffusion coefficient. An equation that takes into account all of these parameters is desired.

3 Correlation of Field Data by Hanna and Franzese

The 12 field and wind tunnel tests mentioned prior were analyzed by Hanna and Franzese to determine a suitable along-wind dispersion coefficient that resulted in an equation comparable to Chatwin's theoretical equation[5]**Error! Bookmark not defined.**. Results of the data and the predicted equation are shown below in Figure 1.

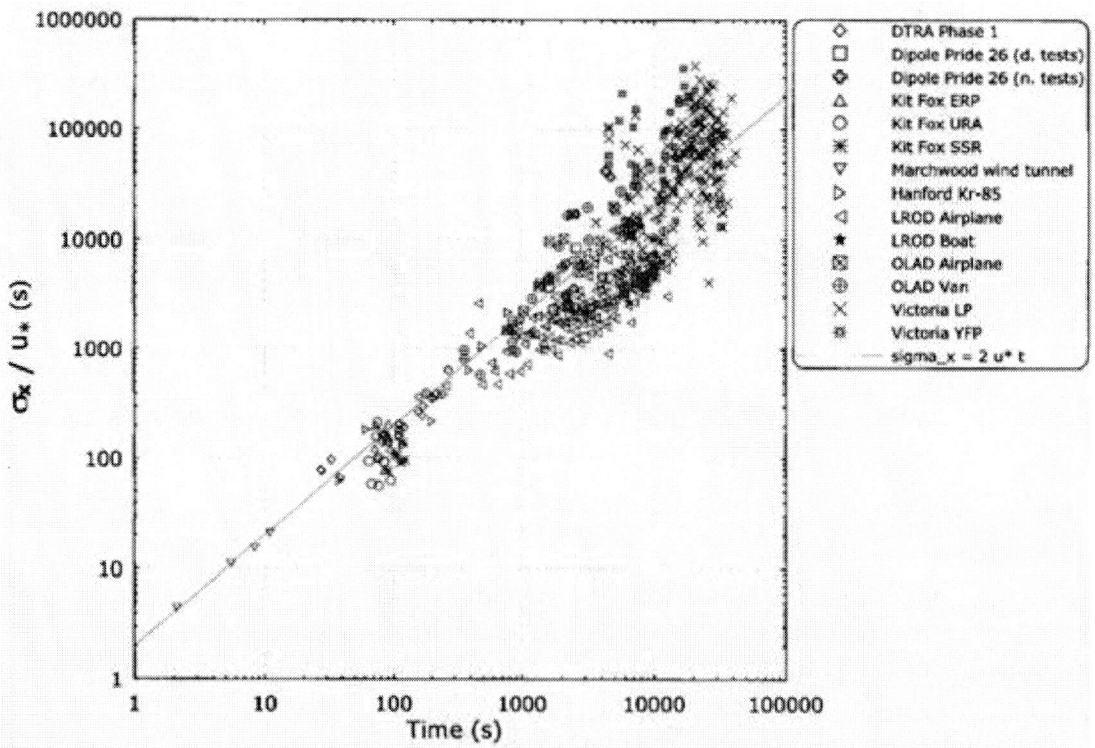

Figure 1. Hanna and Franzese results of along-wind dispersion coefficient divided by friction velocity and plotted vs time. Line represents $\sigma_x = 2u_*t$.[8]

4 Reconsideration of Field Data Using Dimensionless Groups

Figure 1 shows a correlation of the along-wind dispersion coefficient with the friction velocity as a function of time. While this is linear, their data can be up to a factor of 10 away from the predicted equation. There is also a noticeable trend of data beneath the predicted line when time is less than 10,000 seconds and above the predicted line when time is greater than 10,000 seconds. This is believed to be an effect of the near field vs. the far field. This equation for the along-wind dispersion coefficient does not take into effect the near field and far field which results in a very large portion data over predicted from 1,000-10,000 seconds and a large portion of data under predicted from 10,000 seconds and above. An improved equation to improve the correlation accounting for the error and trends is desired.

A non-dimensionalized equation was created to better predict the along-wind dispersion coefficient. The parameters of the non-dimensionalized equation were based off theory and assumptions regarding trends seen with atmospheric dispersion and along-wind dispersion. The non-dimensionalized equation allows for the along-wind dispersion coefficient to be predicted based on wind speed, distance, and stability class for each test in addition to the previous parameters of time and friction velocity.

The non-dimensionalized equation takes into account multiple theories and trends. Batchelor's Lagrangian Similarity Theory predicts σ_x to be a function of friction velocity, u_* , and time, t, while ignoring the release height[4]. Van Ulden predicts separate equations for both neutral and stable conditions[6]. Surface roughness has been shown to have an effect on wind speed[7]. The overall non-dimensionalized equation for along-wind dispersion is shown below

$$\frac{\sigma_x}{x} = \left(\frac{ut}{x}\right)^p \left(\frac{u_*}{u}\right)^q C_1 \qquad (9)$$

where x is the distance in meters and p, q, and C_1 are all dimensionless constants. The $\left(\frac{ut}{x}\right)$ term accounts for the effect of wind speed, u, time, t, and distance, x. The $\left(\frac{u_*}{u}\right)$ term is to take into effect the wind speed, u, and friction velocity, u_*, which ultimately represents the stability class.

Data was accumulated and analyzed by Stephen Hanna and Pasqual Franzese in 1999 from 11 field tests and 1 wind tunnel test[8]. The accumulation of these 12 tests produced 446 data sets over varying conditions. Other tests were accumulated and stored in the Modeler's Data Archive[9] for future reference. The following tests were analyzed in hopes of a better fit for along-wind dispersion in atmospheric modeling. Of the 446 data sets, only six data points were not used in the analysis due to lack of recorded friction velocity and/or wind speed. Reference heights for recording the wind speed from each of the data sets vary. A constant reference height of 10m was implemented across all data sets for a more standardized data analysis. Data sets used in this analysis are described below.

4.1 DTRA Phase 1

The DTRA Phase 1 tests were carried out from the 9th -26th of September in 1996 at the Dugway Proving Ground on Pad 11 producing 23 ensemble averages[10]. Propylene was instantaneous released from air cannons at ground level. Each air cannon produced 0.36kg of propylene in a period of 0.2-0.4 seconds. DigiPIDs and Ultra-Violet Ion Collectors were both used at distances of 200 m, 300 m, 400 m, 800 m, and 1200 m along a fixed sampling line 1.5m above the ground. Increasing distances resulted in an increased number of air cannons to result in an increased mass of propylene for the increased distance. 1 air cannon was used for distances of 200 m and 300 m, 2 air cannons for 400 m, 4 air cannons for 800 m, and 8 air cannons for 1200 m. Wind speed is referenced at a height of 2m which ranged from 2-7 m/s. The Great Salt Lake Desert produced a surface roughness of less than 1mm.

4.2 Dipole Pride

Dipole Pride tests were grouped into day and night tests from the 4th -20th of November in 1996 at the Department of Energy Nevada Test Site – Yucca flat. SF_6 was released from 2 vertically mounted, pressurized cylinders on the back of a 5-ton flatbed truck. These point sources released 8-21.6 kg of mass in a duration of 1-3 seconds[11]. 90 whole air samplers, with 15 minute time averaging, and 6 continuous TGA-4000 SF_6 analyzers were placed in 3 lines at distances of 1 km, 10 km and 20 km[12]. Wind speeds were referenced at a height of 10 m resulting in wind speeds of 3-5 m/s during the day and 2-5 m/s during the night. Surface roughness at the Yucca flat was around 32 mm. The 23 total tests produced 10 adequate day releases and 5 adequate night releases.

4.3 Kit Fox (ERP, URA, SSR)

All Kit Fox releases took place at the National Spill Test Facility at the Nevada Test Site. The Equivalent Roughness Pattern (ERP) and Uniform Roughness Array (URA) tests took place from the 24th -31st of August in 1995 whereas the Smooth Surface Rough (SSR) tests took place the 11th -15th of September in 1995.[13] The tests were similar in all conditions other than surface roughness. The Kit Fox ERP tests had a surface roughness of 0.2 m, the Kit Fox URA tests had a surface roughness of 0.02 m, and the Kit Fox SSR tests had a surface roughness of 2 mm[14]. Carbon Dioxide was released from a ground level area source with dimensions of 1.5 m x 1.5 m x 1 m at a rate of 1 m/s resulting in 1.5 kg/s – 4 kg/s being released[15]. Concentrations were measured at distances of 25 m, 50 m, 100 m, and 225 m. 84 high and low infrared concentration sensors, solid state infrared CO_2 sensors, and sample bag Nova sensors, were all used to detect concentrations of CO_2 with an averaging time of around 20 seconds[16]. Wind speeds were referenced at a height of 2 m which resulted in wind speeds of 2-3 m/s for the ERP and SSR test and 3-4 m/s for the URA tests. All tests took place right before sunset which resulted in lower wind speeds with neutral to stable stability classes. The ERP and SSR tests resulted in 6 sufficient trials each and the URA tests resulted in 9 sufficient trials to be used in this analysis.

4.4 Hanford Kr-85

The Hanford Fr-85[17] tests took place from 18 August to 8 November 1967 at the Battelle Pacific Northwest Laboratory. A vial containing Krypton 85 at ambient pressure was crushed by a guillotine-like device at ground level, instantaneously releasing a 1 m puff of the radioactive gas. Concentration was detected by 64 halogen-quenched Geiger-Muller tubes at a distance of 200 m and an elevation of 1.5 m above the ground. 24 halogen-quenched Geiger-Muller tubes were positioned on 6 towers at a distance of 800 m at elevations of 0.8 m, 4.6 m, 10.7 m and 21.3 m. The time averaging associated with the Geiger-Muller tubes was 4.8 seconds. 8 trials were performed, but only 6 releases remained within the boundary of the array of sensors. The wind speed was taken at a reference height of 1.5 m which resulted in wind speeds of 1-8 m/s. The roughness was around 30 mm due to the sagebrush and steppe grass.

4.5 LROD

The Long Range Overwater Dispersion (LROD)[18] tests took place from 13th -26th of July in 1993 at the Pacific Missile Range Facility in Kauai, Hawaii. A crosswind line source of pressurized, liquid SF_6 was released from a high pressure hose from the back of a plane at an elevation of 90 m at a rate of 9.6 g/meter and a speed of 129 m/s over water. A total duration of the release of SF_6 was around 13 minutes for the 100 km distance. Concentration was measured at ground level from boats along with planes at various elevations using electron capture detectors (ECD) and TGA-4000 whole air samplers. The averaging time was around 30-120 seconds for the sequential bag samples. Noise associated with the continuous analyzer resulted in limited use. Concentration was detected from the plane by flying parallel to the dissemination line and 45 km south of the sampling line (5 km, 15 km, 30 km, 60 km, and 100 km). The plane, equipped with a continuous SF_6 analyzer, would fly to the dissemination line then back to 25 m. It would slowly ascend from 25 m to 2500 m for various heights. Testing over water resulted in a surface roughness of 0.02-0.86 mm. A reference height of 10 m was used for wind speed which resulted in speeds of 3-11 m/s. The LROD tests produced 11 ensemble averages.

4.6 OLAD

The Over Land Along-wind Dispersion (OLAD)[19] tests were similar to the LROD tests, except over land. These tests took place from the 8th -25th of September in 1997 at the Dugway Proving Ground, West Desert Test Center. SF_6 was released from both plane and truck line sources. The plane released 100 kg of SF_6 at an elevation of 100 m from a N_2 pressurized tank at a rate of 33.3 kg/min for a distance of 20 km. The truck released 15 kg of SF_6 at an elevation of 3 m from a mounted tank at a rate of 1.5 kg/min for a distance of 10 km. Concentrations were recorded with TGA-4000 whole air samplers, with time averaging of 15 minutes, and 2 continuous electron capture detector (ECD) analyzers at the end of each line at distances of 2 km, 5 km, and 10 km for the truck releases and 10 km, 15 km, and 20 km for the plane releases. Wind speed was measured at a height of 2 m which resulted in speeds of 0.5 – 7 m/s. The surface roughness was measured around 30 mm. These tests resulted in 3 ensemble averages for the aircraft release and 8 surface ensemble averages for the truck release.

4.7 Marchwood wind-tunnel

The Marchwood[20] wind tunnel tests are the only tests that took place in a wind tunnel. A mixture of propane and helium, resulting in neutral buoyancy, was released from a 15 mm ground level point source. The mixture was released for a duration of 0.5 – 12 seconds. 150-300 releases were summarized into 4 ensemble averages. A single FID measured the concentration with a 5 ms sampling interval at locations of 1 m, 3 m, 5 m, and 7 m. Wind speeds ranged from 1-4 m/s at neutral stability class. Surface roughness was measured at 0.3 mm for the wind tunnel.

4.8 Oceanside

The Oceanside[21] tests took place at the shoreline of Oceanside and Del Mar in Southern California. 20 μm particles of green and yellow fluorescent pigment (FP) were released by plane, boat, and land with standardized Dugway Proving Ground equipment. For the plane releases, the DPG model D-1 released 1500 g/min of FP for a duration of 10 minutes at an elevations of 61 m, 122 m, and 152 m over a 40 km distance. For the boat releases, a DPG Mark IV was used to release 110 g/min for a duration of 2 hours at sea level over 40 km. For instantaneous land releases, the DPG Mark IX was used at three different inland locations to release 10-50 g of FP instantaneously at ground level. Concentration of FP were recorded from rotorod samplers, sequential samplers, vertical samplers, 100 m square grids, dense sampling lines, and metronics drum impact samplers. Time averaging ranged from 5 minutes at coastal regions to 15 minutes at inland conditions. Surface roughness ranged from 30-50 m due to the coastal shoreline. The wind speed reference height varied from 8-120 m which resulted in wind speeds ranging from 1-5 m/s.

4.9 Fort Wayne

Fort Wayne[22] releases took place in the Summer and Fall of 1963 in Fort Wayne, Indiana. An aerosol of fluorescent pigment (FP) was released via 2 aircrafts in the form of an elevated, 91-244 m, line source. Dry green and yellow FP were released from the planes via FP dispensers supplied by the Dugway Proving Ground. The amount of release was measured by the difference in weight, before and after. The release was upwind of the 180,000 population urban setting with a river running through the city and agriculture surrounding. Rotorods, Gelman throw-away plastic filter holders, and Gelman paper tape samplers were used for concentration measurements of the FP. The tape provided an averaging time of 5 minutes where the filter holders had an averaging time of 30 minutes. Wind speed was measured at the release height and resulted in speeds ranging from 5-16 m/s.

4.10 Short Range Diffusion Experiment Series

The Short Range Diffusion Experiment Series[23] took place in 1979 in Japan. A 200 m x 200 m vacant lot covered in turf and surrounded by trees was used for the releases. NO_x was released instantaneously from a balloon at an elevation of 1.5 m. Chemiluminescence NO_x meters were used to detect concentration with a sampling time of <2 minutes. Sample lines were chosen at distances of 60 m and 100 m. Wind speeds were referenced at heights of 0.5 m, 3 m, and 9 m which resulted in speeds of 1-2 m/s.

4.11 Results and Discussion

A multivariable regression analysis, also known as multiple regression, is an extension of linear regression[24]. A regression analysis determines the relationships among variables.

With a multivariable regression analysis, the relationships between multiple independent variables can be determined. A regression analysis is a useful tool in understanding the changes of one variable with respect to others. The form of a typical multi-variable regression analysis is shown below

$$Y = a + b_1x_1 + b_2x_2 + \cdots + b_nx_n \qquad (10)$$

A multi-variable regression analysis can be applied to the non-dimensionalized equation. The non-dimensionalized equation is rearranged with the natural log of each variable to perform the multi-variable regression analysis. This equation is shown below

$$ln\left(\frac{\sigma_x}{x}\right) = p\,ln\left(\frac{ut}{x}\right) + q\,ln\left(\frac{u_*}{u}\right) + C_2 \qquad (11)$$

where $C_2 = \ln(C_1)$. A multivariable regression analysis with known data is used with this equation to determine the constants for p, q, and C_1. An analysis is done for the overall data and categorized into stability classes using macros in Excel[25].

Multiple parameters were analyzed for optimal results using the multivariable regression analysis. Data was quantified and analyzed based on individual tests, surface roughness, friction velocity, Monin-Obukhov length, distance, time, near field, and far field.

Data from all 12 experimental data sets referenced above were analyzed with a multi-variable regression analysis to determine appropriate exponents and a constant for an optimum non-dimensionalized equation. While many parameters were originally analyzed for the optimum non-dimensionalized equation, the quantification of near field and far field as a function of time was chosen as the most prominent parameter. The optimum non-dimensionalized equation was parameterized by near field with time less than 10,000 seconds and far field with time greater than 10,000 seconds. These equations are shown below in Table 1. The ratio, α, of the along-wind dispersion coefficient, σ_x, for the data vs σ_x for the predicted non-dimensionalized equation is used for comparison of the difference in equations. A non-parametric confidence interval was determined for the non-Gaussian distribution of α.[26] Hanna and Franzese's along-wind dispersion coefficient was analyzed in the same manner in Table 1.

Table 1. Comparison of along-wind dispersion equations with 90% confidence intervals

Model	Equation	90% Confidence Interval of $\alpha = \frac{\sigma_x\ (predicted)}{\sigma_x\ (data)}$	Range of Confidence Interval
Hanna and Franzese	$\sigma_x = 2u_* t$	1.63 < α < 1.88	26%
Near Field Non-Dimensionalized	$\frac{\sigma_x}{x} = 1.72\left(\frac{ut}{x}\right)^{1.18}\left(\frac{u_*}{u}\right)^{1.10}$	1.16 < α < 1.31	16%
Far Field Non-Dimensionalized	$\frac{\sigma_x}{x} = 0.05\left(\frac{ut}{x}\right)^{0.67}\left(\frac{u_*}{u}\right)^{-0.27}$	0.85 < α < 1.05	20%

From Table 1, the range of the 90% confidence intervals for each of the equations is shown. The non-dimensionalized equations have a tighter correlation of the predicted values. The 90% confidence intervals shows that Hanna's predicted along-wind dispersion equation has a trend over predict the σ_x values. An interesting result from the multi-variable regression analysis is the determination of the coefficient for the $\left(\frac{u_*}{u}\right)$ term. The -0.27 exponential value essentially confirms that the stability class has no effect on the along-wind dispersion coefficient after a specific amount of time. A discontinuity is understood between the near field and far field non-dimensionalized equation and is currently being evaluated.

The results of the non-dimensionalized predicted σ_x values were divided by the friction velocity and plotted against the original data to determine their relationship. This relationship is shown below in Figure 2.

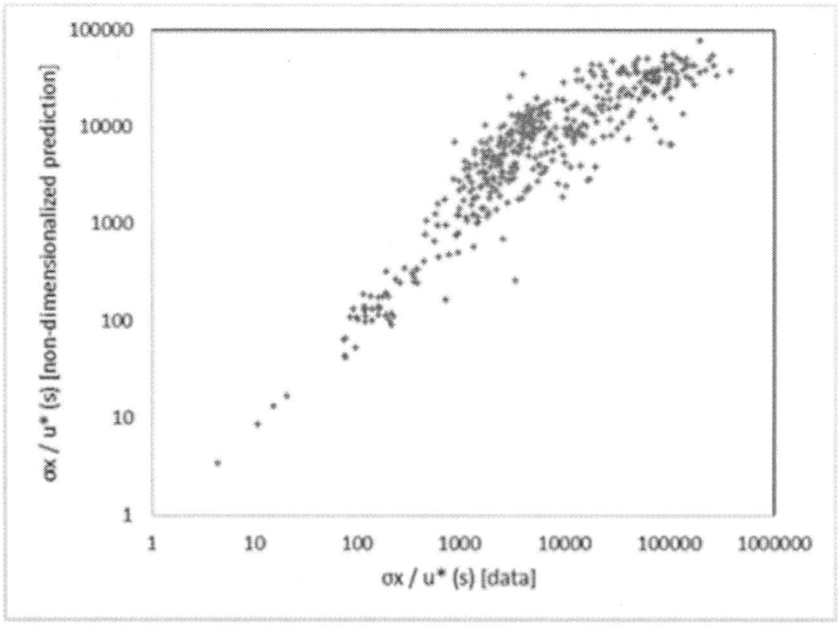

Figure 2. Along-wind dispersion coefficient divided by the friction velocity of the non-dimensionalized modeled values vs the original data.

As shown in Figure 2, the correlation between the data and non-dimensionalized predicted values shows a 45° line. This correlates a good fit for the predicted σ_x values when compared to the original σ_x values.

A visual representation of the non-dimensionalized predicted σ_x values, Hanna and Franzese's σ_x values, and the original σ_x values as a function of time are all shown below in Figure 3.

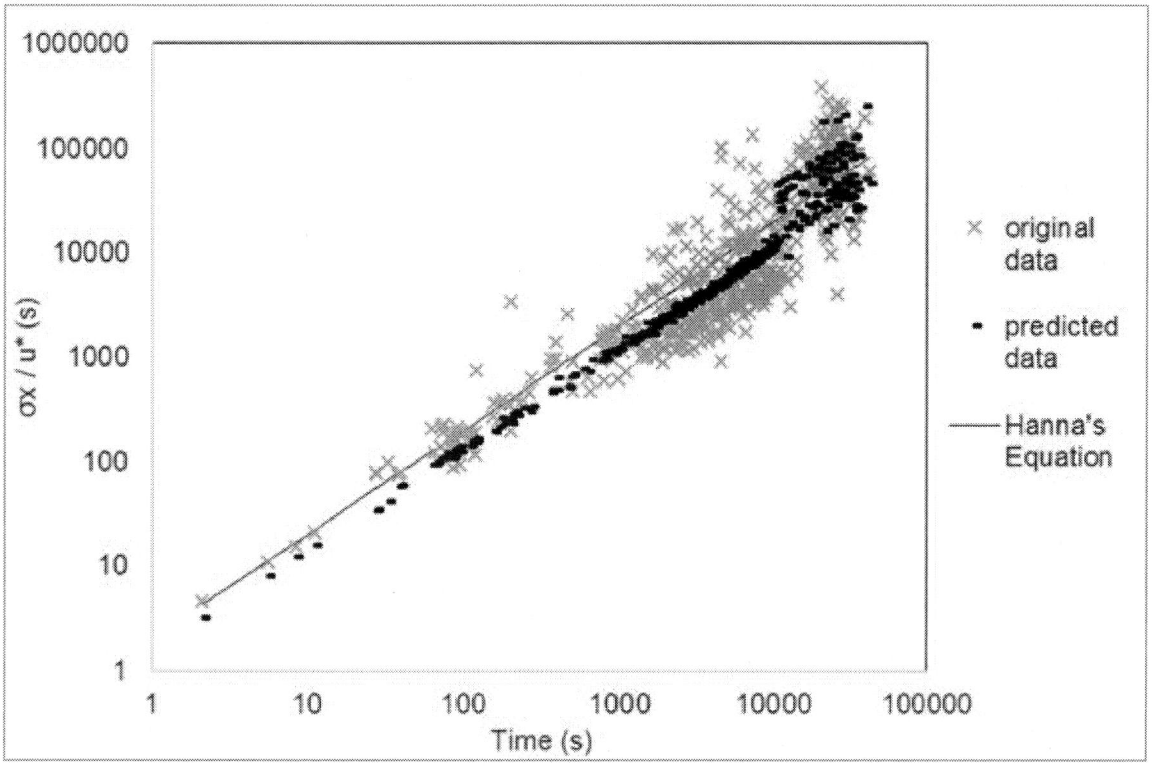

Figure 3. Time vs. the along-wind dispersion coefficient divided by the friction velocity as a function of the original data, non-dimensionalized predicted data, and Hanna's predicted data.

From this figure, the predicted values from both Hanna and Franzese's predicted data and the non-dimensionalized predicted data are shown in comparison with the original data. The non-dimensionalized equation takes into account additional parameters, other than time, which results in the spread of σ_x values as a function of time and additional parameters in comparison to the single line by the $\sigma_x = 2u_*t$ along-wind dispersion equation. The non-dimensionalized equation also takes into effect the near field and far field. This is clearly seen at times greater than 10,000 seconds where the σ_x values are spread over a wider range. Overall, the non-dimensionalized equation for along-wind dispersion proves better than the previous $\sigma_x = 2u_*t$ correlation.

5 Summary

Along-wind dispersion is critical when modelling toxic, flammable, or hazardous time-limited or finite-duration releases into the atmosphere. While many models exist from a theoretical standpoint or parameterization of experimental data, there is no overarching correlation to predict the along-wind dispersion coefficient. This paper took into effect established theoretical equations validated with experimental data to determine a better equation to model the effects of along-wind dispersion. A successful non-dimensionalized equation parameterized by the near field and far field was determined and validated against available field-scale data which showed improvement in comparison to previous equations. The non-dimensionalized equation for along-wind dispersion solved for in this paper will improve the existing atmospheric modeling with along-wind dispersion.

6 Acknowledgments

We gratefully thank our colleagues from DNV GL, Henk Witlox and Maria Fernandez, who provided insight and expertise that greatly assisted the research. We also thank Steven R. Hanna and Pasquale Franzese for sharing their accumulated data for further analysis.

7 References

[1] Boybeyi, Zafer, Sethu Raman, and Paolo Zannetti. "Numerical investigation of possible role of local meteorology in Bhopal gas accident." *Atmospheric Environment* 29.4 (1995): 479-496.
[2] Høiset, S., et al. "Flixborough revisited—an explosion simulation approach." *Journal of hazardous materials* 77.1 (2000): 1-9.
[3] De Marchi, Bruna, Silvio Funtowicz, and Jerome Ravetz. "Seveso: A paradoxical classic disaster." *The long road to recovery: Community responses to industrial disaster* (1996): 86-120.
[4] Batchelor, G. K. "Diffusion from sources in a turbulent boundary layer." *Arch. Mech. Stosow* 3.6 (1964): 661-670.
[5] Chatwin, P. C. "The dispersion of a puff of passive contaminant in the constant stress region." Quarterly Journal of the Royal Meteorological Society 94.401 (1968): 350-360.
[6] Van Ulden, A. P. "A surface-layer similarity model for the dispersion of a skewed passive puff near the ground." *Atmospheric Environment. Part A. General Topics* 26.4 (1992): 681-692.
[7] DNV GL. DNV GL Theory Manual for Unified Dispersion Model. London UK 2015.
[8] Hanna, Steven R., and Pasquale Franzese. "Alongwind dispersion—a simple similarity formula compared with observations at 11 field sites and in one wind tunnel." *Journal of Applied Meteorology* 39.10 (2000): 1700-1714.
[9] Chang, Joseph C., and Steven R. Hanna. "Modelers' Data Archive – A Collection of Atmospheric Transport and Dispersion Data Sets." (2010).
[10] Biltoft, C., Phase I of DSWA transport and dispersion model validation study. DPG-FR-97-058, (1997): 156.

[11] Biltoft, C., Dipole Pride 26: Phase II of DSWA transport and dispersion model validation study. DPG-FR-98-001, (1998): 130.
[12] Watson, T. B, R. E. Keislar, B. Reese, D. H. George, and C. A. Biltoft, The DSWA Dipole Pride 26 field experiment. NOAA Tech. Memo. ERL ARL-225, (1998): 99.
[13] King, S. Bruce. The Kit Fox Field Demonstration Project and Data Set. *International Conference and Workshop on Modeling the Consequences of Accidental Releases of Hazardous Materials.* (1999).
[14] Spicer, T. and Havens, J. Description and Analysis of Atmospheric Dispersion Tests Conducted by EPA. *International Conference and Workshop on Modeling the Consequences of Accidental Releases of Hazardous Materials.* (1999).
[15] Sheesley, D. et al.. Chemical Dispersion Modeling Project, 1995 Kit Fox Field Demonstration DOE Nevada Hazardous Spill Center. Volume I – Experiment Description and Data Processing. (1998).
[16] Hanna, S. R., and K. W. Steinberg, Studies of dense gas dispersion from short-duration transient releases over rough surfaces during stable conditions. *Air Pollution Modeling and Its Applications XI,* Plenum Press, (1996): 481–490.
[17] Nickola, P. W., J. D. Ludwick, and J. V. Ramsdell Jr., An inert gas tracer system for monitoring the real-time history of a diffusing plume or puff. *J. Appl. Meteor.*,9, (1970): 621–626.
[18] Bowers, J. F., G. E. Start, R. G. Carter, T. B. Watson, K. L. Clawson, and T. L. Crawford, Experimental design and results for the Long-Range Overwater Diffusion (LROD) Experiment. DPG/JCP-94/012, (1994): 531.
[19] Biltoft, C., S. D. Turley, T. B. Watson, G. H. Crescenti, and R. G. Carter, Over-Land Atmospheric Dispersion (OLAD) test summary and analysis. WDTC Doc. WDTC/JCP-99/048, (1999): 38.
[20] Robins, A. G., and J. E. Fackrell, An experimental study of the dispersion of short-duration emissions in a turbulent boundary layer. *Air Pollution VI,* C. A. Brebbia, C. F. Ratto, and H. Powers, Eds., WIT Press, (1998): 697–707.
[21] Smith, and B. L. Niemann, Shoreline diffusion program, Oceanside, CA. Vols. I and II. MRI-64-FR-860, (1969): 230 and 456.
[22] Hilst, G. R., and N. E. Bowne, A study of the diffusion of aerosols released from aerial line sources upwind of an urban complex. Vols. I and II. Rep. DA-42-007-AMC-38®, (1966): 231 and 583.
[23] Sato, J., F. Kimura, and T. Yoshikawa, The longitudinal spread of puff in the short range diffusion experiment. *Pap. Meteor. Geophys.*,32, (1981): 155–162.
[24] Draper, Norman Richard, Harry Smith, and Elizabeth Pownell. *Applied regression analysis.* Vol. 3. New York: Wiley, (1966).
[25] Davis, Richard A. "Practical Numerical Methods for Chemical Engineers: Using Excel with VBA". Lexington, KY, (2015).
[26] Conver, W. J. "Practical Nonparametric Statistics". Wiley, New York, (1971).

478500 Review and Analyze the Effects of Parameters for H_2s Gas Mixture Dispersion Models: Degadis and CFD Models

Abstract:

Due to the increasing demands of energy, natural gas explorers and developers are interested in exploring sour gas fields, which have a high concentration of hydrogen sulfide (H_2S). Because H_2S is highly corrosive and could reduce the heating values of natural gas, H_2S needs to be removed from natural gas before transfer. Then the removed H_2S gas will be converted to sulfide as the products in the Claus process. The H_2S mixture (*e.g.,* H_2S and CO_2) may cause corrosion to pipelines and thus, could lead to H_2S gas mixture release.

H_2S mixture behaves as a heavy gas, which travels down to the ground causing harm to the local population and the environment. In this paper, the DEnse GAs DISpersion (DEGADIS) model and the Computational Fluid Dynamics (CFD) model (*e.g.,* ANSYS Fluent) are implemented for the H_2S mixture simulation, and assumptions in the models are also studied. The effects of topography and meteorological parameters such as atmospheric temperature, wind speed, and humidity are investigated. Finally, the simulation results from the DEGADIS model and the CFD model for flat terrain with constant wind velocity are compared.

The simulation results show that the DEGADIS model, which is suitable for flat terrain and constant wind conditions, can be applied to simple terrain geography and stable atmospheric conditions. Whereas the CFD model, which considers unsteady wind speed and turbulence near the obstacles, can be utilized for complex topography. The series of simulation results can provide guidance for the risk assessment of H_2S mixture releases under different topographies and meteorological conditions.

478998 Considerations When Justifying the Purchase of an Operator Training Simulator

Abstract:

Highly skilled plant operators ensure higher levels of plant safety, reliability and profitability. An Operator Training Simulator accelerates both the time to competence and raises the competency level of an operator. The ARC Advisory Group recently surveyed companies who have invested in Operator Training Simulators (OTS). Two critical results from the survey are that over sixty five percent of those companies report fewer abnormal situations, fewer trips, and incidence avoidance attributable to the use of OTS. Over sixty percent of these companies report improved safety and reduced risk and lower liability attributable to the use of the OTS.

Many leading companies have embraced the use of Operator Training Simulators across their Enterprise however many facilities and companies have yet to embrace the idea. The most common reason given for not investing in such a solution is that the cost of an OTS cannot be justified. The nuclear industry mandates that a nuclear power plant must have an OTS whereas the American Petroleum Institute only recommends the use of Training Simulators. The Process industries overall will benefit by the greater adoption of training simulators and this paper discusses the different ways that can be used to justify the investment in an OTS.

479500 Consequence Analysis of Gas Explosion in Underground Common Utility Tunnels

Abstract:

The purpose of this study is to figure out the effects of blast overpressure in gas piping leakage in underground common utility tunnels through the explosion simulation tool FLACS (Flame Acceleration Simulator) and find and apply methods that can reduce the risk.

Common utility tunnels are facilities to jointly bury pipes in underground spaces. Placing gas piping in common utility tunnels has the advantages of reducing the risk of corrosion and preventing the physical damage due to civil engineering works and excavating work on the ground. However, when an explosion has occurred in a common utility tunnel, the consequences could be larger in that it is a closed space. Nevertheless, studies on underground common utility tunnels are currently limited to fire accidents. Therefore, risks due to explosions should be quantitatively studied.

Therefore, in this study, the effects of explosions were analyzed assuming cases where gases leaked from gas piping in underground common utility tunnels exploded. In addition, blast overpressures were evaluated for diverse scenarios regarding ignition sources, leakage locations, and discharge rates and the results were statistically derived. Blast overpressures high enough to cause dangerous situations of the underground common utility tunnels were observed. Methods that can reduce the risk, that is, methods that can reduce the frequency of explosions, and methods for designs of internal structures of common utility tunnels that can mitigate the consequences will be presented.

479158 Air Cooler Explosion Analysis Using CFD

Abstract:

Air coolers or fin-fans are located in most process plants. They cool process fluids using forced air flow over radiators containing the fluid. Air coolers initially appear to offer many features that would result in a significant explosion source should a flammable cloud reach the cooler. The radiator coils are closely spaced pipes which will accelerate flames. The metal box represents confinement and the turbulent air flow produced by the fans will accelerate flames similar to how congestion produces turbulence ahead of a flame front. Also, the flow of air can draw rich flammables near the ground up into the air cooler.

This study examines the severity of an air cooler explosion using computational fluid dynamics (CFD). The commercial software FLACS is used to examine the parameters mentioned above (congestion, confinement, and turbulence) and their effect on explosion severity. Comparisons between FLACS and the Multi-Energy Method are made and recommendations provided for assessing air coolers in non-CFD facility siting studies.

13TH GLOBAL CONGRESS ON PROCESS SAFETY

Revisions to the QMEFS Vapor Cloud Explosion Model

Jeffrey D. Marx and Benjamin R. Ishii
Quest Consultants Inc.®
908 26th Avenue N.W.
Norman, OK 73069
jdm@questconsult.com / bri@questconsult.com

Prepared for Presentation at
American Institute of Chemical Engineers
2017 Spring Meeting and 13th Global Congress on Process Safety
San Antonio, TX
March 26 – 29, 2017

Revisions to the QMEFS Vapor Cloud Explosion Model

Jeffrey D. Marx and Benjamin R. Ishii
Quest Consultants Inc.®
908 26th Avenue N.W.
Norman, OK 73069
jdm@questconsult.com / bri@questconsult.com

Keywords: vapor cloud explosion (VCE), Baker-Strehlow-Tang (BST) model, estimating flame speeds

Abstract

The Quest model for estimating flame speeds (QMEFS) was introduced in 2007, has since been published, and has been used by Quest and other companies to provide a detailed method for vapor cloud explosion (VCE) predictions. QMEFS provides a set of parameters that predict a flame speed, which then is used with the Baker-Strehlow-Tang (BST) blast curves to predict overpressure and impulse. The original intent of the QMEFS model was to provide better characterization of confined or congested areas, as well as fuel reactivities, which fit between the low, medium, and high categories offered by the BST model, so that the large differences in predictions between categories could be eliminated. In this way it provided a more refined system for predicting the consequences of a VCE, while remaining anchored within the flame speeds provided by the BST matrix.

In the years since QMEFS' publication, Baker has added to the BST flame speed matrix. In addition, use of the model has highlighted some of the areas where the model could be improved. This paper shows several ways in which the model has been enhanced to encompass a wider range of conditions than are covered by the BST model.

1 Introduction

Any release of a flammable fluid in a petrochemical facility has the potential to generate a flammable vapor cloud that, if ignited, could produce a vapor cloud explosion (VCE). Most VCEs that generate damaging levels of overpressure originate in what is called a potential explosion site (PES), where confinement and congestion contribute to flame acceleration and the generation of overpressure. When flammable vapors enter the PES and are ignited, the possibility of human injury/death, asset damage, or event escalation becomes a concern. The concern for human injury or death is most often addressed in the form of a building siting study. Because people are somewhat less likely to be injured or killed when outside, as compared to when inside a building, the siting study focuses on the potential impacts to buildings within and around petrochemical facilities. It then becomes the task of process safety professionals to estimate the potential for VCE events, and their resulting overpressure impacts on buildings.

The Baker-Strehlow-Tang (BST) model and the TNO Multi-energy (ME) model use blast curves to estimate the overpressure impacts generated by an exploding vapor cloud. Both approaches require the user to estimate the strength of the explosion as a function of the reactivity of the flammable material and the degree of confinement or congestion present in the cloud. This information is then used to determine which strength curve (in the case of the TNO model) or flame speed curve (in the case of the BST model) is used to calculate the overpressure of the explosion as a function of distance from the center of the explosion. It is not clear, however, how best to choose the correct curve when modeling an explosion.

There have been several papers published regarding blast curve selection for the ME model. These methods are generally presented as a qualitative ruleset which acts as an index into which of the 10 curves should be used. The BST model provides a simple set of guidelines that result in a selection of a flame speed, which corresponds to an overpressure vs. distance relationship. Three parameters are used to determine the flame speed: reactivity of the flammable gas, the degree of confinement of the flammable cloud, and the degree of obstruction due to obstacles within the flammable cloud. These parameters are represented by a set of what are effectively low, medium, and high choices for each parameter.

While the BST model provides reasonable guidance on choosing the explosion parameters, there are still areas that could be improved. One problem is that the BST flame speed matrix includes large jumps in flame speed between categories of confinement, reactivity, or congestion. This was first addressed by Baker in a 1998 paper [1] where a new confinement category, 2½-D, was added for those situations that were more confined than the 3-D case, but less so than the 2-D case. The flame speeds used for the 2½-D confinement were simply the arithmetic average of the flame speeds for the 2-D and 3-D cases for a given reactivity and congestion class. Even with this extension, large discontinuities remained among the categories for flame speed according to the prescribed methodology. These discontinuities drive the need for a new method to determine flame speed based upon quantifiable properties of the flammable gas and its surroundings that varies smoothly across the range of conditions that are found in actual process plants.

The creation of the QMEFS model [2] improved upon the basic BST scheme by combining it with the GAMES correlations [3] and providing a numerical resolution to the BST matrix. Integration of the GAMES model allows the strength of the explosion to be defined by numeric input parameters and avoids the low/medium/high categorizations.

2 Parameters Affecting Flame Speed

With the incorporation of the GAMES correlations, the resulting explosion flame speed is characterized by five parameters:

- Laminar burning velocity (LBV) is used as the characteristic property for fuel reactivity

- Volume blockage ratio (VBR) and average obstacle diameter define the level of congestion
- Confinement is characterized as the number of confining planes, or walls, that are available to confine the expanding gases.
- Flame run-up distance is used to characterize the flame acceleration path

The last part of explosion characterization is the explosive energy of the fuel/air mixture, which is defined by the gas concentration, the combustion energy of the fuel, and the volume occupied by the fuel/air mixture.

2.1 Reactivity

Fuel reactivity is a measure of the propensity of the flame front to accelerate and create overpressures or potentially undergo a deflagration-to-detonation transition (DDT). The BST model originally classified reactivity as high, medium, or low based on the fuel in the fuel-air mixture. The original rule set [1] defined propylene oxide, ethylene oxide, and hydrogen (or materials with a 0.8 m/s LBV or larger) as high reactivity; methane and carbon monoxide were classified as low reactivity, and everything else as medium reactivity. According to the current BST model, materials having a laminar flame speed greater than 0.75 m/s are considered high reactivity (to accommodate ethylene being high reactivity) while those having a laminar flame speed below 0.4 m/s are considered low reactivity [1], although some BST publications list the low boundary as 0.45 m/s. In 2016, the BST categories were expanded to include a very-low category for fuels like ammonia, which has an LBV in the range of 0.07 m/s to 0.15 m/s [4].

In the experimental studies that form the basis of the BST model, ammonia, methane, propane, and ethylene were used to define the very low, low, medium, and high reactivity categories. These categories produce confusion due to how they create boundaries between the BST categories. For example, the LBV of ethylene, depending on the cited reference, ranges between 0.64 and 0.83 m/s. With the current definition of 0.75 m/s as the high reactivity boundary, this makes ethylene a "borderline" fuel – one that may either be defined as medium or high reactivity. However, ethylene is explicitly categorized as a high reactivity material, as their experimental work used ethylene as the high reactivity material [5]. This then begs the question: what about ethylene oxide (LBV 1.0 m/s) or acetylene (LBV 1.6 m/s)? These two materials are clearly more reactive than ethylene, but are still categorized as high reactivity. The BST model effectively says that an ethylene explosion will be as severe as an acetylene explosion, when all other parameters are held constant.

2.2 Congestion

In the original BST model, low obstacle density was defined as having an area blockage ratio (ABR) of less than 10%, while high obstacle density was defined as an ABR of 40% or greater, and everything in between is considered medium [6]. Later BST guidelines have used the volume blockage ratio (VBR) as the parameter for classifying obstruction [5]. In practice both blockage ratios will affect how fast a flame accelerates, but, for a uniform obstacle field, the two are related simply by the pitch-to-diameter ratio of the obstacles. If the obstacles included in an ABR are not repeated quasi-uniformly throughout the obstacle field, then their effect would be more accurately portrayed by VBR.

The classifications of low, medium, and high are often too coarse for proper assignment of congestion to a particular PES. In the experimental programs that helped to define the BST model, VBRs of 1.5%, 4.3%, and 5.7% were used to represent the low, medium, and high categories. As with other VCE parameters, the differences between these categories leaves significant room for variation between the categories. In addition, no definition of 0% VBR is provided, nor is any relationship to VBRs higher than 5.7%.

2.3 Confinement

The effect of the confinement of the flammable cloud is taken into account by determining the number of dimensions in which the burning gas may expand. 3-D expansion allows the burning cloud to expand freely in all directions and results in the slowest flame acceleration and lowest overpressures. 2-D expansion, such as a flame between two flat plates, generates higher overpressures because the combustion gases have fewer directions in which to expand resulting is a higher flame acceleration.

Confinement is included in the BST model by identifying the number of dimensions that are available to the products of combustion for expansion. Expansion into free space (except for the plane of the ground) is considered 3-D expansion, expansion between two parallel planes is considered 2-D expansion, and expansion in a pipe is considered 1-D expansion. The 1-D case was removed in the more recent publications discussing the BST model [5]. To handle the case of a frangible or partially-confining plane, such as a very closely spaced pipe rack, the BST model added a 2.5-D classification which simply averaged the 2-D and 3-D flame speed results [5].

While the classification is fairly straightforward, especially after the 2.5-D category was added, some situations are still difficult to define. These include partial walls, pieces of equipment that may form more confinement than congestion, or surfaces that are porous such as grated decks.

2.4 Other Factors

Several researchers [5, 7, 8] have acknowledged that the overall dimensions of the flammable vapor cloud before ignition directly affect the final flame speed and consequent overpressures of the vapor cloud explosion. Since a flame will accelerate until it reaches a maximum sustainable value (or undergoes a deflagration to detonation transition, a DDT), the maximum dimension of a flammable cloud is expected to be an important variable in the creation of overpressure. As a point of reference, the length available for flame acceleration in the more recent BST tests was 15 meters [5].

In addition, research [7, 8] suggests that the scale of obstacles within the congested area also affects the maximum flame speed that may be achieved. As a flame front moves around and past obstacles, the turbulence that is generated can increase the flame surface area and accelerate combustion. The characteristic obstacle diameter for the BST tests was 2 inches (0.0508 meters); other test series (including the MERGE/EMERGE tests, varied the average obstacle diameter).

2.5 *Parameters Comparison*

The BST parameters discussed above are listed in Table 1, along with the corresponding numerical values that are equivalent to those applied in the QMEFS model. When the numerical values in Table 1 are applied in the QMEFS model, the results will match the predictions of the BST model.

Table 1. BST Flame Speed Parameters

BST Parameter	BST Category	Model Parameter	Corresponding Parameter Value
Reactivity	Very-Low (e.g., Ammonia)	S_l	0.08 m/s
	Low (e.g., Methane)		0.37 m/s
	Medium (e.g., Propane)		0.43 m/s
	High (e.g., Ethylene)		0.75 m/s
Obstacle Density	Low	*VBR*	0.015
	Medium		0.043
	High		0.057
Expansion	3-D	*nPlanes*	1
	2.5-D		1.5
	2-D		2

3 Model for Estimating Flame Speed

The QMEFS model was created to overcome some of the limitations described above. By incorporation of the GAMES correlations and the overpressure/impulse curves from the BST method, a model that provides the capability for more detailed descriptions of explosion scenarios has been created. Thus, QMEFS incorporates the modeling methodologies and experimental data sets that have been used in both the BST model and in the GAMES project.

3.1 *Fundamental Correlations*

The fundamental correlation is based on correlations developed from the GAMES project [3]. These correlations related overpressure to parameters based on the MERGE and EMERGE experiments [7,8]. These correlations can be combined with acoustic theory to relate flame speed to explosion parameters [2] to produce the equations below.

For no confining surfaces other than the ground (3D expansion):

$$2.4 * \frac{M_f^2}{1+M_f} * p_0 = 0.84 * \left(\frac{VBR*L_f}{D}\right)^{2.75} * S_l^{2.7} * D^{0.7} \quad (1)$$

For confinement between two flat planes (2D expansion):

$$2.4 * \frac{M_f^2}{1+M_f} * p_0 = 3.38 * \left(\frac{VBR*L_f}{D}\right)^{2.25} * S_l^{2.7} * D^{0.7} \quad (2)$$

where:

- VBR = volume blockage ratio
- L_f = maximum distance a flame may propagate in the obstructed region (i.e., the run-up distance), m
- D = average obstacle diameter, m
- S_l = laminar burning velocity (LBV) of the flammable gas, m/s
- p_0 = the ambient pressure, bara
- M_f = the flame speed relative to a fixed observer, expressed as a Mach number

To correct the flame speed to account for confinement in spaces other than one or two planes, a curve fit of the ratio of burned cloud to unburned cloud radii versus the number of confining planes was made. A curve fit was generated with an expansion ratio (α) equal to 7, a typical expansion ratio for common hydrocarbon fuels in air. This is also the value assumed in the derivation of portions of the BST model [9]. The equation resulting from this curve fit is:

$$NPF = \frac{1}{0.5035 - 0.0757 \cdot nPlanes} - 1.338 \quad (3)$$

where:

- NPF = flame speed correction factor
- $nPlanes$ = number of confining planes

For true 3-D spherical expansion, *nPlanes* = 0. For a typical explosion at grade level, *nPlanes* = 1. For 2-D expansion, such as a flame propagating between a floor and a ceiling, *nPlanes* = 2. For 1-D expansion, such as a flame front in a pipe, *nPlanes* = 4. Planes that are not completely solid or rigid may be accounted for using a fraction of a plane.

The flame speed correction factor, *NPF*, does not fully account for increases in flame speed as the number of confining planes increases. This is accomplished by applying the following factor to the flame speed when *nPlanes* is equal to 2:

$$\beta = \frac{M_{f,2D}}{M_{f,3D} \cdot NPF(nPlanes = 2)} \quad (4)$$

This can be implemented for nPlanes =1 to nPlanes = 4:

$nPlanes = 1$

$$M_f = M_{f\,3D} * NPF \tag{5}$$

$1 > nPlanes < 2$

$$M_f = M_{f\,3D} * NPF * \beta \tag{6}$$

$nPlanes >= 2$

$$M_f = M_{f\,3D} * NPF * [1 + (\beta - 1) * (nPlanes - 1)] \tag{7}$$

3.2 *Flame Speed Limits*

The equations above are implemented in the QMEFS model to define the flame speed (M_f) and then coupled with the latest BST flame speed matrix [4]. One limitation to the GAMES correlations is that they can predict unrealistic flame speeds. This is a direct result of the correlations fitting the data within the scope of the MERGE and EMERGE tests, and not being designed to extend beyond that scope. By coupling with the BST model, this problem is partially resolved because the BST matrix provides a set of upper limits to flame speeds (and by extension, the explosion strength).

Upon examination, it can be seen the GAMES correlations will predict higher and higher flame speed values as parameters such as L_f or S_l, increase. In reality, for a given obstacle configuration, geometry, and flammable gas mixture, there is a limit to the flame speed that can be achieved. This flame speed may be either subsonic or the cloud may undergo a deflagration-to-detonation transition (DDT) in which the flame front becomes a detonation propagating at roughly the Chapman-Jouguet (C-J) detonation velocity for the flammable mixture. For most common flammable hydrocarbons in air, this speed is roughly 1800 m/s or M_f= 5.2. For purposes of this model, any flame that accelerates to a calculated velocity greater than M_f = 3.0 is considered to have undergone DDT [10, 11] and the flame speed is set to the C-J velocity, M_f= 5.2.

The flame speeds suggested by the latest BST model are applied as the upper limit flame speeds to be used in this model, as they have been "scaled up" to account for industrial scale flammable clouds [5] and account for DDTs.

The result of these correlations is a model that estimates the flame speed based on *VBR*, *D*, S_l, L_f and the number of confining planes. The resulting flame speed is compared to the published BST model, which provides a matrix of flame speeds as a function of reactivity, obstacle density, and flame expansion. The BST matrix values are used as maximum flame speed values for the set of parameters given in Table 1. Values between the BST matrix elements are calculated by linear interpolation (as was done for Baker's 2½-D flame expansion category). This methodology provides an extended, systematic approach for estimating flame speeds resulting from the combustion of a flammable cloud in an obstructed and/or confined region. Predicted flame speeds are used with the existing BST blast curves to produce estimates of overpressure at a distance from the explosion source.

3.3 Expansion of the BST Matrix

The range of the QMEFS model is limited by the upper and lower bounds of the BST matrix elements for congestion, confinement, and fuel reactivity. Through application of this model to various petrochemical projects, the need to expand the matrix became obvious. As described above, the model is built to provide predictions for the *nPlanes* (confinement) parameter in the range of 1.0 to 4.0 (the model is not validated nor trusted above this value). Values of average obstacle diameter and flame run-up distance are built into the GAMES correlations, so naturally are accounted for within reasonable limits. This leaves the congestion (VBR) and fuel reactivity parameters as places to potentially expand the model.

The intent of model expansion is to widen the range of congestion and reactivity inputs so that it removes questions about its use in the "below low" and "above high" regions. With proper expansion of the model described above, the predicted flame speed following ignition of a flammable cloud may then be used with the BST blast curves to estimate the overpressure impacts in the area surrounding a vapor cloud explosion.

To accomplish this expansion, additional categories were added to the BST matrix for congestion and reactivity. This follows Baker's own expansion into the **very-low reactivity** category when ammonia explosion experimental results were published [4].

For a **very-high reactivity** category, hydrogen was assumed as the characteristic fuel, and represents a true upper bound for LBV. Expansion to the very-high category, for parts of the matrix with high flame speeds is simple because DDTs are expected in these conditions already. For other areas, published data was required for the basis of the expansion.

In the case of the **very-low congestion** category, open air (3-D flame expansion) explosion tests were performed by Harris and Wickens [12] as well as Sato and Iwabuchi [13] for a variety of materials. The **very-high congestion** category, several of the BFETS series of tests were used [14] due to their use of a VBR in the 10% range.

The expanded matrix applied in the new QMEFS model is presented in Table 2. The latest BST flame speed matrix values are outlined by bold lines in Table 2. Values outside these areas have been added as part of the QMEFS expansion.

While a 1-D confinement is not expected to be applied in large-scale industrial scenarios, the *nPlanes* = 5 (1-D) confinement category is included in this model to provide an interpolation point beyond *nPlanes* = 2 (2-D flame expansion).

Table 2. Expanded Flame Speed Matrix

Confinement	Reactivity	Congestion				
		Very-Low	Low	Medium	High	Very-High
1D	Very-High	DDT	DDT	DDT	DDT	DDT
	High	DDT	DDT	DDT	DDT	DDT
	Medium	1.029	1.029	1.765	2.265	DDT
	Low	0.5	0.5 *	1.029	2 *	DDT
	Very-Low	0.023	0.023	0.124	0.168	0.303
2D	Very-High	DDT	DDT	DDT	DDT	DDT
	High	0.588	0.588	DDT	DDT	DDT
	Medium	0.47	0.47	0.66	1.6	DDT
	Low	0.079	0.079	0.47	0.66	1.244
	Very-Low	0.008	0.008	0.049	0.069	0.130
3D	Very-High	0.5	DDT	DDT	DDT	DDT
	High	0.06	0.36	DDT	DDT	DDT
	Medium	0.03	0.11	0.44	0.5	1.084
	Low	0.025	0.026	0.23	0.34	0.678
	Very-Low	0.003	0.003	0.024	0.036	0.073

* Values adjusted from BST publication for model consistency

4 Application

It is helpful to test the QMEFS model outside of the published BST matrix, in the areas where the QMEFS matrix has been expanded. Large scale hydrogen experiments were performed in a hemispherical test rig with no congestion. The test volume was 300 m^3 with a base of 10.4 meters diameter [13] and center ignition. The tests were modeled with the latest BST model and the expanded BST model. The model parameters listed in Table 3 were applied to these calculations. Figure 1 presents the overpressure in relation to distance from center of the fuel-air mixture, for the test results and both BST and QMEFS predictions. The BST model does not have a category for a hydrogen vapor cloud with no congestion, thus the scenario falls into the high fuel reactivity and low congestion portion of the matrix. The result is an underprediction of the overpressure. The BST high fuel reactivity works well for fuel such as ethylene (LBV = 0.76 m/s), but will often underpredict the impacts of a hydrogen (LBV = 3 m/s) explosion. The expanded matrix allows a higher fuel reactivity category and a lower congestion category, and consequently a better prediction of the explosion impacts. As seen in Figure 1, the QMEFS result is a slight overprediction of the overpressure.

Table 3. Modeling Parameters for an Unconfined Hydrogen Explosion

Parameter	BST	Expanded QMEFS
Volume [m^3]	300	300
nPlanes	1	1
Reactivity	High	Very-High
VBR	N/A	0
Congestion	low	N/A
Run-up distance [m]	N/A	5.2

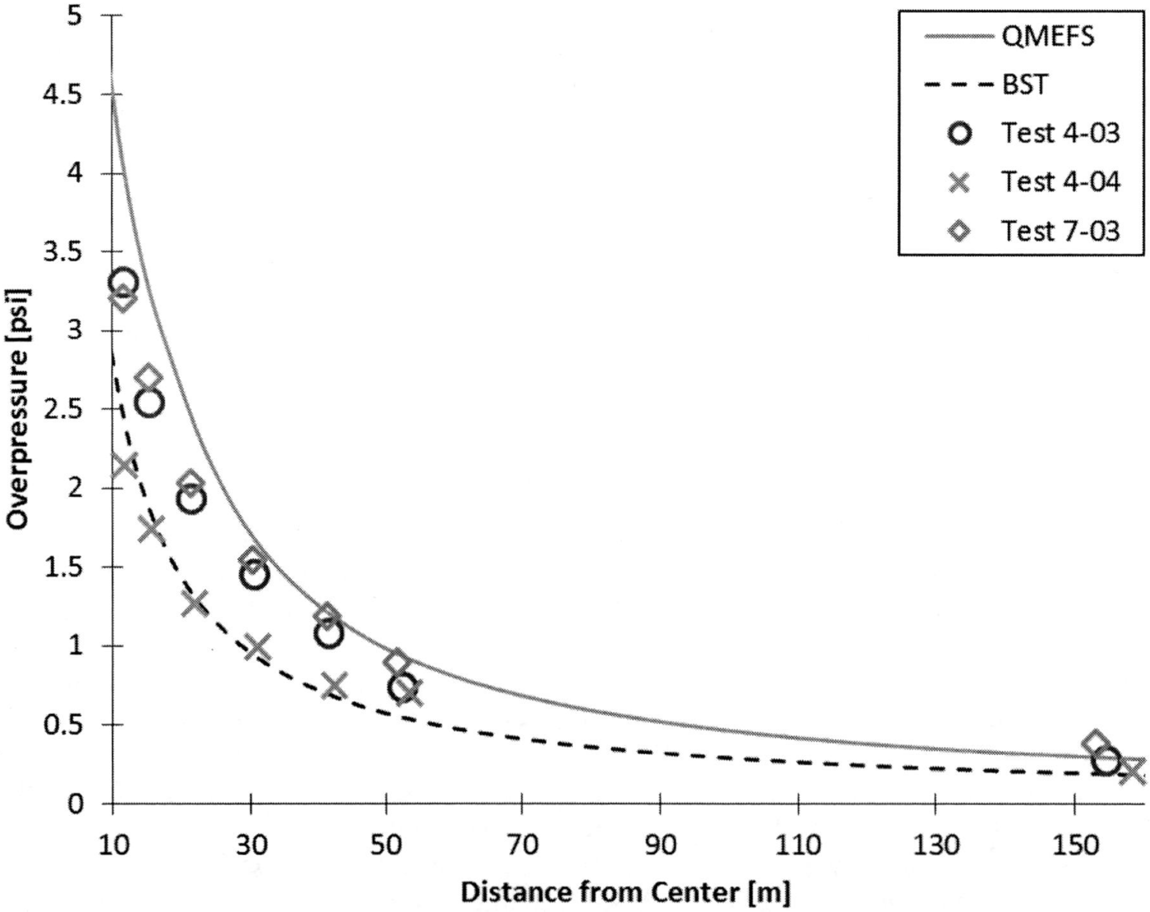

Figure 1. Overpressure vs. Distance for Open-Air Hydrogen Tests

As discussed in the paper, the distance that is available for the flame to accelerate, i.e. the run-up distance, can have a significant effect on the flame speed attained in a flammable gas cloud. The flame speeds presented in the original Baker-Strehlow (BS) model [6] were generated in a test rig whose largest dimension was less than 6 feet (1.8 meters) [5]. The flame speeds presented in the 2005 BST paper are based on tests where the largest dimension was 48 feet (14.6 meters) with the published flame speed results "scaled up" to account for the maximum size of a typical industrial plant [5]. One test of the QMEFS model is to determine the flame speed versus run-up distance for the test configurations used in the newest BST model when specific parameters are varied.

Figure 2 shows the results for the 3-D flame expansion (*nPlanes* = 1), medium reactivity (S_l = 0.43 m/s) case. The curves plotted in Figure 2 show six VBR values – the high (5.7%), medium (4.3%), and low (1.5%) congestion values used in the BST experiments, one intermediate value to show how the model behaves as the VBR is adjusted, a fifth value (1%) below the BST low category to demonstrate the behavior in the "less-than-low" congestion range, and the sixth value (10%) above the BST high category to demonstrate the behavior in the "higher-than-high" congestion range.

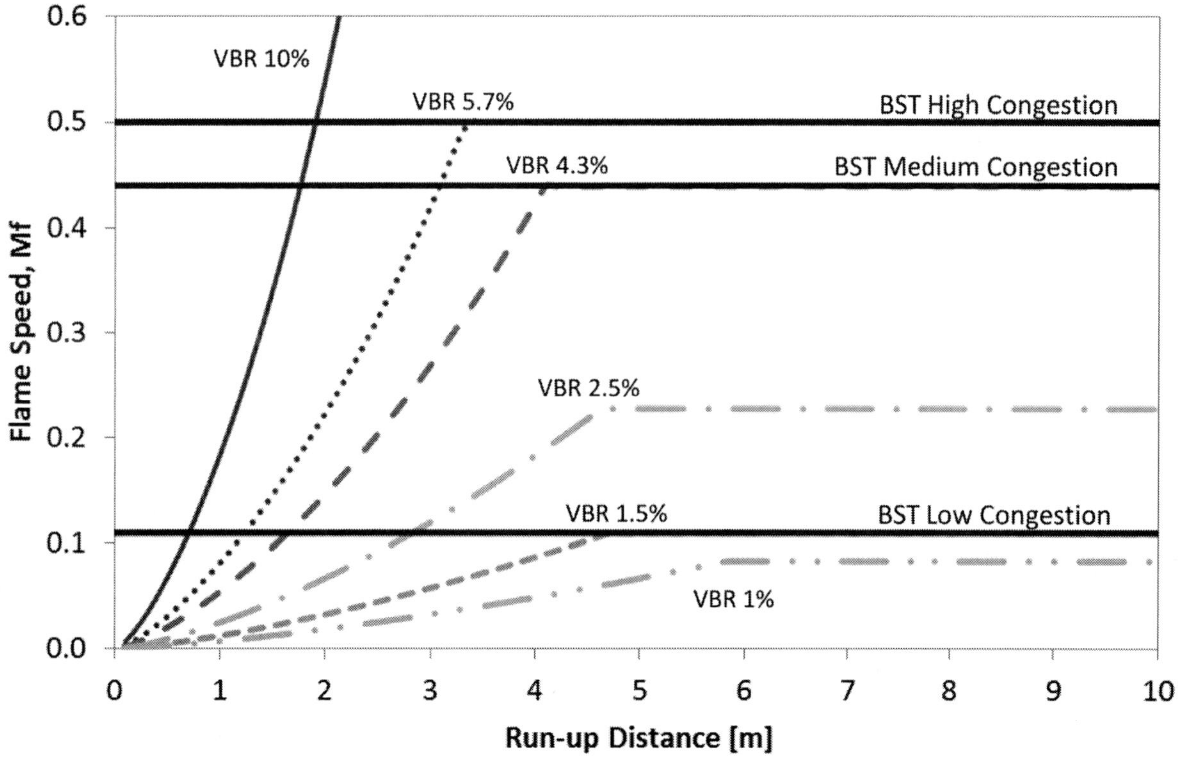

Figure 2. Flame Speed vs. Run-Up Distance; 3-D, Propane, D = 2"

Figure 3 demonstrates how flame speed varies with *nPlanes* and LBV. The basis of the figure is an explosion involving a medium VBR (0.043) and a 4-inch average obstacle diameter. The *nPlanes* parameter is varied between 1.0 (3-D flame expansion) and 4.0 (the maximum value recommend for use in the QMEFS model). The curves plotted in Figure 3 show five materials with various LBV values – ammonia (0.08 m/s), methane (0.37 m/s), propane (0.43 m/s), ethylene (0.76

m/s) and ethylene oxide (0.91 m/s). In the BST model ethylene oxide and ethylene are grouped together into the high category, but these two fuels have different burning properties as can be seen with their LBV values. Thus, by accounting for the LBV of the fuel, the new model provides a finer resolution for flame speed. In addition, as seen in Figure 3, the QMEFS model provides more resolution for the flame expansion parameter through the use of the *nPlanes* variable.

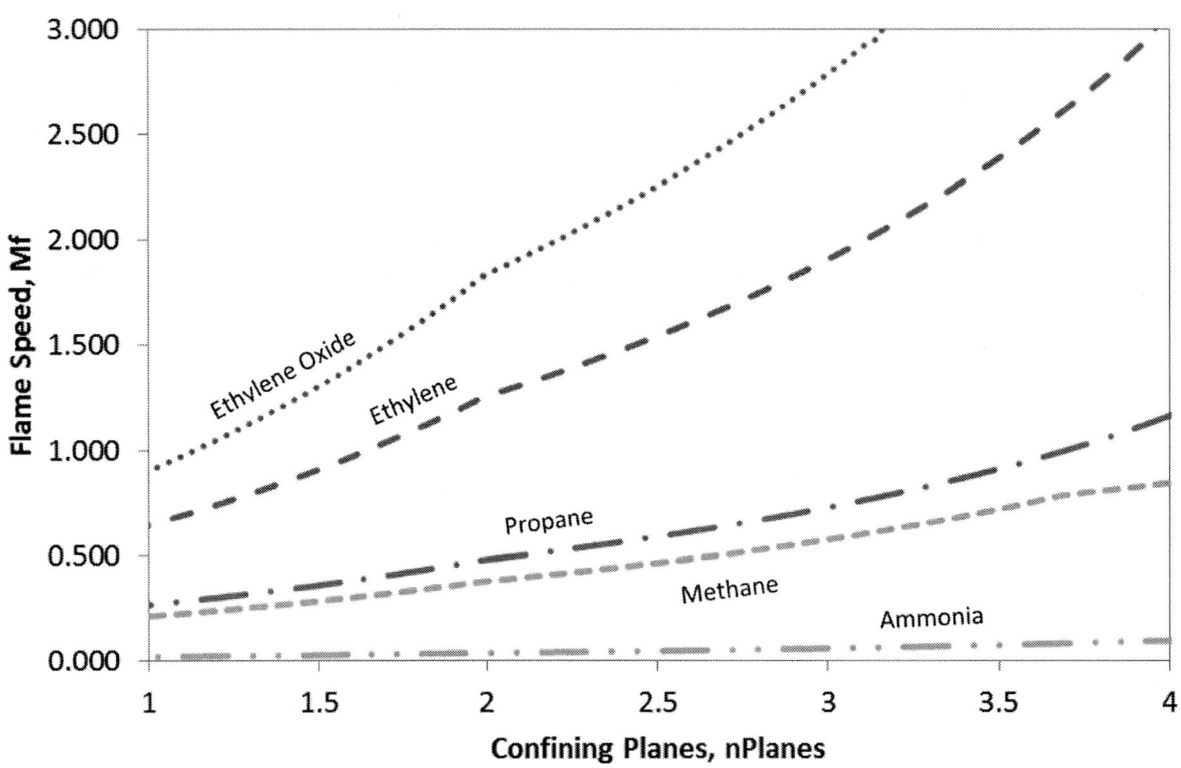

Figure 3. Flame Speed vs. Confining Planes; Medium VBR, D = 4"

5 Conclusions

The model described in this paper provides an improved method to describe the characteristics of a vapor cloud explosion over a wide range of conditions. It combines elements of the work conducted by Baker to develop the BST model, as well as the European MERGE/EMERGE projects; thus incorporating data from the two largest, modern vapor cloud explosion test projects.

Because the BST categories are resolved into numeric ranges, a higher resolution solution space and a clearer set of measureable inputs are achieved. With the ability to model the reactivity of a flammable gas cloud based on laminar flame speed instead of a low, medium, or high classification, the model is able to more accurately describe a wide range of flammable gases and mixtures. Expansion of the constraining flame speed categories to very-low and very-high then provides an even wider range of possibilities.

With the QMEFS matrix expanded to congestion (VBR) categories of very-low and very-high, the model provides a wider range of predictive capability for explosion scenarios. These expansions and enhancements are still confined by the flame speeds in the BST system, and are applied within the set of curves for overpressure and impulse. Thus, with the QMEFS model presented here, process safety experts have a more versatile tool to predict VCE impacts in petrochemical plants.

6 References

[1] Baker, Q.A., C.M. Doolittle, G.A., Fitzgerald, and M.J.Tang, "Recent Developments in the Baker-Strehlow VCE Analysis Methodology," *Process Safety Progress*, v.17, No.4 (1998).

[2] Melton. T.A. and Marx J.D., "Estimating Flame Speeds for Use with BST Blast Curves," Process Safety Progress, v. 28, March, (2009).

[3] Mercx, W.P.M., A.C. van den Berg, and D. van Leeuwen, A*pplication of correlations to quantify the source strength of vapour cloud explosions in realistic situations*. Final report for the project: 'GAMES', TNO report PML 1998-C53, Rijswijk, The Netherlands (1998)

[4] Thomas, K.J., Malik. D.R., Downes, A.M., Rodges, S., "Ammonia Vapor Cloud Explosion Testing and Implications for Facility Siting Studies, Consequence Analyses and Risk Assessments," *50th Loss Prevention Symposium*, AIChE, April, (2016).

[5] Pierorazio, A.J., J.K.Thomas, Q.A.Baker, and D.E.Ketchum, "An Update to the Baker-Strehlow-Tang Vapor Cloud Explosion Prediction Methodology Flame Speed Table," *Process Safety Progress*, v.24, no.1 (2005).

[6] Baker, Q.A., M.J. Tang, E. Scheier, and G.J. Silva, "Vapor Cloud Explosion Analysis," *28th Loss Prevention Symposium*, AIChE, April, (1994).

[7] Mercx, W.P.M., editor, *Extended Modelling and Experimental Research into Gas Explosions*, Final summary report of the project EMERGE, CEC contract EV5VCT930274, European Commission, Directorate General XII, Brussels, Belgium (1997).

[8] Mercx, W.P.M., A.C. van den Berg, C.J. Hayhurst, N.J. Robertson, and K.C. Moran (2000), *Developments in Vapour Cloud Explosion Blast Modeling*, Journal of Hazardous Materials 71 (2000) 301-319.

[9] Tang, M.J. and Q.A.Baker, "A New Set of Blast Curves from Vapor Cloud Explosion," *Process Safety Progress*, v.18, no.4 (1999).

[10] Oran, E.S. and V.M. Gamezo, "Origins of the deflagration-to-detonation transition in gas-phase combustion," *Combustion and Flame*, 148 (2007).

[11] Vasil'ev, A.A., "Estimation of Critical Conditions for the Detonation-to-Deflagration Transition," *Combustion, Explosion, and Shock Waves*, v.42, no. 2 (2006).

[12] Harris, R.J. Wickens, M.J., "Understanding Vapor Cloud Explosions – An Experimental Study," The Institute of Gas Engineers 55th Autumn Meeting, (1989).

[13] Sato, Y., Iwabuchi, H., Groethe, M., Merilo, E., Chiba, S., "Experiments on Hydrogen Deflagration," Journal of Power Sources v.159 (2006).

[14] Health & Safety Executive, "Explosions in Full Scale Offshore Module Geometries Main Report," Offshore Technology Report – OTO 1999 043. Prepared by BG Technology, April, (2000).

479444 Do We Need a Quantitative Risk Assessment (QRA) If My Consequence Analysis Facility Siting Presents a Problem?

Abstract:

There are three steps that a regular facility siting study (FSS) can go through. The first step is the basic separation table approach descrived in API 752. The second step is the consequence analysis facility siting and the third and final one is the quantitative risk assessment. It turns out, however, that the vast majority of the consequence analysis FSS studies quickly demonstrate that the impact to buildings can be more significant that what have been defined as acceptable. For such cases, the analysis normally turns into the QRA, exceedance curves, building structural assessment, etc. The idea of this paper is to cover a intermediate step in early stages to use computational fluid dynamics (CFD) technique in the early stage of the analysis without compromising the time schedule and the budget, this would include dispersion, radiation and explosion modeling. The more accurate results obtained with CFD modelling lead to cloud sizes which are mainly less conservative than traditional methods. The objective of this paper is to go a setp beyond a previous analysis for the dispersion using CFD, but rather make a full evaluation of a FSS using CFD.

Proposed Method for Investigating the Reactivity of Chlorine with Environmental Materials under Relevant and Controlled Conditions

Audrey Feuvrier, PhD.
Ralph E. Martin Department of Chemical Engineering
University of Arkansas, Fayetteville, AR
audrey.feuvrier@uark.edu

Tom Spicer, Ph.D., P.E.
Ralph E. Martin Department of Chemical Engineering
University of Arkansas, Fayetteville, AR
tos@uark.edu

Prepared for Poster Presentation at
American Institute of Chemical Engineers
2017 Spring Meeting and 13th Global Congress on Process Safety
San Antonio, TX
March 26 – 29, 2017

Proposed Method for Investigating the Reactivity of Chlorine with Environmental Materials under Relevant and Controlled Conditions

Audrey Feuvrier, PhD.
Ralph E. Martin Department of Chemical Engineering
University of Arkansas, Fayetteville, AR
audrey.feuvrier@uark.edu

Tom Spicer, Ph.D., P.E.
Ralph E. Martin Department of Chemical Engineering
University of Arkansas, Fayetteville, AR

Keywords: Chlorine reactivity, chlorine release, dry deposition, Jack Rabbit II

Abstract

Many common toxic industrial chemicals (TICs) such as chlorine and ammonia are very reactive with commonly encountered materials in the environment. The amount of TIC material reacted in the environment is an important factor in the impact of a release. Typically characterized with a deposition rate in atmospheric dispersion models, the reaction rate and total deposition quantities have predominately been studied in field scale experiments. Laboratory experiments have also been undertaken, but past investigations have inadequately quantified the available turbulence which can importantly impact surface reaction rates. The state of the art description of surface deposition represents the process as a series of resistances including the effect of atmospheric turbulence and surface reaction(s) depending on the surface material (substrate). Sponsored by the Chemical Security Analysis Center (CSAC) of the U.S. Department of Homeland Security and Transport Canada, an experimental program is underway to test potential substrates under known inlet gas concentrations with levels of turbulence that are comparable to those of the atmosphere. The objective is to develop a test methodology for subjecting a substrate to a reactive gas flow of known TIH concentration with turbulence levels consistent with those in the atmosphere. With data from repeat experiments, deposition of a reactive TIH chemical can be quantified with the proposed experimental protocol.

1 Introduction

Estimating the quantity of a Toxic Industrial Chemical (TIC) removed from the environment through chemical processes is critical to proper consequence assessment of accidental (or intentional) releases, particularly in urban areas [1]. TIC removal mechanisms play an important role in the rational assessment of risk, as these phenomena may significantly and quickly reduce the TIC concentration near a release [2,3], particularly for a reactive chemical such as chlorine. Typically characterized as dry deposition, reactions of chlorine with the environment was not included in the computational models used to predict the consequences of large releases of chlorine in the past [2], but new efforts are underway to characterize chlorine reactivity and incorporate such effects in consequence and risk assessments.

While species like sulfur dioxide and iodine have been the subject of extensive experimental studies [3], studies about deposition involving chlorine are scarce and show a great deal of variability. The high reactivity of chlorine as well as its toxicity make it difficult to study in a laboratory environment, and consequently, many of the studies are conducted as field studies [4]–[6], which restricts test conditions to a limited set of atmospheric conditions. In one set of laboratory experiments, Hill [7] studied the effect of deposition of several TICs on an alfalfa canopy in a chamber where humidity, light, and wind speed were controlled. He showed that chlorine deposition is rapid on alfalfa and that chamber air velocity has a significant effect on the chlorine uptake, but the level of turbulence in the chamber is unknown. Other experimental studies gave promising results such as [8]–[10], but these studies involved no other movement of the flow than the one produced by the mixing device or the circuit used to monitor the concentration level. The state of the art description of surface deposition, [11], represents the process as a series of resistances including the effect of atmospheric turbulence and surface reaction(s) depending on the surface material (substrate).

In this context, an experimental program, sponsored by the Chemical Security Analysis Center (CSAC) of the U.S. Department of Homeland Security and Transport Canada is underway to test potential substrates in a controlled environment. The project includes the design and characterization of a mixing chamber as well as development of a protocol for the measurement of chlorine removal rate for different samples of soil and vegetation. Concentration, velocity, and turbulence measurements will be made in the working section to confirm that the flow is uniform and consistent. It is also desired that the flow in the working section of the apparatus is consistent with the atmospheric flow with target turbulence intensities on the order of 10%. Working section flow speed will be variable, with a maximum in the section of 8 m/s, and expected velocities between 0.5 to 3 m/s. Passive flow conditioning will be used to eliminate non-uniformities in the flow. Recognizing that chlorine gas concentrations as high as 10% or 100,000 ppm may be used, apparatus materials of construction will be carefully chosen to resist damage and also to not remove significant amounts of chlorine during a test. Finally, to limit the hazards involved with the use of chlorine at high concentration level, the apparatus will need to fit under a 6 ft long (about 1.8 m) fume hood.

In the following, the proposed designs for the apparatus and its main elements are introduced. The choice of the mixer, fan, and materials will be discussed. In the second part of the article, the future protocol of measurement will be presented. Focus will be put on the means of measurement used and the characterization of the chamber. The steps to determine chlorine removal rates for several samples types under different environmental conditions will be considered.

2 Proposed design

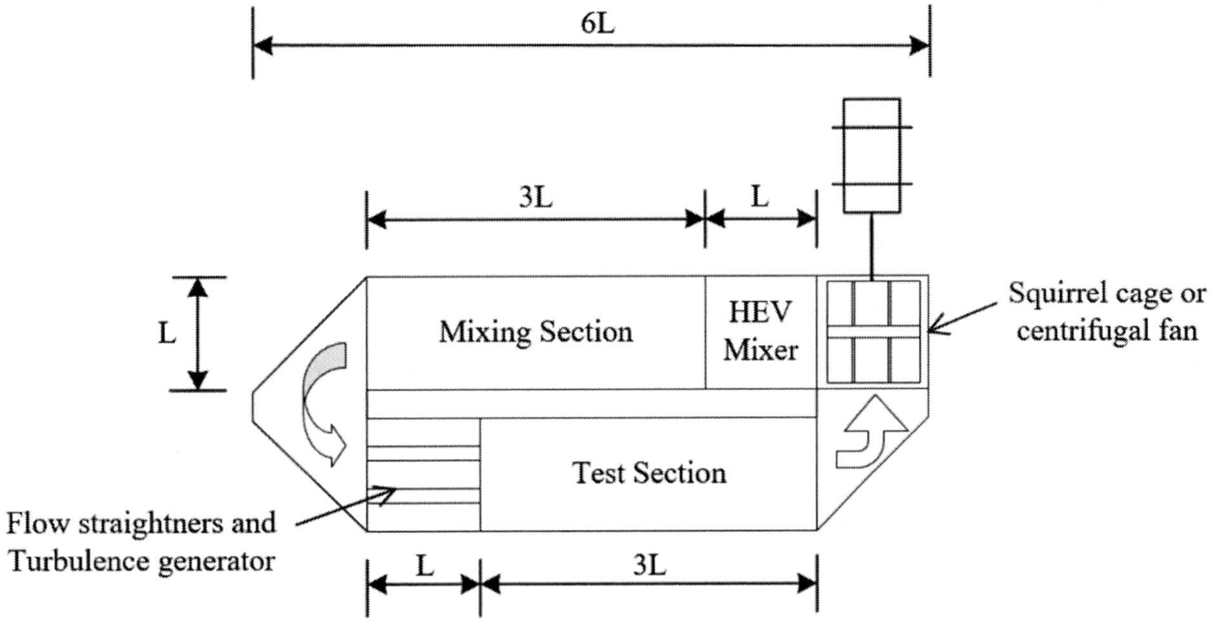

Figure 1. Preliminary apparatus design

Figure 2 shows the preliminary design of the apparatus following the criteria laid down previously using a simple closed return design. The test or working section is placed under the mixing section to facilitate access to the test section through a lab hood sash. The characteristic length, L, chosen to fit in a standard 6 ft hood, will be between 0.25 and 0.3 m, giving a total volume for the apparatus of between 0.16 and 0.28 m^3. A standard lecture bottle contains 454 g of chlorine, which would be sufficient to provide chlorine for 8 to 14 tests depending on the length L based on a maximum concentration of 100,000 ppm. The use of lecture bottles limits the quantity of chlorine required in the lab during the experiments and effectively eliminates the hazard associated with the storage of large quantities of chlorine on site. These volume values are approximate because it may be convenient to use a circular conduit for mixing and a square cross-section for the working section. A circular conduit reduces the required volume of the mixing section (with some minor reduction in power consumption), and a working test section with flat surfaces facilitates optical measurements during characterization and sample placement in the testing phase.

2.1 Mixer and flow conditioner

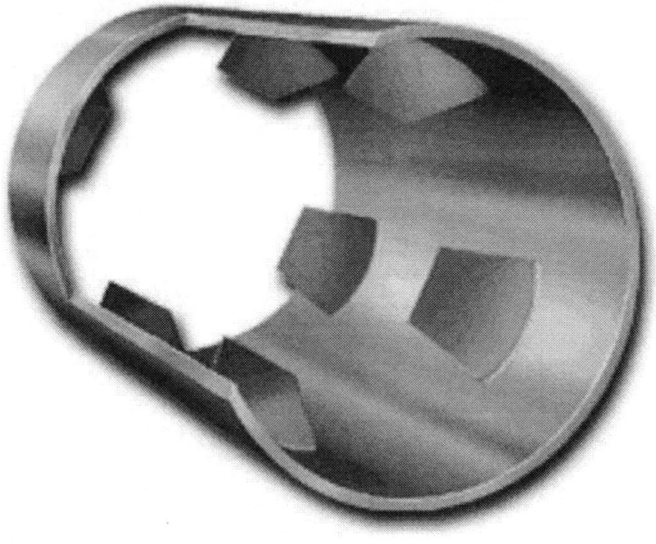

Figure 2. Typical High Efficiency Vortex (HEV) mixer [12].

Figure 3 shows a typical High Efficiency Vortex (HEV) mixer which is placed downstream of the fan. This type of mixer was selected because it provides efficient mixing with turbulent flow with minimal pressure drop. For a Kenics™ HEV mixer, Figure 4 shows the relative coefficient of variation, COV_R , defined as the quotient between the outlet and inlet coefficient of variation, as a function of the Reynolds number (defined as $U_\infty L/\upsilon$ where U_∞ represent the characteristic velocity in the tunnel, L is the characteristic length and υ is the kinematic viscosity of the gas). The coefficient of variation for the mixing of two fluid represents the quotient of the standard deviation of the concentration and the mean concentration. Therefore, the lower the COV_R, the better the mixture quality. The gray zone on Figure 4 represents the range of Reynolds number of this study. Based on Myers et al [13], the pressure drop for our (custom made) apparatus was estimated to be between 1.7 and 17.3 Pa, making it negligible. The HEV mixing process rely on the production of hairpin vortices downstream of the blades of the mixer combined with a pair of counter rotating longitudinal vortices [14]. A mixing section downstream of the device and equal to three times the length of the mixer is recommended to provide for sufficient time for the vortices to complete the mixing process and consequently obtain the predicted COV_R. Additional turbulence generated between the mixing section and the test section will only improve the COV_R before the flow enters the test section..

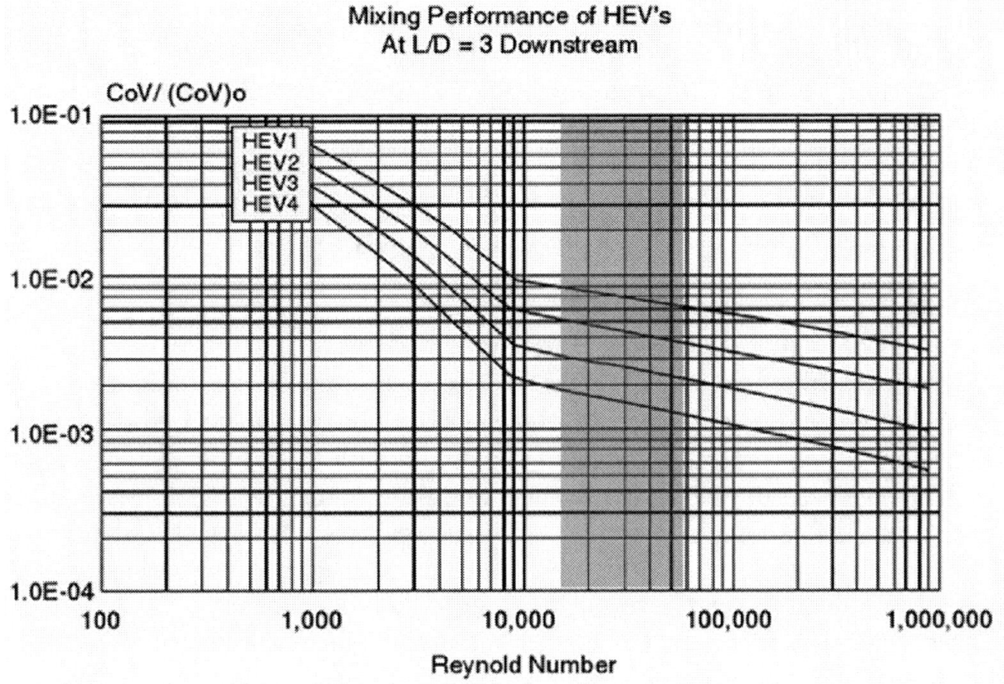

Figure 3. Relative coefficient of variation (COV_R) vs. Reynolds Number and number of tabs row for the Kenics HEV, [12].

The corners currently consist of two 45° bends with a short duct between them, and the 45° bends are preferred to 90° bends to prevent the accumulation of gases in the dead space. To help maintain good flow distribution in the channel, sets of vanes will be placed in each bend. Operating at low wind speed allows us to consider simple circular cross sectional shapes for the vanes [15].

After the two first corners, the flow conditioner section is planned to eliminate large eddies that might be present in the flow ensuring uniformity of the flow and simultaneously raising the turbulence intensity up to 10%. To achieve this, we plan to use a combination of screens or perforated plates. Samples will be placed in the test section through a removable side with provision for a tray in the floor of the test section to accommodate soil and/or plant roots. After experiments, ventilation of the apparatus will be achieved through two hatches on each bend between the mixing and flow conditioner sections.

2.2 The fan

On Figure 2, the motor of the fan is represented outside of the apparatus because chlorine would be expected to react and corrode any of its exposed surfaces. The motor shaft will be supported by Nylon or Teflon bearings located in the side of the apparatus to isolate the motor from chlorine exposure. A centrifugal (squirrel cage) fan is being considered to allow placement of the fan above the apparatus. The fan blades and shaft exposed to chlorine will be treated to limit damage due to chlorine exposure as discussed below.

2.3 Building materials

During the first phase of the experimentation, the apparatus will only use a mixture of air and an inert gases will be used for optic velocity and turbulence measurements and concentration measurements (including a propylene tracer). During this preliminary phase, the structure of the test section will be made with plywood, but two of its walls will be made with polycarbonate to allow visualization and optical measurements of the velocity in the section.

In the second phase of the experimentation involving chlorine, all elements in contact with the test mixture must be made with a material or coated with a material capable of resisting high chlorine concentrations. In addition to extending the workable lifetime of the apparatus, reducing the potential for chlorine to react with materials in the test chamber is also desired so that the reduction in chlorine gas concentration can be attributed to reaction with the sample tested as opposed to the chamber itself. In the case of the fan blades and shaft, a commercial coating is planned (such as Cerakote or titanium). Titanium is known to be highly reactive with dry chlorine, however 1% of water may be enough for passivation after mechanical damage to titanium under normal conditions [16]. For the mixing section, a metal or plastic (PVC) duct will be used since no optical access is required. The structure of the corners and the test section will be made with plywood. In the second phase, polycarbonate may be replaced by plywood. The flow straighteners will likely be made of a suitable metal. All surfaces except the fan blades and shaft are planned to be coated with Viton™, a coating material widely used in applications with chlorine in the chemical industry. We are currently evaluating whether the Viton™ coating under consideration will adhere sufficiently to plywood, and if it does not, then the entire apparatus may need to be fabricated from sheet metal with all internal surfaces painted with Viton™.

3 Protocol for future experimentations

3.1 Experimental resources and characterization of the mixing chamber

Several instruments will be used during the characterization of the chamber and concentration measurement. During both phases of the experimental measurements, the temperature, pressure and humidity of the ambient air will be monitored.

For the characterization of the chamber, the first step will be verifying that the flow entering the test section is uniform and fully mixed through visualization. The working section will be equipped with LED lights and a ROSCO 1500 fogger machine will provide the fog. The fog will be introduce downstream of the fan, at the same point planned for injection of the gases. Video recordings of the fog in the working section will be made with several velocities. After visualization experiments, velocity measurements will be made with a 2D-LDV (Laser Doppler Velocimetry) system using an Argon Ion 300 mW air-cooled laser. This single point technique is non-intrusive and allows simultaneous measurement of two components of the velocity with a good spatial and temporal resolution. Finally, the concentration measurements will be made using an Aurora Scientific mini-PID (photo ionization detectors) The PID measurement method relies on photoionization of a volatile organic compound (in this case a propylene tracer). Small access points will have to be made in the test section for the PID. In the characterization phase, a simulant gas mixture (using sulfur hexafluoride) with be used to reproduce the molecular weight of chlorine. Sulfur hexafluoride is non-toxic, non-flammable, non-reactive and can be used in the PID system.

In conclusion, the characterization of the mixing chamber will assess the quality of the mixing and the flow in the working section, the steps are listed below :

1. Visualization of the flow going in the test section.
2. Velocity profiles taken in the test section will ensure the uniformity of the flow and that the turbulence intensity level is respected.
3. Concentration profiles will provide information on the uniformity of the concentration of the simulant gases in the chamber.

3.2 Concentration measurements and evaluation of the reactivity of chlorine with different samples

After the characterization of the working section, the apparatus will be modified to undergo tests with high concentration of chlorine. The changes have been described in section 2.3. Furthermore, the tunnel will have to allow the access of the JAZ sensors in order to measure the concentration of chlorine. As with characterization of the chamber, the evaluation of the reactivity of chlorine with different samples will request several steps listed hereunder.

1. Characterization of the response of the apparatus to chlorine : Tests will be done without samples, so that the chlorine concentration decay due to the apparatus itself can be measured. A correction factor will be calculated and implemented in the following concentration measurements. Multiple tests will be done, in order to ensure of the repeatability of the measurements.
2. Removal rate estimation : Samples will be placed in the tray, allowing the removal rate of these samples to be estimated through concentration measurements.
3. Analysis of the results : The measurement of the concentration decay in the chamber and the study of the samples put in contact with chlorine will provide valid information in order to calculate the removal rate of chlorine for these different samples. Analytical techniques for determining the amount of chlorine uptake by the test material will depend on the material.

4 Conclusion

In this article, the design of an innovative apparatus or mixing chamber and its main elements have been presented in details. One of the main difficulties encountered was to scale the device to fit under a six feet long fume hood to limit the hazards involved with the use of chlorine at high concentration level in the laboratory. One other obstacle was to find materials and coating suitable and able to resist chlorine in the long run. The future means and procedures of characterization and measurement have been introduced as well. Once build and characterized, this apparatus will provide new data sets for chlorine deposition on vegetal samples and small objects in conditions similar to the atmospheric environment. In the end, this study aim at the improvement of the understanding and computational modeling of toxic inhalation hazards and the consequences to affected population and infrastructure.

5 Aknowledgment

This project is supported by the Chemical Security Analysis Center (CSAC) of the U.S. Department of Homeland Security and Transport Canada. The authors wishes to thank Katherine Elmore for her assistance with this project.

6 References

[1] M. B. Dillon, "The role of deposition in limiting the hazard extent of dense-gas plumes," *J. Hazard. Mater.*, vol. 164, no. 2–3, pp. 1293–1303, 2009.

[2] S. Hanna, S. Dharmavaram, I. Zhang, J., Sykes, H. Witlox, S. Khajehnajafi, and K. Koslan, "Comparison of Six Widely-Used Dense Gas Dispersion Models for Three Actual Railcar Accident," in *Process Safety Progress*, 2008, vol. 27, no. 3, p. 248–259.

[3] S. P. Arya, *Air pollution meteorology and dispersion*, 6th ed. 1999.

[4] J. P. Dimbour, A. Dandrieux, and G. Dusserre, "Reduction of chlorine concentrations by using a greenbelt," *J. Loss Prev. Process Ind.*, vol. 15, no. 5, pp. 329–334, 2002.

[5] G. Dusserre and J. Ollivier, "Small scale field experiments of chlorine dispersion," *J. Loss Prev. Process Ind.*, vol. 15, pp. 5–10, 2002.

[6] J. D. Hearn, R. Weber, R. Nichols, M. V. Henley, and S. Fox, "Deposition of Cl_2 on soils during outdoor releases," *J. Hazard. Mater.*, vol. 252, pp. 107–114, 2013.

[7] A. C. Hill, "Vegetation: A Sink for Atmospheric Pollutants," *J. Air Pollut. Control Assoc.*, vol. 21, no. 6, pp. 341–346, Jun. 1971.

[8] J. Hearn, J. Eichler, C. Hare, and M. Henley, "Cl_2 deposition on soil matrices," *J. Hazard. Mater.*, vol. 237–238, pp. 307–314, 2012.

[9] J. Hearn, J. Eichler, C. Hare, and M. Henley, "Effect of soil moisture on chlorine deposition," *J. Hazard. Mater.*, vol. 267, pp. 81–87, 2014.

[10] D. Brown, "The removal of toxic gases from the air by water, plants and soils.," 2006.

[11] M. L. Wesely and B. B. Hicks, "A review of the current status of knowledge on dry deposition," *Atmos. Environ.*, vol. 34, pp. 2261–2282, 2000.

[12] J. R. Couper, W. R. Penney, and J. R. Fair, *Chemical process equipment revised 2E: selection and design.* Gulf Professional Publishing, 2009.

[13] K. J. Myers, A. Bakker, and D. Ryan, "Avoid agitation by selecting static mixers," *Chem. Eng. Prog.*, vol. 93, no. 6, pp. 28–38, 1997.

[14] W. J. Gretta and C. R. Smith, "The Flow Structure and Statistics of a Passive Mixing Tab," *J. Fluids Eng.*, vol. 115, no. 2, p. 255, 1993.

[15] R. H. Perry, D. W. Green, and J. O. Maloney, *Perry's chemical engineers' handbook.* McGraw-Hill, 2008.

[16] Timet corporation, "Corrosion resistance of titanium," 1997.

13TH GLOBAL CONGRESS ON PROCESS SAFETY

Fire Proofing Cost Reduction Measures for LNG Plant

Raminaidu Girada
Senior Process Safety Engineer, Bechtel Oil, Gas & Chemicals, Inc.
3000 Post Oak Boulevard, Houston, TX 77056
rgirada@bechtel.com

Jegan Babu Arumugakan Thuraiswamy
Senior Process Safety Engineer, Bechtel Oil, Gas & Chemicals, Inc.
3000 Post Oak Boulevard, Houston, TX 77056
jbarumug@bechtel.com

Nino Nicotra
Process Safety Engineering Chief, Bechtel Oil, Gas & Chemicals, Inc.
3000 Post Oak Boulevard, Houston, TX 77056
anicotra@bechtel.com

Prepared for Presentation at
American Institute of Chemical Engineers
2017 Spring Meeting and 13th Global Congress on Process Safety
San Antonio, TX
March 26 – 29, 2017

Fire Proofing Cost Reduction Measures for LNG Plant

Raminaidu Girada
Senior Process Safety Engineer, Bechtel Oil, Gas & Chemicals, Inc.
3000 Post Oak Boulevard, Houston, TX 77056
rgirada@bechtel.com

Jegan Babu Arumugakan Thuraiswamy
Bechtel Oil, Gas & Chemicals, Inc.
3000 Post Oak Boulevard, Houston, TX 77056
jbarumug@bechtel.com

Nino Nicotra
Process Safety Engineering Chief, Bechtel Oil, Gas & Chemicals, Inc.
3000 Post Oak Boulevard, Houston, TX 77056
anicotra@bechtel.com

Keywords: Consequence Modeling, Liquefied Natural Gas (LNG) & Liquefied Petroleum Gas (LPG) and Safe Design

Abstract

The current industry practice in designing fireproofing requirements for LNG facilities follows the traditional approach of "Liquid Pool Fire". The common industry practice for fire proofing design criteria defines a fire potential equipment (FPE) as the equipment that handles flammable liquid above a specific threshold volume. The FPE generates fire scenario envelope whose dimension ranges are recommended by API 2218. Equipment and structural supports that falls inside the fire envelope are considered for fire proofing.

LNG plants are handling products with very low boiling point, and any leak to atmosphere results in an instant flashing into vapor phase or aerosols. The results from our recent works on pool fire consequence modelling has proved that most of the currently considered fire potential equipment seldom create pool fire. This consequence modeling results clearly demonstrates that the current fire proofing design approach is overdesign, that is not effective. The apparent saving in man-hours for not modeling the liquid release will result in a considerable amount of engineering efforts and material cost being spent on the design and installation of fire proofing for equipment and structure with potentially no real benefits to the inherent safety of the process plant. Considering the current economic trend, a change in approach using consequence modelling will effectively result in considerable cost savings in major LNG projects.

1 Introduction

Currently in LNG projects, fireproofing is considered for equipment supports, pipe rack support, structural supports and other auxiliary critical hydrocarbon carrying pipe supports within a fire proofing zone to improve its structural integrity during a fire. The principal value of fire proofing is realized during the early stages of a fire by providing additional time to fire fighters to control or extinguish a fire thus by delaying additional loss of containment which may escalate the incident. The fireproofing requirement is determined by identifying the fire potential equipment and the associated fireproofing zones. Fire potential equipment is defined as an equipment holding flammable liquid above a specific threshold volume, which will have the potential to create a liquid pool and subsequent pool fire if ignited. Per API 2218, the fireproofing zone is defined for every fire potential equipment, that extends 20 ft. to 40 ft. (6 m to 12 m) horizontally and 20 ft. to 40 ft. (6 m to 12 m) vertically from the source of liquid fuel. As a typical approach most projects consider fireproofing zones that extend 30 ft. (9m) horizontally and 30 ft. (9m) vertically from the source.

The above mentioned qualitative approach does not reflect the realistic possibility of pool fire scenario. Therefore, a quantitative analysis may be performed to analyze whether a pool fire is realistic for equipment handling multiple refrigerants and LNG. This analysis has been performed on some process equipment handling multiple refrigerants and LNG using Canary software by Quest Consultants.

2 General Technical Guidance

2.1 Scope of Fire Proofing

Fire scenario envelopes for an LNG plant are based on pool fire scenarios. Jet fires are typically not considered as credible for defining fire scenario envelops for fireproofing as described below:

- Jet fires are handled by Prevention, Intervention and Protection. Prevention is by designing the facility in compliance with general fire prevention activities such as those included in NFPA 59A and in the broad application of Process Safety Management(PSM). Once a pressurized jet fire occurs, the intervention and protection can be achieved by a high flow rate firewater stream directed on the point of impingement. This has been proved as an effective mitigation measure in preventing escalation with Jet fires. This will be achieved by strategically locating the fire water monitors around high risk process area, where jet fire is foreseen as a credible scenario.

- Emergency isolation and depressurization system will be provided to lower the pressure of specific isolated sections of the plant that may be involved in a localized jet fire scenario and to de-inventory of gas to flare.

- Efforts on similar LNG projects to extend the fire proofing systems for jet fire scenarios led to enormous impractical level of fireproofing requirement. This could also introduce other hazards such as corrosion.

- Jet fire represents a significant risk or concern for personnel life, working on offshore installations whereas property protection is the basis of pool fire for on-shore LNG facility or petroleum refinery or petrochemical facilities.

2.2 Pool Fire Formations

A pool fire may result via a number of scenarios and it begins typically with the release of flammable material from process equipment. If the material is liquid, stored at a temperature below its normal boiling point, the liquid will collect in a pool. The geometry of the pool is dictated by the surroundings (i.e., diking), but an unconstrained pool in an open, flat area is possible, particularly if the liquid quantity spilled is inadequate to completely fill the diked area. If the liquid is stored under pressure above its normal boiling point, then a fraction of the liquid will flash into vapor, with un-flashed liquid remaining to form a pool in the vicinity of the release.

The pool of flammable liquid may be ignited creating a pool fire. If the release of flammable material from a pressurized process equipment continues, then a jet fire is also likely. If the ignition source occurs at the very beginning of the release, then inadequate time is available for the liquid to form a pool and only a jet fire will result.

Pressurized liquefied gases such as LNG, methane, ethylene and propane will in general disperse as mists or vapor, but may possibly form a small pool if the source operates at low pressure conditions. An exception has to be made for butanes, pentanes and heavier hydrocarbons, which always create pool.

Refer to the following **Figure 1** for pictorial view of the pool fire formation.

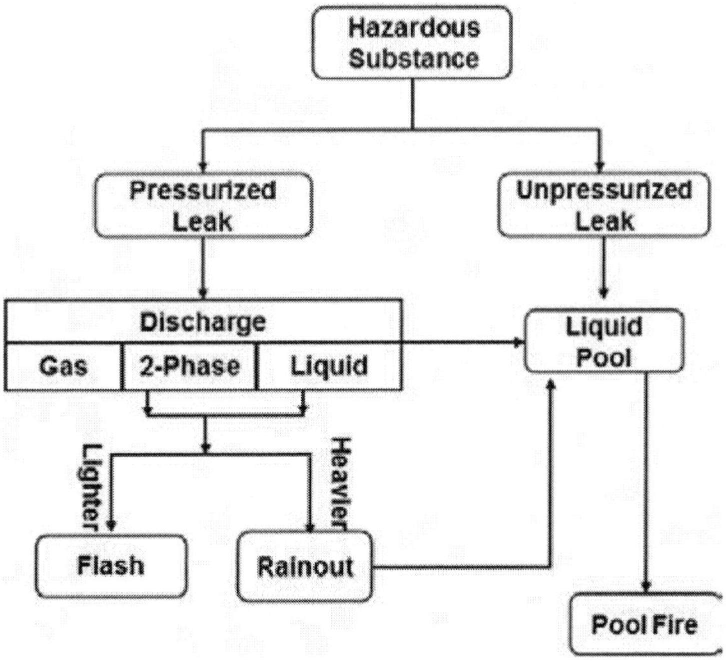

Figure 1: Pool Fire Formation

2.3 Determination of Pool Fire scenario

This section provides an overview of the methodology for determining pool fire existence.

In case of accidental release of flammable materials such as Pentanes and heavier hydrocarbons create liquid pool irrespective of their operating conditions. However, the lighter hydrocarbons such as butane and propane may create liquid pool only under specific operating conditions. The potential for creating liquid pool is based on certain parameters as described below:

- **Butanes**

  Butane releases potentially create liquid pool except for high pressure releases and slight sub cooling of release. Further consequence analysis is required to check if a liquid pool is created from the release.

- **Propane and Ethylene**

  Ethylene and propane liquids may or may not pool, depending on conditions. It has been found that a pool formation can be correlated on several parameters: (1) vessel operating pressure, (2) size of rupture hole for self-leakage vessel, (3) the amount of subcooling, which is also important but the assumption is the liquid will be slightly subcooled, and (4) confinement of *fire hazard area*. Together, these variables establish the size of the pool.

  Higher pressure release are more likely to vaporize and mist, and low pressure releases are more likely to pool. Based on a slightly subcooled liquid at pressure, the analysis is divided into three regimes (1) no liquid pools, (2) highly likely to form a significant pool fire, and (3) possible pool fire. Depending on the size of the release, slope, temperature and composition, a consequence model may be necessary to determine if a pool fire is warranted.

- **LNG**

  Similar to propane and ethylene liquids, LNG may or may not pool depending on several parameters: (1) vessel operating pressure, (2) size of rupture hole, (3) the amount of subcooling, and (4) confinement.

Therefore, a quantitative approach was performed in the following section.

3 Pool Fire Case Study

In this case study, the following light flammable fluids are modelled: methane, propane, ethylene and LNG. Based on the current fire proofing design approach a total of 28 equipment of the LNG plant that handles the above fluids are qualified as Fire Potential Equipment. All those leakage scenarios are modelled using Canary modelling tool, based on its operating condition.

3.1 Leak Scenarios

To provide a spill rate using CANARY, one has to postulate a leak scenario.

Typical hole size choices range from small (3-10mm) to full bore ruptures (diameter of pipe). Hole size selection can define the general level of risk (larger, high consequence releases are significantly less frequent than small, low consequence release).

CCPS book of "Guideline for Fire Protection in Chemical, Petrochemical, and Hydrocarbon Processing Facilities" defined hole sizes as small (3-10 mm), medium (10-50 mm), large (50-150 mm), and full-bore ruptures. Discussion around these hole sizes also points out that the large and rupture releases are not always the worst-case events due to release duration and dosage effects that should be discounted for.

The Canadian Standard Association (CSA) standards Z276 specifies design spills which define the potential extent of hazard zones that are not allowed to impact the public. Z276 defines, among other (smaller) release scenarios, that the design spill should be "The maximum flow of LNG that could be spilled, from the largest branch connection, 60 mm (2-inch) or smaller..."

Based on the above references, 2-inch is considered as an appropriate hole size for consequence modeling. Therefore, the pool fire modeling has been performed for 2-inch hole size. However, sensitivity analysis has been done considering 6-inch hole to check if larger hole size can result in pool formation. The modeling results for both 2-inch and 6-inch are presented below.

The release elevation was considered to be constant at 1 m height (the lowest height is deemed to result in higher consequences). The release directions in all scenarios were assumed to be horizontal. The consequence modeling input data shown below were used for the calculations. All scenarios were modeled using CANARY software developed by Quest. The software is widely used in industrial consequence modelling studies.

3.2 Atmospheric Conditions

The following more stringent atmospheric conditions that favors/sustains the pool scenario are considered:

Wind Speed: 2 m/s

Stability Class: F

Ambient temperature: 80 °F

3.3 Liquid Methane Release at High Operating Pressure

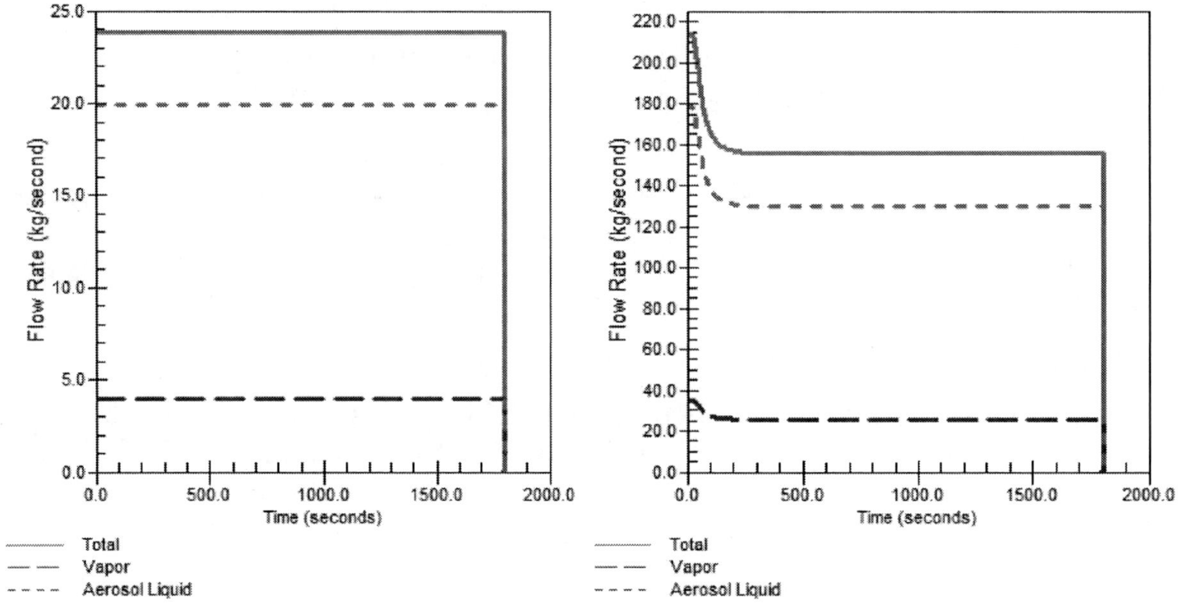

Figure 2: Liquid Methane Release – 2-inch (left figure) & 6-inch (right figure)

Consequence modeling for a liquid methane release was performed at specific process conditions for 2-inch and 6-inch hole size. As shown in the above figures, most of the release is flashing into aerosol liquid and remaining release is in vapor phase. No liquid pool is generated from the liquid methane release.

3.4 Liquid Propane Release

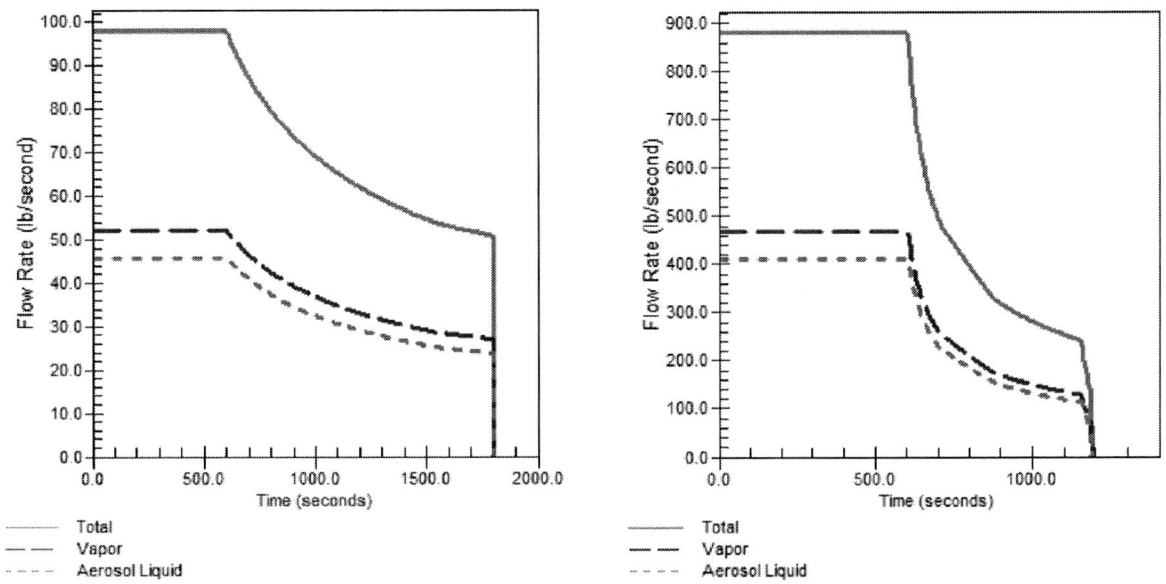

Figure 3: Propane Release – 2-inch (left figure) & 6-inch (right figure)

Consequence modeling for a propane release was performed at specific process conditions for 2-inch and 6-inch hole size. As shown in the above figures, the release is flashing into aerosol liquid and vapor phase. No liquid pool is generated from the propane release.

3.5 Liquid Ethylene Release

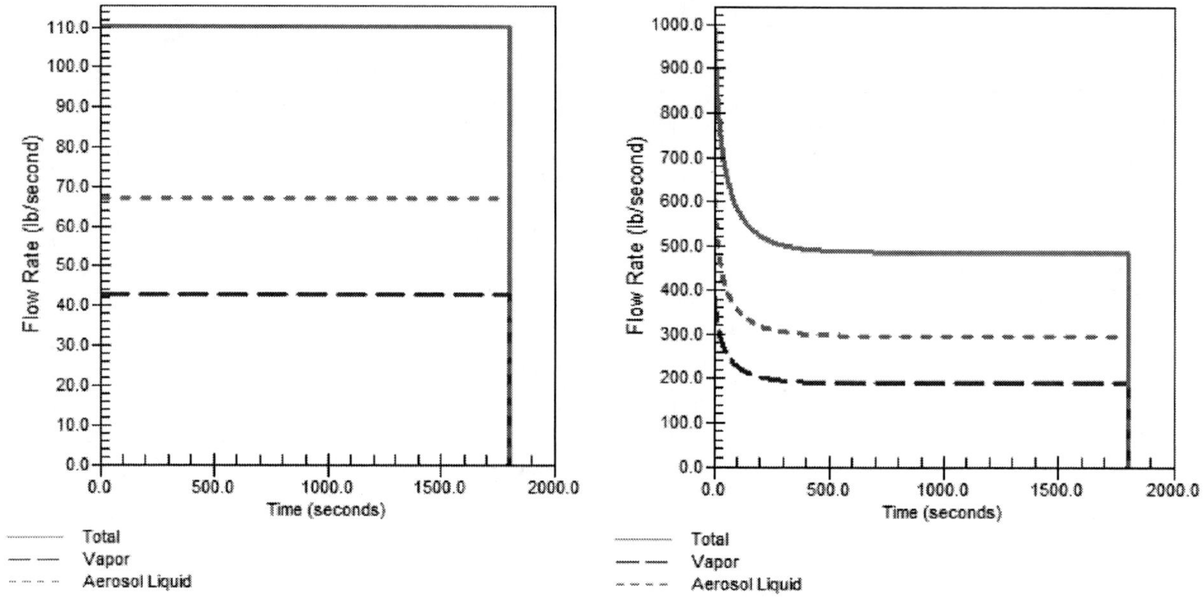

Figure 4: Ethylene Release – 2-inch (left figure) & 6-inch (right figure)

Consequence modeling for ethylene release was performed at specific operating process conditions for 2-inch and 6-inch hole size. As shown in the above figures, the release is flashing into aerosol liquid and vapor phase. No liquid pool is generated from the refrigerant release i.e. Ethylene release.

3.6 LNG Release at Low Operating Pressure condition

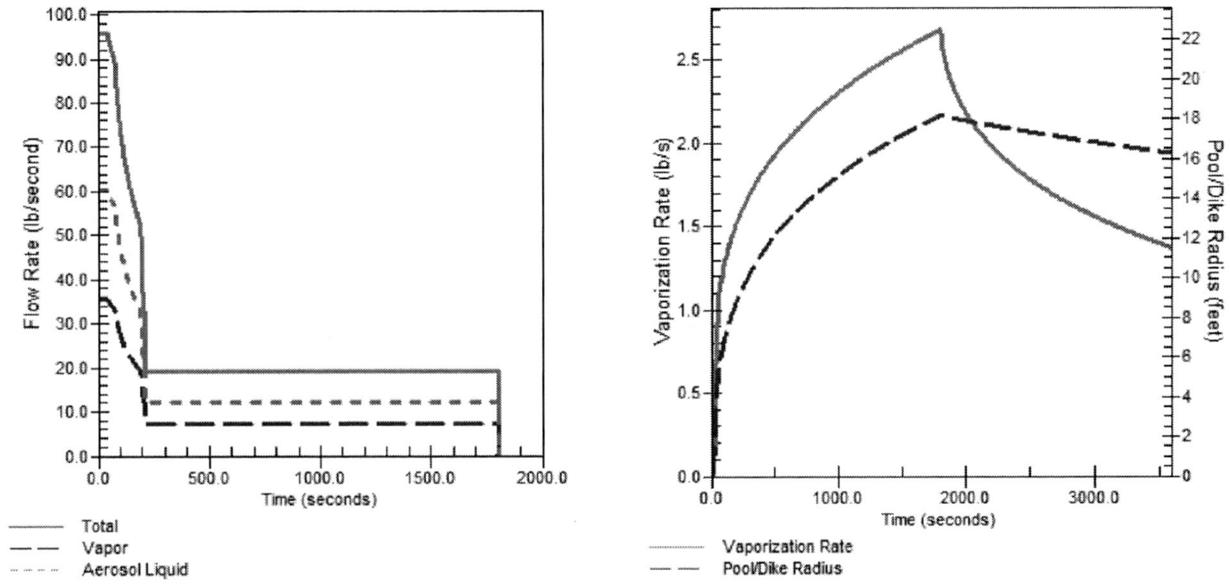

Figure 5: Liquefied Natural Gas Release – 2-inch (left figure) & 6-inch (right figure)

Consequence modeling for a liquefied methane was performed at the process conditions of an equipment for 2-inch. and 6-inch hole size. As shown in the above left side figure, the release for 2-inch hole size is flashing into aerosol liquid and vapor phase, without creating any pool. However, as shown in the right side figure, a release from 6in. hole size from the same equipment is creating a liquid pool of 16 ft. radius.

4 Conclusions

Consequence modeling results showed that out of the total twenty-eight (28) equipment of LNG liquefaction plant that were normally considered as Fire Potential Equipment (FPE) based on specific threshold volume, thirteen (13) equipment that handles methane, propane, ethylene at some specific operating conditions (e.g. low or medium or high pressure) did not create any liquid pool. Instead they flash along forming a large fraction of aerosol liquids, which never creates liquid pool.

The pool fire modeling was performed for a 2-inch release and a sensitivity analysis was carried out for 6-inch release to check the possibility of liquid pool formation for larger size releases for all the 28 equipment. The 6-inch release did not make a major difference in pool fire modeling results except for one equipment which contain LNG at low operating pressure condition. As stated earlier, thirteen (13) out of twenty-eight (28) equipment had not produced a liquid pool irrespective of the release size.

One (1) out of the remaining fifteen (15) equipment, that handles LNG at lower pressure condition did not generate liquid pool at 2-inch release, however it has generated liquid pool for a larger size (6-inch) release. Therefore, it is recommended to perform the consequence modeling for larger hole size to check the liquid pool formation.

The consequence modeling proved that 46% of the total equipment containing liquid such as methane or propane or ethylene or LNG had not produced a liquid pool. Therefore, considering the current economic trend, consequence modeling based approach for determining fireproofing requirements in conjunction with API 2218 requirement can be used in major LNG projects to justify considerable cost savings.

5 References

[1] J.D. Marx and A. Nicotra, "Is a Two-Inch Hole Adequate for Siting Study?", 12th Global Congress on Process Safety, American Institute of Chemical Engineers, Houston, Texas, 2016.

[2] API. "Fireproofing Practices in Petroleum and Petrochemical Processing Plants", American Petroleum Institute, 2013.

[3] CCPS (2003), "Guideline for Fire Protection in Chemical, Petrochemical, and Hydrocarbon Processing Facilities", Center for Chemical Process Safety of the American Institute of Chemical Engineers, 3 Park Avenue, New York, 10016-5991. (ISBN 0-8169-0898-2).

[4] CAN/CSA Z276, "Liquefied natural gas (LNG) - Production, Storage, and Handling", National Standard of Canada, 2015.

479423 Consideration of Escalation Risks in Facility Siting

Abstract:

A small fire when uncontrolled, can potentially lead to failure of an equipment containing large inventory of hydrocarbon or a structure holding hydrocarbon bearing equipment within the same process unit, resulting in an escalation of a minor event to a much larger event with catastrophic consequences. Similarly, a major fire or an explosion in a process unit may escalate to cause failure of equipment and piping in adjoining units or facilities, thereby causing more severe damage. Identifying and quantifying the escalation risk is the subject of this paper in terms of summarizing the current approaches and how adopting this in a rigorous manner can help minimize losses to asset and production in addition to saving lives. A few case studies are presented to illustrate the benefits of incorporating mitigation measures to minimize escalation risks.

Various accidents in the past such as the Formosa Plastics propylene explosion in 2005 show that small leaks or fires can escalate to cause damage to the unit and also escalate to cause damage to adjoining units. Inadequate separation between units, inadequate active and passive protection of structures and equipment, particularly for jet fires, inadequate isolation and blowdown measures, inadequate specification of safety critical elements such as ESDVs and flare headers are some of the causes for escalation. To minimize asset loss and injuries, facility siting and plant layout need to incorporate a systematic assessment of escalation risk to ensure mitigation measures are identified and incorporated.

Escalation risks can be assessed following a consequence-based approach or a risk-based approach incorporating both frequency and consequence estimation. While consequence-based analysis is a simple and easy tool to use, the selection of credible failure scenarios can become a subject of debate - many projects have adopted a generic 1" or 2" hole size as the basis for analysis whereas some have considered larger hole sizes. Given the increased size of piping and equipment due to larger throughputs, a 1" or 2" hole size may be considered as insufficient. The alternative is to follow a risk based approach wherein all leak sizes are modelled together with mitigation system failure, directionality and ignition/explosion probability to determine the cumulative likelihood of a damaging fire or explosion event. Such an approach would typically be based on a target frequency threshold of 1E-4 per year. This method is subject to the usual uncertainties associated with a frequency assessment including those associated with failure frequency data and the various event probabilities.

Case studies for onshore installations handling hazardous substances are presented considering both risk-based and consequence-based approaches for comparison. The case studies also illustrate how mitigation measures can be designed based on such an analysis to reduce the risk of escalation. Examples of mitigation measures considered include increased spacing between process units, increased spacing between sections of a large process unit by segregated fire zones, spill containment, fire and gas detection, isolation and blowdown, and active and passive fire protection. The case study results are also compared with typical deterministic spacing tables, which are commonly used for plot plan development, whereby recommendations are provided to minimize the risk of escalation early on during the initial plot plan development stage of the project.

The Facility Siting Cycle (Revalidation)

Michael Weinberg
ABS Group
140 Heimer Road, Ste. 300, San Antonio, TX 78232
MWeinberg@absconsulting.com

Prepared for Presentation at
American Institute of Chemical Engineers
2017 Spring Meeting and 13th Global Congress on Process Safety
San Antonio, TX
March 26 – 29, 2017

The Facility Siting Cycle (Revalidation)

Michael Weinberg
ABS Group
140 Heimer Road, Ste. 300, San Antonio, TX 78232
MWeinberg@absconsulting.com

Keywords: Facility Siting, Revalidation, Process Safety

Abstract

Facility siting is an assessment of occupied buildings given potential exposure to explosion, fire, and toxic hazards. As stated in 29 CFR Part 1910, it is requirement for Process Safety Management covered processes [1]. Sites are always changing and, as a part of the management of change, the facility siting should be maintained in-line with these changes. As a minimum, the facility siting study should be updated and revalidated at least every five years, per 29 CFR Part 1910 section 119.e.6. While maintaining a current facility siting study can be a challenge, revalidation has many benefits aside from ensuring a safer working environment.

Revalidation studies are initially seen as opportunities for sites to address large site changes (e.g., new process units, changing process conditions, changes in occupied buildings). However, other benefits from revalidation include: 1) Consolidation of various preliminary or detailed studies into a single revised facility siting study generating a coherent picture enabling the owner to optimize mitigation plans, 2) Taking credit for operational changes and mitigations implemented, 3) Benchmarking the progress of the facility siting remediation plan, 4) Evaluate and revise existing hazard criteria, 5) Take advantage of modeling, software, and methodology improvements to illustrate reduced hazards, and 6) Assessing the effectiveness of the Management Of Change (MOC) system to identify changes to the site that influence facility siting. This paper discusses the different ways to approach revalidations studies, what sites should expect during revalidation, and the additional opportunities/insights that can be recognized beyond the study itself.

1 Introduction

A facility siting study is an assessment of occupied buildings for potential exposure to explosion, fire, and toxic hazards as required in 29 CFR Part 1910.119 under the PHA element. Sites and processes are always changing and updating a facility siting study in-line with these changes is required by OSHA at least every five years as part of the PHA umbrella. While maintaining a current facility siting study can be a challenge, revalidation has many benefits aside from promoting a safer working environment. Revalidation studies are typically viewed as solely as an exercise to address all major changes to site operations (e.g., new process units, changing process conditions, changes in occupied buildings). However, other there are a variety other direct and indirect benefits and opportunities available during revalidation as discussed in this paper.

1.1 Acronyms

Table 1. Common Acronyms

API	American Petroleum Institute
BLEVE	Boiling Liquid Expanding Vapor Explosion
CCPS	Center for Chemical Process Safety
CFD	Computational Fluid Dynamics
CFR	Code of Federal Regulations
MOC	Management of Change
OSHA	Occupational Safety and Health Administration
PHA	Process Hazard Analysis
PSM	Process Safety Management
PVB	Pressure Vessel Burst
TNO	TNO Prins Maurits Laboratorium
VCE	Vapor Cloud Explosion

2 Facility Siting

2.1 Who Requires Facility Siting and Its Revalidation?

Facility siting is required for onshore owner/operators with PSM covered process (e.g., refineries, chemical operations, gas liquefaction, specialty chemical, and pharmaceutical manufacturing). The PHA element of OSHA PSM indicates that facility siting must be performed and shall be updated and revalidated on a maximum 5-year cycle.

2.2 What is the Scope of Facility Siting?

Facility siting is intended to address hazards for personnel inside occupied buildings within the plant boundary. Facility siting does not address hazards directly for personnel in-the-field during an event or potential community impacts. The three hazard types typically evaluated in facility siting include explosions (e.g., VCE, PVB, BLEVE), fire (e.g., jet, pool, flash), and toxic releases.

2.3 What are Facility Siting Recommended Practices?

Resources, guidance, and recommended practices for facility siting can be found in several sources:

- API
    - API RP 752, Management of Hazards Associated with Location of Process Plant Permanent Buildings, Third Edition, 2009. [2]
    - API RP 753, Management of Hazards Associated with Location of Process Plant Portable Buildings, First Edition, 2007. [3]
    - API RP 756, Management of Hazards Associated with Location of Process Plant Tents, First Edition, 2014. [4]
- CCPS
    - Guidelines for Facility Siting and Layout; First Edition, 2008. [5]
    - Guidelines for Evaluating Process Plant Buildings for External Explosions, Fires, and Toxic Releases; Second Edition, 2012. [6]
    - Guidelines-for-Chemical-Process-Quantitative Risk Assessment, Second Edition, 2000. [7]
- UK - Chemical Industries Association
    - CIA Guidance for the Location and Design of Occupied Buildings on Chemical Manufacturing Sites, Third Edition, 2010. [8]

2.4 What are Acceptable Technical Approaches to Facility Siting?

There are a variety of technical approaches to facility siting. These approaches vary in their levels of applicability, conservatism, and technical difficulty. Both consequence-based and risk-based methods can benefit from these approaches. For example, VCE hazards can be evaluated with widely-used screening/traditional methods (Gaussian dispersion and VCE blast curves) or can be modeled using (CFD) methods for both dispersion and explosion.

Fire hazards can be assessed (in order of complexity) using spacing tables, empirical modeling, geometric analysis (using view factor and shading from obstacles), and/or CFD fire modeling. Fire evaluation criteria can use heat flux values for certain consequences, such as thermal radiation intensity tolerable for evacuating personnel. A more detailed approach would be to perform a thermal dose calculation that integrates thermal radiation intensity exposure over time to predict personnel vulnerability.

Toxic release hazards can be assessed using traditional dispersion models or with CFD. Evaluation criteria can be based on endpoints of concentration levels, such as Immediately Dangerous to Life or Health (IDLH), or on a dose calculation that integrates exposure level over time to predict personnel vulnerability

3 The Facility Siting Cycle

It is not uncommon for the initial facility siting study to be considered "the end" or the "goal." However, facility siting is an ongoing cycle. It is intended to be revised and changed as processes and facilities change. This cycle can be broken down into 6 steps:

1. Corporate Practice
2. Pre-Study/Initial Siting Study
3. Detailed Assessments
4. Remediation Planning
5. Implement Remediation
6. Management of Change

3.1 Corporate Practice

This is the first stage in the cycle and commonly includes policies and procedures. The policies are developed by the site's corporate owner or operator. For example, these practices may specify requirements for facility siting to be consequence-based or risk-based assessment. They may also define credible and non-credible scenarios (e.g., VCEs, BLEVE) or exceedance criteria (e.g., toxic exposure limits during evacuation). In the event owner/operators do not have set policies and procedures, facility siting providers should be able to provide guidance based on reference documents, industry practice, experience, and past events.

3.2 Pre-Study/Initial Siting Study

The pre-study or initial facility siting study is more aptly referred to as a screening study as this better represents its purpose. Initial studies typically use traditional/ simple methods (e.g., filling a unit with flammable gas, assuming an explosion severity level) with the purpose of screening out those occupied buildings that clearly meet the acceptance criteria and identifying buildings which require closer inspection (e.g., re-evaluate use case, detailed analysis). The pre-study is typically a part of a facility siting study and includes identifying occupied buildings, locating/generating building data (e.g., structural design), and compiling process information (e.g., pressure, composition, line size). The initial study itself includes a site survey (e.g., verify pre-study information, document congestion) and a screening level evaluation of occupied buildings vs. the exceedance criteria for explosion, fire, and toxic hazards. The conclusion of this step should identify which buildings require additional attention through remediation or more detailed study.

3.3 Detailed Assessments

Detailed assessments can be very effective and cost efficient approaches for addressing issues identified in a screening level study. In some cases, the generalized methods for evaluating unique hazard situations may be conservative or cannot take into account specific considerations (e.g., topography, complex geometry). A variety of analysis options can be used with increasing detail and technical knowledge (e.g., TNO blast curves, CFD, evacuation routes). Multiple assessment approaches may be necessary to identify the safest, most cost-efficient approach, and most practical for remediation (e.g., building retrofit, new design, process changes).

3.4 Remediation Planning/Implement Remediation

Once the necessary detailed assessments have been completed, owner/operators should develop and implement a remediation plan and schedule as identified in the following industry guidance.

- Guidelines for Evaluating Process Plant Buildings for External Explosions, Fires, and Toxic Releases [6].
- API RP 752, Management of Hazards Associated with Location of Process Plant Permanent Buildings [2].

The schedule and prioritization of these efforts will be based on the overall remediation strategy and estimated options and costs. Implementation of this plan would include finalizing designs, interim changes, and long term solutions.

3.5 Management of Change

Over time sites may undergo a variety of changes which will test the ability of their MOC system to account for changes that could affect facility siting. Common changes include:

- Change in existing building occupancy levels (e.g., new occupancy, de-occupancy),
- new occupied building construction, and
- plant expansion (e.g., new process units, process changes).

These changes are commonly addressed by directly applying previous results (e.g., new construction in a previously evaluated area) or assessing specific new hazards (e.g., adding a new unit). These cases should be analyzed and the results, recommended actions, and past remediation actions should be documented and tracked.

4 Revalidation

As stated in 29 CFR Part 1910.119 section e.6, revalidation is required every 5-years (based on the original study's completion date). In its simplest form revalidation is a verification of the hazard assessments (e.g., scenario conditions, congestion assessment) and occupied building details while also accounting for any large site changes (e.g., new process units). In addition to providing a safe working environment, there are several other benefits with revalidation.

4.1 Revalidation Benefits

Owner/operators may have performed a variety of detailed assessments and/or preliminary hazard assessments for potential process or site changes. Revalidation provides an opportunity to combine these various studies into a single document. This can make it easier for owner/operators to maintain a clear picture of the site's current facility siting status and avoid conflicting or overly burdensome conclusions based on separate documents that lead to unjustified or unnecessary mitigation costs.

Revalidation also provides the ability to take into account risk or hazard mitigation implementations since the previous study was issued. This also minimizes the need to maintain multiple documents to capture the site's current facility siting status. A single consolidated document makes it easier to respond to and explain current site and remediation status to regulatory bodies (e.g., OSHA).

After the screening level study, sites may perform hazard assessments for planned new units or even a variety of preliminary units. These studies are frequently based on general layouts and process information. Even for assessments based on 90% complete models, revalidation provides an excellent opportunity to ensure that assumptions and idealized processes properly match the as-built product.

A specific timeframe for implementation of a mitigation plan is not specified in the API or CCPS guidance documents; although, CCPS points out that such implementations can take "several years" but should be implemented "promptly" like other PHA findings. So, remediation timelines can extend beyond the required 5-year revalidation cycle. Revalidation serves as a valuable tool to benchmark progress of the remediation effort and to determine if the remediation plan needs a course-correction based on changing conditions.

Over time a site's ownership may change. New management may have different policies and standards regarding performance of facility siting, such as different evaluation criteria or differences in adopted methodology, such as risk-based vs consequence-based. Revalidation should address such changes.

Industry accidents provide lessons learned that may influence or have a direct impact on facility siting. A specific example would be the generation of API 753 for portable buildings after the 2005 BP Texas City incident. Likewise, the 2005 Buncefield accident highlighted the potential for surrounding foliage to support an unexpectedly severe VCE. All of these conditions may be reason to evaluate and revise the existing hazard criteria.

Apart from the simplest hazard assessment methods, most facility siting studies utilize a wide variety of software packages (commercial and in-house) to evaluate hazards with varying levels of complexity. As with any software package, technology is upgraded and additional capabilities are implemented. As more advanced software packages are utilized for this type of work, they become more prevalent in industry. This along with hardware improvements can lower the cost of analysis. This improves the cost-benefit for screening level studies and detailed assessments like CFD.

Another value gained by revalidation is assessing the effectiveness of the MOC system to identify changes to the site that influence facility siting. MOC is a challenge for most sites to effectively keep track of changes to personnel, buildings, and processes. Effective

MOC necessitates that personnel be knowledgeable of the site's facility siting criteria and the limitations of the methods employed. Training should be refreshed to ensure current personnel can address MOC issues and/or identify potential issues for further assessment.

Finally, revalidation can also incorporate lessons learned from onsite explosions, fires toxic releases, or near-misses. If an accident has occurred, it is likely that a safety system has been implemented. The safety system should be evaluated to see if can address other previously identified scenarios. Near-misses can also be evaluated as accidents to quantify their hazard potential if not properly mitigated.

4.2 What should be expected during Revalidation?

Revalidation should start with a reassessment of current owner/operator's policies and procedures. This is followed by consideration of historical events in industry as well as onsite recorded incidents and near-misses. MOC events including new process units and new buildings should be evaluated. Any studies based on preliminary or final design should be verified to match the as-built/operating elements. Remaining buildings and onsite congestion will be verified to ensure changes were not missed by the MOC system. Detailed assessments should ensure that they are still applicable, accounting for any site changes that may affect the results, prior to inclusion in the revalidation study. Any implemented migration efforts should also be verified and ensure they are maintained over the life cycle needed. Scenarios and models can then be re-evaluated to address any software improvements or updated technical approaches (e.g., TNO blast curves to CFD). Finally, the results of the revalidated study will either continue with the existing remediation plan or make adjustments to account for new hazards.

5 Conclusion

Facility siting is a cycle. It is intended to be revised and updated as processes and facilities change. One of the key required components of the cycle is revalidation. In its simplest form revalidation is a verification of the hazards assessments (e.g., scenario conditions, congestion assessment) and the occupied buildings, accounting for any large site changes (e.g., new process units). Benefits from revalidation include: 1) Consolidation of various preliminary or detailed studies into a single revised facility siting study generating a coherent picture enabling the owner to optimize mitigation plans, 2) Taking credit for operational changes and mitigations implemented, 3) Benchmarking the progress of the facility siting remediation plan, 4) Evaluate and revise existing hazard criteria, 5) Take advantage of modeling, software, and methodology improvements to illustrate reduced hazards, and 6) Assessing the effectiveness of the Management Of Change (MOC) system to identify changes to the site that influence facility siting. While revalidation can be treated in a very cursory manner, this paper was intended to highlight the many opportunities and advantages available to owner/operators during this process.

6 References

[1] OSHA, Process safety management of highly hazardous chemicals, https://www.osha.gov/pls/oshaweb/owadisp.show_document?p_table=STANDARDS&p_id=9760, Accessed on 03/04/2017.

[2] API RP 752, Management of Hazards Associated with Location of Process Plant Permanent Buildings, Third Edition, 2009.

[3] API RP 753, Management of Hazards Associated with Location of Process Plant Portable Buildings, First Edition, 2007.

[4] API RP 756, Management of Hazards Associated with Location of Process Plant Tents, First Edition, 2014.

[5] CCPS, *Guidelines for Facility Siting and Layout*, First Edition, 2008.

[6] CCPS, *Guidelines for Evaluating Process Plant Buildings for External Explosions, Fires, and Toxic Releases*, Second Edition, 2012.

[7] CCPS, *Guidelines-for-Chemical-Process-Quantitative Risk Assessment*, Second Edition, 2000.

[8] Chemical Industries Association, *Guidance for the Location and Design of Occupied Buildings on Chemical Manufacturing Sites*, Third Edition, 2010.

13TH GLOBAL CONGRESS ON PROCESS SAFETY

Strategic Facility Siting – A Different Perspective

Steve Clepper
Jacobs | Buildings + Infrastructure | Advance Planning Group
10 Tenth Street Suite 1400
Atlanta, GA 30309
Steve.clepper@jacobs.com

Hollie Schmidt
Jacobs | Buildings + Infrastructure | Advance Planning Group
201 N Franklin Street Suite 1400
Tampa, FL 33602
Hollie.schmidt@jacobs.com

Prepared for Presentation at
American Institute of Chemical Engineers
2017 Spring Meeting and 13th Global Congress on Process Safety
San Antonio, TX
March 26 – 29, 2017

Strategic Facility Siting – A Different Perspective

Steve Clepper
Jacobs | Buildings + Infrastructure | Advance Planning Group
10 Tenth Street Suite 1400
Atlanta, GA 30309
Steve.clepper@jacobs.com

Hollie Schmidt
Jacobs | Buildings + Infrastructure | Advance Planning Group
201 N Franklin Street Suite 1400
Tampa, FL 33602
Hollie.schmidt@jacobs.com

Keywords: facility siting, building hardening, master planning, risk assessment, risk mitigation, risk reduction, non-essential personnel, API 752

Abstract

The overall purpose of this presentation is to highlight the human and financial benefits of strategic facility siting versus tactical facility siting for the purposes of the reduction of risk exposure to workers in a petro-chemical environment. Jacobs has been the leader in conducting facility siting driven by API 752 compliance for buildings in petro-chemical plants for over a decade. Our comprehensive analysis of the entire portfolio of buildings in conjunction with risk assessment conclusions can lead to significant risk metric reductions, cost saving and improved worker welfare through new construction of consolidated building functions in lower risk exposure areas of the plant. Jacobs has developed a unique methodology called the "Circles of Safety" which is a tool used in conjunction with the risk assessment findings to locate occupied buildings by proximity which the worker should incur by nature of their job function.

1 Introduction and Background

Very few plants are constructed entirely in one phase at one time. In fact, most chemical plants are the result of cumulative growth through decades of infrastructure investment, migrating operations, changing management or even ownership, surrounding community encroachment and evolving regulations. The initial phase of most chemical plants consists of a few processing units with nearby support buildings and enabling infrastructure. Typical support buildings include shops, field offices, shipping/receiving facilities and other buildings to house people whose job functions do not require them to work directly on the processing units. Incremental process growth and the addition of new risk generating infrastructure often "embeds" these occupied buildings in high-risk exposure areas. These occupied buildings are prime candidates for a facility siting project. Poor facility siting is often to the result of reactive site selection that are short-term tactical decisions versus long-term strategic decisions.

The challenge is often exacerbated by the behavioral culture of legacy personnel who have been working in these embedded support buildings for a long period of time. Occupants tend to believe that the building's location is acceptable because "it's the way it's always been" or "nothing bad has ever happened" As a result, these personnel become somewhat, or entirely, de-sensitized to the risk. A safety-centered facility siting strategy evaluates the cumulative results of a risk assessment, translates the results into easily understood consequences, then identifies and locates these personnel away from the most severe hazards if their job function does not require them to be immediately in the core of the plant. The cultural changes in behavior that are necessary for successful facility relocation are addressed with the client leadership team as a part of the future master plan and must be endorsed and championed at the highest level of plant management.

Recent petro-chemical plant events and resultant tragedies have focused greater attention on the danger and liability of locating non-essential personnel near high-risk generating processing units and infrastructure. Regulatory requirements and industry best practices, such as API 752 and 753, are currently evolving to respond to the inherent safety challenges at refineries and chemical plants. OSHA and the American Petroleum Institute (API) are leading this evolution by updating key documents related to occupied buildings within petrochemical plants.

Mitigating risk exposure to existing workers in occupied buildings in petro-chemical plants present significant opportunities for loss prevention. It is a common occurrence, and therefore a ripe opportunity for risk reduction, to find high-risk occupied buildings dispersed across most plants. In some cases these high-risk facilities house people who do not need to be exposed to risk by nature of their job function. Modern risk assessments highlight the non-compliance to As Low As Reasonably Practicable (ALARP) corporate standards of worker risk exposure in those buildings.

2 Strategic Facility Siting verses Tactical Facility Siting

The nature of the petro-chemical business is focused on revenue-generating assets which are the units, tanks and enabling support infrastructure to keep them running. These assets get the most attention and funding to keep them in prime operational condition. Alternatively, the non-revenue-generating assets (i.e. buildings) are not considered a priority for maintenance, funding and optimization. This leads to an environment and culture where tactical versus strategic facility siting decisions are made. The definition of the terms strategic and tactical facility siting for the purposes of this paper is:

- Strategic: forming an integral part of a stratagem (attain a goal or to gain an advantage)
- Tactical: plan of action designed as an expedient toward gaining a desired end or temporary advantage

The differential is that a tactical approach tends to be for a dire immediate need with focused problem solving for a singular objective. A strategic approach is more holistic problem solving for a wider range of issues, achieved over a longer timeframe addressing multiple objectives. The variables in these two approaches are urgency, scope, scale and overall, long-term benefit.

2.1 Tactical Facility Siting Application

Tactical facility siting projects are often organized and managed as reactive "silo" projects, with small, defined boundaries, incomplete information about the surrounding area and potential strategic options for better siting decisions.

Tactical facility siting focuses on one or two high risk occupied buildings with the goal to build hardened replacement buildings to a "nearby" vacant site. Only the worst rated, high priority buildings that do not meet ALARP standards are addressed. Other buildings are left unaddressed. Organizational units that exist in the high risk building are relocated to the new building but are not consolidated and optimized. This limited view misses opportunities to improve operational efficiencies or consolidate fragmented operations. These opportunities would be apparent with a strategic approach to facility siting.

This strategy does not always recognize that there are multiple ways in which to mitigate risk to buildings. Often the first, and only, option considered is to "harden" the building in place. The perception is that this is the quickest, easiest and cheapest solution. The option of relocating the function to a new "fit for purpose" constructed building is seen as too complex with a lot of resultant logistics to be to solved and is therefore rejected. In actuality, the inverse is almost always true. New buildings are more efficient, cost less to maintain, are constructed to withstand the risk exposure where it is located and can solve other plant challenges such as vehicle congestion, increased security and sufficient parking.

Finally, tactical facility siting does not acknowledge the overall condition of the entire portfolio of buildings and the added operations and maintenance costs of a "run to fail" model for buildings. Most buildings in petro-chemical plants are older construction that do not comply with current safety standards and risk mitigation measures and are often being used for a different purpose than they were intended (i.e. old control rooms used for office space or laboratories in old warehouse buildings). The attitude is that "buildings do not make money" and therefore are not maintained to the same standard as revenue generating infrastructure.

2.2 Tactical Facility Siting Limitations

The limitations of tactical facility siting is that it focuses only on select occupied buildings and therefore leaves personnel in dangerous but less than critical, non-ALARP-compliant, occupied buildings. The "easy button" action is typically to relocate functions in the critical buildings to a nearby vacant site which is often an expensive solution requiring building hardening. These solutions typically are complicated and expense due to their presence inside the battery limits of an operating plant. (security, safety, liability, etc.). The silo approach offers limited information to the team and limits big thinking on potentially better, holistic solutions.

2.3 Strategic Facility Siting Application

Strategic Facility Siting projects are organized as "holistic land planning" projects with a plant manager's view of the entire plant. In addition to the typical knowledge of a tactical facility siting approach, this strategic view also has knowledge of other high risk buildings, building condition and fragmented operations that could be consolidated and relocated to create a safer and more productive plant.

Strategic facility siting first makes a collective and comprehensive high level assessment of all occupied buildings. It examines the populations and job positions of those building occupants to determine their need to work near risk generators as a result of their daily responsibilities. Also, it is common to find organizations in fragmented locations simply because space was available in a certain building at their time of need. A strategic facility siting exercise examines the best siting for different types of buildings balancing safety with productivity. It also seeks to consolidate fragmented operations for improved operational efficiencies and increased time on tools.

A successful strategic facility siting is a comprehensive process that includes sponsorship of plant leadership to "tear down the silos" to create more opportunities for success. The strategic approach to facility siting should consider future growth and change management at the plant and propose a wide range of strategies to mitigate the risk.

2.4 Challenges to a Strategic Facility Siting Process

Typical issues to overcome for a strategic facility siting process can include:

- Lack of a Clear Framework for Growth from Plant Leadership – Lack of a comprehensive plant-wide business/production output plan, designation of land uses and asset management strategy.

- Lack of a Future Vision - Uncertain land acquisition strategy that accommodates growth and creates a buffer zone where it is needed.
- Knee Jerk Reactions – Reactive decision making about facility siting without considering the holistic picture. Often associated with the response to a recent incident on site or within the company.
- Unsafe Embedded Buildings – Existing occupied facilities too close to hazards.
- Emergency Response Equipment at Risk – Many plants have firehouses near processing units where a large event could render firefighting resources unusable.
- Workforce vs. Workplace Imbalance – Modular wooden trailers are often used as permanent facilities to accommodate overflow.
- Security and Traffic Enhancements – A traffic plan is needed to accommodate a large number of vehicles within the secured perimeter due to location of warehouse and shops inside perimeter. Gate traffic congestion sometimes results when vehicles are not segregated by type.
- Inadequate Road and Parking System – Poorly defined roads within the plant and lack of extra parking cannot accommodate a large turnaround event.
- Problems Attracting and Retaining Talent - Lack of investment in employee facilities and workplaces affect recruitment and retention of good employees.
- Community Issues – A community outreach program may be needed to address complains of risk, noise, odors, flares, etc.

2.5 Strategic Facility Siting Benefits

There are numerous benefits to strategic facility siting. The main benefits are:

- Balance Safety with Operations – Balance safety with excellence in productivity, operational efficiency and sustainability by utilizing a "close but safe" philosophy. Often, the perception that all workers need to be in the high risk process areas to be operationally efficient. This is simply not true. Valid solutions can be identified and quantified to achieve a good balance of these two critical factors.
- Consolidate Fragmented Operations – Incremental growth in processing capability over time can result in fragmented multiple shops, control rooms, etc. Through the consolidation of multiple smaller buildings into a single building can eliminate wasted production time caused by unnecessary movement between fragmented facility locations. Development of a comprehensive strategic facility siting strategy can be used to quickly identify a migration strategy, avoiding a separate, time consuming facility siting study for each building.
- Improve Operational Efficiency – Increase operational efficiency is achieved through optimized functional relationships that locates all workers in the same division together and co-locates divisions that frequently work together in the same building or location. Replacement facilities can be constructed in phases as high risk buildings are removed.
- Limit Liability – Decrease the potential liability for loss of life, lost production and fines for non-compliance.
- Reduce Overall Risk – Reduced risk metrics through good facility siting and other mitigation measures such as "fit for purpose" new construction. Facilities housing people

now house job function does not require risk exposure can be built to normal standards (no hardening, etc.)

- Comply With Applicable Regulations –Comply with current facility siting regulations, including corporate ALARP, OSHA, API 752, and company-specific HSSE policies.

Figure 1 illustrates the typical conditions for a petro-chemical plant in terms of number of people, number of buildings, building risk exposure (red is worst and green is best) and square footage. Thresholds and tolerance of risk vary greatly between plants and between corporations. Generally, the highest risk is represented in red and lowest risk is represented in green. It is clear to understand the potential for key safety improvements as shown in Figure 2. This existing condition, which is from a Jacobs project, had 111 out of 130 buildings and 1,744 out of 1,863 (94%) of workers exposed to risk where the building did not mitigate the varying degrees of risk.

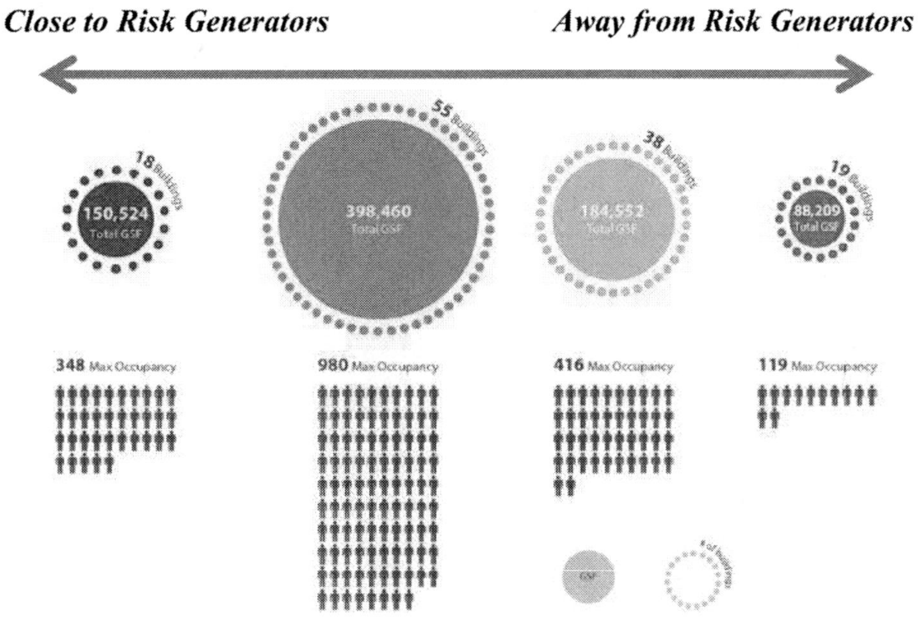

Figure 1 – Existing Conditions before Strategic Facility Siting Plan

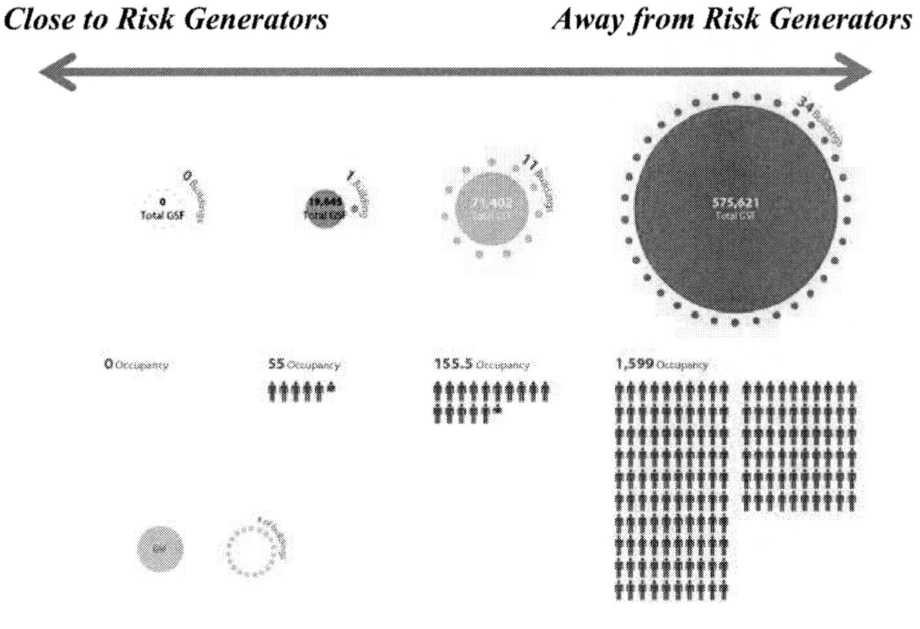

Figure 2 – Effect of Strategic Facility Siting Implementation

2.6 Additional Opportunities (Beyond Facility Siting Risk Reduction) Presented by A Strategic Approach

The strategic benefits outlined in Section 2.5 are a primary justification to engage in the recommended analysis and action plan, however, there are unexpected opportunities that are presented by this methodology that go beyond the scope of facility siting.

- Opens to New Paradigms – Achieve the best outcome by encouraging plant leadership to explore new ways of thinking about the potential of the plant and embrace change.
- Be a Good Neighbor and a Good Place to Work - Promote "Industry Best Practices" related to security, reduction of societal risk exposure, environmental protection, quality of workplace and worker welfare issues.
- An opportunity to discover process expansion sites inside the fence when high risk buildings are demolished and the land is cleared.
- Ability to decongest the core of the plant by removing buildings that attract high truck volumes, such as warehouses and shops, generate a lot of foot and gator/forklift traffic and can be conflict areas between vehicles and workers. By eliminating truck traffic you also eliminate sources of ignition near the units.
- Ability to provide more room for laydown, turnaround/shutdown logistics area and the ability to make roadway improvements for lane widening and turning radii compliance for larger vehicles.
- Ability to better secure the core of the plant where the units are located by eliminating a large portion of vehicles and workers that do not need to be at the units as part of their primary job function.
- Solve More Than One Problem – Greater Bang for Buck

3 Strategic Facility Siting Process

3.1 Identify Industry Best Practices

A first step in the strategic process is to understand the latest best practices on locating non-essential personnel (those with job functions not requiring close proximity to the units/core risk areas). API 752, states that "Consideration should be given to locating non-essential personnel as far as practicable from the hazard and discouraging congregation of personnel in buildings close to process areas."

3.2 Circles of Safety by Job Function

In order to determine essential versus non-essential personnel, or direct support versus in-direct support of the units, the planning team utilizes the "Circles of Safety" tool, a tool unique to Jacobs, which our team has developed specifically for industrial clients with employees exposed to risk. This tool categorizes personnel by job function in rings of progressive proximity to risk exposure. These Circles of Safety have a direct correlation and influence on where workers sit at their "desk", or perform their job duties, which aim to site future operational support facilities in the safest location possible while still optimizing operations.

In the Circles of Safety shown in Figure 3, employee job functions are shown in the circle appropriate for the job's proximity to the process units and high risk areas. Those employees working directly at the process units are located in the center circle in red in the High Risk (HR) zone. Those that must be near but safe are located in the middle circle in the yellow Moderate Risk (MR) zone and those that do not need to be located near the process are in the green outermost circle in the Low Risk (LR) zone. Working with plant management to identify the job position placements provides key data on who is considered essential to the process and it also provides clues on how to group personnel of similar risks in a facility strategy.

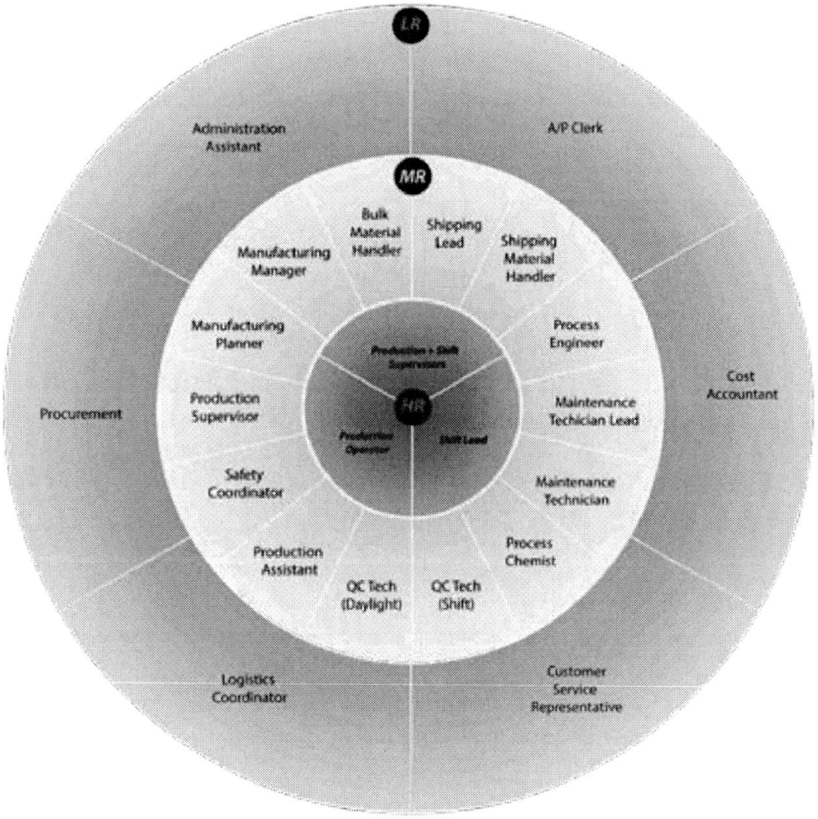

Figure 3: Circles of Safety

3.3 Building Retention or Disposal Strategy.

The results of the Circles of Safety tool, combined with the results of the risk assessment, determine the buildings retention or disposal strategy. There are multiple outcomes of the retention or disposal strategy:

a) <u>Retain</u>

   i. Retain "As Is": No mitigating measures are needed.
   ii. Retain and Harden/Improve: Retain the building but undergo improvements to bring it into compliance with risk exposure, current building codes, corporate standards or improved operations.
   iii. Retain, De-occupy and Repurpose: Retain the building but due to risk exposure de-occupy but keep for repurposing such as storage.

b) <u>Vacate and Demolish</u>

   i. In a high risk exposure area
   ii. Have reached their usable life span (can determine with a Facility Condition Assessment)
   iii. Are small and inefficient (can determine through programming interviews with departments)
   iv. Demolition will create resultant valuable new land space near the core processing area that can be used for optimized hydrocarbon development, increased mobility, safer vehicle movement, laydown or process expansion

c) <u>Relocate</u>

   i. These are small modular buildings that can be easily transported to a new location.

Typical non-process facilities that could be impacted include:

- Administration buildings to house office based workers
- Training and conferencing
- Shops to house maintenance personnel
- Maintenance support and their offices
- Warehouses / shipping + receiving to house inventory of materials and shipping /receiving loading and unloading docks for material trucks
- Laboratories to house lab personnel for QC of plant products.
- Changing rooms for plant employees that work in the processing areas of the plant
- Firehouse / Emergency Operations Centers house specialized fire response equipment and provide a central control room for emergency operations
- Guard houses / security buildings house security personnel responsible for gates and perimeter security. These personnel also play a key role in policing the plant and responding to emergencies

- Field offices house office based workers that spend a portion of their time in the process area daily but predominately work in a field office. These personnel typically include contractors and maintenance and operational personnel
- Control Rooms house process control systems and personnel. These facilities can be centralized using contemporary digital control systems and can be located outside the high risk area

3.4 Non-Process Facility Siting + Typical Plant Layout.

Each facility and function identified to be demolished and consolidate into new construction have ideal locations on a well-planned site that includes clearly defined roads, non-process area for location of support facilities, clear entry exit and handling of traffic all considered for their proximity to risk. Figure 4 illustrates ideal relationships between plant elements.

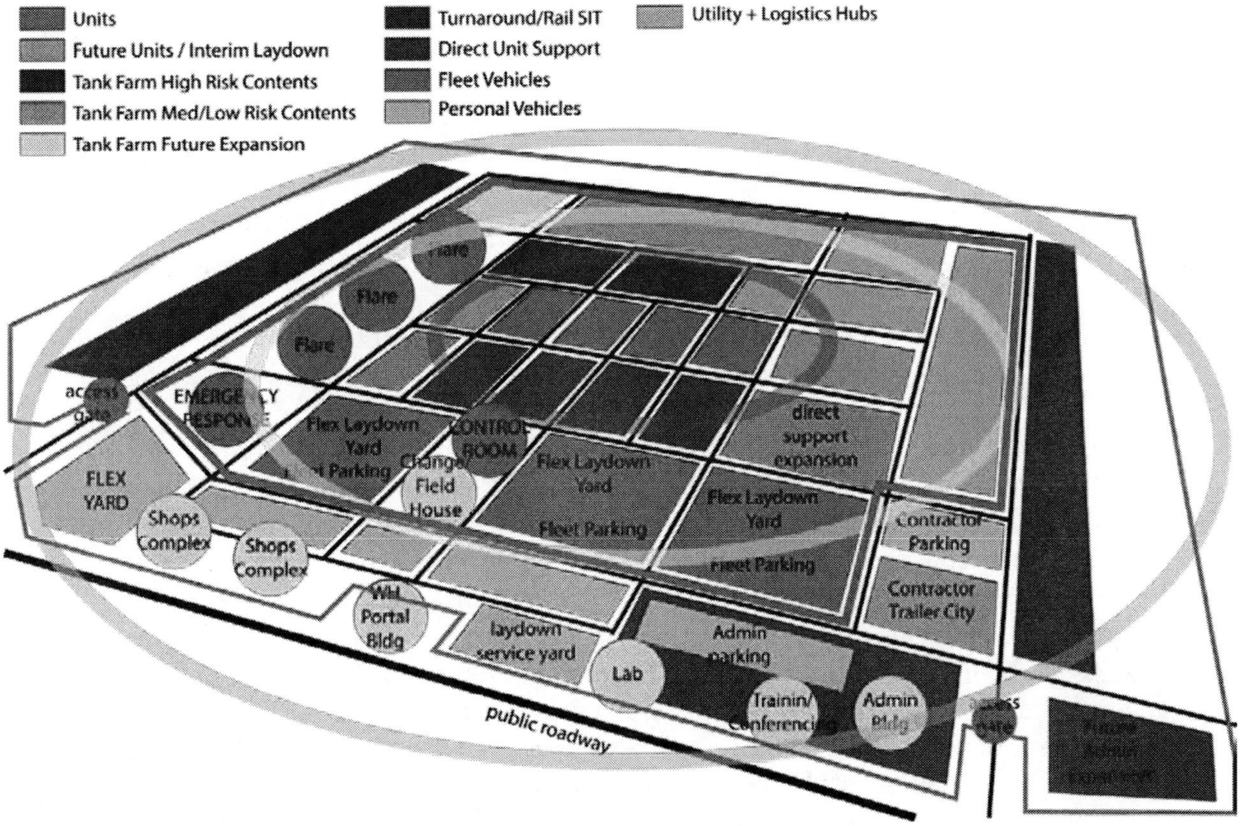

Figure 4 – Ideal Typical Plant Relationships

The guiding principles that factor into the plant layout and facility siting are:

- Have a Complete Understanding of the Risk – It is important to have a clear and thorough understanding of the major risk factors of overpressure, toxic release and thermal as well as security regulations. The planning team must understand the conclusions of the risk assessment and translate the data into a plan of action to mitigate the risk.
- Identify Situational Safety Issues – Challenge work cultures that place people in unnecessary risk. Often employees are located haphazardly in the plant as a result of individual moves over time, placing process non-essential workers in risk areas. Sometimes employees are moved to a certain location to take advantage of available space, such as an administrative assistant in a shop or an engineer in a control room. Use the Circles of Safety exercise to locate those employees relative to the risk.
- Identify the Most Economical Way To Move People Away From Risk - Relocate people from harm's way wherever possible, especially those non-essential to the process. Develop a people migration and backfill strategy for the movement of workers from high risk exposure buildings to new development.
- Construct Fit-For-Purpose Buildings - For those employees that must work near processing units, construct appropriate facilities that mitigate risk (overpressure, toxic release, thermal). Almost any situation can be mitigated with high-cost, blast-resistant design and shelter-in-place facilities. The real question is whether it is better to construct buildings outside the risk zone at a lower cost, especially for personnel who are not directly involved in the process.
- Identify High Value Facility Siting Issues for Priority – In addition to the facility siting drivers related to risk, the plant must also be evaluated to understand development opportunities and constraints posed by the land form, environmental considerations, security requirements, mobility and circulation, utility and enabling support infrastructure and adjacent surrounding development.
- Determine Best Long-Term Strategic Location for Facilities – A comprehensive analysis will not only consider the current state at the plant for facility siting but will also project likely process expansion and the resultant change to the risk footprint and plan for the full buildout of the plant.
- Seek Strategic Opportunities to Consolidate Functions – The benefits of building consolidation are significant and include improved security monitoring, optimized time on tools, floor area reductions due to shared spaces such as break and conferencing, reduced utility costs, economies of scale, reduced O+M, increased knowledge share with increased worker interaction which results in more flexible "hybrid" workers that have an expanded job knowledge base. Figure 5 highlights the benefits of facility consolidation.

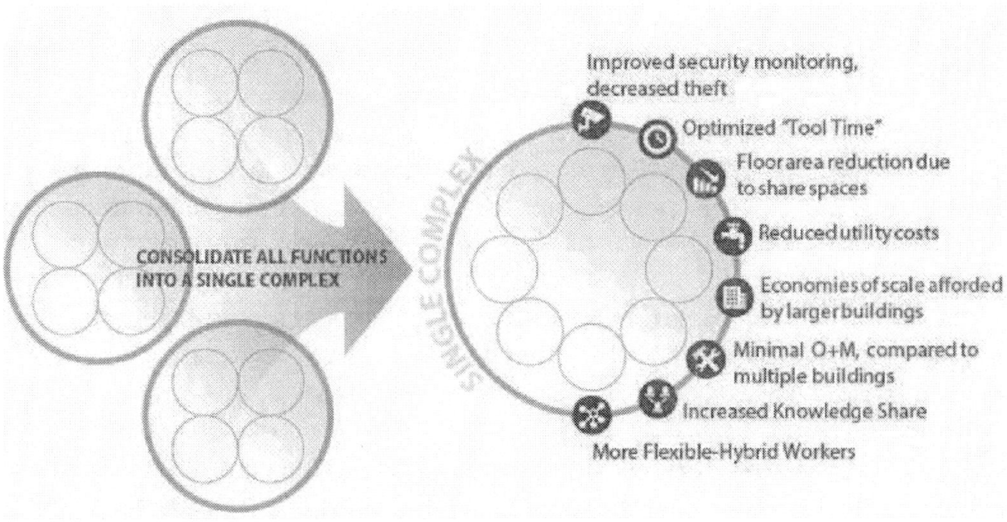

Figure 5 – The benefits of Facility Consolidation

4 The Business Case for Safety

4.1 Safety Savings – Risk Cost Avoidance or Cost of a Safety Incident

Investment in a replacement facility located in a safe area of the plant will help avoid the high cost of future risk related incidents and harm to workers. Prudent business leaders know that, as with other investments, there should be a positive return on investment in facility siting. Much information on the return on investment for improved safety is available on the internet from reputable sources such as OSHA and The Society of Safety Engineers.

- Some key facts from 2014 gathered from these sources include:
  - The total 2014 cost to society of occupational injuries and deaths in the United States was $198.2billion. Of that total, $92.2 billion were in medical, wage, and productivity losses.
  - The 2014 cost per death, ranges between $1.42 million to $3 million per fatality.
  - One dollar invested in 2014 for injury prevention returns between $2 and $6.
- A presentation from CCPS, an AICHE Industry Technology Alliance, titled "The Business Case for Process Safety" stated:
  - A robust process safety program will help your company reduce risk and avoid loss
    - Save lives and reduce injuries
    - Reduce property damage costs
    - Reduce business interruptions
    - Protect market share
    - Reduce litigation costs
    - Reduce regulatory penalties from accidents
    - Reduce regulatory attention

4.2 Safety + Health Management Programs

Safety and health management programs have consistently shown that their implementation has a positive effect on the company's bottom line. Worker's compensation claims, civil liability damages, regulatory fines, and litigation expenses are savings that are easy to see. Other less visible, but nonetheless, significant returns include improved employee morale and productivity. The costs of a process safety accident can be catastrophic, especially with loss of life. The cost of doing the right safety investments reduces the effect of these accidents on people and the company's bottom line. Specific examples of results from petrochemical plants as reported by OSHA include:

- Monsanto Chemical Company – In their Pensacola, FL plant, which employs 1,600 workers, the Lost Workday Case Rates steadily declined during the period the worksite was implementing effective safety investments. Lost Case Workday Rates fell from 2.7 in 1986 to 0.1 in 1994.
- Mobil Chemical Company – When Mobil brought all its existing plants into VPP, OSHA reported that the company's recordable injuries were reduced by 32%.
- Occidental Chemical Company – This chemical company reported that investment in safety has reduced their company-wide injury/illness rate from 6.84 in 1987 to 1.84 in 1993, a %73 decline.
- Kerr-McGee Chemicals – After instituting a VPP program, OSHA noted that production hit an all-time high that is 35% above pre-VPP levels.
  - The business case for safety is compelling. Investment in safety provides significant returns and encourages goodwill from employees, contractors, visitors, and neighbors. The avoidance of catastrophic losses combined with improving employee morale and productivity can only result in improved financial performance.
  - Some companies may look narrowly at only the cost of relocating a group of employees from an area of high risk to lower risk. This is a restricted view that does not account for the full return on investment. Both investment and return should be properly represented to inform educated decisions by management.

5 Strategic Facility Siting Success Stories

Figures 6 and 7 on the following pages are infographics that highlight the success stories of strategic facility siting for two of Jacobs actual clients

US GULF COAST POLY CHEMICAL PLANT

Date Plan Prepared: 2009 | Total Acres: 575 | Total Population: 1,100 | Client Satisfaction Score: 90.0%

Number of Occupied Buildings

- Total Number of Occupied Buildings: 250
- Total Number of High Risk Buildings before SCMP: 60
- Total Number of High Risk Buildings after SCMP: 2

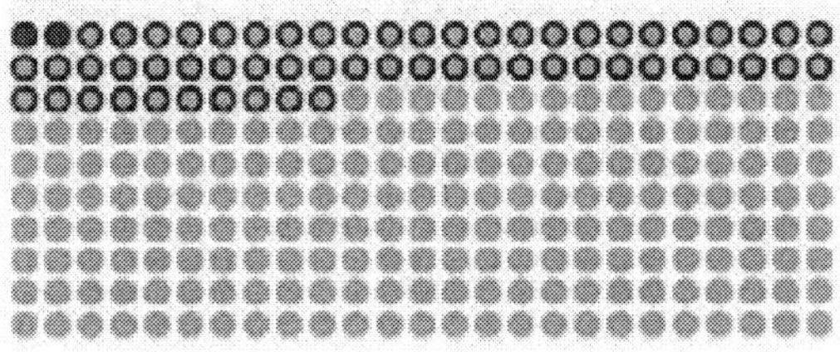

Gross Square Feet (GSF)

This US Gulf Coast Poly Chemical Plant produces petrochemicals and polyethelene. The plant was looking to expand due to advantageous reductions in the cost of feedstock. This price reduction created motivation for new processing units.

The SCMP planned for future process expansion capability. Key roads around the plant were studied and recommendations were made for improvements to handle the large number of personnel on site during capital construction of two major units. The corporate evaluation of the potential sites pointed out access to this site was limited and may not be able to handle the flow of vehicles.

The follow-up planning efforts included showing a phased construction approach and road improvements to overcome any corporate objections to expanding at this site.

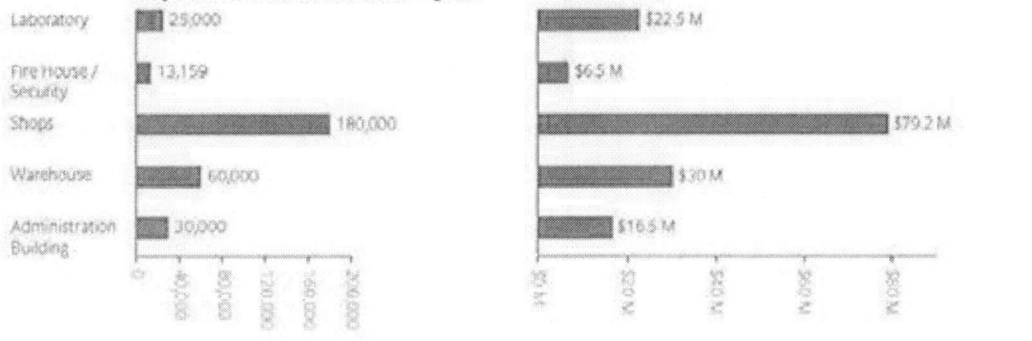

Occupants

68% of 190 Occupants Relocated to Lower Risk

"The planning process made it possible for our plant to attract corporate investment in our major process expansion."

Figure 6

US MIDWEST SPECIALTY CHEMICAL PLANT

Date Plan Prepared: 2009 | Total Acres: 300 | Total Population: 295 | Client Satisfaction Score: 94.8%

Number of Occupied Buildings

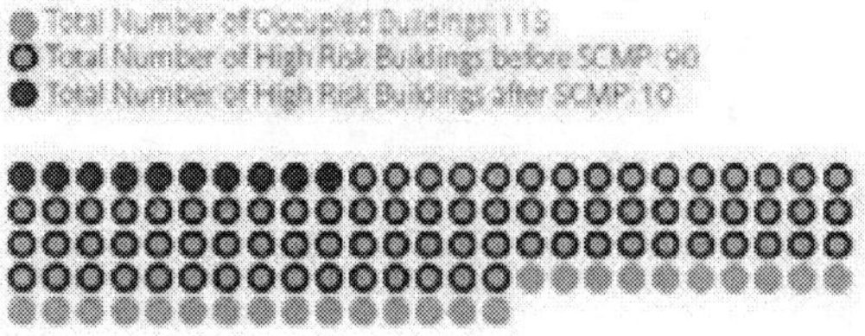

Gross Square Feet (GSF)

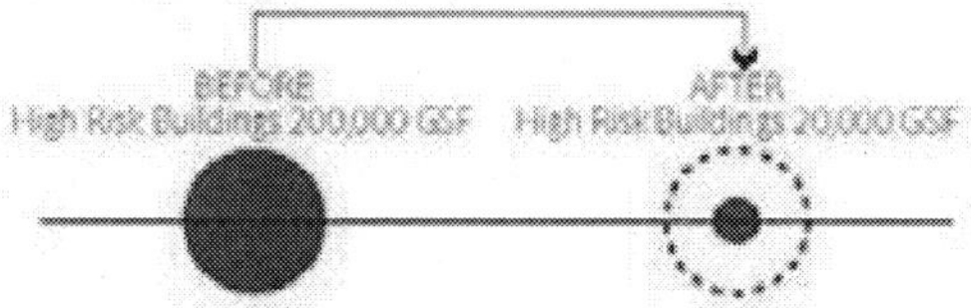

This US Midwest Speciality Chemical Plant was a small chemical plant and had very limited land resources. Both site and corporate management believed the site had been built out to the site's fullest extent.

SCMP revealed a large opportunity to create space inside the fence for future growth by removing embedded non-process buildings. The company owned nearby land that could be connected to the plant, but was located in a low risk zone. This low risk zone was perfect for non-process buildings.

A total of 25% of the processing land could be freed by relocation of non-process buildings to a safer location.

Replacement or Harden Existing GSF

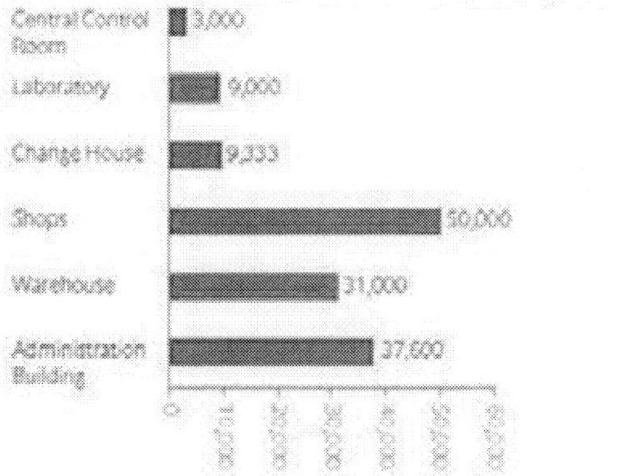

TIC Estimate

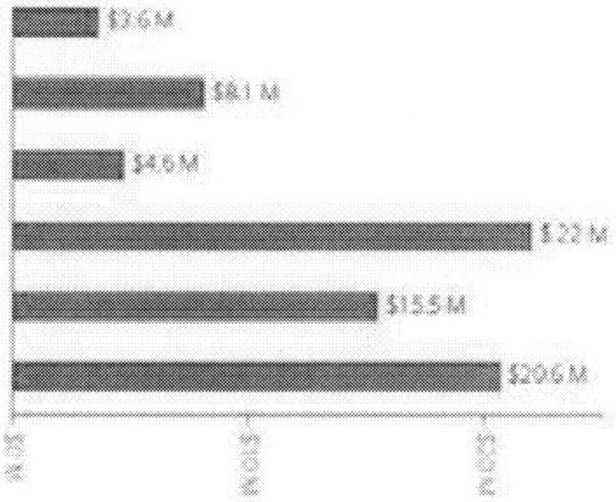

Occupants

100% of 250 Occupants Relocated to Lower Risk

"You have given us a new way to look at our plant. We did not imagine we could attract new processing units with our limited site."

Figure 7

6 Conclusion

In conclusion, the application of a strategic process to facility siting provides the ability to envision and aim for enduring long-term business goals despite short-term priorities of the business environment. A strategic view of facility siting avoids the potential mistakes of a tactical approach.

It creates safer sites for non-process facilities and building consolidation breaks down work culture silos and improves productivity. Clients realize cost avoidance of well performed facility siting and the benefits of a well performed strategic facility siting plan can be unexpectedly beneficial and comprehensive given the plan uses a holistic, multiple disciplined approach.

Large Scale Testing Confirms: A Deflagration to Detonation Transition Needs to be Considered When Facility Siting

Scott Davis
Gexcon US
Bethesda, MD USA
Scott.G.Davis@gexon.com

John Pagliaro, Drew Botwinick, Derek Engel
Gexcon US
Bethesda, MD USA

Kees van Wingerden
Gexcon AS
Bergen, Norway

Erik Merilo, Adam Ziemba, Mark Groethe
SRI International
Menlo Park, CA, USA

Prepared for Presentation at
American Institute of Chemical Engineers
2017 Spring Meeting and 13th Global Congress on Process Safety
San Antonio, TX
March 26 – 29, 2017

Large Scale Testing Confirms: A Deflagration to Detonation Transition Needs to be Considered When Facility Siting

Scott Davis
Gexcon US
Bethesda, MD USA
Scott.G.Davis@gexon.com

John Pagliaro, Drew Botwinick, Derek Engel
Gexcon US
Bethesda, MD USA

Kees van Wingerden
Gexcon AS
Bergen, Norway

Erik Merilo, Adam Ziemba, Mark Groethe
SRI International
Menlo Park, CA, USA

Keywords: *vapor cloud explosions, DDTs, industrial explosions, mitigation measures*

Abstract

A large vapor cloud explosion (VCE) followed by a fire is one of the most dangerous and high-consequence events that can occur at petrochemical facilities. As the size and complexity of facilities increase designs must consider the potential adverse effects associated with vapor cloud explosions in large congested areas and understand the potential for more devastating deflagration-to-detonation transitions (DDTs) on these facilities. While the likelihood of DDTs is lower than deflagrations, they have been identified in some of the most recent large-scale explosion incidents including: 2005 Buncefield explosion, 2009 San Juan explosion, and 2009 Jaipur event. The consequences of DDTs can be orders of magnitude larger than deflagrations because they have the ability to self-propagate outside the region of high congestion/confinement. Hence, it is critical to understand how a facility's geometry or equipment layout can affect explosion consequences and assist in their mitigation and/or prevention. However, most tools at present cannot accurately predict DDTs largely because they have been developed using experiments conducted at small scales.

Due to the inability to predict such devastating phenomena on the large scale, owners and designers cannot evaluate installations for risk of DDTs and provide "inherently safer" layout or mitigation measures to significantly reduce or eliminate such hazards. This paper will present the results of large scale testing that has been conducted in a newly developed test rig of over 50,000 ft^3 (1,500 m^3) gross volume to validate the necessary design tools used to predict the risk of DDT. The facility was designed in a modular-type fashion, whereby the congestion levels can be modified and adjusted to ensure detonations for fuels with different reactivities. These tests determined under what conditions two different reactivity fuels (propane and methane) transition from deflagrations to detonations at stoichiometric, lean and rich mixtures. The results indicate that DDTs are more likely to occur than originally anticipated at the large scale, and that current tools developed using smaller scale experiments require modifications. In addition, preliminary mitigation tests demonstrate that inhibitors are effective at slowing the propagation of a fast deflagration and help mitigate the transition to detonation.

1 Background and Introduction

A large vapor cloud explosion (VCE) followed by a fire is one of the most dangerous and high-consequence events that can occur. As the size and complexity of facilities increase, designs must consider the potential adverse effects associated with vapor cloud explosions in large congested areas. However, one of the most devastating explosions is when the subsonic deflagration transitions to a supersonic flame front, which can travel at speeds greater than 1700 m/s (3,800 miles per hour) and generate pressures greater than 17 barg (250 psig). This phenomenon is called a deflagration-to-detonation transition (DDT), whereby the resulting detonation velocities are of the same order as high order explosives. In addition, the resulting damage from such events can be quite devastating to plants and personnel.

While DDTs are less likely to occur than deflagrations, they have been identified in some of the most recent large-scale explosion incidents including: 2004 Skikda explosion, 2005 Buncefield explosion [1][2], 2009 San Juan explosion [3], and 2009 Jaipur event [4]. The consequences of DDTs can be orders of magnitude larger than deflagrations because they have the ability to self-propagate outside the region of high congestion/confinement and produce blast overpressures up to two orders of magnitude higher. For example, Figure 1.1 and Figure 1.2 show evidence of extremely high overpressures from an explosion at a LNG liquefaction plant in the Skikda, Algeria. Figure 1.3 and Figure 1.4 show the devastating damage caused by the 2005 explosion at the Buncefield fuel depot. Due to the inability to predict such devastating phenomena on the large scale, owners and designers cannot evaluate installations for risk of DDTs and provide "inherently safer" layout or mitigation measures to significantly reduce or eliminate the hazards associated with these events.

Figure 1.1: Images of the 2004 explosion and ensuing fire at the Skikda LNG liquefaction facility.

Figure 1.2: Images of the damage from the 2004 Skikda explosion: left image – truck, right image – maintenance building.

Figure 1.3: Images of the 2005 explosion and ensuing fire at the Buncefield tank farm.

Figure 1.4: Images of the damage from the 2005 Buncefield explosion: left image – automobile, right image – Fuji building.

The inability to predict DDT phenomena at the large scale becomes increasingly problematic as the size and complexity of facilities increases. Designs must consider the potential adverse effects associated with vapor cloud explosions in large congested areas and understand the potential for more devastating deflagration-to-detonation transitions (DDTs) within these facilities. Hence, it is critical to understand how a facility's geometry or equipment layout can affect explosion consequences and assist in their mitigation and/or prevention.

In order to achieve such goals, Gexcon and SRI International have performed large-scale experiments to validate DDT onset prediction and active mitigation at scales relevant to large-scale designs. This work provides the bases for validation and refinement of industry-accepted CFD tools and simplified methods, which are currently used to assess the consequences of explosions in the oil/gas industry [5][6][7]. Gexcon, who has teamed with SRI, was awarded Subcontract 12121-6403-01 under the Research Partnership to Secure Energy for America (RPSEA) The objectives of this project were to: (1) improve, adapt, and validate the tools necessary to predict high consequence events early in the design phase of facilities and (2) provide guidance and recommended practices to facility owners and designers which, when utilized, will help minimize and design against the consequences of fire/explosion incidents.

Gexcon and SRI have developed new advanced flame tracking techniques to identify the onset of transition for DDTs and present the new results in this paper. In addition, Gexcon is currently developing shock front tracking procedures to not only evaluate DDT events, but also the rarefaction waves that propagate back towards the ignition location. Lastly, preliminary results are presented on the predictions using the current version of FLACS as well as novel models being developed to improve prediction of such events.

2 DDT Test Data and Results

2.1 Testing Setup

Gexcon teamed with SRI to develop a joint large-scale explosion test site at SRI's 480 acre CHES facility in California. This test pad was designed for continued use after the RPSEA test matrix. As part of the testing plan, Gexcon and SRI developed a modular test assembly. An example of a single model can be seen in the left image in Figure 2.1. This allowed for extensive flexibility in the testing, whereby congestion levels and congestion orientations could easily be modified within the rig by installing addition pipes (e.g., approximately 7 and 2 inches) as shown in Figure 2.1. The modules were placed in an array for testing with 2 modules across, and between 12-15 modules deep and 1 module high as shown in Figure 2.2.

Two module configurations were used: 24 modules with a gross volume for these tests of approximately 1,200 m^3 (42,000 ft^3) with dimensions of approximately 8 m wide by 48 m long and 4 m tall; and 30 modules with a gross volume of approximately 1,500 m^3. No pipes were added to the "low" configuration and numerous pipes were added for the final tests outlined in the test matrix with a "high" congestion configuration.

Figure 2.1: Single module (left) and additional piping added to increase congestion (right).

Figure 2.2: Finalized modules on the test site.

A single polyethylene sheet of either 0.05 or 0.1 mm (0.002 or 0.004 in) thickness covering the modules was used to retain the gas/air mixture prior to ignition. The polyethylene sheet was secured directly to the modules. Some tests were performed with a tent cutting apparatus to account for any variation in results due to the initial confinement the sheeting provided. An electric spark was used as the ignition source and was located approximately half a module length within the structure, centered and at mid-module height.

Gas was introduced into the rig at four separate locations and mixed thoroughly with numerous box fans. Samples were extracted at 8 locations throughout the rig, with half the sample points in the upper section and half near the ground. The extracted samples were pumped to an IR gas analyzer (IR4800D three channel), which was calibrated prior to testing. Tests included commercial grade propane (HD-5, >95% purity), experimental grade propane (>99.5% purity) and ultra-high purity methane (>99.97%). Once uniform concentrations were established in the rig, the fans were turned off and the mixture was ignited 1 minute later (with the exception of Test #8, during which the fans were left on up until ignition).

For Test #5, the flame progression was filmed with two high-speed video cameras and two normal speed cameras. Test #6 and beyond included an additional high-speed camera setup for a close-up view of the initial flame development in the first few rows of modules. The high-speed video data is used to provide a very accurate flame-front location as a function of time from which the flame speed can be determined and will be discussed later.

Ionization pin sensors were placed in the rig and used to determine the flame arrival times within the modules for Tests #12-17. The overpressure development within the test rig was measured using 13 pressure transducers located along the major axis, nominally flush with ground level and coupled with the ion pins within the rig. There are four other pressure transducers outside the rig, two along the major axis and two perpendicular to the anticipated DDT location. Two onsite weather stations recorded local temperature, wind speed, wind direction and humidity at the time of each test.

Table 2.1 provides details for each test, including the fuel, equivalence ratio, thickness of the plastic containment sheet, rig size, and level of congestion. The last column in the table includes test-specific notes to be discussed below, such as whether mitigation methods were tested. Tests #1-4 were practice and calibration tests and therefore test numbering starts at #5 in the table.

Table 2.1: Test Matrix.

Test #	Fuel	Equivalence Ratio	Plastic Thickness (mils)	Rig Size (modules)	Congestion	Notes
5	Propane (94.5%)	1.10	4	24	Low	
6	Propane (95.2%)	1.10	4	24	Low	
7	Propane (95.2%)	1.05	2 on first three rows, 4 on rest	24	Low	
8	Propane (99.5%)	1.05	2	24	Low	Circulator fans left on
9	UHP Methane	0.95	2	30	Low	
10	Propane (99.5%)	0.80	2	30	Low	
11	Propane (99.5%)	1.35	2	30	Low	
12	Propane (99.5%)	1.10	2	24	Low	Cut tent
13	UHP Methane	1.05	4 on top and edges, 2 on sides	30	High	Cut tent
14	Propane (99.5%)	0.90	4 on top and edges, 2 on sides	30	High	Cut tent
15	Propane (99.5%)	1.35	4 on top and edges, 2 on sides	30	High	Cut tent
16	UHP Methane	1.05	4 on top and edges, 2 on sides	30	High	Mitigation agent added
17	UHP Methane	1.05	4 on top and edges, 2 on sides	14	High	Mitigation agent added

2.2 Test Results

2.2.1 Low congestion

Tests #5-12 were performed in the "low" congestion rig with no pipes added to the rig. The volume blockage ratio (e.g., volume occupied by the obstacles divided by the volume of the rig) was approximately 5.8% for this configuration.

Propane – near stoichiometric conditions

Commercial grade HD-5 propane at near stoichiometric conditions was the first fuel tested with 24 modules and a DDT was observed approximately half way down the rig. (see Figure 2.3 and Figure 2.4). A repeat of this test, using a different batch of similar commercial propane resulted in a DDT; however, the DDT occurred approximately two modules farther down the rig (see Figure 2.5).

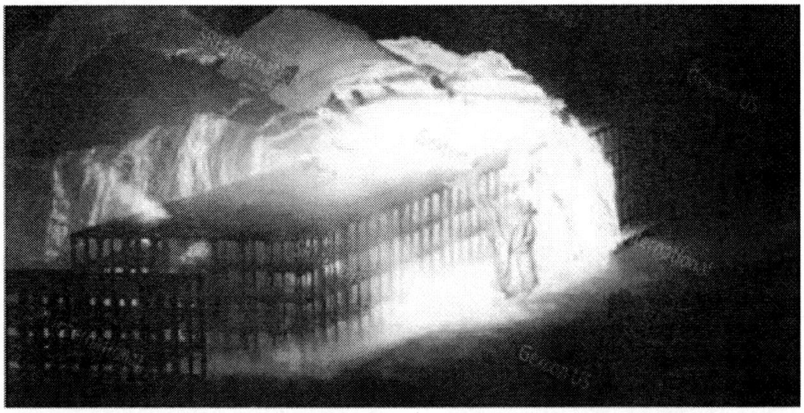

Figure 2.3: Image of flame propagation just prior to detonation in Test #5.

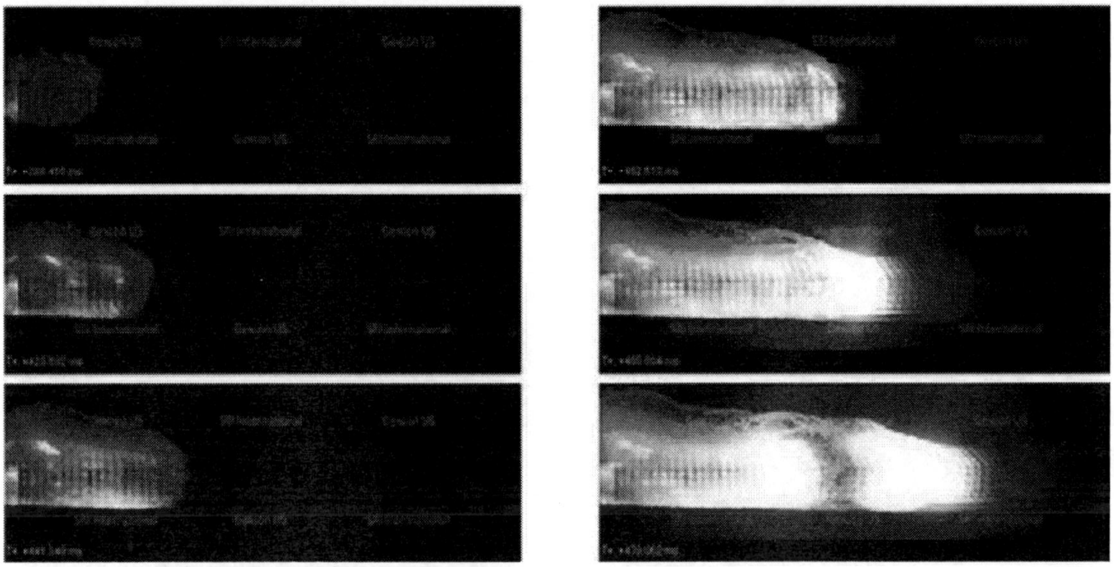

Figure 2.4: Images of flame propagation before and after transition to detonation in Test #5.

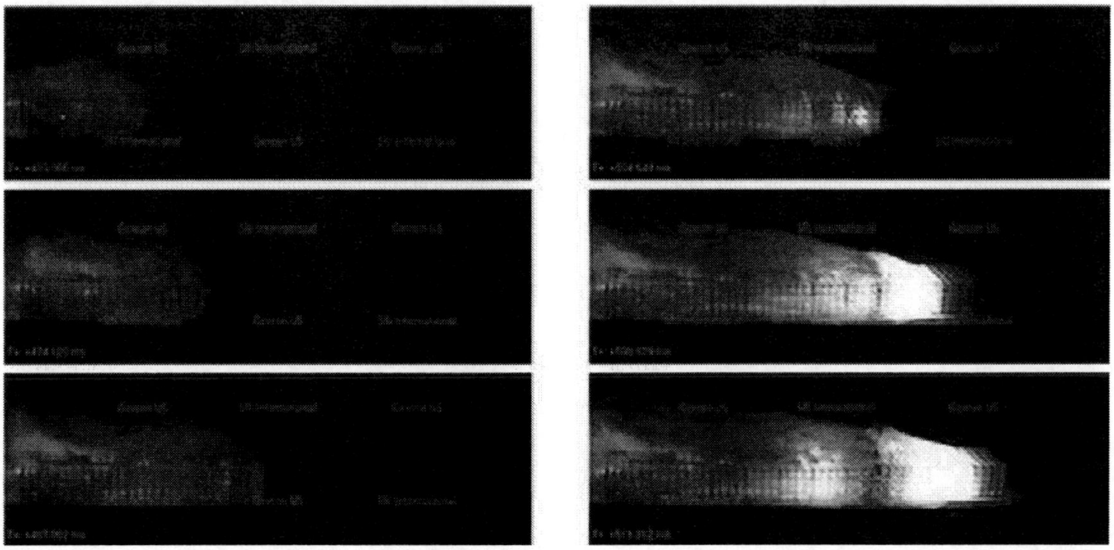

Figure 2.5: Images of flame propagation before and after transition to detonation in Test #6.

Additional tests were performed in order to better understand the potential reasons that propane exhibited a deflagration-to-detonation transition for "low" congestion conditions. More specifically, the effect of the plastic sheet and propane purity were evaluated. The next test conducted (Test #7) was with thinner plastic sheets (2 mils instead of 4 mils) on the first 3 rows. This would alleviate some initial confinement during initial flame development. The end result was nearly identical to that of Test #6 as shown in Figure 2.6 and Figure 2.7. Test #8 was conducted with a higher purity propane (~99.5%); however, the fans were inadvertently left in operation during the test. As expected, the initial turbulence created by the fans allowed the flame to transition to detonation much early than the tests without initial turbulence.

One final test (#12) with the higher purity propane was conducted in order to eliminate the effect of the containment plastic. In this test, a mortar was launched connected with wire to the upper corners along the entire length of the congestion rig. The timing of the system was set to cut the plastic ahead of the flame front and eliminate the "weight" of the plastic. With the mortar system in place, the higher purity propane again had a deflagration-to-detonation transition, indicating that neither the purity of the propane nor the confinement of the plastic had an appreciable effect on the results for propane.

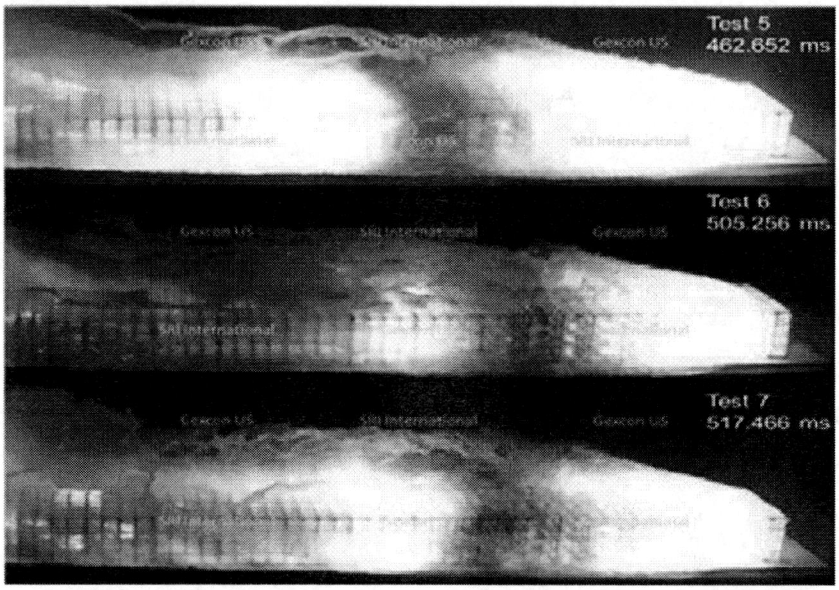

Figure 2.6: Comparison of the detonation fronts as they reach the end of the rig, along with the elapsed time from ignition.

Figure 2.7: Damage comparisons for Tests #5-7.

Propane – lean and rich conditions

Tests #10 and #11 were conducted with lean and rich mixtures of propane respectively. The equivalence ratios at ignition were 0.8 and 1.35. They were conducted in the full 30 module uncongested array. These tests did not result in a DDT.

Methane – near stoichiometric conditions

Test #9 was conducted with a near stoichiometric concentration of methane ($\varphi = 0.95$). This was the first test conducted with the full 30 module array. The thinner 2 mils plastic was used for this test. This tests did not result in a DDT and the resulting flame speed reached a constant value of 75 m/s.

2.2.2 High congestion

Beginning with Test #13, the congestion in the modules was changed to the "high" configuration. This configuration added numerous 7" and 2" pipes into each module with different horizontal and vertical orientations for two-dimensional congestion. The resulting volume blockage ratio was approximately 10.9%. Tests were performed with near stoichiometric methane, lean and rich propane.

Methane – near stoichiometric conditions

Test #13 was performed with near stoichiometric methane ($\varphi = 1.05$). The results indicated that for this level of congestion and methane/air mixture, the velocity of the flame front reached fast deflagration speeds of approximately 250 m/s (discussed next), and no transition to detonation was observed.

Propane – near stoichiometric conditions

When tests using lean ($\varphi = 0.90$) and rich mixtures ($\varphi = 1.35$) of propane were performed on the same high obstacle density rig, they both resulted in detonation.

2.2.3 Mitigation Tests

The "high" congestion test with near stoichiometric methane (Test #13) using 30 modules did not result in a detonation, this configuration was chosen as the "base" case to test the mitigation system. In contrast to the base case, the plastic sheet was secured in place for test series. In fact, the winds at the time of testing were quite strong and the plastic was held in place with a stronger mechanism at the bottom. The idea of the mitigation was to effectively apply a "mitigation" curtain using a solid inhibitor as shown in Figure 2.8 to slow down the propagating flame front. While the width of the inhibitor curtain was not measured, visual inspection showed that the width was nearly uniform from top to bottom of the rig for approximately 3.7 m (12 ft), with the lower region of the curtain spanning approximately 6 m (20 ft).

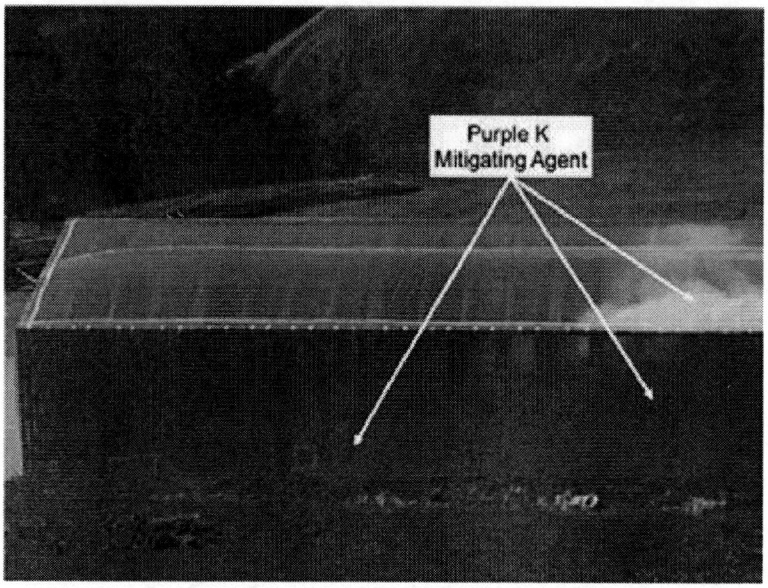

Figure 2.8: Solid inhibitor "curtain" released within the modules.

The first test (Test #16) applied the inhibitor in the last third of the rig. This test resulted in methane propagating at nearly 800 m/s as an overdriven deflagration after only 25 meters, with a shock wave traveling with the flame front. Once the deflagration impacted the inhibitor curtain, the flame velocity decreased considerably (~400 m/s) yet the shock wave continued into the unburned mixture ahead of the flame (see Figure 2.9). The flame was not fully extinguished and when it reached the flammable mixture behind the shock front, it accelerated up again to near overdriven conditions.

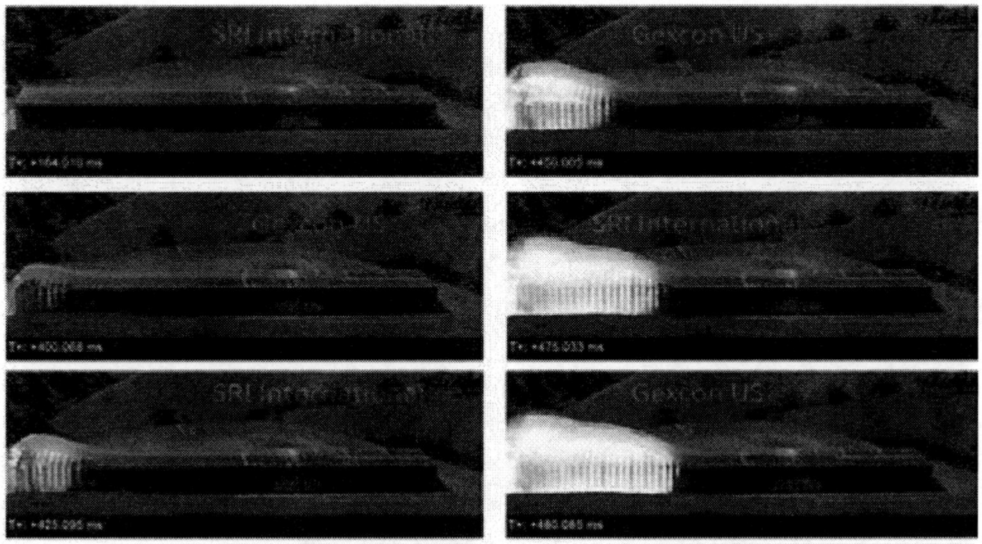

Test 16 – Deflagration with Flame Acceleration-Part 1 Screenshots

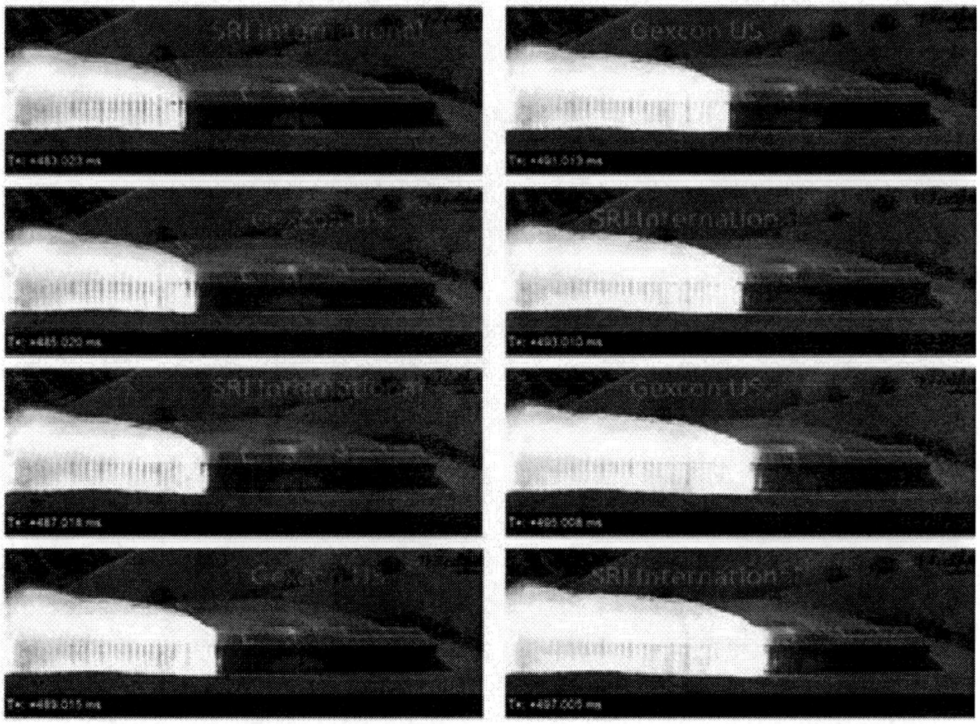

Test 16 – High Speed Constant Velocity Deflagration - Screenshots

Figure 2.9: High congestion stoichiometric methane with mitigation system.

The extra reinforcement of the plastic sheet apparently created initial confinement that helped accelerate the flame front to speeds not observed in the previous test where the plastic was cut. Therefore, in the second mitigation experiment (Test #17), a shorter test rig was used and the mitigation region was positioned considerably closer to the ignition point, keeping the same extra reinforcement of the plastic sheet. The flame front hit the mitigation region at approximately 200 m/s and drastically decelerated to below 50 m/s after the mitigation region. When the front left the mitigation region, it was able to redevelop into an approximately planar wave in the rig. The curtain effectively prevents the overdriven deflagration from occurring, which was shown to occur in the absence of that mitigation for the gas mixture used.

2.2.4 Transition to Detonation Summary

The following table summarizes the transition to detonation test results. All tests conducted with near stoichiometric propane transitioned to detonation. Tests with both lean and rich propane mixtures, with low congestion, did not experience a transition to detonation. When the congestion was increased to the high configuration, lean and rich propane did transition. Detailed comparisons to other studies will be provided in the Analysis section.

Table 2.2: Transition to detonation test results.

Test	Fuel	Equivalence Ratio	Congestion	DDT?
5	Propane (94.5%)	1.1	Low	Yes
6	Propane (95.2%)	1.1	Low	Yes
7	Propane (95.2%)	1.05	Low	Yes
8	Propane (99.5%)	1.05	Low	Yes
9	Methane	0.95	Low	No
10	Propane (99.5%)	0.8	Low	No
11	Propane (99.5%)	1.35	Low	No
12	Propane (99.5%)	1.1	Low	Yes
13	Methane	1.05	High	No
14	Propane (99.5%)	0.9	High	Yes
15	Propane (99.5%)	1.35	High	Yes
16	Methane	1.05	High	Mitigation Test
17	Methane	1.05	High	Mitigation Test

3 Analysis

3.1 Flame front visualization and tracking

Gexcon and SRI determined flame front position versus time from the high-speed video footage of each test. We tried several different methods to see if the results varied and to determine which method yielded results closest to ion pin data. Flame positions versus time at specified fixed heights within the rig were determined with commercially available motion-tracking software (referred herein as Method 1). The flame front was identified in each frame based on a pre-defined intensity threshold that required adjustment as the flame progressed through the rig. Adjustment was necessary because the flame intensity varied during the deflagration and detonation periods of propagation, and because during the detonation period, significant light was projected out in front of the flame front. The top image of Figure 3.1 shows an example of the flame front tracking at fixed heights and the bottom figure shows the flame front position versus time data extracted using Method 1. As seen in the provided example, the measured flame front position varies only slightly with tracking height (i.e. flame propagation is planar).

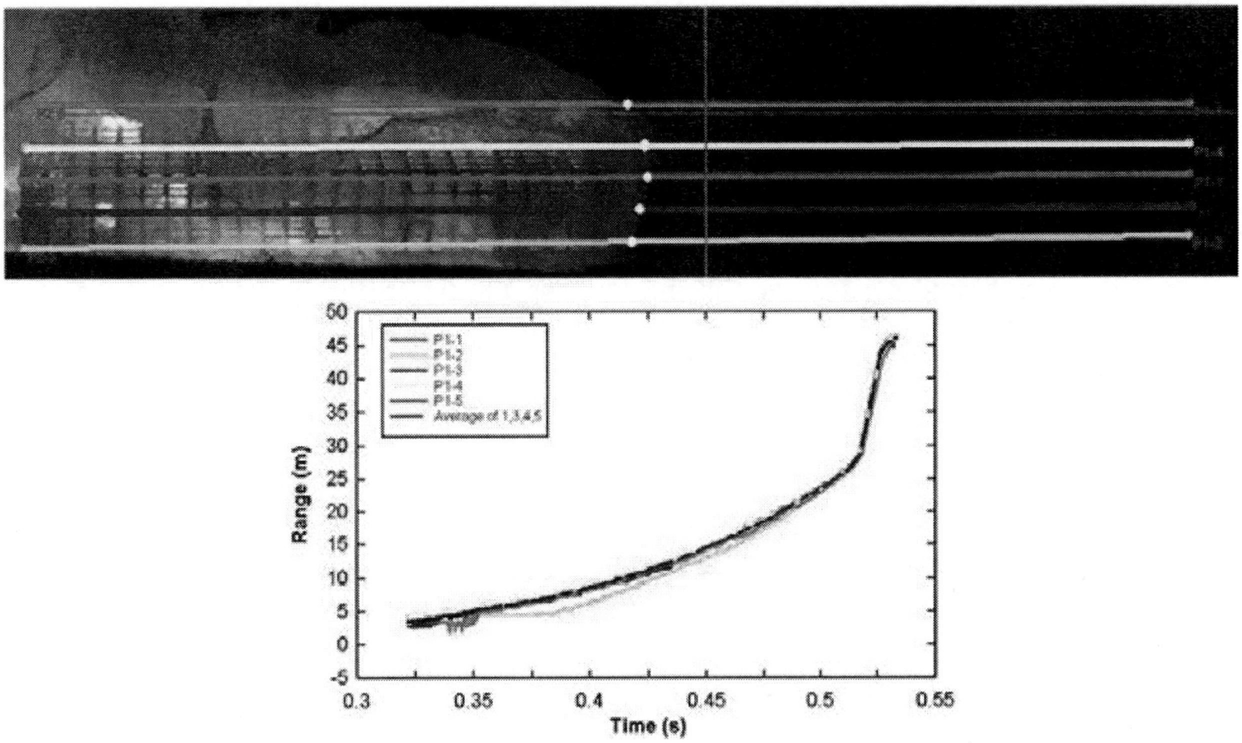

Figure 3.1: Flame front tracking at various heights using Method 1. The top image shows the tracking heights and the bottom figure provides the flame front position versus time at the different tracking heights.

Recently developed in-house software provided alternative flame front tracking methods. In the software, the user defines a specific row of pixels at any fixed height and the software extracts the row from each frame. Each row is then rotated 90° and stacked sequentially. Pixel widths (before rotation) are determined from the specified start and end points of the rig and its length. The height of each pixel (before rotation) represents a fixed duration based on the video frame rate. Hence once the pixel rows are rotated and combined they form a position versus time plot (see Figure 3.2 for examples), which we refer to further as an x-t diagram. The software can generate figures in true color, gray scale, or with a color map. The software can also take the gradient of the light intensity to highlight areas with sharp light intensity changes such as the interface between the flame front and darker background.

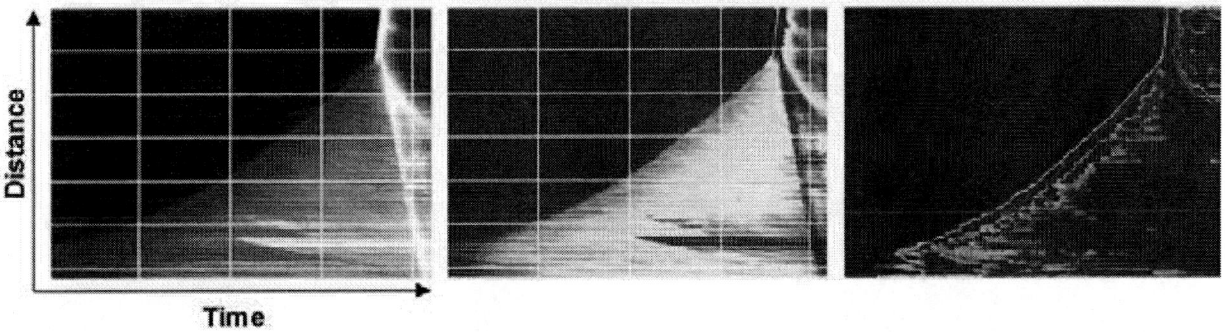

Figure 3.2: Example x-t diagrams: full color output in the left frame, color map intensity output in the center frame, and color map gradient output in the right frame.

The color map x-t diagram of light intensity and the color map x-t diagram of the gradient of light intensity were digitized and the leading edge of the flame front was identified near the mid-height of the rig (referred herein as Method 2 and Method 3). These later techniques proved to be the most accurate methods as the user could view the video simultaneously and manually filter out noise such as the light reflecting off of the containment sheet ahead of the flame. Another major source of noise, which could be accounted for using this method, was the influence of the test rig structure. For certain tests, depending on the ambient lighting conditions, the flame front would clearly disappear behind the structural elements along the rig creating a step-like flame front position in the x-t diagrams. With this method, the flame front position could be identified when it was in between structural elements and interpolated when behind structural elements. The red circles in the left frame (a) of Figure 3.3 shows the points extracted as the flame front. The image in the right frame (b) shows an example of the step-like flame front position caused by structural interference.

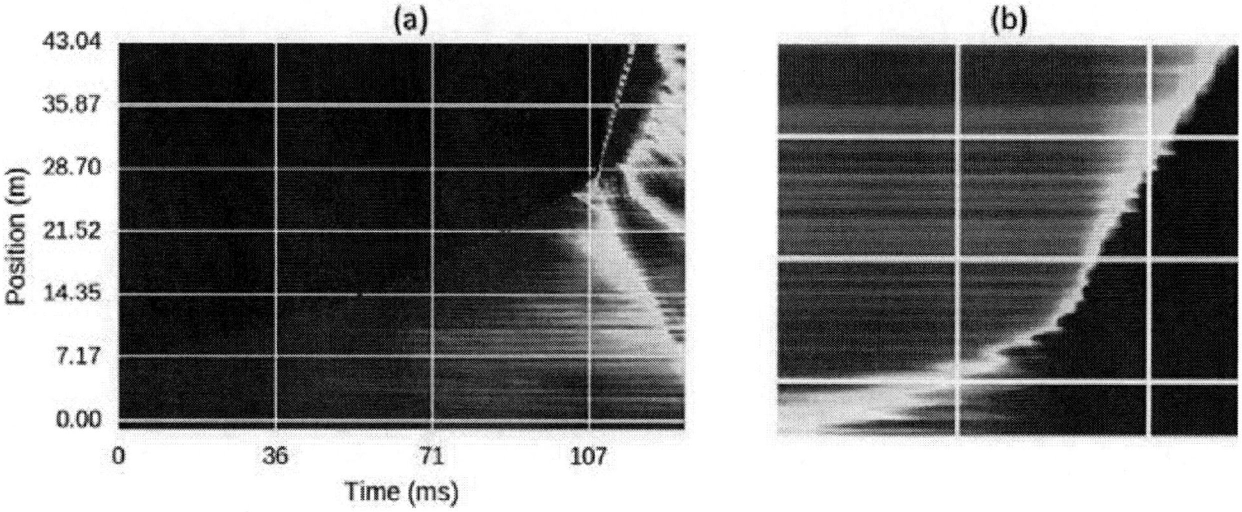

Figure 3.3: (a) Example extraction of flame front position from a color map intensity x-t diagram and (b) example of step-like flame front caused by structural interference.

Extracted positon versus time using Methods 1, 2, and 3 are compared to ion pin data in Figure 3.4 for Tests #13 and #15. The data for these tests were chosen to demonstrate tracking accuracy during both deflagration and detonation (Test #13 – no detonation; Test #15 – DDT). Figure 3.4 shows that Methods 2 and 3 (x-t diagram methods) yield very similar results to each other and are closer to the ion pin data than results from Method 1. Flame front positions from Method 1 (tracked at the high position shown in Figure 3.1) are noisier and exhibit the step-like propagation since there is no correction for when the flame disappears behind structural elements.

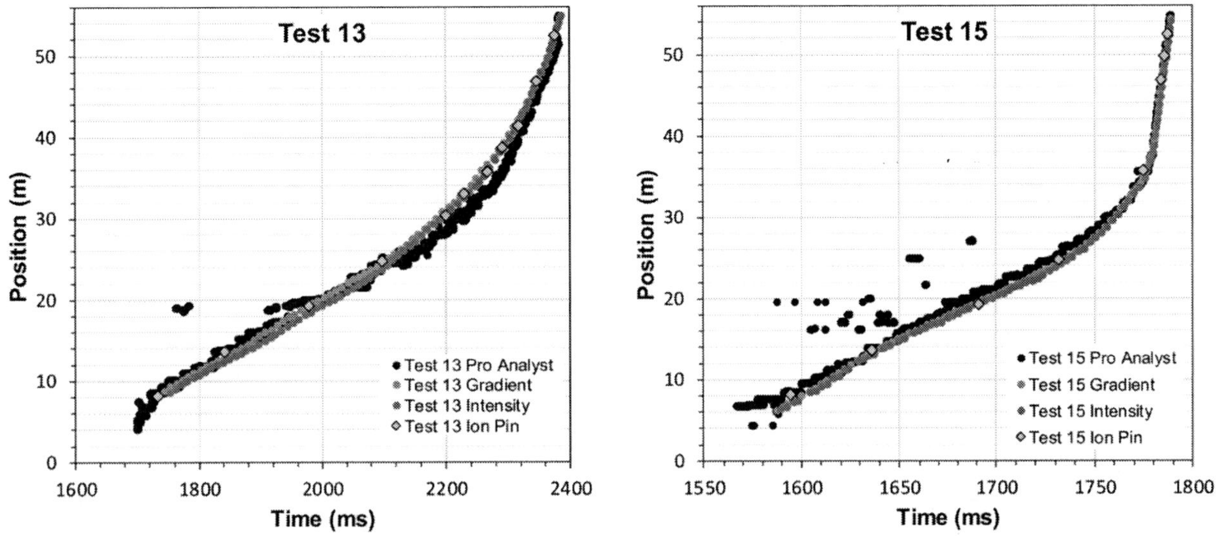

Figure 3.4: Examples of extracted flame front position versus time from videos using Methods 1, 2, and 3 compared to ion pin data.

Figure 3.5 provides flame position as a function of time for tests that did not detonate (frame a), tests with mitigation (frame b), and tests that transitioned to detonation (frame c) using Method 2 described above.

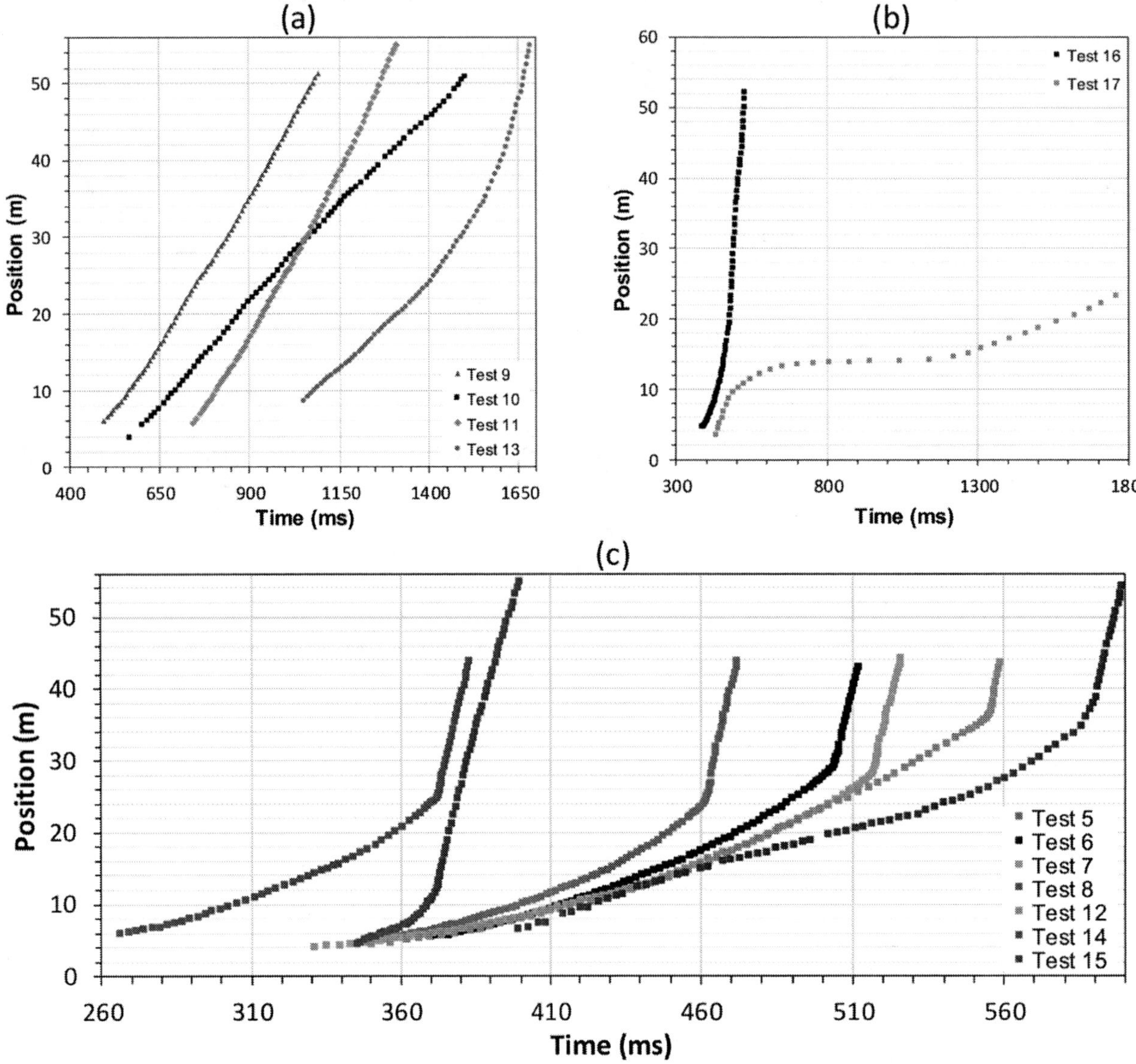

Figure 3.5: Flame front position as a function of time for (a) tests that did not detonate (b) tests with mitigation measures, and (c) tests that transitioned to detonation.

3.2 Flame front velocity

Flame position versus time can be differentiated to yield flame front velocity as a function of time and position. The preliminary approach taken (and used to generate results presented in refs. [8][9]) was to fit the position versus time data with polynomials and then take the analytical derivate of the fits to yield flame front velocity. This approach provided continuous flame speed curves for the deflagration and detonation portions (if present) of each test, but after further analysis, yielded inaccurate flame front velocities around the point of transition.

Previously reported flame front velocities were determined by fitting the deflagration and detonation (if occurring) portions of the data separately. Flame positions during deflagrations were fit with a 2nd order polynomial. For tests that transitioned to detonation, flame positions after detonation were fit with a linear function. For tests that did not detonate, the flame velocity was determined as the derivate of the fit to the data during deflagration. For tests that did detonate, the fits to the data during deflagration and detonation were combined to yield the flame speed for the full duration of the test. The point when the first fit ended and the second fit started was varied until the sum of the R^2 values of each fit was maximized. This point was considered the point of transition from deflagration to detonation. Note that the flame speed just prior to transition to detonation may be very sensitive to this selected point.

For the present dataset, initial results using this method (presented in refs. [8][9]) indicated transition to detonation once flame speeds approached the speed of sound in the unburned reactant mixture. For tests that transitioned, the average transition velocity originally reported was around 250 m/s and the resulting detonation velocities were in the range of 1600-1800 m/s.

Figure 3.6 illustrates why the previously reported transition flame speeds were too low. Frame (a) shows the position versus time data and the fits to the data from the initial flame development up to when the flame reaches the end of the rig. Frame (b) focuses on the data and fits near the point of transition. Frame (c) shows the flame front velocity versus position determined based on the fits shown in frames (a) and (b). The change in flame position near the transition region is no longer captured by the 2nd order fit and appears to increase exponentially. This method therefore provided reasonable flame speeds for the majority of the deflagration and detonation stages, but less accurate flame speeds around the point of transition.

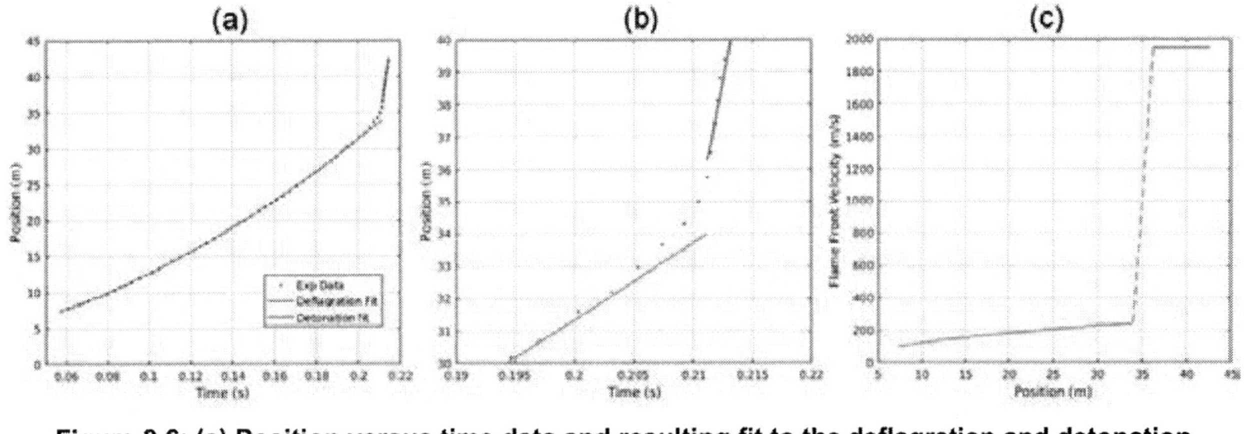

Figure 3.6: (a) Position versus time data and resulting fit to the deflagration and detonation portions of the data; (b) zoomed-in view of the transition region; (c) resulting flame front velocity versus position.

Flame front velocity was alternatively determined at each flame position as the derivate of a linear fit through the local data (i.e. a moving linear regression). For the present results, two data points before and after were considered in the linear fits providing flame speed at each position. This approach does not yield continuously increasing and smooth flame velocities (due to uncertainty in the extracted flame fronts from video), but instead yields a more accurate instantaneous flame speed for the entire flame propagation, including the transition region. Figure 3.7 shows the flame speeds determined using this method. For clarity, dashed lines connect square symbols representing flame speeds at the flame positions extracted from the videos. In Figure 3.7, frame (a) shows flame speeds for tests that did not detonate, frame (b) shows flame speeds for tests with mitigation, and frame (c) shows flame speeds for tests that transitioned to detonation.

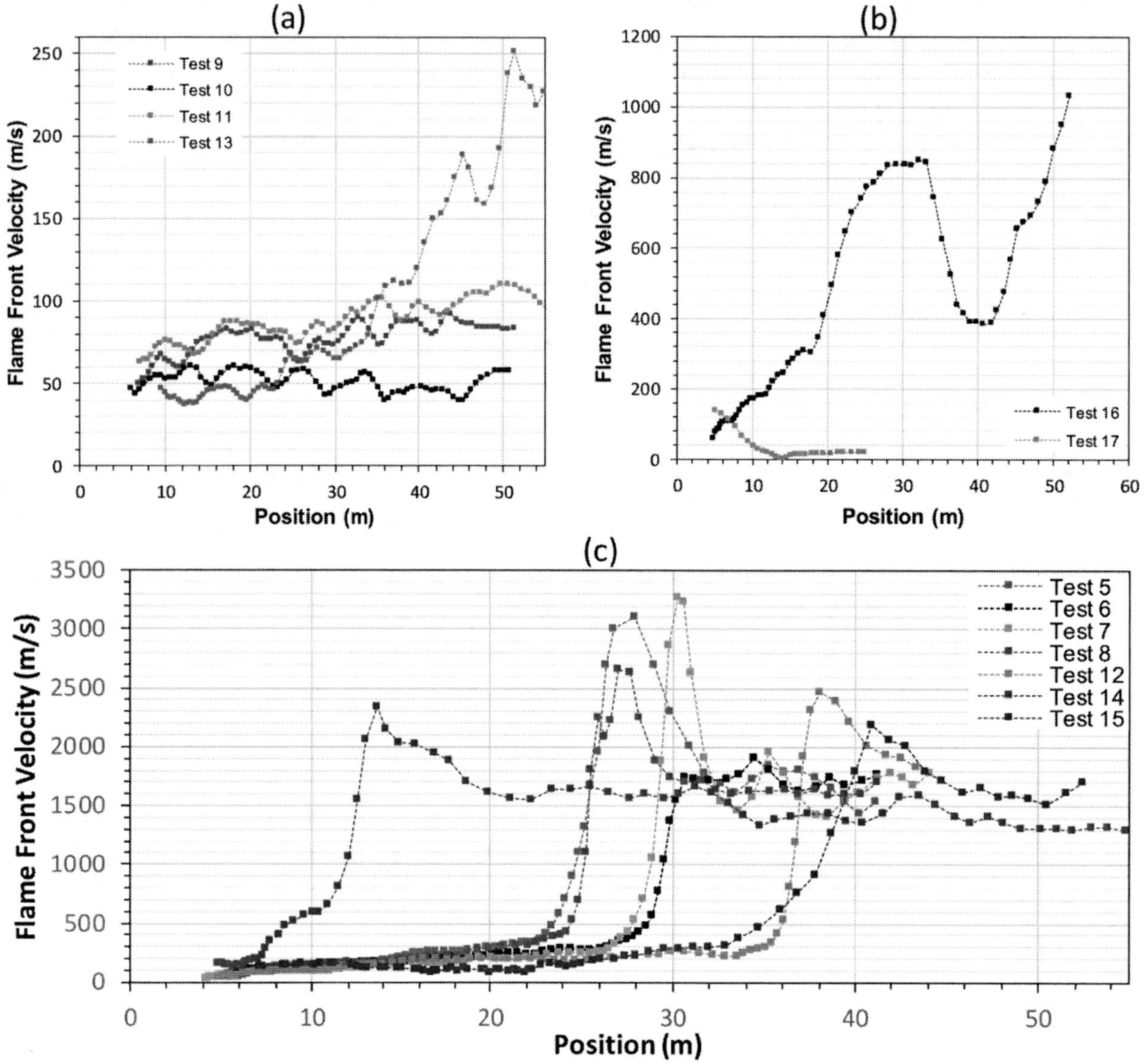

Figure 3.7: Flame front velocities for tests that (a) did not detonation, (b) involved mitigation, and (c) transitioned to detonation.

For the low congestion deflagration results of near stoichiometric methane (Test #9) and lean/rich propane (Tests #10 and #11), the flame speeds were generally near 100 m/s or lower. For the high congestion deflagration of near stoichiometric methane (Test #13), flame speeds were continuously increasing to values approximately 250 m/s at the end of the rig. For the mitigation tests, Test #16 with methane resulted in the flame front propagating at approximately 800 m/s as an overdriven deflagration after only 25 meters, with a shock wave traveling with the flame front. Once the deflagration impacted the inhibitor curtain, the flame velocity decreased considerably down to 400 m/s and the flame was not fully extinguished upon reaching the end of the inhibitor region. Once it exited this region, it accelerated again to near overdriven conditions by the time it reached the end of the rig. When the inhibitor in Test #17 was moved closer to the ignition location and prior to the region of overdriven conditions observed in Test #16, it successfully reduced that flame speed to very small values and there was no evidence of the overdriven condition.

For the DDT propane dataset, transition to detonation occurs closer to 500 m/s as opposed to the originally reported value of 250 m/s. These results are more consistent with previous studies that reported necessary flame speeds of approximately 600 m/s for DDTs to occur (see refs [11] [12] [13]). Consistent with the previous study, the present data shows that detonation velocities from almost all tests were on average between 1600-1840 m/s, except for the rich propane case (phi = 1.35) in high congestion where the average detonation velocity was approximately 1500 m/s. Theoretical values of the Chapman-Jouguet (CJ) velocities for the propane mixtures is approximately 1800 m/s. The measured values are fairly close to the theoretical values and may be slightly lower due to the flow resistance caused by the obstacles and agree well with previous studies.

The experimental flame velocities exhibit a brief overshoot before reaching the stable detonation velocity for each mixture, reaching 2200-3300 m/s before dropping down to near CJ velocity values of 1600-1840 m/s. This overshoot is likely the result of the pre-compression of the unburned gases followed by the shock wave passing through this pre-compressed region during the transition to detonation, leading to higher pressures and velocities than the stable CJ point. Once the shock front establishes and auto-ignites the mixtures behind it the stable CJ velocities are observed. Other investigations have also observed flame velocities much higher than the CJ velocity right after the transition to detonation occurred. Shell has previously presented overdriven detonation velocities with ethane as a fuel. Shell's congested rig, of dimensions, 20 by 6 by 3 m, with a volume blockage ratio of approximately 12% produced flame speeds of 3300 m/s just outside the congested region with ethane as a fuel, over the open space the flame speeds were still over the 1825 m/s CJ velocity [11].

The main uncertainties in the present calculations involve the interpretation of the flame front at each point in the rig and how the flame speed is calculated from the flame position data. The resolution in the flame front depends on the frame rate used in the recording, the resolution of the images and the post-processing of how the flame front is defined. Resolution of the present study was more than ample and in fact averaging over small distances needed to be conducted to avoid local "noise" in the flame velocities. Various techniques were utilized and resulted in very similar flame front positions versus time. The main uncertainty was how to determine the slope of the flame position versus distance, and accurately predict the flame speed, especially at locations where high transients occur (e.g., transition to detonation and overshoot of the CJ velocity right during transition to DDT). Preliminary calculations showed a high sensitivity to the value of the flame speed velocity at transition, and further study is currently being conducted to compare the given method with those reported in previous studies.

3.3 Shock fronts

Different image filters and processing techniques can be applied to highlight the shockwave resulting from detonation. A discontinuity in gas density is created across a shockwave as it travels. Since density and refractive index are closely related, the shock changes the refractive index of the air it occupies. This in turn alters the pixel intensities by a small amount in the image. This change may be difficult to detect directly, but after filtering and taking spatial gradients, it becomes more clear due to the discontinuous nature of the shock front.

In addition to the flame front, shock waves are also visible in the x-t diagram. Figure 3.8 shows an example of a color map x-t diagram showing the gradient of light intensity. As mentioned earlier, the flame front and transition to the detonation shock front are clearly seen in this image. An additional piece of information measurable with this analysis is the formation and propagation of another pressure wave following detonation. The wave is seen to propagate back towards the ignition source after detonation occurs, and is labeled in Figure 3.8. This second shock wave or rarefaction wave originates at the same point that the detonation wave starts, and propagates against the direction of flame travel.

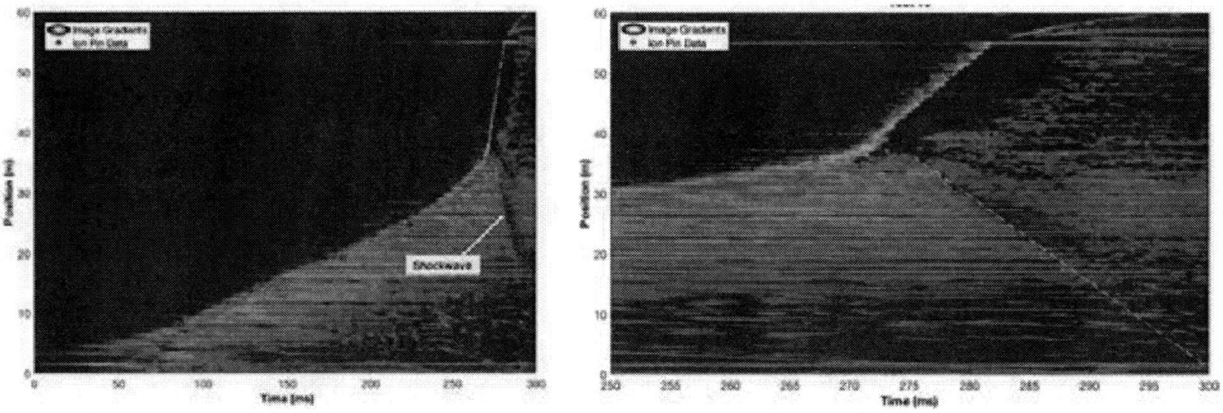

Figure 3.8: x-t diagram highlighting rarefaction shockwave in Test #15.

This shock wave can also be seen using another type of filtering. The image sequence in Figure 3.9 shows the propagation of the pressure front.

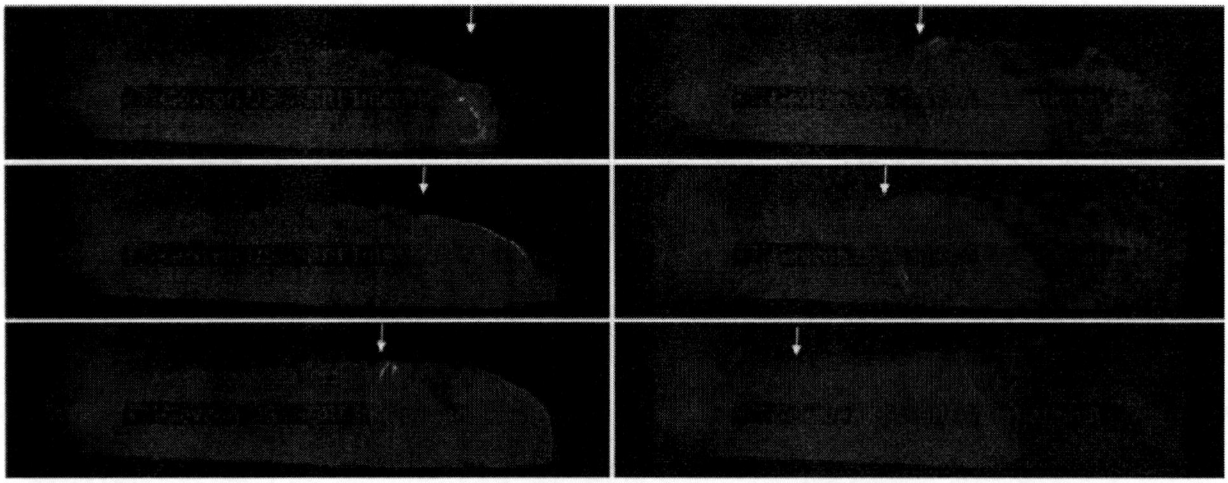

Figure 3.9: Propagation of rarefaction wave.

The rarefaction wave velocities were determined as the derivative of a linear fit to the rarefaction position versus time in the videos. The steady propagation speed of this shockwave is lower than the observed detonation speeds by about 200-600 m/s and ranged in values from approximately 1000 m/s in Test #7 to 1450 m/s in Test #15.

Another powerful result obtained from the x-t diagram analysis was observed in the mitigation case, Test 16, are shown together in Figure 3.10. Four distinct flame propagation events occur: 1) a deflagration ensues after ignition (left image, 0 to 130 ms); 2) the flame front accelerates to near-constant velocities close to the speed of sound in the burned gases (approximately 1/2 the CJ velocity) shown as constant velocity position versus time slope from 100 to 130 ms on the left image where the shock and flame are together; 3) the flame front encounters the mitigation region and significantly decelerates (beginning at 155 ms, labeled as “Shock Detaches”), where the flame detaches from the shock wave; and 4) the flame front passes the mitigation region and accelerates back to velocities near the speed of sound in the burned gases and catches the shock wave as it approaches the end of the rig (185 ms and beyond).

The image gradient analysis clearly illustrates the four different events and shows excellent agreement with the ion pin data. Furthermore, it shows that after the flame front traveling in excess of 800 m/s encounters the mitigation region, it slows down considerably to near 400 m/s, but the shockwave generated by the choked flow does not. Hence, the shock detaches from the flame front and outpaces it in this region. The right frame of Figure 3.10 shows a close-up of this. The shock continues at approximately the same velocity while the flame front decelerates, causing the separation between the two. The flame front then accelerates again, catching up to the shock near the end of the rig, at which point it propagates again at a rate in excess of 800 m/s, or at choked conditions. This information (i.e. the detachment of the shock, speed of detached shock, and rejoining with flame front) was not easily extracted from the ion pin data or traditional high-speed video, showing the value and power of the gradient x-t diagram technique.

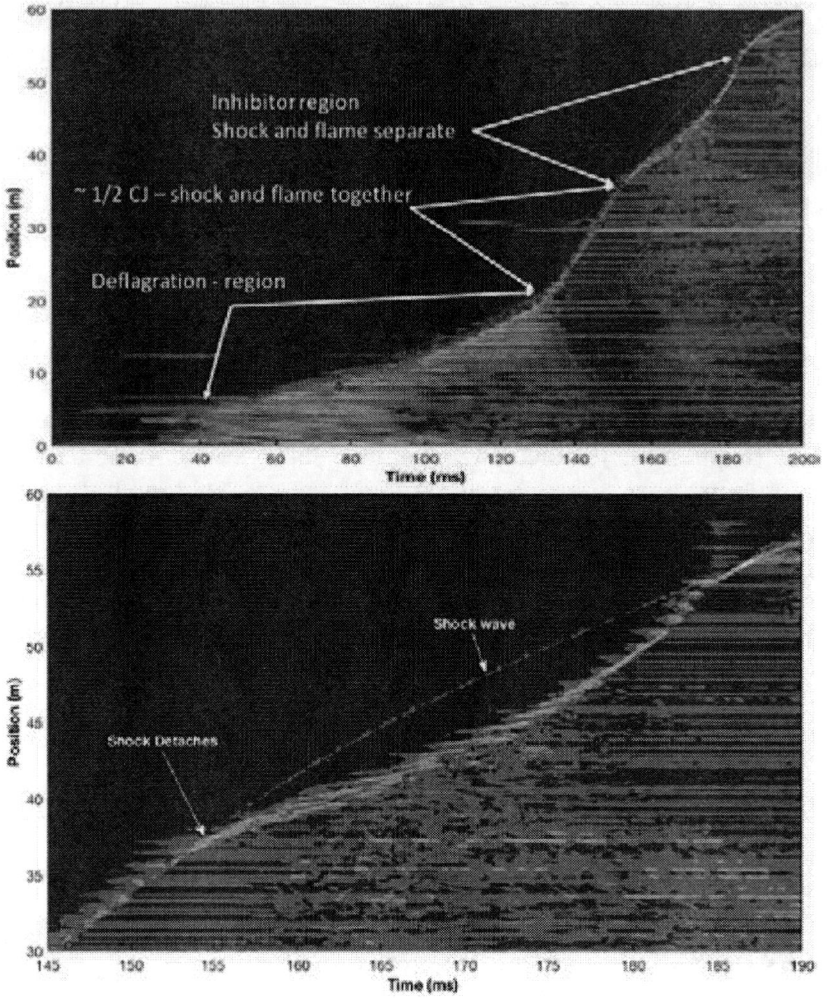

Figure 3.10. x-t diagram showing shock detachment from flame front when it reaches mitigation curtain in Test 16. The upper image shows entire flame propagation (deflagration, DDT, inhibitor, end of inhibitor) and right image is focused in on the inhibitor zone.

3.4 Preliminary FLACS results

To facilitate the design of "inherently safer" facilities, we need to be able to predict high consequence events for vapor cloud explosions and evaluate how design changes affect the possible consequences, especially at realistic large scales. Determining whether a deflagration-to-detonation transition (DDT) can occur is imperative for accurate calculation of such events. Currently, there is a lack of adequate tools that can accurately predict DDT. Work is continuously ongoing to validate FLACS and the capabilities of other analytical tools in predicting DDTs for scales and geometries relevant to petrochemical installations.

FLACS is a commercially available code that is currently used by over 100 companies, and each of the tests being performed has been replicated in the FLACS CFD software package in an effort to further validate the solver. Future software package releases will include updates as a result of validation against the tests performed here.

Flame speed and experimental pressures are one set of parameters that can be used to validate FLACS predictions as compared to the experimental data. Recall that FLACS will not reproduce an actual transition to detonation in the CFD model, so the target was to match the flame speed and overpressure predictions from the test prior to transition to detonation. FLACS also monitors the spatial pressure gradient across the flame front at any location and any moment in order to judge the likelihood of a DDT occurring in the deflagration. The value of this spatial pressure gradient is normalized using the grid resolution and initial pressure:

$$\left.\frac{dP}{dx}\right|_{normalized} = \left.\frac{dP}{dx}\right|_{actual} \frac{X_{CV}}{P_0} \qquad \text{Equation 1}$$

where P_0 is the initial pressure and X_{CV} represents the grid resolution. A magnitude of the pressure gradients of order 1 indicates that DDT is possible, with values around 10 or larger indicating that DDT is very likely [14][15].

3.4.1 Preliminary FLACS predictions of DDT

Efforts were initially placed into comparing the FLACS predicted flame speeds against experimental test results. While many cases were simulated, the base case DDT test shown here is Test #12 (high purity propane, φ=1.1, 24 modules with the tent cut). The exact configuration of the modules was built into FLACS and can be seen in Figure 3.11. As the experimental results have previously shown minimal sensitivity to the initial confinement of the plastic sheeting for propane, FLACS simulations were only performed not considering the effect of the plastic sheeting. The equivalence ratio and components of the fuel mixture (greater than 99.5% propane) were matched with the recorded values. The simulation was performed with a uniform 20 cm grid cell size. Simulations were performed as experimental results became available. The simulation versus the experimental deflagration flame front can be seen in Figure 3.11.

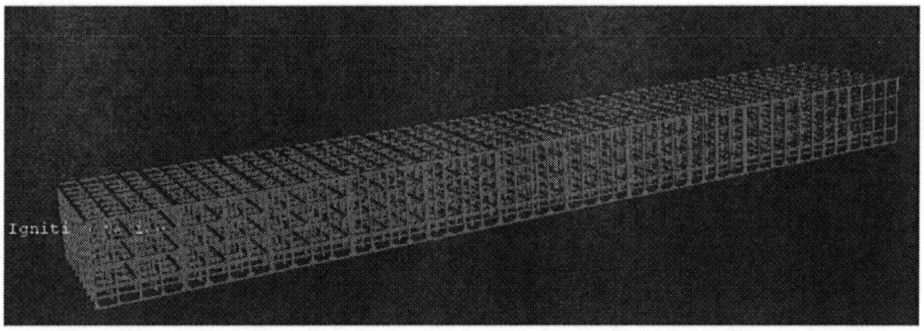

Figure 3.11: CASD image of geometry setup for use with FLACS.

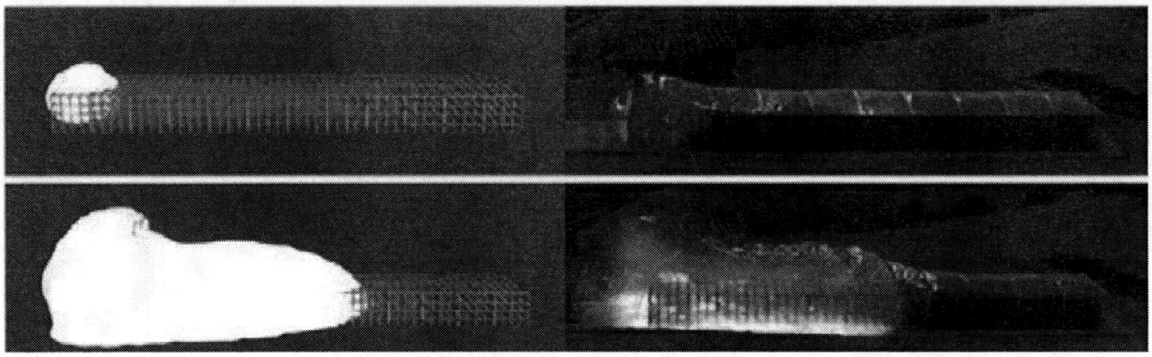

Figure 3.12: FLACS visualization of flame development prior to detonation.

The flame speed of the CFD results was calculated by determining at what time the concentration of products of combustion was greater than 20% or the temperature of the gaseous mixture reached over 1000 K. This was a calculated value along similar visual lines of monitor points as outlined in Section 3.1. The results for the standard issue version of FLACS compared to the experimental results prior to detonation can be found in Figure 3.12. Reasonable agreement was observed for the observed and predicted deflagration flame speed for all cases. For example, while FLACS predictions are very close for Test #12, they would slightly under-predict Tests #5 and #6 as they transitioned earlier. Work is ongoing to improve the accuracy of the predictions. In fact, new sub-grid and combustion models are currently being evaluated in FLACS to improve agreement.

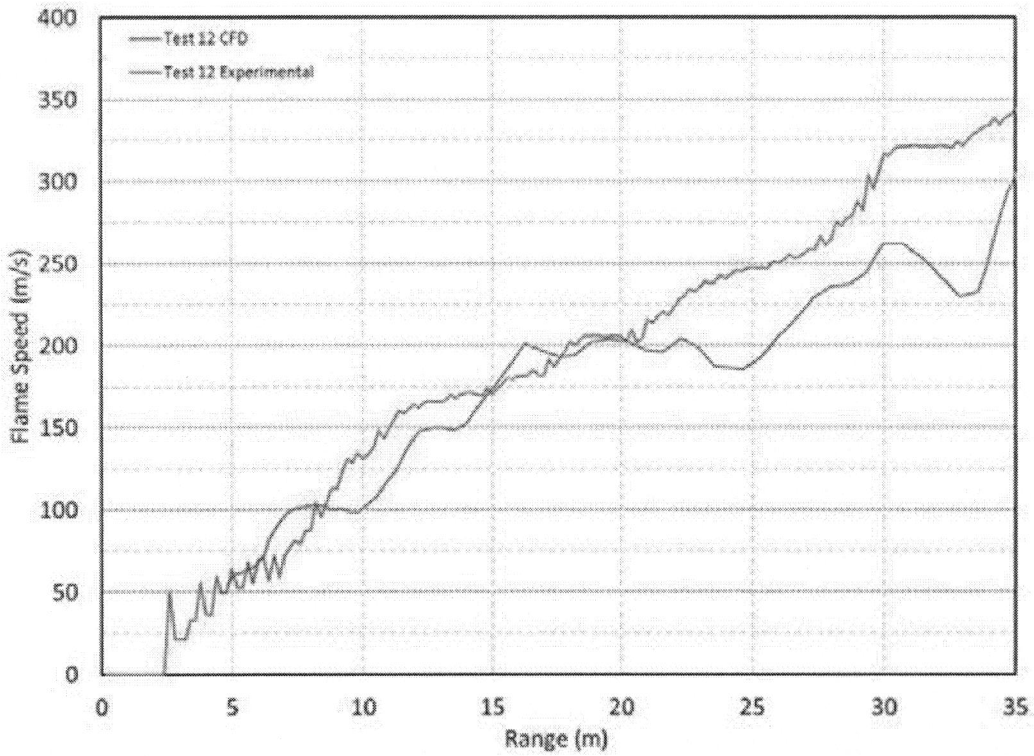

Figure 3.13: CFD calculated flame speed for Test #12 along with experimentally measured flame speed prior to DDT.

In addition to flame speed, the DDT parameter was evaluated for the current tests. While this functionality is being continuously improved, the current version of FLACS does predict DDT onset for several of the tests executed in the test matrix, however, not always in the correct location. Figure 3.14 shows a 2D plot of the DPDX parameter as output from the DDT module of the FLACS solver. This plot shows the results of the CFD simulation of Test #12 (high purity propane, φ=1.1, 24 modules with the tent cut). This plot shows that DPDX values greater than 1 were observed in the 10th row of modules with values over 5 (up to a maximum of 14) in the last row of modules. The transition to DDT actually occurred in the 9th row of modules in the experiments. The results show that FLACS predicts DDT is "possible" near the region in which the experiment transitions to detonation. For other tests, such as Test #7 (HD-5 propane, φ=1.05, 24 modules with 2 and 4 mils plastic), FLACS did not predict the location where one was observed. Details of each of the tests has been provided to the developers for refinement of the CFD tool.

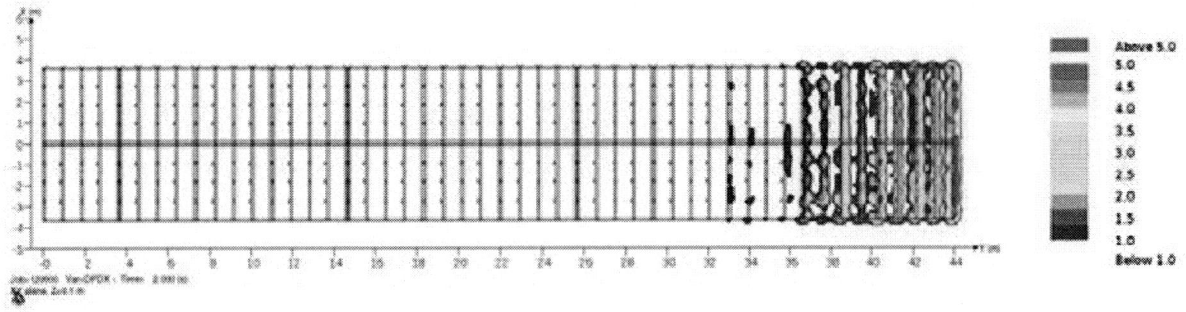

Figure 3.14: FLACS DPDX prediction for Test #12.

The overpressure predictions from the FLACS solver, were challenging as FLACS does not actually model the transition to detonation, and the detonation event. Therefore, no direct comparison for Test #12 pre- and post-transition to detonation is available and more work is being conducted to properly evaluate these results.

3.4.2 Preliminary FLACS Predictions of Deflagrations

In addition to propane, initial FLACS modeling was also conducted for Methane. Test #9 was chosen as the test case for methane, which was conducted at an equivalence ratio of 0.9 and performed in 30 modules with low congestion. Again, given the minimal influence of the plastic tent, the simulations were only performed without including its influence of the sheet. The predicted and experimental flame speed results for Test #9 can be found in Figure 3.15. Agreement between the predicted and experimental results was reasonably close for methane. More work is currently being performed on the lean and rich propane results.

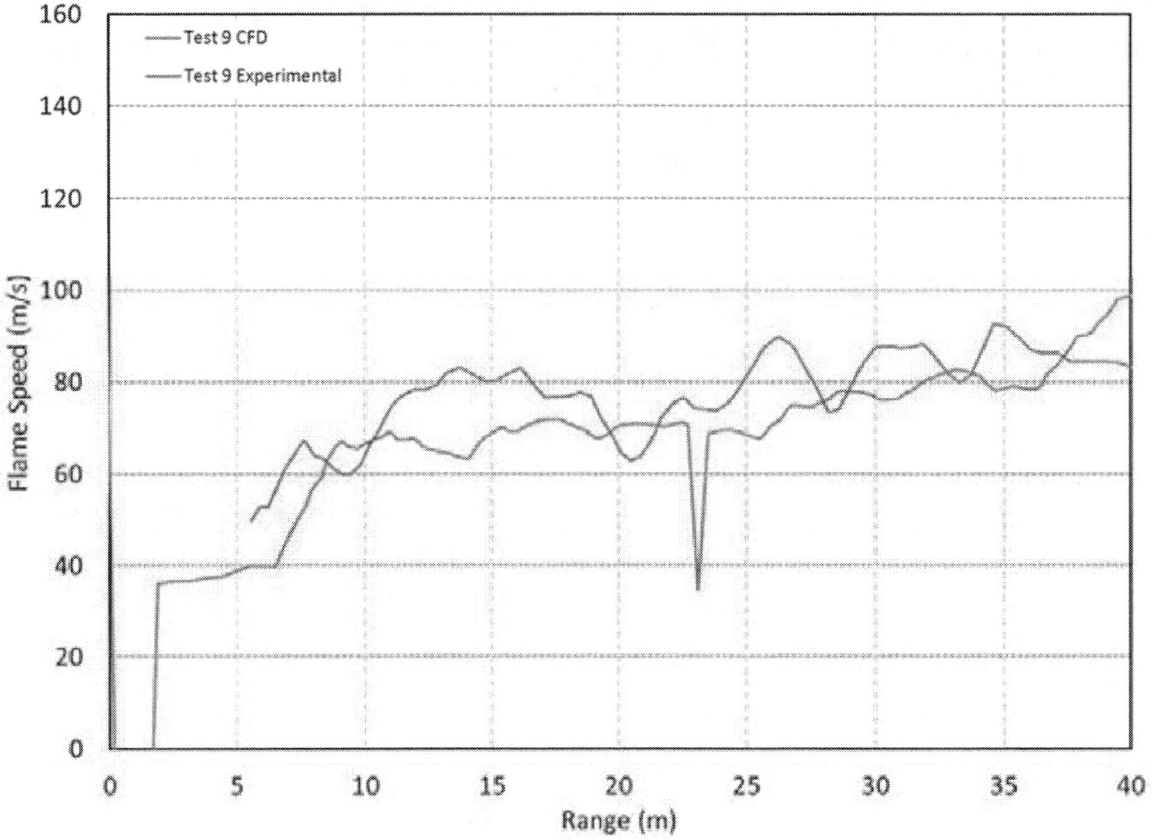

Figure 3.15: CFD calculated flame speed and experimentally measured flame speed for the length of the test rig for Test #9.

The CFD predictions of overpressure show reasonable agreement with the experimental results (see Table 3.1). For the methane deflagration case, Test #9, the results show that the overpressure prediction was, in some cases, about twice of the experimental results. While this seems extreme when presented as a percentage difference, the largest actual pressure difference was less than 100 mbar and on average only 32 mbar.

Table 3.1: Overpressure results from Test #9 and CFD replication of test.

Gauge	Distance from Ignition (m)	Peak Deflagration Pressure (barg)	FLACS CFD Prediction (barg)
P2	11.9	0.037	0.073
P3	17.4	0.045	0.058
P4	22.9	0.040	0.062
P5	25.7	0.043	0.064
P6	28.5	0.061	0.056
P7	31.3	0.049	0.078
P8	34.0	0.060	0.081
P9	36.7	0.113	0.093
P10	39.6	0.063	0.105
P11	42.4	0.059	0.131
P12	45.0	0.072	0.171
P13	47.8	0.072	0.129

The results have been provided to the developers of FLACS to compare with as-yet unreleased solvers with updated turbulence and combustion sub-models. Updates and comparisons to the updated version will be made once a new version of FLACS is released.

4 Conclusions

Deflagration to detonation transitions were observed during testing in a large-scale (1200-1500 m^3) adjustable length rig with different levels of congestion. Transitions to detonation occurred in the low congestion configuration with near-stoichiometric propane/air mixtures and in the high congestion configuration with lean and rich propane/air mixtures. Additional tests were performed to determine whether the plastic confinement sheet or the purity of the propane fuel influenced or facilitated the transition to detonation. These tests yielded similar transitions showing that, at large scales, DDTs are more likely to occur that previously anticipated.

Mitigation systems were considered during two tests and effectively reduced the flame propagation velocity. Experimental results showed that mitigation was more effective when applied closer to the point of ignition.

Analysis of the present experimental results showed that deflagrations transitioned to detonations at flame front velocities near 500 m/s. The present results were somewhat higher than previously reported for the same experimental dataset and the reasons for the differences were discussed. Three flame-front tracking methods (commercially available and developed in-house) were tested to see if results varied between methods and to determined which method yielded results closest to ion pin data. Image filtering methods developed during the present study helped visualize and track flame and shock fronts. Average detonation velocities between 1600-1800 m/s were generally observed for propane/air mixtures.

Preliminary FLACS results are promising, but show that flame speed and overpressure CFD software packages require modification to appropriately model DDTs. There is still more work required to fully understand the onset of DDT and other factors that affect flame and overpressure development in realistic and practical settings.

Further large-scale testing would improve our understanding of DDTs and our ability to predict them. Topics to be considered during future work include (1) the width of the congested test structure, as it has been shown to influence the likelihood of DDTs at large scales; (2) separation gaps between regions of congestion, which can be used to mitigate the flame acceleration and help limit transition to detonation; (3) active mitigation measures such as water deluge and solid inhibitors; (4) realistic congestion tests that more closely resemble inhomogeneous process equipment; (5) realistic gas releases/non-uniform gas concentrations to study the effects of initial turbulence and non-homogeneity of gas mixtures that result from high-pressure releases. Data from additional testing will also help further validate and improve CFD and analytical packages that can lead to safer design of facilities.

5 References

[1] Bakke, J. v. W. K. H. P. &. B. B., 2010. A study on the effect of trees on gas explosions. Journal of Loss Prevention in the Process Industries, pp. 23, 878-884.

[2] Johnson, D., 2010. The potential for vapour cloud explosions. Lessons from the Buncefield accident. Journal of Loss Prevention in the Process Industries, pp. 23, 921-927.

[3] Johnson, D., 2012. Characteristics of the Vapor Cloud Explosion Incident at the IOC Terminal in Jaipur, 29th October 2009. Houston, AIChE.

[4] Middha, P. H. O., 2008. Predicting deflagration to detonation transition in hydrogen explosions. Process Safety Progress, pp. 27 (3), 192-204.

[5] Puttock, J., 2013. Gas explosion modelling with PDRFoam. Cardiff, UK Explosion Liaison Group.

[6] Tang, M. B. Q., 1999. A new set of blast curves from vapour cloud explosions. Process Safety Progress, pp. (18) 235-240.

[7] U.S. Chemical Safety and Hazard Investigation Board, 2015. Caribbean Petroleum Tan Terminal. Explosion and Multiple Tank Fires, Report No. 2010.02.I.PR: s.n.

[8] Davis, SG., Merilo, E., Engel, D., and van Wingerden, K., (2016). Large Scale Detonation Testing: New Findings in the Prediction of DDTs at Large Scales. Proceedings of the 2016 Mary Kay O'Connor Process Safety Center International Symposium, October 2016: 20-39.

[9] Davis, S.G., Engel, D., and van Wingerden, K. (2016). Can Gases Behave Like Explosives – Large Scale Deflagration to Detonation Testing. Eighth International Symposium on Fire Investigation Science and Technology (ISFI 2016). Scottsdale, Arizona, USA, 22-24 September.

[10] MacKay, D., Murray, B., Moen, I., & Thibault, P. (1988). Flame-Jet Ignition of Large Fuel-Air Clounds. The Combustion Institute. Twenty-Second Symposium on Combustion.

[11] Pekalski, A., Puttock, J., Chynoweth, S., (2014), DDT in a vapour cloud explosion in unconfined and congested space: large scale test, Tenth International Symposium on Hazards, Prevention, and Mitigation of Industrial Explosions, Bergen, Norway, 10-14 June 2014.

[12] Harris, R.J. and Wickens, M.J., Understanding vapour cloud explosions – an experimental study, 55th Autumn Meeting of The Institute of Gas Engineers, Communication 1408, 27th and 28th November 1989.

[13] Tomlin, G. and Johnson, D.M., A Large Scale Study of the Venting of Confined Explosions into Unobstructed and Congested Flammable Vapour Clouds, Proc. of the Seventh International Seminar of Fire & Explosion Hazards (ISFEH7), University of Maryland, 2013, pp. 679-688.

[14] Van den Berg, A., 1985. The multi-energy method. A framework for vapour cloud explosion blast prediction. Journal of Hazardous Materials, pp. (12) 1-10.

[15] van Wingerden, K., 2008. On the possibility of DDT in vapor cloud explosions. New Orleans, AIChE. J.A. Klein and B.K. Vaughen. "Implement an Operational Discipline Program to Improve Process Plant Safety," Chemical Engineering Progress, pp. 49-52, June 2011.

479041 Real World Applications of API 753

Abstract:

The first edition of American Petroleum Institute (API) Recommend Practice (RP) 753 - *Management of Hazards Associated with Location of Process Plant Portable Buildings* was released in 2007. Since that time blast resistant modules (BRMs) have become a popular portable building option; however, other portable building options may also satisfy API RP 753 requirements. This paper provides an example of a systematic approach for performing an API 753 blast analysis as well as a discussion of the range of portable building options available to owner/operators seeking to comply with API RP 753. These building options include, but are not limited to, heavy (2×6 or 2×8) wood trailers, retrofitted iso-containers, and commercial BRMs. The paper also discusses the detailed analysis requirements for siting a portable building within the API RP 753 Zone 2 boundary. Finally, the paper presents a brief discussion of the internal debris throw hazard associated with the blast response of two different portable buildings as observed during full scale testing using the BakerRisk Deflagration Load Generator.

Shrapnel–Accounting for Blast Fragments in Facility Siting Studies

Michael S. Schmidt
Bluefield Process Safety
3101 Cherokee St.
St. Louis, MO 63118
bluefieldsafety@gmail.com

Corey Whelehon
Bluefield Process Safety
3101 Cherokee St.
St. Louis, MO 63118
corey.welehon.bluefieldsafety@gmail.com

Prepared for Presentation at
American Institute of Chemical Engineers
2017 Spring Meeting and 13th Global Congress on Process Safety
San Antonio, TX
March 26 – 29, 2017

Shrapnel–Accounting for Blast Fragments in Facility Siting Studies

Michael S. Schmidt
Bluefield Process Safety
3101 Cherokee St.
St. Louis, MO 63118
bluefieldsafety@gmail.com

Corey Whelehon
Bluefield Process Safety
3101 Cherokee St.
St. Louis, MO 63118
corey.welehon.bluefieldsafety@gmail.com

Keywords: Facility sitting, Blast Fragments, Projectiles, Missiles, Explosions Consequences

Abstract

When a pressurized vessel ruptures, stored energy is released. This energy has the potential to cause a shockwave and launch vessel fragments. The need to account for the damage resulting from the shockwave from a physical explosion has long been recognized and methods have been developed and published. There is a general awareness that projectiles are also created during a rupture event, but these blast fragments are typically only addressed anecdotally, in describing the damage done by explosion after the event has occurred, not as part of anticipating the damage done by an event in the course of conducting a facility siting study.

Shrapnel damage by blast fragments is of concern to the extent that shrapnel goes beyond the blast zone caused by the shockwave from an overpressure event. The horizontal range of the fragment field depends on the energy released during the overpressure event. However, the number of fragments that can do damage is finite and the probability that a fragment will do damage is considerably less than 100%.

Understanding how to estimate the horizontal range of a fragment field, to estimate the number of fragments with the potential to do damage, and to estimate the probability that a fragment will cause damage are all essential to accounting for blast fragments in facility siting studies. This paper describes a simplified method for predicting the extent of shrapnel damage by blast fragments resulting from a pressure vessel rupture that staff at a facility can use without special software.

1 Introduction

When considering damages resulting from explosions, it is important to consider the possible types of explosions. The main concerns are boiling liquid expanding vapor explosions (BLEVEs), vapor cloud explosions (VCEs), and pressure vessel explosions (PVEs). BLEVEs and VCEs follow the release of material; without a loss of containment first, the explosion cannot occur. PVEs, on the other hand, result from a sudden and violent failure of a vessel and are the cause of loss of containment.

The major impacts of the various types of explosions have been identified, and methods for estimating those primary consequences are becoming a staple in facility siting studies. When considering the impact of explosions, analysts typically take into account some combination of blast waves, release of contents, and thermal radiation. Another consequence is frequently neglected in most facility sitting studies; projectiles from explosions are commonly considered only in the aftermath of the event. Most documented methods for estimating consequences of explosions mention blast fragments as a concern, but give no definite method for accounting for the effects of the projectiles.

This paper describes an approach for considering damage from blast fragments resulting from a physical vessel explosion and develops a set of calculations for estimating initial damage by the blast fragments from those explosions.

Although the title and the abstract of this paper refer to "shrapnel," that is really for dramatic effect. Strictly speaking, "shrapnel" refers to the fragments of a shell or other small fragments from a projectile weapon specifically designed as an anti-personnel device. Shrapnel takes its name from General Henry Shrapnel, the British officer who invented the fragmenting bomb in the early 19th century. The fragments from an industrial explosion in the chemical process industries are not shrapnel. More correctly, they are referred to by the terms "missiles," "projectiles," and "blast fragments," which are often used interchangeably among sources and professionals in the field.

2 History of Incidents

While BLEVEs and VCEs are frequently the focus of facility siting studies, in part because of the EPA's focus on them in the Risk Management Planning rule (40 CFR-68), physical explosions are on old concern, dating back to the dawn of the industrial revolution. An early review [1] of the blast effects from pressure vessel explosions considered twenty-five previous accidents in its analysis.

A study [2] of the incidents resulting in the largest capital losses in the chemical process industry showed that explosions accounted for thirty of the top one hundred losses. The losses in these explosions were a result of the combination of the initial blast wave damage and the accompanying "domino effect" from blast fragments. In three specific incidents, the damages resulted primarily from projectiles. One of the cases reports the primary damage as being the specific result of blast fragments.

Lee's Loss Prevention [3] also identifies sixteen incidents involving escalation or near misses by missiles resulting from explosions. Five of these incidents are of special interest:

- Texas City, TX, 1969 – A 100 ft high refinery column disintegrated, apparently due to detonation of vinyl acetylene. Fragments were found uniformly over a radius of 1500 ft.
- Bantry Bay, Eire, 1979 – Explosions on an oil tanker at the jetty generated numerous missiles. One fragment weighing 1000 lbs was found at the foot of a large crude oil storage tank 1800 ft from the ship.
- Ponce, Puerto Rico, 1979 – A 13 ft diameter dimerizer vessel suffered massive failure. The 15 ton steel end traveled 1900 ft and struck an adjacent paraxylene plant, setting it on fire.
- Pascagoula, MS, 1986 – Fragments from an explosion in the area of a batch still punctured two atmospheric storage tanks and a pressure storage vessel, releasing flammable material.
- Grangemouth, UK, 1987 – Overpressure due to gas "breakthrough" in a low-pressure separator caused the vessel—10 ft diameter and 30 ft tall—to disintegrate. One piece weighing 3 tons was thrown 3300 ft.

Scilly and Crowther [4] identify seven incidents in their assessment of missile impact on process plants, only one of which is also listed in *Lee's*.

The Chemical Safety Board (CSB) has reported several explosion incidents resulting in consequences from blast fragments:

- Pascagoula, MS, 2002 – An explosion of a distillation column resulted from the decomposition of mononitrotoluene (MNT). Fragments struck a pipe rack above an anhydrous ammonia tank. A large piece of the column sidewall was found 1,100 feet away, within 50 feet of a 250,000-barrel crude oil storage tank. [5]
- Institute, WV, 2008 – A runaway chemical reaction occurred inside a 4,500 gallon pressure vessel causing the vessel to rupture. The explosion resulted in three large fragments traveling nearly 55 miles per hour. A protective steel shield blanket prevented fragments from impacting a 6,700 gallon storage tank that was 70 feet away from the explosion. [6]
- Belvidere, IL, 2009 – Quartz-manufacturing vessel, under 29,000 psig, ruptured, resulting in a fragment weighing approximately 8,600 lbs being thrown through two concrete walls and landing 435 feet away. [7]

Clearly, blast fragments, missiles, and projectiles from PVEs have the potential to cause great damage. The fragments from pressure vessels are not small—they are not shrapnel. Accounting for the damage that these fragments can do and planning accordingly is a reasonable expectation in a facility siting study.

3 Pressure Vessel Explosion

When a vessel ruptures, stored energy is released. While some of the energy is released as a shockwave, some of the energy also goes into launching vessel fragments, which can travel beyond the extent of the shockwave. The energy released is frequently expressed in terms of the equivalent amount of TNT (m_{TNT}) exploded. The distance that vessel fragments travel is a function of the amount of energy released, or in other words, the equivalent amount of TNT exploded. For PVEs, m_{TNT} is a function of vessel volume and burst pressure.

3.1 PVE Overpressure

Pressure vessel explosions were calculated using the TNT equivalency model. Unlike a VCE in an obstructed volume, PVEs behave as point explosions, much like TNT and high explosives. The essence of the TNT equivalency model is that once the amount of energy released in the explosion is determined, that amount of energy is converted to the amount of TNT that would release the same amount of energy, and then the effect of exploding that much TNT (for which there is extensive empirical data) is calculated using empirical scaling laws.

The explosive energy of explosions resulting from the mechanical failure of a pressure vessel depends on the volume of gas released from the vessel when it fails, and the pressure contained by the vessel at the moment it fails. The most conservative method for calculating the explosive energy of a pressure vessel rupture is to assume isothermal expansion. To address the overt conservatism of the isothermal expansion method, Crowl describes in the classic text on chemical process safety a method called "thermodynamic availability" that is similar to the isothermal expansion, but has a correction term that accounts for the energy loss attributable to entropy. [8] It uses the equation

$$E = P_2V \left[\mathit{ln}\,(P_2/P_1) - (1 - P_1/P_2) \right]$$

The term, P_1, refers to atmospheric pressure, at least if the explosion occurs on earth. The term, P_2, refers to the burst pressure, not the maximum allowable working pressure (MAWP). The burst pressure in turn is most closely correlated to the ultimate yield strength of the vessel. The ratio of ultimate yield strength to MAWP is used as a safety factor. The ASME Boiler and Pressure Vessel Code (BPVC) uses a safety factor of four [9], so the burst pressure is 4 times the MAWP. Pressure vessels specified and constructed to European standards will typically have a burst pressure that is 6 times the MAWP. The European standard also requires a safety factor of four, but is based on only two-thirds of the yield strength of the materials used. [10]

Once the explosive energy released by the catastrophic failure of the pressure vessel is calculated using the thermodynamic availability method, the equivalent amount of TNT is calculated by comparing the energies of explosion:

$$m_{TNT} = E/\Delta H_{TNT}$$

The two equations are combined to give:

$$m_{TNT} = \{1/\Delta H_{TNT}\}\{ P_2V \left[\mathit{ln}\,(P_2/P_1) - (1 - P_1/P_2) \right] \}$$

Different values for the heat of explosion of TNT are reported, ranging from around 1,000 kcal/kg to almost 1,200 kcal/kg. Obviously, the lower the value used, the more the equivalent amount of TNT the equation calculates. The method here uses the ΔH_{TNT} used by the EPA RMP rule [11]: 1119 kcal/kg or 4680 kJ/kg. The remainder of the equation depends on the units of the absolute pressures, the volumes, and the mass. The model uses

$$m_{TNT} = 9.21 \times 10^{-5} P_{BURST} V_{BURST} \left[ln\,(P_{BURST}/14.7) - (1 - 14.7/P_{BURST}) \right]$$

where, m_{TNT} is the equivalent amount of TNT, in pounds,

P_{BURST} is the burst pressure of the vessel, in psia,

V_{BURST} is the volume of the vessel bursting, in cubic feet

There is no need for an explosive yield factor; it is built into the thermodynamic availability equation.

4 Blast Fragments

The need to account for the damage resulting from the shockwave from a physical explosion has long been recognized and methods have been developed and published. There is a general awareness that projectiles are also created during a rupture event, but these blast fragments are typically only addressed anecdotally, in describing the damage done by an explosion after the event has occurred, not as part of anticipating the damage done by an event in the course of conducting a facility siting study. Damage by blast fragments is of concern to the extent that it goes beyond the blast zone caused by the shockwave from an overpressure event.

A method for modeling the potential for damage by blast fragments is described here and consists of six steps. The method can be used without depending on complex or expensive modeling software.

The six steps:

1. Determine the maximum horizontal range of blast fragments.

2. Calculate the fractional distance to a potential target, based on the maximum horizontal range of blast fragments. When the fractional distance to a potential target is greater than 1, the blast fragment case may be ignored.

3. When the fractional distance to a potential target is less than 1, use look-up tables to determine the probability of fragments going at least as far as the potential target. There are separate look-up tables for horizontal vessels and for vertical and spherical vessels.

4. Determine the probability that a blast fragment sufficiently large to do damage would travel in the direction of a potential target. The probability of a blast fragment going in the direction of a potential target depends on the number of blast fragments and the area of the target presented to the blast and the distance of the target from the blast.

5. Estimate the surface target density of the process. Not all space presented to a fragmenting pressure vessel would result in a catastrophic release if entered by a fragment.

6. Calculate the probability of damage by blast fragments as the product of surface target density, probability of blast fragments going far enough, and probability of blast fragment going in the right direction.

When the probability of damage is sufficiently low, the blast fragment case may be ignored.

4.1 Horizontal range

The first step in considering damage by projectiles is to determine the maximum horizontal range of blast fragments, d_{MAX}. Crowl and Louvar noted Clancey's empirical relationship [12] between the mass of explosive and the maximum horizontal range of the fragments, as illustrated in Figure 1.

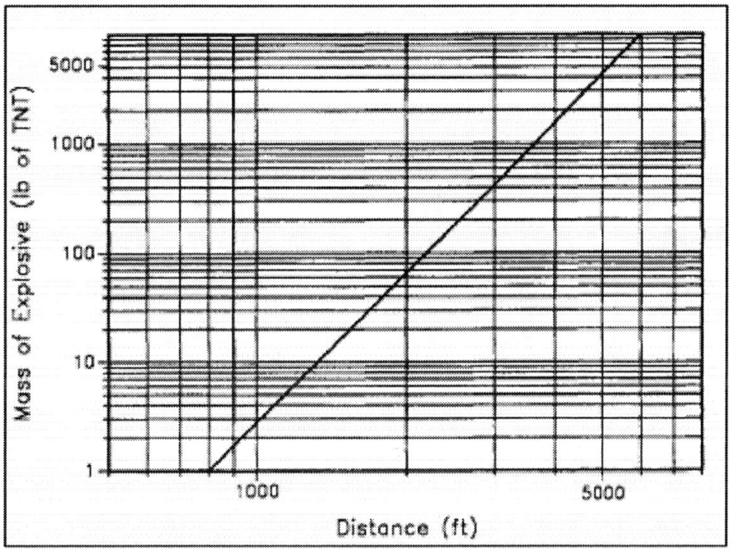

Figure 1. Maximum horizontal range of blast fragments. Data from Clancey, "Diagnostic Features of Explosive Damage"

Derived from the relationship in Figure 1, the maximum horizontal range of blast fragments can be calculated with the following equation

$$d_{MAX} = 800\ m_{TNT}^{0.22}$$

where, d_{MAX} is the maximum horizontal range of blast fragments, in feet
m_{TNT} is the equivalent amount of TNT, in pounds

4.2 Fractional Distance

To determine the impact of fragments hitting other processes, the distance from the initial explosion ,d, is required. Physical measurements, plot plans, or Google Earth can be used to determine this distance. Once the distance from the pressure vessel of interest is known, the fractional distance, d_{FRAC}, to a potential target can be calculated using the equation

$$d_{FRAC}=d/d_{MAX}$$

where, d and d_{MAX} are in the same units.

If d_{FRAC} is greater than 1, then the fragments will not travel far enough to reach the equipment in question and this blast fragment case may be ignored. To quickly identify equipment unaffected by this blast fragment case, d_{MAX} can be used as a radius from the potential event. Equipment outside of this area will have a d_{FRAC} greater than 1 and the blast fragment case may be ignored.

4.3 Blast Fragment Probable Distance

When d_{FRAC} is less than one, then the probable distance that any projectile would travel becomes of interest. Look-up tables are useful for this. Shown below are two tables. The first is for horizontal vessels (Table 1), while the second table is for vertical and spherical vessels (Table 2). With these two tables, it is a straightforward matter to determine the probability, P_{dist}, of a blast fragment flying at least as far as the fractional distance, d_{FRAC}. Interpolation may be performed as necessary.

Table 1. Horizontal Vessel Blast fragment Probable Distance

Horizontal vessel-fractional distance, d_{FRAC}	**Probability of reaching or exceeding fractional distance, P_{dist}**
1.0	0.010
0.9	0.014
0.8	0.018
0.7	0.028
0.6	0.042
0.5	0.063
0.4	0.09
0.3	0.17
0.25	0.22
0.20	0.31
0.16	0.40
0.12	0.51
0.08	0.71
0.04	0.91
0.02	0.99

Table 2. Vertical/Spherical Vessel Blast fragment Probable Distance

Horizontal vessel-fractional distance, d_{FRAC}	**Probability of reaching or exceeding fractional distance, P_{dist}**
1.0	0.010
0.9	0.016
0.8	0.027
0.7	0.045
0.6	0.079
0.5	0.137
0.4	0.23
0.3	0.39
0.25	0.53
0.20	0.69
0.16	0.80
0.12	0.92
0.08	0.985
0.04	0.995
0.02	0.999

Tables 1 and 2 are based on empirical data published by the International Association of Oil & Gas Producers (OGP). [13] These data are based on the results from the study of vessel explosions of different shapes and orientations. The tables were developed using the documented percentage of fragments within a given range for 25 accidental explosions.

4.4 Blast fragment Probable Direction

The blast fragments not only have to travel as far as the equipment distance, but the fragments must also travel in the direction of the process of interest and be of sufficient size to do damage. Therefore, the next step is to calculate P_{dir}, the probability that a blast fragment sufficiently large enough to do damage will travel in the direction of the process of interest.

The probability of a blast fragment going in the direction of the process of interest—a target—depends on the number of blast fragments, the size of the fragment, and the width of the target.

The number of blast fragments is much smaller than most would imagine. This is consistent with the idea that blast fragments are not shrapnel. Pressure vessels are not designed to fragment into very small pieces, as is a hand grenade; instead they fragment at weld seams and other singular failure points.

The OGP report [13] indicates that there are typically a couple of dozen blast fragments created during a PVE. The average number of fragments produced during the seven incidents studied by Scilly and Crowther is 23. [4]

The literature indicates that the number of blast fragments produced in a PVE is very dependent on the cause of the vessel rupture; specifically it depends on whether the cause of the PVE is that the internal pressure of the vessel went well above the pressure the vessel was designed to withstand, or whether the strength of the vessel decreased to well below the normal working pressure. In the latter case, it is common among methods to assume the burst will produce a larger number of fragments—dozens. In the former case, there are typically fewer fragments, as fragmentation occurs along weld seams. The Netherlands Organization for Applied Scientific Research (TNO) reports that the typical number of blast fragments resulting from the PVE of a cylindrical vessel is 2 to 3, while the typical number of blast fragments resulting from the PVE of a spherical vessel, with it much higher number of welds, is 10 to 20. [14]

It may be appropriate to use smaller numbers of fragments in specific cases, but generally, the assumption of 24 fragments is conservative and what this method uses.

The average size of the fragments is directly related to the surface area of the vessel that explodes and the number of pieces into which it fragments:

$$L = (A_{vessel} / N_{Frag})^{0.5}$$

Only half of the fragments formed launch beyond the overpressure zone, those from the top portion of the exploding vessel. The fragments from the bottom portion of the vessel drive into grade in the vicinity of the exploding vessel, causing damage largely indistinguishable from the blast effects. Therefore the number of fragments launched is

$$N = N_{Frag}/2$$

The probability of a blast fragment going in the direction of a particular targeted point at distance, r, is the number of fragments launched times the average size of fragments launched, all divided by the circumference of a circle with a radius equal to the target distance:

$$P_{hit} = N \text{ x } L / \text{Circumference}$$

$$P_{hit} = N_{Frag}/2 \text{ x } (A_{vessel} / N_{Frag})^{0.5} / 2\pi r$$

$$P_{hit} = (A_{vessel} \text{ x } N_{Frag})^{0.5} / 4\pi r$$

This is the probability of hitting a segment of the circle that is the width of the blast fragment. The probability of not hitting that segment is the complement:

$$P_{not\ hit} = 1 - ((A_{vessel} \text{ x } N_{Frag})^{0.5} / 4\pi r)$$

The probability of going in the direction of a target also depends on the width of the target as it presents to the fragmenting pressure vessel. The wider the target, the more likely it is to be hit. Physical measurements, plot plans, or Google Earth, can be used to determine the target width, W, when projected on a plane perpendicular to a line between the fragmenting pressure vessel and process. The probability of not going in the direction of the target depends on not going in the direction of one fragment-width segment, and not going in the direction of the next fragment-width segment, and the next, and the next, until the width of the target has been completely missed. The probability calculation, then, is to take $P_{not\ hit}$, and raise it to the power of the number of sequential segments that must not be hit:

$$P_{target\ not\ hit} = (1 - ((A_{vessel} \text{ x } N_{Frag})^{0.5} / 4\pi r))^n$$

While smallest power must be one, the value of n increases as width increases, because it is expressed as a multiple of fragment size: n = w/L.

The equation for calculating the probability of a blast fragment not going in the direction of a particular process target, then, is

$$P_{target\ not\ hit} = (1 - ((A_{vessel} \times N_{Frag})^{0.5} / 4\pi r))^{W/(A_{vessel} / N_{Frag})^{\wedge}0.5}$$

The equation for calculating the probability of a blast fragment going in the direction of a particular process target is the complement

$$P_{dir} = 1 - (1 - ((A_{vessel} \times N_{Frag})^{0.5} / 4\pi r))^{W/(A_{vessel} / N_{Frag})^{\wedge}0.5}$$

where,

A_{vessel} is the surface area of the exploding vessel
N_{Frag} is the number of fragments generated by the explosion, typically 24 or less
r is the distance from the fragmenting pressure vessel to the process target
W is the width of the process target, as measured perpendicularly to the line from the fragmenting pressure vessel

The area and the two lengths, A_{vessel}, r, and W, can be in any units, but they must be consistent.

4.5 Process Density

Process density, $P_{density}$, is the surface target density that a process area presents to a blast fragment. Not all space presented to a fragmenting pressure vessel will result in a catastrophic release if entered by a fragment. Large, tall storage tanks represent the highest surface target density, approaching 100%, while spaces traversed by a single 2" pipeline represent the lowest surface target density, approaching less than ½%. A typical process area will have a surface target density in the range of 1% to 10%, with the number of lines and vessels capable of a catastrophic release playing a significant role.

Surface target density in process areas can typically be characterized as either low density, medium density, or high density. The methods used for determining obstacle density for multi-energy modeling of VCEs can also be used for characterizing surface target density. While there are several methods for determining obstacle density, three methods are widely used: those developed by Baker-Strehlow, the TNO method, and the approach used by FM Global. For the purpose of this method, the FM Global approach [15] is used for its ease of application to targeted processes:

- Low congestion/obstacle density – coverage less than 2% and more than 6.5 ft between obstacles.

- Medium congestion/obstacle density – coverage between 2% to 6% and between 1.5 to 6.5 ft between obstacles.

- High congestion/obstacle density – coverage greater than 6% and less than 1.5 ft between obstacles.

Commonly, a 5% surface target density is used as the process density for chemical processes.

The targets of concern with respect to facility siting studies in the chemical process industry are vessels and piping. The effect of blast fragments can be mitigated with blast walls, berms, or other shielding. When shielding exists, only the unprotected percentage of the target area is considered. When used, shielding typically is applied to vessels. Piping systems, on the other hand, are frequently neglected since pipeline ruptures can be protected with active measures such as excess flow valves, flow control loops, and emergency shutdowns.

4.6 Probability of Impact

Using the probabilities determined thus far, the probability of impact can be calculated. The probability of impact by blast fragments, P_{impact}, is the product of surface target density, probability of blast fragments traveling far enough, and probability of blast fragment traveling in the right direction. The equation is

$$P_{impact} = P_{dist} \cdot P_{dir} \cdot P_{density}$$

When P_{impact} is less than or equal to 0.01, then the blast fragment case has a 1% or less probability of impacting the process area resulting in a catastrophic release and can be ignored. If P_{impact} is greater than to 0.01, then the blast fragment case is a valid scenario for causing a catastrophic release.

5 Conclusion

While most are quick to consider overpressure waves that result from a BLEVE or VCE as part of the necessary work of a facility siting study, PVEs are often overlooked, especially in terms of the effect of blast fragments. Partly, that is because blast fragment are misunderstood. Blast fragments from a PVE are not shrapnel, as practitioners are fond of calling them, but large pieces. Typically, the number of fragments created during a PVE is less than a couple of dozen.

Estimating the extent and likelihood of damage from blast fragments following a PVE requires estimating three probabilities: the probability that a blast fragment will go as far as a potential target, the probability that a blast fragment will go in the direction of a potential target, and the probability that when a blast fragment enters a potential target, that it will do actually do damage. The last term is related to surface target density, which is similar in concept, though not in use, to obstacle densities used in multi-energy vapor cloud explosion models.

The resulting equation is

$$P_{impact} = P_{dist} \cdot P_{dir} \cdot P_{density}$$

Determining these probabilities does not need to be complicated. An analyst is only required to know the dimensions and orientation of the vessel whose explosion is being considered, the burst pressure of the vessel (typically 4 times the MAWP of the vessel), and the general layout of the vessel in relation to potential targets. With this knowledge, a couple of look-up tables, and arithmetic, an analyst can calculate the probability that blast fragments from a PVE will result in damage.

6 References

[1] Baker, W.E., J.J.Kulesz, R.E.Ricker, P.S.Westline, V.B.Parr, L.M.Vargas, and P.K.Moseley. "Workbook for estimating effects of accidental explosions in propellant ground handling and transport systems," NASA-CR-3023, REPT-02-4778, Southwest Research Institute, San Antonio, TX. 1978. Appendices H-1 and H-2.

[2] Crowl, D.A. *Understanding Explosions*. Center for Chemical Process Safety, American Institue of Chemical Engineers. New York, NY. 2003. p. 1, 11-13.

[3] Mannan, S. "Table 17.85," *Lee's Loss Prevention in the Process Industries (Fourth Edition)* – Volume 1. Elsevier, Inc., Waltham, MA. 2002, p. 1566.

[4] Scilly, N.F. and J.H.Crowther. "Methodologies for Predicting Domino Effects from Pressure Vessel Fragmentation," *International Conference on Hazard Identification and Risk Analysis*, Center for Chemical Process Safety, AIChE. Orlando, FL. 1992.

[5] CSB. *Investigation Report: Explosion and Fire, First Chemical Corporation*. No. 2003-01-I-MS, U.S. Chemical Safety and Hazard Investigation Board, Washington, DC, October 2003. 77 pp.

[6] CSB. *Investigation Report: Pesticide Chemical Runaway Reaction Pressure Vessel Explosion, Bayer CropScience, LP*. No. 2008-08-I-WV, U.S. Chemical Safety and Hazard Investigation Board, Washington, DC, January 2011. 161 pp.

[7] CSB. *Case Study: NDK Crystal, Inc., Belvidere, IL, High-Pressure Vessel Rupture*. No. 2010-04-I-IL, U.S. Chemical Safety and Hazard Investigation Board, Washington, DC, November 2013. 27 pp.

[8] Crowl, D.A. and J.F.Louvar, *Chemical Process Safety—Fundamentals with Applications, (Third Edition)*. Prentice Hall, Upper Saddle River, NJ. 2011. pp. 298-302.

[9] ASME, "Section VIII-Pressure Vessels," *2015 ASME Boiler and Pressure Vessel Code. American Society of Mechanical Engineers*. New York, NY. 2015.

[10] UNM, *EN13445 Unfired pressure vessels*. European Committee for Standardization, Brussels, Belgium. 2014.

[11] USEPA, "Risk Management Program Guidance For Offsite Consequence Analysis", *Chemical Emergency Preparedness and Prevention Office RMP Series*. United States Environmental Protection Agency, April 15, 1999. 256 p.

[12] Clancey, V.J., "Diagnostic Features of Explosion Damage," *6th International Meeting on Forensic Sciences*, Edinburgh, Scotland, 1972.

[13] OGP, "Vulnerability of plant/structure," Risk Assessment Data Directory, Report No. 434-15, International Association of Oil & Gas Producers, March 2010. pp. 14-16.

[14] TNO, *Methods for the calculation of physical effects due to releases of hazardous materials (liquids and gases)* (The "yellow book"). The Netherlands Organization of Applied Scientific Research. The Hague, The Netherlands. 2005. pp. 742-746.

[15] FM Global. "Evaluating Vapor Cloud Explosions Using a Flame Acceleration Method." *Property Loss Prevention Data Sheets 7-42.* Factory Mutural Insurance Company. West Glocester, RI. 2013. pp. 14-16.

479147 A Risk-Based Approach for Predicting Domino Effects Due to Fires - Combining Exceedance Curves with Thermal Stress Dynamic Analysis

Abstract:

This paper proposes a risk-based method for characterization of domino effects and potential escalation for process equipment affected by thermal radiation (i.e. fires). This approach intends to answer two key questions: (1) which process equipment is impacted by a heat flux capable of resulting in escalation due to equipment failure?; and (2) how long until a process equipment is expected to fail; i.e., Time To Failure (TTF)?.

The first phase consists of developing dedicated heat flux exceedance curves for a given location of interest, using a risk-based approach and considering a cumulative frequency threshold based on worldwide recognized risk tolerability criteria. The exceedance curve approach is valuable to graphically identify which heat flux impacts the process equipment at the given frequency of interest.

The second phase involves the prediction of the TTF due to fires (e.g., pool fires, jet fires) impacting the equipment identified in step one. A two-step approach is proposed: (1) vessel wall segmentation to accurately predict heat transfer from fire to defined wall segments, from wall segment to wall segment, and from wall segments to vessel contents. This dynamic simulation is conducted to determine how the Ultimate Tensile Strength (UTS) of the material decreases as a function of temperature. The UTS is then compared with the internal stress (i.e., Hoop stress) by considering the equipment internal pressure combined with the installed overpressure protection performance. This paper defines step-by-step how to conduct a risk-based quantitative assessment and determine the TTF using a case study. It demonstrates the applicability and accuracy of this approach, which can result in a less expensive procedure than, for example, Computation Fluid Dynamics. Furthermore, it helps the decision-making process on how potential mitigation measures can be implemented.

479462 Quantifying the Mass Release Rate for Flashing Two Phase Releases for Consequence Assessment Purposes

Abstract:

Many common toxic industrial chemicals (TICs) such as chlorine and ammonia are stored as liquids at ambient temperature, and these materials will flash when depressurized to atmospheric pressure forming an aerosol. Sponsored by the Chemical Security Analysis Center (CSAC) of the U.S. Department of Homeland Security, the Defense Threat Reduction Agency (DTRA) of the U.S. Department of Defense, and Transport Canada, the Jack Rabbit II tests were designed to release liquid chlorine at ambient temperature in quantities of 5 to 20 T for the purpose of quantifying the behavior and hazards of catastrophic chlorine releases at scales represented by rail and truck transport vessels. In Phase 1 of the two-year campaign, five successful field trials were conducted in which chlorine was released in quantities of 5 to 10 tons through a 6-inch circular breach in the test vessel (disseminator) and directed vertically downward at 1 m elevation over a concrete pad. In Phase 2, 9 to 10 ton quantities of chlorine were released from 6-inch circular breaches at different orientations (vertically downward, 45° below horizontal, and vertically up) from the disseminator along with a final 20 ton release made vertically downward. This paper analyzes disseminator instrumentation (including load cell data) to determine the mass release rate as well as other characteristics such as the mass of chlorine remaining in the vessel for release orientations other than vertically downward. In addition to this data (being the first of its kind to study large openings), previously published data are also considered to demonstrate the impact on release rate of hole size and the superheat limit temperature. The result of this work is the development of a general approach for quantifying the mass release rate of flashing two phase releases for TICs and other flashing liquids such as cryogenics including, for example, liquefied natural gas (LNG) and liquid oxygen.

Effectiveness of Water Sprays in Mitigating Toxic Releases

Samrat Mukherjee - Samrat.Mukherjee@dupont.com
Seshu Dharmavaram - Seshu.Dharmavaram@dupont.com
Stephen Michael Jaskolka - Stephen.M.Jaskolka@dupont.com

E. I. du Pont de Nemours and Company
974 Centre Road, Wilmington, DE-19805

Prepared for Presentation at
American Institute of Chemical Engineers
2017 Spring Meeting and 13th Global Congress on Process Safety
San Antonio, TX
March 26 – 29, 2017

Effectiveness of Water Sprays in Mitigating Toxic Releases

Samrat Mukherjee - Samrat.Mukherjee@dupont.com
Seshu Dharmavaram - Seshu.Dharmavaram@dupont.com
Stephen Michael Jaskolka - Stephen.M.Jaskolka@dupont.com

E. I. du Pont de Nemours and Company
974 Centre Road, Wilmington, DE-19805

Keywords: Toxic Release, Scrubbing, Anhydrous Ammonia, Water Monitors

Abstract

Chemical manufacturing operations must consider and prepare for release of toxic chemicals, such as anhydrous ammonia, anhydrous hydrofluoric acid, etc. from pressurized storage tanks. Water spray curtains are often advertised as means to mitigate the consequences of released chemicals. Spray curtain effectiveness claims by certain vendors are misleading - a curtain placed at the periphery of a tank will only scrub a puff of a release. A spray curtain that can effectively contain a consequential release will look very different and cost much more. Studies that demonstrate spray curtains to be effective assume low gas velocities. However, calculations show that pressurized liquid NH_3 or HF when released from an orifice to the atmosphere comes out at a high velocity and momentum in the form of a two-phase jet. The jet must travel quite a distance before the velocity drops enough to be effectively scrubbed by a water curtain. The water curtain therefore needs to be at this large distance and consequently the diameter of the water curtain manifold ring needs to be quite large to be effective. It also needs to be quite high to accommodate the expanding jet angle, the potentially high point of puncture or upward angle of jet. Sprays lose effectiveness after a short distance (5-6 m) due to coalescence of the droplets. To avoid the toxic gas from escaping above the spray we may need multiple rows of spray nozzles stacked above each other or some form of barrier above the spray manifold. Finally, large quantities of water are needed to feed all these nozzles at a moment's notice. This paper presents engineering calculations to estimate jet velocity, diameter & concentration, calculations to evaluate the efficiency of water sprays and discussion of water spray curtain designs that are more effective.

1 Background

Toxic chemicals, such as anhydrous ammonia, are sometimes stored in chemical plants in large high pressure tanks in liquid form. Potential for leaks in these tanks and associated piping offer a challenge. From 1996-2011, anhydrous ammonia leaks alone have led to 19 deaths and 1,651 injuries [1]. Section 112(r) of the Clean Air Act Amendments (CAAA) in 1990 mandates that facilities that have on site or handle quantities of regulated substances exceeding thresholds submit a Risk Management Plan to federal, state, and local entities and that they have risk management in place.

There are several ways to mitigate the effect of a leak apart from water sprays, such as rapid de-inventory of the tank, immediate isolation of pipes and additives to the chemical that reduces hazard [2]. In this document, we will focus on water curtains as applied to leaks from high pressure tanks, such as that used for ammonia.

Water spray systems are getting adopted by industry. The European Process Safety Centre, an industry-funded organization, conducted a survey and found that 18 out of 22 companies (surveyed) had installed a water curtain as a mitigation tool [3].

Water curtain vendors presented designs to our company that consist of a spray ring around the tank walls. A breach in the tank will lead to high velocity (>80 m/s) of released fluid close to the wall (See Section 2). Yet, the papers that the vendors referenced only considered toxic gases that travelled at a low velocity. In [4], the authors described a field test where ammonia was assumed to be flowing at 3-5 m/s and the authors show > 80% removal using a water curtain at 6 m height. Curiously, they claim that "perforated sleeves are used most widely" as water curtain instead of a header with nozzles. In [5], the authors ran wind tunnel experiments where chlorine and ammonia gases were released at 0.1-1 m/s. The removal efficiency was less than 50% with a water curtain. In a white paper written by the software company CHARM [6], the authors report >60% ammonia removal with 264,000 lb/hr of water. They do not mention the ammonia flowrate and velocity assumed for their calculations. Water curtains close to the tank wall may be useful for small (<1/8") continuous release or puffs of gaseous ammonia from low pressure containment in vessels and pipelines and the papers above [4-6] may be applicable in such cases.

Companies must prepare for large releases when there are continuous jets of two-phase and gaseous plumes that are rapidly established. In such cases the fluid initially travels at high velocity, and then slows down with distance as it entrains air and gets diluted. A derivation of velocity, diameter and concentration of the jet is presented in the Section 2.

In Section 3, we discuss what a well-designed water curtain should look like and its estimated efficiency as a function of gas velocity. A simple model is presented for quick estimates of removal efficiency. Jets at high velocity will blow through a water curtain making them ineffective. Water curtains need low gas velocity to be effective.

The analysis in this paper centers around anhydrous ammonia, but can be extended to water soluble chemicals such as anhydrous hydrofluoric acid (HF).

In this paper the words breach, hole, orifice, and spill point have been used interchangeably.

2 Modelling of Two-Phase Release of Ammonia

The model for two phase jet release is presented in [9, 10]. This model assumes that pressurized ammonia is stored in liquid phase at or above saturated vapor pressure. This mimics ammonia tanks commonly found in industry. Storage pressure is typically higher when ammonia withdrawal is in liquid phase. The fluid experiences choked flow at the orifice. This choked flow is two phase if the tank pressure is below the threshold defined in Eq. 3, else not. If the flow through the orifice is single phase use Eq. 4, else use Eq. 6 to calculate the mass flux. The velocity calculation follows. It then flashes to atmospheric pressure in a short depressurization zone (or expansion zone) that is typically less than 0.1 m in length. See Figure 1. Flashing leads to evaporative cooling. Beyond the depressurization zone is the entrainment zone. Here, the vapor-liquid mixture travels as a jet, initially at high velocity, and entrains air. We assume that the vapor and liquid travel at the same speed. The entrained warm air increases vaporization, but also slows down the jet. In the beginning the jet is turbulent and well mixed and there is no radial variation in concentration and temperature. When there is enough entrainment and mixing all the liquid vaporizes. In [9], the authors validated their results against release tests performed in the Frenchman Flat area of the Department of Energy's Nevada Test Site [8]. In these calculations, we neglect the impact of heat generated due to reaction of ammonia with water that is entrained from air.

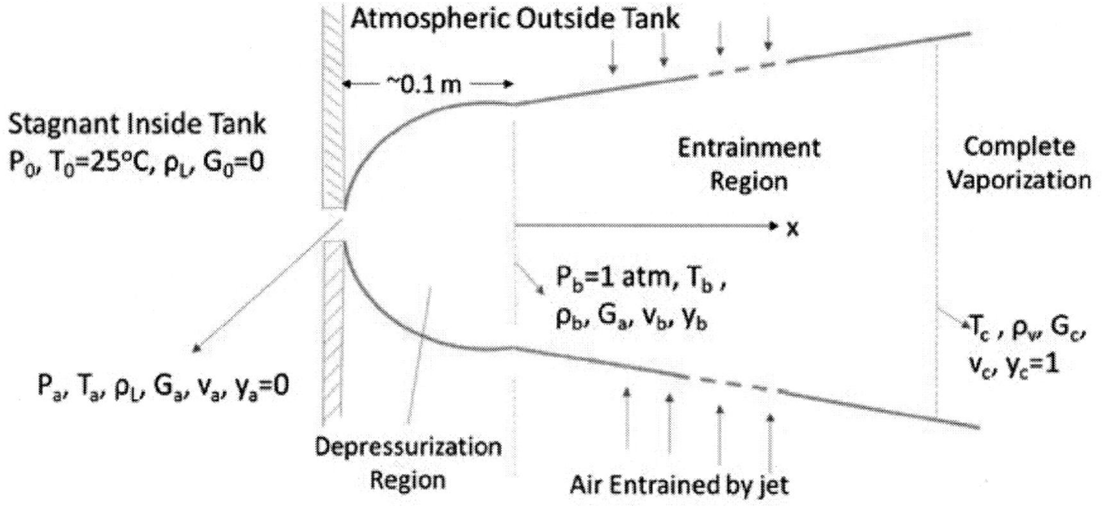

Figure 1: Sketch of pressurized liquid flashing across a hole in the tank wall

For isentropic flashing we can correlate density and pressure using a correlation parameter, ω, as,

$$\frac{\rho_{SL}}{\rho} - 1 = \omega \cdot \left(\frac{P_{SL}}{P} - 1\right) \ldots Eq. 1$$

Where,

ρ_{SL} and P_{SL} are respectively the saturated liquid density and pressure at tank temperature T_0, and

ρ and P are respectively, the density and pressure in the two-phase region.

One can calculate ω by fitting data to Equation 1 or use Equation 2,

$$\omega = \frac{C_{Pl} \cdot T_0 \cdot P_{SL} \cdot \rho_{SL}}{\lambda^2} \cdot \left(\frac{1}{\rho_{SV}} - \frac{1}{\rho_{SL}}\right)^2 \quad ...Eq.2$$

Where,

T_O is the tank temperature

C_{Pl} is the heat capacity of the liquid at T_O,

λ is the latent heat of vaporization at T_O,

ρ_{SV} is the density of saturated vapor at T_O.

For the fluid to be in two phase at the orifice,

$$P_0 \leq \frac{2 \cdot \omega \cdot P_{SL}}{2.\omega - 1} \quad \cdots Eq.3$$

Ammonia in a tank at 25°C has ω of 13.3 and 2.ω/(2.ω-1) of 1.039. If Eq. 3 does not hold, then the flux, G_a, through the hole can be calculated using all-liquid Bernoulli Equation with no phase change,

$$G_a = C_D \cdot \sqrt{2.\rho_{SL} \cdot (P_0 - P_a)} \cdots Eq.4$$

Where,

C_D is the discharge coefficient used for an orifice, typically equal to 0.6

P_O is the stagnant pressure inside the tank

$P_a = P_{SL}$ or the sat. vapor pressure corresponding to the temperature inside the tank

If the tank pressure is at or just above the saturated pressure, then Eq. 3 holds. In this case the pressure at the hole is given by,

$$P_a = P_s \cdot \left(\frac{2 \cdot \omega}{2 \cdot \omega - 1}\right) \cdot \left[1 - \sqrt{1 - \frac{P_0}{P_s}\left(\frac{2 \cdot \omega - 1}{2 \cdot \omega}\right)}\right] \cdots Eq.5$$

And the flux, G_a, is given by,

$$G_a = C_D \cdot P_a \cdot \sqrt{\frac{\rho_{SL}}{\omega \cdot P_s}} \cdots Eq.6$$

The velocity at the hole, $v_a = G_a/\rho_a$.

This is the liquid phase velocity at the hole. As it moves out of the tank the fluid depressurized rapidly and flashes. This is the depressurization region and is typically very short (~ 0.1 m). The flashing leads to an increase in velocity and diameter. From momentum balance equation, we get velocity at the end of depressurization region of,

$$v_b = v_a + \frac{P_a - P_b}{\rho_a v_a} = v_a + \frac{P_a - P_b}{G_a} \cdots Eq.7$$

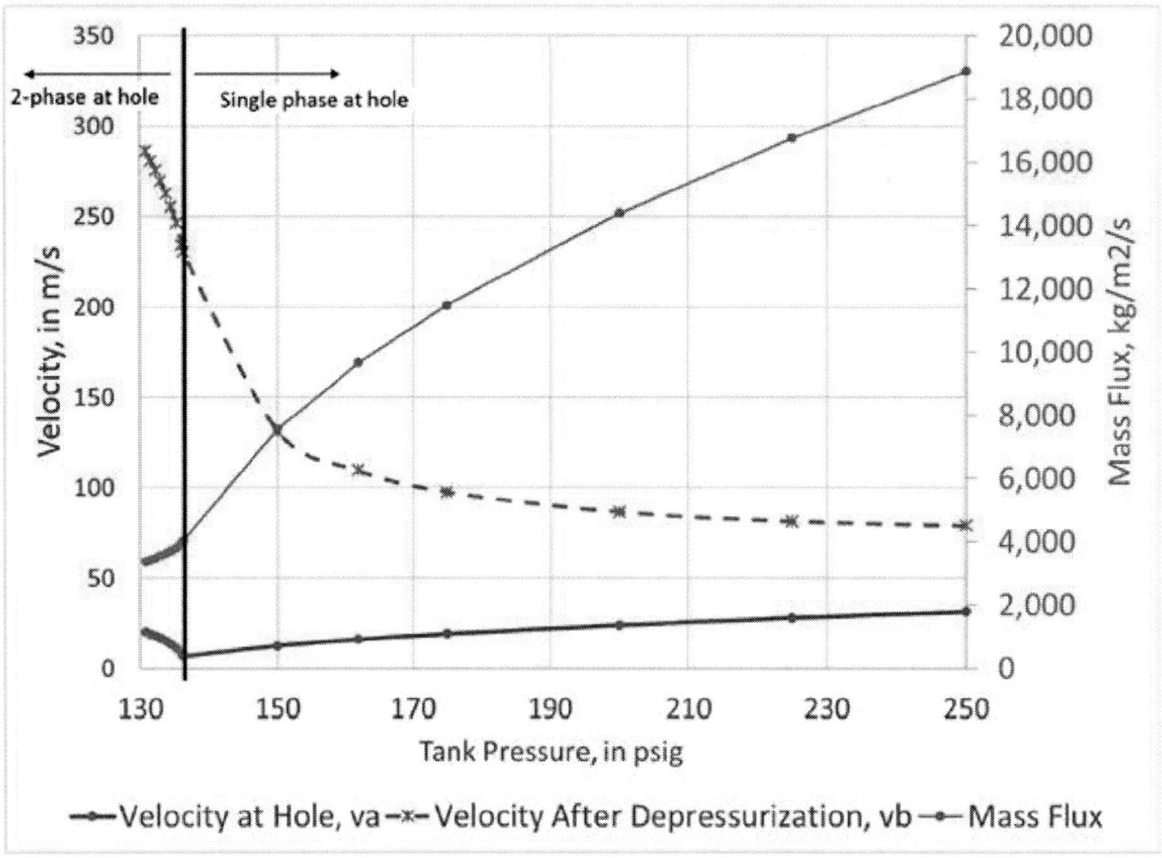

Figure 2: Velocity (v_a & v_b) for an ammonia tank at 25°C. Saturated Vapor Pressure is 130.8 psi.

From energy balance, we can calculate the mass fraction that vaporizes at b using the equation,

$$y_b = \frac{C_P(T_0 - T_{BP})}{\lambda} \cdots Eq.8$$

Where,

C_P is the liquid specific heat

T_{BP} is the boiling point at atmospheric pressure

λ is the latent heat of vaporization

The density of the jet at the start of the entrainment region is given by,

$$\rho_b = \left(\frac{y_b}{\rho_v} + \frac{1 - y_b}{\rho_L}\right)^{-1} \cdots Eq. 9$$

From continuity equation, we calculate the area of the jet at the start of the entrainment zone as,

$$A_b = \frac{A_a \cdot v_a \cdot \rho_L}{v_b \cdot \rho_b} \cdots Eq. 10$$

Where A_a is the hole area.

In the entrainment region, the vapor liquid mixture moves as a jet and entrains air. We assume that the entrainment is adiabatic. Using mass, momentum and energy balance of the jet, authors [12] have derived a correlation for velocity (u), radius (r) and enthalpy (h) as a function of distance (x) from the start of the entrainment zone.

$$\frac{h - h_{air}}{h_b - h_{air}} = \frac{u}{u_b} = \left(\frac{\rho_b}{\rho}\right)^{1/2} \frac{r_b}{r} = \frac{1}{1 + 2 \cdot E_o \cdot \frac{\rho_{air}}{\rho_b} \cdot \frac{x}{r_b}} \cdots Eq. 11$$

Here, density (ρ) is a function of the concentration. E_o is a constant, estimated to be 0.116 [9]. The results from Eq. 11 are plotted in Figures 3-5 for three tank pressures. For each incremental calculation, we assumed a certain mass of air was entrained. Corresponding to this entrained air, we calculate the enthalpy, then the velocity, then the distance travelled and finally the radius of the jet.

Note that all the calculations above, except diameter of jet, are independent of hole area. The total release rate can be calculated by multiplying the velocity, v_a, (shown in Figure 2) with liquid density and hole area. As expected, higher pressure tanks lead to larger rates for the same hole diameter. Non-intuitively, the velocity on depressurization, v_b, is higher with lower tank pressure. The radius of the jet at the start of the entrainment region, r_b, is larger with higher pressure tanks. Tank pressure has little effect on the velocity and jet diameter at larger distances from the spill hole (see Figures 3 and 5). There is, however, some effect on ammonia concentration (see Figure 4).

George Box and Norman Draper [11] wrote that "All models are wrong, but some are useful." This model too is a simplification of reality and is meant to help an engineer bound the problem and calculate approximate solutions with relative ease. As the jet velocity slows down there is a concentration and temperature gradient between the center of the jet and the periphery. Moreover, the effect of wind becomes larger.

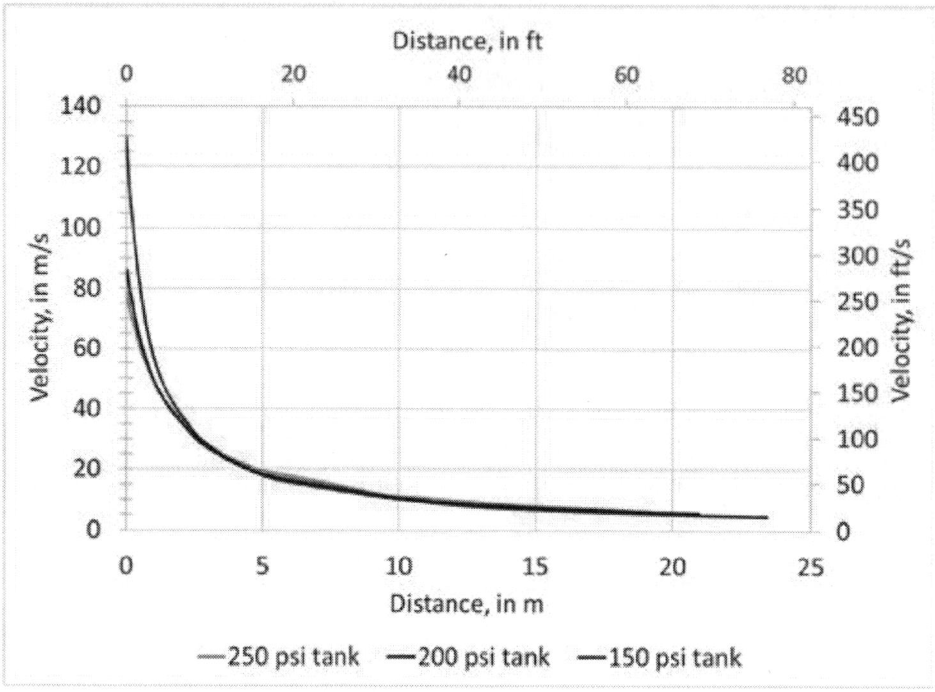

Figure 3: Jet velocity with distance

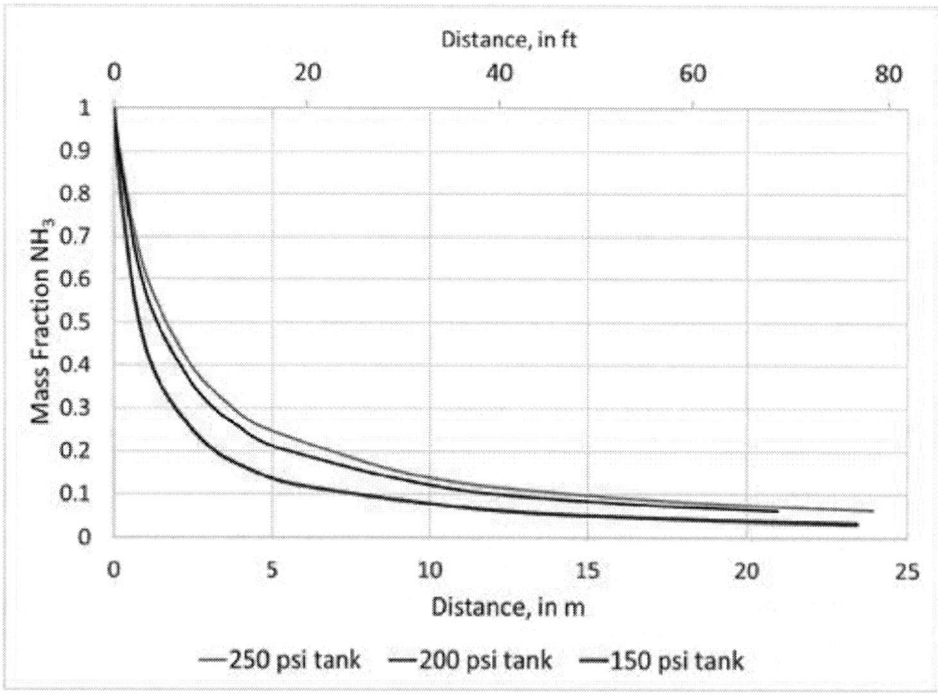

Figure 4: Ammonia Concentration as a function of distance

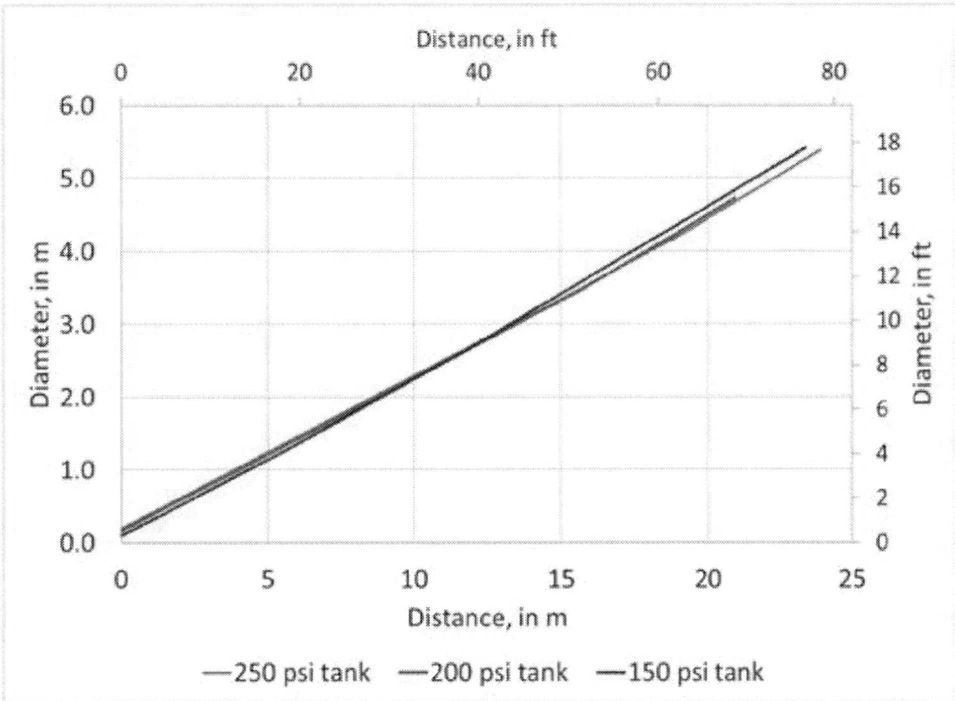

Figure 5: Jet diameter as a function of distance. Assumes a 1" hole in the tank.

3 Water Curtains for Scrubbing

Scrubbing or absorption is an operation in which a gaseous mixture is contacted with a liquid to dissolve one or more components of the gas mixture into the liquid. In our case the gas mixture consists of a toxic gas (NH_3 or HF, etc.), air and small amounts of the toxic material in droplet form. The liquid is water. Gas mixture-liquid contacting can occur in different ways - here we consider liquid spray that is perpendicular to the gas mixture flow.

In this section, we first look at data on scrubbing HF with a water curtain after a large-scale release and details of the configurations used for the tests. Next, we present a simple model for water curtain scrubbing. We validate the model against the limited HF scrubbing data and extend it to NH_3. Finally, we discuss some details of a potential scrubber configuration.

3.1 Experiments with HF Scrubbing

In 1986, six experiments were conducted by Amoco Oil Company and Lawrence Livermore Laboratory, with support from Allied-Signal, DuPont and Mobil, at the Spill Test Facility in which HF was released [13, 14]. This series was dubbed the Goldfish HF Spill Test. These experiments most closely mimic real life high pressure release with scrubbing. The first three tests were without spray mitigation. In the fourth test, compressed air and water spray was used for mitigation. Upflow water curtain spray was employed in test 5. In test 6 downflow water curtain and a combination of upflow and downflow were used. In tests 4-6 a portion of the test was conducted with and a portion without the mitigation system.

Test 4 showed ~ 10% HF removal by the spray and 81% of the water got transported downstream. So, the water helped dilute the HF, but did not help scrub it. Test 5 showed ~ 44% HF removal and 30 % of the water was unaccounted for. It was potentially transported downstream. In test 6, there was 47% HF removal and only 9% of the water was unaccounted for. The H_2O: HF ratio in these tests was 21:1. This matched the results in a flow through chamber run by Allied Corporation. The results from the flow through chamber are in Figure 6. It was later shown in a wind tunnel test (termed the Hawk Tests [15]) that one could achieve ~90% removal with H_2O:HF ratio of 55:1.

The H_2O:HF flow ratio in [13] is misleading. The spill location and direction were predetermined and so the water curtain did not wrap around the tank. It was in a straight line facing the spill point. Some of the HF jet/cloud did not necessarily encounter the curtain. The flowrate per nozzle and nozzle-to-nozzle distance is relevant.

There is no real mention of the HF velocity in the Goldfish test. The spray nozzles were 100 – 105 feet from the spill point. We estimate, using the derivation in Section 2, at that distance the velocity is 5-10 m/s. In Figures 6 and 8, we assume that the velocity is 5 m/s. Furthermore, we assume that HF has completely vaporized before reaching the spray nozzles. In the wind tunnel Hawk test, the spray nozzles were 16' away from the spill point, but the gas velocities was 3 m/s.

The optimal distance of the spray nozzle from the spill point is important – the jet velocity decreases with distance, but the cost of implementation goes up dramatically as the radius of spray nozzle increases. This is discussed in greater detail later.

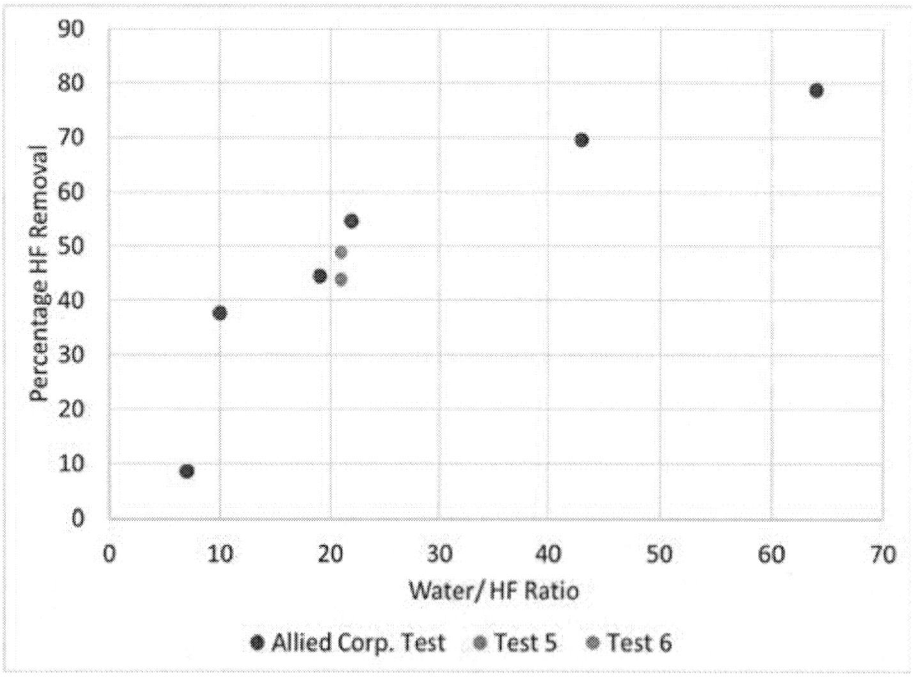

Figure 6: HF removal as a function of Water/HF Ratio. Allied Corp. conducted lab-scale tests.

3.2 Water Spray Model

Researchers [7, 16] have used HGSPRAY, a computer model, to understand the impact of water curtains and validated the model using data from the Hawk Test. This model is shown to be accurate, but is a black box to lay users. It is not easy to understand the impact of various parameters.

Here we borrow from the model used in horizontal cross-flow scrubbers [17] to propose a simplified set of equations that is useful to an engineer evaluating water spray. In horizontal cross-flow scrubbers the gas passes horizontally through a duct and liquid is typically sprayed downwards and perpendicular to the direction of the gas flow. Multiple rows of sprays are often employed. The water curtain is therefore an incarnation of the horizontal cross-flow scrubber, but with a single row of spray nozzles. These are not commonly used commercially because much more water flow is needed than vertical scrubbers, which employ co-current or counter current flow. This methodology does not account for air entrained by the spray and its effect in diluting the toxic component. Therefore, the results are conservative.

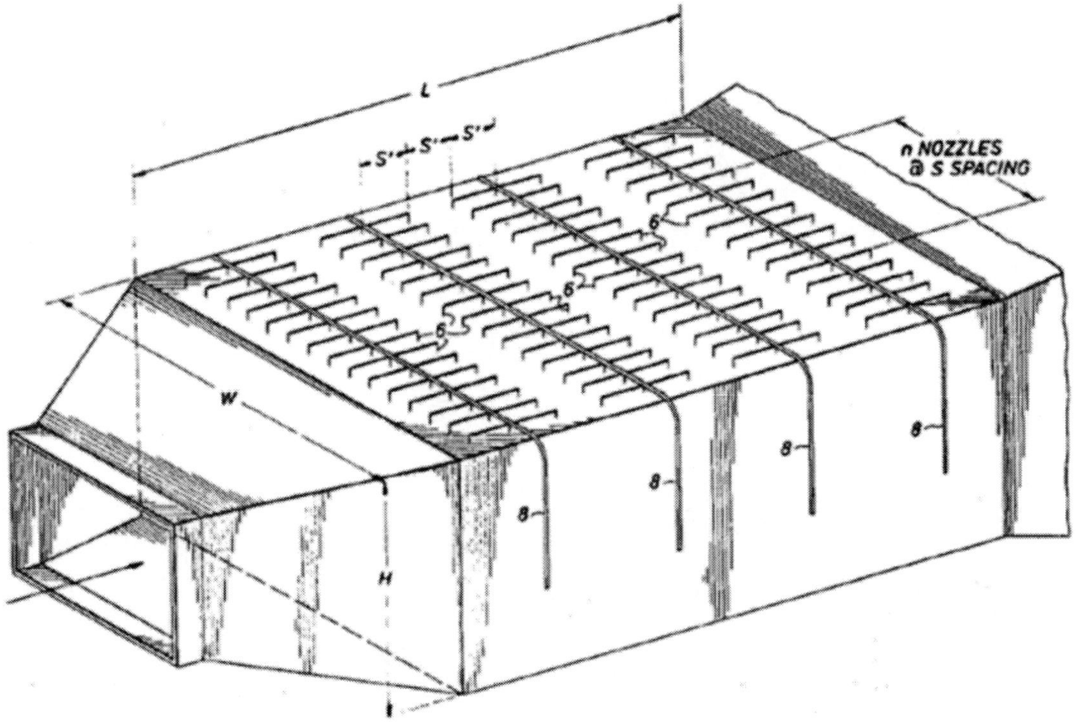

Figure 7: Typical horizontal cross-flow scrubber. From Patent No. CA1124037 A.

In [17], the authors lay out optimal spacing for horizontal scrubbers. If the nozzles are too far apart then fluid being discharged can slip through the spaces in between. If they are too close, then the droplets can collide and combine forming larger – less effective – droplets. The authors assume water flow of 14 l/s (222 gpm) per nozzle, nozzle spacing of 0.47 m (1.5'), 5.5 m (18') high manifold, 1.85mm droplets, 120° cone and droplet velocity of 9 m/s. Scrubbing efficiency decreases with increase in droplet size. To reduce the droplet size (for the same flow) we need higher pressure water upstream of the nozzle. This can increase the droplet velocity, which in turn reduces efficiency.

The applicable equations are as follows:

$$\frac{P}{P_o} = e^{-\phi} \cdots Eq.12$$

$$\phi = 12\sqrt{\frac{D \cdot u_r}{\pi \cdot d}} \cdot \frac{l \cdot N \cdot L}{v \cdot d \cdot G} \cdots Eq.13$$

This set of equations assume that all the mass transfer resistance is in the gas phase and there is no liquid phase mass transfer resistance. A more generalized version of the equation can be found in [17]. Here,

P/P_o is the ratio of the outlet and inlet concentration of the toxic component in the gas

ϕ is the number of gas-phase mass transfer units

D is the diffusivity of the contaminant gas, m2/s. It is ~ $2.2 \cdot 10^{-5}$ m^2/s for NH_3 at 25°C

v is the mean velocity of the spray droplets, m/s

u_r is the relative velocity of the spray droplets and the gas (=$\sqrt{v^2+u^2}$), u is the horizontal gas velocity

d is the (Sauter) mean diameter of the spray droplets, m

l is the mean distance travelled by the spray droplets, m. It is nozzle-height*Cos(spray angle/2)

N is the number of spray stages, one in our case

L is the liquid volume flowrate, m^3/s

G is the gas volume flowrate through the spray area, m^3/s

Note that the model will not extrapolate well to high gas velocities when the momentum of the gas is comparable to the momentum of the liquid droplets. The equations also do not account for scrubbing of liquid entrained with the gas. The equations also do not account for dilution of the gas with air entrained by the spray.

For the Goldfish test the experimenters used 1.76 l/s (28 gpm) per nozzle, spaced 0.91 m (3') apart, 12' high manifold, 0.32 mm droplets and 90° cone. We estimate that the liquid droplet velocity is 33 m/s. In [7] the authors note that the scrubbing efficiencies of NH_3 and HF with water are similar. We could not find similar experimental data for cross-flow ammonia scrubbing.

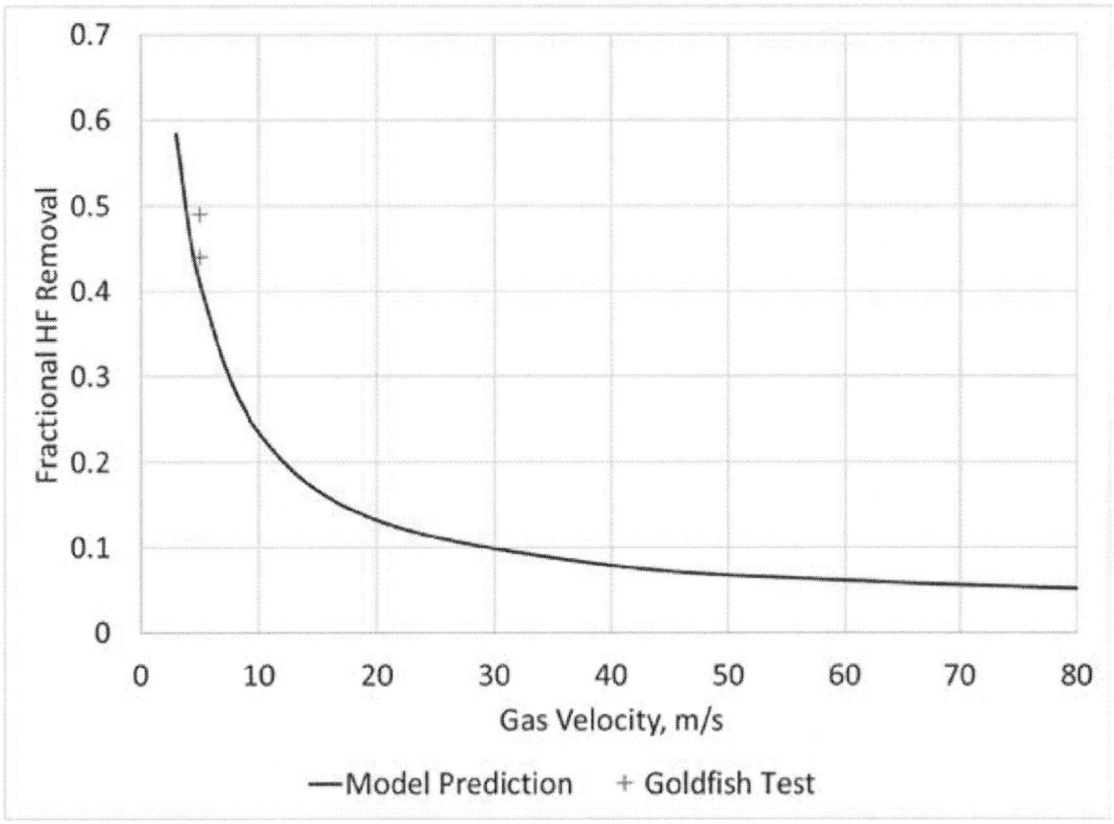

Figure 8: Model versus Goldfish Test Result. In model, nozzle spacing, droplet size, etc. per Goldfish Test. H_2O flow is 28 gpm/nozzle. Model assumes downward flow only and gas at 25°C.

The actual gas velocity at the scrubber in the Goldfish test is unknown, but we can speculate that it is in the 5-10 m/s second range. In Figure 8, we have assumed that the gas velocity is 5 m/s. Also, Test 5 had the water pointing upwards and Test 6 had it both upward and downward. Regardless, the model does a reasonably good job of matching the test data (see Figure 7). Hawk test data shows a substantial drop in efficiency as gas velocity increased from 3 m/s to 6 m/s. A similar drop is seen in Figure 8.

Based on Equations 7-8, we speculate that the efficiency in the Goldfish Test could be higher if the nozzle distance was reduced to 1.5' instead of 3 feet (see Figure 9). This would, however, more water usage.

In [18], the authors undertook outdoor field test and indoor wind-gallery test to understand the effectiveness of water curtains in entraining air into the gas and thus diluting and reducing its concentration. This aspect is not considered in the derivation of Equations 12 & 13, which only considers chemical solubility. The gases used in [18] were Cl_2, CO_2, etc. which have low solubility and the impact of chemical scrubbing was purposely minimized. As in earlier tests, [4, 5 & 15] the authors assumed low gas phase velocity comparable to wind velocity, in the 1-3 m/s range for their experiments and not jet velocity (see Figure 2). The authors proposed a correlation between dilution factor, Df, defined as the ratio of toxin concentration without and with spray and Rm, ratio of momentum of the water-curtain to the representative momentum of the gas cloud. If the gas velocity is 80 m/s and we use the spray curtain structure in the Goldfish test, Rm value is 0.003 and we should not expect any gas dilution based on this calculation methodology – it is outside their charts. On the other hand, at low gas velocities (1-3 m/s), when Rm is 1-10 we should expect the gas dilution effect to be comparable to ammonia removed by scrubbing with water.

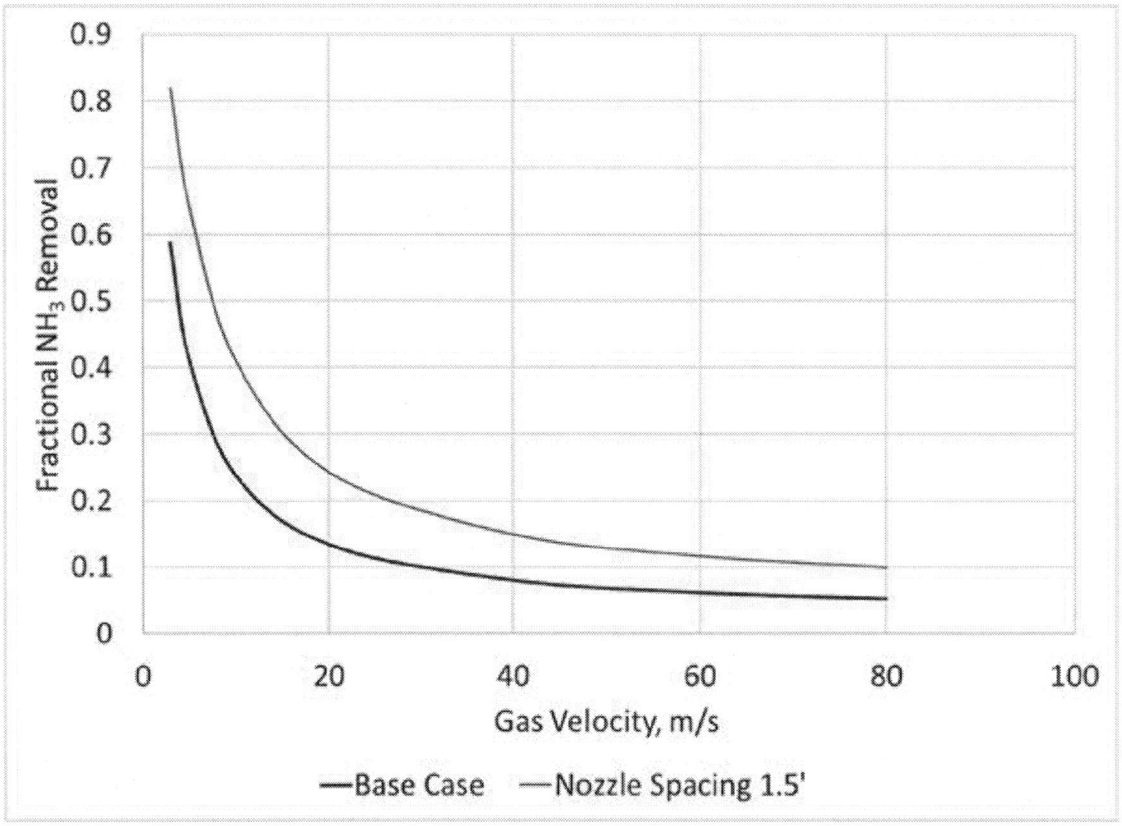

Figure 9: Fractional NH_3 Removal. T=25°C. Base Case has same configuration as Goldfish Test. H_2O flow is 28 gpm/nozzle.

3.3 Initial Water Spray Curtain Design Ideas

Figures 8 and 9 shows that the gas velocity needs to be as low as possible for proper scrubbing. Figure 3 shows that the water curtain needs to be quite far (50-100’) from the spill point to attain the low velocity. Therefore, to scrub a high velocity release the spray curtain needs to form a circular manifold of radius > 50’ (preferably 100’). This corresponds to a manifold perimeter of 314’. If there is a nozzle with 28 gpm every 1.5’on the manifold, we will need ~210 nozzles with ~ 6,000 gpm of water at >60 psi pressure. Note that that this assumes a single row of nozzles is enough. The design would require one to collect and drain the water.

One could reduce the number of nozzle and manifold size by placing a wall/barrier on some of the sides of the tank. Regardless, the kind of area required for such a manifold is often difficult to attain in an industrial setting.

The spill location is not known a priori – it could be in any point around a tank. If the jet points upwards or if the jet expands above the height of the spray nozzles, the gas could travel above the spray. A barrier would then be needed to stop that from occurring.

Maintaining the manifold may be an issue. One needs to ensure that there are no leaks from the water source across the open/close valve as this could lead to corrosion in the spray water manifold. The pump that will deliver the water at a moment's notice needs to be maintained. The standards created for fire water systems could be applicable.

Note that in the event of a leak anyone inside the spray manifold would not be protected.

4 Conclusion

Several vendors approached our company to sell water curtain equipment that will clearly not be effective. There are no accepted engineering standards for such water curtains. Calculating the effectiveness of a scrubber in the event of a release is non-standard and challenging. There are computation fluid dynamic simulators available, but these are typically black box models to most users. This motivated us to present the physics of high pressure release and efficiency of water curtains that is easy to understand and apply to specific applications. The presented equations allow the user to understand the impact of relevant parameters.

The water curtains proposed by some vendors are within a couple of feet of the tank wall. Our calculations show that if there is a breach in a high pressure tank the toxic material released (HF, NH_3, etc.) will have very high velocities near the tank. The fluid velocity will be above 80 m/s for NH_3. Water curtains placed so close will be ineffective (a) as there won't be adequate gas-liquid contacting for scrubbing (as shown in Eq. 12 & 13) and (b) the high momentum of the jet will blow through the water curtain. The water curtain needs to be far from the spill location. This is reflected in the tests done at the STF.

Decreasing the tank pressure will reduce the mass flux. However, calculations show that the velocity near the tank will be higher (see Figure 2). This will make scrubbing more difficult.

Due to the high cost of a water curtain mitigation system that is properly designed and installed it may be beneficial in some instances to use aqueous ammonia solutions that do not need to be stored at high pressure. Other mitigation strategies should be explored.

5. Reference

1. Sofia Plagakis, "Is Anhydrous Ammonia a Risk to Your Community?" http://www.ombwatch.org/node/12472 (accessed on 1/17/2017)
2. John B. Cornwell, and David W. Johnson, "Effectiveness of Mitigation Systems in Reducing Hazards of Hydrogen Fluoride Leaks", In First Risk Control Engineering Seminar Maracaibo, Venezuela October 19-20, 1995.
3. Schoten, H. H., Molag, M., Duffield, J. S., and Powell-Price, M. (2000), "The use of fluid curtains for post-release mitigation of gas dispersion," In Institution of Chemical Engineers Symposium Series (Vol. 147, pp. 287-298). Institution of Chemical Engineers; 1999.
4. G.V. Kotov and S.P. Fisenko, "Efficiency of Contaminated Air-Flow Dispersion by Water Curtains," Development in Earth Science Volume 1 Issue 2, December 2013.
5. E. Palazzi, F. Curro', R. Pastorino and B. Fabiano, "Effectiveness of Reacting Spray Curtains Mitigating Toxic Releases of High Solubility Gases," 2007. http://www.nt.ntnu.no/users/skoge/prost/proceedings/icheap8-pres07/icheap8webpapers/42%20Palazzi.pdf (accessed on 1/18/2017)
6. Estimates of Water Curtain Efficiency for Ammonia, www.charmmodel.com/doc/CurtEff.pdf (accessed on 1/18/17)
7. V.M. Fthenakis, D. N. Blewitt, and W. J. Hague (1995), "Modeling Absorption and Dilution of Unconfined Releases of Hazardous Gases by Water Curtains or Monitors," International Conference and Workshop on Modeling and Mitigating the Consequences of Accidental Releases of Hazardous Materials, New Orleans, Louisiana, September 26-29, pp. 573-592, 1995.
8. H. C. Goldwire, Jr., T.G. McRae, G.W. Johnson, D.L. Hipple, R.P. Koopman, J.W. McClure, L.K. Morris, and R.T. Cederwall. "Desert Tortoise Series Data Report: 1983 Pressurized Ammonia Spills", United States, 1985.
9. Hans K. Fauske and Michael Epstein, "Source term considerations in connection with chemical accidents and vapour cloud modelling", Journal of Loss Prevention in the Process Industries, Volume 1, Issue 2, pp. 75-83, April 1988.
10. J. C. Leung and M. A. Grolmes (1988), A generalized correlation for flashing choked flow of initially subcooled liquid. AIChE J., 34, pp. 688–691, 1988.
11. G. E. P. Box and N. R. Draper, Empirical Model-Building and Response Surfaces, John Wiley & Sons, 1987.
12. J. C. Weimer, G. M. Faeth, and D. R. Olson, Penetration of vapor jets submerged in subcooled liquids. AIChE J., 19, pp. 552–558, 1973.
13. D. N. Blewitt, J. F. Yohn, R. P. Koopman, T. C. Brown, W. J. and Hague, "Effectiveness of Water Sprays on Mitigating Anhydrous Hydrofluoric Acid Releases," Proceedings, International Conference On Vapor Cloud Modeling, American Institute of Chemical Engineers, (1987)
14. D. N. Blewitt, J. F. Yohn, R. P. Koopman, and T. C. Brown, "Conduct of Anhydrous Hydrofluoric Acid Spill Experiments," Proceedings, International Conference On Vapor Cloud Modeling, American Institute of Chemical Engineers, (1987)
15. K.W. Schatz, R.P. Koopman, "Water spray mitigation of hydrofluoric acid releases", Journal of Loss Prevention in the Process Industries, Volume 3, Issue 2, pp. 222-233, 1990.
16. V. M. Fthenakis, "HGSPRAY: A complete model of spraying unconfined gaseous releases", Journal of Loss Prevention in the Process Industries, Volume 6, Issue 5, pp. 327-331, 1993.

17. W.M. Edwards and C.P. Huang, "Horizontal cross-flow scrubber", Patent No. CA1124037 A http://www.google.co.nz/patents/CA1124037A (accessed on 2/2/2017)
18. Karin Hald, Jean-Marie Buchlin, Aurélia Dandrieux, Gilles Dusserre, "Heavy gas dispersion by water spray curtains: A research methodology", Journal of Loss Prevention in the Process Industries, Volume 18, Issues 4–6, pp. 506-511, July–November 2005.

Limiting oxygen concentration of hybrid mixtures

Emmanuel Kwasi Addai
Otto-von-Guericke-Universität,
Institute of Instrumental and Environmental Technology,
Department of Plant Design and Process Safety,
39106, Magdeburg, Germany
emmanueladdai41@yahoo.com

Martin Clouthier
Dalhousie University
Process Engineering & Applied Science,
Halifax, NS, Canada
Martin.Clouthier@Dal.Ca

Jensen Hughes, Inc.,
Halifax, NS, Canada
MClouthier@jensenhughes.com

Paul Amyotte
Dalhousie University
Process Engineering & Applied Science,
Halifax, NS, Canada
Paul.Amyotte@Dal.Ca

Ulrich Krause
Otto-von-Guericke-Universität,
Institute of Instrumental and Environmental Technology,
Department of Plant Design and Process Safety,
39106, Magdeburg, Germany
ulrich.krause@ovgu.de

Prepared for Presentation at
American Institute of Chemical Engineers
2017 Spring Meeting and 13th Global Congress on Process Safety
San Antonio, TX

UNPUBLISHED

Limiting oxygen concentration of hybrid mixtures

Emmanuel Kwasi Addai
Otto-von-Guericke-Universität,
Institute of Instrumental and Environmental Technology,
Department of Plant Design and Process Safety,
39106, Magdeburg, Germany
emmanueladdai41@yahoo.com

Martin Clouthier
Dalhousie University
Process Engineering & Applied Science,
Halifax, NS, Canada
Martin.Clouthier@Dal.Ca

Jensen Hughes, Inc.,
Halifax, NS, Canada
MClouthier@jensenhughes.com

Paul Amyotte
Dalhousie University
Process Engineering & Applied Science,
Halifax, NS, Canada
Paul.Amyotte@Dal.Ca

Ulrich Krause
Otto-von-Guericke-Universität,
Institute of Instrumental and Environmental Technology,
Department of Plant Design and Process Safety,
39106, Magdeburg, Germany
ulrich.krause@ovgu.de

Keywords: limiting oxygen concentration, hybrid mixture explosion, dust explosion, gas/vapor explosion

Abstract

An investigation into the limiting oxygen concentration (LOC) of fifteen combustible dusts and methane mixtures in the standard 20-liter explosion chamber was performed. Three ignition energies (10 J, 2 kJ and 10 kJ) were used. The results show that a 10-J spark leads to significantly higher limiting oxygen concentration values than either 2-kJ or 10-kJ pyrotechnic igniters. This could be due to the "overdriving" effect of the chemical igniters, which produce a hot flame that virtually covers the entire explosion chamber during combustion. With respect to hybrid mixture investigation, the 20-liter sphere was modified to allow the input of methane gas. The limiting oxygen concentrations of the hybrid mixtures were found to be considerably lower than those of dust-air mixtures when the relatively weaker spark igniter was used. There was no significant change in limiting oxygen concentration when the higher-energy chemical igniters were used. Mathematical model to estimate the limiting oxygen concentration of hybrid mixtures is presented. The model predictions show good agreement with experimental results. This research provides safety practitioners with a practicable means to characterize material hazards without the need to perform a large number of tests.

1. Introduction

The Limiting oxygen concentration (LOC) of a fuel-oxidant system is defined as the oxygen concentration at the limit of flammability for the worst-case (most flammable) fuel concentration. It is usually expressed as the percentage of oxygen available in the fuel-oxidant gas mixture. The value of the oxygen concentration is decreased by a step of 1 vol% under the variation of fuel mixture concentration. If no ignition is obtained at any fuel mixture concentration, the oxygen concentration at that point is considered as the LOC. In order to protect process equipment, a safety margin of 2 vol% below the LOC is usually used for cases where the LOC is greater than or equal to 5 %. Where the LOC is less than 5 %, the equipment is operated at no more than 50 % of the LOC.

Much research has been performed on the determination of LOC values for single fuel-air-diluent mixtures [1, 2-4]; however, there are many industrial processes involving fuel mixtures in different states of aggregation (i.e., hybrid mixtures) [5-8]. There has been relatively less research on hybrid mixtures compared to single-fuel-air mixtures, especially for LOC. The current work deals with the investigation of LOC values for hybrid mixtures using fifteen combustible dusts, and one flammable gas (methane). Nitrogen is used as a diluent in the standard 20-liter sphere under the same test conditions. A standard 2-kJ chemical igniter is typically used as an ignition source for the determination of LOC for dusts; however, in this study three different ignition energies (10 J, 2 kJ, and 10 kJ) were employed to measure the LOC of dusts and hybrid mixtures. Comparisons among these three ignition strengths are made in order to estimate the effect of ignition energy on the LOC of different fuel mixtures. Mathematical model to estimate the LOC of dusts and hybrid mixtures is presented with comparisons between the models and the experimental results.

2. Experimental procedure

To obtain a general understanding of the LOC of dust and hybrid mixtures, a wide variety of dusts with different properties and combustion reaction mechanisms were chosen. For dusts, materials as diverse as food substances, metals, plastics, natural products, fuels and artificial substances were chosen. The raw dust samples were received in different forms and conditions. For this reason, both physical and chemical analyses were performed on each dust sample, including particle size distribution, volatile content, moisture content, heat of combustion and elemental analysis as shown in Table 1. The most important (physical, chemical, ignition, thermodynamic) properties of methane gas are shown in Table 2.

Tables 1: Physical and chemical properties of dusts

Dust sample	Median Particle Diameter (μm)	Volatile Content (% wt.)	Moisture Content (% wt.)	Heat of combustion (MJ/kg)	Elemental analysis (% wt.)				
					C	H	O	S	N
Starch	14	93.77	0.50	15.30	44.3	6.3	48.9	0.4	0
Lycopodium	32	91.06	0.35	28.44	69.3	9.6	19.6	0.4	1.3
Toner	14	90.18	0.92	35.79	86.1	7.7	5.2	1.0	0.0
Wood	61	84.38	0.20	16.44	50.3	6.3	43.2	0.1	0.0
Wheat flour	52	79.60	0.38	15.64	45.6	6.6	45.9	1.8	0.0
Protein	46	81.47	1.93	20.49	53.0	7.2	23.5	0.3	16
Peat	45	68.76	1.43	17.46	57.4	5.8	35.3	1.1	0.6
Charcoal	79	14.99	0.77	29.36	90.2	2.8	6.7	0.2	0.2
Antioxidant	18	94.56	0.58	32.16	72.8	6.7	12.5	5.3	0.7
Cork	78	80.79	0.86	22.99	63.8	6.9	28.4	0.4	0.4
Polyamide	52	97.98	0.97	29.46	63.7	6.0	14.3	0.0	15
Alfalfa	32	76.00	1.96	16.27	50.6	6.6	39	0.2	3.5
Ascorbic acid	32	83.14	0.42	11.46	41.8	4.7	53.9	0.3	0.0
Aluminium	54	—	1.21	—	—	—	—	—	—
Sulphur	42	—	0.67	—	—	—	—	100	—

Table 2: Properties of methane [9, 10].

Properties	Methane
Molecular formula	CH_4
Purity (%)	99.87
Density (g/cm^3)	6.6E-4
Molecular weight (g/mol)	16
Explosible range (vol. %)	4.4-17
Melting point (°C)	−161
Specific heat capacity (J/mol. K)	35.69
Boiling point (°C)	−182.5
Heat of vaporization (kJ/mol)	−74.87
Maximum explosion pressure (barg)	8.1
MESG (mm)	1.14
Temperature class	T1
Explosion Group	IIA
Heat of combustion: (MJ/kg) : (kJ/mol)	55 -286
Diffusivity (m^2/sec)	1.60
Adiabatic flame temperature (K)	2226
Burning velocity (cm/s)	39

2.1. Experimental setup

Experiments were performed in a 20-liter sphere (Siwek apparatus) to determine the limiting oxygen concentration of dusts, gases and hybrid mixtures. Measurements were performed in accordance with EN 14034-4 [11]. Figure 1 presents a technical diagram of the 20-L sphere. The commercially-available test apparatus consists of a hollow sphere made of stainless steel with an internal volume of 20 L, jacketed cooling system, vacuum pump for evacuating the chamber, distribution system for gases and dusts, ignition source at the center of the sphere, pressure transducers, and data acquisition system (Kuhner KSEP-Software). The apparatus was modified to include a distribution system for methane.

An ignition source was placed at the center of the sphere through a flange located on the top of the combustion chamber. International standards [12] require ignition energies of 10 kJ (2 x 5 kJ) for testing dusts, 10 J for gases and 2 kJ (2 x 1 kJ) for hybrid mixtures. The igniter assembly is shown in Figure 2.

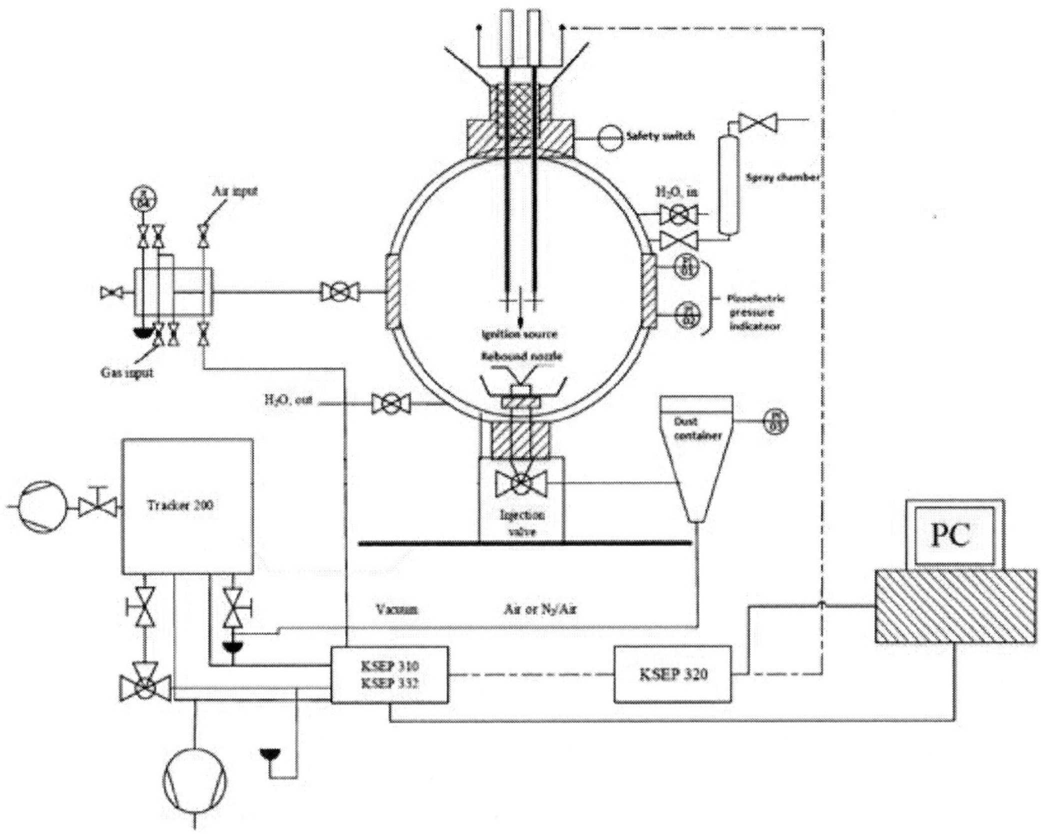

Figure 1: Technical diagram of the 20-liter sphere apparatus

The electrical spark electrodes are two round tungsten rods with ends ground to a point. The electrodes are spaced 6 mm apart and are supplied with power from a high-voltage transformer (KSEP 320) which produces a continuous spark (15 kV, 30 mA) [13]. The chemical igniters (40 wt.% zirconium, 30 wt.% barium nitrate, 30 wt.% barium peroxide) are activated electrically by a low-voltage electrical signal and provide a densely-dispersed cloud of hot particles with very little gas by-product. All experiments in this study were performed with an ignition delay of 60 ms.

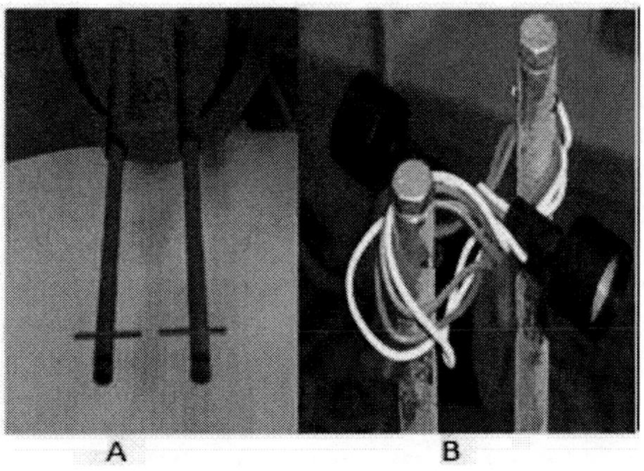

Figure 2: (A) electrical ignition assembly (B) chemical ignition assembly

2.2. Limiting oxygen concentration testing

The tests to determine LOC were performed in the 20-liter sphere according to the procedure stated in the European Standard EN 14034-4 [14]. Equipment modification were made to allow for the addition of nitrogen and methane gas into the sphere. To test for the LOC of a specific fuel-air-diluent mixture, the mixture was initially tested at 21 vol% oxygen. Further tests were performed by systematically reducing the oxygen concentration until ignition was no longer observed after three successive attempts. The concentration of oxygen which will just not allow an explosion of the fuel/oxygen/inert gas mixture in three successive tests was considered to be the limiting oxygen concentration. Different ignition energies such as spark ignitor (10 J) and chemical ignitors (2 kJ and 10 kJ) were used for this study and the results were compared. To assess ignition, the explosion overpressure value recorded by the piezoelectric pressure transducers must be greater than 0.2 bar.

3. Results and Discussion

3.1. Limiting oxygen concentration of dusts

Initially, the LOC values of single-fuel mixtures were determined. Figure 3 presents experimentally-determined LOC values for dusts using different ignition energies. It can be seen that the ignition energy has a significant influence on the LOC of dusts. The results obtained using the 10-J electrical igniter are much higher than those from the chemical igniters, which are in the range of 11 vol% to 17 vol% for aluminium and peat, respectively. The LOC values decrease when the ignition energy is increased to 2 kJ. A further decrease is also observed when the ignition energy is increased by a factor of five (10 kJ). For example, the LOC obtained for brown coal with a 10-J ignition strength is 16 vol%. The LOC decreases to 8 vol% when the ignition energy is increased to 2 kJ, and to 7 vol% when the ignition energy is increased to 10 kJ. The reason for the pronounced effect of ignition energy on the LOC of dusts may be attributed to the energy released from the ignition source during combustion. The chemical igniters provide a dense cloud of hot dispersed particles that occupy the entire volume of the combustion chamber. Complete fuel consumption could be assumed since the entire fuel mixture ignites almost simultaneously. Unlike chemical igniters, an electrical igniter produces a continuous spark in a small localized region at the center of the combustion chamber. Ignition occurs at this point and flame propagates outward from this point. So, if the energy produced during the energy transfer process is not enough for flame propagation, the overpressure criterion will not be met.

These results confirm the findings of Going et al. [15], who reported a reduction in LOC with increased ignition energy. The authors also explained that the LOC values obtained from the electrical igniter (10 J) appeared to be between 4 vol% and 8 vol% higher than the 2-kJ chemical igniter, thus indicating that the explosion may have been underdriven by the weak ignition source. ASTM-E2931-13 [16] indicates that LOC decreases with increasing ignition energy. For example, the LOC values obtained for Pittsburgh pulverized coal were 13.5 vol%, 11 vol% and 9.5 vol% when 1-kJ, 2.5-kJ and 5-kJ igniters were used, respectively.

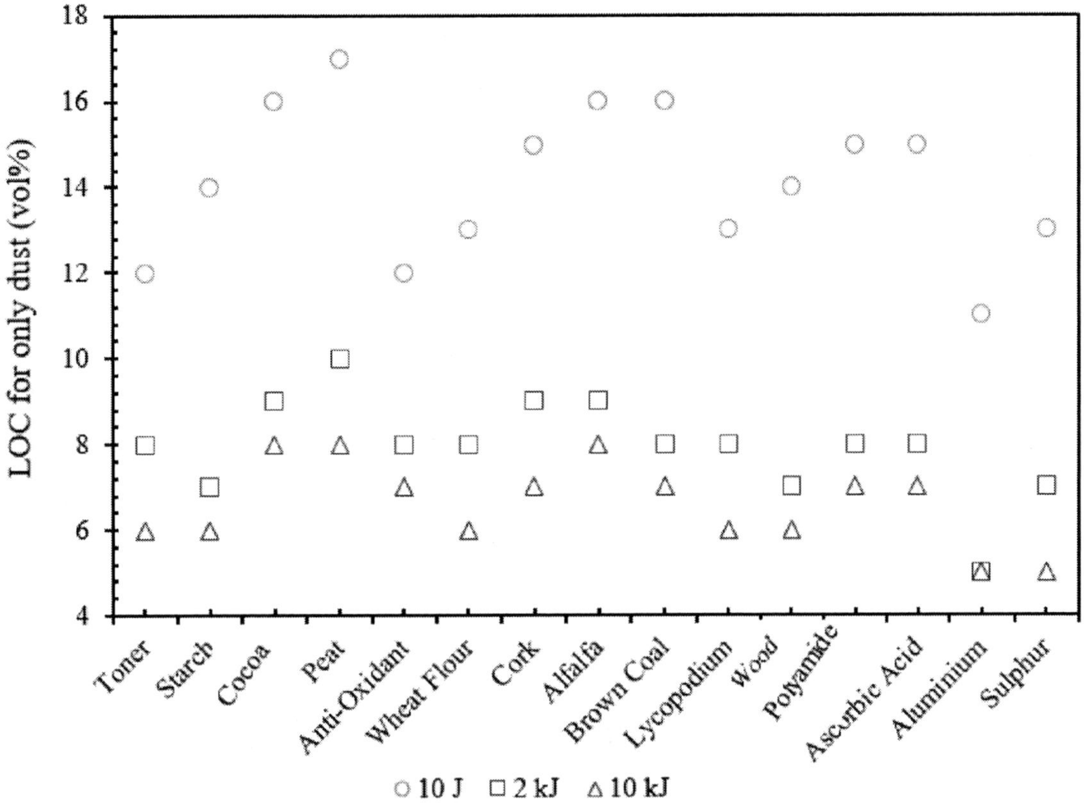

Figure 3: Limiting oxygen concentration of dusts with different ignition energies.

Furthermore, the differences in the LOC values obtained with ignition energies of 2 kJ and 10 kJ did not exceed 2 vol% of oxygen. This suggests that, taking into account the effect of overdriving as reported by Going et al. [15] for the 10-kJ igniters, an ignition energy of 2 kJ would perhaps be the most appropriate choice for LOC tests in the 20-liter sphere.

Going et al. [15] also provided justification for the lower LOC values in the 20-litre sphere using a 10-kJ igniter compared to a 2-kJ igniter. The authors provided a comparison of LOC data obtained from the 20-liter chamber and the 1-m³ explosion vessel, concluding that a 10-kJ ignition source is appropriate for the 1 m³ vessel, but using the same ignition source for the 20-liter tests produced overdriven explosions.

These results also confirm the work done by Mittal [17] who undertook experimental investigations of LOC of coal with different ignition energies and noted that the LOC of the coal dust decreased from 11 vol% at 1 kJ to 7 vol% at 2.5 kJ, 6 vol% at 5 kJ and 5 vol% at 10 kJ.

3.2. Limiting oxygen concentration of hybrid mixtures

In the next step, the LOC of hybrid mixtures of fifteen dusts with methane were investigated. Tests were performed using three different ignition energies which included the 10-J electrical igniter and both 2-kJ and 10-kJ chemical igniters. Figure 4 presents the LOC data for hybrid mixtures of methane gas and dusts using three ignition energies. It can be seen that the LOC of hybrid mixtures decreased as the ignition energy increased. For example, the LOC of a hybrid mixture of methane and sulphur dust decreased from 11 vol% at 10 J to 8 vol%

at 2 kJ and 6 vol% at 10 kJ. A similar explanation as provided for the effect of ignition energy on the LOC of single substance presented in the previous section, can be applied in this case.

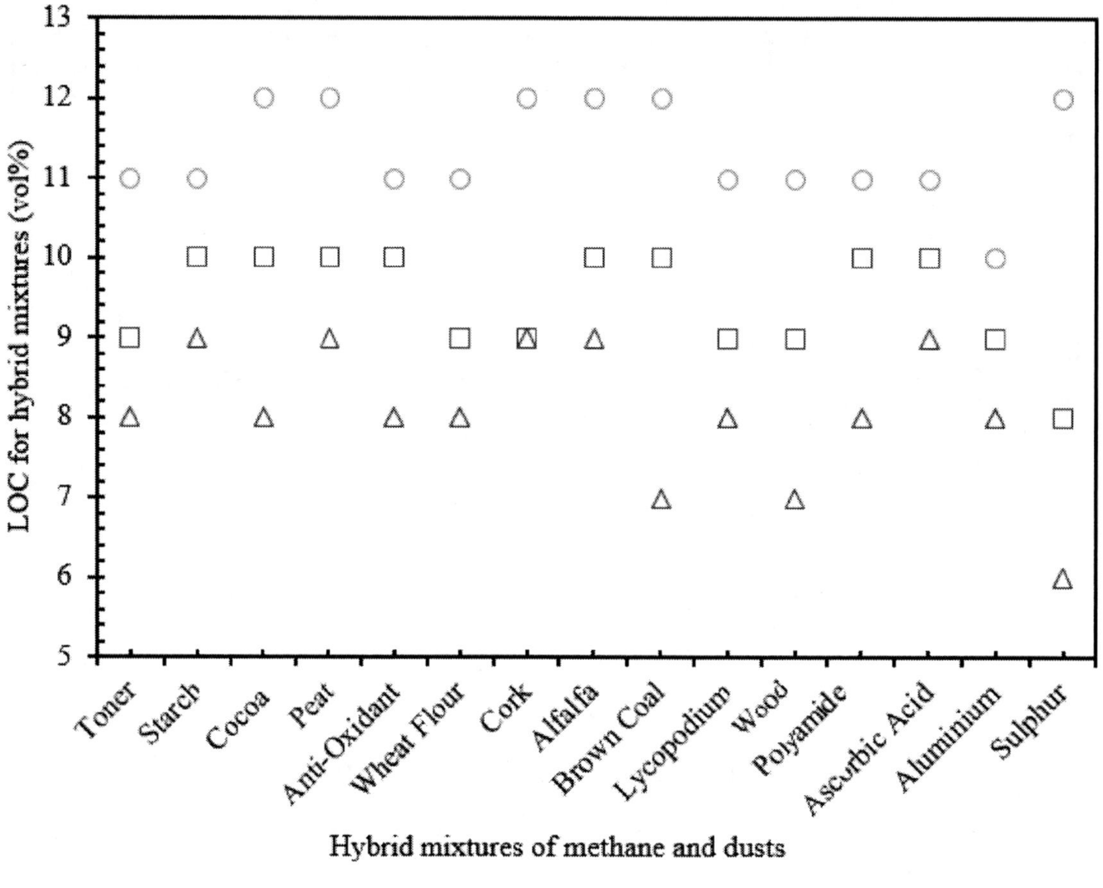

Figure 4: Limiting oxygen concentration of hybrid mixtures of methane and dusts with different ignition energies.

Table 3 shows the experimental results for the LOC of hybrid mixtures. In all cases, a comparison was made between dusts and the hybrid mixtures results. Regarding the 10-J electrical igniter, the LOCs for hybrid mixtures were comparatively lower than for single dusts. For example, the LOC of peat decreased from 17 vol % to 12 vol% when 39 g/m^3 of methane was added.

Table 3 further presents experimentally-determined LOC values for hybrid mixtures using both 2-kJ and 10-kJ chemical igniters. Unlike the 10-J electrical igniter for which the dusts recorded the highest LOC, in the case of 2-kJ chemical igniters, the dusts recorded lower LOC values. A similar trend was noticed for the tests with 10-kJ igniters.

Table 3: LOC of hybrid mixtures of methane and dusts with different ignition energies.

Dusts Samples	10 J			2 kJ			10 kJ		
	LOC (vol %)	Gas conc. (g/m^3)	Dust conc (g/m^3)	LOC (vol %)	Gas conc. (g/m^3)	Dust conc (g/m^3)	LOC (vol %)	Gas conc (g/m^3)	Dust conc. (g/m^3)
Toner	11	39	250	9	39	500	8	46	375
Anti-Oxidant	11	53	500	10	39	500	8	33	500
Lycopodium	11	46	250	9	33	250	8	39	250
Starch	11	33	500	10	33	250	9	33	200
Wheat Flour	11	39	500	9	33	500	8	33	500
Wood	11	33	750	9	39	250	7	33	250
Cocoa	12	59	500	10	39	500	8	33	375
Cork	12	46	750	9	39	500	9	33	500
Polyamide	11	53	500	10	33	375	8	33	375
Peat	12	46	375	10	39	250	9	33	250
Alfalfa	12	59	750	10	39	500	9	33	500
Ascorbic Acid	11	59	500	10	33	250	9	33	500
Aluminium	10	39	375	9	33	375	8	33	375
Brown Coal	12	53	500	10	39	500	7	33	500
Sulphur	12	33	250	8	39	250	6	33	175

3.3. Model to estimate the limiting oxygen concentration of hybrid mixtures

A model to estimate the φ_{LOC} of hybrid mixture is presented by considering the assumption that the φ_{LOC} of hybrid mixtures is the summation of the φ_{LOC} of dust and gas multiplied by the respective mass fractions of the individual components in the mixture, as shown in the following equation. Details of this model are provided in [18, 19].

$$\varphi_{LOC\_hybrid} = 100.\ [(\sigma_{fu-g}.M_{O2}.\ (0.55\frac{100}{1+1.193k}\ .\ \beta_g)) + (\sigma_{fu-d}.\frac{M_{O2}}{\rho_{O2}}\ .\ (\frac{a}{H_0} - b)\ .\ \beta_d)]$$
$$k = 4C + H + 4S - 2O - N - 2Cl - 3F - 5Br$$

Where C, H, S, O, N, Cl, F, Br are number of atoms of carbon, hydrogen, sulphur, oxygen, nitrogen, chlorine, fluorine, bromine in the molecule, M_{O2} is the molecular weight of oxygen, ρ_{O2} is the density of oxygen, σ_{fu-g} is the fuel number of gas [mol/g], φLOC is the limiting oxygen, σ_{fu-d} indicates the fuel number of dust [mol/g], β_d represent the mass fraction of gas in the mixture, β_d is the mass fraction of dust in the mixture, subscripts g and d are used to indicate gas or dust. Possible values for constants a and b are shown in Table 4, values for mass fraction of dusts (β_d) and gases (βd) are shown in Table 3, values for σ_{fu-g} and σ_{fu-d} are shown in Table 5.

Table 4: Constants for Schönewald model [20].

Constants	Coating Powder	Industrial Dusts	Fuel Dusts	Metal Dusts
a in J/g	1.235 x 10^6	1.194 x 10^6	1.390 x 10^6	1.132 x 10^6
b in g/m^3	2.532	0.604	7.952	1.540

Table 5: Mass fractions of elements, MEC, heat of reaction and estimated fuel numbers for the fifteen dusts.

Samples	μ_c	μ_H	μ_0	MEC (g/m^3)	σ_{fu}(mol/g)	H_o (J/g)
Wheat	0.42	0.061	0.476	64.7	0.035	18360
Starch	0.433	0.064	0.477	63.9	0.035	17310
Brown coal	0.69	0.060	0.22	69.5	0.066	17148
Wood	0.491	0.063	0.440	77.7	0.043	15395
Lycopodium	0.695	0.094	0.186	37.9	0.075	30554
Polyamide	0.637	0.097	0.142	46.4	0.073	29461
Cork	0.42	0.06	0.51	83.90	0.035	14230
Alfalfa	0.44	0.05	0.34	74.6	0.038	15990
Peat	0.50	0.05	0.3	64.8	0.044	18420
Antioxidant	0.72	0.08	0.12	37.35	0.076	31960
Ascorbic acid	0.40	0.04	0.53	104	0.024	11469
Toner	0.86	0.07	0.05	32.75	0.089	35792
Cacao	0.52	0.06	0.28	57.13	0.049	20678
Aluminium	-	-	-	-	-	-
Sulphur	-	-	-	-	-	-
Methane	0.75	0.25	0	29.5	0.116	55 E6

Figure 5 presents a comparison between the experimental and computational results for hybrid mixtures with different ignition energies. An error bar based on the uncertainty analysis from the experimental procedure is also plotted with the experimental results. See [18] for detailed information on the error and uncertainty analyses. With respect to the results obtained for methane and dust mixtures with different ignition energies as presented in Figure 5, it could be seen that the computed values were always lower than, or equal to, the experimental ones for both 10-J and 2-kJ ignition energies. However, for 10-kJ chemical igniters, the computed values were below the experimental values with only one deviation, where the experimental value was 1 vol% higher than the estimated value. For industrial applications, in order to protect process equipment, usually a safety margin of 2 vol% below the determined LOC is used.

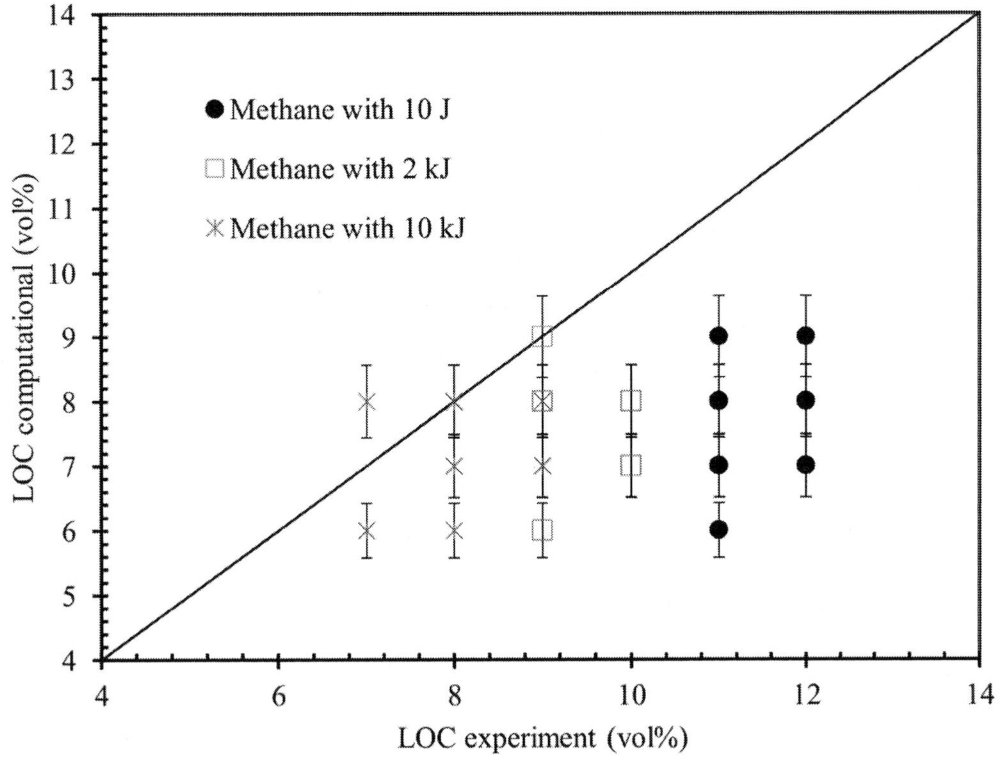

Figure 5: A comparison between computational and experimental results for the LOC of hybrid mixture of methane and dusts with different ignition energies.

4. Conclusion

Both experimental and theoretical limiting oxygen concentration values for fifteen combustible dusts and methane hybrid mixtures in the standard 20-liter explosion chamber have been determined. Three different ignition energies (10 J, 2 kJ and 10 kJ) were used as ignition sources. The results show that an electrical igniter providing approximately 10 J of energy led to significantly higher values for LOC than pyrotechnical igniters of either 2 kJ or 10 kJ. This is due to the fact that the chemical igniters overdrive the explosion by producing a hot flame which covers the entire explosion chamber during combustion. The LOC values for the hybrid mixtures were significantly lower than those of dust-air mixtures, when the relatively weak electrical igniter was used; however, by using chemical igniters, no significant change in the limiting oxygen concentration could be observed. Furthermore, a mathematical model to estimate the limiting oxygen concentration of hybrid mixtures was presented. A comparison between this estimated values and experimental results show good agreement between for all mixtures, with the exception of the results obtained using a 10-kJ chemical igniter, where some deviations were recorded within the safety margin commonly used in industry.

References

[1] Cashdollar, K.L., (1996). Coal dust explosibility. Journal of Loss Prevention in the Process Industries, 9, 65-76.
[2] Continillo, G., Crescitelli, S., Fumo, E., Napolitano, F., & Russo, G. (1991). Coal dust explosions in a spherical bomb. Journal of Loss Prevention in the Process Industries, 4, 223-229.
[3] Wilén, C., Moilanen, A., Rautalin, A., Torrent, J., Conde, E., Lodel, R., & Brehm, K. (1999). Safe handling of renewable fuels and fuel mixtures. Fuel, 68, 1123-1133.
[4] Woskoboenko, F. (1988). Explosibility of Victorian brown coal dust. Fuel, 67, 1062-1068.
[5] Addai, E. K., Gabel, D., & Krause, U. (2016). Experimental investigation on the minimum ignition temperature of hybrid mixtures of dusts and gases or solvents. Journal of Hazardous Materials, 301, 314-326.
[6] Addai, E. K., Gabel, D., & Krause, U. (2015). Explosion characteristics of three component hybrid mixtures. Process Safety and Environmental Protection, 98, 72-81.
[7] Addai, E. K., Gabel, D., & Krause, U. (2015). Lower explosion limit of hybrid mixtures of burnable gas and dust. Journal of Loss Prevention in the Process Industries, 36, 497-504.
[8] Addai, E. K., Gabel, D., Kamal, M., & Krause, U. (2016). Minimum ignition energy of hybrid mixtures of combustible dusts and gases. Process Safety and Environmental Protection, 102, 503-512.
[9] Brandes, E., Möller, W., Molnárné, M., Schendler, T., & Schröder, V. (2003). Sicherheitstechnische Kenngrößen. Wirtschaftsverlag NW, Verlag für neue Wissenschaft, 2, 20-40
[10] Rossini, F.D. (1952). Selected values of chemical thermodynamic properties.US Government Printing Office Washington, DC., 500, 128-137
[11] National Fire Protection Association. (2005). NFPA 654: Standard for the Prevention of Fire and Dust Explosions from the Manufacturing, Processing, and Handling of Combustible Particulate Solids. National Fire Protection Association.
[12] European Standard, EN 14034 1-4, (2004). Potentially Explosible Atmospheres, Explosion Prevention and Protection. Determination of explosibility and severity of dust clouds Dust/air Mixtures in the 20-liters sphere. ECS, Brussels, Belgium.
[13] Cesana, C., Kühner, A. G., & Siwek, R. B000_071 23.03. 11 CC.
[14] European Standardization, EN 50281-2-1, (2007). Potentially Explosible Atmospheres, Explosion Prevention and Protection, Determination of limiting oxygen concentration for Dust/air Mixtures. ECS, Brussels, Belgium.
[15] Going, J. E., Chatrathi, K., & Cashdollar, K. L. (2000). Flammability limit measurements for dusts in 20-L and 1-m^3 vessels. Journal of Loss Prevention in the Process Industries, 13, 209-219.
[16] Dastidar, A. G. (2016). A new standard for the limiting oxygen concentration of combustible dusts. Process Safety Progress, 2, 159-164.
[17] Mittal, M. (2013). Limiting oxygen concentration for coal dusts for explosion hazard analysis and safety. Journal of Loss Prevention in the Process Industries, 26, 1106-1112.
[18] Addai, E. K. (2016). Investigation of Explosion Characteristics of Multiphase Fuel Mixtures with Air. Western Engineering, Inc., 1, 100-120
[19] Krause, U. Addai, E. K., & Gabel, D. (2016). Determination of the limiting oxygen concentration of dust/air and hybrid mixtures based on thermochemical properties, 11th International Symposium on Hazards, Prevention, and Mitigation of Industrial Explosion (ISHPMIE), Dalian, China, Volume: 11.

[20] Schönewald, I. (1971). Vereinfachte Methode zur Berechnung der unteren Zündgrenze von Staub/Luft-Gemischen," Staub- Reinhaltung der Luft, Nr. 31, 376-378.

.

13TH GLOBAL CONGRESS ON PROCESS SAFETY

Ignitability of Combustible Dust Fueled Flash Fires with Industrial Ignition Sources

Michael C. Stern, Ph.D., PE, CFEI
Exponent, Inc.
Natick, MA 01760
mstern@exponent.com

Sean C. O'Hern, Ph.D., PE, CFEI
sohern@exponent.com

Alfonso F. Ibarreta, Ph.D., PE, CFEI
aibarreta@exponent.com

Russell A. Ogle, Ph.D., PE, CSP, CFEI
rogle@exponent.com

Timothy J. Myers, Ph.D., PE, CFI, CFEI
tmyers@exponent.com

Prepared for Presentation at
American Institute of Chemical Engineers
2017 Spring Meeting and 13th Global Congress on Process Safety
San Antonio, TX
March 26 – 29, 2017

Ignitability of Combustible Dust Fueled Flash Fires with Industrial Ignition Sources

Michael C. Stern, Ph.D., PE, CFEI
Exponent, Inc.
Natick, MA 01760
mstern@exponent.com

Sean C. O'Hern, Alfonso F. Ibarreta, Russell A. Ogle, Timothy J. Myers
Exponent, Inc.

Keywords: Combustible Dust, Flash Fire, Ignitability

Abstract

Combustible dust flash fires represents a serious hazard in industrial settings and can require significant mitigation efforts. Increasingly, standards require the use of Dust Hazard Analyses (DHAs) to determine the appropriate level of protection for specific processes. However, assessing the likelihood of ignition can be difficult to assess for hard-to-ignite dusts. Common ASTM dust cloud testing methods either use capacitive sparks up to about 1-Joule, which are typically too weak to ignite hard-to-ignite dust, or pyrotechnic igniters, which are difficult to correlate to real-world industrial ignition sources. Based on previous testing by the authors, many dusts that are explosible in confined 20-L tests using the ASTM E1226 method do not appear to propagate flames under unconfined flash-fire conditions. This paper extends the previous testing to include the use of relevant ignition sources such as diffusion flames, pre-mixed propane flames, and oxy-acetylene torch flames to further develop techniques that can be used in practical hazard analyses and risk assessments.

1 Introduction

The current hazard assessment and mitigation methodologies for combustible dust flash-fire hazards focus heavily on limiting the availability of the fuel source, and put less emphasis on prevention through the elimination of ignition sources. When protecting against flammable gas or vapor-fueled fires and explosions, the elimination of all ignition sources can be challenging due to the low ignition energies of many flammable gases and vapors. Combustible dust ignition energies, however, can vary by orders of magnitude based on particle size, morphology, and chemical composition [1]. While some dusts are susceptible to ignition from electrostatic discharges, as will be shown in this paper, other dusts that are combustible in confined 20-L testing are not susceptible even to premixed flames at temperatures over 1000°C. In this paper, carbonaceous dusts with varying volatility that are explosible as tested in a 20-L confined apparatus are tested against a variety of flames to evaluate the differences between the 20-L confined testing methods and real world ignition sources.

ASTM standards for ignitability of combustible dust due to sparks [2], hot surfaces [3], and high dust cloud temperatures [4] are available. These tests, however, often cannot evaluate high energy sparks (> 1 J) and do not evaluate diffusion or premixed flames. Therefore, a practitioner could not use the result of a Minimum Ignition Energy (MIE) test to determine if the dust is ignitable by a spark caused by an electrical short, welding, or a pilot light.

Electrical arcs in 20-ampere protected residential circuits can release several Coulombs of charge capable of igniting nearby materials or generating molten metal sparks [5]. Standard MIE tests, even if performed at higher energies, cannot simulate the potential of ignition due to molten metal ejecta or small flames resulting from the spark. If arcing occurs with an aluminum conductor, the molten aluminum ejecta can be an effective ignition source due to its exothermic reaction with air.

Through a more complete understanding of the dusts that are ignitable by common industrial ignition sources hazard analysis practitioners can more effectively evaluate the likelihood of an ignition scenario and, therefore, more accurately assess the risk of a process. In this paper, we use our previously described unconfined dust testing apparatus to expose carbonaceous dusts at varying volatilities to a variety of flames that might exist in industrial settings.

2 Experimental Method

2.1 Materials

To create dusts with varying combustion properties, blends of Pittsburgh Pulverized Coal (PPC) and graphite were mixed at varying ratios. PPC is a common and well-characterized dust standard used by testing laboratories to calibrate their equipment. PPC is derived from bituminous coal and obtained from NIOSH. The artificially grown graphite was acquired from Sigma-Aldrich. Relevant properties of the materials are shown below in Table 1.

Table 1. Properties of materials used in testing

Property	Pittsburgh Pulverized Coal	Artificial Graphite
Mass percent below a 200 mesh sieve (~75 microns)	58%	100%
Moisture Content (wt %)	1.7%	0.1%
Volatility (wt %)	36%	0%
Heat of Combustion (MJ/kg)	25-33	33

Mixing PPC with graphite allows for the creation of carbonaceous dust mixtures with varied volatilities. The volatiles content of the combustible dust governs the relative contributions of homogeneous and heterogeneous chemical reactions in the combustion process [6-9]. High volatiles content favors homogeneous reactions and larger fireball volumes. Low volatiles content favors heterogeneous reactions and smaller fireball volumes.

2.2 Test Equipment

Both a standard Siwek 20-L confined combustion sphere made by Kühner and a custom unconfined flash fire apparatus were used. Significant literature regarding the Siwek 20-L vessel is available. The unconfined flash fire apparatus used was previously described in articles by Stern, et al. [10-12]. A photograph of the test equipment can be seen in Figure 1.

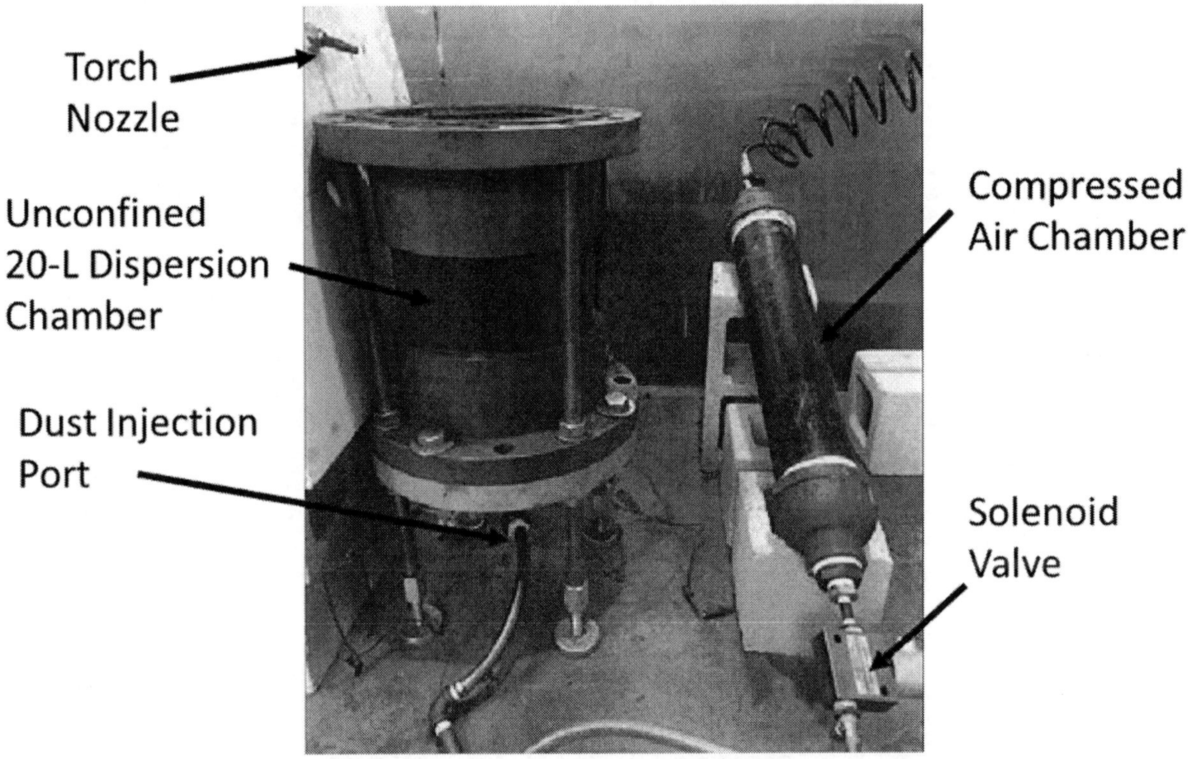

Figure 1. Photograph of unconfined test apparatus with a premixed MAPP-Pro flame.

The flash fire apparatus creates an unconfined combustible dust cloud within and above the apparatus which is then subjected to an ignition source. A nominal 5-L air reservoir is pressurized to 3.3 barg and used to disperse 30 grams of the dust mixture for each test. This produces a nominal concentration (before dilution into the local environmental) of 1200 g/m^3. The ignition source is located approximately 5 cm above the rim of the apparatus. The diffusion flames are created with a propane torch without allowing air entrainment in the nozzle. Both small (1-inch) and large (5-inch) flames were tested. Premixed 5-inch flames of propane and MAPP-Pro are created using the nozzles with air entrainment. Finally, an oxyacetylene flame is tested using an oxy-acetylene torch. An open tube apparatus with an oxy-acetylene torch has previously been used in Norway as a dust cloud ignition screening test [13]. Figure 2 shows examples of all of the flames tested.

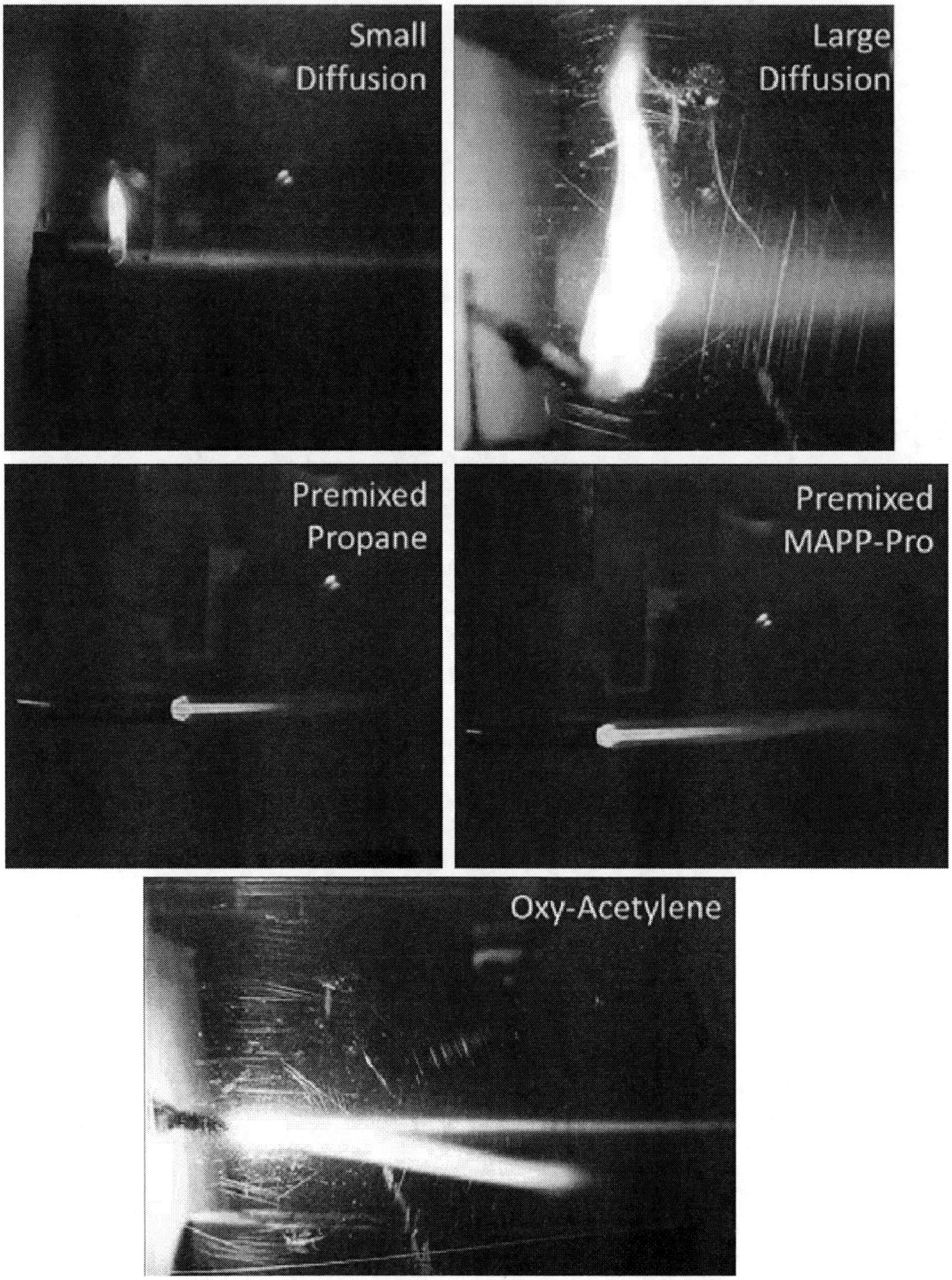

Figure 2. Images of the different flames tested with the unconfined dust testing apparatus.

The confined 20-L Siwek chamber allows the creation of a repeatable dust cloud through injection of dust from a pressurized chamber containing both the dust and air through a high-flow solenoid valve and a dispersion nozzle. After a specified delay time, usually 60 milliseconds, the dust is ignited using a small quantity of pyrotechnics. ASTM test methods often recommend the use of ignition energies from 2.5 kJ to 10 kJ depending on the type of test. In the "Go/No-Go" test used to assess the explosibility of a dust, 5-10 kJ of ignition energy is recommended. 2.5 to 5 kJ ignition energies are recommended in minimum explosible concentration and limiting oxidant concentration (MEC and LOC) testing. A schematic diagram and photograph of the 20-L system are shown in Figure 3.

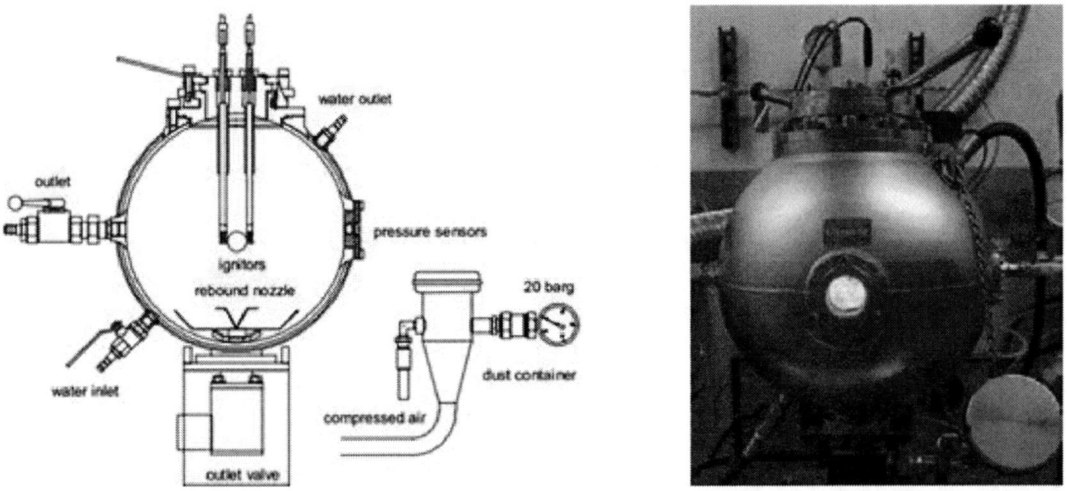

Figure 3. Schematic (left) and photograph (right) of the 20-L Chamber

3 Results and Discussion

3.1 Confined 20-L Testing

The minimum pyrotechnic ignition energies (MPIE) for many combustible dusts are significantly less than the 2.5-10 kJ used for most tests. To evaluate the MPIE of mixtures of the PPC with graphite, tests were performed with 250 J, 500 J, and 1000 J pyrotechnic igniters at 500 g/m^3. A threshold maximum overpressure of 1 bar-g calculated based upon the P_m values as described in ASTM E1226 is used to determine if a sample is explosible under a given set of conditions. The results of the confined 20-L testing performed in the Siwek chamber can be found in Table 2.

Table 2. MPIE test results in 20-L confined Siwek combustion sphere with three different graphite and PPC mixtures. All tests performed at 500 g/m^3.

PPC Concentration	Ignites at 250 J	Ignites at 500 J	Ignites at 1,000 J	P_m at 10 kJ (barg)	dP/dt at 10 kJ (bar/s)
100%	Yes			7.8	461
75%	No	Yes		7.2	314
50%	No	No	Yes	6.9	213
0%	No	No	No	5.9	194

3.2 Unconfined Flash Fire Testing

The three mixtures tested in the 20-L confined Siwek chamber were tested using the unconfined apparatus with the three ignition sources described in the methodology section. The results of the testing can be seen in Table 3. An ignition is determined to occur when the flame propagates away from the areas either immediately adjacent or upstream of the torch flame.

Table 3. Ignition results of unconfined flash fire testing with spark and flame ignition sources

PPC Concentration	MPIE[1] (J)	1 in. Diffusion Flame	5 in. Diffusion Flame	5 in. Premixed Propane Flame	5 in. Premixed MAPP-Pro Flame	Oxy-Acetylene Flame
100%	250	Ignition				Ignition
75%	500	No Ignition	No Ignition	No Ignition	No Ignition	Ignition
50%	1000				No Ignition	No Ignition

[1] Based on results shown in Table 2.

Based on the results summarized in Table 2 and Table 3, the dust ignitable with a MPIE of 250 J is ignitable with the small 1-inch pilot flame, the dust ignitable with a MPIE of 500 J is ignitable with a oxy-acetylene torch only, and the mixture with a MPIE of 1,000 J is not ignitable with any of the flames. The ability of the oxy-acetylene flame to ignite 75% PPC sample is likely attributed to its high heat output and possible excess oxygen released by the torch. It should be noted that while the ignition of the 100% PPC propagated in both directions (with the direction of air flow and counter to the direction of air flow back into the open chamber) the flame of the 75% PPC sample only propagated with the direction of air flow. Still photographs of the extent of propagation for the three mixtures with the oxy-acetylene torch are shown in Figure 4.

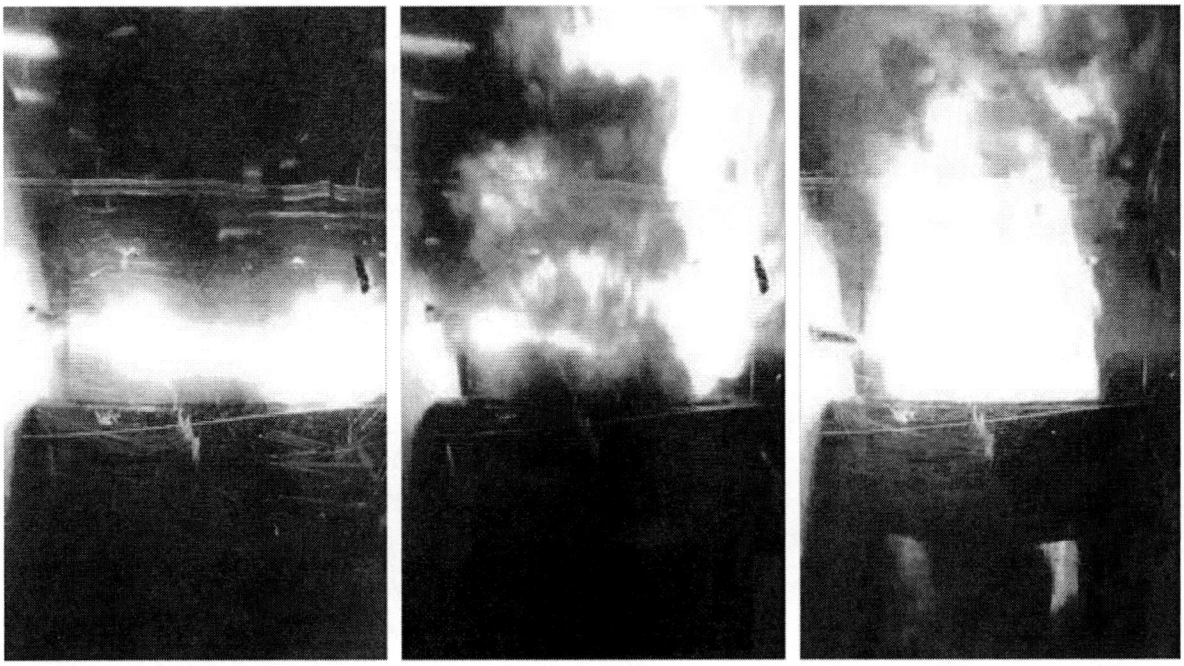

50% PPC 75% PPC 100% PPC

Figure 4. Still photographs of oxy-acetylene torch tests with the three dust mixtures tested. Note the propagation of the flame into the lower chamber for the 100% mixture.

These results are similar to the results of previous investigators evaluating the inerting of coal dusts [15-17]. These investigators found that significantly higher concentrations of inert dusts are required to inert a given combustible dust when measured in a 20-L confined chamber versus a 1-m^3 chamber. This discrepancy, and the general tendency of the overestimation of explosibility parameters when testing in a 20-L chamber as opposed to a 1-m^3 chamber, occurs due to overdriving [18]. Overdriving is caused by a combination of two potential factors:

1. The energy of the igniters can preheat and pressurize the chamber causing an increase in the reaction kinetics and devolatization of the dusts.
2. The thermal energy of the igniters occupies the majority of the 20-L volume, which causes confined dust to burn in the flame of the igniter.

While testing in a 1-m^3 chamber alleviates the problem of overdriving, it can be challenging due to the large quantities of dust required (50 times that of 20-L testing), the significant capital expense of the testing equipment, and the potential disconnect that still exists between a standardized confined experiment and real-world hazards in unconfined geometries. Additionally, it is difficult to correlate the pyrotechnic igniters to other real world ignition sources. The advantages of the 1-m^3 testing are that it is standardized, can detect explosibility in difficult to ignite dusts, and can be used for more quantitative assessments of explosion severity parameters.

It should also be noted that with some dusts, predominantly metal dusts, the 20-L confined chamber can under-predict the explosibility of a dust when compared to 1-m^3 testing [21].

Therefore, the choice of test scale should be considered carefully when expanding these methodologies to metal dusts.

The flash fire testing performed in this work does not currently provide a quantitative measure, but for a qualitative measurement (can support flash fire propagation or cannot support flash fire propagation), the unconfined testing apparatus provides an option that requires less than one kilogram of dust, low capital cost, and is not subject to overdriving. However, because this does not simulate a confined system, it may not be appropriate for evaluating explosion hazards, only flash-fire hazards. Other ignition sources, such as higher energy electric sparks, could be evaluated with this flash-fire apparatus.

Measurement of quantitative properties through unconfined experiments has been previously achieved [19-20]. These additional methods of performing unconfined testing of combustible dust clouds in recent years are also the result of concerns with confined chambers with limited optical access.

4 Conclusions

The results presented in this paper indicate that common industrial ignition sources, flames, in unconfined environments provide significantly different results from confined 20-L testing with 5-10 kJ chemical igniters used in standard test methods and referenced in several NFPA standards. These results, however, may correspond well with 1-m^3 testing, where overdriving is not as significant. This indicates that in addition to ignitability, the ability of a cloud to sustain ignition in areas that are not preheated should be considered when assessing combustible dust flash fire hazards. Unconfined flash fire tests using either diffusion or premixed flames offer the potential for a test that does not require significant quantities of dust, is inexpensive, does not require handling of pyrotechnics, and is directly relatable to the potential hazard. In our preliminary assessment of this methodology, we find that an oxy-acetylene torch, one of the most potent ignition sources likely to exist outside of process equipment, yields comparable results to a 500 J ignition source in a confined 20-L chamber. Further testing is necessary to determine if that correlation is appropriate for other dusts.

5 References

[1] R. Ogle, “Dust Explosion Dynamics,” *Elsevier*, Chapter 2, 2017.

[2] ASTM E2019-03(2013) *Standard Test Method for Minimum Ignition Energy of a Dust Cloud in Air.*

[3] ASTM E2021-15, *Standard Test Method for Hot-Surface Ignition Temperature of Dust Layers.*

[4] ASTM E1491-06(2012), *Standard Test Method for Minimum Autoignition Temperature of Dust Clouds.*

[5] M. Stern, O'Hern S., Morse T., Bishop J., and Kytomaa H.. "Fire Risks Due to Unintentionally Energized Metal Structures." *Proceedings, International Symposium of Fire Investigators*, Scottsdale, AZ, 2016.

[6] C. Proust and Veyssiere B.. "Fundamental Properties of Flames Propagating in Starch Dust-Air Mixtures," *Combustion Science and Technology*, 62: 149-172, 1988.

[7] C. Proust. "Flame propagation and combustion in some dust-air mixtures," *Journal of Loss Prevention in the Process Industries*, 19: 89-100, 2006.

[8] W. Gao, Dobashi R., Mogi T., Sun J., and Shen X.. "Effects of Particle Characteristics on Flame Propagation Behavior During Organic Dust Explosions in a Half-Closed Chamber," *Journal of Loss Prevention in the Process Industries*, 25: 993-999, 2012.

[9] W. Gao, Mogi T., Yu J., Yan X., Sun J., and Dobashi R.. "Flame Propagation Mechanisms in Dust Explosions," *Journal of Loss Prevention in the Process Industries*, 36:186-194, 2015.

[10] M. Stern, Rosen J., Ibarreta A., Ogle R., and Myers T.. "Quantification of the Thermal Hazard From Metallic and Organic Dust Flash Fires." *Journal of Loss Prevention in the Process Industries*, 44: 528–537, 2016.

[11] M. Stern, Rosen J., Ibarreta A., Myers T., and Ogle R. "Quantification of the Thermal Hazard of Metallic and Organic Dust Flash Fires." *Proceedings Mary K O'Connor Process Safety Symposium*, College Station, TX, 2015.

[12] M. Stern, J. Rosen, A. Ibarreta, T. Myers and R. Ogle. "Unconfined Deflagration Testing for the Assessment of Combustible Dust Flash Fire Hazards." *Proceedings, 11th Global Congress on Process Safety*, Austin, TX, 2015.

[13] R. Echkhoff, Dust Explosions in the Process Industries, Third Edition, Gulf Professional Publishing, p. 489, 2003.

[14] ASTM E1226-12a, *Standard Test Method for Explosibility of Dust Clouds*.

[15] P. Amyotte. "Solid inertants and their use in dust explosion prevention and mitigation," *Journal of Loss Prevention in the Process Industries*, 19:161–173, 2006.

[16] A. Dastidar and Amyotte P., "Determination of Minimum Inerting Concentrations for Combustible Dusts in a Laboratory-Scale Chamber," *Trans IChemE*, 80B:287-297, 2002.

[17] A. Dastidar and Amyotte P., "Flammability Limits of Dusts: Minimum Inerting Concentrations," *Process Safety Progress*, 18:56-63, 1999.

[18] T. Myers, A. Ibarreta, J. Bucher and K. Marr. "Assessing the hazard of marginally explosible dusts. *Proceedings, 47th Loss Prevention Symposium,*" *American Institute of Chemical Engineers Spring Meeting,* San Antonio, TX, 2013.

[19] T. Skjold, K. Olsen, and D. Castellanos. "A constant pressure dust explosion experiment." *Journal of Loss Prevention in the Process Industries*, 26:562-570, 2013.

[20] P. Julien, Vickery J., Whiteley S., Wright A., Goroshin S., Bergthorson J., and Frost D.. "Effect of Scale on Freely Propagating Flames in Aluminum Dust Clouds." *Journal of Loss Prevention in the Process Industries*, 36:230-236, 2015.

[21] J. Taveau and R. Seidel. "Explosibility and Large-Scale Venting Tests with 4 Metal Powders," *NFPA 484 Committee Meeting,* Knoxville, TN, 2015.

A New Methodology to Evaluate System-Level Performance of Explosion Suppression Systems

Jenny Chao
FM Global, Chemical Operations
1175 Boston-Providence Turnpike
Norwood, MA 02062
jenny.chao@fmglobal.com

Sergey Dorofeev
FM Global, Research Division
1151 Boston-Providence Turnpike
Norwood, MA 02062
sergey.dorofeev@fmglobal.com

Prepared for Presentation at
American Institute of Chemical Engineers
2017 Spring Meeting and 13th Global Congress on Process Safety
San Antonio, TX
March 26 – 29, 2017

A New Methodology to Evaluate System-Level Performance of Explosion Suppression Systems

Jenny Chao
FM Global, Chemical Operations
1175 Boston-Providence Turnpike
Norwood, MA 02062
jenny.chao@fmglobal.com

Sergey Dorofeev
FM Global, Research Division

Keywords: explosion suppression, dust explosions, gas explosions

Abstract

A parametric experimental study was performed in order to understand the effect of various factors on the explosion suppression phenomenon. Full-scale suppression experiments were conducted in FM Global's 2.5 and 25 m^3 vessels. Either quiescent propane-air or cornstarch-air explosions were suppressed using a 2.5, 5, 10, or 50 L suppressant bottle, filled with sodium bicarbonate and pressurized with nitrogen to approximately 62 bar. The growth of the suppressant cloud in the open atmosphere was also measured in conjunction with the pressure drop in each size of suppressant bottle. From these results, it was found that an expanding flame must be entirely enveloped by suppressant in order for successful suppression of the flame to occur. Based on this concept, a simple physics-based model was developed to determine the maximum protected throw distance, surface area, and volume of a single suppressant bottle (as part of a suppression system). This can be easily scaled to larger volumes than what was used in the experiments.

1 Introduction

An accidental explosion that occurs inside an enclosure will typically generate an overpressure of about 6 to 8 bar, which far exceeds the strength of almost all enclosures in industrial locations (e.g., spray dryers, dust collectors, and rooms with typical wall strengths). Needless to say, an accidental explosion can result in serious property damage, long periods of business interruption, injury to personnel, and loss of life. Although explosion venting has been established as a very effective means of reducing damaging explosion overpressures (e.g., [1]), it is not always possible or desired for various reasons (e.g., not enough surface area of the enclosure is available for venting, space is limited, or there are safety concerns for adjacent personnel). As a result, explosion suppression systems as alternate means of explosion protection have become fairly popular.

An explosion suppression system is essentially designed to detect and quench an incipient explosion inside a protected volume. Once an explosion is detected by a pressure transducer or a visual flame sensor (ideally at the initial stages of its development), a control unit is then prompted to actuate suppressant storage containers that rapidly inject suppressant (such as sodium bicarbonate or monoammonium phosphate) into the enclosure. Successful operation of a system occurs when the injected suppressant quenches the expanding flame and reduces the overall pressure rise to a level below the enclosure strength.

Each explosion suppression system is unique and designed by a manufacturer for a specific application. The type and location of the equipment (i.e., size, number, and placement of suppressant bottles) needed to protect a given volume is determined by using a design methodology that a manufacturer has developed, which is generally an empirical correlation based on many large-scale experiments (e.g., [2, 3]). The models and computers codes that have been developed by manufacturers are proprietary information, and it is not straightforward to independently evaluate the system-level performance of a given explosion suppression system, particularly at scales larger than those used in tests.

In order to address this, a multi-year parametric experimental investigation was undertaken in order to understand explosion suppression phenomena and to develop a methodology that can be used to evaluate the effectiveness of any given system. Open-air discharge tests and full-scale suppression tests were conducted at FM Global's Research Campus in West Glocester, RI. The open-air discharge tests were conducted to measure the growth of a suppressant cloud as the high-pressure contents inside a suppressant bottle were released into the open atmosphere. The full-scale suppression experiments were conducted in 2.5 and 25 m^3 explosion vessels using propane-air or cornstarch-air mixtures in order to understand how the suppression phenomenon is affected by different factors such as the amount of suppressant used, the location of suppressant injection relative to ignition, system-activation pressure, amount of suppression, distribution of suppressant or suppressant bottles, and scale.

A simple physics-based model was developed to synthesize the experimental data and to help elucidate the dynamics of the suppression phenomenon. It was found that this basic model can be used in conjunction with performance tests to determine maximum throw distance, protected volume, and surface area limits for a desired reduced pressure and suppressant bottle (with a given nozzle configuration and system-activation pressure). As a result, any system that is evaluated using this new methodology can be installed with confidence at larger scales – provided that the maximum performance limits are satisfied. An overview of the development of this new methodology is discussed in the present paper, along with important experimental results. For the sake of brevity, only the cornstarch-air results are shown; however, this new methodology applies to both dust gas explosions.

2 Experimental Details

2.1 Open-Air Discharge Experiments

Four different suppressant bottles or HRDs (High-Rate Discharge) were used in the present study; the HRDs were 2.5, 5, 10, and 50 L in nominal volume. Each HRD was filled with sodium bicarbonate according to the manufacturer's specifications and mounted on a support structure with a dispersion nozzle that was fabricated in-house; see Fig. 2 for a photograph of the experimental setup. A dispersion nozzle with twelve equidistantly-spaced holes was made for each HRD; the holes were 38.1 mm in diameter for the 50 L HRD and 25.4 mm in diameter for all other HRDs

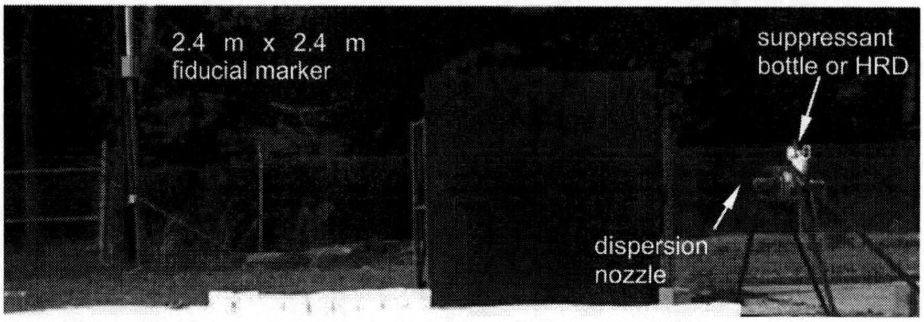

Figure 1. Photograph of the experimental setup.

Once activated, the high-pressure discharge of suppressant into the open atmosphere was recorded by a high-speed camera at 1,000 fps. A 2.4 m x 2.4 m black plywood sheet was used as a fiducial marker. The pressure drop inside the suppressant bottle was monitored by a Kistler piezoresistive pressure transducer. A 5-V signal and an electric match placed in the field of view of the camera were triggered simultaneously with the suppressant bottle in order to synchronize the high-speed video with the pressure data. All the data were recorded using a high-speed 32-channel data acquisition system with a sampling rate of 25 kHz.

2.2 Full-Scale Suppression Experiments

All full-scale suppression experiments were conducted in FM Global's 2.5 and 25 m^3 (nominal) steel explosion vessels. The smaller vessel is 1.4 m in diameter and 2.0 m in height; the larger vessel is 2.9 m in diameter and 4.2 m in height. The same HRDs that were used in the open-air discharge experiments were used in the full-scale suppression experiments. The HRDs were filled with suppressant and nitrogen according to the manufacturer's specifications and then mounted onto an explosion vessel at the desired location. See Fig. 2 for a schematic of the two vessels along with the various injection locations.

For gas explosions, a stoichiometric propane-air mixture was prepared via the method of partial pressures and recirculated to ensure a homogenous composition. Ignition was effected in the center of the vessel via a weak spark. For dust explosions, 6 or 60 kg of cornstarch was placed into a dust container on either the 2.5 or 25 m^3 vessel, respectively. The vessel was partially evacuated to a pressure of about 0.85 bar. Cornstarch and air were then dispersed into the vessel via an air cannon, pressurized to 9 bar, such that the initial pressure prior to ignition was 1 bar. Ignition was effected in the center of the vessel by a 10 kJ Sobbe igniter.

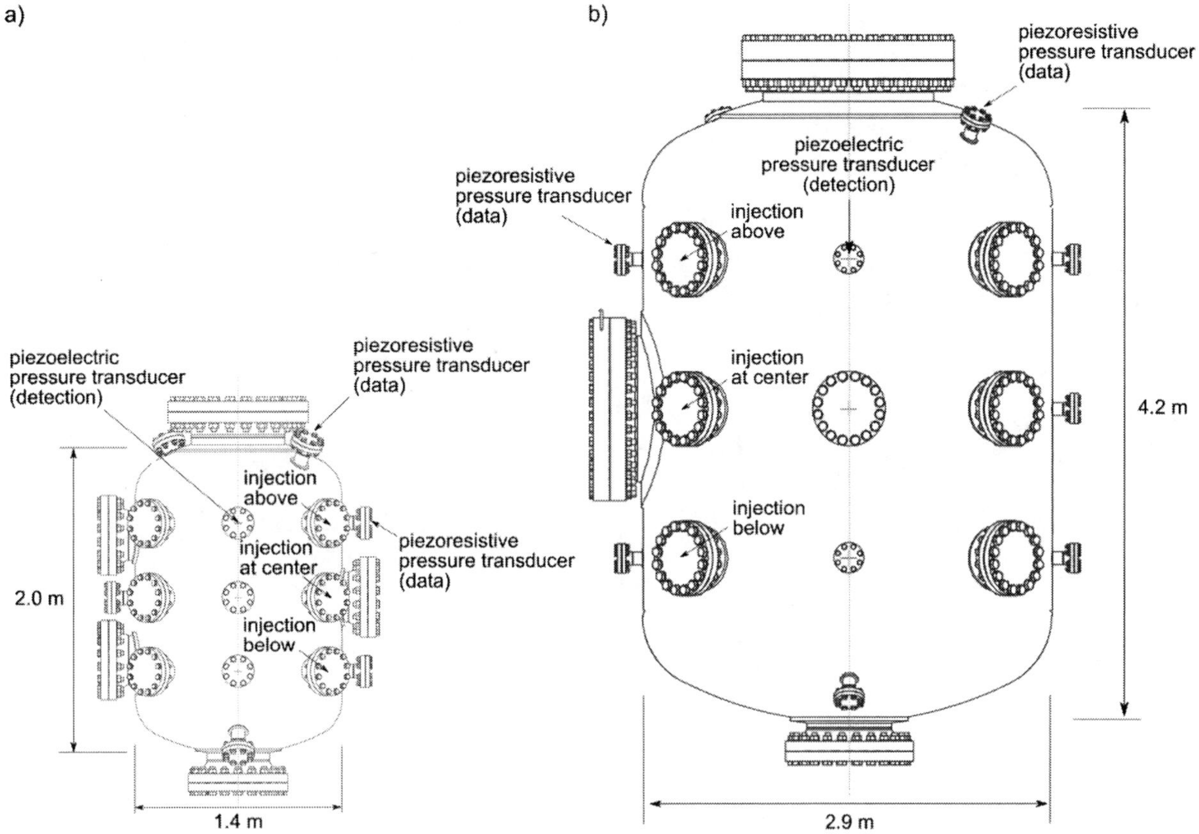

Figure 2. Schematic of FM Global's a) 2.5 and b) 25 m^3 explosion vessels with pressure transducer and suppressant-injection locations.

An in-house detection system/controller was used, which included a Kistler piezoelectric pressure transducer to detect the incipient explosion (and always remained fixed; see Fig. 2 for its location). A programmable microcontroller was used to monitor the pressure inside the vessel and to trigger the suppression system at an activation pressure of $\Delta p_a = 0.1$ or 0.2 bar.

In each experiment, the overall pressure development inside the vessel was monitored by two piezoresistive pressure transducers with a frequency response of 3 kHz. All data were recorded using a high-speed 32-channel data acquisition system with a sampling rate of 25 kHz.

3 Experimental Results

3.1 Open-air discharge

A series of photographs from the high-speed camera is shown in Fig. 3 to illustrate a typical discharge of sodium bicarbonate into the open atmosphere. At $t = 0$, the HRD was activated and the contents of the suppressant bottle were released. The formation of the suppressant cloud can be seen at the dispersion nozzle in the first frame at $t = 1$ ms. The subsequent growth of the suppressant cloud occurred fairly quickly over the next three frames (at $t = 5$, 10, and 20 ms) as the pressure dropped inside the suppressant bottle.

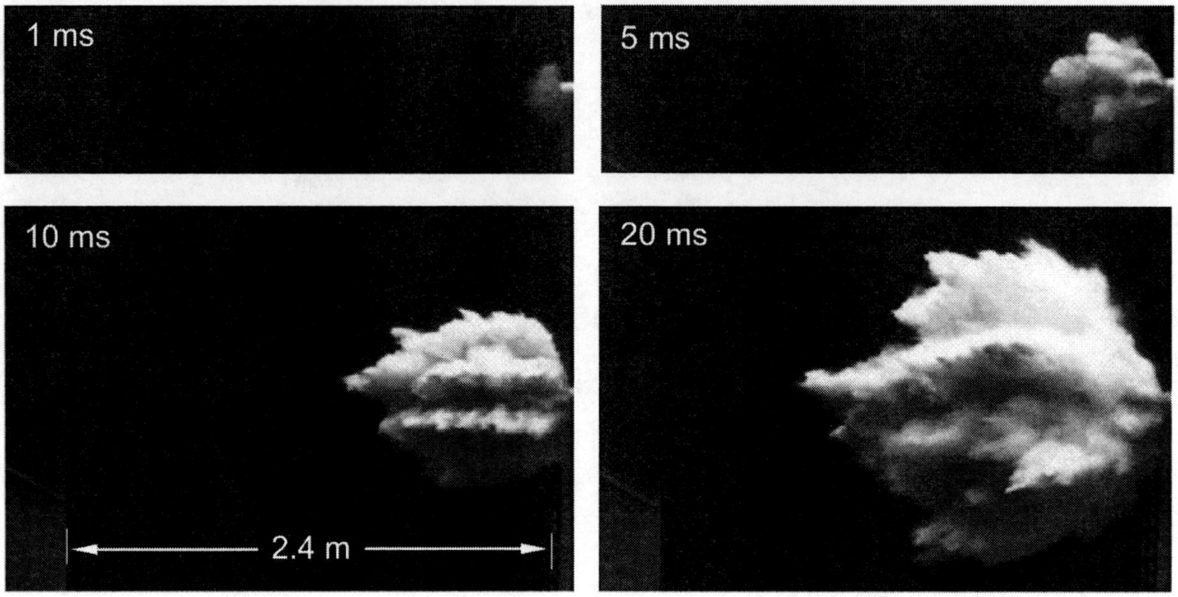

Figure 3. Photographs of discharge of sodium bicarbonate from a 5-L HRD.

3.2 Full-Scale Suppression

Examples of typical pressure histories are shown for cornstarch-air in the 2.5 m^3 vessel in Fig. 4.

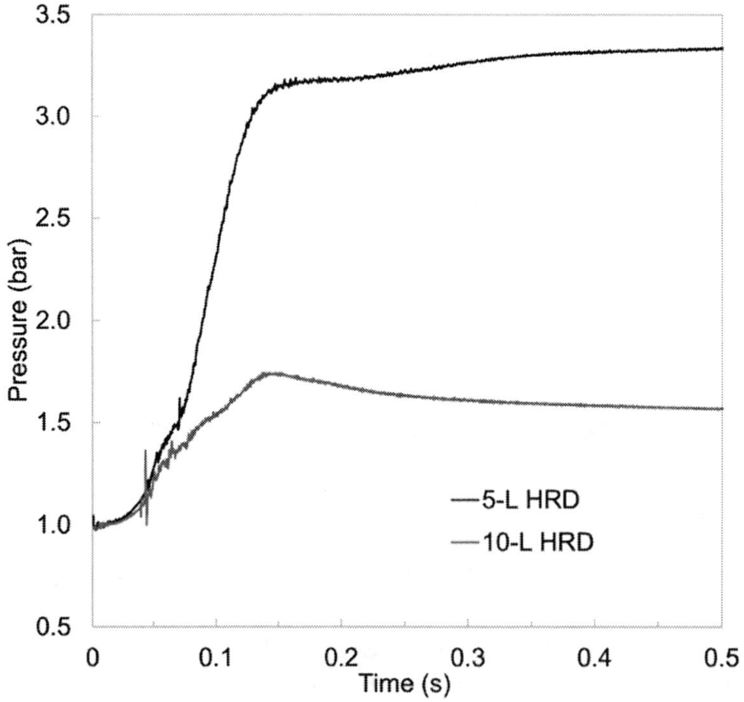

Figure 4. Pressure histories of suppressed ST2 cornstarch-air explosion using a single HRD in the 2.5 m^3 vessel (Δp_a = 0.1 bar).

It should be noted that, for dust explosions, unsuppressed closed-volume tests were first conducted prior to any suppression test in order to establish the required ignition delay time that emulates an ST1 or ST2 reactivity class (see [4] for details). For the pressure histories shown in Fig. 4, an ignition delay time of 1.25 s was used to emulate the ST2 reactivity class (i.e., $200 < K_{ST} < 300$ bar m/s). Ignition occurred at $t = 0$, and the detection pressure was set at $\Delta p_a = 0.1$ bar. The HRDs were located above the ignition location.

After the suppression system was triggered at the detection pressure of $\Delta p_a = 0.1$ bar, the actual injection of suppressant occurred after a small increase in pressure at $p_{inj} = 1.16$ and 1.12 bar for the 5-L and 10-L HRDs, respectively. The injection of suppressant into the vessel can be seen in the pressure histories as sharp pressure fluctuations. For the case where the 5-L HRD was used, the flame continued to propagate until a maximum reduced pressure of $p_{red} = 3.37$ bar was achieved. For the 10-L HRD, a lower reduced pressure of $p_{red} = 1.69$ bar was observed.

4 Basic Suppression Model

When ignition occurs inside an enclosure that is filled with an explosive mixture, a flame will begin to expand spherically and cause the pressure inside the enclosure to increase. The pressure continues to increase due to flame propagation as the suppression system becomes activated (at a certain threshold) and suppressant is discharged into the enclosure.

Suppression of the explosion only begins when the suppressant cloud reaches the flame ball and interacts with it; at this point, the pressure ceases to increase as the flame surface area becomes quenched (see Fig. 5 for a sketch of the phenomenon where the shape of the suppressant cloud is approximated as a hemisphere with radius, r_s).

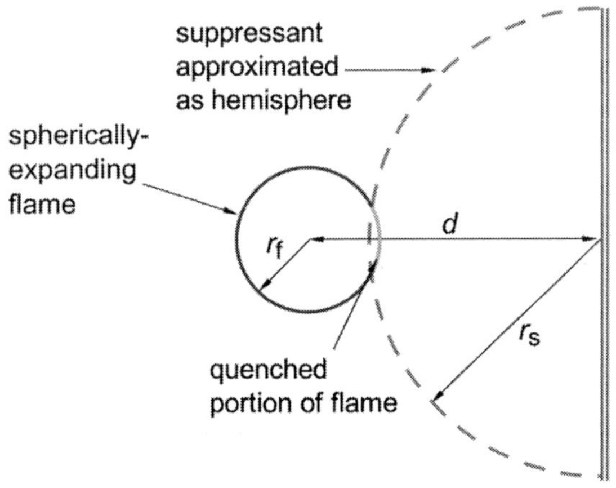

Figure 5. Sketch of the suppression phenomenon where a spherically-expanding flame is quenched by a hemispherical suppressant cloud.

The pressure continues to decrease as the suppressant cloud grows larger and quenches more of the flame surface area. The final reduced pressure inside the enclosure can be considered a result of the increase in pressure due to flame propagation which then ceases to increase due to suppression.

Although an expression can be developed to describe the change in pressure with time due to flame propagation and suppression (which can then be solved numerically to determine the final reduced pressure), it suffices to approximate the phenomenon in a simpler manner. Using the same concept originally proposed by Lewis and von Elbe [5], the reduced overpressure can be estimated by assuming that the fraction of unburned mixture remaining in the vessel is given by:

$$\frac{m_u}{m_0} = \frac{p_{max} - p_{red}}{p_{max} - p_0} \qquad 1$$

where m_u is the unburned mass, m_0 is the total mass inside the vessel, p_0 is initial pressure, p_{max} is the maximum constant-volume explosion pressure, and p_{red} is the reduced pressure.

Assuming isentropic compression, Eq. 1 can be written as:

$$\left(\frac{p_{red}}{p_0}\right)^{\frac{1}{\gamma}} \left(1 - \frac{V_f}{V_0}\right) = \frac{p_{max} - p_{red}}{p_{max} - p_0} \qquad 2$$

which can be used to estimate the value of p_{red} due to explosion suppression once the flame volume at suppression, V_f, is determined.

4.1 Model parameters

4.1.2 Flame volume

The suppression phenomenon is complete when the flame no longer propagates because either:

1. all chemical reactions are quenched by the suppressant, or
2. the chemical reactions are not completely quenched and the flame is able to propagate until it touches the enclosure wall.

In the first case, it can be assumed that chemical reactions are quenched when the suppressant cloud (of sufficient concentration) expands quickly enough to traverse the distance between the injection and ignition locations, *d*, and then to reach the flame front propagating away from the suppressant cloud; *i.e.*,

$$\frac{r_s}{\alpha} = r_f + d \qquad 3$$

where r_s is the radius of the hemispherical suppressant cloud, r_f is the flame radius, and α is a fitted parameter specific to a particular experimental condition. This expression is developed assuming a single suppressant bottle. Although the suppressant concentration is not explicitly taken into consideration, the parameter α can be considered a measure of the effectiveness of the suppressant concentration for a given experiment. This can be visualized in the sketch shown in Fig. 6 of the trajectories of the suppressant and of the flame front moving away from the suppressant.

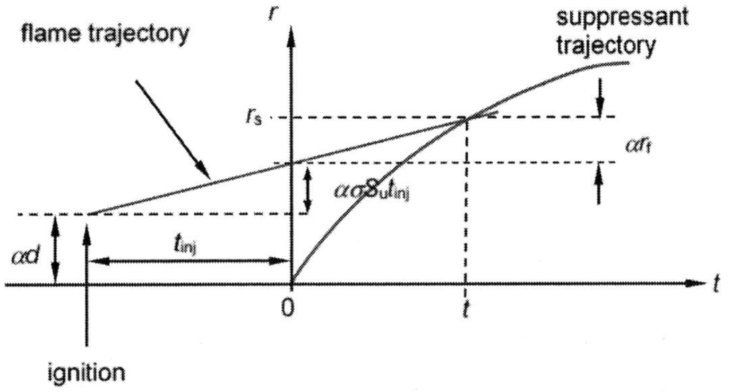

Figure 6. Sketch of the trajectories of the suppressant and of the flame front moving away from the suppressant where the suppressant cloud has penetrated a sufficient distance into the flame ball beyond its centroid.

Equation 3 can be rewritten as:

$$r_s = \alpha\left(\sigma S_u\left[t + t_{inj}\right] + d\right) \quad 4$$

where σ is the expansion ratio of the explosive mixture, S_u is a burning velocity, t is the time for suppression to occur, and t_{inj} is the time between ignition and actual suppressant injection. Equation 4 is satisfied when the flame trajectory intersects the suppressant trajectory, and the flame volume can be calculated from:

$$V_f = \frac{4}{3}\pi r_f^3\left(1 - \frac{\beta}{d^3}\frac{2}{3}\pi r_s^3\right) \quad 5$$

where $r_f = \sigma S_u(t + t_{inj})$ and β is a fitted parameter that is normalized with d^3 and represents the volume of the flame ball where the suppressant cloud has effectively quenched the chemical reactions. Equation 5 can then be substituted into Eq. 2 to determine the final reduced pressure.

In the second case, when the chemical reactions are not completely quenched (*i.e.*, Eq. 3 is not satisfied), the flame front moves away from the suppressant more quickly than the expansion of the suppressant cloud and consumes the remaining mixture inside the vessel. This is illustrated in the sketch in Fig. 7 where the two trajectories do not intersect at any point.

In this case, the flame volume in Eq. 2 is essentially the total vessel volume less the volume of the suppressant cloud:

$$V_f = V_0\left(1 - \frac{\beta}{d^3}\frac{2}{3}\pi r_s^3\right) \quad 6$$

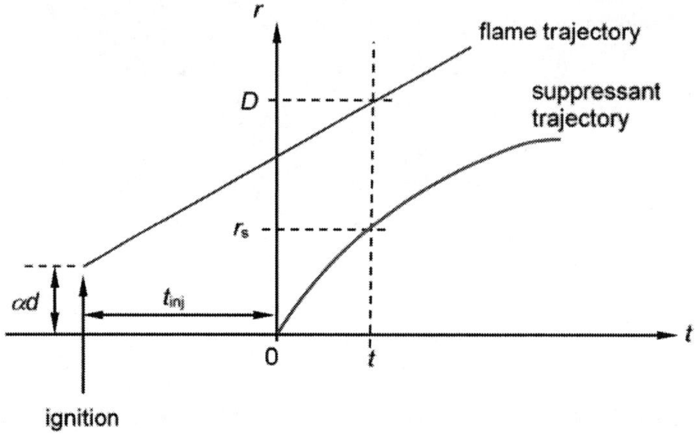

Figure 7. Sketch of the trajectories of the suppressant and of the flame front moving away from the suppressant where the suppressant cloud is not able to penetrate a sufficient distance into the expanding flame ball beyond its centroid.

4.1.3 Burning velocity

The burning velocity is required in Eq. 4 to determine the increase in radius of a spherically expanding flame with time. For a gaseous fuel-air mixture, like propane-air, the laminar burning velocity is either well known or can be easily measured. For a dust-air mixture, on the other hand, K_{ST} (or K_{eff}) is used to describe the reactivity of a dust explosion; however, its applicability to the basic suppression model is fairly limited as it cannot be directly measured in suppression experiments and does not describe the growth of the flame ball at early times before suppression begins.

The basic suppression model requires a burning velocity to describe the growth rate of a cornstarch-air flame, which is generally not known – particularly for concentrations of cornstarch of 750 g/m^3. Furthermore, the conditions under which a cornstarch-air mixture is ignited in the present investigation is very turbulent, and a burning velocity that is obtained from lab-scale experiments may not accurately describe the actual experimental conditions. Therefore, in order to evaluate a burning velocity that can describe the early stages of flame propagation in an explosive dust-air mixture, the initial portion of the pressure history of each experiment is used and an effective burning velocity is determined using the same procedures of Chao and Dorofeev [4]. In the present investigation, burning velocities vary from test to test and range from 0.5 to 2.0 m/s.

4.1.4 Expansion of a suppressant cloud

From the high-speed camera images (e.g., Fig. 3), the axial and radial dimensions of the suppressant cloud were measured using an image-processing algorithm. They are shown in Fig. 8 along with the pressure drop inside the suppressant bottle as functions of time.

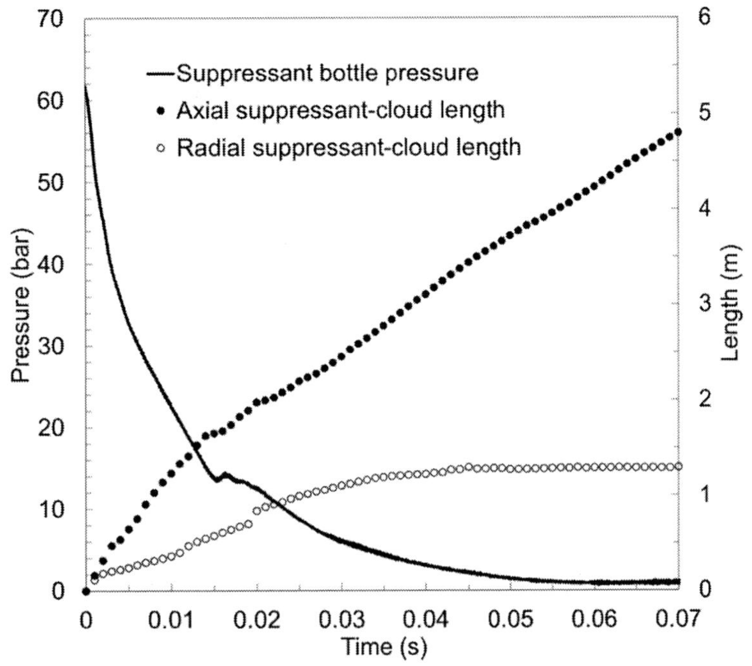

Figure 8. Pressure drop inside a 5-L HRD and corresponding axial and radial lengths of the suppressant cloud as functions of time.

It can be seen in the figure that the axial length of the suppressant cloud grew continuously, even after the entire contents of the HRD were expelled at roughly $t = 0.060$ s. The radial length, on the other hand, increased until it reached an asymptotic value of roughly 1.3 m. This suggests that the growth of the suppressant cloud is driven by the expansion of the high-pressure contents inside the HRD and that the apparent continued growth in the axial direction (once the suppressant bottle is depleted of its contents) is simply a result of the suppressant cloud getting convected downstream by the momentum of the high-velocity jet. Therefore, the throw distance can be considered the asymptotic radial length, which marks the transition from expansion growth to translation of the suppressant cloud.

The radius of the suppressant cloud, r_s, in Eq. 4 can be determined as a function of time directly from the experimental results shown in Fig. 9 by fitting a constant expansion velocity and a throw distance to the radial growth of the suppressant cloud; *i.e.*,

$$r_s = \min(u_s t, l_{throw}) \quad 7$$

where u_s it the expansion velocity of the suppressant cloud and l_{throw} is the throw distance. The least-squares fit is shown in comparison to the radial suppressant-cloud length in the figure, which yields values of $u_s = 38.1$ m/s and $l_{throw} = 1.27$ m.

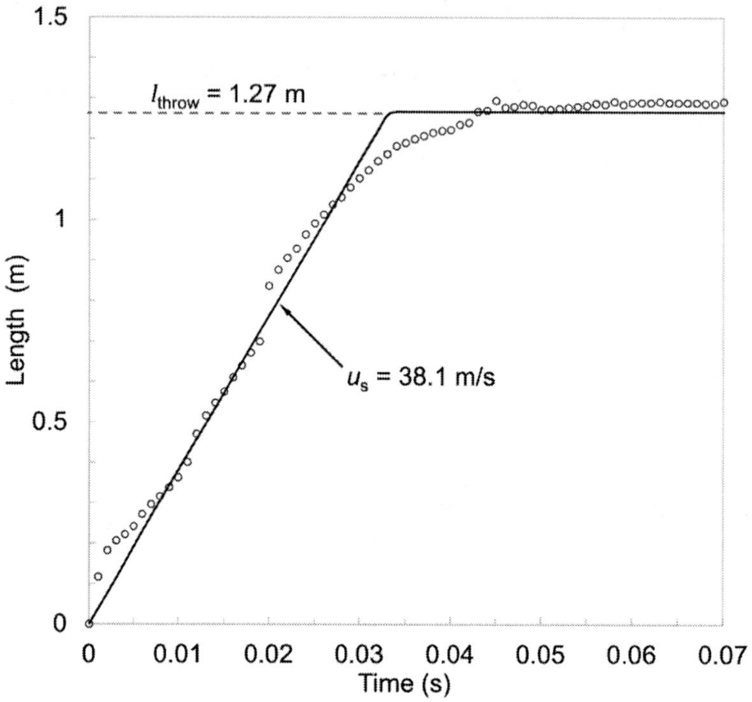

Figure 9. Least-squares fit of the radial suppressant-cloud length as a function of time from the discharge of a 5-L HRD.

4.2 Application of the basic suppression model

The basic suppression model can only be applied to a single suppressant bottle. In order to determine the appropriate form of Eq. 5 or 6 (for V_f) to be used to calculate p_{red} from Eq. 2, the time, t, for successful suppression is first solved from Eq. 4 (and knowing that $r_s = u_s t$):

$$t = \frac{\sigma S_u t_{inj} + d}{\frac{u_s}{\alpha} - \sigma S_u} \qquad 8$$

It is then compared to the maximum time that is allowed for suppression, $t_{s,max}$, and the time at which the flame surface touches the vessel wall, t^*.

The maximum time that is allowed for suppression, $t_{s,max}$, depends on the expansion velocity and maximum radius of the suppressant cloud (where $r_{s,max}$ is given by l_{throw} shown in Fig. 9):

$$t_{s,max} = \frac{r_{s,max}}{u_s} \qquad 9$$

The time, t^*, when the flame reaches the vessel wall is:

$$t_s^* = \frac{D}{2\sigma S_u} - t_{inj} \qquad 10$$

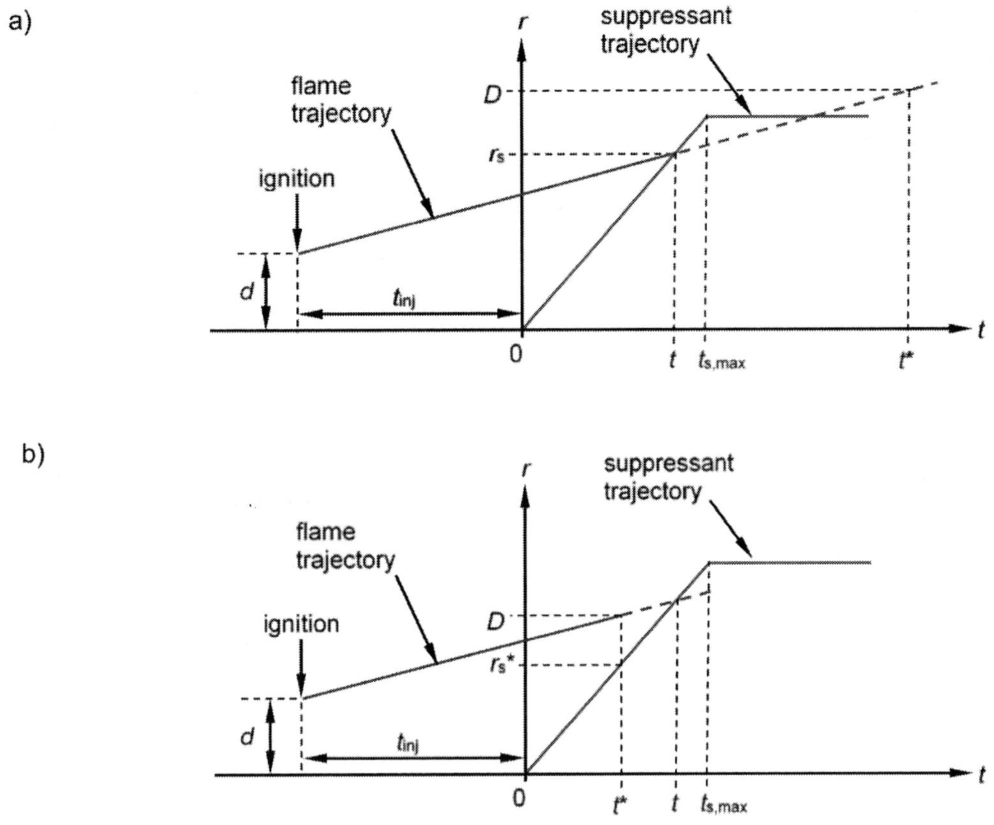

Figure 10. Flame and suppressant trajectories for $t \leq t_{s,max}$ illustrating the two possible scenarios of a) $t \leq t^*$ and b) $t > t^*$.

If $t \leq t_{s,max}$, two possible scenarios can occur, depending on the relative magnitude of t^*:

1. $t \leq t^*$ (see Fig. 10a)

Here, suppression is successful as the suppressant cloud expands sufficiently quickly to reach the flame front before the flame touches the vessel wall. Therefore, the flame volume from Eq. 5 becomes:

$$V_f = \frac{4}{3}\pi\left(\sigma S_u\left[t + t_{inj}\right]\right)^3 \left(1 - \frac{\beta}{d^3}\frac{2}{3}\pi[u_s t]^3\right) \qquad 11$$

2. $t > t^*$ (see Fig. 10b)

In this case, suppression is not successful because the flame touches the vessel wall before the suppressant cloud can reach the flame front, and the flame volume (from Eq. 6) becomes:

$$V_f = V_0\left(1 - \frac{\beta}{d^3}\frac{2}{3}\pi[u_s t^*]^3\right) \qquad 12$$

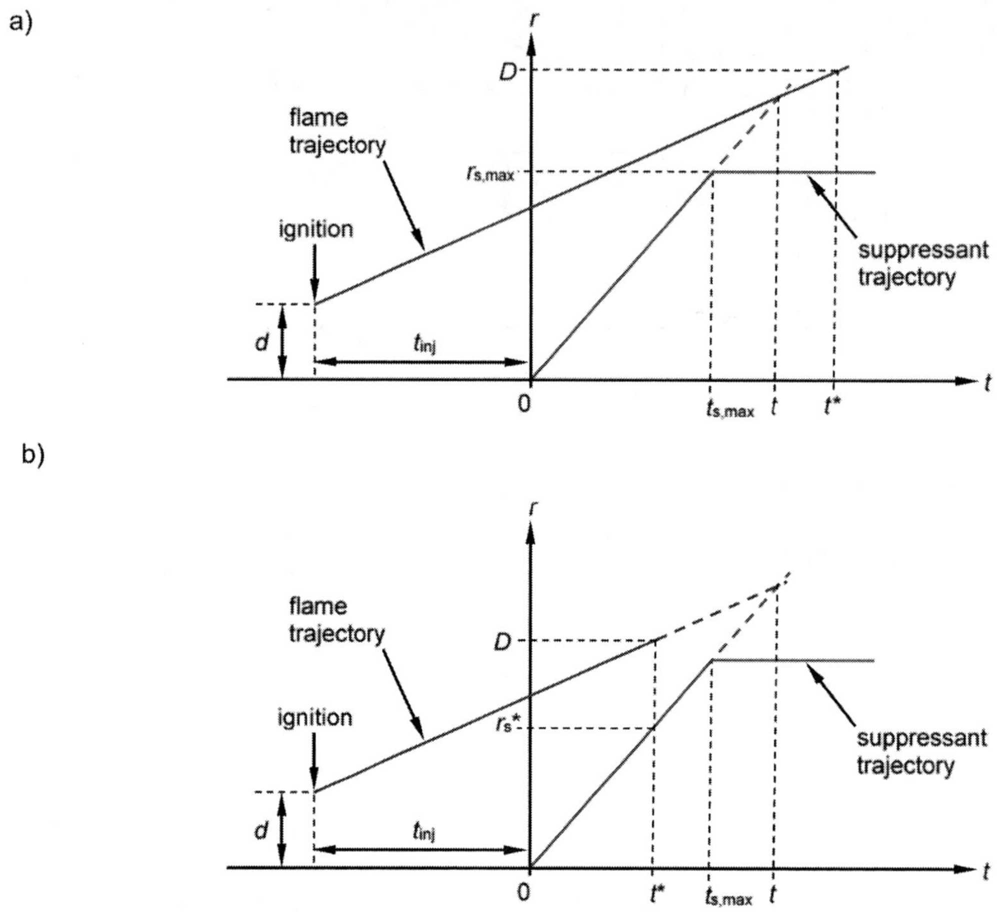

Figure 11. Flame and suppressant trajectories for $t > t_{s,max}$ illustrating the two possible scenarios of a) $t_{s,max} \leq t^*$ and b) $t_{s,max} > t^*$.

On the other hand, if $t > t_{s,max}$, the suppressant cloud simply does not expand sufficiently quickly compared to the flame speed (see Fig. 11). As a result, suppression is not successful at all, and the functional form of Eq. 6 to be used in the basic suppression model will depend on the relative magnitude of t^*:

3. <u>$t_{s,max} \leq t^*$ (see Fig. 11a)</u>

In this case, the suppressant cloud expands to its maximum radius, $r_{s,max}$ (or l_{max}), which is substituted into Eq. 6 to determine the flame volume:

$$V_f = V_0 \left(1 - \frac{\beta}{d^3} \frac{2}{3} \pi {r_{s,max}}^3\right) \qquad 13$$

4. <u>$t_{s,max} > t^*$ (see Fig. 11b)</u>

In this case, the flame touches the vessel wall before the suppressant cloud reaches its maximum radius; thus, the flame volume is calculated from Eq. 12.

Once V_f is determined, p_{red} can be calculated from Eq. 2 by fitting the parameters α and β such that the least-squares difference between the calculated and measured values of p_{red} is minimized. Because suppressant-cloud trajectories are unique to a suppressant bottle and nozzle configuration, α and β are fit in the present study for each HRD that was used. The parameters are shown in Table 1.

Table 1. Fitted parameters, α and β, for cornstarch-air from the basic suppression model.

HRD size (L)	α	β
2.5	0.90	0.01
5	0.94	0.03
10	0.76	0.06
50	0.83	0.09

In general, it can be seen from the table that α is roughly equal to unity, which suggests that the suppressant cloud from each HRD is of sufficient concentration to quench the growing flame ball by the time it has enveloped the entire flame. It can also be seen in the table that the fitted values of β are very small. The value of β for the 50-L HRD is slightly greater, which may be because a different dispersion nozzle was used with the 50-L HRD.

As a general test of the model, the calculated values of p_{red} (using the values of α and β from Table 1) are shown in Fig. 12 in comparison to the measured values. It can be seen that the agreement between the model and experimental results is fairly reasonable for the different HRDs.

Despite some data scatter in Fig, 12, it appears that the basic suppression model is a useful tool to understand the dynamics of successful and unsuccessful suppression. It can be seen from Eq. 4 that the velocity at which suppressant is discharged from a suppressant bottle must be sufficiently high in order for the suppressant cloud (of sufficient concentration) to traverse the distance from the injection site to where ignition occurred and then catch up to the flame front that is propagating away from the suppressant. Essentially, the suppressant cloud must envelop the entire flame ball and quench all chemical reactions. If this occurs, then it is likely that the reduced pressure will be kept to a low value. If, on the other hand, the expansion velocity of the suppressant cloud is too slow to traverse the distance between the injection and ignition locations and cannot reach the flame front that is propagating away, the flame will continue to propagate until all of the explosive mixture not engulfed by the suppressant cloud is consumed. This will generally result in high values of reduced pressure.

A given suppressant bottle filled with a certain amount of suppressant has a fixed expansion velocity that is determined by the initial pressure inside the bottle. Assuming that there is a sufficient concentration of suppressant, this suggests that the suppressant agent is only able to travel a certain distance in a certain time frame before it cannot successfully suppress an expanding flame. Therefore, for a given system-activation pressure and nozzle configuration, it is reasonable to assume that there exists a maximum distance from ignition to injection, d_{max}, beyond which the expansion velocity of the suppressant is too slow to envelop a propagating flame. Once d_{max} is determined for a given suppressant bottle, limits on the maximum volume, surface area, and throw distance can be easily determined via geometric considerations.

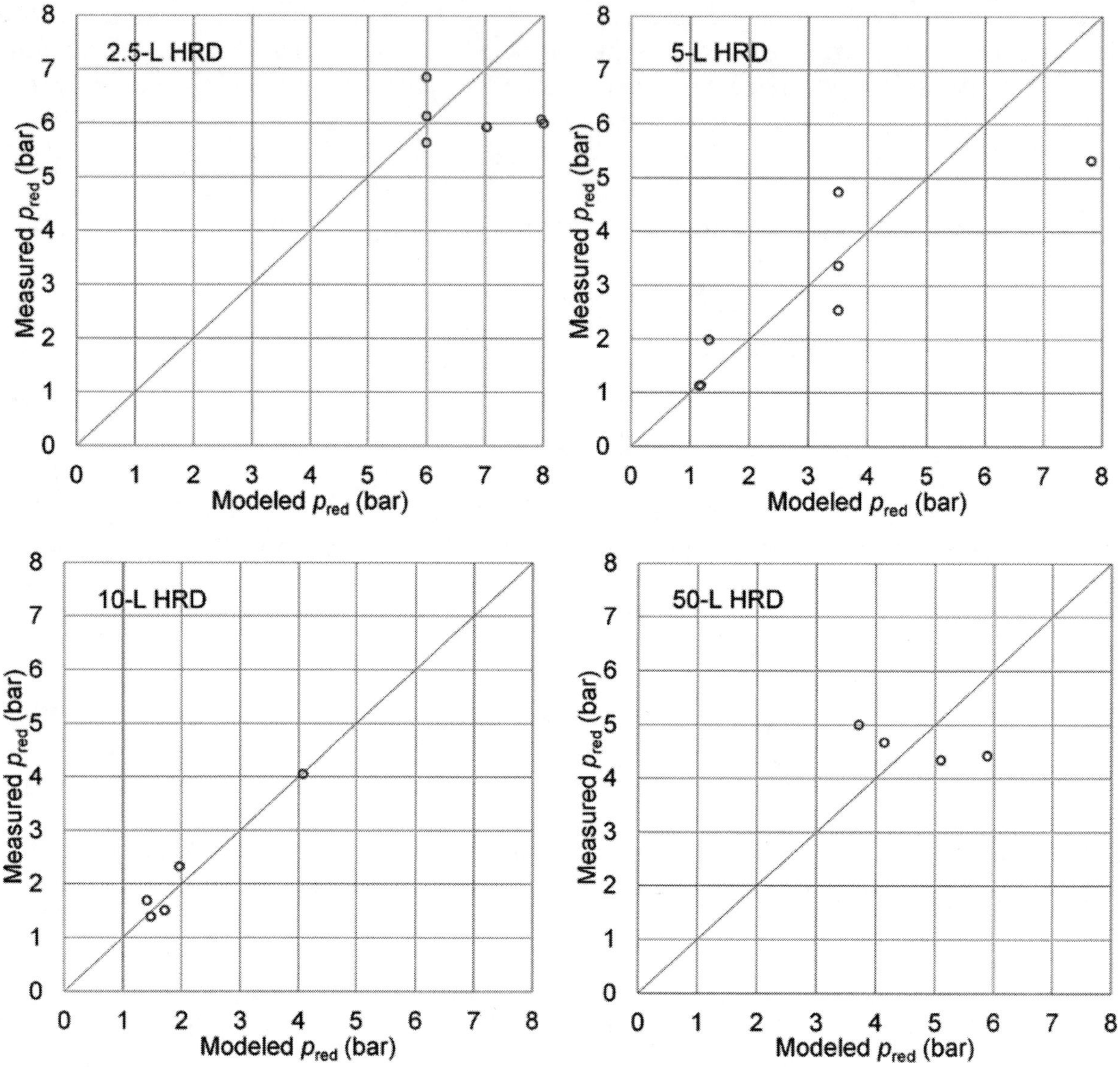

Figure 12. Comparison of calculated and measured values of p_{red} for cornstarch-air experiments using a single 2.5, 5, 10, or 50-L HRD in the 2.5 and 25 m^3 vessels.

4.3 Evaluation of performance limits

Equation 4 can be manipulated to determine the maximum distance, d_{max}, between the injection location and the ignition location:

$$r_{\mathrm{s,max}} = \alpha\sigma S_{\mathrm{u}} t_{\mathrm{s,max}} + \alpha\left(d_{\mathrm{max}} + \sigma S_{\mathrm{u}} t_{\mathrm{inj}}\right) \quad 14$$

Knowing that $t_{s,max} = r_{s,max}/u_s$, Eq. 14 becomes:

$$d_{\max} = \frac{r_{\mathrm{s,max}}}{\alpha}\left(1 - \frac{\alpha\sigma S_{\mathrm{u}}}{u_{\mathrm{s}}}\right) - \sigma S_{\mathrm{u}} t_{\mathrm{inj}} \qquad 15$$

Because the exact location where ignition can occur in a real application is unknown, the suppressant bottles should be arranged such that there is sufficient coverage of the protected enclosure regardless of where ignition occurs. This can be achieved by allowing a portion of the volume to remain unprotected; the size of the unprotected volume will depend on the strength of the enclosure or the desired reduced pressure. This is shown in the plan-view projection of Fig. 13a where the locus of possible ignition locations $d_{\max}$ away from an injection point is represented by a red semicircle for each suppressant bottle. Four suppressant bottles (denoted by the blue arrows in the figure) are arranged at each corner of a protected cube of side length, s. This is conservative, non-optimized configuration (because opposite suppressant bottles are not staggered) is preferred as final installation details can be difficult to regulate and control.

In the configuration shown in Fig. 13a, there is sufficient overlap of the suppressant clouds in the protected cube with the exception of ignitions occurring in the yellow region in the center of the figure. This region represents the volume where ignition would occur beyond a distance of $d_{\max}$ away from each injection location, and the suppressant clouds would not expand quickly enough (or far enough) to envelop a flame initiating in that region. This, therefore, is an unprotected volume that is bound by the overlapping suppressant clouds and is roughly approximated by a cube with length ($\sqrt{3}s - 2d_{\max}$) for values of $s/2$ that are roughly equal to $0.7d_{\max}$.

In order to visualize and calculate the exact unprotected volume, the orthogonal projection of the volume protected by a single suppressant container can be used (shown in Fig. 13b). It can be seen in the figure that the volume of the yellow region can be found by subtracting the volume of the sphere and four spherical caps of height ($d_{\max} - s/2$) from the volume of the cube; i.e.,

$$V_{\mathrm{unprotected}} = s^3 - \left(\frac{4}{3}\pi d_{\max}^3 - 4V_{\mathrm{cap}}\right) \qquad 16$$

Because $V_{\mathrm{cap}} = \frac{1}{3}\pi\left(d_{\max} - \frac{s}{2}\right)^2\left(2d_{\max} + \frac{s}{2}\right)$, Eq. 16 becomes:

$$V_{\mathrm{unprotected}} = s^3 - \frac{4}{3}\pi\left(d_{\max}^3 - \left(d_{\max} - \frac{s}{2}\right)^2\left(2d_{\max} + \frac{s}{2}\right)\right) \qquad 17$$

The ratio between the unprotected volume and the total volume of the protected cube can then be given by:

$$\frac{V_{\mathrm{unprotected}}}{V_{\mathrm{total}}} = 1 - \frac{4}{3}\pi\frac{d_{\max}^3 - \left(d_{\max} - \frac{s}{2}\right)^2\left(2d_{\max} + \frac{s}{2}\right)}{s^3} \qquad 18$$

The ratio of $V_{\mathrm{unprotected}}/V_{\mathrm{total}}$ corresponds to $\Delta p_{\mathrm{red}}/\Delta p_{\max}$ and can be defined as an effectiveness parameter, ζ; a value of $\zeta = 0$ corresponds to complete protection, whereas $\zeta = 1$ indicates no protection at all. The side length, s, of the protected cube can now be determined from Eq. 18 via iteration once a value of ζ is selected that corresponds to the strength of the protected enclosure.

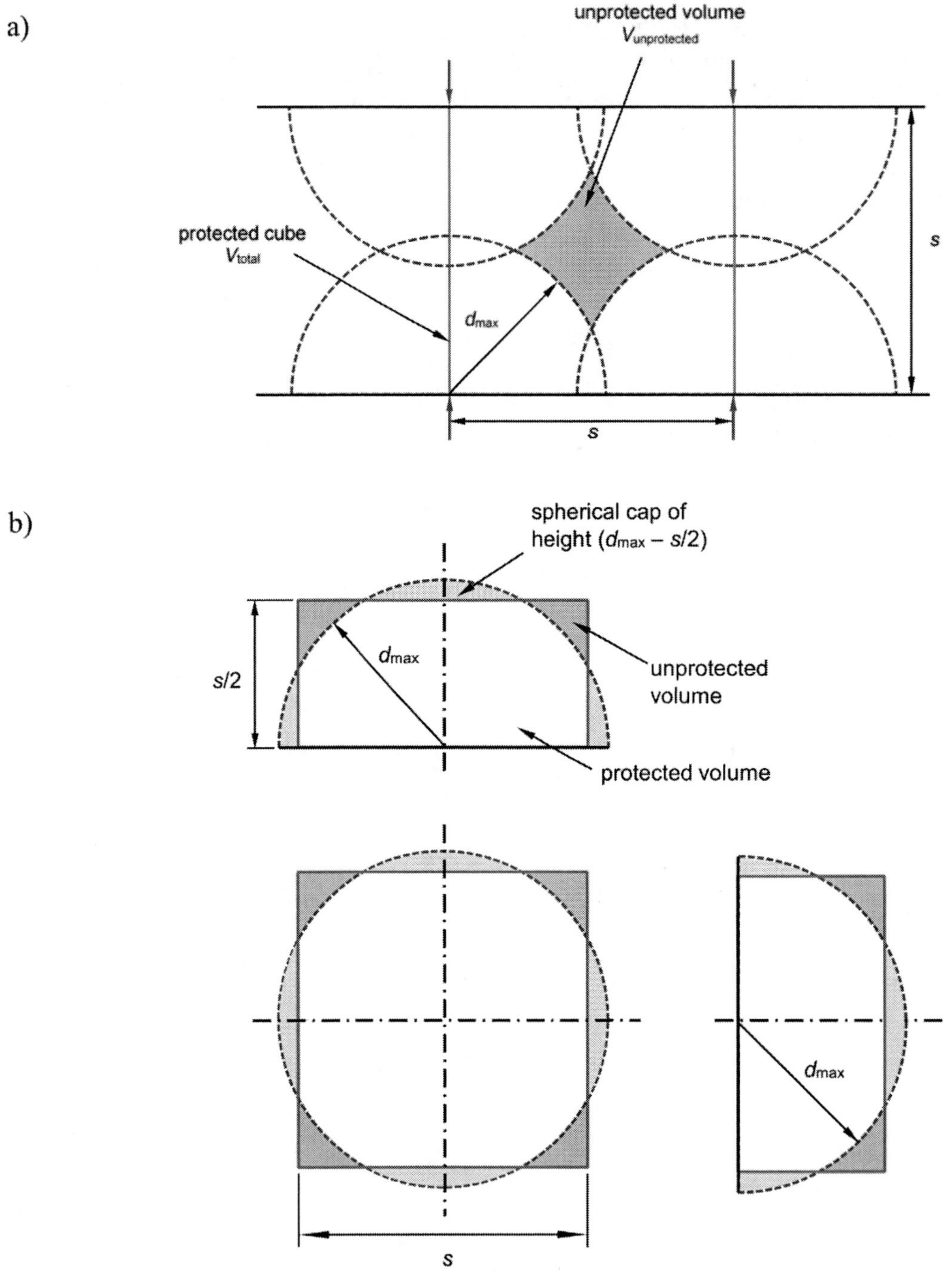

Figure 13. a) Locus of ignition locations a distance of dmax from the injection point of each suppressant bottle, and b) an orthogonal projection of protected and unprotected volumes by a single suppressant container; each bottle location is denoted by a blue arrow.

It follows from geometry that the limits on maximum protected throw distance, surface area, and volume for a single suppressant bottle are given by:

$$r_{max} = \frac{s}{2}, A_{max} = \frac{s^2}{2}, \text{and } V_{max} = \frac{s^3}{2} \quad 19$$

To determine these performance limits for a given suppressant bottle, d_{max} must first be calculated from Eq. 15 (using values of $r_{s,max}$ and u_s from open-air discharge tests and the fitted parameter α from full-scale suppression tests).

As an example, the maximum protected distance, surface area, and volume are shown in Table 2 for the 5-L HRD that was used in the present study. The performance limits are shown for different suppressant injection pressures and two different values of the effectiveness parameter (*i.e.*, $\zeta = 0.1$ and 0.2, which correspond to reduced overpressures of about 0.8 and 1.6 bar, respectively, assuming a maximum unsuppressed overpressure of 8 bar). For cases where more than one test was conducted at a given injection pressure, the maximum limits were calculated for each test; only the averaged values along with the standard deviation are shown in the table.

Table 2. Maximum protected distance, surface area, and volume for a 5-L HRD with current nozzle configuration for ζ = 0.1 and 0.2.

		$\zeta = 0.1$			$\zeta = 0.2$		
p_{inj} (bar)	d_{max} (m)	r_{max} (m)	A_{max} (m^2)	V_{max} (m^3)	r_{max} (m)	A_{max} (m^2)	V_{max} (m^3)
1.1	0.92 ± 0.05	0.73 ± 0.04	1.06 ± 0.12	1.55 ± 0.25	0.80 ± 0.04	1.30 ± 0.14	2.09 ± 0.34
1.2	0.56 ± 0.24	0.44 ± 0.19	0.44 ± 0.28	0.47 ± 0.34	0.49 ± 0.21	0.54 ± 0.34	0.64 ± 0.45
> 1.3	0.24	0.19	0.07	0.03	0.21	0.09	0.04

Therefore, using the present methodology, the maximum performance limits of the 5-L HRD and nozzle configuration used in the present study can be determined for the particular value of ζ (or reduced pressure) that corresponds to the strength of the desired application. For example, if the strength of an enclosure to be protected is 0.8 bar (assuming the maximum overpressure of the combustible in the protected enclosure is 8 bar), then the maximum limits corresponding to $\zeta = 0.1$ should be selected.

4.4 Assessment of the approach

To evaluate the suitability of these prescribed limits, they are applied to the experimental results where multiple 5-L HRDs were used. In Table 3, reduced pressures are shown for each test along with the vessel volume and surface area (given by $V^{2/3}$ for simplicity) normalized by the number of 5-L HRDs that were used in each test. The values of V/n and A/n are simply the volume and surface area protected by each suppressant bottle. If the maximum limits are indeed suitable, then the ratios of $(V/n)/V_{max}$ and $(A/n)/A_{max}$ should be less than one for successful suppression to occur. For the tests conducted in the 2.5 m^3 vessel, the 5-L HRDs were located on the same side of the vessel; therefore, the protected distance, in this case, should be at least equal to the diameter of the vessel. For the tests conducted in the 25 m^3 vessel, the HRDs were mounted diametrically opposite to each other, and the vessel radius is used as the protected distance.

Table 3. Distance, surface area, and volume protected by each suppressant bottle in experiments conducted using multiple 5-L HRDs.

Test #	D (m)	A (m^2)	V (m^3)	n	V/n	A/n	p_{inj} (bar)	p_{red} (bar)
LgVesPROP012	1.45	8.5	25	4	6.25	2.14	1.2	5.71
LgVesPROP013	1.45	8.5	25	4	6.25	2.14	1.2	6.52
LgVesPROP014	1.45	8.5	25	8	3.125	1.07	1.3	4.98
LgVesCORN034	1.45	8.5	25	4	6.25	2.14	1.2	1.24
LgVesCORN035	1.45	8.5	25	4	6.25	2.14	1.2	1.40
LgVesCORN039	1.45	8.5	25	3	8.33	2.85	1.1	1.65
LgVesCORN040	1.45	8.5	25	3	8.33	2.85	1.1	5.92
LgVesCORN041	1.45	8.5	25	3	8.33	2.85	1.1	5.59
LgVesCORN042	1.45	8.5	25	3	8.33	2.85	1.3	5.48
LgVesCORN043	1.45	8.5	25	3	8.33	2.85	1.2	6.06

The reduced pressures from Table 3 are shown in Figs. 14a to c as functions of the maximum limits in distance, surface area, and volume, respectively. The dashed vertical line in each figure indicates the required distance, surface area, and volume that satisfies the maximum limits in Table 2. The region to the left of the vertical line indicates where the actual distance, surface area, and volume are within the maximum limits and should result in successful suppression. The region to the right represents configurations where the maximum limits are not satisfied; in this case, suppression is expected to be unsuccessful.

The two cases of $\zeta = 0.1$ and 0.2 are also shown in the figure (as open and closed symbols, respectively), representing desired reduced pressures of 1.8 and 2.6 bar (absolute). Although the limits for the case of $\zeta = 0.2$ are larger than those for $\zeta = 0.1$, it can be seen that suppression was successful when the prescribed maximum limits of $D/r_{max} = V/V_{max} = A/A_{max} = 1$ were satisfied; in fact, suppression was successful for cases where the distance, surface area, and volume were larger than the maximum limit, which suggests that the prescribed limits are quite reasonable. It is not known if the ignition locations used in the present investigation are realistic worst cases; therefore, by selecting the maximum limits to be $D/r_{max} = V/V_{max} = A/A_{max} = 1$ ensures that there is a sufficient safety factor that account for the uncertainty of where ignition actually occurs in a real application. Furthermore, the safety factor also takes the uncertainty of the actual shape of the suppressant cloud, as well as the burning velocity of dust-air flames, into consideration.

These results indicate that the present methodology of evaluating a single suppressant bottle can be used successfully to prescribe a configuration that yields adequate protection in a larger volume than what was used in the performance test (*i.e.*, 25 vs. 2.5 m^3). Therefore, it seems that the maximum limits on throw distance, surface area, and volume determined by the present methodology – for a given value of ζ (or desired reduced pressure) – can be confidently applied to real installations that span a wide range of conditions that cannot be known *a priori*.

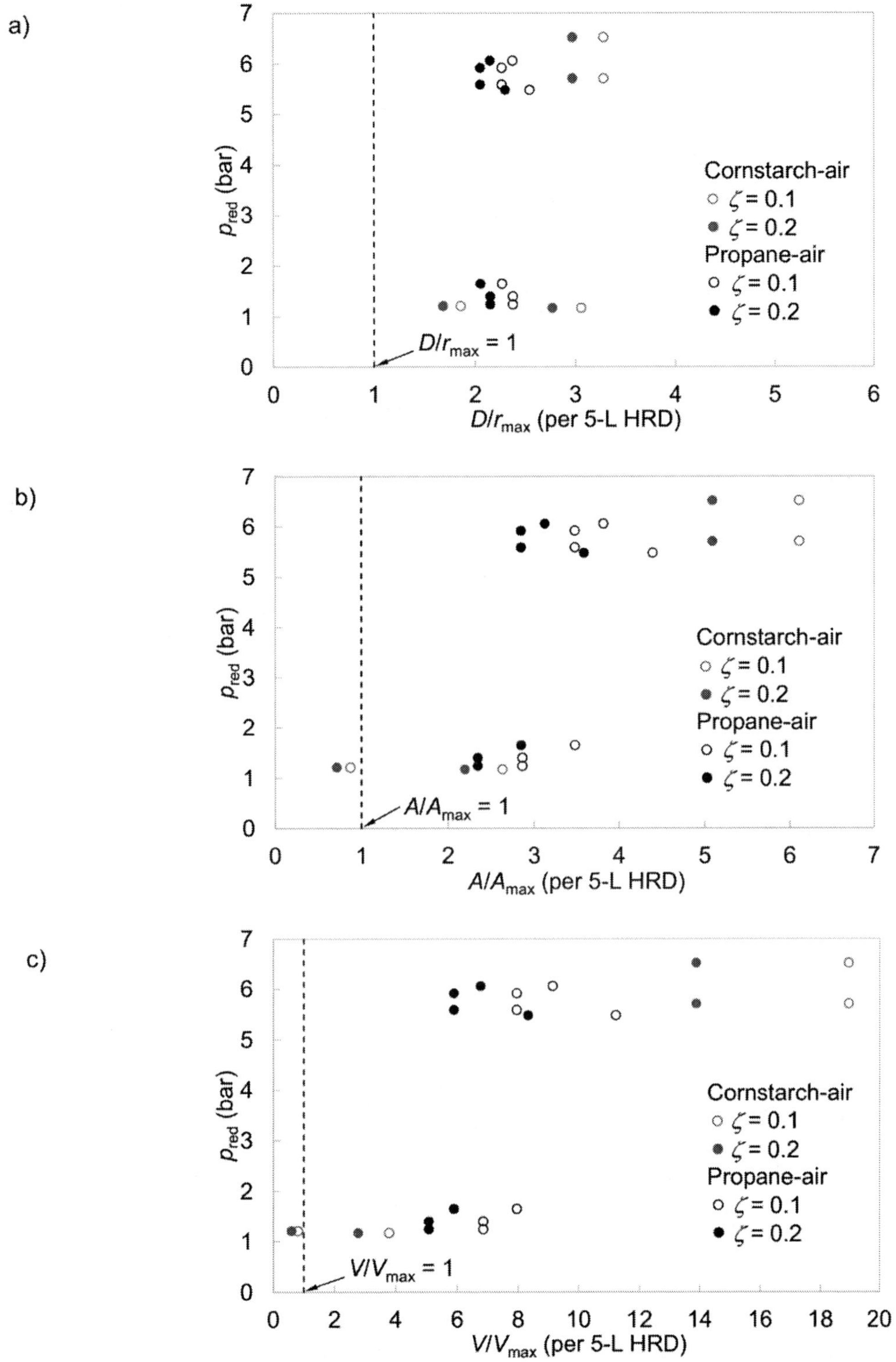

Figure 14. **Measured p_{red} as a function of maximum protected a) throw distance, b) surface area, and c) volume for $\zeta = 0.1$ and 0.2.**

5 Concluding Remarks

It has been shown that the methodology presented in this study can be used to evaluate system-level performance of an explosion suppression system. The approach that is used is based on performance testing of a single suppressant bottle within a system; limitations to the maximum protected distance, surface area, and volume are determined for a single suppressant bottle from the tests and can then be used to evaluate a system that is comprised of multiple suppressant bottles.

Therefore, the evaluation of a single suppressant bottle should be performed for a particular suppressant, dispersion-nozzle system, system-activation threshold, and class of dust or gas to be protected. Each suppressant bottle should undergo two different sets of tests:

1. An open-air discharge tests in order to characterize the growth of the suppressant cloud, and
2. Full-scale suppression tests where the distance from ignition to injection, *d*, is varied in order to maximize the range of reduced pressures that is achieved.

From the test results, the expansion velocity, throw distance (or, equivalently, maximum suppressant radius in the basic suppressant model), burning velocity, and time between ignition and suppressant injection can be calculated for each test. These results can be used in conjunction with the basic suppression model in order to determine maximum protected limits for throw distance, surface area, and volume.

6 References

[1] F. Tamanini, Scaling Parameters for Vented Gas and Dust Explosion, J. Loss Prev. Process, 14 (2011) 455-461.

[2] K. Chatrathi and J. Going, Dust Deflagration Extinction, Proc. Saf. Prog., 19 (2000) 146-153.

[3] P.E. Moore, Suppressant for the Control of Industrial Explosions, J. Loss Prev. Process 9 (1996) 119-123.

[4] J. Chao and S.B. Dorofeev, Evaluating the Overall Efficiency of a Flameless Venting Device for Dust Explosions, J. Loss Prev. Process, 36 (2015) 63-71.

481543 Understanding vapor cloud explosions

Abstract:

Accidental releases of flammable gases can result in vapor cloud explosions, with potentially severe consequences. This tutorial explains the circumstances that can lead to such an event, how the magnitude of the explosion can be estimated, how severe the consequences could be and potential mitigation strategies.

481544 Running the Gamut of Hazard Identification and Risk Analysis Methods

Abstract:

Hazard identification and risk analysis (HIRA) methodologies, including those used for process hazard analyses (PHAs), run the gamut from a basic hazard identification exercise all the way to a fully quantitative risk analysis. This tutorial will give a bird's-eye view of these techniques, highlighting similarities, differences, inter-relationships, and discussing the appropriate application of each. The tutorial will show how identification and analysis of incident scenarios is at the heart of all risk analysis techniques, whether the method of choice is a What-If Analysis, a HAZOP Study, a LOPA, or a quantitative risk analysis.

481545 Q&A on QRA

Abstract:

What exactly is a QRA? How do you do one? What tools are needed? What can a QRA tell you? This tutorial session will help to answer these questions and others, with the goal of providing a basic understanding of quantitative risk analysis (QRA). The nature of consequences and probabilities, and their combination, within the context of chemical process safety, will be explored and explained. The tutorial will provide introductions to consequence modeling tools, probability assessment methods, various risk measures, the computational requirements for developing a QRA, and ways that QRA can be applied to process safety. After the tutorial session, the attendee will be better equipped to know what a QRA is and how it can be used to evaluate the risks inherent to chemical processing.

479519 The Top Ten Reasons for Fixed Equipment Failures in the Hydrocarbon Process Industries

Abstract:

This presentation summarizes the top ten reasons why fixed equipment fails in-service, thus causing significant reliability and process safety events e.g. explosions, fires, toxic releases, environmental damage, shutdowns, LPO, etc. This top ten list does is not based on the physical causes of failure like, corrosion, cracking, embrittlement or the other 65+ damage mechanisms and failure modes that afflict fixed equipment in the petroleum petrochemical industry. Those damage mechanisms and failure modes are just the final result after several other layers of protection fail first. Rather, this top-ten list is based on human system failures associated with inadequate or poorly implemented management systems for fixed equipment mechanical integrity (FEMI). Being aware of these human system failures and mitigating the reasons for them can substantially reduce process safety and reliability events caused by fixed equipment failures.

The hydrocarbon process industry as a whole continues to have numerous repeat failures of pressure equipment and piping, and too many of them involve big process safety events with major asset losses, environmental damage, injuries, and fatalities, even though most of the reasons for those failures and the necessary mitigation steps are well known and documented in industry codes and standards. Having been involved in numerous failure investigations and having read about hundreds more of them, I have come to the conclusion that there are ten primary reasons why they continue to occur. Certainly there are more than ten reasons and there are lots of different contributing as well as root causes for each major failure, but it is my opinion that if the necessary management systems and mitigation steps for these top ten reasons were adequately implemented and sustained at each hydrocarbon process site, the vast majority of the these failures could be avoided. This presentation will address those top ten reasons for equipment and piping failures in the hydrocarbon process industries and suggest ways to improve upon each.

13TH GLOBAL CONGRESS ON PROCESS SAFETY

Advancing Asset Integrity Management in Chevron Upstream

Vincent Attanucci
Manager of Process Safety
Chevron Upstream, a division of Chevron U.S.A Inc.
1400 Smith Street
Houston, TX 77002
vatt@chevron.com

Philip Delpero
Asset Integrity Advisor
Chevron Upstream, a division of Chevron U.S.A Inc.
1400 Smith Street
Houston, TX 77002
Philip.Delpero@chevron.com

Elizabeth Leaverton
Process Safety Specialist
Chevron Upstream, a division of Chevron U.S.A Inc.
1400 Smith Street
Houston, TX 77002
eleaverton@chevron.com

Prepared for Presentation at
American Institute of Chemical Engineers
2017 Spring Meeting and 13th Global Congress on Process Safety
San Antonio, TX
March 26 – 29, 2017

Advancing Asset Integrity Management in Chevron Upstream

Vincent Attanucci
Process Safety Manager
Chevron Upstream, a division of Chevron U.S.A Inc.
1400 Smith Street
Houston, TX 77002
vatt@chevron.com

Philip Delpero
Asset Integrity Advisor
Chevron Upstream, a division of Chevron U.S.A Inc.
1400 Smith Street
Houston, TX 77002
Philip.Delpero@chevron.com

Elizabeth Leaverton
Process Safety Specialist
Chevron Upstream, a division of Chevron U.S.A Inc.
1400 Smith Street
Houston, TX 77002
eleaverton@chevron.com

Keywords: Asset Integrity, safeguard verification

Abstract

The presentation describes efforts Chevron Upstream has taken over the past decade to improve performance on Asset Integrity Management (AIM). The AIM program within Chevron goes beyond more traditional fixed equipment[1] integrity programs and covers all equipment types and equipment related layers of protection. Quantitative risk analyses and hazard and operability (HAZOP)/layers of protection analysis (LOPA) studies identify key layers of protection or safeguards that must be in place in order to reduce the risk in an identified high consequence scenario. Both preventive and mitigating layers of protection are incorporated into the program using bow tie analysis techniques and leveraging existing process hazard analysis studies. The AIM program implements various inspection and maintenance strategies to maintain the functionality of safeguards throughout the operational life cycle.

An important element of an AIM Program is for frontline operations personnel to have a clear understanding of the identified equipment-related safeguards and to have an active role in ongoing verifications that the identified safeguards are in place and functional. Verification and audit programs related to Asset Integrity Management must be targeted and focused to ensure that the safeguards are in place, functioning and effective for Loss of Containment (LOC) prevention and mitigation. The methodology describes a systematic approach to develop and operationalize plant and field based AIM verification programs, to effectively manage the key equipment related safeguards through their full lifecycle.

1 Introduction

From 2000 to 2009, the equipment care method within Chevron Upstream was focused on regulatory compliance and preventing frequent failures that impacted production This compliance and reliability-based view drove business (operating) units to create the systems needed to track failures and conduct necessary inspections, tests, and preventive maintenance. However, this view of equipment did not utilize available information regarding potential high-consequence, low-likelihood events. During this period, industry and Chevron Upstream risk assessment processes were growing more robust in design and implementation. From macroscopic Integrated Hazard Identification Studies (IHAZID) to more detailed Hazard and Operability (HAZOP) studies, risk studies grew in their detail and maturity, and increased the importance and visibility of equipment related safeguards. However, the risk assessment outcomes were not widely communicated to maintenance, inspection, and operational personnel, or understood by those who most interacted with the equipment.

Beginning in 2010, there was a focused shift toward equipment care with the purpose of preventing high-consequence, low-likelihood events. Comprehensive risk studies were providing increased knowledge about equipment as safeguards, and this knowledge was leveraged to support the execution of asset care programs. Maintenance and inspection processes moved beyond a focus on only fixed and rotating equipment with recurring failures, and utilized the detailed risk studies to identify the equipment that was both preventive and mitigating for potential high-consequence scenarios at each facility. The methods and structures within the Asset Integrity Management program helped ensure that this equipment received preventive maintenance and inspections, and that required repairs were performed in a timely manner.

A greater understanding of the safeguards that prevent or mitigate potential high-consequence scenarios revealed the importance of ongoing verification efforts for these equipment layers of protection, to ensure that the integrity and effectiveness are maintained. Verification programs are structured from the frontline workers through management with appropriate scope and frequency of task to create multiple layers of verification for key safeguards. Ongoing management and verification of key safeguards, along with other efforts, has helped reduce the number of Loss of Containment (LOC) events over the past several years by 75%.

2 SERIP Foundation

The Surface Equipment Reliability and Integrity Process (SERIP) within Chevron Upstream provides a systematic, staged approach to deploy and execute standardized procedures, that enable operation and maintenance of facilities to sustain integrity and prevent incidents. SERIP procedures are executed in a prescribed order referred to as *stages*. **Figure 1** outlines the SERIP stages and what is included in each. All upstream units follow this standardized approach.

SERIP stage 1 provides the foundation needed to manage work and materials effectively by building systems that enable maintenance planning and tracking. Stages 1 and 2 share a focus on equipment reliability, in order to keep process systems in operation. Stages 3 and 4 build on the basics in Stage 1 and 2 and add an emphasis on Asset Integrity and alarm management.

The intent of the staged approach is to build upon the foundation laid in prior SERIP stages. Acceptable performance levels must be achieved before moving to the next stage. Performance is first verified through self-assessment and then through formal validation. An assessment of performance in prior stages will be included in any stage validation or assessment since the expectation is that the business units maintain fluency in prior stages while proceeding to later stages. Timing of business unit progress through the SERIP stages is based on complexity and potential consequences associated with the facilities.

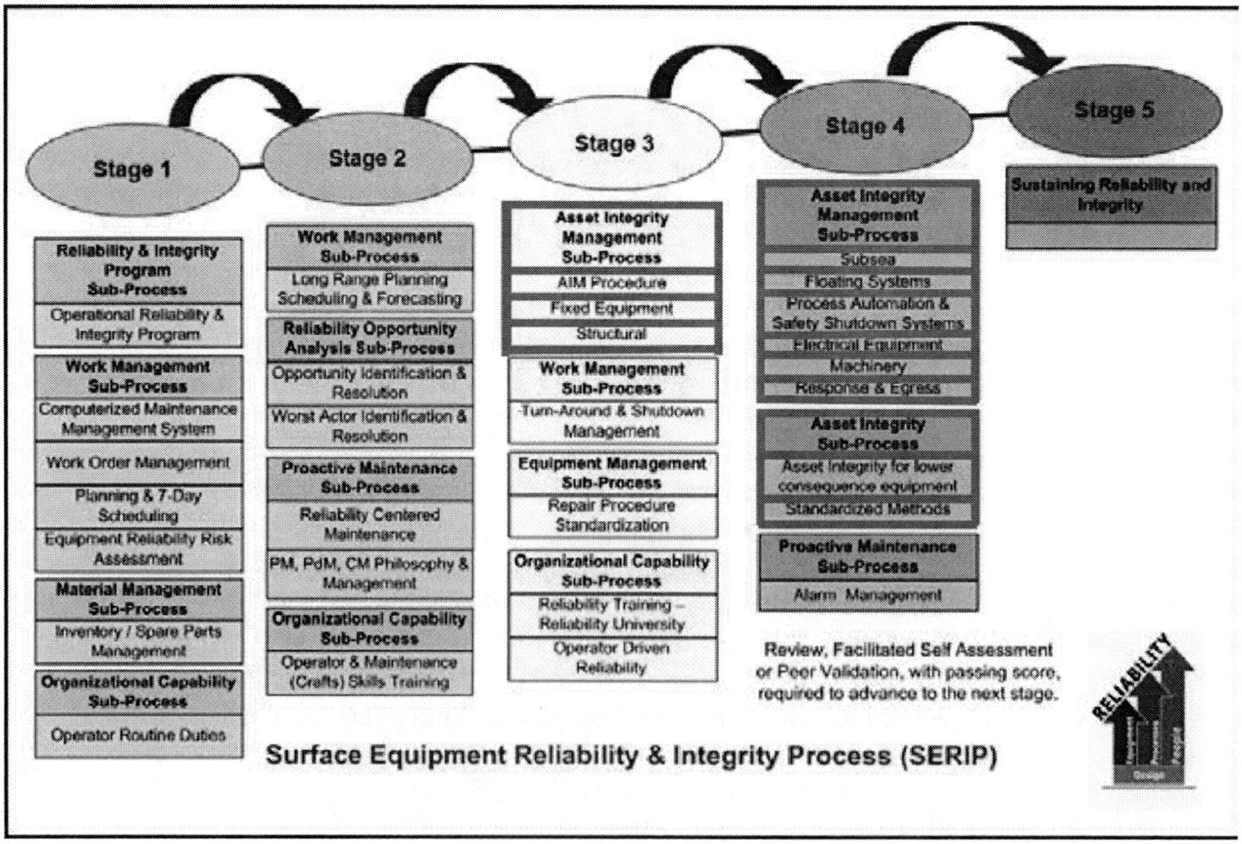

Figure 1: SERIP Process, red outline indicating Asset Integrity Management

3 Risk Assessment

The Risk Management process matured in parallel with the SERIP process in Upstream. High-level IHAZIDs were performed to identify risks across the whole facility or activity, obtain a preliminary understanding of these risks, determine where further risk assessments must be conducted, and identify opportunities for initial risk reduction. Additional detailed risk studies were performed including HAZOP studies and quantitative risk assessments (QRAs) for facilities with a potential for high consequence scenarios. More detailed studies provided more information on the layers of protection that are in place to prevent and mitigate the high-consequence scenarios and determine whether they are robust enough to prevent the hazards from becoming consequences. The deeper understanding of the equipment that serves as an independent barrier against severe incidents led to the greater efforts to prioritize inspections, testing, and preventive maintenance of key equipment safeguards.

4 Asset Integrity Management Design

While risk and safeguard fluency were growing across the industry and within Chevron Upstream, it was recognized that most severe incidents are not caused by recurring equipment failures, but, rather, are high-consequence, low-likelihood rare events that occur when multiple barriers fail simultaneously and allow a catastrophic consequence. In addition to identifying and managing the equipment critical to reliable operations, it is necessary to identify and prioritize the management of equipment that protects against high-consequence, low-likelihood incidents.

Figure 2: Four basic steps to Asset Integrity Management implementation

Based on the foundation the previous SERIP stages built, the Asset Integrity Management procedures were released in 2009-2011 as a part of SERIP Stages 3-4, as shown in **Figure 1** in the red-outlined boxes. Asset Integrity Management is the systematic implementation of activities such as inspections, tests, and maintenance tasks necessary to ensure that important equipment will be suitable for its intended application throughout its service life and prevent or mitigate severe or catastrophic health, safety, environment, or asset incidents. The Asset Integrity Management program went beyond more traditional mechanical integrity programs and covers all equipment types and equipment related layers of protection. The objective of the AIM procedures was to overlay onto the existing SERIP procedures an emphasis on activities that are proactive and obtain the required level of performance (safe and dependable) of assets to prevent or mitigate the effects of incidents.

The basic steps of the Upstream Asset Integrity Management program are shown in **Figure 2**. The first step is to identify the "important equipment" or the equipment acting as preventive or mitigating safeguards against a high-consequence, low-likelihood event. The AIM program takes a consequence based approach. The equipment within the scope of the AIM program is considered integrity critical equipment and is identified by leveraging the detailed risk studies performed for the facility and assessing consequence and risk for individual equipment. If equipment is associated with a potential high-consequence scenario (consequence indices 1 or 2, indicated by the arrows in **Figure 3**) at a facility, it is managed by the Asset Integrity Management program. **Figure 3** shows the Chevron Integrated Risk Matrix to indicate which equipment is included in the AIM program and which is managed through other inspection, maintenance, and reliability programs. The intent is to have robust and industry-compliant maintenance and inspection plans in place for all equipment (indicated by the bottom blue box in Figure 3), but to add an extra layer of rigor for the integrity critical equipment. In-scope equipment is flagged as integrity critical in the maintenance management system.

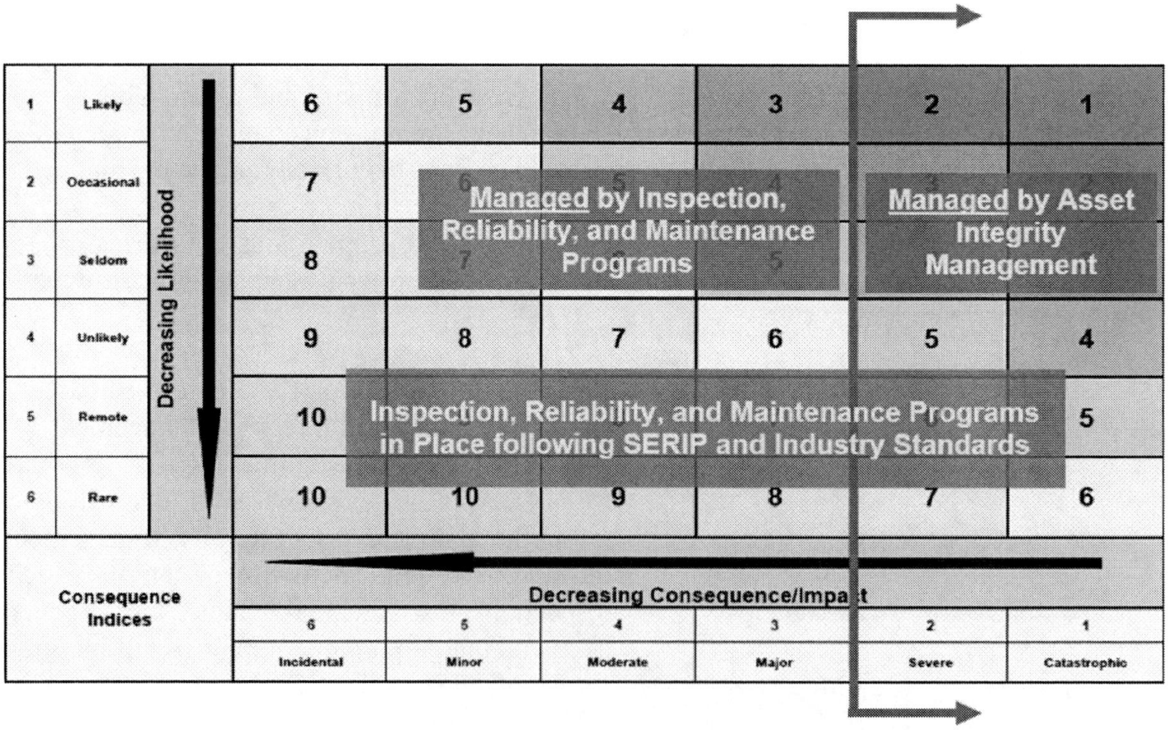

Figure 3: Chevron Integrated Risk Matrix and Equipment Management

The Chevron Barrier Model is shown in **Figure 4**. Chemical hazards are present in the flammable, corrosive, and toxic materials contained within process equipment. Preventive safeguards keep the hazards under control. An "event" occurs when control of a hazard is lost. An example of an event is a release of process material. Mitigating safeguards prevent the event from escalating to a high-consequence impact on health, safety, environmental or asset loss. Because there are other administrative safeguards that help prevent an event or mitigate a consequence, such as design, procedures, and training, the Asset Integrity Management program focuses on the equipment safeguards.

Preventive equipment safeguards are primarily focused on process containment and functionality. Mitigating equipment safeguards may detect an event (warning devices such as alarms), mitigate and control an event (engineered safety devices such as emergency shutdown valves) or respond to an event (safety equipment such as fire suppression systems). Asset Integrity Management extends through many layers of protection to help ensure that equipment safeguards will function as intended.

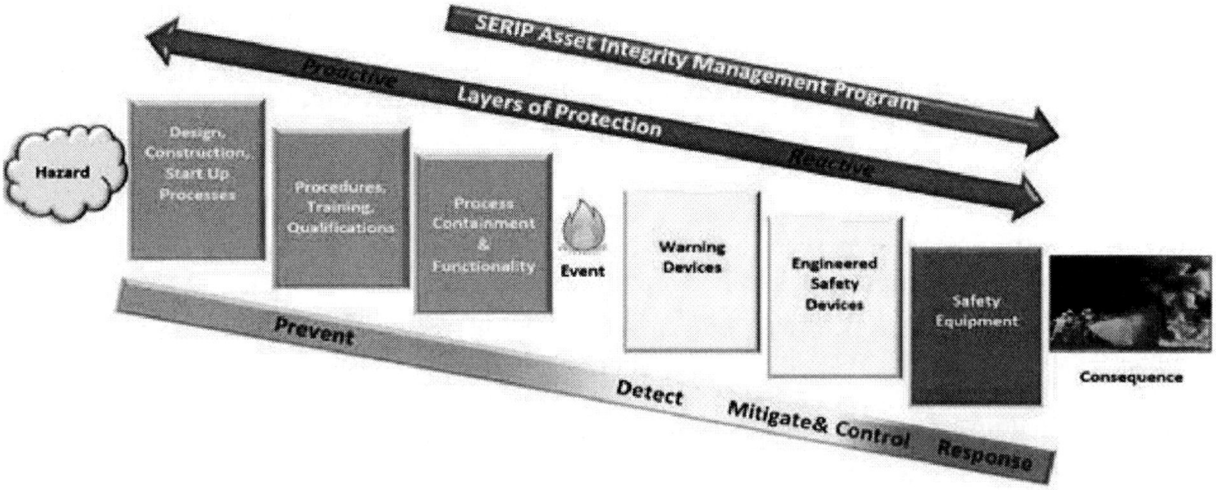

Figure 4: Chevron Barrier Model

The second step of the Upstream Asset Integrity Management program is to develop industry-compliant asset integrity plans that specify the frequency (or condition) and type of activities to be accomplished to provide reasonable assurance of ongoing equipment integrity. These activities are based on maintaining and verifying the integrity of the barriers identified during the initial consequence assessment. The integrity tasks developed as part of the integrity plan are to be identified as integrity critical. However, not all tasks on a given piece of equipment are integrity critical. Only the tasks that support the mandatory functionality of the asset in scenarios described through associated barrier models are integrity critical. Detailed procedures are developed for integrity critical tasks and they contain requirements for appropriate acceptance criteria for each task.

The third step of the Upstream Asset Integrity Management program is to execute the planned tasks. A forward-looking planning process showing planned tasks that are coming due enables the business to plan work such that asset integrity tasks are performed on or ahead of schedule.

The fourth step of the Upstream Asset Integrity Management program is to monitor and manage deviations. There are two types of deviations: schedule or date deferrals, and discrepant material which involves operating with known deviations from performance requirements such as metal thickness or valve closure times. The nonconformance process is used to highlight overdue integrity critical tasks, deviations in methods or scope from the asset integrity plan, or integrity critical equipment operating with discrepant materials that affect the ability of an integrity critical asset to perform its integrity critical functions. A nonconformance report (NCR) is created whenever a piece of equipment within the AIM program is operated or planned to be operated with a nonconformance. For example, an NCR is required to be approved:

i. Before an under-repair or out-of-service asset with a discrepant material condition left uncorrected or with overdue integrity tasks is brought back into service.
ii. Before any temporary repairs are performed and the asset brought back into service.
iii. When delaying the performance of an integrity critical task past its due date

When nonconformances are identified, the risk is assessed, an action plan is generated, and a NCR is sent to senior operations leadership for approval to continue operating. Management involvement makes leaders aware of the risk involved in continuing operation when the integrity of integrity critical equipment is threatened.

An additional upstream tool for supervisors and inspection groups to monitor and manage inspection plans has been deployed in the form of a performance dashboard for process safety leading indicators. This dashboard pulls information on a daily basis directly from Computerized Maintenance Management Systems (CMMS) and Integrity Data Management Systems (IDMS). It provides visualization of open and scheduled inspection and testing tasks. These are projected ahead for the next 12 months, in order to facilitate planning and resource discussions between operations and maintenance groups. The system also allows drill down into the system of record to pull up detailed information on individual inspections within a given facility. Visibility of planned schedules to frontline and senior leaders has helped drive down the number of overdue inspections to near zero across all business units.

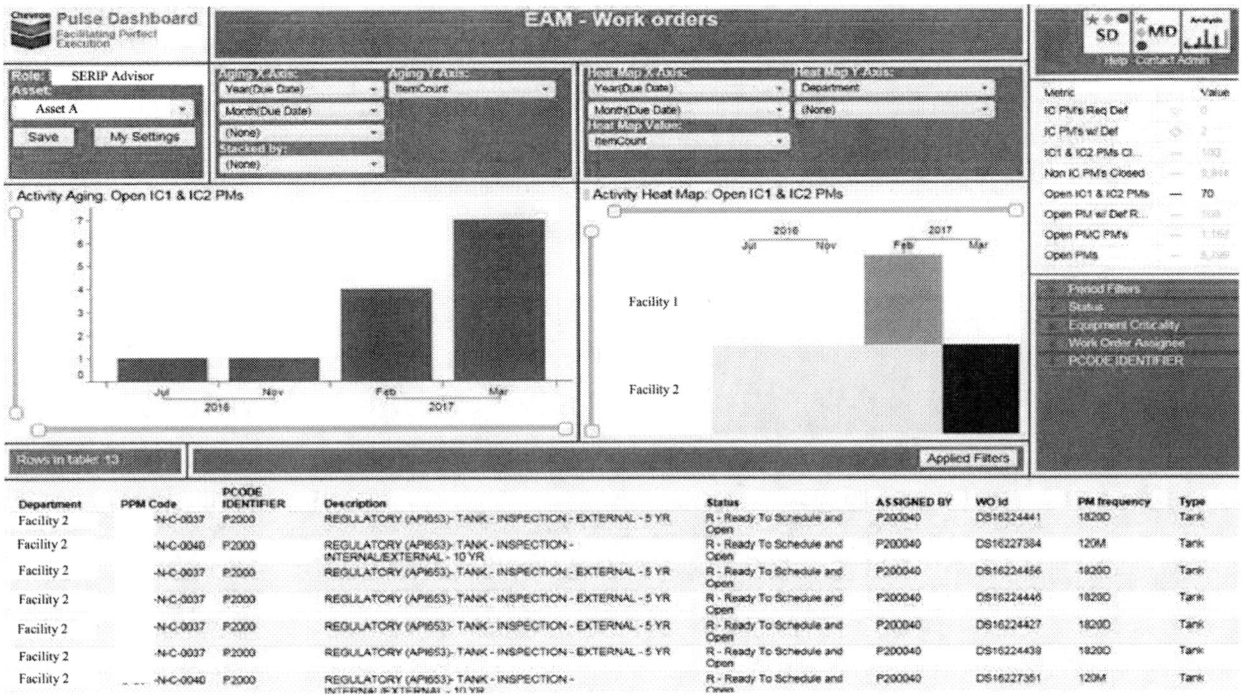

Figure 5: View of process safety dashboard illustrating the status of workorders specific to Asset Integrity inspections.

5 Asset Integrity Management Implementation

The Upstream Asset Integrity Management program was designed with significant input from the business units, so there was ownership and commitment when the Asset Integrity Management programs were implemented. The AIM team utilized in-person reviews with the business units to conduct baseline assessments to determine current state and worked with the business unit to develop shaping plans. This “soft start” allowed time for understanding and planning for what was required. After the framework was in place, more rigor was added to the reviews to determine if requirements were being met.

As shown in **Figure 6**, the Upstream Asset Integrity Management program included a step-wise implementation phase-in of Asset Integrity Management for equipment that is part of the layers of protection preventing events with health, safety, environment, or asset consequences of severe or catastrophic (i.e., consequence indices of 1 or 2 in **Figure 3**) from occurring. The initial focus in implementing Asset Integrity Management was on establishing a solid program foundation and providing specific requirements on the initial containment barriers of fixed equipment, structures and well systems. The second focus area expands to address preventative barriers associated with floating systems and subsea equipment. It also includes safety shutdown systems and primary mitigating barriers including response and egress. The first and second focus areas are aligned with the barrier model to focus first on the equipment barriers that can prevent an event and then on those that can detect, mitigate and control, and respond. The final focus area includes machinery and electrical equipment. This focus area adds incremental rigor to the existing reliability program developed for these equipment types in the early stages of SERIP. The approach of phased-in focus areas ensures the entire scope of integrity critical equipment safeguards are included in the Asset Integrity Management program and allows time for effective implementation within the Business Units.

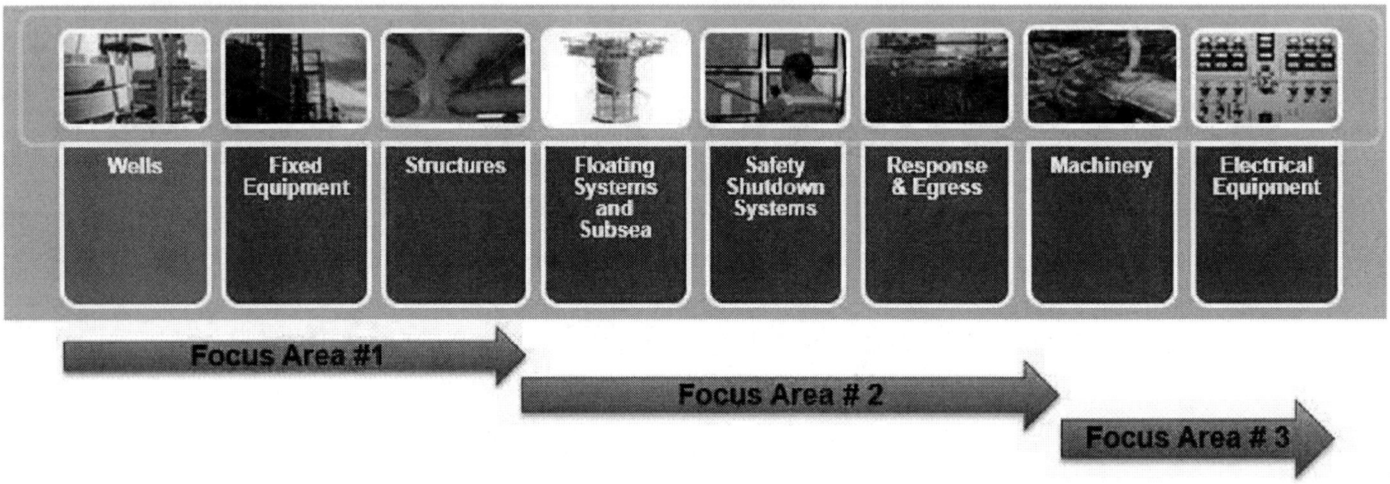

Figure 6: Phased-In Focus Areas of Asset Integrity Management Implementation

6 Assurance of Asset Integrity Management

When safeguards are critical to preventing and mitigating high consequence scenarios, it is imperative to verify that they are in place and effective. The rigor and frequency of verifications, including audits, should reflect what could occur in the event of a potential safeguard failure. If the safeguard is critical in protecting against a potential high consequence event, multiple layers of verification may be warranted. Verifications may be conducted by different levels in the organization. Possible methods of verification may include inquiry, observation, inspection or audit.

Chevron's Operational Excellence (OE) Assurance process provides a structure for assessing the verifications of safeguards associated with prevention of a high consequence scenario. Using available risk studies, incident and near-miss learnings, the high consequence scenarios and safeguards in place to prevent them are identified. Safeguard failure vulnerabilities are assessed (e.g. how could this safeguard fail or not work as intended?). Existing verifications for key safeguards are assessed to determine if they cover potential safeguard vulnerabilities, or if additional layers of assurance are needed. The OE Assurance program does not duplicate existing inspection and verification efforts, but rather leverages them to provide a cross-functional, multi-layer view on the assurance of key safeguards. In addition to the equipment safeguards, OE Assurance also considers verifications for administrative safeguards critical to preventing serious incidents such as operating procedures, a management of change program, and work permits for activities in high consequence facilities.

The verification activities span from the operator level, through frontline supervisors, up to leaders and can be in the form of inspections, audits, work observation, record sampling, and other related activities. It is important to consider the appropriate frequency and the appropriate role and subject matter expertise for the verification activity. **Figure 7** shows an example of what different layers of verification could look like for the Asset Integrity Management program. The involvement of leaders of various levels in verifications increases awareness, visibility, and ownership of key safeguards. Multiple layers of verification also emphasize the criticality of key safeguards to the operations workforce. A deeper understanding of how the safeguard functions and the consequences it is preventing or mitigating also increases knowledge of the operations workforce. Outcomes of verification activities are tracked and action items are generated to close any observed gaps.

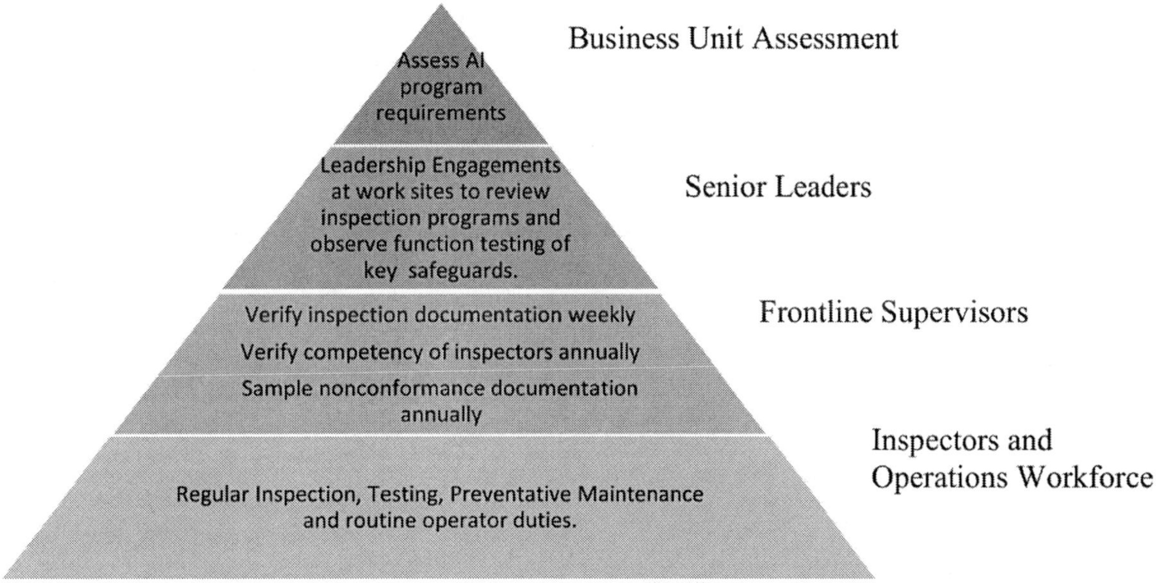

Figure 7: Example of Layers of Verification added to Asset Integrity

A specific approach to operationalizing the Asset Integrity Management program and verification of key safeguards is outlined as follows:

i. Confirm the scope of a high consequence scenario related to a plant or a field system. Confirm (or include) initiating events that have been identified in risk management studies that could trigger a high consequence scenario.
ii. Confirm full suite of key safeguards (including engineered, administrative and applicable personal safeguards) utilizing the safeguards identified in risk management

studies, and document improvement opportunities as well as applicable verifications (including responsibility, frequency, documentation location) associated with the scenario.

iii. Cross-reference these verifications with any current or planned equipment focused verification programs related to Asset Integrity Management and associated administrative safeguards; the goal is to avoid overlap of efforts.

iv. Determine if/how these scenario verifications will be further checked through higher level independent audit programs.

v. Address how specific verifications will be owned and managed through operations leadership and workforce, as well as how they will support inspection and maintenance groups.

vi. Provide visibility of the scenario and verification programs to the frontline workforce. Document this effort as part of the greater business unit assurance plan.

vii. On a regular basis, review progress and resolve findings of associated verification programs.

The results of this approach can be visualized through the use of block flow diagrams for the system described in the high consequence scenario, and annotated with key safeguards (**Figure 8**). There are also additional tables listing the safeguards, safeguard function and stewardship strategy (typically inspections, tests or operator routine duties). These materials are maintained at the facility site for communication and ongoing reference.

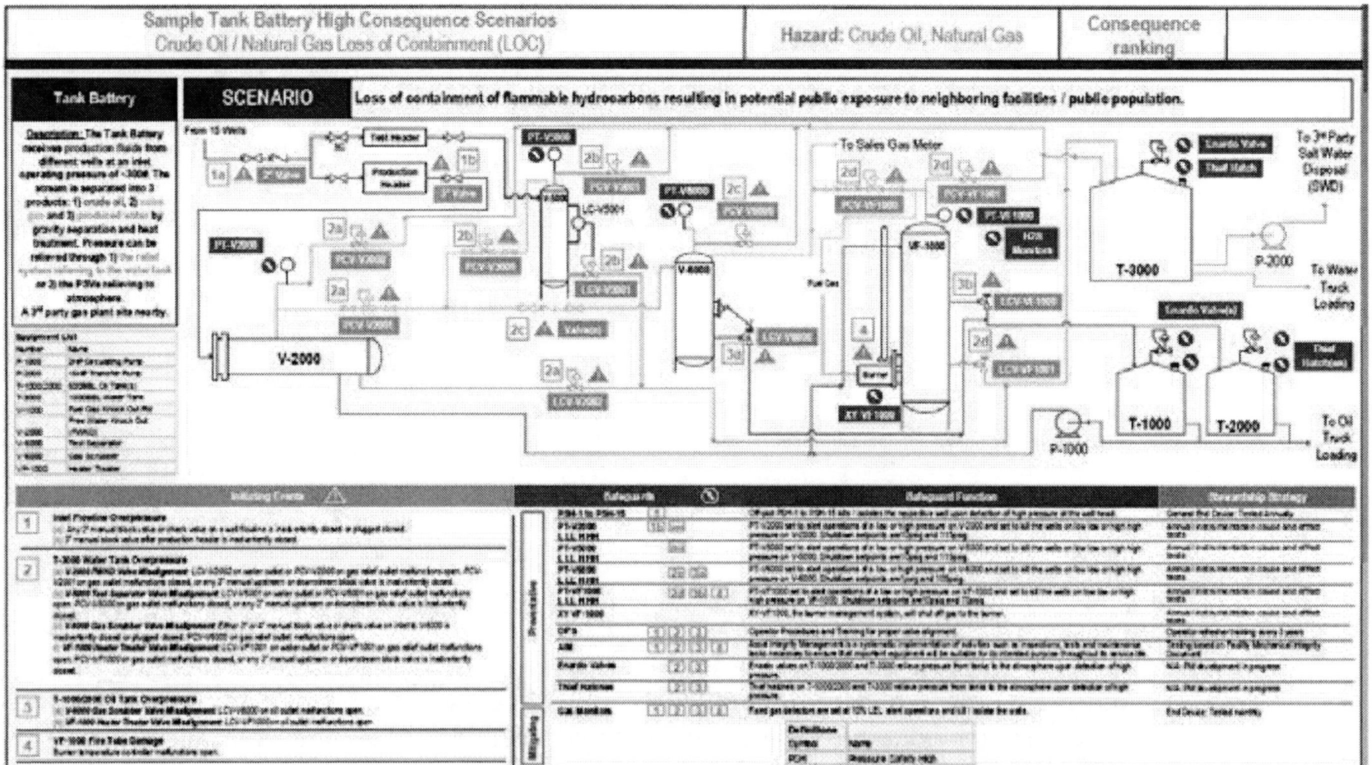

Figure 8: Example of key equipment safeguards and stewardship strategies for a high consequence scenario

7 Conclusions

Asset Integrity Management is the method Chevron Upstream uses to ensure the functionality of integrity critical equipment. Over time, this process has increased in both scope and depth to help ensure that equipment safeguards that prevent or mitigate severe consequences are well-maintained and are effective. This program was built on the foundation of a strong reliability program and started with a focus of specific equipment types. The business units developed their AIM programs and were assessed to determine whether their current program scope met requirements, before adding in additional equipment safeguards. Through a phased-in implementation, all integrity critical safeguards are covered by the AIM program.

The OE Assurance process helps ensure that the equipment and administrative safeguards that help prevent or mitigate severe consequences have the appropriate verification structure to ensure effectiveness. This cross-functional initiative helps fill in any gaps that may exist in providing assurance for key safeguards.

OE Assurance and Asset Integrity Management are two of many initiatives designed to reduce serious incidents by managing safeguards to ensure effectiveness. These efforts and others have been influencing factors in the 75% reduction of Tier 1 and Tier 2 Loss of Containment events over the past several years. Continued dedication to management of key safeguards can help drive down the frequency and severity of these incidents even further.

8 References

1] CCPS. Asset Integrity Management. Center for Chemical Process Safety, American Institute of Chemical Engineers, New York, NY, 2017.

2] CCPS. Guidelines for Risk Based Process Safety. Center for Chemical Process Safety, American Institute of Chemical Engineers, New York, NY, 2007.

9 Acknowledgement

1] Tracey R. Driver, OE Assurance Team Leader, Chevron OE/HES Technical Services, Houston Texas, has been instrumental in developing and deploying the assurance programs described above, and provided content for this paper.

Using HAZOP/LOPA to Create an Effective Mechanical Integrity Program

Steven T. Maher, PE CSP
Risk Management Professionals
Steve.Maher@RMPCorp.com

David J. Childs
Risk Management Professionals
David.Childs@RMPCorp.com

Prepared for Presentation at
American Institute of Chemical Engineers
2017 Spring Meeting and 13th Global Congress on Process Safety
San Antonio, Texas
March 26-29, 2017

Using HAZOP/LOPA to Create an Effective Mechanical Integrity Program

Steven T. Maher, PE CSP
David J. Childs
Risk Management Professionals

Keywords: PSM, RMP, CalARP, Mechanical Integrity, OSHA, EPA, Process Safety

Abstract

Many people view the conduct of a HAZOP/LOPA to address regulatory requirements as a chore, and stop there. However, the implementation of a quality HAZOP/LOPA has the potential to provide a framework for addressing numerous safety and operational optimization issues at plants, including the formulation/refinement of the Mechanical Integrity Program. The purpose of this paper is to focus on the mechanical integrity program, illustrate how a quality HAZOP/LOPA can support the effective implementation of some of the new Damage Mechanism Review requirements for California Refineries (e.g., 5189.1(k)), and optimize key elements of an effective Mechanical Integrity Program, e.g.:

- Inspection/testing methods
- Testing intervals
- Maintenance outage periods
- Repair prioritization and allowable outage
- Identification of low priority equipment

1. Mechanical Integrity Defined

When you look at the parallel evolution of modern Safety Management Systems (SMS) (Figure 1.1), such as OSHA's Process Safety Management (PSM) Program[1], U.S. EPA's Risk Management Program (RMP)[2], and the Bureau of Safety and Environmental Enforcement's (BSEE's) Safety and Environmental Management Systems (SEMS) Program[3], **<u>the same key Safety Management System elements are at the core of PSM, RMP, and SEMS, spanning an entire spectrum of facility</u>**

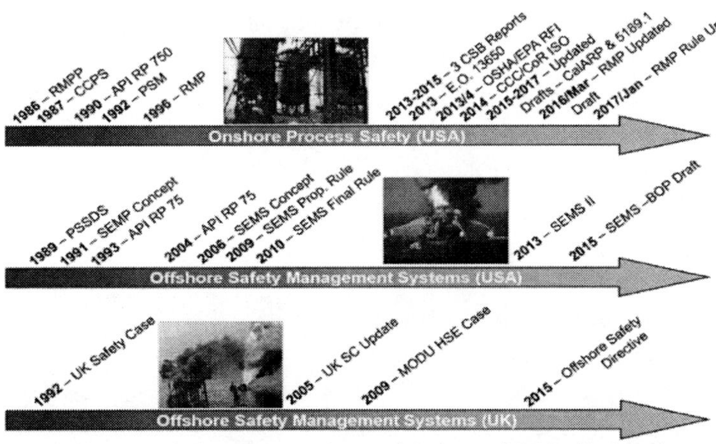

FIGURE 1.1 – Evolution of Select SMS Guidelines

types and geographic application. Although these regulatory programs were developed independently, at different times, and in different locations, industry and the regulatory community noted the importance of SMS application, and a fundamental part of this has always been maintaining the integrity of the process and functionality of equipment. As can be seen in Figure 1.2, "Mechanical Integrity" (MI) is a critical part of any SMS application.

FIGURE 1.2 – Key PSM Elements (2016)

The core objective of MI is "to maintain the on-going integrity of process equipment." This includes the integrity of the process boundaries as well as the reliability of operating/standby equipment. 29 CFR 1910.110(j) lays a foundation for:

- Typical process equipment to be included in the MI Program
- Written procedures to allow the program to function
- Training for process maintenance activities, with a focus on safety
- Inspection and testing, including procedures and definition of frequency
- Documentation of inspections and tests
- Correction of equipment deficiencies
- Quality assurance

Now that we have identified what MI is and what the requirements are, let's take a look at another key element of PSM.

2. Why do a Process Hazard Analysis (PHA)?

PSM is a performance-based standard, and as such, it is designed to focus on key objectives such as minimizing potential hazards and maintaining the desired level of safety at the plant site. PHA is a key early step in minimizing potential hazards by first identifying and understanding them in order to focus management systems (e.g., MI) on equipment/characteristics of importance. There are numerous PHA tools (see Figure 2.1) that have various advantages / disadvantages for different applications[4].

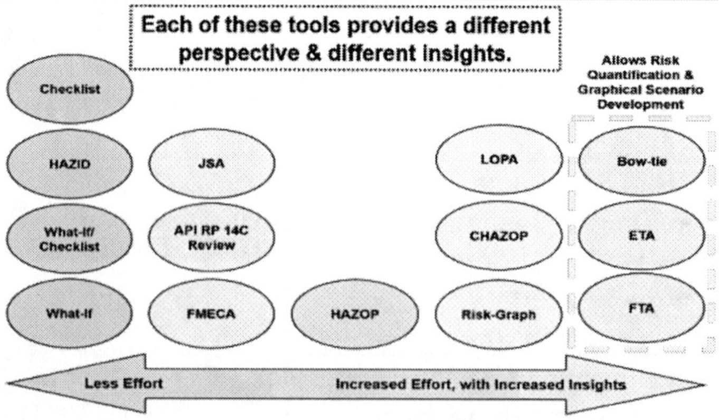

FIGURE 2.1 – Hazard Analysis Tool Spectrum

However, one of the more broad-spectrum PHA techniques is the Hazard and Operability (HAZOP) Study.

The guideword HAZOP technique is based on the premise that hazards and operability problems originate from deviations from design intent when a process is running under normal operating conditions. For example, adding the guideword "NO" to the parameter "FLOW" to get the deviation "NO FLOW" would prompt the leader to ask the Team, "What causes could result in no flow in this node or line segment?" The potential hazard scenarios that include possible "Causes" and potential "Consequences" are documented in the report worksheets. The possible "Safeguards" in place to reduce the risk associated with the specific cause/consequence scenario are then discussed and documented.

The HAZOP Study proceeds sequentially, studying each piece of equipment contained in the process. Thus, if applied comprehensively, HAZOP systematically creates a roadmap of key paths that lead to undesired events (hazards or operability issues, depending on the study objectives). Because this roadmap provides a framework for assessing the likelihood and severity of each path to an undesired event, the importance of the contribution of causal events and safeguards can be assessed, as well as the need to prioritize reliable equipment function.

FIGURE 2.2 – HAZOP/LOPA Requires a Multidisciplinary Approach

Since HAZOP is a scenario-based method that explicitly identifies the failure of equipment that can potentially lead to a hazardous condition (cause), explicitly identifies and illustrates the importance of active protection features (safeguards), and applies a measure of importance (consequences) to their failure, it is a helpful platform for identifying important equipment requiring prioritization of reliability. This information, derived from the contributions of diverse technical disciplines (e.g., engineering, operations, maintenance) is fundamental to the establishment of a balanced MI Program. Reference 5 is a very good source of pragmatic tips for the implementation of HAZOP, and Reference 15 provides some general background on the HAZOP method and its application during the design process.

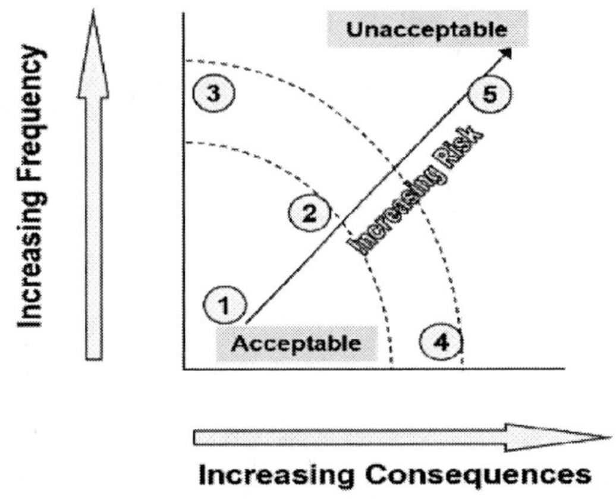

FIGURE 3.1
Scenario-Based Analysis Objectives

3. Using Layer of Protection Analysis (LOPA) to Dig Further

Section 2 describes the essence of a PHA, which is the identification of scenarios with sufficient detail to balance likelihood and severity to understand their risk contribution, and thus, the importance of the scenario and associated equipment. Figure 3.1 graphically illustrates how the clarity provided by assessing both the likelihood and severity for different scenarios (1-5 for this example) provides an improved perspective on the risk contribution of the scenario, and thus, the importance of associated equipment reliability.

Since the 1980s, advances in electronics (see Figure 3.2) facilitated the application of more reliable control/protection equipment that provided a platform for improved levels of safety and reliability. For most facilities subject to PSM/RMP, these improvements are implemented in a "phased-approach," as-needed, typically as part of capital projects. So, at any point in time, a facility has a wide-spectrum of equipment applied to control/protection systems (Figure 3.3). The challenge is applying a tool to evaluate their reliability contribution that can be scaled up/down, depending on the level-of-detail needed, and that can build on all of the work done during a HAZOP. LOPA is a tool that is well-suited for this challenge.

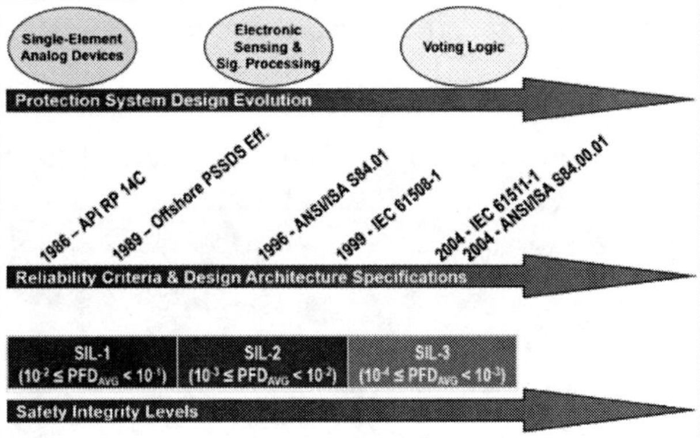

FIGURE 3.2
Tandem Advances in Protection System Design Architectures & Analysis

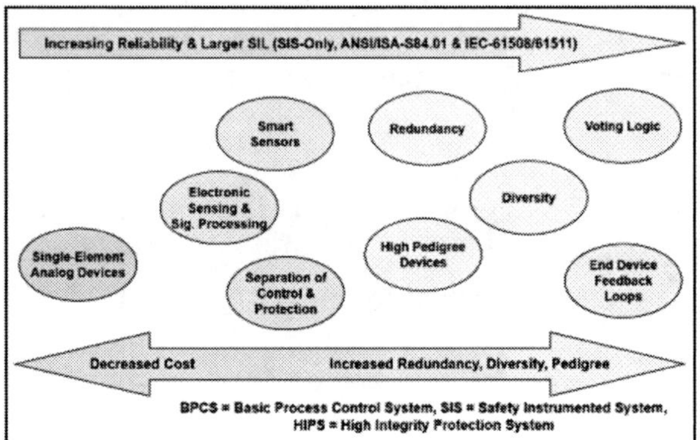

FIGURE 3.3
Control/Protection System Spectrum – BPCS & SIS/HIPS

Like a HAZOP, LOPA is also a scenario-based tool that is often coupled with a HAZOP. The primary difference is depth, specificity, and the ability to infuse more complex quantitative information (see Table 3.1). References 6, 7 and 8 are very good sources of pragmatic tips for the implementation of LOPA.

TABLE 3.1 - Defining the Scenario and Equipment Importance (Contrasting HAZOP & LOPA)

Likelihood		Severity	
HAZOP	**LOPA**	**HAZOP**	**LOPA**
Cause	Initiating Cause		
Safeguards	IPL & non-IPL		
Likelihood Ranking from a Risk-Ranking Matrix	Product of Initiating Cause Frequency, Enabling Condition Probability, Conditional Modifiers, and the IPL PFD	The severity value used for the HAZOP and LOPA is typically the same, but an opportunity exists for LOPA to apply more quantitative differentiation.	

LOPA typically applies representative order-of-magnitude quantitative values to the frequency of causal events and the Probability of Failure on Demand (PFD) to safeguards to provide a frequency of reaching an undesired consequence that can be compared to the company target value to assess acceptability (see Figure 3.4). LOPA also drills a little deeper with respect to understanding if a safeguard is an "Independent Protection Layer" (IPL) and the potential for common-mode failure. LOPA can also apply various Enabling Event Probabilities and Conditional Modifiers to better characterize the potential for reaching the Ultimate Consequences (see Figure 3.5).

- Risk Framework
  - $Risk(R) = \Sigma\ F_i * C_i$
- Scenario Frequency Assessment as Absolute Value
  - $f_i^C = f_i^I * \prod P_{ij}^{EC} * \prod PFD_{ij} * \prod P_{ij}^{CM}$
- Scenario Frequency Assessment as a Ratio

$$LOPA\ Ratio\ (Safety) = \frac{TF_{Safety}}{ICL * PFD_1 * PFD_2 * PFD_3 ... * ECi * CMi}$$

- Where:
  - $ICL(f_i^I)$ – Initiating Cause Likelihood (Frequency)
  - PFD – Probability of Failure on Demand
  - TF – Target Frequency
  - EC – Enabling Condition
  - CM – Conditional Modifier

FIGURE 3.4 – LOPA Snapshot

LOPA's primary purpose is to determine the adequacy of existing IPLs and determine if additional protection features are needed. LOPA is also used to assign a target Safety Integrity Level (SIL) value for a Safety Instrumented System (SIS)[9, 10]. SIL assignment is based on an instrument's likelihood to function upon demand. A higher SIL level device has more "value" in risk reduction and is determined based on the specifications the instrument is manufactured to meet. These applications identify one of the other very useful functions for LOPA. It is able to identify reliability targets for equipment that might "cause" a potential hazard and identify reliability targets for equipment that can function as a protective feature (safeguard). One can capitalize on these characteristics to fortify the structure of a MI Program.

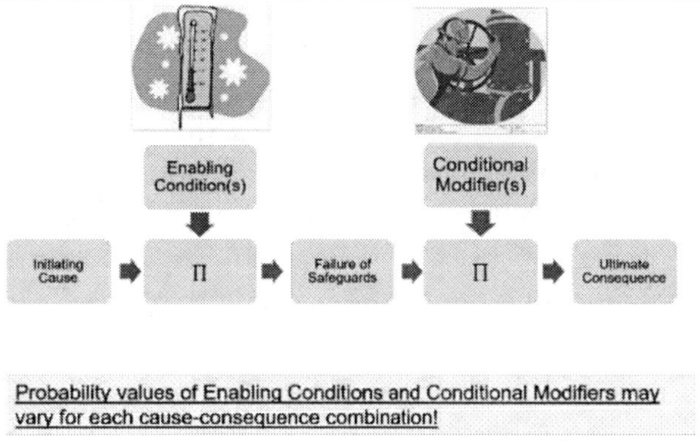

FIGURE 3.5
Addressing Enabling Conditions & Conditional Modifiers in LOPA

4. Pulling It Together

4.1 Basics

Section 1 defined MI and identified relevant regulatory requirements. Both MI and PHA are key elements of PSM/RMP, and as such, properly structured, they can be mutually supportive. Critical to effective implementation is an understanding of key MI Program Elements (see Figure 4.1). If one were to create a wish list that could provide a basis for a MI Program, it might include:

- Accommodating both safety and operational issues

- Identifying when a safety feature is needed
- Being able to scale up/down
- Provide optional quantification
- Scenario-based

When we look at these needs, it clearly points towards PHA, specifically HAZOP/LOPA as having the ability to provide the information needed to define a good MI Program. Being a performance-based standard, PSM doesn't provide an exact prescription for defining a MI Program or its elements. Therefore, as long as performance-based objectives are met, any number of ways to define the program and the various key elements, such as inspection frequencies may be acceptable. However, diligent implementation of various elements of the MI Program and HAZOP/LOPA can greatly increase effectiveness.

FIGURE 4.1
MI Program Elements

4.2 Desirable MI Program Characteristics

Figure 4.2 illustrates that there can be a very wide range of acceptable approaches to the implementation of a performance-based standards like PSM and RMP. However, certain characteristics facilitate the effective implementation of a MI Program, as well as allowing constructive interface with other PSM/RMP elements such as PHA:

- Configuration of a Computerized Maintenance Management System (CMMS) to allow for trending
- Programmatic checks/balances that allow for consistent trending
- Assign of allowable outage times
- Communications with Operations, Safety, and other stakeholders if equipment is out-of-service for maintenance, inspection, testing, or repair
- Assignment of maintenance, inspection, testing, or repair priorities
- Application of consistent equipment tag number patterning and utilization that matches with other Process Safety Information (PSI)

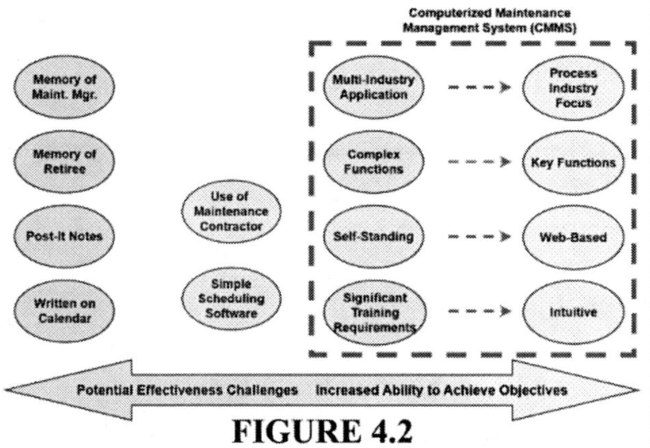

FIGURE 4.2
MI Implementation Spectrum

4.3 Desirable HAZOP/LOPA Characteristics

The ability to utilize the results of a HAZOP/LOPA is greatly dependent on the quality of the study and documentation, which is often linked to the experience and diligence of the Facility/Scribe Team heading the effort. For this reason, inconsistencies in the HAZOP/LOPA results have often created a challenge. However, certain characteristics can facilitate the effective utilization of the HAZOP/LOPA in support of the MI Program:

- Availability of a high quality HAZOP/LOPA (Reference 4 provides tips on the implementation of high quality HAZOP/LOPA Studies)
- Documentation that consistently, accurately, and comprehensively applies equipment tag numbers that match with other Process Safety Information (PSI)
- Clear documentation of safeguard functions
- Ready access to machine-readable HAZOP/LOPA outputs, for searching

4.4 Using HAZOP/LOPA to Formulate the MI Program

Many companies/individuals seem to struggle with identifying equipment to be encompassed by the MI Program and frequency/scope of testing, inspection, and preventive maintenance to be applied. Although there are a number of different ways to approach MI, since the purpose of the MI Program is to support safe and reliable plant operation, using a high quality HAZOP/LOPA is one straightforward way that can at least offer a good starting point and a defensible basis:

- If an active component is a safeguard identified by HAZOP/LOPA, then there is an implicit or explicit reliability assumed by the Team. The MI Program needs to be designed to support that reliability.
- If the failure of a piece of equipment is a causal event, there is an implicit/explicit assumption of failure frequency. The MI Program needs to be designed to support that reliability.

Thus, if a piece of equipment that is a safeguard in a HAZOP/LOPA is not at least defined in the MI Program with a reasonable testing, inspection, and preventive maintenance assignment, this would seem to be a deficiency and difficult to justify its absence. At the other end of the spectrum, the plant maintenance department needs to be able to justify not tracking, testing, inspecting, and maintaining every subcomponent. Again, the HAZOP/LOPA can help clarify that the objective is

TABLE 4.1 – Example Values Used for LOPA	
Initiating Cause Likelihoods	
Initiating Cause	**Events / Year**
BPCS instrument loop failure	1×10^{-1}
Regulator failure	1×10^{-1}
Pumps and other rotating equipment failure	1×10^{-1}
Safety valve opens spuriously	1×10^{-2}
Pump seal failure	1×10^{-1}
Independent Protection Layer (IPL) Probability of Failure on Demand (PFD)	
IPL	**PFD**
Basic process control system, if not associated with the initiating event being considered	1×10^{-1}
Safety valve fails to open on demand	1×10^{-2}
Rupture disc fails to open on demand	1×10^{-2}
SIL-1 IPL	$> 1 \times 10^{-2}$ & $\leq 1 \times 10^{-1}$
SIL-2 IPL	$> 1 \times 10^{-3}$ & $\leq 1 \times 10^{-2}$
SIL-3 IPL	$> 1 \times 10^{-4}$ & $\leq 1 \times 10^{-3}$

to achieve the desired reliability of the equipment referenced in the HAZOP/LOPA (see Table 4.1), and if the subcomponent in question is implicit in that reliability, it does not need to be independently tracked in the PSM MI Program.

In addition to defining the universe of components to be encompassed by the MI Program, HAZOP/LOPA can be used to support prioritization. Equipment (and key failure modes) encompassed by the MI Program can be divided into four main classes:

- Safety Instrumented Functions (SIF)
- Safety – High Priority
- Safety – Low Priority
- Operational

Although some expert judgment and experience can be used when classifying equipment (and failure modes) into these categories, as a starting point, the results of the HAZOP/LOPA can be helpful and provide a complimentary perspective to the expert judgement classically used:

- SIF – If a facility has committed to IEC 61508/61511, these are typically treated as the highest priority with well-defined testing, inspection, and preventive maintenance requirements.
- "Safety – High Priority" Equipment Considerations
    - Equipment failure modes that can initiate a high consequence HAZOP/LOPA scenario (if unmitigated)
    - IPL Safeguards that could mitigate a high consequence HAZOP/LOPA event
    - IPL Safeguards that could mitigate a HAZOP/LOPA event with a safety consequence, and where that is the only protection feature for that safety scenario
    - IPL Safeguards that could mitigate multiple scenarios associated with lower consequence HAZOP/LOPA events
- "Safety – Low Priority" Equipment Considerations
    - Other equipment failure modes that could result in a safety consequence (if unmitigated) identified by the HAZOP/LOPA
    - IPL Safeguards that could mitigate a lower consequence HAZOP/LOPA event
    - Non-IPL Safeguards credited by the HAZOP/LOPA
- Operational Considerations for the MI Program

Binning equipment and the key failure modes of concern support meaningful prioritization by the Plant Maintenance Department to ensure that the SIF and "Safety – High Priority" equipment and failure modes receive the proper support and application of testing, inspection, and preventive maintenance that meets or exceeds industry standards and best practices.
Other Tips:

- During the HAZOP/LOPA, avoid including safeguards that aren't important IPLs, as their inclusion into the MI Program, even as low priority items, can dilute the Plant Maintenance Department's efforts on more critical equipment.

- Testing (functional) and inspection activities in the MI Program should focus on the failure modes identified in the HAZOP/LOPA as important.
- Without the perspective of the HAZOP/LOPA, instrumentation designers can often overdesign the protection features and include SIF where they may not be necessary. A good use for the HAZOP/LOPA is to identify where a SIF could be converted to a BPCS, so that the Plant Maintenance Department can focus resources in other, more critical, areas.
- Tracking and trending of failure data as part of the MI Program can be geared to the level-of-resolution of the failure mode in the HAZOP/LOPA.

4.5 Using HAZOP/LOPA to Support the MI Program During Plant Operation

Whereas the previous subsections focus on the ability to utilize the HAZOP/LOPA to initially formulate the MI Program, interaction between the MI Program and the HAZOP/LOPA models can be useful during plant operation. Plant operations can be a quite dynamic environment with priorities continually shifting as new challenges arise. If HAZOP/LOPA information is readily available during plant operation, more effective decision-making and prioritization can be accomplished:

- If diligently documented, the HAZOP/LOPA can be used to determine if out-of-service equipment has a potentially critical safety impact.
- In a similar way, allowable outage time and repair priorities can to be geared towards an understanding of the role equipment may play as a safeguard.

5. Complementary Methodologies

The approaches discussed in Section 4 address the majority of the needs of a PSM MI Program; however, for some equipment and process configurations, especially those associated with high-consequence potential hazards, additional tools may be required to define the associated inspection, testing, and maintenance frequencies and activities.

5.1 API RP 581[11]

In 1993, the American Petroleum Institute (API) released Recommended Practice 581 which provides guidance on performing a risk based, quantitative analysis to develop an inspection program tailor-made to a facility based on facility conditions and company expectations of risk at the facility. The practice includes calculations of probability of failure (POF) and the consequences of failure (COF) similar to the methodology used in a HAZOP Study when looking at potential consequences and likelihoods of failure within a process. By assigning a risk rank to equipment individually, inspections and mechanical integrity programs can be tuned to provide the level of attention necessary to equipment. In generalized or standardized programs, some equipment may be serviced or inspected too infrequently resulting in higher risk whereas other, lower risk equipment may be serviced or inspected at a rate above what would be necessary to meet a company's risk target.

API RP 581 provides a comprehensive structure for analyzing equipment in the following groups:

- Pressure Vessels and Piping
- Atmospheric Storage Tank
- Pressure Relief Devices
- Heat Exchanger Tube Bundles

For each equipment group, specific methods for determining probability of failure, consequences of failure and inspection planning guidelines are available. This process also allows for differing levels of inspection which would facilitate effective implementation based on the size and resources available at a facility.

5.2 Damage Mechanism Review (DMR)

The Richmond Refinery fire on August 6, 2012 triggered a fresh look at several SMS programs, the application of hazards identification techniques (as applied to hazardous material containment integrity), and resulted in several proposals for the modernization of PSM and RMP, including the performance of a "Damage Mechanism Review."[12,13] A key focus of DMR requirements is piping systems, even though 29 CFR 1910.110(j)(1)(ii) identifies "Piping systems" as types of process equipment that for which a MI Program should be applied.

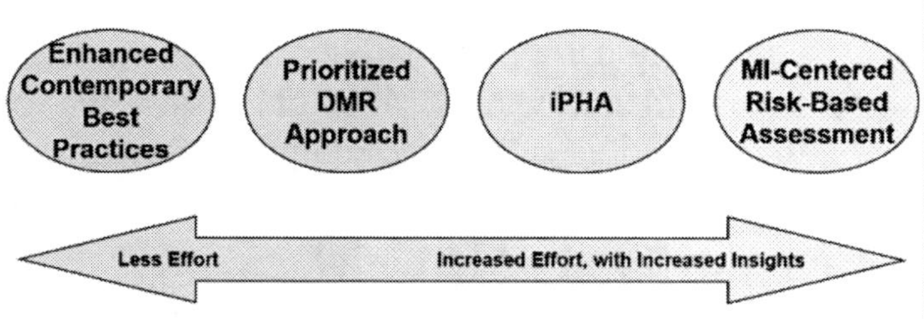

FIGURE 5.1 – DMR Implementation Spectrum

The complete implementation of DMR can require extensive resources, and Figure 5.1 depicts the range of approaches that can be used to address DMR requirements. In short, one of the most effective approaches is to encompass DMR by the PHA and treat the failure of select piping as a causal event, thus capitalizing on the insights from similar types of releases considered by the PHA Team.

The following resources clarify the challenge and provide some focused/practical approaches for implementation:

- Maher, Nour, Schultz, "Using PHA as a Framework for Effectively Addressing Evolving PSM/RMP Guidelines, Such As Damage Mechanism Hazard Reviews," Global Congress on Process Safety 2015[17].
- RMP/PSM Series Educational Webinars (March 26, 2015 and August 27, 2015)[14]

5.3 Effective Use of Standardized Maintenance Schedules

The aforementioned methods will provide a robust and focused MI Program for a facility. Based on the size, complexity and level of risks at a given facility, these methods may be more or less important. In many cases, facilities will use recognized standards within industry for maintenance intervals as a baseline. There are multiple groups that provide recommended maintenance and inspection intervals. Some of the more commonly referenced ones are listed below:

- OSHA (Occupational Safety and Health Organization)
- Cal/OSHA (California Occupational Safety and Health Organization)
- ANSI (American National Standard Institute)
- IIAR (International Institute of Ammonia Refrigeration)
- IEC (International Electrochemical Commission)
- API (American Petroleum Institute)
- NBIC (National Board Inspection Code)
- CCPS (Center for Chemical Process Safety)
- Department of The Army Technical Bulletin

These organizations offer guidance on various equipment groups with information regarding frequencies of maintenance and the types of actions that are to be taken within a time interval. These actions will be independent of facility conditions (in some cases corrosion is taken into consideration) and offer a standard for all facilities to follow. If a facility chooses to opt for a more robust methodology (such as API 581), the recommended actions by these organizations can be used as a "litmus test" to ensure the advanced methodology is achieving its goal. Table 1 shows some examples of commonly-referenced standards for specific equipment groups:

TABLE 5.1 – Examples of Commonly-Referenced MI Standards

Maintenance Standard	Description
API 510	Multiple equipment groups including pressure vessels and PRVs
API 570 & ASME B31.1-2007	Piping Inspection, Repair and Corrosion Examination
IEC 61508	Functional safety of electrical /electronic/programmable electronic safety-related systems
API 653	Tank Inspection, Repair, Alteration, and Reconstruction
IIAR 110	Shutoff and control valve maintenance, daily inspection recording,

Some of these standards such as API and IIAR are associated with a specific industry, however they can act as a starting point for all facilities. These standards can also be used in conjunction with manufacturer recommendations of maintenance intervals. A conservative method would be to compare the manufacturers proposed actions and intervals to those offered by the organizations and taking the more involved of the two.

6. Select Statistics to Optimize Your MI Program

The implementation of a real MI Program can be quite dynamic, and various issues may materialize:

- Variance of inspection/testing intervals
- Variance of inspection/testing methods

- Impact of maintenance outage time on equipment reliability
- Repair prioritization and allowable outage time
- Feedback of reliability observations back into the MI Program

Every component has a certain degree of uniqueness, and theoretical application of the bathtub curve concept never exactly echoes component-specific performance; however, equipment in a process facility is generally utilized during a period of its existence where it is not subject to burn-in or wear-out failures, and the failure rates is generally constant (see Figure 6.1). However, during this period, the inspection, testing, and preventive maintenance features of the PM Program impact various categories of equipment differently, e.g.:

- Monitored-Repairable Components
- Unmonitored-Repairable Components
- Standby Components

Understanding these differences can provide useful insights to optimize the PM Program with respect to cost and equipment reliability. This section is designed to convey basic concepts behind the driving forces of equipment reliability.

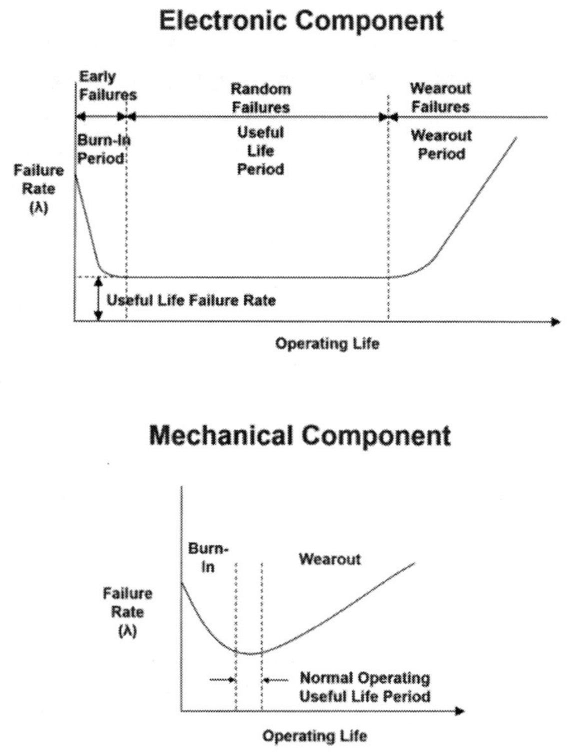

FIGURE 6.1
General Component Life Cycles

6.1 Monitored-Repairable Components

Examples in this category include active valves, where a failure would be noticed, or contemporary electronics with high-pedigree self-diagnostics. In these cases, the failure mode of interest would be revealed and can then undergo repair. Note that not all failure modes associated with a piece of equipment may be able to be monitored. A fundamental issue for any MI Program is the choice of what failure modes can be monitored and what failure modes can be functionally tested. A brief review of some key definitions is in order:

- Reliability – Probability that the component experiences no failures during time (0,t)
- Availability (A(t)) – Probability that the component is normal (available) at time "t" = $\frac{Total\ Operating\ Time}{Total\ Time\ of\ Interest}$
- Unavailability (Q(t)) = $\frac{Total\ Down\ Time}{Total\ Time\ of\ Interest}$
- Mean-Time-To-Failure (MTTF) – Average time interval between failures

- Mean-Time-To-Repair (MTTR, $^1/_\mu$) – Average time to repair a failed component
- Failure Rate (λ) = $^1/_{MTTF}$

Figure 6.2 illustrates the time periods that might contribute to the overall availability/unavailability of a piece of equipment and identifies the associated calculations that can provide insights into equipment availability/unavailability. Based on the criticality of the equipment with respect to its reliability and contribution to plant safety via the HAZOP/LOPA, the Plant Maintenance Department can use these concepts, as well as MTTR and MTTF to judge the need to invest in resources to minimize MTTR (e.g., warehoused spares) or to maximize MTTF (e.g., higher reliability equipment replacements).

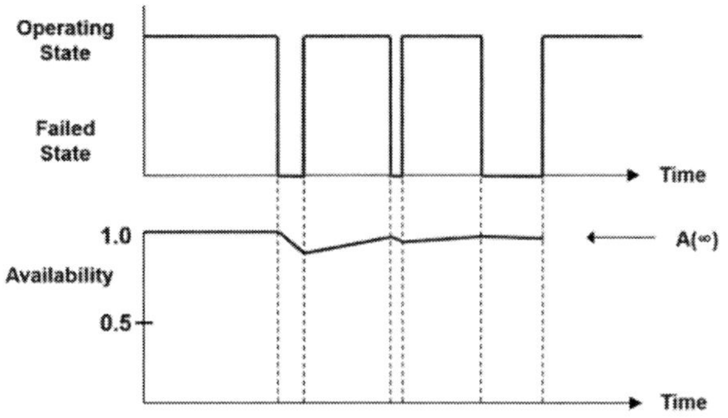

- $A(\infty) = \frac{MTTF}{MTTF+MTTR}$
- $Q(\infty) = \frac{MTTR}{MTTF+MTTR}$
- $A(\infty) + Q(\infty) = 1$
- Example
  - For λ = 1E-6/hr, MTTR = 10 hr
  - Q = 1E-5

FIGURE 6.2
Monitored-Repairable Components

6.2 Unmonitored-Repairable Components

Examples in this category include pressure safety valves (PSVs). Unmonitored components are subject to a similar relatively-uniform failure rate during the active life of the equipment; however, it would be a covert failure, or unrevealed, until such time as a planned test would identify that the component has failed. This is illustrated by Figure 6.3 and covers a wide range of safeguards in a typical process unit. Based on the importance of equipment function and functionality needed (e.g., from the HAZOP/LOPA), the PM Program can be tuned to optimize testing/inspection intervals (i.e., cost-benefit) and testing/inspection methods (i.e., to address the failure mode and functionality needed).

- Component Unavailability = $\frac{Mean\ Time\ of\ Unavailability}{Mean\ Time\ of\ Interest}$
- Mean Time of Interest (τ) = Time Between Tests
- Mean Time of Unavailability =
  $$\int_0^\tau F(t) * dt = \int_0^\tau \lambda t * dt = \frac{\lambda\tau^2}{2}$$
- $Q = \frac{\lambda\tau}{2}$

FIGURE 6.3
Unmonitored-Repairable Components

6.3 Standby Components

Standby components typically do not behave with only the simple parameters identified in Section 6.2. Figure 6.4 illustrates the contributions of testing/inspection intervals, testing/inspection durations, repair duration, and preventive maintenance duration on the unavailability of a standby component. To add to the complexity, different failure modes or piece of equipment may be unrevealed (covert) or revealed failures, and the different failure modes may have a different importance with respect to plant safety/operability, as identified via the HAZOP/LOPA. The challenge of the PM Program is to optimize equipment reliability and associated costs or achieving that reliability. Whereas, there is no perfect solution, a clear understanding of the need stemming from the HAZOP/LOPA and understanding fundamental reliability concepts can help tune the PM Program to achieve the desired degree of optimization.

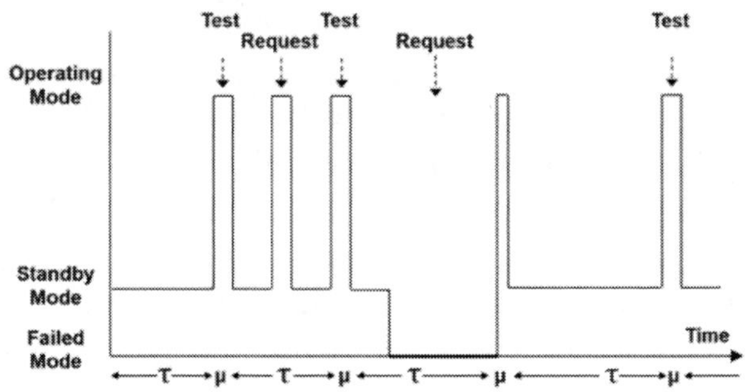

- Unavailability Due to Hardware Failure (q_1)
  - Unrevealed Failures = $\lambda_s \tau / 2$
  - Failure to Start on Demand = λ_d
- Unavailability Due to Test (q_2) = μ / τ
- Unavailability Due to Maintenance (q_3) = μ_m / τ_m
- Unavailability Due to Repair (q_4) = $\frac{MTTR}{MTTF+MTTR}$
- $Q_T = q_1 + q_2 + q_3 + q_4$

FIGURE 6.4
Standby Components

6.4 Feedback of Reliability Observations into the MI Program

Most CMMS provide an ability to log equipment failures and support data trending. There is a fundamental challenge associated with carefully logging the information and correlating the specific failure mode of the equipment to a failure mode of importance to the HAZOP/LOPA. Assuming that this has been done diligently, various approaches[11] (e.g., Bayesian statistics) can be used to update manufacturer reliability data with the specific experiences at the plant site. This information can be fed back into the MI Program to further optimize testing, inspection, and preventive maintenance practices (see Figure 4.1) to optimize its cost-effectiveness. This feedback mechanism can often result in re-focusing limited Plant Maintenance resources towards areas of greater importance.

7. Conclusion

Because they are core elements of PSM/RMP, the ties between the MI Program and HAZOP/LOPA are very strong, but are typically underutilized. When formulating the MI Program, there is a wealth of information that can be drawn from HAZOP/LOPA to focus and enhance the effectiveness of the MI Program. This effectiveness can manifest itself in many ways, e.g.:

- Ensuring that high-priority equipment gets the attention needed
- Optimizing inspection, testing, and preventive maintenance frequencies
- Identification of low-priority equipment, so that Plant Maintenance Department can focus on high-priority equipment
- Identification of over-application of SIS, where a BPCS component can provide adequate reliability with much lower recurring MI costs

Similarly, during the course of plant operations, when the inevitable challenges occur that compromise planned inspection, testing, and preventive maintenance activities, HAZOP/LOPA can provide insight regarding importance and may identify desirable options.

8. References

[1] PSM – 29 CFR 1910.119, "Process Safety Management (PSM) of Highly Hazardous Chemicals, Explosives and Blasting Agents," 1992.

[2] RMP – 40 CFR Part 68, "Risk Management Programs (RMP) for Chemical Accidental Release Prevention," 1996.

[3] SEMS Final Rule – Federal Register – Title 30, Code of Federal Regulations (CFR) Part 250 – "Oil and Gas and Sulphur Operations in the Outer Continental Shelf – Safety and Environmental Management Systems," Federal Register, Vol. 78, No. 66, April 5, 2013.

[4] http://www.RMPCorp.com/HAZOP-Study-series-module, HAZOP/LOPA Facilitation Best Practices Webinar Series.

[5] CCPS "Guidelines for Hazard Evaluation Procedures, 3rd Edition, 2008.

[6] CCPS "Layer of Protection Analysis – Simplified Process Risk Assessment," 2001.

[7] CCPS "Guidelines for Initiating Events and Independent Protection Layers in Layer of Protection Analysis," 2015.

[8] CCPS "Guidelines for Enabling Conditions and Conditional Modifiers in Layer of Protection Analysis," 2013.

[9] IEC 61508, "Functional Safety of Electrical/Electronic/Programmable Electronic Safety-Related Systems."

[10] IEC 61511, "Functional Safety - Safety Instrumented Systems for the Process Industry Sector."

[11] API Recommended Practice 581, "Risk-Based Inspection Technology."

[12] http://www.caloes.ca.gov/cal-oes-divisions/fire-rescue/hazardous-materials/california-accidental-release-prevention, California Accidental Release Prevention (CalARP) Program Proposed Updates, February 14, 2017.

[13] http://www.rmpcorp.com/wp-content/uploads/2014/08/15-day-Notice-Process-Safety-Management-for-Petroleum-Refin.pdf, Proposed General Industry Safety Order (GISO) §5189.1, Process Safety Management for Petroleum Refineries, February 10, 2017.

[14] http://www.RMPCorp.com/rmppsm-series/ - RMP/PSM Series Educational Webinars.

[15] Maher, Reyes, Vasudevan, "Assimilating Design Formulation and Design Review into a HAZOP," Global Congress on Process Safety 2012.

[16] "Relief Valve Testing Interval Optimization Program for the Cost-Effective Control of Major Hazards," Second Symposium on Preventing Major Chemical Accidents, Oslo, May 1988.

[17] Maher, Nour, Schultz, "Using PHA as a Framework for Effectively Addressing Evolving PSM/RMP Guidelines, Such As Damage Mechanism Hazard Reviews," Global Congress on Process Safety 2015.

[18] Clean Air Act (CAA) Section 112(r)(1) – General Duty Clause.

[19] http://www.CSB.gov – Source website for the Chemical Safety Board.

[20] http://www.CalEPA.CA.gov/Refinery/ – Source website for the Interagency Refinery Task Force.

[21] http://www.calepa.ca.gov/publications/Reports/2014/RefineryRpt.pdf, “Improving Public and Worker Safety at Oil Refineries,” February 2014.
[22] http://www.RMPCorp.com/SMS_Regulatory_Updates/ - Website Tracking Safety Management Systems U.S. Regulatory Updates.
[23] CCPS "Guidelines for Process Equipment Reliability Data with Data Tables," 2010.
[24] OREDA Handbook 2015, 6th edition – Volume I and II.
[25] IEEE-500-1984 - IEEE “Guide To The Collection And Presentation Of Electrical, Electronic, Sensing Component, And Mechanical Equipment Reliability Data for Nuclear-Power Generating Stations”.
[26] SINTEF “Reliability Data for Safety Instrumented Systems,” 2010.
[27] SINTEF “Reliability Data for Control and Safety Systems,” 1998.

13TH GLOBAL CONGRESS ON PROCESS SAFETY

NEW GUIDELINES FOR ASSET (MECHANIAL) INTEGRITY MANAGEMENT

Eric Freiburger, P.E.
Praxair, Inc.
Danbury, Connecticut
eric_freiburger@praxair.com

Robert W. Johnson
Unwin Company
Columbus, Ohio
rjohnson@unwin-co.com

Michael Broadribb
Baker Engineering and Risk Consultants, Inc.
Genoa, Nevada
mbroadribb@bakerrisk.com

John F. Murphy, P.E.
Process Safety Services
Punta Gorda, Florida
hamjfm@embarqmail.com

Prepared for Presentation at
American Institute of Chemical Engineers
2017 Spring Meeting
9th Global Congress on Process Safety
San Antonio, Texas
March 27-29, 2017

UNPUBLISHED

NEW GUIDELINES FOR ASSET (MECHANIAL) INTEGRITY MANAGEMENT

Eric Freiburger, P.E.
Praxair, Inc.
Danbury, Connecticut
eric_freiburger@praxair.com

Robert W. Johnson
Unwin Company
Columbus, Ohio
rjohnson@unwin-co.com

Michael Broadribb
Baker Engineering and Risk Consultants, Inc.
Genoa, Nevada
mbroadribb@bakerrisk.com

John F. Murphy, P.E.
Process Safety Services
Punta Gorda, Florida
hamjfm@embarqmail.com

Keywords: Asset integrity management, mechanical integrity, reliability, damage mechanisms, safety critical, inspection and testing, preventive maintenance, quality management, RAGAGEP

Abstract

Ensuring the initial and ongoing integrity of process equipment, including instrumentation and safety systems, has become known as ***asset integrity management*** or ***AIM***. Asset integrity is a fundamental component of successful, risk-based process safety programs. However, facilities continue to be challenged to develop and maintain successful AIM programs. CCPS' Technical Steering Committee commissioned *Guidelines for Asset Integrity Management* as an update and expansion of the previous CCPS document *Guidelines for Mechanical Integrity Systems*, to continue assisting facilities in meeting this challenge. The change in terms from *mechanical integrity* to *asset integrity* reflects international usage, consistent with the elements in CCPS' more recent *Guidelines for Risk Based Process Safety*, and a recognition that a much broader set of assets needs to be properly designed, installed and maintained at process facilities than those requiring "mechanical" integrity.

In addition, the term *equipment* is often associated with an individual piece of equipment such as a pump, compressor or vessel. The term *asset* better reflects how equipment and its associated instrumentation, utilities and connections function together as a system and the integrity of which must be maintained holistically as a system.

The new, expanded *Guidelines* document includes updated information on failure modes, damage mechanisms, and good practices for managing asset integrity throughout the life cycle of a facility. Guidance and examples are provided for selecting and maintaining critical safety systems. This presentation summarizes the new *Guidelines* and highlights changes from the original document.

1. Introduction

This paper introduces to the process safety community the new Center for Chemical Process Safety book *Guidelines for Asset Integrity Management* (CCPS 2017), emphasizing changes from the earlier CCPS book *Guidelines for Mechanical Integrity Systems* (CCPS 2006). The major changes include terminology (e.g., asset integrity vs. mechanical integrity, systems vs. equipment), safety-critical systems, definition of recognized and generally accepted good engineering practice (RAGAGEP), metrics, and other asset management tools. The new, expanded Guidelines document also includes updated information on failure modes and damage mechanisms as well as life cycle considerations. Some of the key changes are discussed in the following sections, including some excerpts from the AIM Guidelines. The Appendix to this paper gives a side-by-side comparison of the 2006 and 2017 *Guidelines* tables of contents.

2. Background and Terminology

Asset integrity management (AIM) is a management system for ensuring the integrity of assets throughout the life cycle of the assets. In this context, an *asset* is a process or facility that is involved in the use, storage, manufacturing, handling or transport of chemicals, or the equipment comprising such a process or facility. Examples of assets include off-shore and on-shore extraction and processing equipment; process and auxiliary tanks, vessels and piping systems including their internal components; control systems; safety systems; buildings and other structures; and transport containers. The selection of which assets are "important" is discussed in Section 1.3 and in Chapter 5 of the AIM Guidelines.

AIM is a product of many activities, usually performed by many people. When these activities are done well, AIM can provide the foundation for a safe, reliable facility that minimizes threats to the workforce, the public and the environment. Effective AIM is also consistent with good business practices.

Mechanical integrity (MI) is a term found in the U.S. Occupational Safety and Health Administration (OSHA) Process Safety Management Standard, 29 Code of Federal Regulations 1910.119. The OSHA definition of mechanical integrity is primarily concerned with the loss of containment of hazardous materials as defined by the standard. The CCPS asset integrity definition includes the concern of loss of containment but also the broader concern of suitability for its intendent purpose throughout its life cycle.

Integrity needs to be built into a facility's assets and planned operation before the facility is started up. This asset integrity is then maintained over time by proper facility operation; by inspection, testing, and preventive maintenance (ITPM) activities; by managing facility changes; and by learning from experience. When failures do occur or deficiencies are detected, integrity is re-established by proper repairs or replacements. Thus, managing asset integrity begins well before a new process/facility is started up and extends throughout the facility lifetime until final decommissioning. It recognizes that the functional integrity of equipment can degrade over time by corrosion, erosion, fatigue and various other mechanisms (i.e., "aging plants"), so these mechanisms need to be understood, detected, and corrected before containment, control, or the ability to respond to an abnormal or emergency situation is lost.

3. Safety-Critical Systems

The underlying philosophy of managing some assets within an AIM program differently than others, considering them to be "critical assets," is generally a recognition of the realities that resources are limited and the best maintenance planning is thwarted by unforeseen circumstances, resulting in situations that may not allow all inspection, testing, preventive maintenance and deficiency management tasks to be done on time and according to plan. Even so, because some assets are critical to the operation of a process, the priority on the maintenance of critical equipment needs to be elevated such that, if not all AIM tasks can be accomplished on time for whatever reason, maintenance of the critical equipment will not be deferred. Although personnel safety, public safety and environmental protection is the primary AIM focus from a process safety perspective, business considerations might also be important when considering equipment criticality.

While initially the main focus of equipment criticality ranking was on production and quality issues, increasingly plants have also used an equipment ranking approach for occupational safety and environmental performance (emissions, ISO 14001, regulatory compliance). More recently, some companies in the process industries have started to apply criticality ranking for process safety purposes, including regulatory compliance (in Europe), and to make decisions regarding which ITPM tasks may be rescheduled and which must be done on the specified date.

In the case of a criticality determination for process safety reasons, equipment falling into this prioritized realm has been called ***safety-critical equipment (SCE)*** or ***safety-critical elements*** *(also abbreviated* ***SCE****)*. The term "safety-critical" does not fully reflect the objective of identifying and managing critical equipment. The term ***major-incident critical*** would be more appropriate. There are not one but two basic ways to define these terms. Suggested definitions given in the *Guidelines* are as follows:

Safety-critical equipment/element
(consequence-based definition)

Equipment, the malfunction or failure of which could cause or contribute to a major incident, or the purpose of which is to prevent a major incident or mitigate its effects

Safety-critical equipment/element
(risk-based definition)

Equipment, the malfunction or failure of which is likely to cause or contribute to a major incident, or the purpose of which is to prevent a major incident or mitigate its effects

Using the risk-based definition to illustrate this concept, Figure 1 (which is Figure 5-1 in the AIM Guidelines) shows an example flowchart for helping to decide which equipment is safety critical from a risk-based perspective. Suggestions are given in the AIM Guidelines for the several terms needing to be defined by the user in order to use this figure, as noted in the caption.

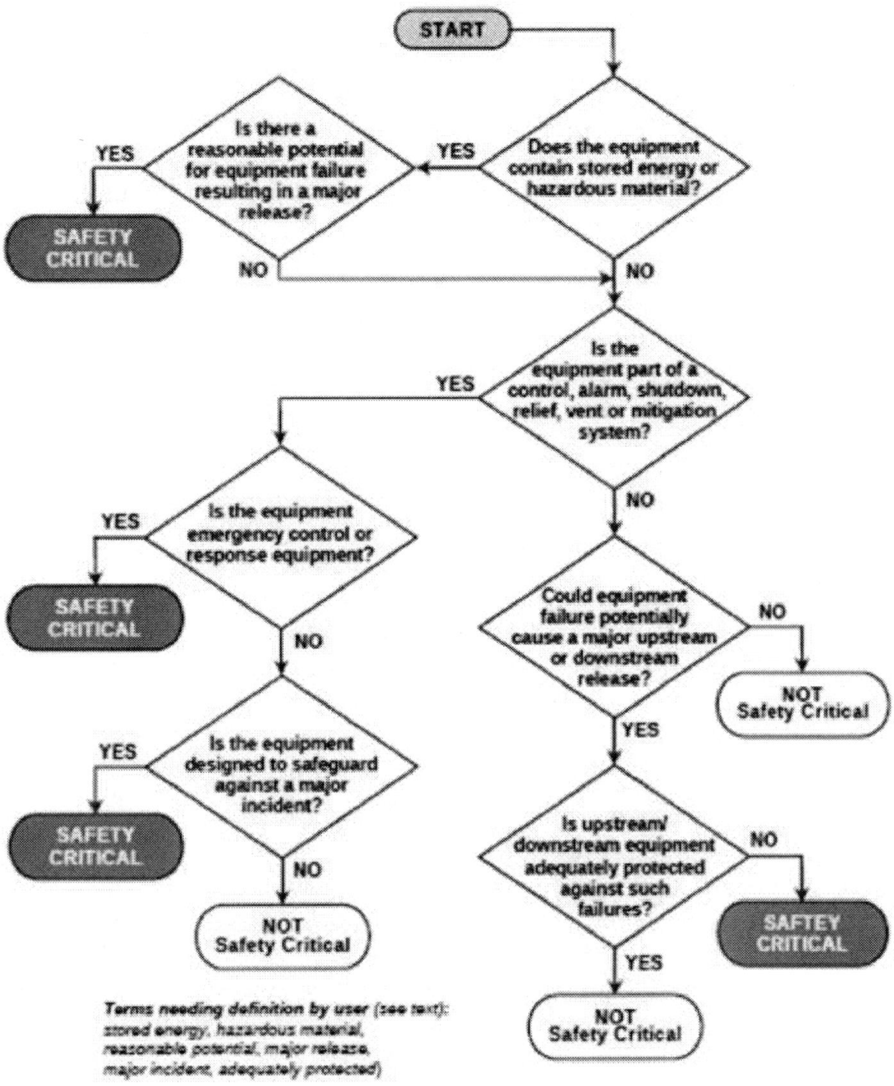

Figure 1. Risk-based determination of safety-critical equipment (CCPS 2017, Figure 5-1)

4. Recognized and Generally Accepted Good Engineering Practice (RAGAGEP)

Codes, standards, and practices, which are sometimes termed "recognized and generally accepted good engineering practice" (RAGAGEP), are an important resource for an AIM program. RAGAGEP stems from the selection and application of appropriate engineering, operating, and maintenance knowledge when designing, operating, and maintaining process facilities with the purpose of ensuring safety and preventing process safety incidents. RAGAGEP involves the application of engineering, operating or maintenance activities derived from engineering knowledge and industry experience based upon the evaluation and analyses of appropriate internal and external standards.

Consensus RAGAGEP Definition. The new CCPS definition of RAGAGEP (presented below) was prepared by the CCPS Asset Integrity Management (AIM) Subcommittee. The subcommittee openly solicited additional input from OSHA, ACC, EPA, API, CGA, and others at the 2014 Global Congress on Process Safety, so that a wide range of stakeholders would have a chance for input to the AIM subcommittee. The final subcommittee definition drew from many published articles, case studies, and reviews of legislation and regulation and represents a substantial review of many years of industry and regulatory experience. The subcommittee definition also included legal review by several companies and the American Chemistry Council to ensure its long-term validity with regulations, legislation, and pending cases. One of the more important revisions to the definition is the inclusion of appropriate internal company guidance into a company's derivation of a selected practice.

The revised definition in the new AIM Guidelines book is now as follows. It can also be found in the CCPS on-line Process Safety Glossary (https://www.aiche.org/ccps/resources/glossary).

> *"Recognized and generally accepted good engineering practice", a term originally used by OSHA, stems from the selection and application of appropriate engineering, operating and maintenance knowledge when designing, operating and maintaining chemical facilities with the purpose of ensuring safety and preventing process safety incidents.*
>
> *It involves the application of engineering, operating or maintenance activities derived from engineering knowledge and industry experience based upon the evaluation and analyses of appropriate internal and external standards, applicable codes, technical reports, guidance, or recommended practices or documents of a similar nature. RAGAGEP can be derived from singular or multiple sources and will vary based upon individual facility processes, materials, service, and other engineering considerations.*

5. Metrics, Audits and Continuous Improvement: Learning from Experience

While much of the AIM efforts at a facility are generally directed at developing and implementing the AIM program, successful facilities also have an objective and desire to learn from experience and continuously improve the program. This often takes the form of the following activities.

Establishing Performance Measurement Systems. Effective performance measures can help organizations evaluate the ongoing performance of the overall AIM program and key AIM program activities such as compliance with the ITPM task schedule. The metrics can include direct measurement of AIM program performance as well as leading indicators (predictors) of program performance.

Conducting Periodic Audits and Assessments of Program Activities. Audits and assessments are typically used to evaluate how the management systems of the AIM program are functioning, such as the ITPM program and the quality management system, and how they satisfy relevant requirements such as process safety regulations. Periodic AIM program audits are generally a mandatory requirement in process safety regulations.

Learning Lessons from Equipment Failures, Process Safety Incidents and Near Misses. Another improvement activity is to systematically evaluate equipment failures, including near misses and excursions outside of integrity operating windows, using structured evaluation processes such as root cause analysis (RCA). Using systematic analyses provides an effective means to identify the contributing and root causes of failures. From the results of these processes, personnel can make recommendations for eliminating underlying causes, then management can develop and implement appropriate actions to eliminate or reduce the likelihood of future failures and excursions.

Each of these means of learning from experience are discussed in the AIM Guidelines document.

6. Other Asset Management Tools

In recent years, facilities have applied risk-based analytical techniques in an effort to develop more performance-based AIM programs. Some inspection standards, such as American Petroleum Institute (API) 510 and API 570, now include provisions for determining inspection requirements based on risk. Also, API and other organizations have developed standards and recommended practices that encourage the use of risk-based techniques to define inspection and testing requirements, including:

- API Recommended Practice 580, *Risk-Based Inspection*
- API Recommended Practice 581, *Risk-Based Inspection Technology*
- IEC 61511, *Functional Safety: Safety Instrumented Systems for the Process Industry Sector - Part 1: Framework, Definitions, System, Hardware and Software Requirements*. Also available as ANSI/ISA-84.00.01 Part 1 (IEC 61511-1 Mod).

Many organizations find that using risk analysis techniques to aid in the management of process safety provides several benefits, such as:

- Assurance that a structured, systematic and technically defensible approach is used to make decisions
- Improved knowledge of the system operation and the cause/effect relationships that result from specific asset failures
- Confidence that AIM resources are being focused on the most important failures by explicitly assessing the risk and then assigning resources to those areas for which they will be the most effective.

This chapter summarizes some of the important attributes of risk-based analytical techniques. Various analytical techniques can assist decision making in the AIM program. However, facilities need to recognize when these techniques are best applied, what decisions can be made with the results, and other related issues using the techniques such as timing, advantages and resources. In general, the following are typical uses of analytical techniques in an AIM program:

- ***FMEA/FMECA*** to identify and prioritize potential equipment failure modes that need to be addressed by ITPM tasks
- ***RCM*** to optimize proactive maintenance tasks such as predictive maintenance, preventive maintenance, and failure-finding tasks; usually applied to functional failures of active components (e.g., pumps failing off, erratic control)
- ***RBI*** to optimize inspection tasks and frequencies for fixed equipment (e.g., vessels, tanks, and piping) and pressure relief devices
- ***LOPA*** and similar analysis approaches to define performance requirements for independent protection layers (IPLs), including safety instrumented functions (SIFs)
- ***Fault Tree Analysis and Markov Analyses*** to verify that SIF and IPL designs meet performance requirements (i.e., targeted PFD).

Incorporating Risk into AIM Decisions. A primary objective of an AIM program is to reduce unreliable performance resulting from asset failures. An AIM program can accomplish this by identifying and prescribing tasks to prevent asset failures, detecting the onset of asset failures, and/or discovering hidden asset failures before they affect system performance, safety and/or the environment. Risk-based analytical techniques can be used to assess unreliable performance by identifying potential loss exposures and then estimating their risks. Based on this analysis, facility staff can determine the most effective AIM tasks for reducing those loss events.

Risk is the combination of the severity of consequences and the frequency (likelihood) of occurrence of an undesired event. Consequences may include safety, health, economic, environmental and/or other types of loss and harm impacts. The likelihood aspect addresses the frequency of a single occurrence of the undesired event (e.g., asset failure) or the average frequency of occurrence of the undesired event. For example, an equipment item that fails once a year causing $100,000 in lost production each time it fails has the same risk (annualized loss rate) as another equipment item that fails 10 times a year causing $10,000 in lost production each time it fails. Every year, the failures of each item results in $100,000 in lost production. The economic risks associated with operating these two equipment items are equivalent. Using this example, the risk of operating the two equipment items is measured as an annual cost (dollars per year). This example expresses risk in terms of dollars to illustrate the calculation of risk. Other adverse effects (e.g., safety, environmental) can be measured with units such as injuries per year, affected area of fires per year, or mass of a chemical released per year.

Understanding system risk can help a facility develop a performance-based AIM strategy. Identifying higher-risk areas provides better opportunities for risk reduction. Priorities can be established by comparing systems, system functions and component failure modes. Also, risk characterization tools can be used during AIM task development to predict the effectiveness of the tasks being assigned. When assigning tasks, the objective is to reduce the risk of a failure to a tolerable risk threshold, which can be considered as "balancing the risk." Risk is considered to be in a balanced state when the assigned maintenance tasks are effective in reducing the risk to a tolerable level. A tolerable-risk threshold is known as a *risk tolerance criterion*—the rate of loss that decision makers are willing to tolerate for a given consequence. This tolerable rate of loss will vary with many factors, including the type of loss, the magnitude of loss, and the individuals or groups affected by the loss. Risks that exceed the risk tolerance criterion require actions to lower the risk.

Reliability-Centered Maintenance. Unlike the four approaches outlined in Chapter 7 (code/standard, regulatory, company-specific, and RBI approaches) that are used to develop AIM inspection and test plans, Reliability-Centered Maintenance (RCM) is an established technique that is used for defining maintenance tasks, including ITPM tasks. RCM deals only with specific preventive maintenance and inspection tasks for an asset and may not include sufficient detail to be called a true inspection plan that meets code/standard or regulatory requirements.

Originally developed for the aviation industry and later expanded to other industries, RCM can be used in an AIM program to evaluate critical and complex systems to determine (1) which potential failures pose the highest risk and (2) what ITPM tasks and task frequencies are needed to address the potential failures; i.e., what is best strategy for managing the potential failures. It is a useful tool to determine which tasks are important to meet the operating context of an asset (i.e., how the asset will be employed and what its intended function will be). The RCM method is most commonly applied to an asset with a goal of increasing equipment availability (uptime) rather than meeting code/standard, regulatory or corporate requirements from an AIM perspective.

RCM provides a systematic approach for identifying potential failures that can affect process or system performance. The characteristics of the potential failures are then evaluated to determine (1) appropriate maintenance tasks or (2) potential design or operational changes (referred to as *one-time changes* in RCM).

Protection Layer Analysis Techniques. As defined in IEC 61511, *Functional Safety: Safety Instrumented Systems for the Process Industry Sector - Part 1: Framework, Definitions, System, Hardware and Software Requirements*, a safety instrumented system (SIS) is a system composed of sensors, logic solvers, and final control elements for the purpose of taking the process to a safe state when predetermined conditions are violated. (Other terms used for SIS include emergency shutdown system, safety shutdown system, and safety interlock system.) IEC 61511 also defines a safety life cycle that includes the application of two analyses:

1. ***Layer of Protection Analysis (LOPA)*** or an alternate tool to determine if additional independent protection layers (IPLs), which may include one or more safety instrumented functions (SIFs), are needed to achieve tolerable risk levels and, if tolerable risk is not achieved, to define the required risk reduction. Other techniques including Hazard and Operability (HAZOP) Studies can be used for this purpose.
2. ***Reliability analysis*** (unavailability analysis) to determine the protection layer configuration (including redundancy, if required) and functional test frequency needed to achieve the required risk reduction for SIFs in terms of the overall probability of failure on demand

(PFD), based on specified system components and their failure rates. Tools include Fault Tree Analysis (FTA), simplified equations, and Markov analysis.

Asset Failure And Root Cause Analyses. *Failure analysis* is designed to help personnel understand how a failure occurred and to determine and implement needed corrections and improvements to assets. These corrections typically focus on redesigning the asset item, altering its ITPM plan and/or changing the process conditions. This analysis may result in (1) changes that affect only the asset item that failed or (2) changes that affect the failed asset item as well as similar asset items.

Usually, failure analysis is based on forensic engineering principles. On the other hand, *root cause analysis* (RCA) is designed to discover why the failure occurred so that corrections and improvements can be made to the management systems that promote asset integrity (e.g., AIM program, operating procedures, engineering design practices, personnel training). Because RCA focuses on changes to management systems, the results tend to affect the asset item that failed, similar assets, and seemingly unrelated assets that are managed by the same management systems. For example, if a pressure vessel leaks because of corrosion and RCA finds that nondestructive testing (NDT) activities were not being performed as planned, the RCA might result in improvements in the ITPM program that not only affect pressure vessel inspections, but also instrument calibration, rotating equipment lubrication and so forth.

7. AIM Activity Matrices

The AIM Guidelines incorporates guidance on AIM activities for specific equipment types in Chapter 12. The chapter discusses codes/standards/practices for each equipment type, then includes in an appendix four-page activity matrices for specific equipment types, such as pumps and relief valves. Shown below are the first two of the four pages for one specific equipment type; namely, safety instrumented systems and emergency shutdowns. The first page gives activities for the design, fabrication and installation of new equipment. The second page gives inspection and testing activities. (The third and fourth pages give preventive maintenance and repair activities, respectively.) Chapter 12 has a total of 96 pages of equipment-specific AIM activities guidance.

New Equipment Activities for SISs and Emergency Shutdowns (from CCPS 2017)

New Equipment Design, Fabrication and Installation		
Example activities and typical frequencies	**Activity**	**Frequency**
	Identification of materials of construction	When received
	Calibration of field devices (e.g., transmitter, switches)	Before or at initial installation
	Visual inspection of field devices and installation	Initial installation
	SIS loop check	Initial installation
	SIS functional test	Initial installation
	Manufacturer tests of logic solver, operator interface and engineering interface	During fabrication
	Factory acceptance tests of logic solver, operator interface and engineering interface	Before delivery and confirmation tests after delivery
Technical basis for activity and frequency	• API RP-554, API RP-551 • Manufacturers' recommendations • Industrial insurers' recommendations • Common industry practices	
Sources of acceptance criteria	• Device specifications • SIS specifications • Manufacturers' recommendations • Industrial insurers' recommendations • Company engineering and/or maintenance standards	
Typical failures of interest	• Leakage resulting from improper installation or materials of construction (e.g., incorrect gasket, incorrect metallurgy of wetted parts) • Failure to operate on demand caused by improper installation (e.g., incorrectly wired), incorrect configuration of control system and/or incorrect calibration of the device • Potential ignition source or electrical shock due to improper installation • Improper rating for the electrical classification of the installed location • Improper management of intervening valves between the instrument and the equipment being protected	
Personnel qualifications	• Craft skills and knowledge required by the individual procedures • Training on the use and operation of special tools (e.g., signal simulators) required by procedures	
Procedure requirements	• Procurement and receiving procedures to ensure proper materials of construction • Device-specific testing, calibration and installation procedures • Manufacturers' manuals • Special tool (e.g., signal simulator) use and operation procedures	
Documentation requirements	• Vendor material of construction reports • Calibration record • Loop check sheet • Functional test record • Manufacturers' test report • Factory acceptance test report • Facility acceptance test report	Special notes: • Records retained for the life of the equipment • Installation documentation to support PSSR requirements

Inspection and Testing Activities for SISs and Emergency Shutdowns (from CCPS 2017)

Inspection and Testing		
Example activities and typical frequencies	**Activity**	**Frequency**
	• Calibration of field devices (e.g., transmitters, switches) • Loop check • Functional test • Checking/running of logic solver diagnostics	As specified to meet the safety performance requirements
Technical basis for activity and frequency	• API RP-554 • API RP-551 • IEC 61508 • Manufacturers' recommendations • Industrial insurers' recommendations • Common industry practices	
Sources of acceptance criteria	• Device specifications • SIS specifications • Manufacturers' recommendations • Industrial insurers' recommendations • Company engineering and/or maintenance standards	
Typical failures of interest	• Failure to operate on demand or spurious trip of system as a result of wiring failure (e.g., loose connection, short), failure of input device (e.g., sensor electronic failure), failure of controller/local solver (e.g., I/O card failure), or unauthorized change to controller/logic solver configuration and/or bypassing/forcing of the interlock/alarm • Failure to operate on demand caused by isolation/plugging of input device connection or other process conditions (e.g., buildup on a temperature probe) that render input device inoperable or incapable of accurately measuring process conditions (e.g., pressure, temperature) • Potential ignition source if the device or connection shorts • Leakage at process connection as a result of overpressurization of the joint • Improper management of intervening valves between the instrument and the equipment being protected	
Personnel qualifications	• Craft skills and knowledge required by the individual procedures • Training on the specific procedures for the inspection and testing activities • Training on the use and operation of special tools (e.g., signal simulators) required by procedures	
Procedure requirements	Written procedures describing the inspection or test activity including: • The manner, the extent, the location and the timing for the inspection or test and by whom • The documentation and analysis of results • The resolution of functions or conditions not meeting acceptance criteria	
Documentation requirements	• Calibration record • Loop check sheet • Functional test record • Diagnostic check sheet	Special notes: • Records retained long enough to establish trends • Documentation by exception may be acceptable (for certain tasks and selected electric motors)

8. Conclusions

As with all CCPS projects, this book was a team effort. It is hoped that these new AIM Guidelines will prove useful to the process safety community for continuous improvement of asset integrity and reliability throughout the life cycle of process facilities.

Contributors

The chair of the Asset Integrity Management Subcommittee was Eric Freiburger of Praxair. Mike Broadribb of Baker Risk was co-chair. The CCPS staff consultant was John F. Murphy. The AIM Subcommittee had the following contributing members:

Andrew Basler, Mallinckrodt
Matt Hedlund, Eastman Chemical
Kevin Blackwell, Honeywell
Tom Sandbrook, Chemours
Russ Davis, Mistras Group
John Traynor, Evonik
Jonas Duarte, Chemtura
Chris Urbanowich, Petrobras

Robert W. Johnson of Unwin Company prepared this document under contract with CCPS, with input from Steven W. Rudy.

Robert W. Ormsby was the CCPS staff consultant at the beginning of this project, and Tom Folk prepared the detailed outline for the AIM Subcommittee. ABSG Consulting Inc. prepared CCPS' precursor publication, *Guidelines for Mechanical Integrity Systems.*

Insights and suggestions from the peer reviewers helped ensure a balanced perspective:

Robert A. Bartlett, Pareto Engineering & Management Consulting
James M. Broadribb, Wood Group
Robert A. Bartlett, Pareto Engineering & Management Consulting
Hugh Hemphill, Chevron
Mark Jackson, FM Global
Adrian Sepeda, Process Safety & Risk Management
Razzack Syed, Praxair Inc.
David Thaman, PPG
Terry A. Waldrop, AIG

Please address any questions on this paper to the authors.

References

CCPS 2006. Center for Chemical Process Safety, *Guidelines for Mechanical Integrity Systems*, New York: American Institute of Chemical Engineers.

CCPS 2017. Center for Chemical Process Safety, *Guidelines for Asset Integrity Management*, New York: American Institute of Chemical Engineers.

Appendix: Tables of Contents Comparison *(Notes: Some titles abbreviated. Only first and second level headings shown. Section numbers ending with "A" are chapter appendices.)*

	Guidelines for Asset Integrity Management		***Guidelines for Mechanical Integrity Systems***
			ITEMS ON CD ACCOMPANYING THIS BOOK
			ACRONYMS AND ABBREVIATIONS
	PREFACE		GLOSSARY
	ACKNOWLEDGEMENTS		ACKNOWLEDGEMENTS
	FILES ON THE WEB		PREFACE
			MANAGEMENT OVERVIEW OF THE GUIDELINES
1	**INTRODUCTION**	**1**	**INTRODUCTION**
1.1	Background and Scope		
1.2	What is Asset Integrity Management?	1.1	What is Mechanical Integrity?
1.3	What Assets are Included?		
1.4	AIM Life Cycle		
1.5	Relationship to Other Programs	1.2	Relationship to Other Programs
		1.3	Expectations for the MI Program
1.6	Relationship to RAGAGEP	1.4	The Effect of RAGAGEPs
1.7	Structure of this Document	1.5	Structure of this Guidelines Book
2	**MANAGEMENT RESPONSIBILITY**	**2**	**MANAGEMENT RESPONSIBILITY**
2.1	Leadership Roles and Responsibilities	2.1	Facility Leadership's Roles and Responsibilities
2.2	Technical Assurance Responsibilities	2.2	Technical Assurance Responsibilities
3	**AIM LIFE CYCLE**		
3.1	Overview		
3.2	Research Through Process Development		
3.3	Process Design		
3.4	Engineering, Procurement and Construction		
3.5	Commissioning		
3.6	Operation and Maintenance		
3.7	Decommissioning		
3.8	RAGAGEP Selection & Application at Each Stage		
3A	Design Review Suggestions		
4	**FAILURE MODES AND MECHANISMS**		
4.1	Introduction		
4.2	Equipment Functions and Functional Failure		
4.3	Failure Modes		
4.4	Damage Mechanisms		
4.5	Failure Effects		
4.6	Risk		
4.7	Analysis		
4.8	ITPM Task Assignments		
4.9	Operational Issues		
4.1	Other Related Activities		
4A	Risk Concepts Related to AIM		
5	**ASSET SELECTION, CRITICALITY DETERMINATION**	**3**	**EQUIPMENT SELECTION**
5.1	Program Objectives and Philosophy	3.1	Reviewing Program Objectives
5.2	Asset Selection Criteria and Principles	3.2	Establishing Equipment Selection Criteria
5.3	Level of Detail	3.3	Defining Level of Detail
5.4	Asset Criticality Determination		
5.5	Documentation	3.4	Documenting the Equipment Selection
5.6	Roles and Responsibilities	3.5	Equipment Selection Roles and Responsibilities
5A	Sample Guidelines for Selecting Program Assets	3A	Sample Guidelines for Selecting MI Equipment

Guidelines for Asset Integrity Management		*Guidelines for Mechanical Integrity Systems*	
6	**INSPECTION, TESTING AND PREVENTIVE MAINT**	**4**	**INSPECTION, TESTING AND PREVENTIVE MAINT**
6.1	ITPM Task Planning	4.1	ITPM Task Planning
6.2	ITPM Task Execution and Monitoring	4.2	Task Execution and Monitoring
6.3	ITPM Program Roles and Responsibilities	4.3	ITPM Program Roles and Responsibilities
6A	PdM, NDT Techniques for Mechanical Equip	4A	Common PdM and NDT Techniques
7	**APPROACHES FOR DEVELOPING ITPM PLANS**		
7.1	Code/Standard Approaches		
7.2	Regulatory Authority Approaches		
7.3	Company-Specific Approaches		
7.4	Risk-Based Inspection (RBI)		
7.5	FMECA Approaches		
7.6	Safety Instrumented Systems		
8	**AIM TRAINING, PERFORMANCE ASSURANCE**	**5**	**MI TRAINING PROGRAM**
8.1	Skills and Knowledge Assessment	5.1	Skills/Knowledge Assessment
8.2	Training For New and Current Workers	5.2	Training For New and Current Workers
8.3	Verification, Document'n of Performance Assurance	5.3	Verification, Documentation of Training
8.4	Certifications	5.4	Certification
8.5	Ongoing and Refresher Training	5.5	Ongoing and Refresher Training
8.6	Training Techs, Operators Performing Maint Tasks		
8.7	Training for Technical Personnel	5.6	Training for Technical Personnel
8.8	Contractor Issues	5.7	Contractor Issues
8.9	Roles and Responsibilities	5.8	Roles and Responsibilities
8A	Sample Training Survey	5A	Sample Training Survey
8B	Sample Training Guide		
9	**ASSET INTEGRITY PROCEDURES**	**6**	**MI PROGRAM PROCEDURES**
9.1	Procedure Types Supporting the AIM Program	6.1	Procedure Types Supporting the MI Program
9.2	Identification of Needs	6.2	Identification of MI Procedure Needs
9.3	Procedure Development Process	6.3	Procedure Development Process
9.4	Format and Content	6.4	MI Procedure Format and Content
9.5	Other Sources of AIM Procedures	6.5	Other Sources of MI Procedures
9.6	Implementing and Maintaining AIM Procedures	6.6	Implementing and Maintaining MI Procedures
9.7	AIM Procedure Program Roles, Responsibilities	6.7	Procedure Program Roles and Responsibilities
9A	Example AIM Procedure		
10	**QUALITY MANAGEMENT**	**7**	**QUALITY ASSURANCE**
10.1	Design	7.1	Design
10.2	Procurement	7.2	Procurement
10.3	Fabrication	7.3	Fabrication
10.4	Receiving	7.4	Receiving
10.5	Storage and Retrieval	7.5	Storage and Retrieval
10.6	Construction and Installation	7.6	Construction and Installation
10.7	In-service Repairs, Alterations and Rerating	7.7	In-service Repairs, Alterations, and Rerating
10.8	Temporary Installations and Temporary Repairs	7.8	Temporary Installations and Temporary Repairs
10.9	Decommissioning / Re-use	7.9	Decommissioning/Reuse
10.1	Used Assets	7.1	Used Equipment
10.11	Spare Parts	7.11	Spare Parts
10.12	Contractor-Supplied Assets and Materials	7.12	Contractor-Supplied Equipment and Materials
10.13	QA Program Roles and Responsibilities	7.13	QA Program Roles and Responsibilities
		7A	Design Review Suggestions
10A	Sample Vendor QA Plan	7B	Sample Vendor QA Plan
10B	Positive Material Identification	7C	Positive Material Identification
10C	Sample Service Contractor QA Plan	7D	Sample Service QA Plan

Guidelines for Asset Integrity Management		*Guidelines for Mechanical Integrity Systems*	
11	**EQUIPMENT DEFICIENCY MANAGEMENT**	**8**	**EQUIPMENT DEFICIENCY MANAGEMENT**
11.1	Equipment Deficiency Management Process	8.1	Equipment Deficiency Management Process
11.2	Acceptance Criteria	8.2	Acceptance Criteria
11.3	Equipment Deficiency Identification	8.3	Equipment Deficiency Identification
11.4	Responding to Equipment Deficiencies	8.4	Responding to Equipment Deficiencies
11.5	Equipment Deficiency Communication	8.5	Equipment Deficiency Communication
11.6	Tracking of Temporary Repairs	8.6	Permanent Correction of Equipment Deficiencies
11.7	Deficiency Management Roles and Responsibilities	8.7	Deficiency Management Roles and Responsibilities
		8A	Fitness for Service
12	**EQUIPMENT-SPECIFIC INTEGRITY MANAGEMENT**	**9**	**EQUIPMENT-SPECIFIC INTEGRITY MANAGEMENT**
12.1	Vessels, Tanks and Piping	9.1	Fixed Equipment
12.2	Relief and Vent Systems	9.2	Relief and Vent Systems
12.3	Instrumentation and Controls	9.3	Instrumentation and Controls
12.4	Rotating Equipment	9.4	Rotating Equipment
12.5	Fired Equipment	9.5	Fired Equipment
12.6	Electrical Systems	9.6	Electrical Systems
12.7	Fire Protection and Suppression Systems	9.7	Fire Protection Systems
12.8	Ventilation and Purge Systems	9.8	Miscellaneous Equipment
12.9	Protective Systems		
12.1	Passive Mitigation Systems		
12.11	Solids-Handling Systems		
12.12	Refrigeration Systems		
12.13	Utilities		
12.14	Safety Equipment		
12A	Asset Integrity Activities by Equipment Type	9.9	Equipment-specific MI Activity Matrices
13	**AIM PROGRAM IMPLEMENTATION**	**10**	**MI PROGRAM IMPLEMENTATION**
13.1	Budgeting and Resources	10.1	Budgeting and Resources
13.2	Use of Data Management Systems	10.2	Use of Software in MI Programs
13.3	AIM Benefits and Return on Investment	10.3	Return on Investment (ROI)
13A	AIM Program Design Activity Worksheets		
14	**METRICS, AUDITS, CONTINUOUS IMPROVEMENT**		
14.1	Performance Measurement and Monitoring		
14.2	AIM Program and Implementation Audits		
14.3	Continuous Improvement		
14A	AIM-Related Regulatory Citations		
15	**OTHER ASSET MANAGEMENT TOOLS**	**11**	**RISK MANAGMENT TOOLS**
15.1	Introduction to Risk-based Analytical Techniques	11.1	Introduction to Risk-based Analytical Techniques
15.2	Incorporating Risk into AIM Decisions	11.2	Incorporating Risk into MI Decisions
		11.3	FMEA/FMECA
15.3	Reliability-Centered Maintenance	11.4	RCM
		11.5	Risk-based Inspection
15.4	Protection Layer Analysis Techniques	11.6	Protection Layer Analysis Techniques
15.5	Asset Failure and Root Cause Analyses		
	ACRONYMS AND ABBREVIATIONS		
	GLOSSARY		
	INDEX		INDEX

Fracture Toughness and Brittle Failure – A Pressure Vessel Case Study

John Puryear, PE
ABSG Consulting Inc. ("ABS Consulting")
140 Heimer Rd. San Antonio, TX
jpuryear@absconsulting.com

Guillermo Ramirez, PE PhD
gramirez@absconsulting.com

Clint Botard
cbotard@absconsulting.com

Kollin Kenady
kkenady@absconsulting.com

Prepared for Presentation at
American Institute of Chemical Engineers
2017 Spring Meeting and 13th Global Congress on Process Safety
San Antonio, TX
March 26 – 29, 2017

Fracture Toughness and Brittle Failure – A Pressure Vessel Case Study

John Puryear, PE
ABS Consulting
140 Heimer Rd. San Antonio, TX
jpuryear@absconsulting.com

Guillermo Ramirez, PE PhD
gramirez@absconsulting.com

Clint Botard
cbotard@absconsulting.com

Kollin Kenady
kkenady@absconsulting.com

Keywords: pressure vessel, fracture toughness, normalization, stress relief, heat treatment

Abstract

That the manufacturing and fabrication process can introduce low fracture toughness and cause brittle failure in steel has been documented in well-known studies. These include failure of Liberty ships during World War II and, more recently, weld failures in steel moment frames during the 1994 Northridge earthquake. In both of these cases, the manufacturing and fabrication process introduced stress states that reduced fracture toughness and caused brittle failures.

Control of the manufacturing and fabrication process to maintain sufficient fracture toughness remains a challenge for the Oil & Gas Industry. In this paper, we will review an example of a brittle failure of a pressure vessel. The head of the pressure vessel, which was operating as a low-pressure separator, detached about its circumference at a pressure much less than the vessel's maximum operating pressure. A root cause analysis of the incident identified the performance gaps and root causes from the vessel's service conditions, manufacture, and fabrication that combined to cause the brittle failure.

This paper examines the performance gaps that lead to the failure and their root causes. The effect of the root causes on the vessel's mechanical properties is discussed. Further, the performance gaps are related to material and fabrication guidelines in the *ASME Boiler & Pressure Vessel Code (BPVC)*. Finally, recommendations for correcting the performance gaps are offered.

1 Introduction

That the manufacturing and fabrication process can introduce low fracture toughness and cause brittle failure in steel has been documented in well-known studies. One example is failure of the Liberty ships during World War II. The hulls of several early ships broke apart in high seas. Subsequent investigation found one factor to be the weld process introducing residual stresses that reduced hull weld's fracture toughness. When this factor was combined with others, such as low temperatures in the North Atlantic, the consequence was brittle, catastrophic failure [1].

A second example is damage to buildings during the 1994 Northridge earthquakes in California. Post-event investigation found instances of fractured connecting welds in steel moment frames. These fractures were unexpected because structural analysis suggested steel moment frames would perform well in a seismic event, compared to other structural systems. As was observed in fabrication of the Liberty ships, the welding process used to connect moment frames in Northridge introduced residual stresses that reduced the fracture toughness and caused brittle failure [2]. This result led to revision of seismic design guidelines as discussed in *FEMA 356 Prestandard and Commentary for the Seismic Rehabilitation of Buildings* [3].

In a more related example, an investigation lead by Harry McHenry was performed to determine the cause of a pressure vessel burst which resulted in 17 deaths at the Union Oil Plant in 1984 [4]. A catastrophic failure occurred after the vessel had undergone repair. The repair welding process introduced high residual stress and material embrittlement, but a post-weld heat treatment was not performed. Metallography revealed pre-existing cracks which initiated adjacent to the weld repair. The presence of hydrogen from a sour environment and a susceptibility of the material near the welds to hydrogen cracking resulted in the formation of a crack in weak material which ultimately led to the final fracture of the vessel under normal operating conditions.

From these examples, it is clear that controlling the manufacturing and fabrication process to maintain sufficient fracture toughness has been a challenge historically. It remains a challenge to this day, particularly in the Oil & Gas Industry, given its global scope and correspondingly long supply chains. This paper presents a case study of a pressure vessel failure to illustrate how manufacturing and fabrication process affect fracture toughness and can introduce brittle failure.

It is particularly important that fracture toughness be controlled during manufacture and fabrication because, after the unit is installed, operators commonly do not have the means to specifically verify fracture toughness through non-destructive means during routine inspections. Operators that lack third-party verification are particularly vulnerable to inadvertently putting equipment with insufficient fracture toughness into service.

In the motivating, incident for this study, the head of the pressure vessel, which was operating as a separator, detached about its circumference at a pressure that was much less than the vessel's maximum operating pressure. A root cause analysis of the failure determined that factors from the vessel's service conditions, manufacture, and fabrication had combined to cause the brittle failure. Performance gaps in prevention and detection were identified, and root causes of these gaps were determined.

After summarizing the failure incident, this paper examines the effects of the vessel's manufacture and fabrication on its mechanical properties. The vessel's manufacturing process is further related to guidelines in the *ASME Boiler & Pressure Vessel Code (BPVC)*. The performance gaps that led to the failure and their root causes are then discussed. Finally, recommendations for correcting the performance gaps are offered.

2 Pressure Vessel Failure

2.1 Incident Summary

A full circumferential failure occurred in one head of a low-pressure condensate separator resulting in the total release of gas and condensate. The separator had operated offshore in sour service for over 30 years.

Ten months before the failure, the separator was taken out of service. Industry standard internal and external inspections were performed on it. No relevant findings were identified indicating a problem that would have prevented the start of operation. In fact, it was determined that the remaining operating life was at least ten years. Operations were accordingly restarted, and the separator was put back into service two months before the failure.

2.2 Failure Sequence

A discontinuity in the interior surface of the head, near the weld at about the 7 o'clock position (Figure 1), either existed when the separator was delivered or occurred during operation of the separator. Sulfide stress corrosion cracking began at the discontinuity and progressed slowly over years. This corrosion mechanism is consistent with the sour service conditions and age of the separator.

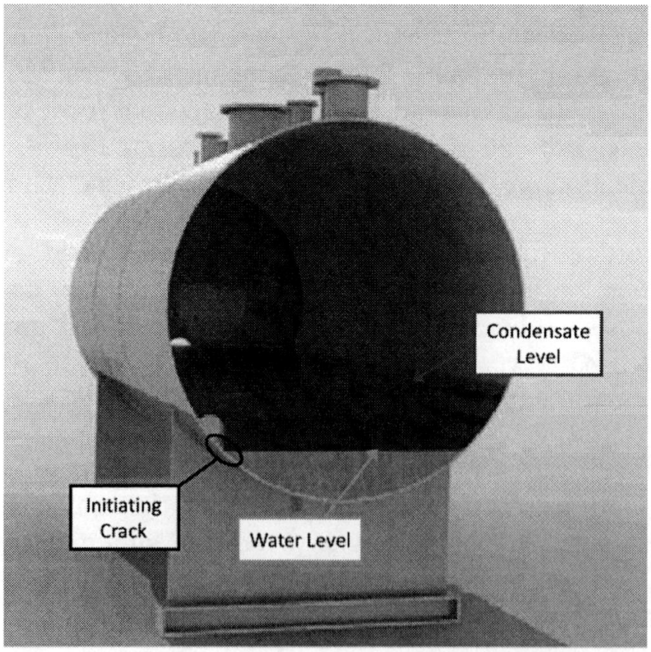

Figure 1. Location of Crack Initiation

The initial discontinuity and the small, tightly closed cracks were not detected by standard inspection methods before, or after, the accident. Only subsequent laboratory examination was able to identify the cracks. Standard inspection methods before and after the accident did identify thinning from general corrosion; nevertheless, the thickness of the head was above the retirement thickness. The cracking progressed through the thickness and caused a small release that was detected by gas detectors. Within minutes, the crack grew to the critical length that caused the complete circumferential failure of the head. The progress of the crack is illustrated in Figure 2.

(a) Initial Crack (b) Crack Pop-Through (c) Circumferential Propagation

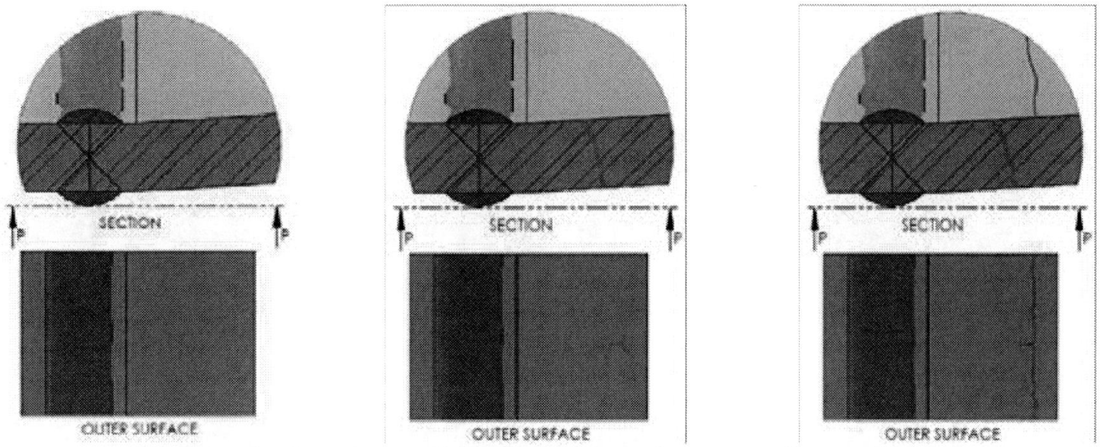

Figure 2. Crack Propagation from Inner to Outer Surface

The integrity performance of the separator—catastrophic failure following development of a through crack—was inadequate in terms of ASME BPVC Section VIII design philosophy. A vessel designed to the BPVC Section VIII is intended to exhibit leak-before-failure performance. Specifically, the vessel's material must be adequate to develop a through crack and leak product in a controlled manner without catastrophic failure. The integrity performance of the separator did not satisfy this requirement.

The failure sequence described above was developed and verified using onsite personnel interviews, line operation data, post-incident evidence gathering, gas detector data, analytical calculations, numerical analysis and laboratory examination and analysis. Output from a simulation of the failure sequence is shown in Figure 3. That simulation was performed using the Arbitrary-Lagrangian Eulerian (ALE) method in LS-DYNA, a multi-physics solver developed by Livermore Software Technology Corporation (LSTC).

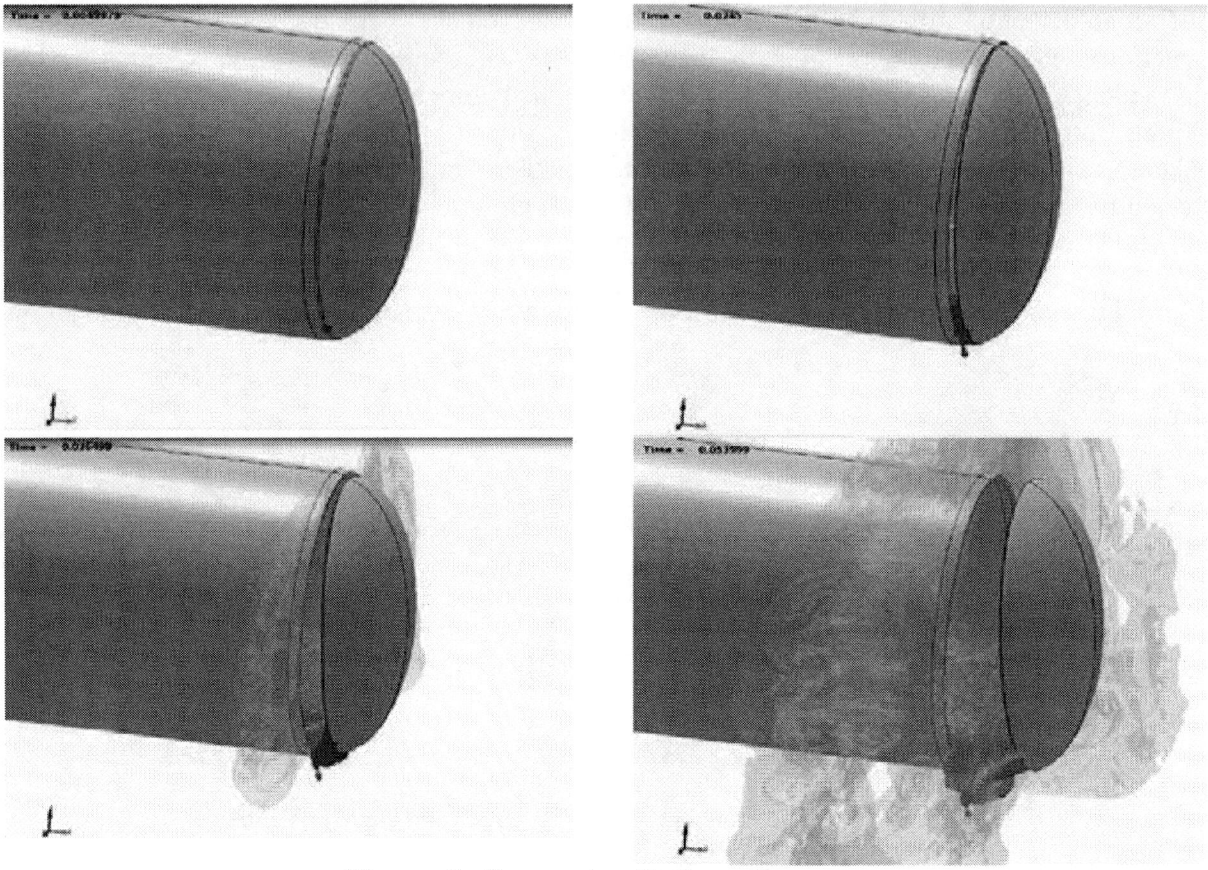

Figure 3. Separator Failure Sequence

3 Manufacturing Process

3.1 Mechanical Properties

Laboratory analyses performed following the incident revealed the effects of the manufacturing process on the separator's mechanical properties with the results showing that the head metal was not normalized steel. Normalization is a heat treatment that produces a more uniform microstructure in the steel; the process is illustrated in Figure 4. In terms of mechanical properties, the mill plate used to manufacture the heads had relatively low fracture toughness and was thus vulnerable to brittle failure. In practical terms, lack of normalization meant that the heads did not have the heat treatment that would reduce the potential for crack growth and propagation.

Figure 4. Normalization Process [5], [6]

The head manufacturing process further increased this vulnerability to brittle failure. The cold working during dishing and flanging (Figure 5) introduced high residual stresses, increased the ultimate strength of the steel, decreased its ductility and reduced its fracture toughness. These observations came from laboratory tests comparing mechanical properties at the apex of the head to properties at the knuckle.

The residual stresses introduced during manufacture remained because of a lack of heat treatment and/or insufficient post-weld heat treatment. These processes are shown in Figure 6 and Figure 7. Residual stresses in the vicinity of the failure were further increased by welding the head to the vessel shell. The reduction in residual stress concentrations due to stress relief and post-weld heat treatment is illustrated in Figure 8. In the figure, T refers to tensile stress, and C refers to compressive stress.

Figure 5. Head Dishing and Flanging Process [7]

Figure 6. Stress Relief Process [8]

Figure 7. Post-Weld Heat Treatment Process [9]

(a)

(b)

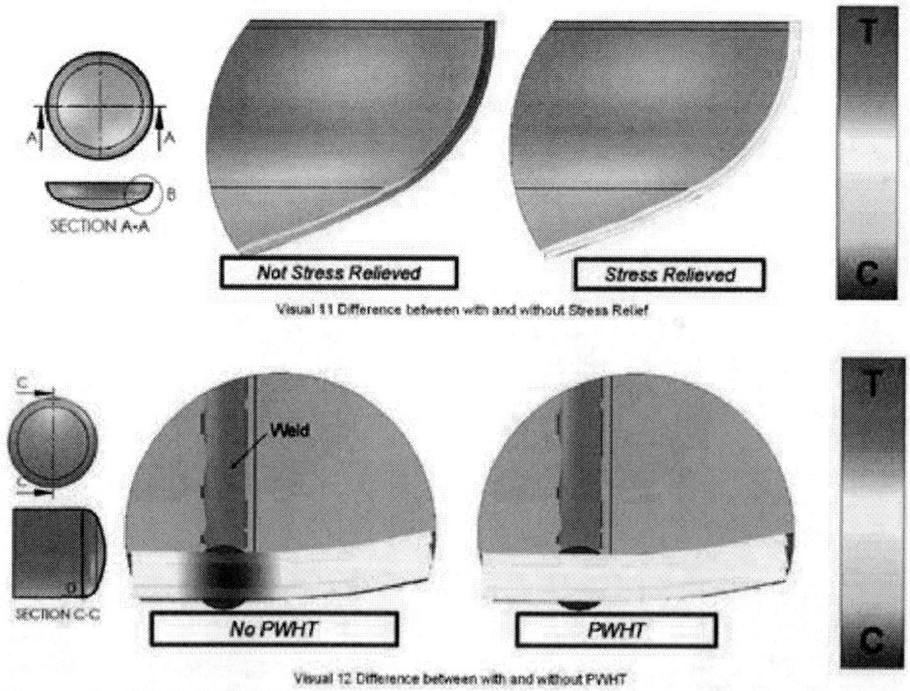

Figure 8. Reduction in Residual Stress Concentration due to (a) Stress Relief and (b) Post-Weld Heat Treatment

3.2 *Manufacturing Standards*

A normalized steel provides additional material ductility to reduce the possibility of a complete circumferential failure when a through-thickness crack is produced. Stress relieving allows a significant reduction in residual stresses that occur from the cold-forming processes. However, the BPVC accepted at the time of the separator's manufacture did not require that the metal used in forming the sour service separator heads be a normalized steel or that stress-relief after cold forming be applied to the head.

The post-weld heat treatment allows a significant reduction in residual stresses that occur from the welding process. The BPVC at the time of the separator's manufacture did require post-weld heat treatment, but the operator accepted the separator, presumably because the manufacturer had indicated that all requirements had been met. Pressure vessels for sour service do not require verification of sufficient heat treatment. Therefore, the inspections performed during maintenance of the pressure vessels were not focused to identify insufficient heat treatments in the separator's manufacture or the small, tightly closed cracks that might form under these conditions.

4 Performance Gaps and Failure Root Causes

The root cause analysis process involved first focusing on the identification of equipment failures or front-line personnel performance gaps that are the causal factors contributing to the incident. A single equipment-related causal factor for the failure was identified: the separator lost mechanical integrity.

Once the causal factor was identified, the root cause analysis focused on the identification of performance gaps that contributed to the occurrence of the causal factor and the full development of the relevant root cause paths. The investigation team used a root cause map to help direct the asking of why-why-why until the management system weaknesses representing the root causes contributing to the incident were identified. Specifically, associated with the causal factor were two performance gap categories:

- Prevention gaps - the separator was initially put into service with the heads having insufficient heat treatment during fabrication and assembly.
- Detection gaps - insufficient or non-existent heat treatment of the separator heads was not detected after the separator's installation.

The root causes of these performance gaps are discussed in the following subsections.

4.1 *Prevention Gaps*

Five root cause paths related to preventing resulted in the separator being put into service with the heads having insufficient heat treatment.

Two of the root cause paths were related to design. First, the mill plates used in the fabrication of the heads were not made of normalized steel. The use of normalized steel is recommended in the technical literature for construction of vessels operating in corrosive environments [10], [11]. Evidence suggests that there was no requirement for steel normalization of sour service pressure vessel heads less than 1.5-in. thick in internal design standards, nor was there such a requirement in international design standards (e.g., BPVC) at the time of design. Second, the separator heads did not receive effective stress relief after cold forming. Evidence suggests that there was no requirement for stress relief of pressure vessel heads with less than 5% fiber elongation during forming in internal design standards, nor was there such a requirement in international design standards (e.g., BPVC) at the time of design. Post-weld heat treatment after fabrication was, however, required by the BPVC at the time of design.

Three additional causes were related to the user's procurement process for the separator. It was inferred that purchasing documentation for the separator (which were not available for review) did not include requirements for normalization or stress relief because the design standards did not have such a requirement. In addition, the separator was designed to the ASME BPVC, but the operator had no requirement that an ASME stamp be on the separator. Finally, it was deduced that the separator was accepted from the manufacturer (receipt documents were not available for review) either without, or with insufficient, post-weld heat treatment. Accordingly, a measure to verify that the heat treatment had occurred was either not in place or not exercised.

4.2 Detection Gaps

Three root cause paths related to detection allowed the separator to remain in service over 30 years vulnerable to brittle failure. The root cause paths were limitations in hazard/defect identification and analysis through inspection. First, the pre-startup review, which was performed prior to initial start-up of the separator, did not identify the insufficient heat treatment. Verification of heat treatment documentation was not required as a part of the pre-startup review. Second, the inspections performed throughout the life of the separator did not identify the non-normalized steel. Pressure vessel inspection plans did not include a requirement to verify steel normalization.

Third, the inspections performed throughout the life of the separator did not detect the cracks that had formed and grown. There was no requirement for the determination of manufacture adequacy of operating sour service pressure vessels which would have identified this vessel as inadequate (due to lack of post weld heat treatment) and prompted mitigation or removal from service. The root causes for these three root cause paths all involved the lack of requirements to support detection.

It is noted that the material condition created a situation where, quite likely, inspections may not have been able to detect the initial damage, and even if the inspections did detect the initial damage, the inspections would be under the thresholds of criticality. In this way, inspections serve as a limited second line of defense against manufacture and fabrication process control.

5 Conclusions and Recommendations

Loss of mechanical integrity of a separator operating at low pressure (the head fractured about its circumference in the knuckle region) had eight root causes. These causes resulted in prevention gaps in design and procurement and detection gaps in inspections. These performance gaps combined to result in the loss of mechanical integrity.

Several recommendations were proposed to address the performance gaps. First, with regard to the prevention gaps, it was recommended that the operator design standard add a requirement that all carbon steel plate and low-alloy steel plate used in the manufacturing of the shell and heads of pressure vessels in sour service be normalized. A second recommendation was to add a requirement that any cold-formed shell and head should be heat-treated for stress relieving.

It remains the operator's responsibility to verify that the stress relief is performed. To that end, it was recommended that the heat-treatment requirements (normalization, stress relief and post-weld heat treatment) for pressure vessels in sour service in offshore installations be included in the contracts for purchasing, design, or construction. In addition, establishing a requirement that all new pressure vessels in critical sour service in offshore installations have an ASME stamp was recommended; such a requirement had not existed previously for this international operator.

With regard to the detection gaps, it was recommended that the pre-startup safety review verify the evidences that the required heat treatment has been adequately performed, and, if applicable, that the ASME stamp for new pressure vessels in critical sour service in offshore installations is the correct. A final recommendation was that the operator verify the required heat treatment was performed on other vessels that are currently in service.

6 References

[1] Frank, Karl H. Class Notes from "CE 397L Advanced Structural Metals." University of Texas at Austin, Fall 2005.

[2] *FEMA 343 Case Studies: An Assessment of the NEHRP Guidelines for the Seismic Rehabilitation of Buildings*. Federal Emergency Management Agency, Washington, D.C. September 1999.

[3] *FEMA 356 Prestandard and Commentary for the Seismic Rehabilitation of Buildings*. Federal Emergency Management Agency, Washington, D.C. November 2000

[4] McHenry, Harry I. *Examination of a pressure vessel that ruptured at the Chicago refinery of the Union Oil Company on July 23, 1984*. U.S. Dept. of Commerce, National Bureau of Standard, Boulder, CO. 1986

[5] Image courtesy of Prisma Impianti, www.prismagroup.it

[6] Image courtesy of Wesman Group, www.wesman.com

[7] Image courtesy of Slawinski, www.slawinski.de

[8] Image courtesy of Myungjjn Heat-Treatment, Inc., www.mhti.co.kr

[9] Image courtesy of Indiamart, http://dir.indiamart.com/

[10] *NACE Standard MR0103-2012 Item 21305 Standard Material Requirements Materials Resistant to Sulfide Stress Cracking in Corrosive Petroleum Refining Environments*. NACE International, Houston, TX. 2012. p. 6.

WHC-SD-W236A-ES-003 Rev. 0 Material Selection for Multi-Function Waste Tank Facility Tanks. U.S. Department of Energy, Washington, D.C. November 1994. p. 1.

An Integrated Prediction Model for H_2S/CO_2 Corrosion in the Pipe of Refinery

Lihan Zeng
Mary Kay O'Connor Process Safety Center
Artie McFerrin Department of Chemical Engineering
Texas A&M University
3122 TAMU, JEB 403, College Station, TX, USA
Lihan_zeng@tamu.edu

Yan-Ru Lin, Yan Cui, M. Sam Mannan
Mary Kay O'Connor Process Safety Center
Artie McFerrin Department of Chemical Engineering
Texas A&M University
3122 TAMU, College Station, TX, USA Email address

Prepared for Presentation at
American Institute of Chemical Engineers
2017 Spring Meeting and 13th Global Congress on Process Safety
San Antonio, TX
March 26 – 29, 2017

An Integrated Prediction Model for H_2S/CO_2 Corrosion in the Pipe of Refinery

Mary Kay O'Connor Process Safety Center
Artie McFerrin Department of Chemical Engineering
Texas A&M University
3122 TAMU, JEB 403, College Station, TX, USA
Lihan_zeng@tamu.edu

Yan-Ru Lin, Yan Cui, M. Sam Mannan
Mary Kay O'Connor Process Safety Center
Artie McFerrin Department of Chemical Engineering
Texas A&M University
3122 TAMU, College Station, TX, USA Email address

Keywords: corrosion, pipe, H_2S, CO_2, 90° elbow, turbulent flow, wall shear stress, mass transport, electrochemical kinetics

Abstract

Corrosion of equipment is one of the most vital factors that results in serious process safety incidents. Though various types of equipment are subject to corrosion issues to different extents depending on the process conditions, the pipe and the piping network connecting units and equipment are relatively more vulnerable to corrosion. The vulnerability of a pipe to internal corrosion is predominantly due to the process conditions. Among the factors contributing to the internal corrosion, substances creating corrosive conditions in the pipe, such as CO_2 and H_2S, are the most common factors, followed by the flow condition of processes (flow rate and temperature). In this paper, a single-phase integrated prediction model for H_2S/CO_2 corrosion is developed to study a holistic effect of the most important variables. The model investigates the electrochemical kinetics of corrosive substances, the scale formation conditions, and the flow conditions that have an impact on the mass transport of corrosive species. The COMSOL software based on the finite element method is used to simulate all these processes and the prediction results will be compared with the previously published model and the field data.

1 Introduction

The oil refining industry plays a significant role in the global economy and development. Refined products are not only the source of energy for transportations, but also the important feedstocks in petrochemical and other manufacturing industries that produce a multitude of products such as resins, plastics, engineering materials and fibers. Therefore, the existence of the refinery sector is a strategic advantage for a country.

Corrosion is one of the most critical factors that results in the serious process safety incidents in the oil refining industry. The reliability of equipment can be reduced gradually until a severe failure occurs as a result of uncontrolled corrosion, which could eventually cause the release of hazardous materials. Although different types of equipment are subject to corrosion issues to different extents depending on the process conditions, the pipe and the piping network connecting units and equipment are relatively more vulnerable to corrosion. Corrosion failure was reported as a contributing factor to 20% of the investigated 137 major refinery incidents in EU countries since 1984 [1], while in the U.S, the cost related to corrosion in the oil and gas industry has been estimated at $3.7 billion per year[1]. Oil and gas products can carry many impurities that are inherently corrosive. Among those corrosive impurities, the most common ones are CO_2 and H_2S, giving rise to sweet and sour corrosion, respectively, with the presence of water in the pipe.

Because of the changes in process conditions such as the increase in flow rate and temperature that can facilitate the corrosion rate over time, the equipment failure due to corrosion sometimes can take place before the increased corrosion rate is detected. Thus, many corrosion prediction models have been developed to help corrosion engineers and scientists make decisions on scheduling mechanical integrity inspections. One major type of these corrosion prediction models is the mechanistic model, which is able to comprehensively describe chemical reactions, electrochemical kinetics and mass transport involved in the corrosion process.

One of the first mechanistic prediction models for CO_2 corrosion was developed by Turgoose *et al.*, in the1990s [2], where an electrochemical model and a diffusion model were coupled to estimate the corrosion rate, based on the difference in concentrations between steel surface and bulk solution. Sundaram *et al.,* constructed a more comprehensive model to describe corrosion behavior [3]. The key difference from Turgoose's model is the extension of the mass transport process that includes the study of porous diffusion and turbulent layers instead of simply examining the molecular diffusion layer. Different from Sundaram's work, the model published by Dayalan *et al.*, applied a correlation to treat the mass transport processes in different layers as a whole, by introducing an effective diffusion coefficient, D_{eff} [4]. In the early 2000s, Nesic *et al.*, established the Kjeller Sweet Corrosion (KSC) model, which incorporated the concentration profile of corrosive species with the Nernst-Planck equation [5-8]. Recently, an advanced H_2S/CO_2 mechanistic model was published by Zheng and Nesic *et al.*, based on the principles of FREECORP and MULTICORP corrosion models [9]. The advantage of this model lies in the fact that the computation time and the complexity were effectively reduced by solving the mass transport in only two concentration nodes: one in the bulk and the other at the steel surface.

In common engineering practice, the flow condition within a pipe is always turbulent. Furthermore, pipes with elbows are associated with more complex flow features. However, most of these mechanistic models, which only focused on the one-dimensional mass transport process, were not practical to describe the complex mass transport process in a pipe elbow under turbulent conditions.

The flow condition determines the Mass transfer coefficient. Meanwhile, the Mass transfer coefficient affects the concentrations of corrosive species at the steel surface that is proportional to the corrosion rate. Thus, a better description of the flow condition and mass transport is very helpful for predicting corrosion rate. In fact, many prediction models for corrosion in a bend or a 90° elbow have been developed to correlate the Mass transfer coefficient with corrosion rate [10-13]. However, all these models calculated the mass transfer coefficient with a numerical method where the concentrations of corrosive species were assumed to be zero at the steel surface. In this approach, the corrosion rate is solely mass-transfer controlled, leaving it difficult to incorporate electrochemical reaction and scale formation information into the model.

In the present paper, an electrochemistry model is constructed to describe the electrochemical kinetics and corrosion rate at the steel surface. Also, a mass transport model describing mass transport together with equilibrium reactions in the boundary layer is developed, which is then combined into the electrochemistry model. Moreover, a flow model in the pipe and elbow is set up to obtain the flow parameters such as wall shear stress, boundary layer thickness, and turbulent diffusivity. After this, the flow model is linked with the mass transport model by the wall shear stress data.

2 Model description

2.1 Water chemistry model

H_2S and CO_2 existing in the pipe can be treated as weak acids in the aqueous solution and can partially dissociate to generate hydrogen ions. If the concentrations of H_2S and CO_2 in aqueous phase are low enough, Henry's law can be applied to calculate the solubility. The equilibrium chemical reactions involving H_2S and CO_2 in bulk solution and boundary layer are listed in Table 1, and major equations for the equilibrium constant calculation that will be used in the model are listed in Table 2.

Table 1. Chemical reactions of H_2S and CO_2 in aqueous solution and relevant reaction equilibrium equations

Reaction name	Reaction equation	Equilibrium equation
H_2S dissolution	$H_2S_{(g)} \rightleftharpoons H_2S_{(aq)}$ (1)	$K_{sol} = \frac{c_{H_2S}}{p_{H_2S}}$ (2)
H_2S dissociation	$H_2S_{(aq)} \rightleftharpoons H^+_{(aq)} + HS^-_{(aq)}$ (3)	$K_{hs} = \frac{c_{HS^-}c_{H^+}}{c_{H_2S}}$ (4)

HS^- dissociation	$HS^-_{(aq)} \rightleftharpoons H^+_{(aq)} + S^{2-}_{(aq)}$ (5)	$K_{bi} = \frac{C_{S^{2-}}C_{H^+}}{C_{HS^-}}$ (6)
CO_2 dissolution	$CO_{2(g)} \rightleftharpoons CO_{2(aq)}$ (7)	$K_{sol} = \frac{C_{CO_2}}{p_{CO_2}}$ (8)
CO_2 hydration	$CO_{2(aq)} + H_2O \rightleftharpoons H_2CO_{3(aq)}$ (9)	$K_{hyd} = \frac{C_{H_2CO_3}}{C_{CO_2}}$ (10)
H_2CO_3 dissociation	$H_2CO_{3(aq)} \rightleftharpoons H^+_{(aq)} + HCO^-_{3(aq)}$ (11)	$K_{ca} = \frac{C_{HCO_3^-}C_{H^+}}{C_{H_2CO_3}}$ (12)
HCO_3^- dissociation	$HCO_3^- \rightleftharpoons H^+_{(aq)} + CO^{2-}_{3(aq)}$ (13)	$K_{bi} = \frac{C_{CO_3^{2-}}C_{H^+}}{C_{HCO_3^-}}$ (14)
H_2O dissociation	$H_2O \rightleftharpoons H^+_{(aq)} + OH^-_{(aq)}$ (15)	$K_{wa} = C_{H^+}C_{OH^-}$ (16)

Table 2. Equations for the equilibrium constant calculation for H_2S and CO_2

Equilibrium constant			Source
$K_{sol_{(H_2S)}} = 10^{-(634.27+0.2709T_k-0.00011132T_k^2-16719/T_k-261.9log(T_k))}$	(mol/bar)	(17)	[14]
$K_{hs} = 10^{(782.43945-0.36126T_k-0.00016722T_k^2-20565.7315T_k-142.7417222log(T_k))}$	(mol)	(18)	[14]
$K_{bs} = 10^{(-23.93+0.030446T_k-2.4831\times10^{-5}T_k^2)}$	(mol)	(19)	[15]
$K_{sol(CO_2)} = \frac{14.5}{1.00258} \times 10^{-\left(2.27+5.65\times10^{-3}T_f-8.06\times10^{-6}T_f^2+0.075I\right)}$	(mol/bar)	(20)	[16]
$K_{hyd} = 2.58 \times 10^{-3}$		(21)	[17]
$K_{ca} = 387.6 \times 10^{-\left(6.41-1.594\times10^{-3}T_f+3.52\times10^{-6}T_f^2-3.07\times10^{-5}p-0.47721I^{0.5}+0.11807I\right)}$	(mol)	(22)	[16]
$K_{bi} = 10^{-\left(10.61-4.97\times10^{-3}T_f+1.331\times10^{-5}T_f^2-2.624\times10^{-5}p-1.66I^{0.5}+0.34661I\right)}$	(mol)	(23)	[16]
$K_{wa} = 10^{-(29.3868-0.0737549T_k+7.47881\times10^{-5}T_k^2)}$	(mol^2)	(24)	[15]

*T_f is temperature in °F, T_k is absolute temperature in K, I is ionic strength in mol: $I = \frac{1}{2}\sum_i C_i Z_i^2$, and p is the pressure in bar

In the aqueous solution, the concentrations of nine species need to be determined, which are H_2S, HS^-, S^{2-}, CO_2, H_2CO_3, HCO_3^-, CO_3^{2-}, OH^- and H^+, while there are only 8 equations available (see Table 1). Due to the electro-neutrality of solution, the number of negative charges should be equal to the number of positive charges. Therefore, the ninth equation (Equation (25)) is found to solve the concentration of the nine species.

$$C_{H^+} = C_{HS^-} + 2C_{S^{2-}} + C_{HCO_3^-} + 2C_{CO_3^{2-}} + C_{OH^-} \quad (25)$$

By comparing the equations in Table 1 and Table 2, it can be indicated that one major difference of H_2S water chemistry from CO_2 is that H_2S can directly form the weak acid in the aqueous solution with a single step, while the formation of H_2CO_3 involves two steps: dissolution and hydration. Thus, the corresponding calculations for concentrations of H_2S and H_2CO_3 are $C_{H_2S} = K_{sol_{(H_2S)}}p_{H_2S}$ and $C_{H_2CO_3} = K_{sol(CO_2)}K_{hyd}p_{CO_2}$, respectively.

2.2 *Electrochemistry model*

2.2.1 Cathodic reaction

Hydrogen ions dissociated from either H_2S or H_2CO_3 are the most readily reduced species due to its large diffusion coefficient compared with other species. As expressed in Equation (26), The total current density of the reduction of H^+ includes two parts: the charge-transfer current density and the diffusion-limit current density,

$$\frac{1}{i_{H^+}} = \frac{1}{i_{\alpha,H^+}} + \frac{1}{i^d_{lim,H^+}} \tag{26}$$

where, i_{H^+} is the total current density of H^+ reduction, i_{α,H^+} is the charge-transfer current density and i^d_{lim,H^+} is the diffusion-limit current density. For the charge-transfer current density, it can be expressed by Equation (27):

$$i_{\alpha,H^+} = i_{0,H^+} \times 10^{-\frac{\eta}{\beta_c}} \tag{27}$$

$$\beta_c = \frac{2.303RT}{\alpha_c F} \tag{28}$$

where, β_c is the Tafel slope, which is a function of temperature. According to Bockris [18], for H^+ reduction, the apparent symmetry coefficient (α_c) is 0.5. F is Faraday constant (F = 96,485.33 s·A/mol). η is the difference between the applied potential E and the reversible potential E_{rev}. The reversible potential of hydrogen ion reduction is related with the pH value of the solution. Due to the extremely low partial pressure of H_2, the second term in Equation (29) can be ignored for calculation.

$$E_{rev(H^+)} = \frac{2.303RT}{F}\text{pH} - \frac{2.303RT}{2F} log P_{H_2} \tag{29}$$

The calculation of exchange current density (i_{0,H^+}) is shown in Equation (30), where the reference exchange current density (i_0^{ref}) is chosen to be 0.03 A/m^2 at the reference temperature (T_{ref} = 298.15 K) and the reference concentration of hydrogen ion (C_{H^+ref} = 10^{-4} M). 30 kJ/mol is taken for the enthalpy of activation (ΔH) as indicated by Nordsveen, et al., [7]. $C_{H^+\_s}$ represents the concentration at the steel surface.

$$i_{0,H^+} = i_0^{ref}\left(\frac{C_{H^+\_s}}{C_{H^+ref}}\right)^{0.5} \times e^{-\frac{\Delta H}{R}\left(\frac{1}{T}-\frac{1}{T_{ref}}\right)} \tag{30}$$

For diffusion-limit current density, it can be calculated by Equation (31). k_{m,H^+} is Mass transfer coefficient of hydrogen ion (m/s), which can be determined from the flow model discussed in the later section. And $C_{H^+\_b}$ is the concentration in the bulk solution.

$$i^d_{lim,H^+} = k_{m,H^+} F C_{H^+\_b} \tag{31}$$

H_2S reduction

Besides the reduction of hydrogen ions, the dissolved H_2S can also be reduced to contribute to the corrosion process. The total current density of H_2S reduction is also consist of the charge-transfer current density and the diffusion-limit current density, which is given by Equation (32):

$$\frac{1}{i_{H_2S}} = \frac{1}{i_{\alpha,H_2S}} + \frac{1}{i^{d}_{lim,H_2S}} \tag{32}$$

For the calculation of charge-transfer current density, it is same with the hydrogen ions. And $E_{rev(H_2S)}$ is calculated by Equation (34), where the partial pressure of H_2S (P_{H_2S}) is considered.

$$i_{\alpha,H_2S} = i_{0,H_2S} \times 10^{-\frac{\eta}{\beta_c}} \tag{33}$$

$$E_{rev(H_2S)} = \frac{2.303RT}{F} pH - \frac{2.303RT}{2F} logP_{H_2S} \tag{34}$$

For the calculation of exchange current density, i_0^{ref} = 0.00015 A/m^2, T_{ref} = 293.15 K, $C_{H_2S,ref}$ = 10^{-4} M, and C_{H^+ref} = 10^{-4} M. The enthalpy of activation is taken to be 60 kJ/mol, which was in good agreement with the experimental results[19].

$$i_{0,H_2S} = i_0^{ref} \left(\frac{C_{H_2S\_S}}{C_{H_2Sref}}\right)^{0.5} \left(\frac{C_{H^+\_S}}{C_{H^+ref}}\right)^{-0.5} \times e^{-\frac{\Delta H}{R}\left(\frac{1}{T} - \frac{1}{T_{ref}}\right)} \tag{35}$$

Similar with the calculation in Equation (31), the diffusion-limit current density of H_2S is expressed in Equation (36):

$$i^{d}_{lim,H_2S} = k_{m,H_2S} F C_{H_2S\_b} \tag{36}$$

H_2CO_3 reduction

Similar with H_2S, H_2CO_3 formed from the dissolved CO_2 can also be directly reduced with the electrochemical reaction at steel surface. As shown in Equation (37), the charge-transfer current density and the hydration-limit current density consist of the total current density of H_2CO_3 reduction.

$$\frac{1}{i_{H_2CO_3}} = \frac{1}{i_{\alpha,H_2CO_3}} + \frac{1}{i^{r}_{lim,H_2CO_3}} \tag{37}$$

$$i_{\alpha,H_2CO_3} = i_{0,H_2CO_3} \times 10^{-\frac{\eta}{\beta_c}} \tag{38}$$

$$\beta_c = \frac{2.303RT}{\alpha_c F} \tag{39}$$

For charge-transfer current density of H_2CO_3 reduction, β_c is approximate to 120 mV/decade, when α_c is equal to 0.5. Similar with H_2S, the reversible potential of H_2CO_3 reduction is a function of both pH and CO_2 partial pressure (see Equation (40)).

$$E_{rev(H_2S)} = \frac{2.303RT}{F}pH - \frac{2.303RT}{2F}logP_{CO_2} \quad (40)$$

For the calculation of exchange current, i_0^{ref}= 0.018 A/m^2 at T_{ref} = 293.15 K, $C_{H_2CO_3ref}$ = 10^{-4} M, and C_{H^+ref} = 10^{-4} M. The enthalpy of activation is taken to be 50 kJ/mol [7, 20].

$$i_{0,H_2CO_3} = i_0^{ref}\left(\frac{C_{H_2CO_3\_S}}{C_{H_2CO_3ref}}\right)^{0.5}\left(\frac{C_{H^+\_S}}{C_{H^+ref}}\right)^{-0.5} \times e^{-\frac{\Delta H}{R}\left(\frac{1}{T}-\frac{1}{T_{ref}}\right)} \quad (41)$$

The hydration-limit current density is calculated by Equation (42) [20],

$$i_{lim,H_2CO_3}^{r} = F \times C_{CO_2\_b} \times \left(D_{H_2CO_3}K_{hyd}k_{hyd}^{f}\right)^{0.5} \quad (42)$$

where, K_{hyd} is the CO_2 hydration constant equal to 0.00258 and the k_{hyd}^f is the forward hydration reaction constant which is calculated in Equation (43) [21]:

$$k_{hyd}^{f} = 10^{329.85-110.541\times logT_k-\frac{17265.4}{T_k}} \quad (43)$$

H_2O reduction

Since the corrosion model in this paper considers single-phase condition with H_2O as solvent, the total current density of H_2O reduction is equal to its charge-transfer current density. Therefore:

$$i_{H_2O} = i_{\alpha,H_2O} = i_{0,H_2O} \times 10^{-\frac{\eta}{\beta_c}} \quad (44)$$

where, β_c can be calculated by Equation (28) [22]. Since the thermodynamics of reduction of H_2O and H^+ are similar, the reversible potential of H_2O can also be calculated by Equation (29) [19].

According to Zheng, *et al.*, for the calculation of exchange current density of H_2O in the presence of H_2S (Equation (45)), the reference exchange current density (i_0^{ref}) and the enthalpy of activation are 10^{-6} A/m^2 and 90 kJ/mol, respectively, while the C_{H_2Sref} and C_{H^+ref} are same with the value taken for H_2S reduction [19].

$$i_{0,H_2O} = i_0^{ref}\left(\frac{C_{H_2S}}{C_{H_2Sref}}\right)^{-0.1}\left(\frac{C_{H^+}}{C_{H^+ref}}\right)^{-0.5} \times e^{-\frac{\Delta H}{R}\left(\frac{1}{T}-\frac{1}{T_{ref}}\right)} \quad (45)$$

HS^- and HCO_3^- reduction

Since both HS^- and HCO_3^- are negatively charged, the reductions of these two species are more thermodynamically difficult than H^+, H_2S, H_2CO_3 and H_2O. Thus, the reduction of H^+, H_2S, H_2CO_3 and H_2O is dominant in corrosion process, while the reduction of HS^- and HCO_3^- can be ignored.

2.2.2 Anodic reaction

The current density at anode due to the oxidation of iron is expressed by Equation (46), which is only related to the charge-transfer current density, since the diffusion process of the generated Fe^{2+} at the steel surface can be ignored as compared with the other species.

$$i_{Fe} = i_{0,Fe} \times 10^{-\frac{\eta}{\beta a}} \quad (46)$$

When in the acidic solution with the presence of H_2S, the anodic adsorption is dominated by HS^-. The expressions of exchange current density for anodic reaction controlled by HS^- concentration is expressed in Equation (47) and (48).

$$i_{0,Fe} = i_{0,Fe}^{*\prime} \theta_{HS^-} e^{-\frac{\Delta H}{R}\left(\frac{1}{T}-\frac{1}{T_{ref}}\right)} \quad (47)$$

$$\theta_{HS^-} = \frac{K_2 C_{HS^-}}{1+K_2 C_{HS^-}} \quad (48)$$

According to Zheng and Nesic [19, 21], the best-fit values for Equations (47) and (48) are $i_{0,Fe}^{*\prime} = 0.33$, $K_2 = 3.5\times10^6$, ΔH=37.5 kJ/mol and $T_{ref} = 293.15$ K.

2.2.3 The mixed potential theory

The corrosion rate is proportional to either the total cathodic current density or the total anodic current density. In order to calculate the current densities, two unknowns need to be solved: the concentration of corrosive species at the steel surface, and the applied potential (E). The steel surface concentrations can be calculated from the mass transport model, while the applied potential can be obtained by using the mixed potential theory and solving the charge balance equation (Equation (49)). Since the total cathodic current density and the total anodic current density are with opposite signs, the total current density at steel surface is zero.

$$\sum i_{anodic} = \sum i_{cathodic} \quad (49)$$

Once the applied potential is solved, the corrosion rate can be immediately determined by equation (50)

$$CR(mm/yr) = \frac{i_{anode} \times M_{steel}}{2F \times \rho_{steel}} \quad (50)$$

where CR is corrosion rate, i total is current density at the anode (or the cathode) (A/m^2), M_{steel} (g/mol)is the molecular weight of steel (g) and ρ_{steel} is the density of corroding steel (g/m^3).

2.3 Mass transport model

The mass transport process in the boundary layer is demonstrated in Figure 1 and the diffusion coefficients of different corrosive species are listed in Table 3. Here only a one-dimensional mass transport model along the radial direction of the pipe is considered, which will be then coupled with the three-dimensional flow model in the pipe and elbow. Although the diffusion barrier effect of protective scale is not taken into consideration in the present model, whether the protective scales can be formed under the solution condition will be discussed in the later section.

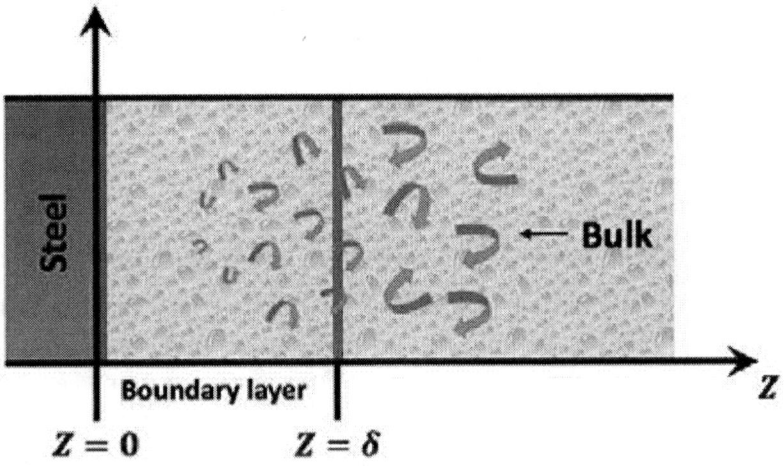

Figure 1. one-dimensional mass transport model for the corrosives species transported from the bulk solution to the steel surface

Table 3. Diffusion coefficient of the corrosive species involved in the present model

Species	Diffusion coefficient (m^2/s)	Source
H^+	9.312×10^{-9}	[23]
OH^-	5.26×10^{-9}	[23]
H_2S	1.61×10^{-9}	[24]
HS^-	2.00×10^{-9}	[7]
S^{2-}	2.00×10^{-9}	[7]
CO_2	1.96×10^{-9}	[24]
H_2CO_3	2.00×10^{-9}	[25]
HCO_3^-	1.105×10^{-9}	[23]
CO_3^{2-}	0.92×10^{-9}	[25]

Fe^{2+}	0.72×10^{-9}	[25]

It is assumed that turbulent mixing is dominant in the bulk solution, so the species concentration in the bulk is considered to be constant. This assumption also applies to the flow model, as a result, the concentration of different species throughout the pipe is also treated as constant.

As for the mass transport in the boundary layer, the mass conservation equation is applied as the governing equation for all species, which is expressed in Equation (51).

$$\frac{\partial c_i}{\partial t} = -\frac{\partial N_i}{\partial z} + R_i \tag{51}$$

where C_i is the concentration of species i, N_i is the flux of species i, and R_i represents the equilibrium chemical reactions that species i are involved in. The flux of species N_i can be expressed by Equation (52):

$$N_i = -D_i \frac{\partial c_i}{\partial z} - z_i u_i F c_i \frac{\partial \phi}{\partial z} + c_i u \tag{52}$$

The first term on the right side of Equation (52) is the flux contributed by molecular diffusion. Because of the diluted solution, the second term, the migration due to charge separation, can be ignored. The third term is the mass transport led by convection in the radial direction. In the boundary layer of a turbulent flow, it is complicated and time-consuming to compute the instantaneous velocity (u) in the convection term. According to Nesic *et al.*, [7], the convection term arising from the presence of eddies can be approximately replaced by $-D_t \frac{\partial c_i}{\partial z}$ [26]. D_t is called turbulent diffusion coefficient, which changes with the distance from steel surface to the bulk solution, and is a function of boundary layer thickness [26]:

$$D_t = 0.18 \left(\frac{z}{\delta}\right)^3 \frac{\mu}{\rho} \qquad (0 \le z \le \delta) \tag{53}$$

In order to obtain the concentration profile in the boundary layer, the boundary layer thickness at different positions of steel surface in the pipe should be determined. According to Cengel [27], the thickness of boundary layer is approximated as a function of wall shear stress in the pipe:

$$\delta = \frac{5v}{\sqrt{\tau_w/\rho}} \tag{54}$$

where v is the kinematic viscosity of water and τ_w is the wall shear stress, which can be obtained from the flow model. If only the steady state mass transport is considered, Equation (51) becomes:

$$0 = (D_i + D_t)\left(\frac{\partial^2 c_i}{\partial z^2}\right) + R_i \tag{55}$$

Therefore, for the ten species, there are ten ordinary differential equations, which are coupled by the equilibrium chemical reactions (R_i). The boundary conditions for solving the OEDs are the constant concentration in bulk solution and zero total current density at the steel surface. Since the electrochemistry model and the mass transport model are coupled by the boundary conditions, the concentration of corrosive species at steel surface and the corresponding current density can be easily solved by the electrochemistry module incorporated in COMSOL software [28].

2.3 Flow model in pipe and pipe elbow

$k-\omega$ model is selected in the present model to simulate the single-phase turbulent flow in the pipe and elbow. As expressed by Equation (56) and (57), the standard $k-\omega$ model is a two-equation model to solve the turbulence kinetic energy (k) and the specific rate of dissipation (ω) (the rate that turbulence kinetic energy k converts into internal thermal energy) [29]. Different from the $k-\varepsilon$ model [30], which is a more commonly used turbulence model, $k-\omega$ solves the specific rate of dissipation of kinetic energy. One of the advantages of the $k-\omega$ model is the near wall treatment, which is more accurate than $k-\varepsilon$ model when simulating internal flows with a strong curvature such as flows in a pipe bend or a 90° elbow.

$$\frac{\partial}{\partial t}(\rho k)+\frac{\partial}{\partial x_j}(\rho k u_i)=\frac{\partial}{\partial x_j}\left[\left(\mu+\frac{\mu_t}{\sigma_k}\right)\frac{\partial k}{\partial x_j}\right]+G_k-Y_k+S_k \qquad (56)$$

$$\frac{\partial}{\partial t}(\rho \omega)+\frac{\partial}{\partial x_j}(\rho \omega u_i)=\frac{\partial}{\partial x_j}\left[\left(\mu+\frac{\mu_t}{\sigma_\omega}\right)\frac{\partial \omega}{\partial x_j}\right]+G_\omega-Y_\omega+S_\omega \qquad (57)$$

where, $G_k=\rho\tau_{ij}\frac{\partial u_i}{\partial x_j}$, $G_\omega=\alpha\frac{\omega}{k}\tau_{ij}\frac{\partial u_i}{\partial x_j}$, $Y_k=\beta^*\rho\omega k$, $Y_\omega=\beta\rho\omega^2$, $\mu_t=\frac{\rho k}{\omega}$, $\alpha=\frac{5}{9}$, $\beta^*=0.09$, $\beta=0.075$, $\sigma_k=\sigma_\omega=2$, S_k and S_ω represents source terms.

In this paper, the model is to simulate the corrosion rate distribution in a 90° elbow connecting two pieces of straight pipe in a refinery facility of Assiut Oil Refining Company (ASORC), with the exception that only single liquid phase flow (water) is considered. The configuration of the pipe and elbow is as shown in Figure 2. In the pipe and elbow, the side with smaller curvature at the elbow is defined as the inner wall, while the side with larger curvature is defined as the outer wall. The corresponding geometry and operation parameters are listed in Table 4.

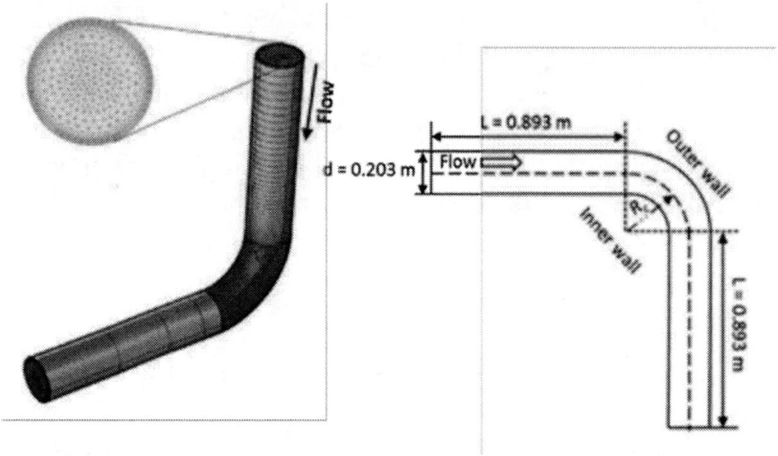

Figure 2. Configuration of the pipe and 90° elbow in the ASORC refinery facility

The results that need to be extracted from the flow model are the shear stress (τ_w) at the steel surface and the mass transfer coefficient (k_m) that is correlated with the diffusion-limit current density as mentioned in the previous section. According to Liu *et al.*, and Rani *et al.*, [31, 32], mass transfer coefficient is proportional to wall shear stress under the turbulent condition:

$$k_m = \left(\frac{\tau_w}{u\rho}\right) Sc^{-2/3} \tag{58}$$

where u is the average flow velocity and Sc is the Schmidt number:

$$Sc = \frac{\mu}{\rho D_i} \tag{59}$$

Table 3. Geometry and operation parameters [33]

d	0.203 [m]	Pipe diameter
L	0.893 [m]	Inlet and outlet pipe length
R_c	0.305 [m]	Radius of curvature
u	0.19 [m/s]	Average flow velocity
P	1.91 [bar]	Operating pressure
μ	0.000355 [kg/(m•s)]	Dynamic viscosity of water
ρ	971.8 [kg/m^3]	Density of water

3. Results and discussion

3.1 Analysis of hydrodynamic parameters

The distribution of wall shear stress in the pipe and 90° elbow is shown Figure 3. The distribution shows the almost uniform wall shear stress in the upstream pipe. The wall shear stress distribution starts to be diversified at the inlet of the elbow in the flow direction, with a sharp increase at the inner wall and a drastic decrease at the outer wall. As the flow goes through the elbow, the wall shear stress increases at the outer wall near the outlet of the elbow, while a static zone is observed at the inner wall near the outlet. This adverse distribution of the wall shear stress continues in the downstream pipe.

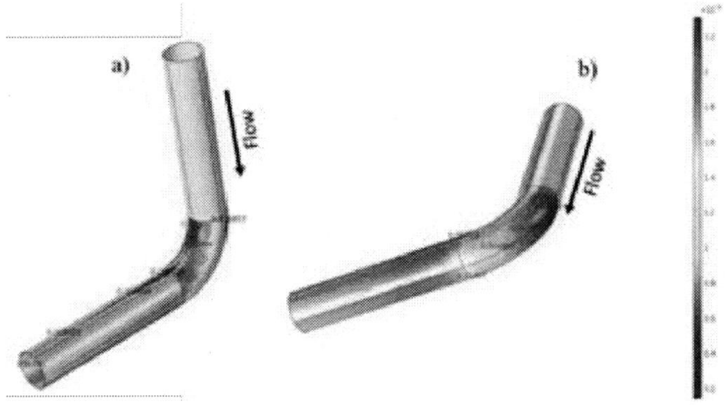

Figure 3. Wall shear stress (τ_w [Pa•s]) distribution in the pipe and 90° elbow with the view on a) inner wall and b) outer wall

As observed in Figure 4, the distribution of the mass transfer coefficient of H^+ follows the same pattern of wall shear stress distribution, the trend of which is in a good agreement with Liu *et al.*, and Rani *et al.*, [31, 32].

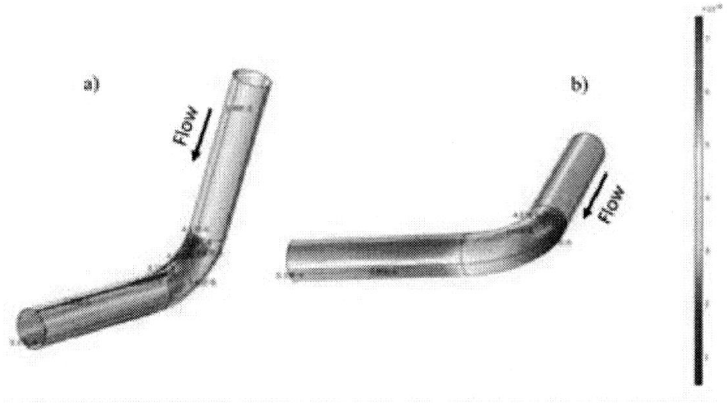

Figure 4. Mass transfer coefficient (k_m [m/s]) distribution of H^+ in the pipe and 90° elbow with the view on a) inner wall and b) outer wall

In general, the increase of wall shear stress would increase the mass transfer coefficient in a flow region, and result in the decrease in boundary layer thickness (see Figure 5), according to Equation (53). As a result, the mass transport from the bulk to the steel surface can be facilitated, which can accelerate the corrosion rate by increasing the diffusion-limit current density and the concentration of corrosive species at the steel surface.

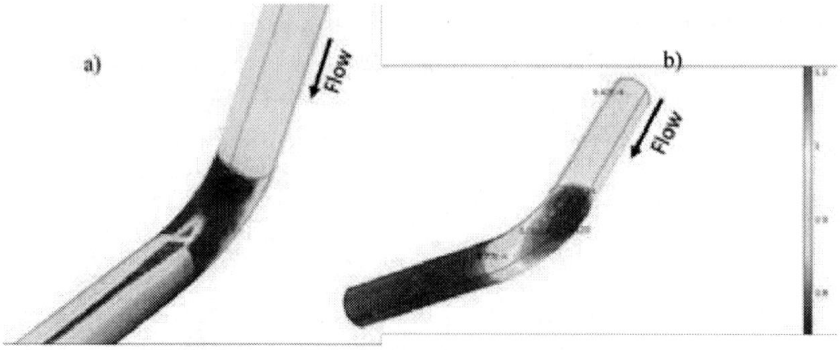

Figure 5. Boundary layer thickness (δ [m]) distribution of H^+ in the pipe and 90° elbow with the view on a) inner wall and b) outer wall

Figure 6 demonstrates the pressure contours in the pipe and elbow. The pressure increases at the outer wall as the flow moves from the inlet of the elbow to about 45° position, while the pressure along the inner wall reduces to negative values. As a result, a secondary flow is developed due to the pressure difference between the inner wall and the outer wall, which affects the motion of the species through the elbow. The mass transport of corrosive species is favored towards the inner wall by means of convection.

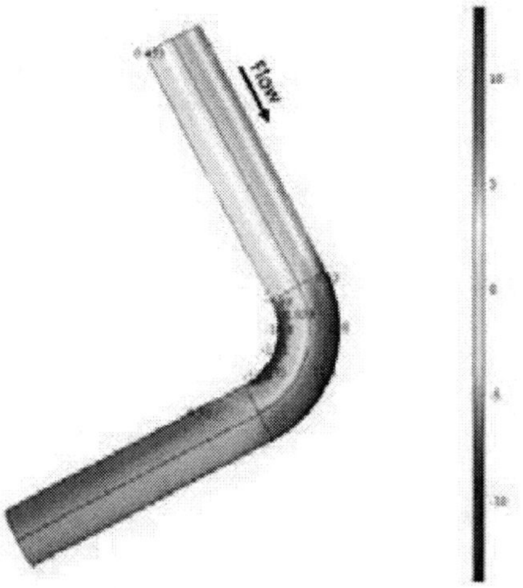

Figure 6. Contours of pressure (Pa) on the wall of pipe and 90° elbow section

Therefore, it is necessary to investigate the hydrodynamic parameters at both the inner wall and outer wall of the pipe as part of the study on corrosion mechanisms, since they are closely correlated with electrochemical reaction and corrosion rates.

3.2 Corrosion in the pipe and elbow

The comparison of the corrosion rate change between the inner wall and the outer wall along the flow direction in the pipe and elbow is demonstrated in Figure 7. The corrosion rate keeps identical in the upstream pipe at both the inner wall and the outer wall until it reaches the inlet of the elbow, where the corrosion rate at the inner and outer wall starts to differentiate. The corrosion rate at the inner wall keeps increasing till the peak value at 25° position of the elbow. Then the corrosion rate decreases along the outer wall to the lowest value at 90° position of the elbow (outlet). After exiting from the elbow, the corrosion rate at inner wall increases again in the downstream pipe and gradually reaches constant. In contrast, the corrosion rate at the outer wall decreases first at the inlet of elbow until the flow arrives at about 10° position in the elbow. Then the corrosion rate raises in the elbow along the flow direction to the outlet. The corrosion rate at the outer wall in the downstream pipe slightly decreases but is still higher than the corrosion rate at the inner wall. Generally, in the elbow region, the corrosion rate at the inner wall is larger than the outer wall, while in the downstream pipe region, the outer wall has larger corrosion rate.

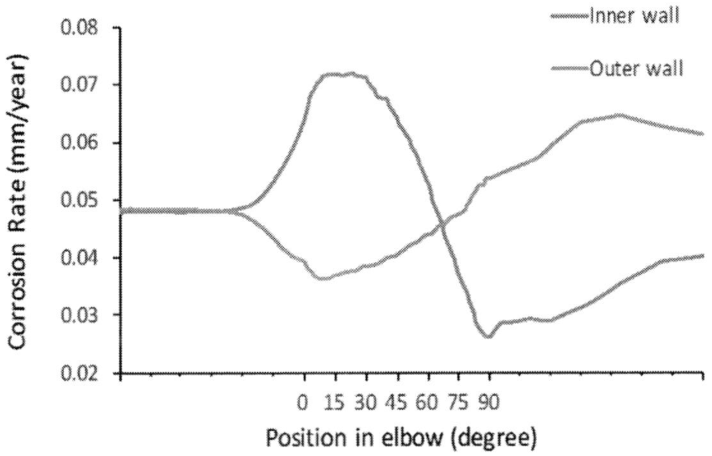

Figure 7. Comparison of corrosion rate between the inner wall and the outer wall in the pipe and elbow with T=80 °C, pH=4, 500 ppm CO_2 and 100 ppm

Comparison with field data of ASORC and previous model

As shown in Figure 8, the predicted corrosion rate from the present model is compared with the measured data from one of the naphtha pipes in ASORC refinery facility and the simulation results from the model by Doheim *et al.*, [11]. The predicted corrosion rate in the elbow decreases along the inner wall while increases along the outer wall in the direction of flow. This trend is in a good agreement with the measured data. The present model's prediction is closer to the measured value as compared with Doheim's model, with the exception that the prediction of the present model at the outer wall is slightly lower than the field data. According to Doheim et al., [10], one of the possible explanations for this difference is that the pipe of the ASORC refinery is also affected by oxygen corrosion, while the present model does not take account for the effect of oxygen corrosion.

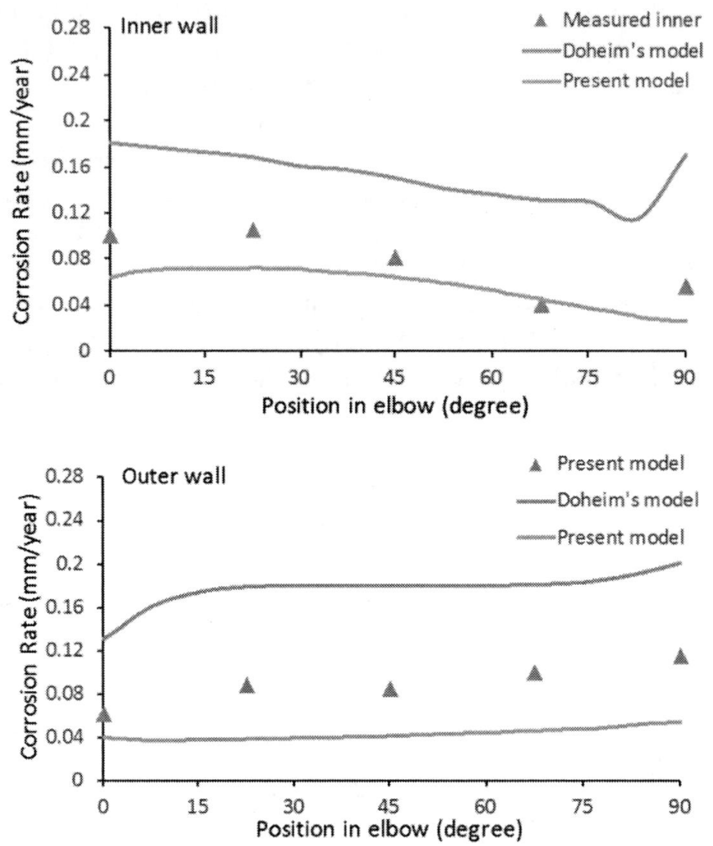

Figure 8. Comparison of present model with the field measurements and the published model under the condition of T=80 °C, pH=4, 500 ppm CO_2, and 100 ppm H_2S. Field data are taken from Doheim et al., [11]

In addition, since the present model includes the contribution of charge-transfer current density to the corrosion rate, the corrosion rate under the influence of both charge current density and diffusion-limit current density is lower than the one only under the control of diffusion-limit current density, which is the case in Doheim's model. From the results of the present model, the charge-transfer current density of H_2S reduction is within the range of 166 ~ 269 A/m^2. In contrast, as shown in Figure 9, the diffusion-limit current density of H_2S is much smaller, ranging from 8.72×10^{-5} to 1.08×10^{-3} A/m^3. According to $\frac{1}{i_{H_2S}} = \frac{1}{i_{\alpha,H_2S}} + \frac{1}{i^d_{lim,H_2S}}$, it is obvious that the total current density of H_2S reduction is controlled by the diffusion-limit current density, and the corrosion rate contributed by H_2S reduction is mass transfer controlled (since the diffusion-limit current density is proportional to mass transfer coefficient). On the other hand, due to the contribution of the charge-transfer current density, the total current density of H_2S is smaller than its diffusion current density (see Figure 9), and the same trend is also observed for the reduction of H^+ and H_2CO_3.

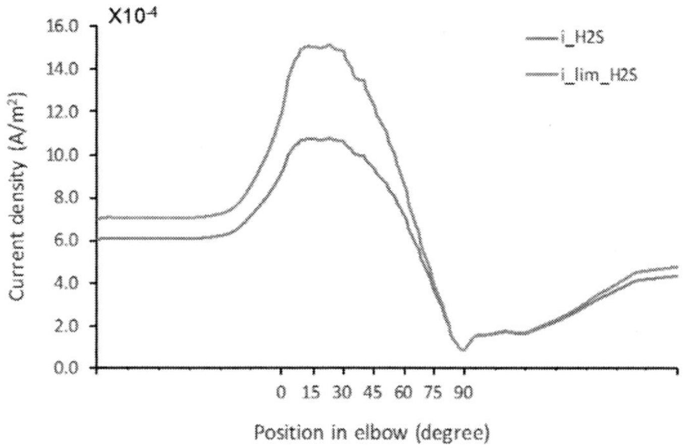

Figure 9. Total current density (i_{H_2S}) and diffusion limit current density (i^d_{lim,H_2S}) of H_2S reduction on inner wall of pipe and elbow under the condition of T=80 °C, pH=4, 500 ppm CO_2, and 100 ppm H_2S

Effect of CO_2 and H_2S

The evaluation of corrosion rate in the pipe and elbow under different concentrations of CO_2 and H_2S environment is also conducted in the present model. As shown in Figure 10, under the condition of 100 ppm H_2S, an increase in corrosion rate at both the inner wall and the outer wall is observed as the concentration of CO_2 increased from 10 ppm to 1000 ppm. However, the magnitude of corrosion rate change is not significant, especially for the difference between 10 ppm and 100 ppm CO_2 conditions.

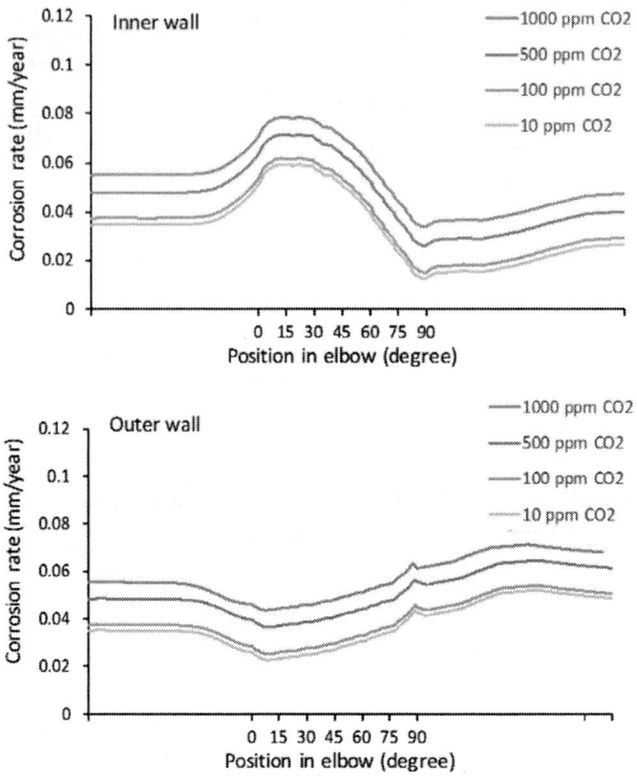

Figure 10. Comparison of corrosion rate in the pipe and elbow under different concentration of CO_2 and with the condition of T=80 °C, pH=4, and 100 ppm H_2S

In contrast, as seen in Figure 11 the corrosion rate under 500 ppm CO_2 environment in the pipe as a function of H_2S concentration, the change of corrosion rate is more sensitive to the change of H_2S concentration from 10 ppm to 1000 ppm. The major reason for the difference in sensitivity of corrosion rate to the concentration change is the extremely slow hydration process of CO_2 to H_2CO_3. For example, for the gas mixture with the same partial pressure of H_2S and CO_2, the concentration of H_2S in the aqueous phase is approximately 1000 times higher than the concentration of H_2CO_3. Therefore, in the presence of both H_2S and CO_2, it is more likely that H_2S will dominate the corrosion process.

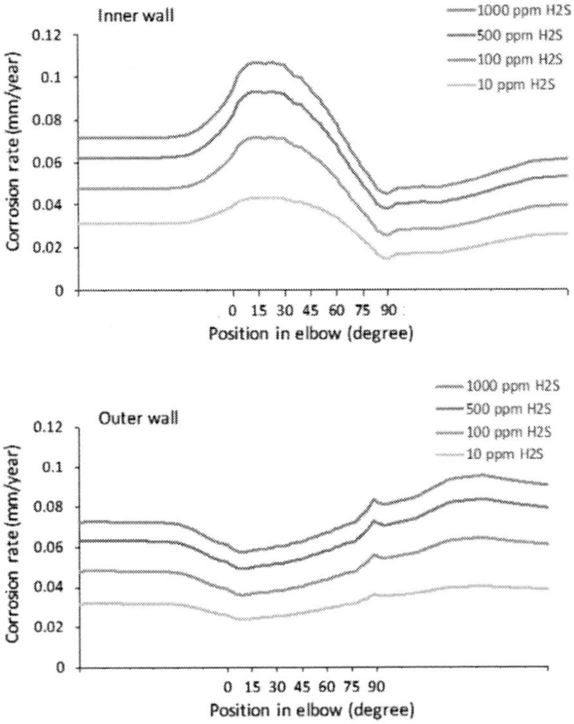

Figure 11. Comparison of corrosion rate in the pipe and elbow under different concentration of H_2S with the condition of T=80 °C, pH=4, and 500 ppm CO_2

Effect of Temperature

In order to investigate the effect of temperature on corrosion rate in the pipe under the environment of H_2S and CO_2, simulation is carried out in the present model under different temperature conditions. As seen in Figure 12, corrosion rate calculated from the model increases by approximately 0.02~0.03 mm/year at both the inner wall and the outer wall with the temperature raising from 20 °C to 80 °C. In general, the temperature is expected to increase the rate of chemical reaction and electrochemical reaction, as well as the mass transport process.

In detail, for the charge-transfer current density, according to $i_0 = i_0^{ref}\left(\frac{C_i}{C_{i,ref}}\right)^{0.5} \times e^{-\frac{\Delta H}{R}\left(\frac{1}{T}-\frac{1}{T_{ref}}\right)}$, the increase of temperature leads to the increase in the value of exponential term of the expression. So the charge-transfer current density is proportional to the temperature. Besides, the increase in temperature facilitates the dissociation reaction of H_2S and H_2CO_3, and leads to the generation of more H^+ ions, which are with higher diffusion coefficient. Thus, the diffusion-limit current density of H^+ ions is favored by higher temperature. As for mass transport process, the mass transfer coefficient is also proportional to the temperature.

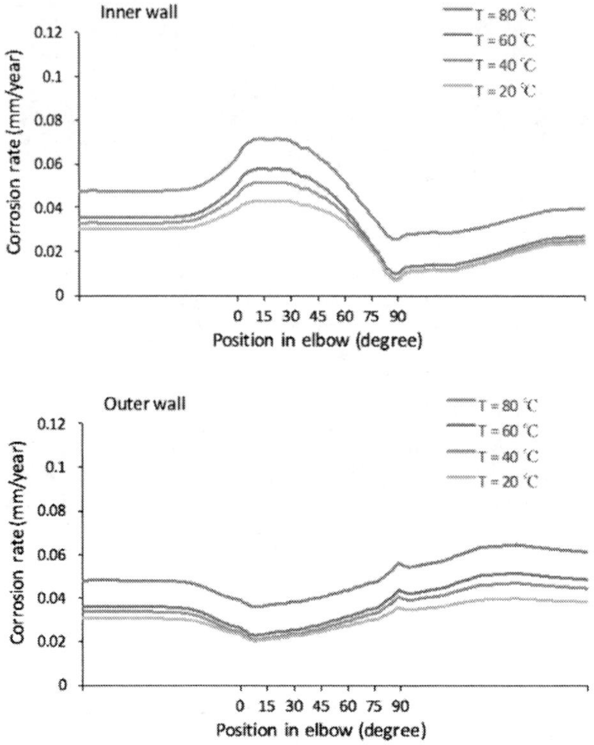

Figure 12. Corrosion rate change in the pipe and elbow as a function of temperature (T=20 °C, 40 °C, 60 °C and 80 °C) under the condition of pH=4, 500 ppm CO_2, and 100 ppm H_2S

Scale formation

One of the advantages of the present model is the capability of calculating not only the concentration of corrosive species at the steel surface, but the equilibrium reactions in the boundary layer. Thus, by obtaining the surface concentration of Fe^{2+}, S^{2-} and CO_3^{2-}, it can be determined if there will be FeS or $FeCO_3$ precipitated and formed as protective scales. According to Equation (59) and (60), if the supersaturation of FeS (or $FeCO_3$) is larger than 1, the protective scales can form.

$$S_{FeS} = \frac{C_{Fe^{2+}} C_{S^{2-}}}{K_{sp\_FeS}} \tag{59}$$

$$S_{FeCO_3} = \frac{C_{Fe^{2+}} C_{CO_3^{2-}}}{K_{sp\_FeCO_3}} \tag{60}$$

where, K_{sp_FeS} and $K_{sp\_FeCO_3}$ are solubility limit of FeS and $FeCO_3$, respectively, which are both the functions of temperature, according to Benning, *et al.*[7], and Sun, *et al.*, [34].

$$K_{sp,FeS}10^{\frac{2848.779}{T}-6.347}K_{hs}K_{bs} \tag{93}$$

$$K_{sp,FeCO_3} = -59.3498 - 0.041377T - \frac{2.1963}{T} + 24.5724logT \tag{94}$$

As shown in Figure 13, the supersaturation of FeS (S_{FeS}) in the pipe and elbow ranges from 1.2 ×10^{-3} to 1.6 ×10^{-3}, while the supersaturation of $FeCO_3$ (S_{FeCO_3}) is 1,000 times smaller. This indicates that, under the operation condition of ASORC pipe, both FeS and $FeCO_3$ scales cannot be formed.

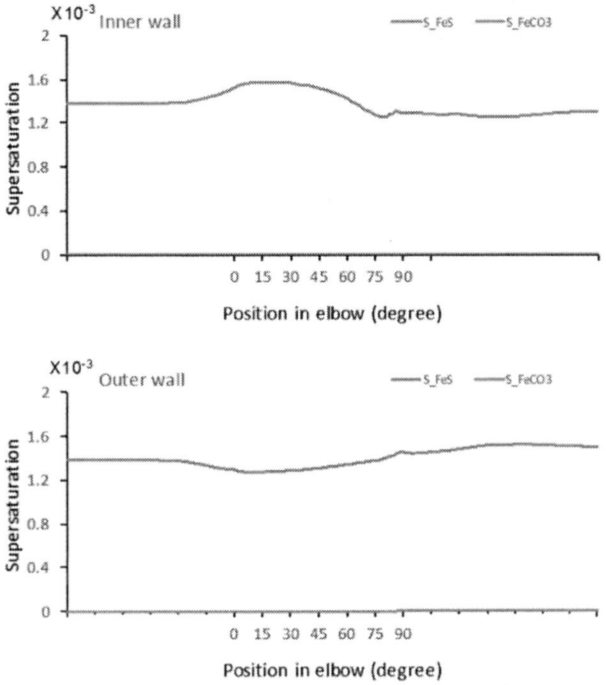

Figure 13. Comparison between the supersaturation of FeS (S_{FeS}) and $FeCO_3$ (S_{FeCO_3}) in the pipe and elbow with the condition of T=80 °C, pH=4, 500 ppm CO_2, and 100 ppm

4. Conclusion

An integrated single-phase prediction model for H_2S/CO_2 corrosion in the pipe, especially for the case of a 90° elbow region, has been developed, which is able to solve the four key facets in the corrosion process by incorporating the flow model, the mass transport model, and the electrochemistry model:

(1) Flow parameters including wall shear stress, boundary layer thickness, and turbulent diffusivity;

(2) Equilibrium reactions and mass transport of corrosive species in the boundary layer, which determines the concentration of corrosive species at the steel surface;

(3) Current density and corrosion rate, which are proportional to the steel surface concentration and the mass transfer coefficient;

(4) Formation of protective scale based on the concentration of Fe^{2+}, S^{2-} and CO_3^{2-} at the steel surface.

The prediction results of the present model are compared with the previously published model and the field data of the pipe in an ASORC refinery facility. This model is in good accordance with the measured field data, and follows the prevalent trend among the existing models for the H_2S/CO_2 corrosion in a pipe and elbow. Additionally, it is found in the present model that the charge-transfer current density has the significant contribution to the corrosion rate, which is ignored in some other models. Apart from that, other comparisons have been made to investigate corrosion rate as a function of flow region, H_2S and CO_2 concentration, and temperature. Finally, the present model is able to provide the information on whether there are FeS (or $FeCO_3$) protective scales formed at the steel surface.

References

[1] M. Wood, A. Vetere Arellano, and L. Van Wijk, "Corrosion-Related Accidents in Petroleum Refineries: lessons Learned from Accidents in EU and OECD Countries," *European Commission Joint Research Centre (JRC), Luxembourg. doi,* vol. 10, p. 379, 2013.

[2] S. Turgoose, R. Cottis, and K. Lawson, "Modeling of electrode processes and surface chemistry in carbon dioxide containing solutions," in *Computer Modeling in Corrosion*: ASTM International, 1992.

[3] M. Sundaram, V. Raman, M. High, D. Tree, and J. Wagner, "Deterministic modeling of corrosion in downhole environments," NACE International, Houston, TX (United States)1996.

[4] E. Dayalan, F. De Moraes, J. R. Shadley, E. F. Rybicki, and S. A. Shirazi, "CO_2 corrosion prediction in pipe flow under $FeCO_3$ scale-forming conditions," in *CORROSION 98*, 1998: Nace International.

[5] S. Nesic, M. Nordsveen, and A. Stangel, "A Mechanistic Model for CO_2 Corrosion with Protective Iron Carbonate Films," 2001.

[6] S. Nešić, M. Nordsveen, R. Nyborg, and A. Stangeland, "A diffusion-based model for uniform corrosion," in *Defect and Diffusion Forum*, 2001, vol. 194, pp. 1661-1674: Trans Tech Publ.

[7] M. Nordsveen, S. Nešic, R. Nyborg, and A. Stangeland, "A mechanistic model for carbon dioxide corrosion of mild steel in the presence of protective iron carbonate films-Part 1: Theory and verification," *Corrosion,* vol. 59, no. 5, pp. 443-456, 2003.

[8] S. Nešic and K.-L. Lee, "A mechanistic model for carbon dioxide corrosion of mild steel in the presence of protective iron carbonate films-Part 3: Film growth model," *Corrosion,* vol. 59, no. 7, pp. 616-628, 2003.

[9] Y. Zheng, J. Ning, B. Brown, and S. Nešić, "Advancement in Predictive Modeling of Mild Steel Corrosion in CO_2-and H_2S-Containing Environments," *Corrosion,* vol. 72, no. 5, pp. 679-691, 2016.

[10] M. A. Doheim, H. M. Elbatsh, and A. F. Hassan, "Prediction for the mass transfer and oxygen corrosion in complex pipelines," *Proceedings of ICFD,* vol. 9, pp. 18-21.

[11] M. A. Doheim, H. M. El-Batsh, and A. F. Hassan, "Prediction of hydrogen sulfide corrosion in petroleum refinery complex pipelines," *Proceedings of ICFD,* vol. 10, pp. 16-19, 2010.

[12] K. Das, D. Basu, and T. Mintz, "Comparative Assessment of Turbulence Models for Prediction of Flow-Induced Corrosion Damages," in *ASME 2011 Pressure Vessels and Piping Conference*, 2011, pp. 51-60: American Society of Mechanical Engineers.

[13] J. Liu, Y. Lin, and X. Li, "Application of numerical simulation to flow-induced corrosion in flowing seawater systems," *Anti-Corrosion Methods and Materials,* vol. 52, no. 5, pp. 276-279, 2005.

[14] O. Suleimenov and R. Krupp, "Solubility of hydrogen sulfide in pure water and in NaCl solutions, from 20 to 320 C and at saturation pressures," *Geochimica et Cosmochimica Acta,* vol. 58, no. 11, pp. 2433-2444, 1994.
[15] Y. K. Kharaka, W. D. Gunter, P. K. Aggarwal, E. H. Perkins, and J. D. DeBraal, "SOLMINEQ. 88: A computer program for geochemical modeling of water-rock interactions," *US geological survey water-resources investigation report,* vol. 88, p. 4227, 1988.
[16] J. E. Oddo and M. B. Tomson, "Simplified calculation of $CaCO_3$ saturation at high temperatures and pressures in brine solutions," *Journal of Petroleum Technology,* vol. 34, no. 07, pp. 1,583-1,590, 1982.
[17] D. A. Palmer and R. Van Eldik, "The chemistry of metal carbonato and carbon dioxide complexes," *Chemical Reviews,* vol. 83, no. 6, pp. 651-731, 1983.
[18] I. Bockris, "'M. The electrode kinetics of the deposition of iron/I. O'M. Bockris, D. Drazic, AR Despic," *Electrochimica Acta,* vol. 4, pp. 325-361, 1961.
[19] Y. Zheng, B. Brown, and S. Nešić, "Electrochemical study and modeling of H2S corrosion of mild steel," *Corrosion,* vol. 70, no. 4, pp. 351-365, 2013.
[20] S. Nesic, J. Postlethwaite, and S. Olsen, "An electrochemical model for prediction of corrosion of mild steel in aqueous carbon dioxide solutions," *Corrosion,* vol. 52, no. 4, pp. 280-294, 1996.
[21] Y. Zheng, J. Ning, B. Brown, and S. Nešić, "Electrochemical Model of Mild Steel Corrosion in a Mixed H_2S/CO_2 Aqueous Environment in the Absence of Protective Corrosion Product Layers," *Corrosion,* vol. 71, no. 3, pp. 316-325, 2014.
[22] K. Videm and J. Kvarekvål, "Corrosion of carbon steel in carbon dioxide-saturated solutions containing small amounts of hydrogen sulfide," *Corrosion,* vol. 51, no. 4, pp. 260-269, 1995.
[23] J. S. Newman, *Electrochemical Systems, 2nd ed.* NJ: Prentice Hall: Englewood Cliffs, 1991.
[24] D. G. R.H. Perry, *Perry's Chemical Engineers' Handbook, 50th ed.* New York, NY: McGraw-Hill, 1984.
[25] J. Kvarekvål, "A kinetic model for calculating concentration profiles and fluxes of CO_2-related species across the Nernst diffusion layer," NACE International, Houston, TX (United States)1997.
[26] J. T. Davies, *Turbulence Phenomena*. London, U.K.: Academic Press, 1992.
[27] Y. A. Cengel, *Fluid mechanics*. Tata McGraw-Hill Education, 2010.
[28] W. B. Zimmerman, *Multiphysics modeling with finite element methods*. World Scientific Publishing Co Inc, 2006.
[29] D. C. Wilcox, "Reassessment of the scale-determining equation for advanced turbulence models," *AIAA journal,* vol. 26, no. 11, pp. 1299-1310, 1988.
[30] D. C. Wilcox, *Turbulence modeling for CFD*. DCW industries La Canada, CA, 1998.
[31] L. Jingjun, L. Yuzhen, and L. Xiaoyu, "Numerical simulation for carbon steel flow-induced corrosion in high-velocity flow seawater," *Anti-Corrosion Methods and Materials,* vol. 55, no. 2, pp. 66-72, 2008.

[32] H. Rani, T. Divya, R. Sahaya, V. Kain, and D. Barua, "Unsteady turbulent flow in 90° bend under the wall thinning degradation environment," *Nuclear Engineering and Design,* vol. 267, pp. 164-171, 2014.

[33] H. El-Batsh, M. Doheim, and A. Hassan, "On the application of mixture model for two-phase flow induced corrosion in a complex pipeline configuration," *Applied Mathematical Modelling,* vol. 36, no. 11, pp. 5686-5699, 2012.

[34] E. Lamy-Pitara and J. Barbier, "The electrocatalytic reactions of oxidation and evolution of hydrogen on iridium electrodes modified by sulphur adsorption," *Journal of Electroanalytical Chemistry,* vol. 416, no. 1-2, pp. 47-51, 1996.

13TH GLOBAL CONGRESS ON PROCESS SAFETY

Waste Solvents to Trash Haulers: Lessons Learned from Hazardous Waste Accidents

Delmar "Trey" Morrison, Ph.D., P.E.
Exponent, Inc.
4580 Weaver Parkway, Suite 100, Warrenville, IL 60555, USA
tmorrison@exponent.com

Michael Stern, Ph.D., P.E.
Exponent, Inc.
Natick, MA, USA
mstern@exponent.com

Carmen H. Osorio-Amado, Ph.D.
Exponent, Inc.
Los Angeles, CA, USA
cosorio@exponent.com

Prepared for Presentation at
American Institute of Chemical Engineers
2017 Spring Meeting and 13th Global Congress on Process Safety
San Antonio, TX
March 26 – 29, 2017

Waste Solvents to Trash Haulers: Lessons Learned from Hazardous Waste Accidents

Delmar "Trey" Morrison, Ph.D., P.E.
Exponent, Inc.
4580 Weaver Parkway, Suite 100, Warrenville, IL 60555, USA
tmorrison@exponent.com

Michael Stern, Ph.D., P.E.
Exponent, Inc.

Carmen H. Osorio, Ph.D.
Exponent, Inc.

Keywords: Hazardous waste, incident investigation, chemical reactivity

Abstract

Process safety management (PSM) of hazardous chemicals is a core tenet of the chemical, oil, and gas industries. Organizations in these industries have increasingly woven the elements of PSM into their strategic vision, day-to-day operations, and interactions with contractors and vendors. This cultural evolution has, in part, been driven by major accidents in these fields. In these industries, the chemicals, their hazardous properties, and their potential interactions, are usually either well-defined or capable of being well-defined. Control and mitigation strategies for those chemicals can be developed based on this available knowledge. In contrast, companies that handle wastes have experienced catastrophic accidents due to the ill-defined nature of the hazards. This industry handles the cast-offs, dregs, and refuse, which makes identification of the hazards and prevention of incidents more challenging. How do you implement a PSM program when your input streams can vary due to processes at a plant you do not control or monitor? Does the EPA (Environmental Protection Agency) Resource Conservation and Recovery Act (RCRA) provide sufficient guidance of safely handling of wastes? The overriding theme in the incidents is that PSM is more complex to apply to unknown, mixed, and uncontrolled chemicals, both solid and liquid. To assist companies that handle wastes, this paper provides several case studies of fires and explosions involving multiple waste material handling processes at different facilities. The incidents involved numerous matrices from solid industrial wastes, EPA RCRA hazardous wastes, to waste solvents. An analysis of the lessons learned will be presented and used to provide realistic guidelines to reduce safety risks based on a pragmatic implementation of process safety management principles. The paper will also compare OSHA (Occupation Safety and Health Administration) PSM with EPA RCRA approaches to characterizing wastes versus managing the hazards, and will emphasize those elements that played critical roles in the accident case studies.

1 Introduction

This paper discusses several case histories of process incidents occurring at waste handling or processing facilities and compiles lessons learned from those accidents. Depending upon the source, the industry, and the regulatory classification of the waste, industrial waste may pass through a municipal waste transfer station to a landfill or be collected, aggregated and then incinerated by a RCRA (Resource Conservation and Recovery Act) [1] permitted hazardous waste Treatment, Storage, and Disposal Facility (TSDF). Managing waste is inherently more challenging than managing typical chemical process hazards where the identity and properties of the process chemicals can be effectively defined. While many types of hazardous waste facilities test for the properties of incoming waste streams, these tests can be challenging if the waste is phase separated, if the waste is still reacting and changing over time, or if the waste is incompletely identified by the generator. Similarly, investigations of accidents are complicated by the fact that wastes involved in the accident could have contained unknown chemicals with unknown properties.

Several case studies are presented in this paper. The accident locations range from waste transfer stations, waste storage facilities, waste treatment facilities, and a cement manufacturing facility using waste as a fuel source. The wastes involved in the accidents ranged from municipal waste, unregulated solid and liquid wastes, to regulated hazardous solid and liquid wastes. Significant fires and explosions occurred in these incidents. The overarching causes of these incidents were due to the gap between the true properties and the recognized properties of the wastes. While the task of characterizing waste streams may be challenging, the overall process safety and incident investigation principles remain similar. In this paper, we start with a typical decision tree analysis for screening chemical reactivity hazards and modify it for use for waste processing facilities. We will then discuss several case studies involving fires and explosions in waste handling and treatment and provide lessons learned.

2 Background

Application of Process Safety Management (PSM) principles in the chemical industry has been codified for highly hazardous chemicals above threshold quantities through Title 29 Part 1910.119 of the Code of Federal Regulation since 1992 [2]. The Center for Chemical Process Safety (CCPS) subsequently expanded on the process safety model presented by the regulations to develop twenty elements of risk-based process safety [3]. The overlap between the CCPS process safety model and the OSHA PSM regulatory elements (bolded) is represented in the list below (Table 1).

Table 1. Overlap between OSHA (bold) and CCPS PSM elements

1. Process Safety Culture	2. Contractor Management **(h) Contractors**
3. Compliance with Standards	4. Training and Performance Assurance **(g) Training**
5. Process Safety Competency	6. Management of Change **(l) Management of change**
7. Workforce Involvement **(c) Employee participation**	8. Operational Readiness
9. Stakeholder Outreach	10. Conduct of Operations
11. Process Knowledge Management **(d) Process safety information**	12. Emergency Management **(n) Emergency planning and response**
13. Hazard Identification and Risk Analysis **(e) Process hazard analysis**	14. Incident Investigation **(m) Incident investigation**
15. Operating Procedures **(f) Operating procedures**	16. Measurement and Metrics
17. Safe Work Practices **(k) Hot work permit**	18. Auditing
19. Asset Integrity and Reliability **(j) Mechanical integrity**	20. Management Review & Continuous Improvement

Deficiencies in many of these elements can be traced back to the root causes of hazardous materials incidents, as will be illustrated in the case studies in this paper. However, it is uncommon for TSDFs and other waste handling facilities to formally apply PSM to their processes. Usually, facilities follow a prescriptive approach to waste characterization and handling that is based upon the EPA (Environmental Protection Agency) regulations and relies heavily upon the waste generator's characterization of the waste properties [4]. This reliance is not unusual when considering the quantity of distinct and unique waste streams that an individual municipal facility or TSDF may handle and dispose.

The primary hazard identified in the case studies is delayed reaction leading to pressure generation, heat generation, and unintended ignition of packaging or the waste material itself. This is commonly referred to as spontaneous ignition or spontaneous combustion since a piloted ignition source is not necessary. Trash and municipal waste handling facilities are frequently susceptible to delayed fires due to spontaneous combustion. Such ignition events may be caused by unintentional mixing of incompatible chemicals or exothermic oxidation of self-heating materials, such as vegetable-derived oils. For example, the National Fire Protection Association (NFPA) estimated that 4,460 annual fires involving spontaneous combustion of trash occurred between 2005 and 2009 [5].

Recent studies have identified several trends in hazardous waste TSDF fires and explosions [6], [7], [8]. Incidents can occur in any stage of the process from initial storage, waste repackaging, or transfer to the final treatment stage (i.e., incinerator, calciner, or kiln). The consequences of a fire or explosion can accelerate rapidly, depending upon the nature of the initial waste and its proximity to surrounding waste materials, surrounding waste materials' properties, and the adequacy of fire protection and emergency response actions. The causes of fires and explosions typically derive from unknown or unrecognized chemical reactivity issues with the waste material in question.

2.1 Waste generation in the United States

According to the US EPA, 26,237 generators produced 33.6 million tons of hazardous waste, of which over 6 million tons were non-wastewater [9]. Of the 6 million tons, 26% is combusted for energy recovery, 19% is put into landfills, and 13% is incinerated. Over 76% of the non-wastewater hazardous waste is treated off-site by a TSDF.

The EPA statistics also show that 258.5 million tons of non-hazardous Municipal Solid Waste (MSW) was generated in the United States in 2013. While the per capita waste has slightly decreased since the year 2000, the total annual generation continues to climb due to population growth [10]. Approximately 89.4 million tons (35%) of the MSW were recycled in 2013. Another 13% of the total MSW was combusted with some amount of energy recovery. The remaining 52.6% was placed into landfills.

2.2 Resource Conservation and Recovery Act

A hazardous waste can be any waste material with a potential to harm human health or the environment, either by itself or through interaction with other factors. However, for regulatory purposes in the U.S., a waste material must meet specific criteria to be considered as a hazardous waste. These conditions, summarized in Figure 1, are defined by the EPA under the RCRA in Title 40 of the Code of Federal Regulations [1].

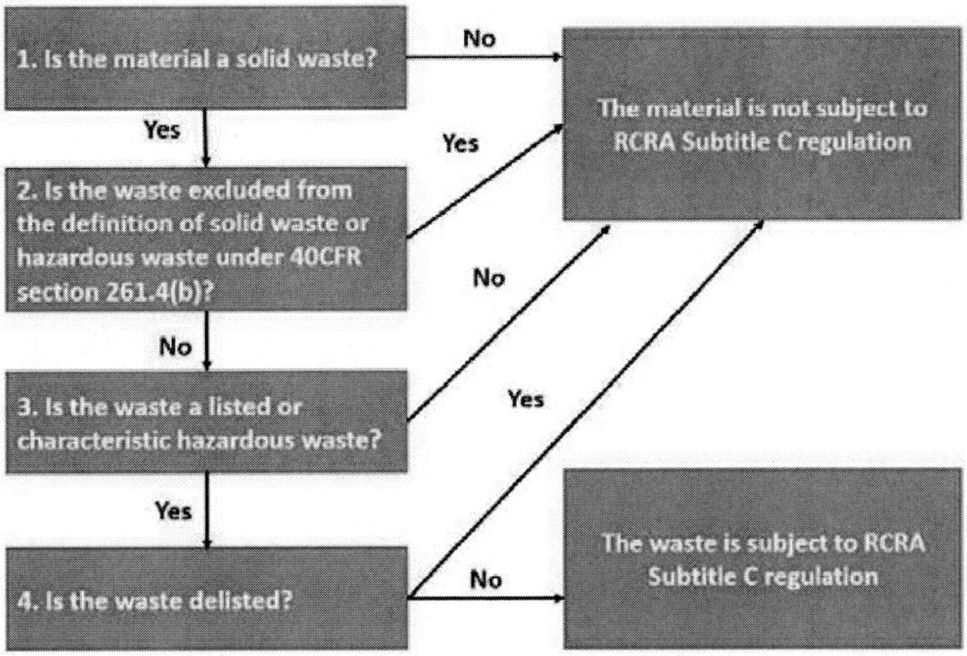

Figure 1. EPA's Hazardous waste identification process [11]

Per the EPA, the first condition of a hazardous waste is to be a solid waste, which paradoxically includes not only wastes that are physically solid, but also wastes that are liquid, semi-solid, or contained gaseous material. A hazardous waste is also required to meet certain characteristics based on its toxicity, ignitability, reactivity, and corrosivity, and needs to be listed in either the F, K, P, or U lists found at the 40CFR section 261 [12].

Some solid wastes are specifically excluded from RCRA regulatory compliance due to different reasons such as impracticability, economy impact, lack of information to justify applying the regulation, regulation by other laws, or they were removed (delisted) from this category. These exclusions refer to conditions (2) and (4) from the hazardous waste identification process shown in Figure 1. For example 40 CFR 261.4(b) excludes household waste and solid waste from commercial or industrial sources (that don't contain solid hazardous waste). It also excludes fly ash waste. These types of waste material may pose physical and chemical hazards, but they are not regulated under RCRA.

2.3 RCRA chemical compatibility testing

As part of the RCRA regulations, hazardous waste generators are required to classify their waste. When TSDFs receive waste, they must perform pre-acceptance analysis to confirm the waste's characteristics and identify incompatible wastes [13]. The EPA states "Compatibility testing may include determining water reactivity (e.g., foaming, heat generating, explosivity), oxidation-reduction potential, spot tests for cyanide and sulfide, among others" [14]. The EPA provides a list of some common potentially incompatible wastes in Appendix V to Part 264. When analysis is required, the EPA has published guidance on how to qualitatively evaluate the compatibility of wastes [15]. The guidance document uses a two-part approach: step-by-step compatibility analysis procedures and then a compatibility chart that considers 41 reactivity groupings. To simplify this analysis when chemical testing is not required, free computer software packages are available for performing quick assessments of chemical compatibility [16].

2.4 Guidelines for evaluating chemical reactivity hazards

According to the U.S. Chemical Safety Board (CSB), a chemical reactive hazard can be defined as "*a sudden event involving an uncontrolled chemical reaction with significant increases in temperature, pressure, or gas evolution that has the potential to, or has caused serious harm to people, property or the environment*" [17]. In order to decide whether a chemical can be considered a reactive hazard or not, the CCPS published a preliminary screening test based on a series of questions summarized in Figure 2 [18].

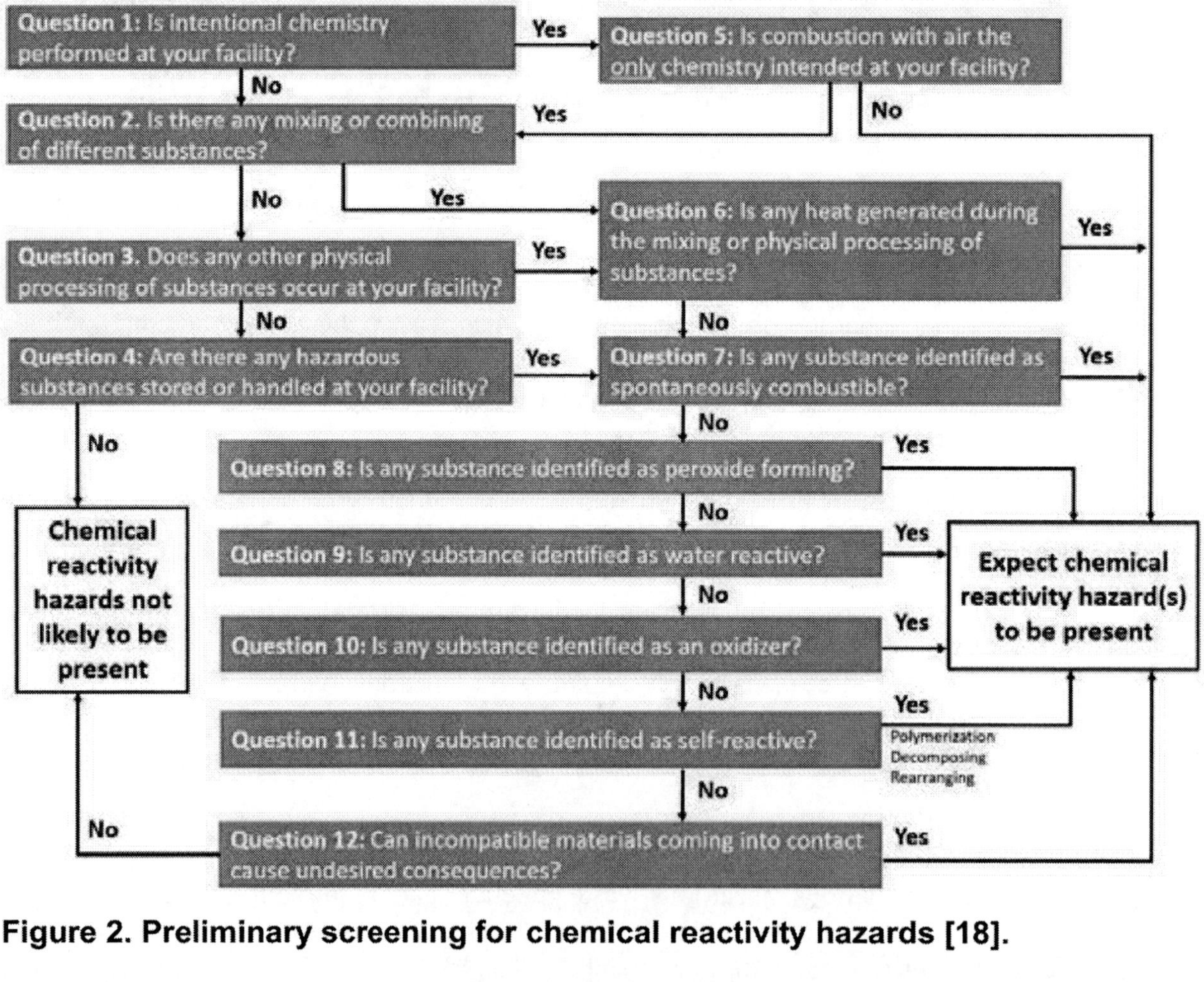

Figure 2. Preliminary screening for chemical reactivity hazards [18].

This test should be conducted on an ongoing basis, in particular when there have been changes in the process/conditions or a new chemical is introduced into the facility. If a chemical reactive hazard is expected to be present at the facility, the CCPS also suggests the implementation of a chemical reactivity management program structured consistent in Figure 3.

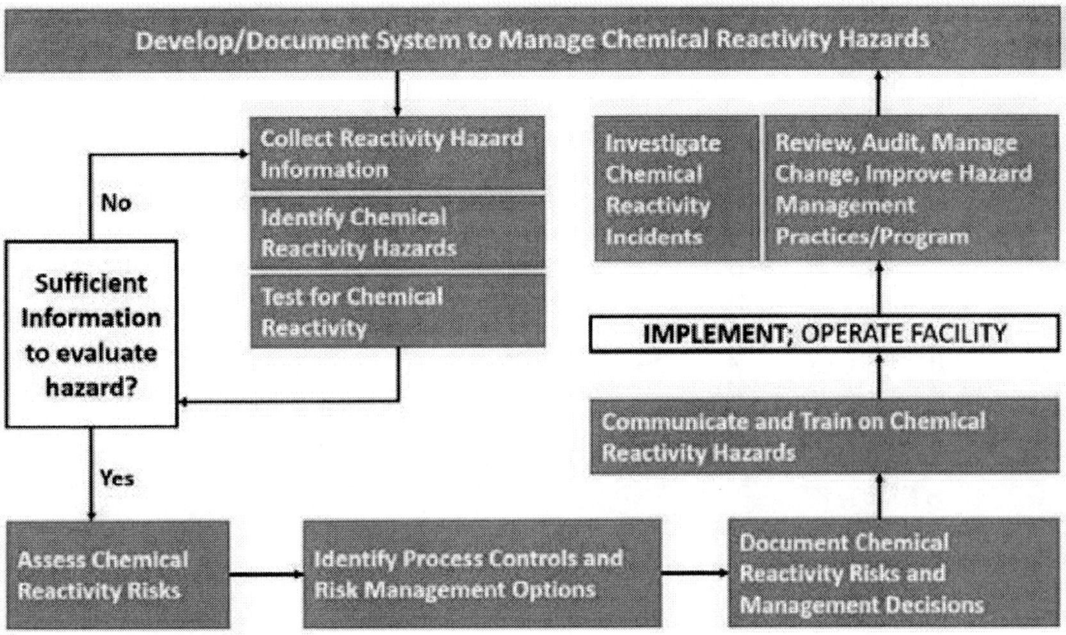

Figure 3. Flowchart for implementing chemical reactivity hazard management [18].

2.5 Why and how an event happened? Protocols for investigating reactive chemical incidents

Given the nature of fires and explosions, especially when involving hazardous waste materials, necessary evidence may be destroyed or lost during the incident or subsequent emergency response. Thus, conclusive determination of the chemistry and proximate causes may be impossible. The investigation process must rely upon the scientific method to develop hypothetical causes, test those causes, and determine the most likely causes and contributing factors.

Two commonly relied upon resources for performing incident investigations per the scientific method are NFPA 921 *Guide for Fire and Explosion* Investigations [19] and the CCPS publication Guidelines for Investigating Chemical Process Incidents [20]. Figure 4 shows the progression of an investigation as described in Chapter 4 of NFPA 921. Two essential concepts, inductive reasoning and deductive reasoning, are used to develop and test hypotheses, respectively. Inductive reasoning is using the analyzed data to develop a consistent hypothesis. Deductive reasoning is used to test the hypothesis by comparing it to all known facts of the accident and scientific literature. NFPA 921 cautions that "[t]he investigator does not have a valid or reliable conclusion unless the hypothesis can stand the test of a careful and serious challenge." Recent editions of NFPA 921 have also added cautionary language regarding the use of *negative corpus* also known as the process of elimination. Investigators should not confuse the lack of an alternative hypothesis for affirmative proof of a hypothesis with no supporting evidence (but that cannot be ruled out with the available information).

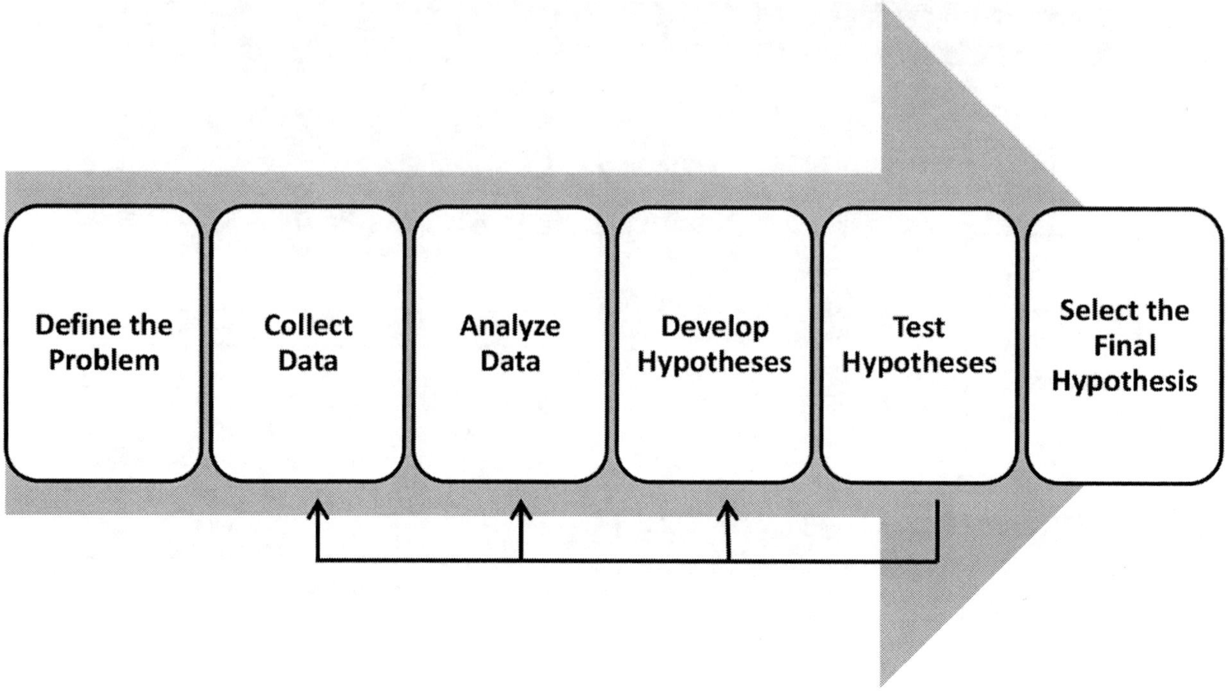

Figure 4. The scientific method of investigation.

The CCPS Guidelines text provides a similar scientific methodology for accident investigation but also offers information specific to industrial accident investigation. For example, information for designing incident management systems within a corporation is provided. Specific investigation and root cause analysis tools commonly employed for investigations, organized by the level of structure, are shown in Figure 5.

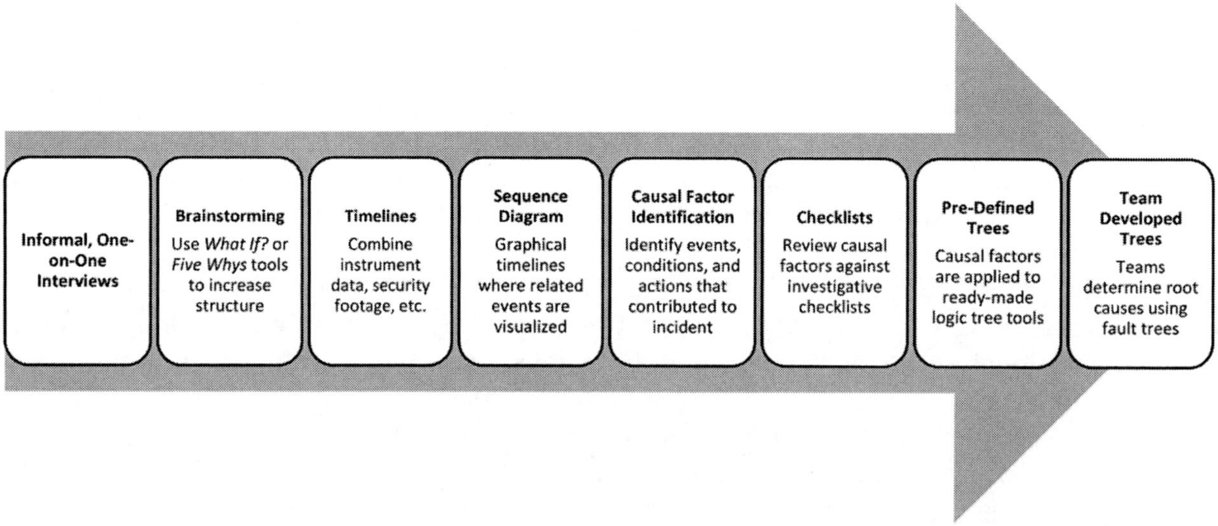

Figure 5. Investigation tools for process incident investigations.

The case studies below describe the causes and lessons learned from MSW handling, solvent reclamation, and TSDF incidents. All of the incidents were characterized by fires or explosions, which caused significant damage and posed potential for catastrophic injuries or fatalities to onsite personnel. Many of the incidents were investigated by external public agencies in addition to the company investigations. The authors also participated firsthand in the investigations of the incidents. Where possible, lessons learned from the investigation process as one element of PSM are shared.

3 MSW Transfer Facility Fires

Two fires, separated by three days, occurred at a MSW transfer station. The first fire involved a trailer of waste awaiting transfer to the landfill, and the second fire involved a different trailer and a large portion of the building and waste contained within. The facility was unsophisticated when considering the element of process safety management, which is understandable because it did not handle reactive materials. As a consequence, early investigation of the causes of the fires relied upon supposition and poor science to conclude that one waste stream was the cause of the fire. The incorrect conclusion adopted by the facility meant that the real cause went unaddressed until a later, thorough engineering investigation occurred.

The waste transfer station was a large, open warehouse structure with two sunken loading bays. The facility handled trash classified as municipal solid waste (MSW) and construction and demolition (C&D) debris. Waste materials were piled on the concrete and moved around using a front end loader. Different materials were commingled by the facility operators, both in piles and in the open trailers located in the loading bays. The commingled waste was periodically hauled to a municipal landfill. The facility did not accept RCRA hazardous waste or liquid wastes. The local regulations and the permit for the facility both acknowledged the reasonable possibility of "hot loads" - in other words, the receipt of loads of waste material that are undergoing self-heating or are already smoldering.

The first fire involved the contents and the rear of a trailer (Figure 6), which was located in a loading bay overnight. The trailer contained a mixture of bottom ash from a fluidized bed coal combustor and other unknown MSW and C&D materials. No formal investigation was conducted to attempt to determine the cause of the trailer fire. The waste material in the trailer was not fully extinguished or overhauled by the fire department, and the facility located the trailer away from the building so that it could eventually burn itself out. The fire witness statements indicated that the contents of the trailer continued to burn for the next several days. The fire damage patterns to the trailer are consistent with a fire originating at the rear wheels, perhaps due to overheated brakes. Three days after the initial fire, a second fire was discovered after a different trailer was loaded with similar waste materials from within the building and held overnight in the undamaged loading bay. The second fire burned to completion and destroyed the building.

Figure 6. Photograph of the trailer damage resulting from the first fire.

An early investigation of the second fire concluded that the combustor bottom ash was the equivalent of pure lime (CaO), caused both fires due to exothermic heat of hydration, and should have been classified as a RCRA hazardous waste. Pure CaO is referred to as "lime" or "quick lime." The reaction of CaO with water vapor results in solid calcium hydroxide, $Ca(OH)_2$, while mixing with excess water creates an aqueous solution, according to the chemical reaction:

$$CaO_{(s)} + H_2O \rightarrow Ca(OH)_2$$

This reaction is exothermic, meaning that it generates heat, with a heat of reaction of -65.27 kJ/mole of CaO reacted [21]. The quantity of heat released is directly proportional to the quantity of CaO and H_2O undergoing the reaction.

The transfer station's incorrect initial conclusion was based on the description of the waste by the generator as "agricultural lime," which is commonly used to describe calcium carbonate ($CaCO_3$), hydroxide ($Ca(OH)_2$), and oxide (CaO) used in farming. Further, the early investigation concluded that the mischaracterization of the waste was a direct cause of the fires. A subsequent engineering investigation analyzed that conclusion and also identified several plausible causes including an electrical fire, spontaneous combustion of MSW or C&D, or re-kindling of the smoldering remains of the first fire after being transported across the facility.

Chemical analysis of the combustor bottom ash and a review of the literature indicated that the ash did not contain sufficient CaO to cause spontaneous combustion through hydration and was excluded under 40 CFR 261.4(b). The residual CaO in combustor tram will undergo hydration by reacting with water (H_2O) in either a vapor (e.g., atmospheric) or liquid form. The hydration reaction for residual CaO would run largely to completion within a few days when exposed to the outdoor atmosphere [22]. Comparison samples of fresh combustor bottom ash revealed that it contained approximately 10% CaO by weight.

If concentrated CaO is mixed with liquid water, the hydration reaction can lead to very high temperatures capable of igniting combustible materials. However, it is unlikely that these conditions existed in the current matter. The maximum temperature reached by the combustor ash via this reaction, if any unreacted CaO is present, will be tempered by the presence of liquid water and surrounding solid material (i.e., $CaCO_3$, $Ca(OH)_2$, $CaSO_4$, sand, and onsite waste materials). The water will absorb the released heat and act to cool the material as the water evaporates; the other solid species will act to absorb the liberated heat and increase in temperature.

The false conclusion that the ash waste was CaO, thus causing the fires, highlights the importance of precise descriptions for waste streams even if they are not bound to the RCRA regulation. Effective description can help investigators, who may be unaware of the precise waste compositions, more effectively assess the important properties of the waste stream after an accident. Similarly, investigators must be careful when examining the properties of non-RCRA waste streams as their descriptions may be more colloquial. Conclusions should not be drawn until a more detailed composition of the waste can be obtained through chemical analysis or discussion with the generator.

4 Solvent Recovery Plant Vapor Cloud Explosion

In 2009, a release of ignitable liquid and vapors, followed by catastrophic vapor cloud explosion and subsequent fires, took place in a solvent recovery plant. The explosion severely injured two employees, destroyed part of the plant (see Figure 7), and broke windows in neighboring facilities and residences. The explosion was investigated by the Chemical Safety and Hazard Investigation Board (CSB) [23].

The solvent recovery plant (Figure 8) was located at a facility that handled hazardous and non-hazardous waste materials that were received via tanker truck, drums, totes, and lab packs. The facility performed fuel blending, solvent recovery, and waste consolidation. Incoming dirty solvents were unloaded into tanks in the dirty tank farm. Clean solvent was recovered from the dirty solvents via one of three distillation units (Unit 1, Unit 3, and Unit 4). The recovery process involved the circulation of distillation bottoms back into the source tank in the dirty tank farm, while purified "clean" solvent was condensed and transferred to the clean solvent tank farm. Final waste liquids and sludge were shipped offsite to a waste incinerator as blended fuels.

At the time of the incident, the three Processing Units were being run by a single outside operator. Unit 1 was used to recycle a paint thinner material whose primary compound was acetone, while Unit 3 was processing a tetrahydrofuran (THF)/water mixture. Unit 4 worked as a closed circuit reflux-distillation unit where a mixture of THF and water near the azeotropic concentration was processed.

Figure 7. Aerial view of the solvent recovery plant after the incident from the CSB report [23].

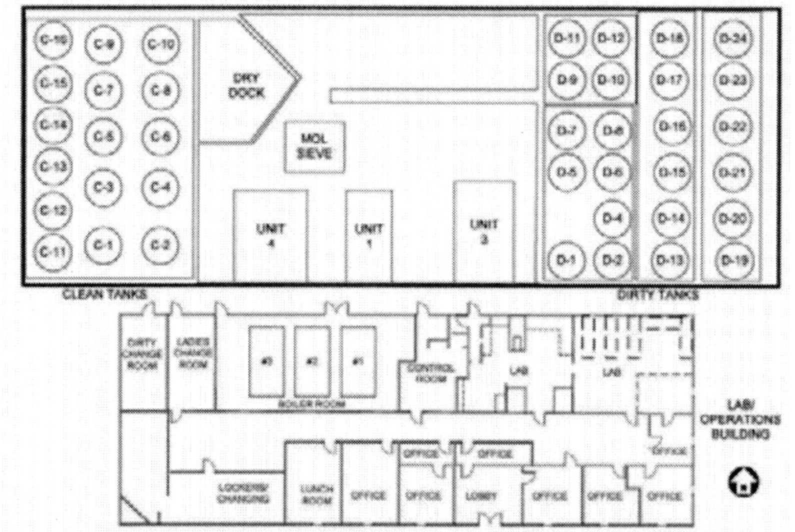

Figure 8. Facility layout (not to scale) [23].

Three of the six workers onsite reported hearing a loud noise like a pressurized release, accompanied by a strong solvent odor and a visible aerosol/or vapor cloud coming from the Dirty Tank Farm prior to ignition. Security camera footage from onsite and a nearby business supported the conclusion of a pressurized release, likely of Tetrahydrofuran (THF) solvent mixture; however, the exact composition and source of the flammable material released could not be determined.

The incident investigation concluded that the vapor cloud spread out from the Dirty Tank Farm, entered the laboratory/Operations Building, and ignited inside the Boiler Room causing, an initial explosion. Consequently, additional flammable material within the Dirty Tank Farm was also ignited causing a blast pressure wave (vapor cloud explosion) and subsequent pool fire that burned for several hours.

The CSB's findings from this incident included the following facility siting concerns:

- Lack of hazard analysis to determine the risks associated to the proximity of occupied buildings to processes: The most critical injured employees were at the Laboratory/operations building at the time of the incident. The nearby building was occupied with non-essential personnel during the day shift.
- Inappropriate area classification: Two boilers were in service in the Laboratory/Operation Building, where the ignition source was identified. The flammable vapor cloud was ingested and ignited by one of the boilers.

Although the facility was covered by the OSHA PSM standard, the CSB found that the TSDFs did not have industry-specific guidance for fire safety and facility siting.

5 Fires and Explosions in Storage Operations at TSDFs

Several fire incidents have occurred during storage of hazardous materials at TSDFs. The ultimate fate of hazardous materials is often destruction through incineration. The type of incinerator and local environmental emission limits dictate handling activities with the hazardous waste. To manage the limits on the heat of combustion for the incinerator, heavy metal emissions, and other factors, TSDFs commonly repackage wastes by either combining similar waste streams in the same container or by breaking up larger containers into smaller container sizes. For example, if the bulk heat of combustion of the 55-gallon drum of waste received by a TSDF is too high for the incinerator's operating limit, then the 55-gallon drum must either be metered (if a liquid) into the incinerator, or the drum must be subdivided into smaller containers (e.g., 30-gallon drums).

Thus, hazardous materials often pass through the following stages at a TSDF:

1. Receipt and characterization
2. Temporary storage
3. Repackaging – either bulk up or break down
4. Storage awaiting disposal
5. Final disposal

The waste generator must provide specific information on the hazardous waste that they are providing to the TSDF. The TSDF oftentimes works with the generator to help characterize the waste stream as part of the initial contracting process. The TSDF typically samples and performs basic characterization of individual wastes as they are received. However, this process is imperfect because the TSDF cannot be guaranteed that a set of samples from distinct locations in a waste container represent the bulk of the waste in that container. The potential deviation of actual properties from the results of this sample may grow significantly for larger-sized containers (e.g., a liquid tote or a 1 m^3 box), wastes that are heterogeneous, wastes with layers or pockets of different composition, or large volume wastes that are not completely sampled (e.g., sampling a subset of containers).

After initial contracting and waste specification, a TSDF will develop a plan for handling that waste stream, which will include sampling, characterization, and verification that as-received waste meets the contract specification. This process falls under Process Knowledge Management in the 20 elements of CCPS PSM. Deviations of as-received waste from the specification should be addressed under a Management of Change element, but the TSDF may not have a robust change management policy. Additionally, changes from the specification may not be identified during the characterization process as discussed above.

Incident investigation after a hazardous materials incident at a TSDF can be very challenging. A TSDF storage area may contain hundreds or even thousands of individual waste containers (see Figure 9). A fire or explosion in the storage area will destroy or damage the waste containers, displace waste containers, and mix the residue from involved containers. All these aspects make determination of the cause of the incident difficult. The likelihood of determination is usually directly related to whether the incipient stages of the fire or explosion were witnessed. The following three case studies provide incident summary and investigation results for a progressively larger TSDF fire incidents in order to characterize the difficulty in cause determination and lessons learned.

Figure 9. Example of TSDF storage warehouse.

5.1 Delayed Ignition - No Repackaging – Cause Undetermined

An investigation was performed on a fire that occurred in a small TSDF's storage bay. The facility handled a wide variety of wastes including flammables materials and acids. The facility included storage bays, laboratories, treatment equipment and office areas. In the storage bay where the fire occurred, several different types of containers were present including:

- Plastic buckets
- Plastic, steel, and paper fiber drums
- Steel encaged plastic totes known as intermediate bulk containers (IBCs)

The 55-gallon drums were placed on wooden pallets in sets of four. The pallets of 55-gallon drums and IBCs were stacked two or three levels high. A photograph of the bay after the fire is shown in Figure 10.

Figure 10. Photograph of storage area after fire.

Based on witness statements, the fire in the storage bay was preceded by a loud hissing sound and a pressurized release of a reddish dust/mist from a container near the center row of the image. The witness did not see the exact container where the release originated, but was able to place it within a few pallet stacks in the area. Immediately after the pressurized release, a fire ensued in the area. Due to fire suppression efforts, the fire damage was contained to the storage bay where it originated. The investigation focused on the adjacent rows containing the waste drums that were damaged. The exact location of each waste container in the neighboring pallet stacks was reconstructed based upon storage records, remaining containers, damage patterns, and witness statements about container placement. The origin was confirmed to be one pallet stack in the center row.

The containers in the center row had not been repackaged or disturbed from the time of receipt until the incident; therefore, commingling of incompatible wastes or mixing of a reactive pocket within the source drum were refuted as potential causes. Delayed decomposition of waste material in one of the 55-gallon drums is the likely source of the pressurization. Delayed decomposition may be the result of chemical incompatibility or thermal instability leading to a temperature rise, gas generating reaction, or both. Eventually, the pressure increased to the level that the source container ruptured. It was most likely that this release occurred from either a steel drum or a plastic drum given their ability to contain some degree of pressurization compared to a fiber drum. After the initial pressurized release and fire, several other containers were damaged and ruptured as a result of the growing fire.

Based on the photo-documentation of the scene, it was possible to identify several potential drums that showed evidence of pressurization and/or rupture. However, due to the extent of the damage, it was not possible to identify the serial number on all of the containers, and therefore, their origin and contents. Examination of the fire damage showed evidence of burned away paint, burned pallets, and melting of non-metal containers. Based on the damage patterns, it was possible to rule out some waste containers including a set of drums containing phosphoric acid. A ground-level area containing four drums was identified as the likely area of origin, but further specificity was not possible due to the extent of the damage and several other drums that could not be conclusively eliminated. In a regular facility, the containers of all the stored containers would be known, providing valuable information to incident investigators. In a waste handling facility, significantly less certainty is available about the exact contents of each container. The ability for many of the waste container to have been the source of the pressure release significantly complicated this specific investigation. The final determination as to which container was the source could not be determined given the degree of damage.

The extent of damage does not always preclude determination of the likely cause of a fire in storage. A similar TSDF facility experienced a fire in a storage bay, which consumed all of the hazardous waste and combustible containers in the area (see Figure 11). The point of origin of the initial fire could be more precisely placed than the incident just discussed because a nearby operator observed the location of the incipient fire at ground level at a specific pallet stack. The source of the fire was most likely a leak of fuming red nitric acid that spilled onto a wood pallet holding the drums. The drums were stacked on combustible wood pallets, and fuming nitric acid is a strong acid capable of igniting such materials. The facility did not identify the potential consequences of a spill of the acid; thus, a noncombustible pallet (i.e., steel) was not selected. Use of a what-if strategy for packaging and storage selection may have prevented the fire (but not necessarily the leak).

Figure 11. Another TSDF storage bay after extensive fire damage.

5.2 EQ Industrial Services Warehouse

A hazardous waste TSDF was complete destroyed due to a fire for which the initial cause was undetermined (see Figure 12). The magnitude of this incident forced thousands of local residents to be evacuated in two days. The CSB investigated the cause of the incident, and identified several hypothetical causes. Their recommendations focused on eliminating those hypothetical causes.

Figure 12. Facility after the incident [24].

Based on witness statements, a strong chlorine odor and plume of smoke came from the TSDF on the night of the incident. Initial firefighters responding to the scene reported that a small fire started in one of the hazardous waste storage bays and rapidly spread to the other locations of the facility, including a bay containing many 55-gallon drums of flammable solvents. The incident investigation revealed that the fire originated in a storage bay where containers of solid chlorine-based pool chemicals and unspent aircraft oxygen generators awaiting for final disposal.

Many incidents have occurred involving storage of chlorine-based pool chemicals (see for example [25]-[28]), and around the time of the incident, the NFPA was pursuing research into fire protection considerations for pool chemicals in warehouse settings [29]. Concentrated calcium hypochlorite or sodium dichloroisocyanurate containing pool chemicals pose potential thermal runaway concerns if contaminated with organic material or if stored at elevated temperature. Given their nature as solid oxidizers, pool chemicals would also serve to accelerate fire growth. The witness observations of a chlorine-odor associated with the smoke likely indicate that these chemicals were present near the fire origin, but the investigators could not conclude that these were cause of the fire.

Although the first material involved could not be determined due to the extent of damage, the CSB investigation findings focused on chemical oxygen generators [30]. The CSB concluded that chemical oxygen generators most likely contributed to the intensification of the fire, which quickly spread to an adjacent bay where flammable materials were stored. The investigation found that the unspent oxygen generators, coming from an aircraft maintenance facility, were not properly packaged by the aircraft maintenance personnel, and were misidentified as general oxidizer waste on shipping paperwork. A lesson learned from this incident was to consider oxygen generators that have exceed their service life as still posing a risk of heat and oxygen generation if accidentally actuated. Therefore, the CSB recommended that unspent oxygen generators should be actuated before they are transported [30].

5.3 Teris El Dorado Warehouse

This incident occurred in a warehouse at a TSDF that contained 4000 to 5000 drums of different wastes. The facility operated two rotary kiln incinerators for solid disposal and thermal oxidation for liquid incineration [31]. The incident was witnessed by an operator who saw a drum rupture, which quickly spread fire throughout the warehouse contents (see Figure 13). The entire contents of the warehouse were consumed, damaged, and/or released during the incident. No injuries or fatalities were reported; however, nearby residents were evacuated during the emergency response.

The investigation found that the fire was caused by spontaneous combustion of sodium chlorate solid waste recovered from a rail car accident site. This material was brought to the facility in $1m^3$ boxes to be incinerated shortly before the incident. However, the waste generator incorrectly profiled this material as containing only sodium chlorate in soil; the waste actually contained vegetative matter, rail car components, and portions of railroad ties. The organic contaminants are incompatible with the strong oxidizer and would act as fuel sources, allowing spontaneous ignition of the mixture. The material was repackaged into smaller drums and a portion was moved into the warehouse prior to the fire. The remainder of the material had not been relocated at the time of the fire. This residual material was isolated in a trailer onsite. The trailer containing the residual material underwent spontaneous ignition several days after the original fire incident, which supported the conclusion that the sodium chlorate waste material was the cause of the fire.

Figure 13. Aerial image during fire event [31].

Given the non-homogeneous nature of the waste material received, grab sampling of a single container was insufficient to fully characterize the mixed material. In contrast to liquid sampling, a thorough visual examination of the mixed, non-homogeneous waste material may have identified the presence of organic materials in the oxidizer and soil waste. Further, the use of "soil" in the waste analysis plan submitted by the generator may have triggered a further review by the TSDF since soil typically contains a significant organic component. It is likely that the repackaging activity mixed reactive sodium chlorate with fresh organic fuel materials that led to the delayed spontaneous combustion event. After the incident, the facility enhanced their material analysis and handling plans for wastes containing solid oxidizers.

6 Fires & Explosions during Incineration Processes

6.1 Explosion during Solid Waste Homogenization

An explosion occurred at an incineration facility that processed hazardous and non-hazardous solid and liquid wastes through thermal destruction in a rotary kiln. Drums of waste were homogenized in a nitrogen-inerted shredder. Drums were scheduled to be incinerated and assembled onto pallets based upon the waste characteristics. The system automatically moved individual pallets into an airlock, purged the airlock with nitrogen, then dumped the pallets into the already inerted shredder chamber. The shredder was used to shred steel drums as well as plastic drums. After shredding, the homogenized waste was conveyed to the rotary kiln via a screw conveyor.

The pallets, packaging materials, and waste materials are combustible materials, and flammable materials may also be processed based on the incinerator waste schedule. Due to the potentially flammable nature of the waste and the likelihood of mechanical sparks and heat generation from the shredder, the system was inerted with nitrogen gas. The nitrogen inerting function was performed by using a fixed timing of nitrogen addition. Although the system did not use an oxygen measurement control loop to dictate the duration of purging, it did use oxygen measurement as a permissive control for the automated conveyor movements and door opening between chambers in the system.

A central incinerator control room at the facility monitored several operating parameters, including the oxygen concentration and temperature within the inerted system. Since one of the primary safety mechanisms for the unit was a reduced oxygen atmosphere, the unit was not intended for processing oxidizing materials including solid oxidizers, oxygen containers/cylinders, explosives, and/or water reactive materials. During one waste load, an explosion occurred in the shredder chamber, which damaged the unit and caused explosion panels to open on the shredder chamber and airlock.

An approximation of the oxygen concentration data and shredder chamber temperature is provided in Figure 14. There was a lag noted in the oxygen sensor response due to the delay for gas to traverse the sampling leg to the sensor. The temperature measurement had less of a response lag; thus, the sudden temperature rise in the shredder chamber actually occurred coincidentally with the explosion and the opening of the explosion vents. On the night of the incident, an additional waste load was fed into the unit, and shredding commenced less than one minute after it was added. Approximately one minute after shredding began, the oxygen concentration started to increase from 2% to over 13% over a period of 20 seconds. According to the recorded data, the temperature in the shredder chamber suddenly increased at approximately the same time as the oxygen concentration increased. This temperature increase is consistent with the heat generated by the explosion within the shredder chamber. An engineering analysis concluded that the most likely source of the rapid oxygen influx was from an item in the waste stream that released oxygen when shredded. All potential sources of air via external sources were examined and rejected.

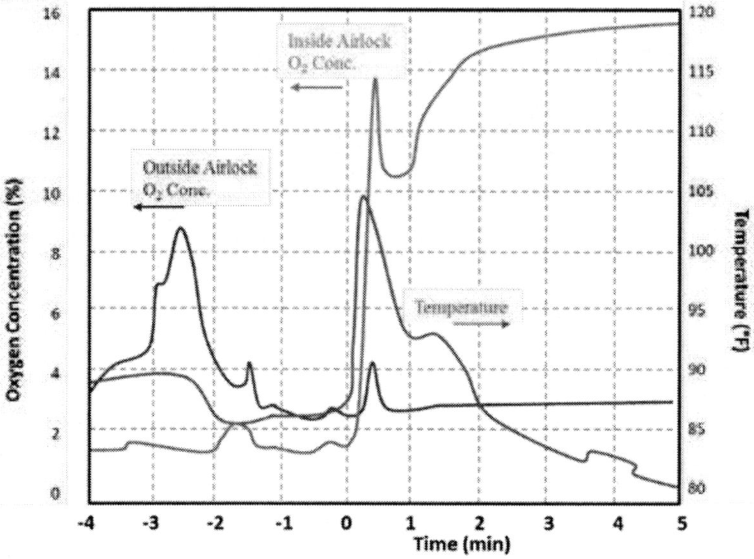

Figure 14. Process oxygen concentration and temperature during incident.

The explosion in the shredding system demonstrated the serious consequences of processing waste with unintended properties. In this case, a system that was intended to operate while inerted inadvertently contained compressed oxygen, resulting in a damaging explosion. While the waste was screened prior being fed into the shredder, the screening process did not identify and divert the oxygen-containing waste. In this case, the explosion mitigation system worked as intended by relieving the explosion pressure through explosion vents. The system was damaged, but the damage was not catastrophic, and there were no injuries.

6.2 Spontaneous Combustion at a Calciner

A fire occurred at a cement production facility that used an engineered solid fuel composited from MSW and other industrial wastes to co-fire in the calciner. The engineered fuel was a blended mixture of solid wastes including paper, plastic, and sawdust as well as non-hazardous waste liquids, slurries, and sludge. Because the waste streams possessed variable energy content, a qualifications process (including laboratory analysis) was used to determine the appropriate mixture for creating a consistent engineered feed. A block flow diagram of the shredding, blending, and mixing process is shown in Figure 15. The engineered fuel had the potential for self-heating due to containing oxidizing materials such as polymerizing agents and vegetable-derived oils. Material handlers inspected the stockpiles for external signs of high temperature due to self-heating and excavated the piles to identify internal high temperatures. Infrared pyrometers were used for surface temperature measurements. The temperatures were monitored with the handheld pyrometers and logged prior to the loading the material into trucks for transport to the cement production facility.

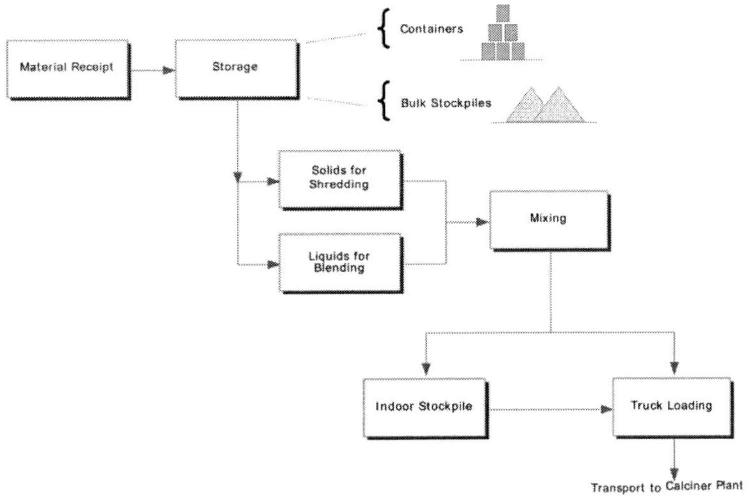

Figure 15. Block flow diagram of the engineered solids feed preparation process.

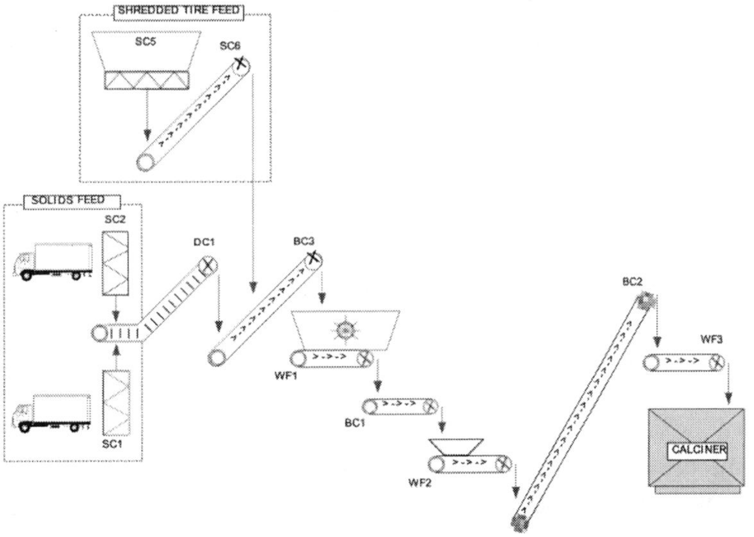

Figure 16. Conveyor flow diagram between truck delivery and the calciner.

The engineered fuel was transported to the cement production facility in tractor trailers for eventual feed into the calciner unit. A conveyor flow diagram is provided in Figure 16. The engineered fuel was unloaded into enclosed hoppers that discharged into horizontal screw conveyors. The horizontal screw conveyors discharged into a central drag conveyor that delivers the fuel to a belt conveyor which also received a separate shredded tire feed. The combined feed then traveled through an additional series of hoppers and conveyors before being deposited on a 1,600 feet long enclosed belt conveyor that climbed from ground level to the upper levels of the calciner structure. The belt of the long conveyor was originally a flame-resistant belt material but had been subsequently patched with non-flame resistant material. The total travel time of the engineered fuel was estimated to be 20-30 minutes by the cement facility's personnel.

A fire was discovered overnight soon after the conveyor system was restarted after a two-day outage. During that two-day period, the belts were left loaded with the engineered fuel and combined feed. When the kiln was restarted, the system was operated for approximately one hour then shut down due to an upstream clog during which a fire was discovered near the head section of the long conveyor. The belt failed and broke as a result of the fire.

An investigation into the fire cause determined that, while various friction-based ignition sources (belts, bearing, etc.) did exist within the conveyor system, the most likely cause of the fire was spontaneous ignition of the engineered feed. This was substantiated because three days after the fire, a stockpile of the engineered feed that had been fed to conveyor prior to the fire enflamed at the storage warehouse. Although staff had monitored stockpiles at the feed mixing facility for external signs of ignition, external methods can be insufficient to detect self-heating inside the pile. After exposing the self-heating pile to oxygen during transport and handling, self-heating may accelerate and lead to spontaneous ignition. The facility implemented a strategy that avoided leaving materials on conveyors for long periods of time. Through the use of conveyor interlock strategies and external dump locations, the system can be operated to empty conveyors in the event of an unintended shutdown.

6.3 TDI Waste Rail Tank Car

An explosion occurred at another TSDF cement plant in the liquid fuels rail tank car unloading area [6], [32]. As consequence, the tank ruptured and was propelled about 750 feet away. Other tank cars, as well as storage tanks located nearby, were also heavily damaged (Figure 17). No injuries or fatalities were reported; however, total property losses were estimated at nearly $8.2 million.

This facility incinerated liquid hazardous waste streams as supplementary fuel in rotary kilns, and used to receive several tank cars containing column bottoms waste from a toluene diisocyanate (TDI) manufacturing process. TDI is a monomer feedstock used in the manufacture of polyurethane polymers. The waste generator had stored tank cars containing this type of waste material for several years, and reprocessed them as the opportunity arose. Prior to the incident, several tank cars were shipped to the TSDF for incineration, and had been successfully emptied, with contents incinerated in the kilns. Despite the TDI waste was partially polymerized in most tank cars, the car's contents were non-homogeneous and did not have identical properties between individual tank cars.

Figure 17. Locations (1) and (2) indicate positions of the tank car's vessel before and after the explosion, respectively [33].

In order to decrease the viscosity of the polymerized TDI, so that it was pumpable to the kilns, the facility had successfully used external steam heating through the built-in steam jackets on the tank cars.. A schematic of the approximately 30,000 gallon tank car is provided in Figure 18. While attempting to unload the incident TDI material, the tank car vessel was steam-heated in two opportunities to ensure the waste mixture was sufficiently fluid to be pumped. After three unsuccessful unloading attempts, the shipment was rejected, but the tank car was left in the unloading location. The explosion incident took place two days after the last heating attempt.

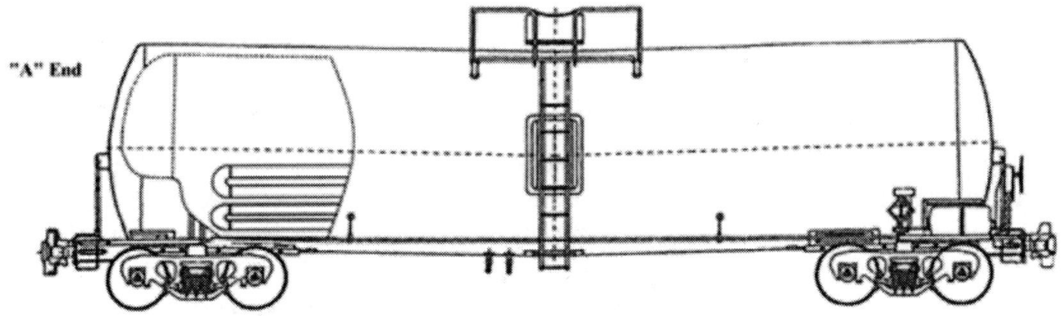

Figure 18. TDI tank car schematic with cutaway showing the external steam heating coils. [20]

Retained samples of the original material and residue ejected during the incident were analyzed, and the presence of self-polymerized TDI, TDI monomer, and ureas were determined. Thermal analysis of the materials indicated that the polymerized TDI and TDI monomer exhibited an onset temperature for thermal runaway that was much higher than possible with the steam heating system. A chemical hazard assessment revealed that the cause of the explosion was the runaway thermal decomposition of a contaminant mixed with the TDI waste. The TSDF operator relied on the waste generator for information regarding the waste composition and safe heating procedures. In this case, the waste generator supplied incomplete waste characterization data and inadequate instructions and warnings for heating and handling the TDI waste.

Four critical factors and lessons learned were identified from this incident [6] [20]:

(1) Use of inappropriate tank car design that was unsuitable for this service, creating difficulties for unloading of TDI solid waste.

(2) Storing of TDI waste for several years in a warm climate

(3) Providing inadequate instructions and warnings for unloading the TDI waste.

(4) Commingling an unknown contaminant with the TDI waste; this last factor was the most serious error in this incident.

7 Discussion and Conclusions

In the case studies presented above, hazardous properties that were not known to exist led to accidents. In several cases, the accidents were the result of a chemical reaction, leading to a release of heat or pressure that caused a storage vessel to fail. Chemical reactions due to the mixing of incompatible chemicals, storage of unstable chemicals, or heating of thermally unstable chemicals occurred in several of the case studies described. While characterization is a significant preventative safeguard for these types of failures, it may not always be possible to determine if a waste stream is completely stable based on testing; therefore, facilities may still need to consider the unlikely event of an exothermic or pressure-generating reaction within their waste streams. This may be accomplished by using separation distance or barriers between certain waste streams, fire protection systems, or fast turnaround from receipt through final processing.

In several incidents, oxidizers including oxygen tank, oxygen generators/tanks and sodium chlorate, are involved in the initiation of the incident or contribute to the severity of the incident. Because many waste streams contain flammable or combustible materials, enriched oxygen atmospheres or the presence of chemical oxidizers increases the likelihood and severity of a fire or explosion. While in many process operations, inerting atmospheres may be an effective and robust protective measure, due to the possibility of oxidizer ingress in mis-identified waste, additional protections such as explosion vents may be required. The following flowchart was created based on the cases presented in this paper, and may serve as a guideline for investigating suspected waste streams and their role on initiating a fire or explosion event.

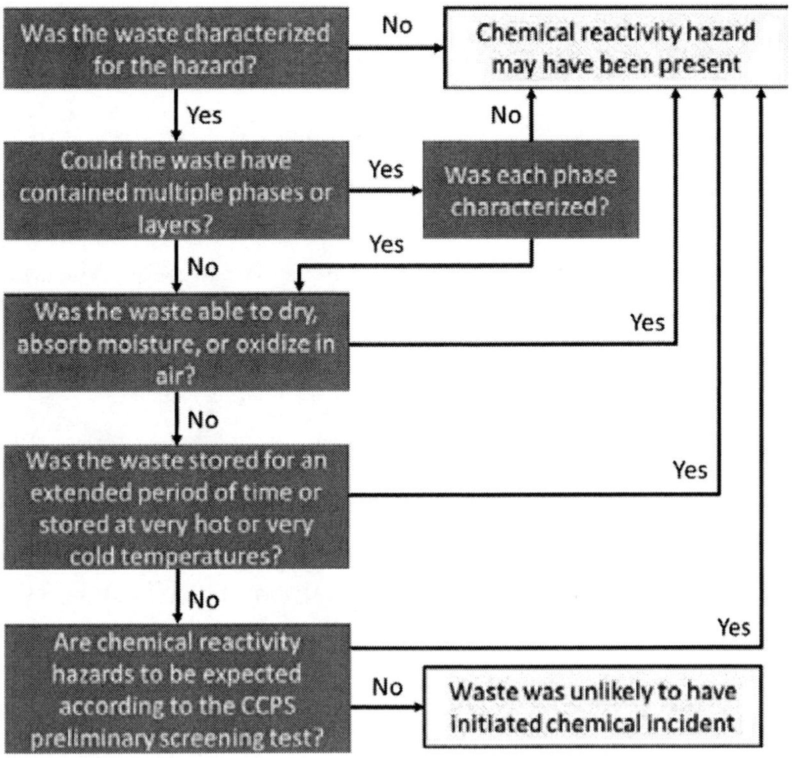

Figure 19. Role of suspected waste stream on initiating a fire or explosion.

8 References

[1] U.S. Environmental Protection Agency (EPA) Resource Conservation and Recovery Act, Title 40 of the Code of Federal Regulations.

[2] 29 CFR §§1910.119

[3] Center for Chemical Process Safety (CCPS). Guidelines for Risk Based Process Safety. Center for Chemical Process Safety, American Institute of Chemical Engineers, New York, NY, 2007.

[4] U.S. Environmental Protection Agency: Resource Conservation and Recovery Act (RCRA) Laws and Regulations, available at https://www.epa.gov/rcra. Accessed on February 15, 2017.

[5] Evarts, B. "Fires Caused by Spontaneous Combustion or Chemical Reactions," National Fire Protection Association, November 2011.

[6] Cox BL, Carpenter AR, Ogle RA. Lessons learned from case studies of hazardous waste/chemical reactivity incidents. Process Safety Progress 2014; 33(4):395–398.

[7] E.C. Buc. "Hazardous Waste Treatment, Storage, and Disposal Facility Fire Code Gap Analysis, Final Report," The Fire Protection Research Foundation, 2015.

[8] Environmental Protection Agency. Report on Emergency Incidents at Hazardous Waste Combustion Facilities and Other Treatment, Storage, and Disposal Facilities (TSDFs). EPA 1999

[9] U.S. Environmental Protection Agency (EPA): 2015 Biennial Report Summary, available at: https://rcrainfo.epa.gov/rcrainfoweb/action/modules/br/national/view. Accessed January 15, 2017.

[10] U.S. Environmental Protection Agency: Advancing Sustainable Materials Management: 2014 Fact Sheet. US EPA, November 2016.

[11] U.S. Environmental Protection Agency, Available at https://www.epa.gov/hw. Accessed on February 15, 2017.

[12] 40 CFR §§261

[13] 40 CFR §§264/265.17

[14] U.S. Environmental Protection Agency. Waste Analysis at Facilities that Generate, Treat, Store, and Dispose of Hazardous Wastes – Final: Guidance Manual (EPA 530-R-12-001). April 2015.

[15] Hatayama H.K., J. J. Chen, E. R. de Vera, R. D. Stephens, and D. L. Storm. A Method for Determining the Compatibility of Hazardous Wastes. U.S. Environmental Protection Agency, Washington, D.C., EPA/600/2-80/076 (NTIS PB80221005), 1980.

[16] National Oceanic and Atmospheric Administration (NOAA), "Chemical Reactive Worksheet," NOAA, available at http://www.aiche.org/ccps/resources/chemical-reactivity-worksheet-40. Accessed January 15, 2017.

[17] Davis E.M., D. Silva, and J. Murphy. The CCPS Chemical Reactivity Evaluation Tool. 8th Global Congress on Process Safety. April 2012.

[18] Center for Chemical Process Safety (CCPS). Essential Practices for Managing Chemical Reactivity Hazards. Center for Chemical Process Safety, American Institute of Chemical Engineers, New York, NY, 2013.

[19] National Fire Protection Agency. NFPA 921: Guide for Fire and Explosion Investigations

[20] Center for Chemical Process Safety (CCPS). Guidelines for Investigating Chemical Process Incidents. 2nd Ed. 2003

[21] Anthony, EJ, L Jia, M Caris, F Preto, and S Burwell, "An examination of the exothermic nature of fluidized bed combustion (FBC) residues," Waste Management Journal, 19 (1999).

[22] Serris, E, L Favergeon, M Pijolat, M Soustelle, P. Nortier, RS Gartner, T Chopin. "Study of the hydration of CaO powder by gas–solid reaction," Cement and Concrete Research 41(2011) 1078-1084.

[23] U.S. Chemical Safety Board (CSB). Veolia ES Technical Solutions, L.L.C. Case Study. Report 2009-10-I-OH. Washington, D.C. 2010

[24] U.S. Chemical Safety Board (CSB). EQ North Carolina Case Study. Report 2007-01-I-NC. Washington, D. C. 2008.

[25] Gray B.F. and B.W. Halliburton. 2000. "The thermal decomposition of hydrated calcium hypochlorite (UN 2880)." *Fire Safety Journal*. 35.3: 223-239.

[26] Gray B.F. 2002. Spontaneous Combustion and Self-Heating. In SFPE Handbook of Fire Protection Engineering, eds. P.J. DiNenno, D. Drysdale, C.L. Beyler, W.D. Walton, R.L.P. Custer, J.R. Hall and J.M. Watts, Quincy, Massachusetts: National Fire Protection Association, 211-228.

[27] Ogle R. A, Carpenter A. R. Toxic gas release caused by the thermal decomposition of a bulk powder blend containing sodium dichloroisocyanurate. Process Safety Progress 2003; 22:75–82.

[28] Ogle RA, Morrison DR, Carpenter AR, Su YS. Missed opportunities in reactive chemical hazard evaluations. Process Safety Progress, March, 2006, 25(1):2-7.

[29] Buc E.C. and D.J. Hoffman, "Development of an Enhanced Hazard Classification System for Oxidizers Research Project," The Fire Protection Research Foundation, One Batterymarch Park, Quincy, Massachusetts, April, 2006

[30] U.S. Chemical Safety Board (CSB). Safety Advisory Dangers of Unspent Aircraft Oxygen Generators (2007-I-NC-01-SA). 2007

[31] U.S. EPA, "Teris El Dorado Explosion," downloaded on March 13, 2017, https://response.epa.gov/site/site_profile.aspx?site_id=NRC745957.

[32] Ogle R.A, Carpenter AR, Morrison DR. Explosion of a railcar containing toluene diisocyanate waste. Process Safety Progress 2004 Dec; 23(4):316-320.

[33] National Transportation Safety Board. Rupture of a Railroad Tank Car Containing Hazardous Waste Near Clymers, Indiana. Hazardous Materials Accident Report NTSB/ HZM-01/01. Washington, DC. 2001

479039 ExxonMobil Torrance Refinery FCC ESP Explosion

Abstract:

On February 18, 2015, an explosion occurred in the electrostatic precipitator (ESP) at the ExxonMobil Refinery in Torrance, California near Los Angeles. The explosion injured four workers caused significant property damage to multiple process units within the refinery, and resulted in an off-site accidental release of catalyst dust. As a result of the incident, the FCC unit was down for over one year. The explosion occurred when hydrocarbons from the FCC unit reached the energized ESP while the unit was idling to perform maintenance on equipment that had failed. Multiple failures contributed to the incident. During the ESP explosion, there was also a near miss release of hydrofluoric acid when a large piece of ESP debris fell within feet of a storage vessel storing thousands of gallons of modified hydrofluoric acid in the neighboring alkylation unit. If the storage vessel had failed due to impact from ESP debris following the explosion, hydrofluoric acid would have been released. This presentation will detail key findings from the CSB investigation into the incident.

13TH GLOBAL CONGRESS ON PROCESS SAFETY

Overfill to Overpressure – Investigation Findings and Contributing Events to the CAPECO Explosion

Scott Davis
GexCon US
Bethesda, MD
sgdavis@gexcon.com

Tom DeBold
Claudio Marsegan
Bryant Hendrickson
GexCon US
Bethesda, MD

Prepared for Presentation at
American Institute of Chemical Engineers
2017 Spring Meeting
13th Global Congress on Process Safety
San Antonio, Texas
March 26-29, 2017

Overfill to Overpressure – Investigation Findings and Contributing Events to the CAPECO Explosion

Scott Davis
GexCon US
Bethesda, MD
sgdavis@gexcon.com

Tom DeBold
Claudio Marsegan
Bryant Hendrickson
GexCon US
Bethesda, MD

Keywords: CAPECO, petroleum, overfill, bulk storage, explosion

Abstract

On October 23, 2009, a vapor cloud explosion occurred after an above ground gasoline storage tank was overfilled at the Caribbean Petroleum Corporation (CAPECO) tank farm facility in Bayamón, Puerto Rico. The blast registered a 2.9 on the Richter scale and sparked several fires throughout the facility that burned for another three days and damaged a total of 21 tanks before emergency responders were able to extinguish the flames.

Operations on the day of the incident included a ship transfer of unleaded gasoline to the CAPECO facility and the transfer was scheduled for four smaller capacity tanks. At approximately 10:00 pm, all of the unleaded gasoline was being transferred to tank 409 and a visible vapor cloud was noted during the walk around at 12:00 am on October 23, 2009. It was determined that tank 409 was overfilled by approximately 5,000 barrels (210,000 gallons) of gasoline, which caused gasoline to flow out of the tank vents near the top of the tank and cascade down the side forming a vapor cloud and liquid pool. This is equivalent to overfilling the tank by over 20 gasoline tanker trucks, which if 10% of the liquid formed the vapor cloud it would cover an area larger than 30 American football fields of a rich flammable mixture. The gasoline vapors mixed with the air forming an ignitable vapor cloud spanning a vast portion of the facility, which subsequently ignited neighboring tanks that burned for three days.

US regulations do not consider CAPECO and other bulk petroleum storage terminals as high hazard facilities, yet a large vapor cloud formed. This paper will present investigation findings and key lessons learned into the massive explosion and fire. This includes: effects of confinement and congestion in a seemingly open space that contributed to the incident; ship offloading and tank filling procedures that revealed significant operational deficiencies; failures to secure the confined drainage system; and level controls.

1 Introduction

On October 23, 2009, a vapor cloud explosion occurred after an above ground gasoline storage tank was overfilled at the Caribbean Petroleum Corporation (CAPECO) tank farm facility in Bayamón, Puerto Rico. The blast registered a 2.9 on the Richter scale and sparked several fires throughout the facility that burned for another three days and damaged a total of 21 tanks before emergency responders were able to extinguish the flames. The explosion and ensuing fires did not cause any fatalities; only a few minor injuries were reported. However, approximately 300 homes and business within 1.25 miles of the site reported damages [1].

GexCon was retained to perform an independent and objective investigation into the cause and origin of the explosion. Due to the nature of the investigation, some information provided to GexCon was confidential and could not be provided. Therefore, information provided in the reports of other agencies were used in the present analysis, and include: the ATF Investigation 764045-10-0011 Report [1]; U.S. Chemical Safety and Hazard Investigation Board Final Investigation Report – Caribbean Petroleum Tank Terminal Explosion and Multiple Tank Fires [2]; and the Puerto Rico Fire Department Fire Marshal Report [3] . This paper will present investigation findings and key lessons learned into the massive explosion and fire that occurred at the CAPECO facility. This includes: effects of confinement and congestion in a seemingly open space that contributed to the incident; ship offloading and tank filling procedures that revealed significant operational deficiencies; failures to secure the confined drainage system; and level controls.

We will first present a brief description of the facility, events leading up to the incident, and details of the explosion. Next, we will evaluate the conditions and contributing factors that lead to the explosion.

CAPECO Facility

The Caribbean Petroleum Corporation (CAPECO) facility was a petroleum storage terminal located in Bayamón, Puerto Rico. At the time of the incident, the petroleum storage terminal employed approximately 65 people. Prior to the gasoline overfill and ensuing explosion, the CAPECO facility (Figure 1.1) consisted of a tank farm of above ground storage tanks, offices, decommissioned refinery, waste water treatment plant and distribution piping that serviced the San Juan Bay. The CAPECO facility stored gasoline, diesel, fuel oil, and jet fuel at the time of the incident. The tank locations along with what fuel was stored in each can be found in Figure 1.2.

The tank farm was split by roads into three areas; the northern, central, and southern tank farms. The tanks were separated by dikes or berms designed to hold 110% of the largest tank capacity in that berm (see Figure 1.2) [3]. The southern tank farm contained eleven tanks, which included diesel and fuel oil tanks, divided by six berms. Jet fuel, diesel, and fuel oil were contained in the central tank farm. The northern tank farm held gasoline, diesel, and fuel oil. CAPECO estimated that 613,816 barrels (25,780,272 gallons) of gasoline, 428,260 barrels (17,986,920 gallons) of diesel, 360,496 barrels (15,140,832 gallons) of fuel oil, and 132,724 barrels (5,574,408) of jet fuel were present at the time of the incident [4].

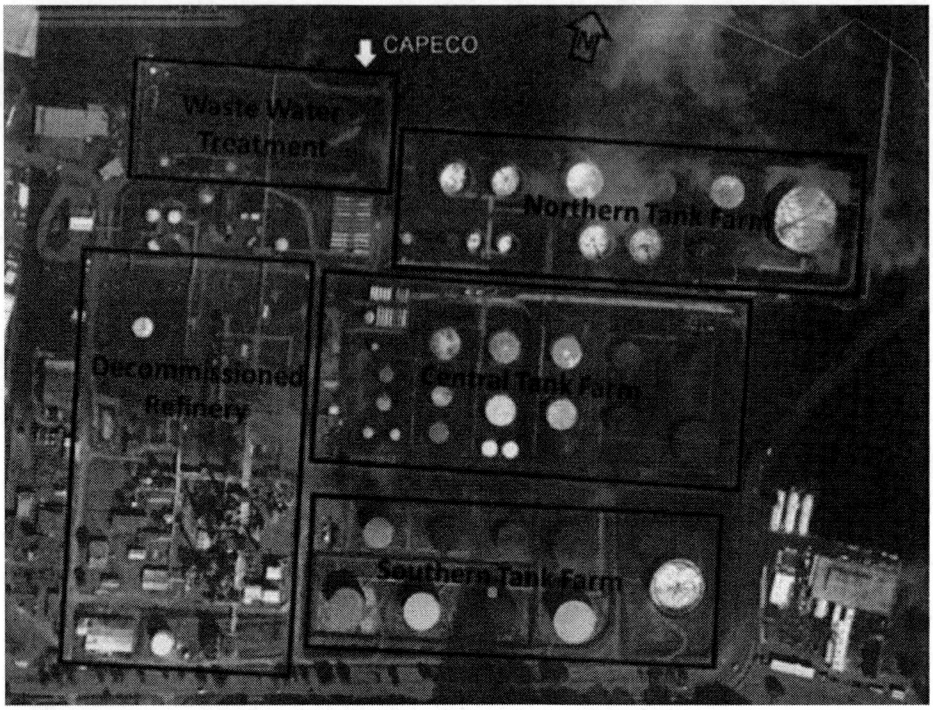

Figure 1.1: Facility Layout.

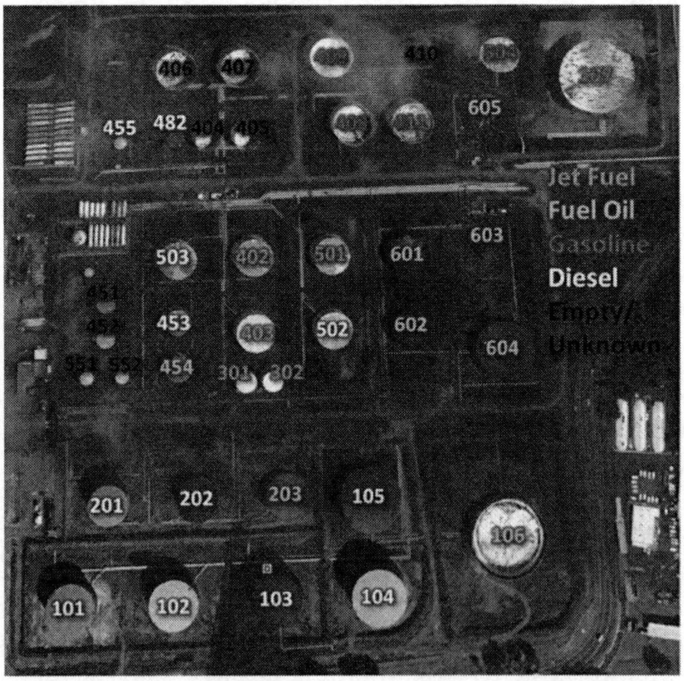

Figure 1.2: CAPECO tank location and inventory.

During fuel transfer operations, vessels would load/unload through the facility's 2 ½ mile pipeline which ended at a loading dock located in the San Juan Bay [2]. The terminal dock could accommodate one tanker of up to 75,000 tons. The transfers were made through six pipelines that connected to the facility transferring crude oil, FCC stock, fuel oil, gasoline, diesel, jet fuel and kerosene. The pipelines were equipped with remote operating valves that shutdown through: an electrical actuator; manually closing the valve; thermal shut-off switch set to 221°F (105°C); five push buttons in the operator's office, guards office, stairway to the main platform, fire pump building and Chiksan platform [5].

Events Prior to the Incident

Operations the day of the incident included: 1) A transfer of diesel fuel to the Puerto Rico Electric Power Authority Plant in Puerto Nuevo from tank 202; 2) A transfer of fuel oil to the Puerto Rico Electric Power Authority Plant in Palo Seco from tank 105; and 3) A transfer of gasoline from the Cape Bruny vessel to the CAPECO facility. Due to CAPECO's largest tank 107 (capacity 500,000 bbls – 21,000,000 gal) being used for a different quality gasoline, the transfer was scheduled for four smaller capacity tanks (see Figure 1.3) [6] .

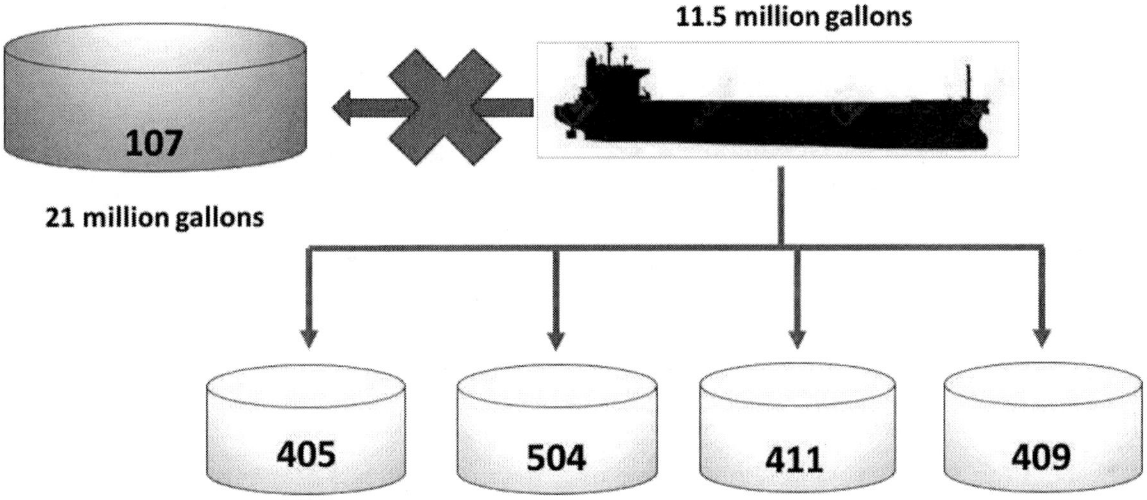

Figure 1.3: Cape Bruny transfer plan.

Pumping from Cape Bruny to the CAPECO facility began on October 21, 2009. Four tanks (Tank 405, 504. 409, and 411) in the northern tank farm received gasoline during the transfer. Initial pumping started at approximately 8:47 p.m. A sequence of operations can be found in Figure 1.4 [2].

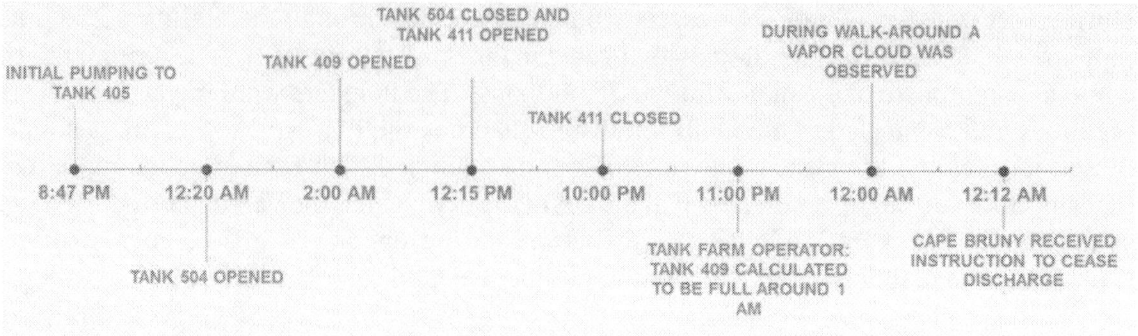

Figure 1.4: Transfer Operations [6].

Shortly after 11:00 p.m., the tank farm operator advised the supervisor in charge of the operations that night that the inspector found a difference between the side gauge, monitor console reading and the physical status of Tank 411 (already closed), and advised that the discrepancy was to be investigated [1].

ATF interviews determined that numerous operators and employees observed a vapor cloud starting at midnight. Tank 409 began to overfill between the 11:00 p.m. and 12:00 a.m. tank checks while the average pumping rate was approximately 11,134 bbls/hr (467,628 gal/hr) [7]. The overfilling of tank 409 caused gasoline to flow out of the tank vents near the top of the tank and cascade down the side forming a vapor cloud and liquid pool (as will be discussed in detail later). Tank 409 was contained in a berm with Tank 410 having a capacity of approximately 132,000 barrels (5,544,000 gallons) based on the 110% capacity requirement [3] . As the overfilling continued, gasoline (liquid and vapor) began to fill the containment area and expand towards the waste water treatment facility resulting in the white mist or vapor cloud that was seen in the facility by employees [6]. It was determined that approximately 5,082 barrels (213,444 gallons) of gasoline overflowed from tank 409 [7]. This is equivalent to overfilling the tank by over 20 gasoline tanker trucks, which if 10% of the liquid formed the vapor cloud it would cover an area larger than 30 American football fields of a rich flammable mixture.

The employees saw a white cloud extending approximately two to three feet above the ground by tank 404, and the white vapor cloud could be seen throughout the area between tanks 453 and 301, and near tanks 201 and 409 (Figure 1.5). The vapor cloud could only be seen in areas that were illuminated and extended approximately 200 yards wide by 400 yards long (183 m by 366 m) [1].

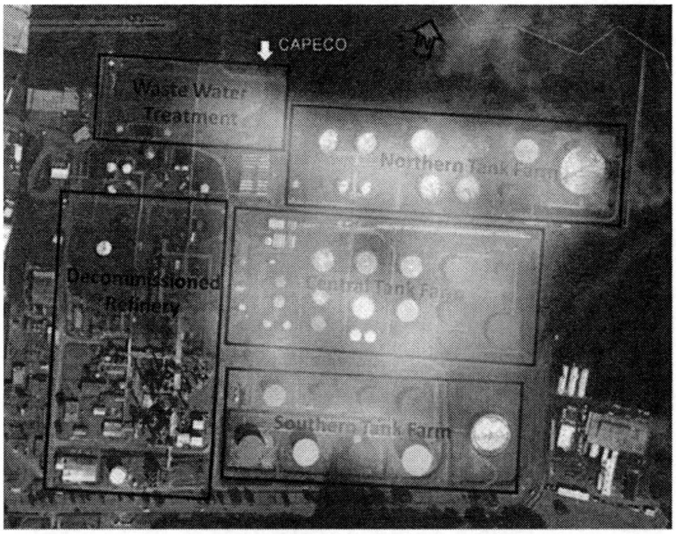

Figure 1.5: Approximate migration area of the vapor cloud.

Incident

At approximately 12:23 a.m. on October 23, 2009, the resulting vapor cloud from the overfill of gasoline ignited on the western side of the facility, away from the tank farm area. As the flame propagated (Figure 1.6 and Figure 1.8) towards the above ground storage tanks, a large explosion (Figure 1.8) occurred in the tank farm area that registered 2.9 on the Richter scale. Due to ignition of the large vapor cloud and ensuing explosion, many of the facility storage tanks were consumed by fire, which burned for another three days before emergency responders extinguished the flames [3].

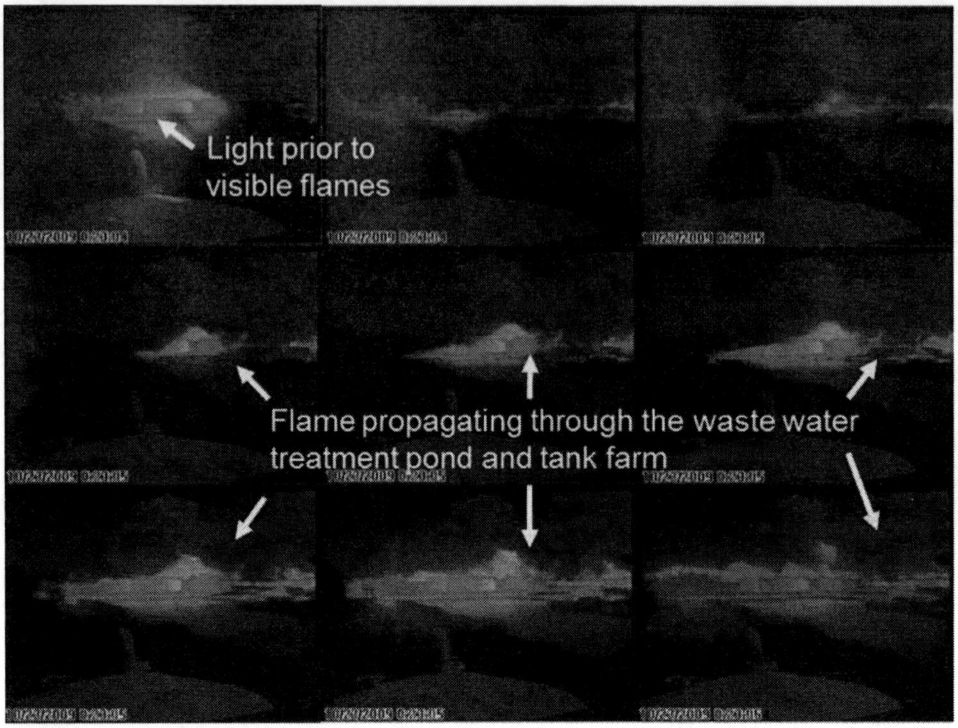

Figure 1.6: CAPECO security footage from camera 4. Visible light is seen prior to the flame propagation through the waste water treatment pond and tank farm.

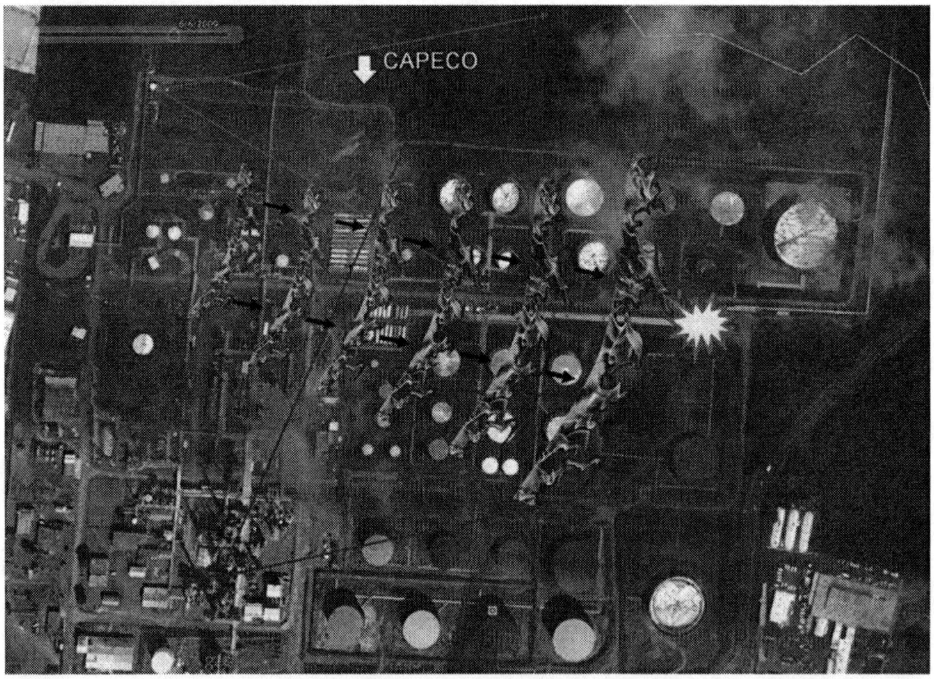

Figure 1.7: Approximate propagation of the flame front prior to the explosion in the underground drainage as indicated by arrows. Camera locations and angles (red-camera 3 and blue-camera 4) at the first sign of ignition.

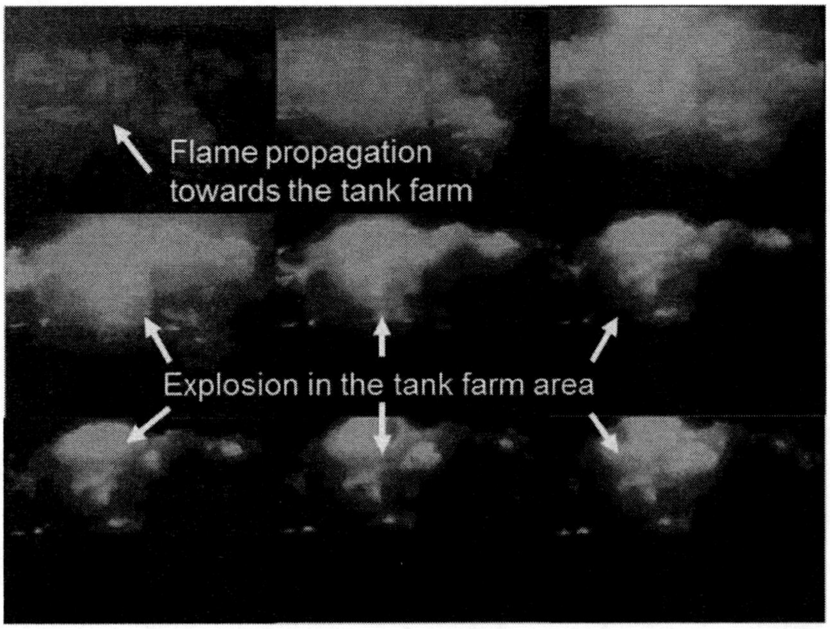

Figure 1.8: CAPECO security footage from camera 3 as the flash fire approached the tank farm from the west, leading to an explosion within the tank farm area (top-left to bottom-right).

Post Incident and Blast Damage

Emergency crews were alerted to the incident when a nearby citizen heard a large explosion and saw flames at the CAPECO facility [1]. At approximately 12:30 a.m. on October 23, 2009 the Bayamón and Cataño Fire Departments arrived at the CAPECO facility where they waited for the site fire chief to arrive approximately 45 minutes later. Eleven of the CAPECO tanks were damaged by the initial fire and explosion.

Firefighting efforts, as shown in Figure 1.9, lasted almost three days until approximately October 25, 2009 at 11:40 a.m. when the fires were extinguished [6]. Additional tanks were damaged by ensuing fires (see Figure 1.10 and Figure 1.11) [3] and when the fire was completely extinguished, the damage caused by the large gasoline vapor cloud explosion and fire included 21 tanks onsite. Both blast and thermal damage were observed on the tanks at the CAPECO facility (see Figure 1.12), consistent with the 11 tanks initially damaged by the large vapor cloud explosion and ensuing gasoline fire, and the others falling victim to the resultant fire [1].

Figure 1.9: Firefighting efforts after the incident [8].

Figure 1.10: Aerial view of the CAPECO facility after the initial gasoline vapor cloud explosion.

Figure 1.11: Secondary tank fires caused by the initial vapor cloud explosion and fire from the gasoline vapors.

Figure 1.12: Post incident damage to the tank farm.

2 Physics of vapor cloud explosions

Prior to the explosion at the CAPECO facility, tank 409 was overfilled and gasoline was being discharged from the overflow vents at the top of the tank. Recall that ignition and burning of liquid hydrocarbons occurs when the hydrocarbon is in the vapor state. A large vapor cloud developed resulting in the 'white cloud' described by CAPECO employees. As the vapor cloud grew from the continuous overflow, the gasoline vapors continued to mix with air forming a flammable mixture throughout the facility. This cloud found a competent ignition source in the western side of the facility and flames propagated towards the tank farm leading up to the main explosion event. This phenomenon is referred to as a vapor cloud explosion (VCE). The specifics on how the vapor cloud was formed and on the VCE will be discussed next.

Formation of vapor clouds

When evaluating the potential for large vapor cloud formation, which may lead to a massive explosion event, it is important to consider the mechanism by which a vapor and vapor clouds are formed. Recent experimental and modeling research has been performed to explore the generation of vapor clouds from overfilling gasoline storage tanks [9], [10]. The research determined that during a tank overfill, the gasoline would flow out of the ventilation holes at the top of the tank and over the deflector plate. The deflector plate channeled some of the escaped fuel onto the tank wall, but the rest ran over the top of the plate, fragmenting into droplets that cascaded through the air. Most of the fuel running down the wall hit a wind girder (a structural stiffening ring) and detached from the tank wall, creating a second cascade of droplets.

These conditions promoted the evaporation of the lighter components of gasoline (e.g., butanes, pentanes and hexanes). The free-fall of droplets leads to entrainment of air and mixing between the air and fuel vapor, and the formation of a rich fuel/air mixture. In addition, the wind conditions at the time of the overfill were calm, and had the wind speeds been higher, the evaporating gasoline vapor would be quickly diluted as it escaped the berm. Therefore, the overfill led to conditions that increased the break-up of gasoline into droplets, which significantly increased the evaporation rate of the lighter components of gasoline as compared to that from the pooled gasoline. Therefore, the overfill rate coupled with low wind speeds generated very large vapor clouds capable of filling a significant region of the CAPECO tank farm as seen in modeling shown in Figure 2.1.

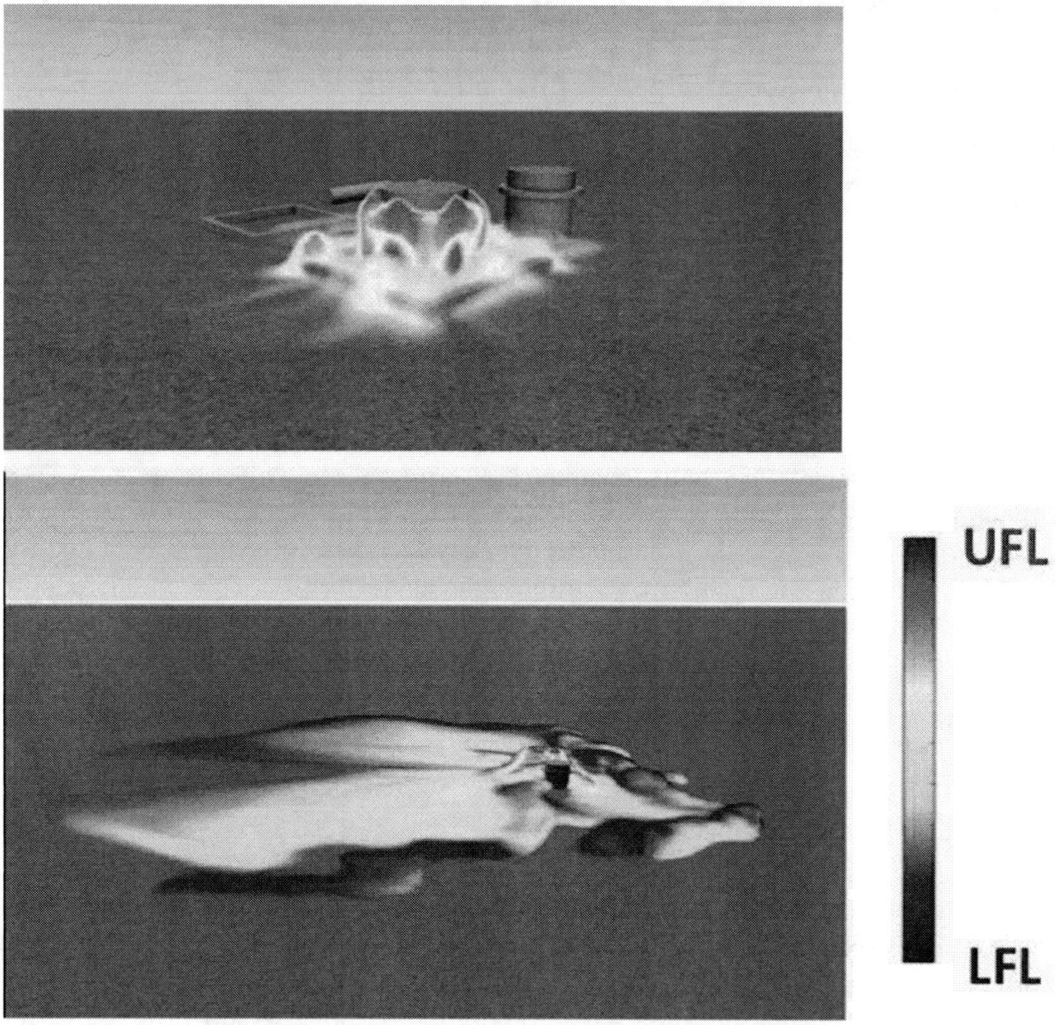

Figure 2.1: Vapor cloud formation and growth as gasoline initially cascades down the side of a storage tank and is dispersed.

Vapor cloud explosions

Vapor cloud explosions (VCEs) are oftentimes understood as "confined explosions", where a flammable mixture of gas (or vapor) and air is ignited within a confined structure or vessel. The heat released from the burning mixture causes the gases to expand and the pressure within the confinement increases above normal atmospheric pressure; this is referred to as "overpressure". This occurs until the envelope of the structure ultimately fails, similar to an ignition of a flammable mixture within an empty tank causing the roof to blow off. The sudden failure of the structure/vessel causes a shock or blast wave to form. However, conditions at CAPECO were mostly unconfined and open, except for the underground drainage system. An open and unconfined environment presents little to no resistance of an expanding flame front when a mixture is ignited; therefore, an overpressure is not expected if the gases are allowed to freely expand. The formation of a shock or blast wave in an open area cannot be explained in the same manner, and in fact, requires a unique set of circumstances.

Overpressures can indeed be created for unconfined vapor cloud explosions and require mechanisms that accelerate the combustion or burning process. The combustion rate for burning vapor/air mixtures is measured by how fast the flame propagates (flame speed) relative to the unburned mixture. For quiescent fuel/air mixtures the flame propagates at approximately 0.5 m/s and minimal overpressure is observed as the hot gases can expand in the unconfined environment. However, the flame propagation rate is extremely sensitive to turbulence in the flammable mixture into which the flame is burning. This turbulence can be caused by repetitive obstacles, (e.g. pipes, equipment, trees) within the vapor cloud itself. This results in a feedback loop, which has been extensively studied by Gexcon (see Figure 2.2), whereby the turbulence causes the flame to burn faster, which in turn pushes the unburned gases past the obstacles even faster creating higher turbulence levels, which in turn causes the flame to burn even faster. This results in the flame accelerating and can lead to very high propagation rates.

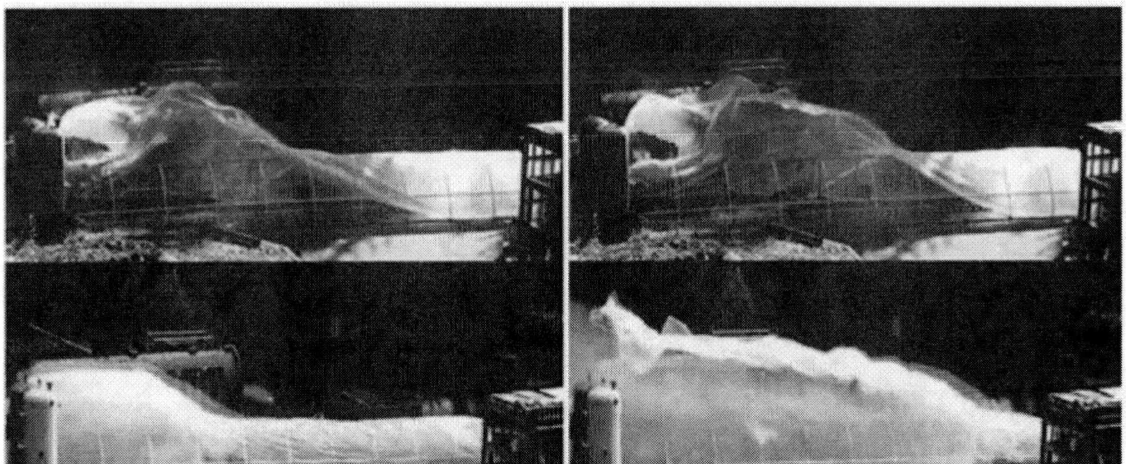

Figure 2.2: Flammable fuel air experiments for "unconfined" vapor clouds with no obstacles (top) and with trees (bottom). Clouds ignited on the left and maximum flame speeds with trees are six times larger than without.

The flame may propagate at such high speeds that the expanding gas ahead of the flame is actually compressed, which can ultimately result in a shock or blast wave. The strength of the blast wave will depend on the flame speeds achieved during the acceleration process.

If the congestion in the cloud is low, the flame speeds will remain low and will not generate a shock or blast wave (i.e., a "flash" or "whoosh" type sound), while relatively higher flame speeds can occur in high congestion areas, resulting in shock wave and "boom" type sound. Flame speeds between 100-150 m/s can generate significant overpressures, and those above 400 m/s can generate pressures well in excess of 1 barg, and can even transition to detonations. This phenomenon was precisely what occurred at the Buncefield explosion in 2005, whereby the flame accelerated into highly congested tree lines that extended over 100 meters. No confinement was found at the Buncefield site and the flame accelerations, and extremely high overpressures, were exclusively caused by the tree density [11].

In order to generate the massive explosion observed at the CAPECO facility, the resulting large vapor cloud needs to migrate into a congested (as described above) or confined region to accelerate the flame. Similar to Buncefield, the CAPECO facility had congestion that included numerous low-lying pipelines connecting the storage tanks, and vegetation to the north and east of the facility. During the post-incident scene investigation, thermal damage was found to extend into the tree line approximately 50 yards past the east property fence [1]. Unlike Buncefield, the CAPECO facility had additional long channels of confined drainage pipes that were not properly closed. Recall that the vapor cloud was ignited western side of the plant propagating towards the waste water treatment (see Figure 1.7) and it propagated east towards the tank farm and vegetation. Video analysis confirmed (see Figure 2.3) that while the main flame propagated above ground through the tank farm, a faster flame front traveled via the underground drainage pipes where a fireball exited the sump pump encasement at the northeast corner of the central tank farm well before the above ground flame.

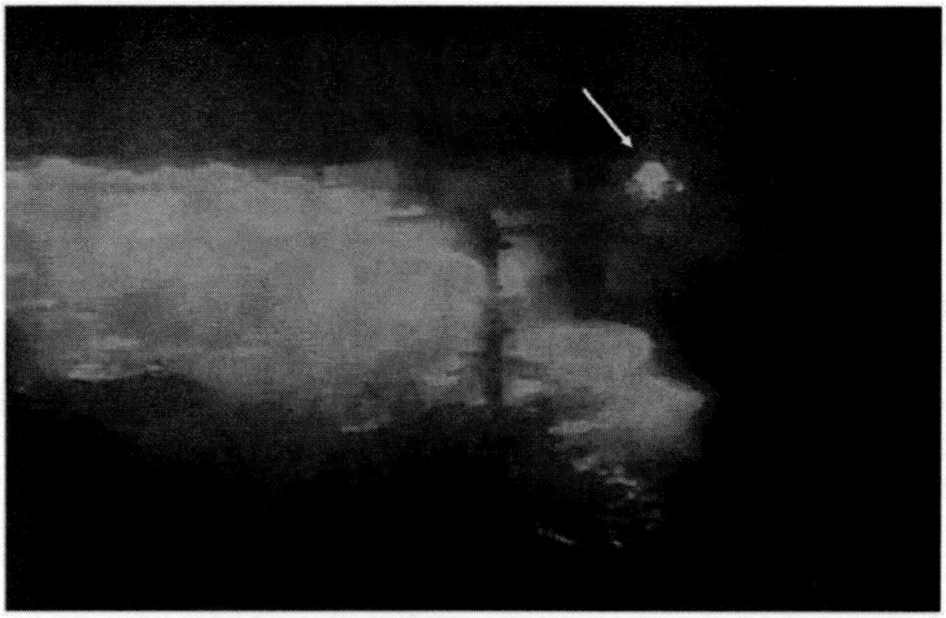

Figure 2.3: View from camera 3 showing the flame front in the underground drainage pipes reaching the northeast corner of the central tank farm before the above ground flame front. Ignition occurred on the western side of the facility resulting in flame propagation from left to right in the image above.

In fact, the 10 ft x 10 ft concrete lid on the sump pump, weighing about 12,000 pounds, was found over five feet from its original location [1]. Furthermore, blast damage to the above ground storage tanks was generally in the direction of the sump pump encasement and eastern vegetation and not the western side of the facility where the cloud was first ignited [1]. This confirms that the explosion pressures were exacerbated due to the confinement caused by the underground drainage pipes, which resulted in higher overpressures and faster accelerating flames that also entered the congested region. To evaluate the effect of the underground drainage system, FLACS simulation were performed to evaluate the resulting overpressures and blast damage both with and without the presence of the drainage system.

The CAPECO tank farm area (both with and without the drainage system) was modeled in FLACS, a computational fluid dynamics (CFD) tool predominantly used in the Oil & Gas and chemical industries to predict the consequences of hydrocarbon deflagrations. A large vapor cloud (e.g., stoichiometric cloud 4 meters tall) was placed throughout the tank farm and water treatment areas, and ignited in the water treatment area. Figure 2.4 shows the tank farm with the large vapor cloud and drainage system.

The flame propagation rates were evaluated both through the above ground cloud and through the drainage system. Figure 2.5 shows that the vapor cloud entering the drainage system resulted in flame speeds of over 400 m/s, significantly higher than above ground flame propagation. In fact, the confinement and acceleration of the flame in the drainage system resulted in overpressures that exceeded 2 barg (see Figure 2.6), which could possibly result in condition where the exiting deflagration transitions to a detonation. In contrast, overpressures generated by the flame burning through the above ground portion of the cloud did not exceed 0.03 barg. Therefore, the confined underground drainage system resulted in flames burning faster than the speed of sound and with pressures exceeding 2 barg, which ignited the remaining unburned vapor cloud in the tank farm. This high-pressure, supersonic flame front resulted in a significantly more devastating explosion than the "flash" or "whoosh" that would have been observed had the drainage system been properly closed prior to the overfill.

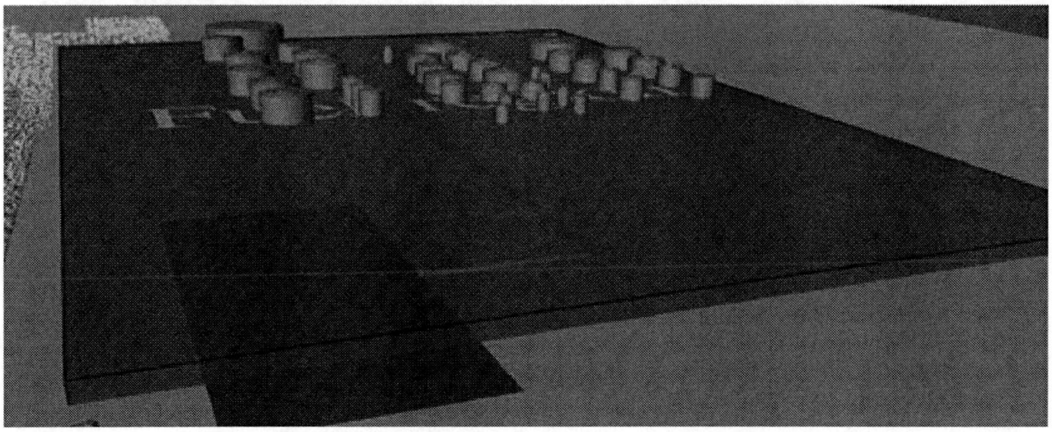

Figure 2.4: FLACS model of the CAPECO underground drainage system.

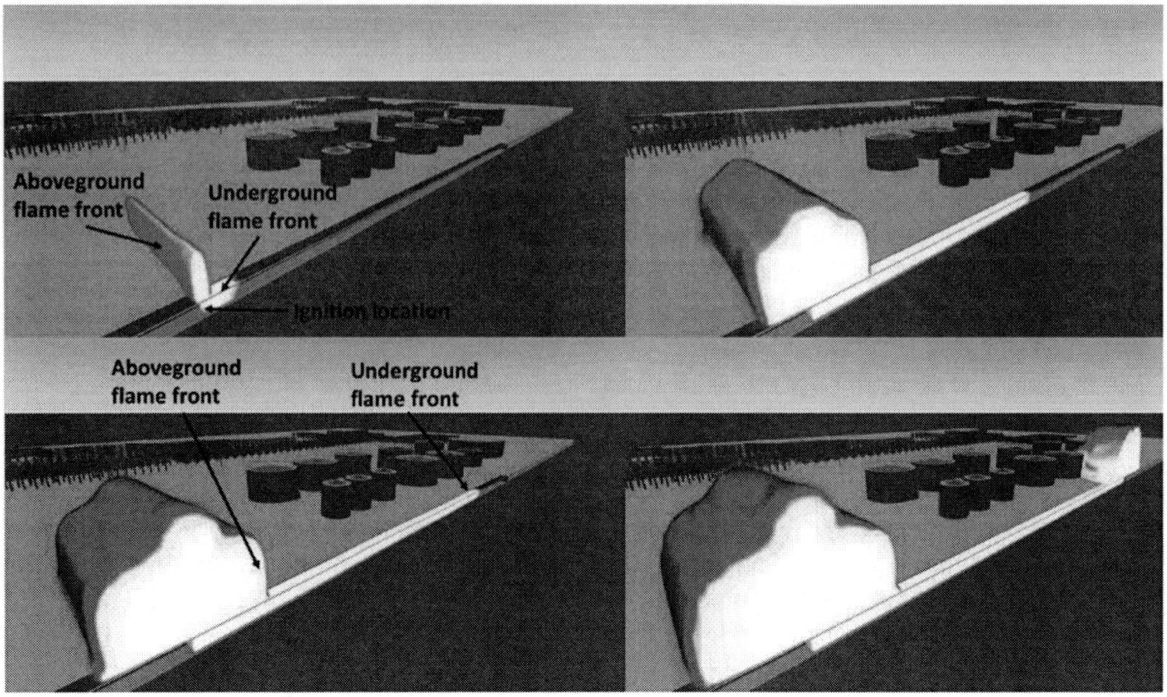

Figure 2.5: Flame front above and underground.

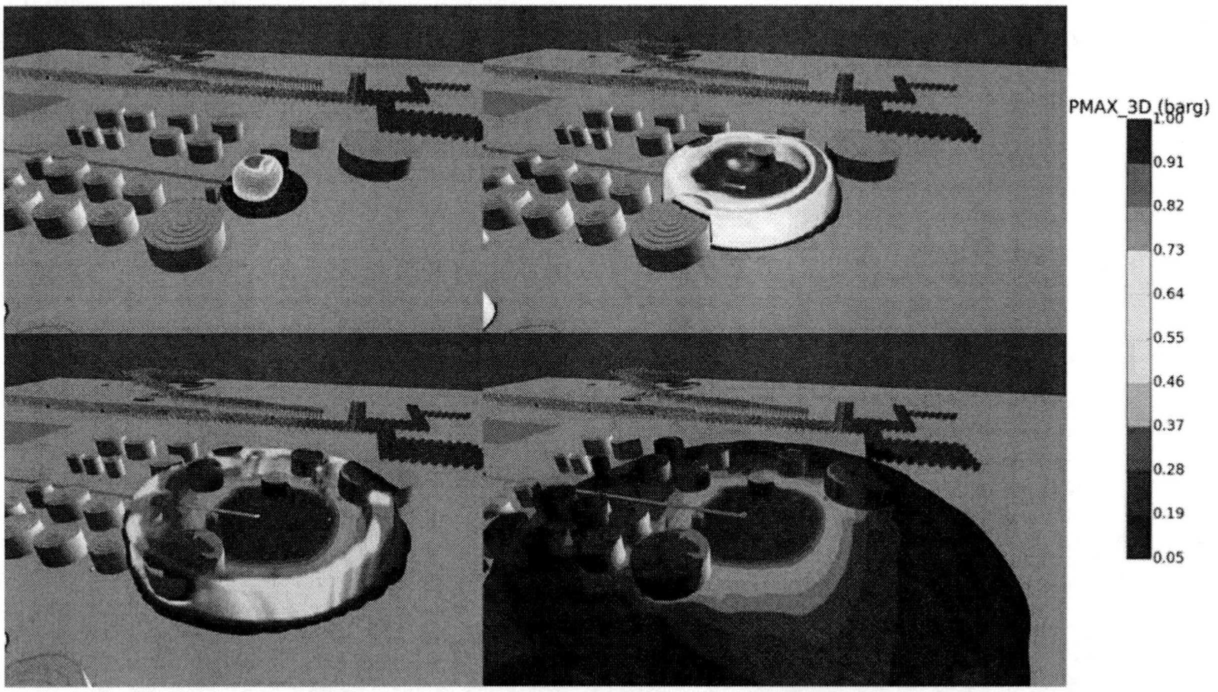

Figure 2.6: Pressure generated from the explosion in the underground drainage system.

3 Investigation findings and key lessons learned

Many of the contributors to the explosive atmosphere present at the CAPECO facility during the early morning hours on October 23, 2009 could have been avoided if the proper precautions were followed. While previous investigation reports have outlined certain contributing factors related to the incident, other major findings and lessons learned were not identified. For example, other groups have determined that the facility did not have automatic level alarms and that the automatic tank gauge system at the facility was not operational, which contributed to the event by not providing key indicators automatically and relying solely on manual measurements. However, key contributors to the incident (in addition to the automatic gauge system) were not identified and included: a failure to accurately monitor and verify the rate at which the gasoline storage tanks were being filled during the vessel offloading; and a failure to ensure "inherent safety" at the facility by not securing the dike drainage systems, which resulted high-pressure, supersonic flame fronts capable of causing the massive explosion as opposed to a "flash" or "whoosh" had the vapor cloud remained above ground. These factors will be discussed next.

Operational deficiencies that led to the Tank 409 Overfill

Recall that automatic controls were down and operations were manually monitored. Using the vessel discharge rates and tank strapping tables, it was estimated that when tanks 504 and 409 were being filled from 4:00 a.m. to approximately 12:00 noon on October 23, 2009, the transfer to each tank was approximately equal (see Figure 3.1). However, when 411 was brought online from 12:00 noon to 10:00 p.m. on October 23, 2009, it was determined that approximately 66% of the fuel was being transferred to tank 411 (see Figure 3.2), with the remaining fuel going to tank 409. This is consistent with operators' observations when preparing to take tank 411 offline: the valve to tank 409 was noted to be only open a little, confirming that tank 411 was receiving a much higher flow rate than tank 409 while they were being filled simultaneously [1].

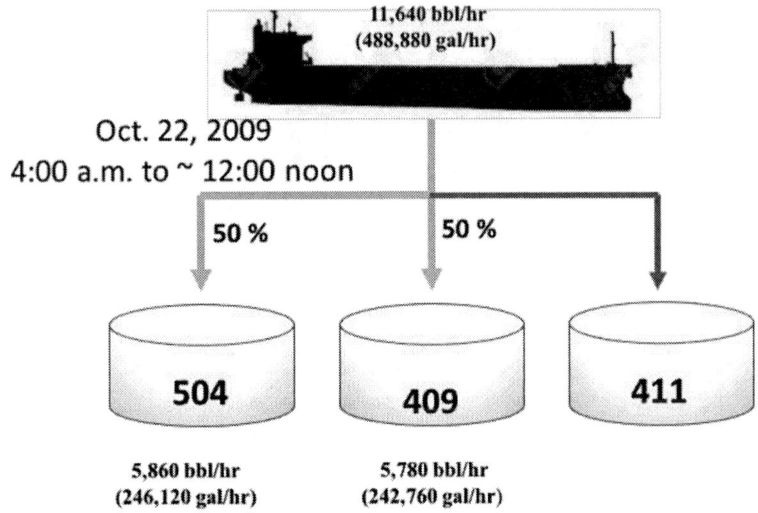

Figure 3.1: Transfer from the Cape Bruny to tanks 504 and 409 from 4:00 a.m. to approximately 12:00 noon on October 22, 2009.

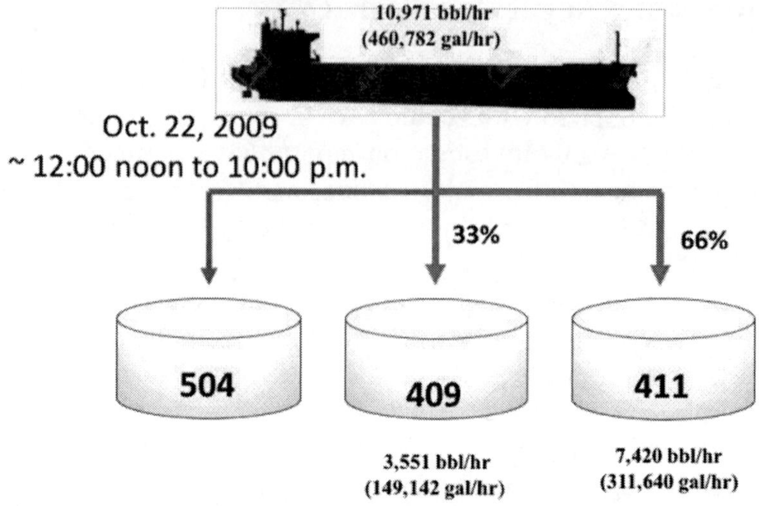

Figure 3.2: .Transfer from the Cape Bruny to tanks 411 and 409 from approximately 12:00 noon to 10:00 p.m. on October 22, 2009.

At 10:00 p.m. on October 22nd, tank 411 was closed and tank 409 was fully opened. Tank 409 contained approximately 98,716 gross barrels (4,146,072 gallons) of gasoline (overfill occurs at approximately 121,062 barrels).[1] The supervisor incorrectly estimated that the tank would be full at 1:00 a.m. by assuming only 33% of the discharge rate was going to tank 409. However, with the full vessel discharge of approximately 11,134 bbl/hr [7] going into tank 409, approximately 5,082 barrels (213,444 gallons) of gasoline overflowed from tank 409 contributing to the generation of a large flammable vapor cloud. This means that overflow occurred for approximately 27 minutes, beginning at approximately 11:45 p.m. on October 22nd until 12:12 a.m. the next morning.

Table 3.1: Estimated amounts of gasoline added to each tank and overflown at the CAPECO facility.

Tank	Estimated gasoline added Gross Barrels (Gallons)
405	4,411 (185,262)
504	62,984 (2,645,328)
411	74,198 (3,116,316)
409	114,325 (4,801,650)
Total	255,918 (255,918)
Total from Cape Bruny	261,000 (10,962,000)
Overfill	**5,082 (213,444)**

Therefore, one major cause of the incident was due to an incorrectly estimated fill time of 1:00 a.m. for tank 409, likely caused by ignorance of the increased flow rate to tank 409 after closing tank 411. Therefore, failure to compare vessel discharge rates to tank fill rates and verify that tank 409 had sufficient space to receive the scheduled amount of product resulted in tank overflow which generated a large flammable vapor cloud [12].

Drainage system

In the post incident investigation, the drainage valve in the berm surrounding tank 409 was found to be in the open position. The open valve drained to the waste water treatment facility, creating a pathway for the liquid gasoline to spread, increasing vapor generation and allowing for flammable gasoline vapor to fill the underground drainage system. The dike drainage valves should be closed during transfer operations so that in the event of an overfill, pool spread is limited. In this case, failure to close the drainage valve led to flammable vapors building up within the underground drainage system, and after ignition, resulted in a deflagration propagating within a completely confined drain and causing significant flame accelerations and overpressures. CFD analyses confirmed that these flame accelerations led to high pressure, supersonic flame fronts capable of causing the massive explosion as opposed to a "flash" or "whoosh" had the vapor cloud remained above ground. Keeping the drainage valves closed during a fuel transfer would have vastly limited the consequences and ensured inherent safety at the facility.

Level controls

The CAPECO facility has an automatic tank gauge system that transmitted the tank liquid depth to the operator's office. It would calculate the fill rate, as well as plot the flow rate and tank liquid depth versus time. Due to issues with the automatic tank gauge system (typical float and tape gauge), manual inspection of the tanks and calculation of the fill time had to occur through hourly readings.

4 Conclusions

Although US regulations do not consider bulk petroleum storage terminals as high hazards, an explosion at the CAPECO facility and ensuing fire destroyed a vast majority of the tank farm facility. This explosion was caused by a vapor cloud resulting from the overfill of a gasoline bulk storage tank, and this vapor cloud entering an improperly closed drainage system. The two main lessons learned from this incident (and previous incidents like Buncefield) include: (1) having systems and procedures in place to limit the likelihood of having such an overfill condition, and (2) providing inherent safety to such facilities to vastly limit the explosion consequences in the event such an overfill occurs (e.g. limiting confinement and congestion).

Regarding reducing the likelihood of such an event, the CAPECO facility was operating as a fully-attended facility and was in compliance with federal regulations at the time of the incident. As of today, fully-attended tank farm facilities do not require independent liquid level alarms and automatic tank gauging which allows other facilities to operate in the same fashion as CAPECO when transferring and storing flammable liquids such as gasoline. Therefore, such facilities should consider redundant controls and procedures to ensure: fill and transfer rates are constantly updated and verified throughout the transfer; and that the automatic tank gauge system is operational and complements manual measurements made at the facilities; and areas of high confinement like dike drainage systems are secured and closed prior to transfer of volatile hydrocarbons.

Inherent safety should also be evaluated at such facilities to ensure significantly lower consequences, even in the event of a significant release. CFD simulations have confirmed that physical layout of both Buncefield and CAPECO resulted in significantly higher consequences that could have resulted for identical overfill conditions. In fact, the densely coppiced tree lines at Buncefield and high confinement of the improperly opened drainage system at CAPECO resulted in explosion pressures and consequences orders of magnitude higher had such congestion/confinement not been present. Inherent safety of such facilities can be enhanced if areas of confinement and congestion are identified and mitigated in standard risk assessments. In fact, the CAPECO and Buncefield incidents would have likely resulted in large flash fires instead of the more devastating VCEs had these congestion and confinement areas been mitigated.

References

[1] ATF Investigation 764045-10-0011 Report.

[2] U.S. Chemical Safety and Hazard Investigation Board Final Investigation Report – Caribbean Petroleum Tank Terminal Explosion and Multiple Tank Fires.

[3] Puerto Rico Fire Department Fire Marshal Report.

[4] Inventory Logs.

[5] Dock Operating Manual.

[6] ATF Investigation 764045-10-0011 Interviews.

[7] Ship Transfer Logs.

[8] Caribnewspr https://www.flickr.com/photos/43578995@N03/.

[9] Davis, SG, Hinze, PC, Hansen, OR, van Wingerden, K (2010). 2005 Buncefield Vapor Cloud Explosion: Unraveling the Mystery of the Blast. Interflame 2010, Nottingham, UK..

[10] Coldrick, S., Atkinson, G.T., and Gant, S.E., Health and Safety Laboratory, Hazards XXII, 2011, pp. 560- 569..

[11] Van Wingerden, M. and Wilkins, B. (2008) Experimental investigation of the effect of flexible obstructions on flame propagation in vapour cloud explosions, GexCon report no. GexCon-08-F44110-RA-1, rev 01.

[12] API RP 2350, Appdx B.3.3 (2005).

Produced Water Tank Explosion Case Study

Presented by:
J. Kelly Thomas, Ph.D.
Baker Engineering and Risk Consultants, Inc.

2017 AIChE
Global Congress on Process Safety

BAKERRISK.

BAKERRISK.

Outline

- Purpose
- Background
- Facility Description
- Incident
  - Overview
  - Fuel
  - Oxidant
  - Ignition
- Related Incidents
- Lessons Learned
- Potential Prevention and Mitigation Approaches

2

Purpose

- Highlight commonly encountered hazard
  - Produced water tank with flammable atmosphere
  - Similar hazard associated with waste (slop) water tanks
    - Incidents involving waste water tanks in industry have occurred
- Identify lessons learned
- Identify potential prevention and mitigation approaches
  - Examples, not exhaustive compilation
  - Other approaches may provide similar level of risk reduction

3

Background

- Produced water (brine water)
  - Water separated from oil / condensate / gas streams
  - Collected in tanks and treated as waste
  - Head space may be in flammable range depending on operating parameters and separator performance (e.g., condensate carryover)
- Not typically viewed as posing significant explosion risk
  - Reasonable on a relative basis if other explosion hazards present which pose much more severe consequences and/or risk
  - Reasonable from a building siting perspective
  - Hazard is local, although debris may be thrown considerable distance
  - However, produced water tanks can, and do, explode
- Should be considered in terms of hazard/risk to personnel

4

Facility Description (1 of 6)

- Incident occurred at gas production site (West Virginia, 2013)
- 6 gas wells (no oil or condensate produced), each with own separator
  - Separators are 2.5 ft diameter by 8 ft long, and operate at 675 psig
  - Gas composition (97.2% CH_4, 2.1% C_2H_6, 0.1% C_3H_8, 0.3 CO_2, 0.3% N_2)
- Produced water tanks
  - Tanks in two groups of 3 (i.e., 6 tanks)
  - Group of 3 tanks share common overflow (3 in.) and vent (4 in.) lines
    - Common vent line not supplied by tank manufacturer
    - No flame arresters on these lines
  - Fixed roof API 12F style, 12 ft. (3.7 m) diameter, 20 feet (6.1 m) tall
  - 400 barrel (16,800 gallons, 64 m^3) & hydrotested at 1.5 psig (0.10 barg)
  - 2 access hatches (Enardo ES-660 and Jayco JT-8); relieve at 0.5 psig

5

Facility Description (2 of 6)

6

Facility Description (3 of 6)

- Fill line and access hatches

Facility Description (4 of 6)

- Common vent and overflow lines

- Common vent line terminates with open end at opposite end of 3-tank group

BAKERRISK

Facility Description (5 of 6)

- Sign at entrance to pad

9

BAKERRISK

Facility Description (6 of 6)

- Warning sign on each produced water tank
- Flammability rating of 3 would be indicative of gasoline

10

Incident (Overview, 1 of 7)

- Approx. 100 bbl removed from each tank every 3 or 4 days
  - Driver goes up to top of tank via catwalk, opens hatch, and gauges using wooden rod before pumping out desired liquid volume
- Driver arrived between 5:30 and 5:45 am, hooked truck up to tank manifold, and (at some point) ascended catwalk to tanks
- Explosion occurs prior to 6:45 am (February, sunrise at 7:17)
- One tank fails at shell-floor joint, ejected, and thrown 110 feet (34 m)
- One tank fails at shell-roof joint, roof thrown 165 feet (50 m)
- Catwalk damaged and displaced, throwing driver from catwalk and outside tank battery, approximately 30 feet (9 m); fatality.

11

Incident (Overview, 2 of 7)

12

Incident (Overview, 3 of 7)

Incident (Overview, 4 of 7)

- Ejected tank

Incident (Overview, 5 of 7)

- Failed tank roof
- Shell-floor joint shows uplift (was approaching joint failure)

15

Incident (Overview, 6 of 7)

- Catwalk displacement (catwalk supported by tank attachment)

16

Incident (Overview, 7 of 7)

- Tank structural capacity roughly 10 to 15 psig (gradually applied internal pressurization) based on past experience
- Trajectory analysis indicates in range of 5 to 15 psig required to achieve observed tank and tank roof throw distances
- Indicates weakly flammable mixture was present in two tanks which failed, and mixture in "middle" tank was either not flammable or barely flammable
- Post-incident flammable gas measurements
  - Tank in adjacent set ≈ 65% LFL days after incident
  - Truck ≈ 15% LFL 2 months after incident
  - Consistent with elevated fuel concentration in tanks at time of incident

17

Incident (Fuel, 1 of 3)

- Separator does not remove "all gas" from the fluid, even when functioning properly and as designed
- Produced water contains gas since at elevated pressure (675 psig), which is released when pressure dropped to atmospheric
  - Gas content function of water pressure, temperature and composition
- Can have gas "carry" over or with stuck/jammed valve in separator
  - Separators were tested post-event and found to work properly
  - Concluded gas source was solely content of produced water transferred to storage tanks from separators

18

Incident (Fuel, 2 of 3)

- Methane solubility data from Duan & Mao (2006) Geochemica, 70: 2269-86.
  - Approximately 0.043 moles/kg at relevant conditions
- See vapor space fuel concentration plot on following slide:
  - Low level to high level of 50 to 150 barrels (1/4 tank capacity)
  - Conditions present (separator pressure, fluid composition, etc.)
  - No air in-leakage (only draw in air when drop liquid level)
- Can easily exceed UFL
  - Fuel concentration would be lower at lower separator pressure
  - Minimum fuel concentration would decrease if either high level or liquid volume extracted were increased (vs. values noted above)
  - > UFL is not a "universal condition" for produced water tanks

19

Incident (Fuel, 3 of 3)

- Fuel concentration for series of 100 bbl fill / extract cycles

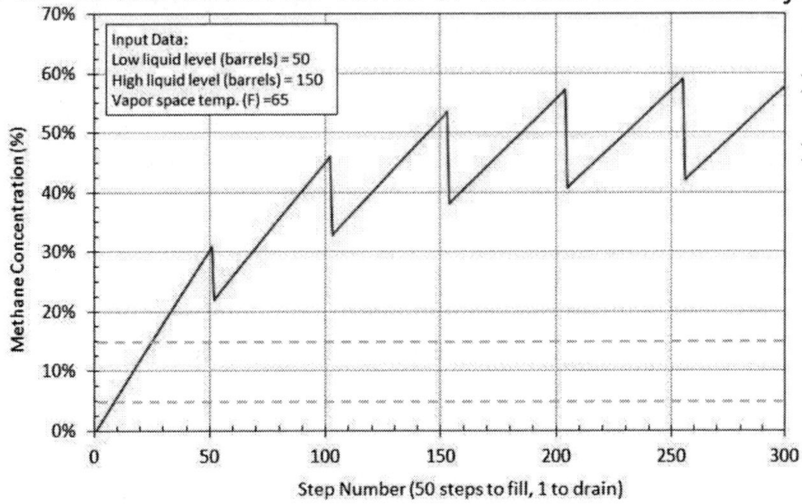

- Can easily exceed the UFL
- Issue is how get enough oxidant to be in flammable range

20

BAKERRISK

Incident (Oxidant, 1 of 4)

- Need additional air to enter flammable range (i.e., < UFL)
- Additional air can be drawn into tank via natural convection
  - Gas within tank lighter than outside air (i.e., unless outside air is hotter)
  - Light gas can flow out common vent line, pulling air in through leak points (e.g., access hatch seals)
    - Analysis subject to uncertainty since inlet flow path not well defined, but leakage rate in range of 125 scfd (2.5 lpm, 22 bbl/day) reasonable
    - Acts to limit maximum fuel concentration
    - Steady decrease in fuel concentration when tank is "stagnant" (i.e., no produced water additions or removals)
  - Much higher air in-leakage rate when hatch is open (2 to 5 scfm)
- Tanks stagnant for period before explosion & hatched opened
- See illustrate plots on following page

21

BAKERRISK

Incident (Oxidant, 2 of 4)

- Fuel concentration for series of 100 bbl fill / extract cycles, followed by stagnant period (125 scfd air in-leakage)

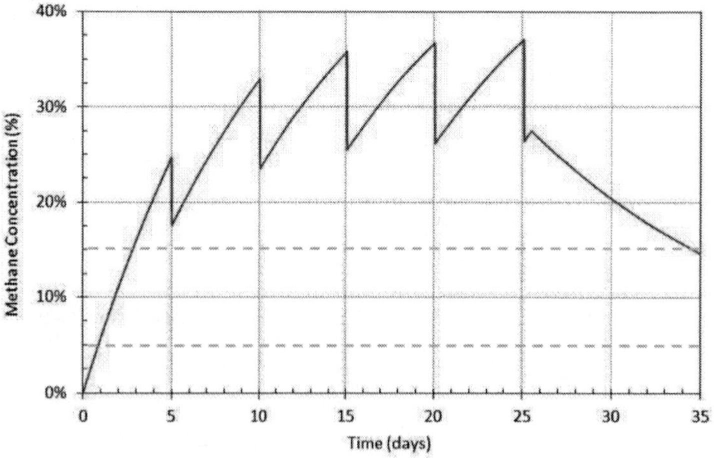

- Above UFL until enter stagnant period
- Air in-leakage during stagnant period allows fuel concentration to drop below UFL and enter flammable range
- Reach flammable condition in about 2 weeks

22

Incident (Oxidant, 3 of 4)

- Fuel concentration for series of 100 bbl fill / extract cycles, followed by stagnant period with hatch open (2 scfm air in-leakage)

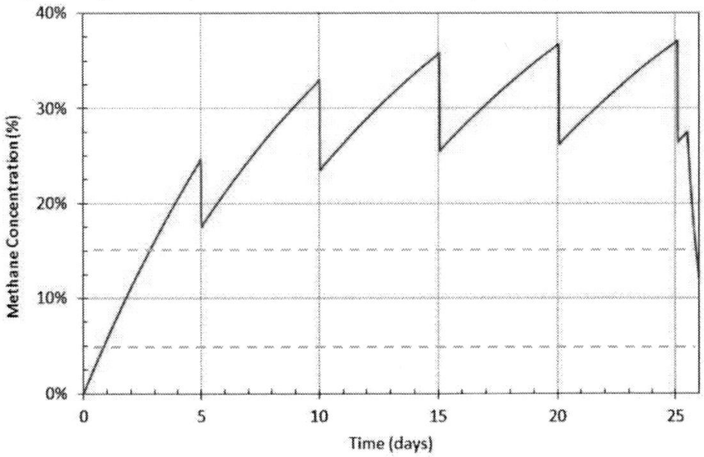

- Above UFL until hatch open for extended period
- Air in-leakage with hatch open for extended period allows fuel concentration to drop below UFL and enter flammable range
- Reach flammable condition in about 8 hours

23

Incident (Oxidant, 4 of 4)

- Natural convection limits maximum fuel concentration and can bring tank vapor space into flammable range, particularly with a tank which is stagnant for an extended period or with a hatch left open
- Note should <u>not</u> conclude that require natural convection in order to form a flammable mixture with a generic produced water tank
  - Lower separator pressure would mean less gas and hence tank could be flammable for portion of time regardless of natural convection
  - Larger volume of produced water withdrawn from tank or higher tank levels could put the tank vapor space into the flammable range
- Should assume a flammable mixture can result if the fuel concentration can exceed the LFL

24

Incident (Ignition Source)

- Driver was performing operation before sunrise
  - No headlamp or flashlight recovered from scene
  - Lighter found 25 feet from body (in same direction as thrown)
  - Driver was a smoker and one glove was off
  - Opened cigarette pack present in truck
  - Reasonable to assume driver used lighter to check liquid level
  - Driver not equipped with an LFL meter
- Static charge cannot be completely ruled out
- Flame propagation to adjacent tanks via connecting lines
- Ignition potential recognized (e.g., API RP 505)

25

Related Incidents

- Produced water tank explosions have happened in the past
- Cursory literature search
  - Texas (2013): Set of six 200 bbl produced water tanks, 2 people smoking on top of tank, tank exploded
  - Canada: Welding on tank which had contained produced water resulted in explosion
- CSB investigation report (2011-H-1):
  - "Public Safety at Oil & Gas Storage Facilities"
  - 26 incidents over 27 year period (44 fatalities and 25 injuries)
  - Includes at least one produced water tank
- Produced water SDS packages vary in flammability warnings (as does the signage on the produced water trucks)

26

Lessons Learned

- Produced water tank explosions can and do happen
  - Lesson already learned in prior incidents .....
- Hazard awareness may not be adequate among all personnel working at oil and gas production sites
- Development of flammable atmosphere in tank depends on separator pressure, tank arrangement and tank ventilation
  - Should assume tank vapor space may be flammable
  - Can also have separator malfunction leading to increased fuel concentration

27

Potential Prevention and Mitigation Approaches

- Increase hazard awareness for all personnel working at site
- Measure level at grade to eliminate routine access to most hazardous location (i.e., tank roof)
- Inert tanks using natural gas
  - Take feed from product gas line
  - Eliminate open vent paths and institute pressure control system (flow as required to maintain positive pressure and preclude air in-leakage),
  - Will result in increased emission of natural gas from facility
- Use a "lead tank" as a secondary separator
  - Inert only this tank with natural gas (limit use of natural gas)
  - Pump or gravity flow from "lead tank" to other produced water tanks
- Monitor tank vapor space

28

Questions

29

481704 Milestone Company Incidents and Their Impact on Air Products Process Safety Journey

Abstract:

Air Products has experienced several significant Process Safety events in our history that have helped define the Process Safety culture that exists at the company today. This presentation describes these events and their root causes, as well as the changes that they launched in the company's approach to Process Safety. The first incident was an explosion at a Nitrous Oxide plant in 1976 that resulted in four on-site fatalities. In the aftermath of this event the need for a formal Hazard Review Process was identified. The second event occurred in 1978 at an air separation plant in which a cryogenic Liquid Oxygen storage tank was ruptured and 800 tons of Liquid Oxygen was released. The Liquid Oxygen vaporized and drifted off-site, resulting in five off-site fatalities. This incident highlighted the necessity of Risk Quantification. While these milestone events served as the impetus for Air Products Process Safety journey, that journey continues to current day, and the presentation will close with a brief update on future direction.

481925 Plymouth Incident

Abstract:

The Williams Company LNG facility on their Northwest Pipeline system located in the state of Washington experience an explosion March 30th 2014. The incident occurred during a routine start-up of the liquefaction process. Directly after the facility was deemed safe the site preservation process was initiated and the investigation began. The various regulatory agencies were involved during the investigation to help determine the cause and to provide recommendations to prevent like kind occurrences elsewhere within the industry. Additional 3rd party firms were called in to review potential structural deficiencies, design flaws, metallurgy testing, and auto ignition testing.

AUTHOR INDEX

AUTHOR INDEX